Lecture Notes in Computer Science   9789

Commenced Publication in 1973
Founding and Former Series Editors:
Gerhard Goos, Juris Hartmanis, and Jan van Leeuwen

More information about this series at http://www.springer.com/series/7407

Osvaldo Gervasi · Beniamino Murgante
Sanjay Misra · Ana Maria A.C. Rocha
Carmelo M. Torre · David Taniar
Bernady O. Apduhan · Elena Stankova
Shangguang Wang (Eds.)

Computational Science and Its Applications – ICCSA 2016

16th International Conference
Beijing, China, July 4–7, 2016
Proceedings, Part IV

 Springer

Editors

Osvaldo Gervasi
University of Perugia
Perugia
Italy

Beniamino Murgante
University of Basilicata
Potenza
Italy

Sanjay Misra
Covenant University
Ota
Nigeria

Ana Maria A.C. Rocha
University of Minho
Braga
Portugal

Carmelo M. Torre
Polytechnic University
Bari
Italy

David Taniar
Monash University
Clayton, VIC
Australia

Bernady O. Apduhan
Kyushu Sangyo University
Fukuoka
Japan

Elena Stankova
Saint Petersburg State University
Saint Petersburg
Russia

Shangguang Wang
Beijing University of Posts
  and Telecommunications
Beijing
China

ISSN 0302-9743   ISSN 1611-3349 (electronic)
Lecture Notes in Computer Science
ISBN 978-3-319-42088-2   ISBN 978-3-319-42089-9 (eBook)
DOI 10.1007/978-3-319-42089-9

Library of Congress Control Number: 2016944355

LNCS Sublibrary: SL1 – Theoretical Computer Science and General Issues

Printed on acid-free paper

This Springer imprint is published by Springer Nature
The registered company is Springer International Publishing AG Switzerland

Preface

These multi-volume proceedings (LNCS volumes 9786, 9787, 9788, 9789, and 9790) consist of the peer-reviewed papers from the 2016 International Conference on Computational Science and Its Applications (ICCSA 2016) held in Beijing, China, during July 4–7, 2016.

ICCSA 2016 was a successful event in the series of conferences, previously held in Banff, Canada (2015), Guimares, Portugal (2014), Ho Chi Minh City, Vietnam (2013), Salvador, Brazil (2012), Santander, Spain (2011), Fukuoka, Japan (2010), Suwon, South Korea (2009), Perugia, Italy (2008), Kuala Lumpur, Malaysia (2007), Glasgow, UK (2006), Singapore (2005), Assisi, Italy (2004), Montreal, Canada (2003), (as ICCS) Amsterdam, The Netherlands (2002), and San Francisco, USA (2001).

Computational science is a main pillar of most present research as well as industrial and commercial activities and it plays a unique role in exploiting ICT innovative technologies. The ICCSA conference series has been providing a venue to researchers and industry practitioners to discuss new ideas, to share complex problems and their solutions, and to shape new trends in computational science.

Apart from the general tracks, ICCSA 2016 also included 33 international workshops, in various areas of computational sciences, ranging from computational science technologies to specific areas of computational sciences, such as computer graphics and virtual reality. The program also featured three keynote speeches and two tutorials.

The success of the ICCSA conference series, in general, and ICCSA 2016, in particular, is due to the support of many people: authors, presenters, participants, keynote speakers, session chairs, Organizing Committee members, student volunteers, Program Committee members, Steering Committee members, and many people in other various roles. We would like to thank them all.

We would also like to thank our sponsors, in particular NVidia and Springer for their very important support and for making the Best Paper Award ceremony so impressive.

We would also like to thank Springer for their continuous support in publishing the ICCSA conference proceedings.

July 2016

Shangguang Wang
Osvaldo Gervasi
Bernady O. Apduhan

Organization

ICCSA 2016 was organized by Beijing University of Post and Telecommunication (China), University of Perugia (Italy), Monash University (Australia), Kyushu Sangyo University (Japan), University of Basilicata (Italy), University of Minho, (Portugal), and the State Key Laboratory of Networking and Switching Technology (China).

Honorary General Chairs

Junliang Chen	Beijing University of Posts and Telecommunications, China
Antonio Laganà	University of Perugia, Italy
Norio Shiratori	Tohoku University, Japan
Kenneth C.J. Tan	Sardina Systems, Estonia

General Chairs

Shangguang Wang	Beijing University of Posts and Telecommunications, China
Osvaldo Gervasi	University of Perugia, Italy
Bernady O. Apduhan	Kyushu Sangyo University, Japan

Program Committee Chairs

Sen Su	Beijing University of Posts and Telecommunications, China
Beniamino Murgante	University of Basilicata, Italy
Ana Maria A.C. Rocha	University of Minho, Portugal
David Taniar	Monash University, Australia

International Advisory Committee

Jemal Abawajy	Deakin University, Australia
Dharma P. Agarwal	University of Cincinnati, USA
Marina L. Gavrilova	University of Calgary, Canada
Claudia Bauzer Medeiros	University of Campinas, Brazil
Manfred M. Fisher	Vienna University of Economics and Business, Austria
Yee Leung	Chinese University of Hong Kong, SAR China

International Liaison Chairs

Ana Carla P. Bitencourt	Universidade Federal do Reconcavo da Bahia, Brazil
Alfredo Cuzzocrea	ICAR-CNR and University of Calabria, Italy
Maria Irene Falcão	University of Minho, Portugal

Robert C.H. Hsu	Chung Hua University, Taiwan
Tai-Hoon Kim	Hannam University, Korea
Sanjay Misra	University of Minna, Nigeria
Takashi Naka	Kyushu Sangyo University, Japan
Rafael D.C. Santos	National Institute for Space Research, Brazil
Maribel Yasmina Santos	University of Minho, Portugal

Workshop and Session Organizing Chairs

Beniamino Murgante	University of Basilicata, Italy
Sanjay Misra	Covenant University, Nigeria
Jorge Gustavo Rocha	University of Minho, Portugal

Award Chair

Wenny Rahayu	La Trobe University, Australia

Publicity Committee Chair

Zibing Zheng	Sun Yat-Sen University, China
Mingdong Tang	Hunan University of Science and Technology, China
Yutao Ma	Wuhan University, China
Ao Zhou	Beijing University of Posts and Telecommunications, China
Ruisheng Shi	Beijing University of Posts and Telecommunications, China

Workshop Organizers

Agricultural and Environment Information and Decision Support Systems (AEIDSS 2016)

Sandro Bimonte	IRSTEA, France
André Miralles	IRSTEA, France
Thérèse Libourel	LIRMM, France
François Pinet	IRSTEA, France

Advances in Information Systems and Technologies for Emergency Preparedness and Risk Assessment (ASTER 2016)

Maurizio Pollino	ENEA, Italy
Marco Vona	University of Basilicata, Italy
Beniamino Murgante	University of Basilicata, Italy

Advances in Web-Based Learning (AWBL 2016)

Mustafa Murat Inceoglu	Ege University, Turkey

Bio- and Neuro-Inspired Computing and Applications (BIOCA 2016)

Nadia Nedjah   State University of Rio de Janeiro, Brazil
Luiza de Macedo Mourell   State University of Rio de Janeiro, Brazil

Computer-Aided Modeling, Simulation, and Analysis (CAMSA 2016)

Jie Shen   University of Michigan, USA and Jilin University,
                        China
Hao Chenina   Shanghai University of Engineering Science, China
Xiaoqiang Liun   Donghua University, China
Weichun Shi   Shanghai Maritime University, China
Yujie Liu   Southeast Jiaotong University, China

Computational and Applied Statistics (CAS 2016)

Ana Cristina Braga   University of Minho, Portugal
Ana Paula Costa Conceicao   University of Minho, Portugal
  Amorim

Computational Geometry and Security Applications (CGSA 2016)

Marina L. Gavrilova   University of Calgary, Canada

Computational Algorithms and Sustainable Assessment (CLASS 2016)

Antonino Marvuglia   Public Research Centre Henri Tudor, Luxembourg
Mikhail Kanevski   Université de Lausanne, Switzerland
Beniamino Murgante   University of Basilicata, Italy

Chemistry and Materials Sciences and Technologies (CMST 2016)

Antonio Laganà   University of Perugia, Italy
Noelia Faginas Lago   University of Perugia, Italy
Leonardo Pacifici   University of Perugia, Italy

Computational Optimization and Applications (COA 2016)

Ana Maria Rocha   University of Minho, Portugal
Humberto Rocha   University of Coimbra, Portugal

Cities, Technologies, and Planning (CTP 2016)

Giuseppe Borruso   University of Trieste, Italy
Beniamino Murgante   University of Basilicata, Italy

Databases and Computerized Information Retrieval Systems (DCIRS 2016)

Sultan Alamri   College of Computing and Informatics, SEU,
                    Saudi Arabia
Adil Fahad   Albaha University, Saudi Arabia
Abdullah Alamri   Jeddah University, Saudi Arabia

Data Science for Intelligent Decision Support (DS4IDS 2016)

Filipe Portela   University of Minho, Portugal
Manuel Filipe Santos   University of Minho, Portugal

Econometrics and Multidimensional Evaluation in the Urban Environment (EMEUE 2016)

Carmelo M. Torre   Polytechnic of Bari, Italy
Maria Cerreta   University of Naples Federico II, Italy
Paola Perchinunno   University of Bari, Italy
Simona Panaro   University of Naples Federico II, Italy
Raffaele Attardi   University of Naples Federico II, Italy

Future Computing Systems, Technologies, and Applications (FISTA 2016)

Bernady O. Apduhan   Kyushu Sangyo University, Japan
Rafael Santos   National Institute for Space Research, Brazil
Jianhua Ma   Hosei University, Japan
Qun Jin   Waseda University, Japan

Geographical Analysis, Urban Modeling, Spatial Statistics (GEO-AND-MOD 2016)

Giuseppe Borruso   University of Trieste, Italy
Beniamino Murgante   University of Basilicata, Italy
Hartmut Asche   University of Potsdam, Germany

GPU Technologies (GPUTech 2016)

Gervasi Osvaldo   University of Perugia, Italy
Sergio Tasso   University of Perugia, Italy
Flavio Vella   University of Rome La Sapienza, Italy

ICT and Remote Sensing for Environmental and Risk Monitoring (RS-Env 2016)

Rosa Lasaponara   Institute of Methodologies for Environmental Analysis,
                          National Research Council, Italy
Weigu Song   University of Science and Technology of China, China
Eufemia Tarantino   Polytechnic of Bari, Italy
Bernd Fichtelmann   DLR, Germany

7th International Symposium on Software Quality (ISSQ 2016)

Sanjay Misra   Covenant University, Nigeria

International Workshop on Biomathematics, Bioinformatics, and Biostatisticss (IBBB 2016)

Unal Ufuktepe   American University of the Middle East, Kuwait

Land Use Monitoring for Soil Consumption Reduction (LUMS 2016)

Carmelo M. Torre	Polytechnic of Bari, Italy
Alessandro Bonifazi	Polytechnic of Bari, Italy
Valentina Sannicandro	University of Naples Federico II, Italy
Massimiliano Bencardino	University of Salerno, Italy
Gianluca di Cugno	Polytechnic of Bari, Italy
Beniamino Murgante	University of Basilicata, Italy

Mobile Communications (MC 2016)

Hyunseung Choo	Sungkyunkwan University, Korea

Mobile Computing, Sensing, and Actuation for Cyber Physical Systems (MSA4IoT 2016)

Saad Qaisar	NUST School of Electrical Engineering and Computer Science, Pakistan
Moonseong Kim	Korean Intellectual Property Office, Korea

Quantum Mechanics: Computational Strategies and Applications (QM-CSA 2016)

Mirco Ragni	Universidad Federal de Bahia, Brazil
Ana Carla Peixoto Bitencourt	Universidade Estadual de Feira de Santana, Brazil
Vincenzo Aquilanti	University of Perugia, Italy
Andrea Lombardi	University of Perugia, Italy
Federico Palazzetti	University of Perugia, Italy

Remote Sensing for Cultural Heritage: Documentation, Management, and Monitoring (RSCH 2016)

Rosa Lasaponara	IRMMA, CNR, Italy
Nicola Masini	IBAM, CNR, Italy Zhengzhou Base, International Center on Space Technologies for Natural and Cultural Heritage, China
Chen Fulong	Institute of Remote Sensing and Digital Earth, Chinese Academy of Sciences, China

Scientific Computing Infrastructure (SCI 2016)

Elena Stankova	Saint Petersburg State University, Russia
Vladimir Korkhov	Saint Petersburg State University, Russia
Alexander Bogdanov	Saint Petersburg State University, Russia

Software Engineering Processes and Applications (SEPA 2016)

Sanjay Misra	Covenant University, Nigeria

Social Networks Research and Applications (SNRA 2016)

Eric Pardede	La Trobe University, Australia
Wenny Rahayu	La Trobe University, Australia
David Taniar	Monash University, Australia

Sustainability Performance Assessment: Models, Approaches, and Applications Toward Interdisciplinarity and Integrated Solutions (SPA 2016)

Francesco Scorza	University of Basilicata, Italy
Valentin Grecu	Lucia Blaga University on Sibiu, Romania

Tools and Techniques in Software Development Processes (TTSDP 2016)

Sanjay Misra	Covenant University, Nigeria

Volunteered Geographic Information: From Open Street Map to Participation (VGI 2016)

Claudia Ceppi	University of Basilicata, Italy
Beniamino Murgante	University of Basilicata, Italy
Francesco Mancini	University of Modena and Reggio Emilia, Italy
Giuseppe Borruso	University of Trieste, Italy

Virtual Reality and Its Applications (VRA 2016)

Osvaldo Gervasi	University of Perugia, Italy
Lucio Depaolis	University of Salento, Italy

Web-Based Collective Evolutionary Systems: Models, Measures, Applications (WCES 2016)

Alfredo Milani	University of Perugia, Italy
Valentina Franzoni	University of Rome La Sapienza, Italy
Yuanxi Li	Hong Kong Baptist University, Hong Kong, SAR China
Clement Leung	United International College, Zhuhai, China
Rajdeep Niyogi	Indian Institute of Technology, Roorkee, India

Program Committee

Jemal Abawajy	Deakin University, Australia
Kenny Adamson	University of Ulster, UK
Hartmut Asche	University of Potsdam, Germany
Michela Bertolotto	University College Dublin, Ireland
Sandro Bimonte	CEMAGREF, TSCF, France
Rod Blais	University of Calgary, Canada
Ivan Blečić	University of Sassari, Italy
Giuseppe Borruso	University of Trieste, Italy
Yves Caniou	Lyon University, France

José A. Cardoso e Cunha	Universidade Nova de Lisboa, Portugal
Carlo Cattani	University of Salerno, Italy
Mete Celik	Erciyes University, Turkey
Alexander Chemeris	National Technical University of Ukraine KPI, Ukraine
Min Young Chung	Sungkyunkwan University, Korea
Elisete Correia	University of Trás os Montes e Alto Douro, Portugal
Gilberto Corso Pereira	Federal University of Bahia, Brazil
M. Fernanda Costa	University of Minho, Portugal
Alfredo Cuzzocrea	ICAR-CNR and University of Calabria, Italy
Florbela Maria da Cruz Domingues Correia	Intituto Politécnico de Viana do Castelo, Portugal
Vanda Marisa da Rosa Milheiro Lourenço	FCT from University Nova de Lisboa, Portugal
Carla Dal Sasso Freitas	Universidade Federal do Rio Grande do Sul, Brazil
Pradesh Debba	The Council for Scientific and Industrial Research (CSIR), South Africa
Hendrik Decker	Instituto Tecnológico de Informática, Spain
Adelaide de Fátima Baptista Valente Freitas	University of Aveiro, Portugal
Carina Soares da Silva Fortes	Escola Superior de Tecnologias da Saúde de Lisboa, Portugal
Frank Devai	London South Bank University, UK
Rodolphe Devillers	Memorial University of Newfoundland, Canada
Joana Dias	University of Coimbra, Portugal
Prabu Dorairaj	NetApp, India/USA
M. Irene Falcao	University of Minho, Portugal
Cherry Liu Fang	U.S. DOE Ames Laboratory, USA
Florbela Fernandes	Polytechnic Institute of Bragança, Portugal
Jose-Jesús Fernandez	National Centre for Biotechnology, CSIS, Spain
Mara Celia Furtado Rocha	PRODEB-Pós Cultura/UFBA, Brazil
Akemi Galvez	University of Cantabria, Spain
Paulino Jose Garcia Nieto	University of Oviedo, Spain
Marina Gavrilova	University of Calgary, Canada
Jerome Gensel	LSR-IMAG, France
Mara Giaoutzi	National Technical University, Athens, Greece
Andrzej M. Goscinski	Deakin University, Australia
Alex Hagen-Zanker	University of Cambridge, UK
Malgorzata Hanzl	Technical University of Lodz, Poland
Shanmugasundaram Hariharan	B.S. Abdur Rahman University, India
Tutut Herawan	Universitas Teknologi Yogyakarta, Indonesia
Hisamoto Hiyoshi	Gunma University, Japan
Fermin Huarte	University of Barcelona, Spain
Andrés Iglesias	University of Cantabria, Spain
Mustafa Inceoglu	Ege University, Turkey
Peter Jimack	University of Leeds, UK

Qun Jin	Waseda University, Japan
Farid Karimipour	Vienna University of Technology, Austria
Baris Kazar	Oracle Corp., USA
Maulana Adhinugraha Kiki	Telkom University, Indonesia
DongSeong Kim	University of Canterbury, New Zealand
Taihoon Kim	Hannam University, Korea
Ivana Kolingerova	University of West Bohemia, Czech Republic
Dieter Kranzlmueller	LMU and LRZ Munich, Germany
Antonio Laganà	University of Perugia, Italy
Rosa Lasaponara	National Research Council, Italy
Maurizio Lazzari	National Research Council, Italy
Cheng Siong Lee	Monash University, Australia
Sangyoun Lee	Yonsei University, Korea
Jongchan Lee	Kunsan National University, Korea
Clement Leung	United International College, Zhuhai, China
Chendong Li	University of Connecticut, USA
Gang Li	Deakin University, Australia
Ming Li	East China Normal University, China
Fang Liu	AMES Laboratories, USA
Xin Liu	University of Calgary, Canada
Savino Longo	University of Bari, Italy
Tinghuai Ma	NanJing University of Information Science and Technology, China
Isabel Cristina Maciel Natário	FCT from University Nova de Lisboa, Portugal
Sergio Maffioletti	University of Zurich, Switzerland
Ernesto Marcheggiani	Katholieke Universiteit Leuven, Belgium
Antonino Marvuglia	Research Centre Henri Tudor, Luxembourg
Nicola Masini	National Research Council, Italy
Nirvana Meratnia	University of Twente, The Netherlands
Alfredo Milani	University of Perugia, Italy
Sanjay Misra	Federal University of Technology Minna, Nigeria
Giuseppe Modica	University of Reggio Calabria, Italy
José Luis Montaña	University of Cantabria, Spain
Beniamino Murgante	University of Basilicata, Italy
Jiri Nedoma	Academy of Sciences of the Czech Republic, Czech Republic
Laszlo Neumann	University of Girona, Spain
Irene Oliveira	University of Trás os Montes e Alto Douro, Portugal
Kok-Leong Ong	Deakin University, Australia
Belen Palop	Universidad de Valladolid, Spain
Marcin Paprzycki	Polish Academy of Sciences, Poland
Eric Pardede	La Trobe University, Australia
Kwangjin Park	Wonkwang University, Korea
Telmo Pinto	University of Minho, Portugal

Reviewers

Abawajy, Jemal	Deakin University, Australia
Abuhelaleh, Mohammed	Univeristy of Bridgeport, USA
Acharjee, Shukla	Dibrugarh University, India
Andrianov, Sergei Nikolaevich	Universitetskii prospekt, Russia
Aguilar, José Alfonso	Universidad Autónoma de Sinaloa, Mexico
Ahmed, Faisal	University of Calgary, Canada
Alberti, Margarita	University of Barcelona, Spain
Amato, Alba	Seconda Universit degli Studi di Napoli, Italy
Amorim, Ana Paula	University of Minho, Portugal
Apduhan, Bernady	Kyushu Sangyo University, Japan
Aquilanti, Vincenzo	University of Perugia, Italy
Asche, Hartmut	Posdam University, Germany
Athayde Maria, Emlia Feijão Queiroz	University of Minho, Portugal
Attardi, Raffaele	University of Napoli Federico II, Italy
Azam, Samiul	United International University, Bangladesh
Azevedo, Ana	Athabasca University, USA
Badard, Thierry	Laval University, Canada
Baioletti, Marco	University of Perugia, Italy
Bartoli, Daniele	University of Perugia, Italy
Bentayeb, Fadila	Université Lyon, France
Bilan, Zhu	Tokyo University of Agriculture and Technology, Japan
Bimonte, Sandro	IRSTEA, France
Blecic, Ivan	Università di Cagliari, Italy
Bogdanov, Alexander	Saint Petersburg State University, Russia
Borruso, Giuseppe	University of Trieste, Italy
Bostenaru, Maria	"Ion Mincu" University of Architecture and Urbanism, Romania
Braga Ana, Cristina	University of Minho, Portugal
Canora, Filomena	University of Basilicata, Italy
Cardoso, Rui	Institute of Telecommunications, Portugal
Ceppi, Claudia	Polytechnic of Bari, Italy
Cerreta, Maria	University Federico II of Naples, Italy
Choo, Hyunseung	Sungkyunkwan University, South Korea
Coletti, Cecilia	University of Chieti, Italy
Correia, Elisete	University of Trás-Os-Montes e Alto Douro, Portugal
Correia Florbela Maria, da Cruz Domingues	Instituto Politécnico de Viana do Castelo, Portugal
Costa, Fernanda	University of Minho, Portugal
Crasso, Marco	National Scientific and Technical Research Council, Argentina
Crawford, Broderick	Universidad Catolica de Valparaiso, Chile

Cuzzocrea, Alfredo	University of Trieste, Italy
Cutini, Valerio	University of Pisa, Italy
Danese, Maria	IBAM, CNR, Italy
Decker, Hendrik	Instituto Tecnológico de Informática, Spain
Degtyarev, Alexander	Saint Petersburg State University, Russia
Demartini, Gianluca	University of Sheffield, UK
Di Leo, Margherita	JRC, European Commission, Belgium
Dias, Joana	University of Coimbra, Portugal
Dilo, Arta	University of Twente, The Netherlands
Dorazio, Laurent	ISIMA, France
Duarte, Júlio	University of Minho, Portugal
El-Zawawy, Mohamed A.	Cairo University, Egypt
Escalona, Maria-Jose	University of Seville, Spain
Falcinelli, Stefano	University of Perugia, Italy
Fernandes, Florbela	Escola Superior de Tecnologia e Gest ão de Bragança, Portugal
Florence, Le Ber	ENGEES, France
Freitas Adelaide, de Fátima Baptista Valente	University of Aveiro, Portugal
Frunzete, Madalin	Polytechnic University of Bucharest, Romania
Gankevich, Ivan	Saint Petersburg State University, Russia
Garau, Chiara	University of Cagliari, Italy
Garcia, Ernesto	University of the Basque Country, Spain
Gavrilova, Marina	University of Calgary, Canada
Gensel, Jerome	IMAG, France
Gervasi, Osvaldo	University of Perugia, Italy
Gizzi, Fabrizio	National Research Council, Italy
Gorbachev, Yuriy	Geolink Technologies, Russia
Grilli, Luca	University of Perugia, Italy
Guerra, Eduardo	National Institute for Space Research, Brazil
Hanzl, Malgorzata	University of Lodz, Poland
Hegedus, Peter	University of Szeged, Hungary
Herawan, Tutut	University of Malaya, Malaysia
Hu, Ya-Han	National Chung Cheng University, Taiwan
Ibrahim, Michael	Cairo University, Egipt
Ifrim, Georgiana	Insight, Ireland
Irrazábal, Emanuel	Universidad Nacional del Nordeste, Argentina
Janana, Loureio	University of Mato Grosso do Sul, Brazil
Jaiswal, Shruti	Delhi Technological University, India
Johnson, Franklin	Universidad de Playa Ancha, Chile
Karimipour, Farid	Vienna University of Technology, Austria
Kapcak, Sinan	American University of the Middle East in Kuwait, Kuwait
Kiki Maulana, Adhinugraha	Telkom University, Indonesia
Kim, Moonseong	KIPO, South Korea
Kobusińska, Anna	Poznan University of Technology, Poland

Korkhov, Vladimir	Saint Petersburg State University, Russia
Koutsomitropoulos, Dimitrios A.	University of Patras, Greece
Krishna Kumar, Chaturvedi	Indian Agricultural Statistics Research Institute (IASRI), India
Kulabukhova, Nataliia	Saint Petersburg State University, Russia
Kumar, Dileep	SR Engineering College, India
Laganà, Antonio	University of Perugia, Italy
Lai, Sen-Tarng	Shih Chien University, Taiwan
Lanza, Viviana	Lombardy Regional Institute for Research, Italy
Lasaponara, Rosa	National Research Council, Italy
Lazzari, Maurizio	National Research Council, Italy
Le Duc, Tai	Sungkyunkwan University, South Korea
Le Duc, Thang	Sungkyunkwan University, South Korea
Lee, KangWoo	Sungkyunkwan University, South Korea
Leung, Clement	United International College, Zhuhai, China
Libourel, Thérèse	LIRMM, France
Lourenço, Vanda Marisa	University Nova de Lisboa, Portugal
Machado, Jose	University of Minho, Portugal
Magni, Riccardo	Pragma Engineering srl, Italy
Mancini Francesco	University of Modena and Reggio Emilia, Italy
Manfreda, Salvatore	University of Basilicata, Italy
Manganelli, Benedetto	Università degli studi della Basilicata, Italy
Marghany, Maged	Universiti Teknologi Malaysia, Malaysia
Marinho, Euler	Federal University of Minas Gerais, Brazil
Martellozzo, Federico	University of Rome "La Sapienza", Italy
Marvuglia, Antonino	Public Research Centre Henri Tudor, Luxembourg
Mateos, Cristian	Universidad Nacional del Centro, Argentina
Matsatsinis, Nikolaos	Technical University of Crete, Greece
Messina, Fabrizio	University of Catania, Italy
Millham, Richard	Durban University of Technoloy, South Africa
Milani, Alfredo	University of Perugia, Italy
Misra, Sanjay	Covenant University, Nigeria
Modica, Giuseppe	Università Mediterranea di Reggio Calabria, Italy
Mohd Helmy, Abd Wahab	Universiti Tun Hussein Onn Malaysia, Malaysia
Murgante, Beniamino	University of Basilicata, Italy
Nagy, Csaba	University of Szeged, Hungary
Napolitano, Maurizio	Center for Information and Communication Technology, Italy
Natário, Isabel Cristina Maciel	University Nova de Lisboa, Portugal
Navarrete Gutierrez, Tomas	Luxembourg Institute of Science and Technology, Luxembourg
Nedjah, Nadia	State University of Rio de Janeiro, Brazil
Nguyen, Tien Dzung	Sungkyunkwan University, South Korea
Niyogi, Rajdeep	Indian Institute of Technology Roorkee, India

Oliveira, Irene	University of Trás-Os-Montes e Alto Douro, Portugal
Panetta, J.B.	Tecnologia Geofísica Petróleo Brasileiro SA, PETROBRAS, Brazil
Papa, Enrica	University of Amsterdam, The Netherlands
Papathanasiou, Jason	University of Macedonia, Greece
Pardede, Eric	La Trobe University, Australia
Pascale, Stefania	University of Basilicata, Italy
Paul, Padma Polash	University of Calgary, Canada
Perchinunno, Paola	University of Bari, Italy
Pereira, Oscar	Universidade de Aveiro, Portugal
Pham, Quoc Trung	HCMC University of Technology, Vietnam
Pinet, Francois	IRSTEA, France
Pirani, Fernando	University of Perugia, Italy
Pollino, Maurizio	ENEA, Italy
Pusatli, Tolga	Cankaya University, Turkey
Qaisar, Saad	NURST, Pakistan
Qian, Junyan	Guilin University of Electronic Technology, China
Raffaeta, Alessandra	University of Venice, Italy
Ragni, Mirco	Universidade Estadual de Feira de Santana, Brazil
Rahman, Wasiur	Technical University Darmstadt, Germany
Rampino, Sergio	Scuola Normale di Pisa, Italy
Rahayu, Wenny	La Trobe University, Australia
Ravat, Franck	IRIT, France
Raza, Syed Muhammad	Sungkyunkwan University, South Korea
Roccatello, Eduard	3DGIS, Italy
Rocha, Ana Maria	University of Minho, Portugal
Rocha, Humberto	University of Coimbra, Portugal
Rocha, Jorge	University of Minho, Portugal
Rocha, Maria Clara	ESTES Coimbra, Portugal
Romano, Bernardino	University of l'Aquila, Italy
Sannicandro, Valentina	Polytechnic of Bari, Italy
Santiago Júnior, Valdivino	Instituto Nacional de Pesquisas Espaciais, Brazil
Sarafian, Haiduke	Pennsylvania State University, USA
Schneider, Michel	ISIMA, France
Selmaoui, Nazha	University of New Caledonia, New Caledonia
Scerri, Simon	University of Bonn, Germany
Shakhov, Vladimir	Institute of Computational Mathematics and Mathematical Geophysics, Russia
Shen, Jie	University of Michigan, USA
Silva-Fortes, Carina	ESTeSL-IPL, Portugal
Singh, Upasana	University of Kwa Zulu-Natal, South Africa
Skarga-Bandurova, Inna	Technological Institute of East Ukrainian National University, Ukraine
Soares, Michel	Federal University of Sergipe, Brazil
Souza, Eric	Universidade Nova de Lisboa, Portugal
Stankova, Elena	Saint Petersburg State University, Russia

Stalidis, George	TEI of Thessaloniki, Greece
Taniar, David	Monash University, Australia
Tasso, Sergio	University of Perugia, Italy
Telmo, Pinto	University of Minho, Portugal
Tengku, Adil	La Trobe University, Australia
Thorat, Pankaj	Sungkyunkwan University, South Korea
Tiago Garcia, de Senna Carneiro	Federal University of Ouro Preto, Brazil
Tilio, Lucia	University of Basilicata, Italy
Torre, Carmelo Maria	Polytechnic of Bari, Italy
Tripathi, Ashish	MNNIT Allahabad, India
Tripp, Barba	Carolina, Universidad Autnoma de Sinaloa, Mexico
Trunfio, Giuseppe A.	University of Sassari, Italy
Upadhyay, Ashish	Indian Institute of Public Health-Gandhinagar, India
Valuev, Ilya	Russian Academy of Sciences, Russia
Varella, Evangelia	Aristotle University of Thessaloniki, Greece
Vasyunin, Dmitry	University of Amsterdam, The Netherlans
Vijaykumar, Nandamudi	INPE, Brazil
Villalba, Maite	Universidad Europea de Madrid, Spain
Walkowiak, Krzysztof	Wroclav University of Technology, Poland
Wanderley, Fernando	FCT/UNL, Portugal
Wei Hoo, Chong	Motorola, USA
Xia, Feng	Dalian University of Technology (DUT), China
Yamauchi, Toshihiro	Okayama University, Japan
Yeoum, Sanggil	Sungkyunkwan University, South Korea
Yirsaw, Ayalew	University of Botswana, Bostwana
Yujie, Liu	Southeast Jiaotong University, China
Zafer, Agacik	American University of the Middle East in Kuwait, Kuwait
Zalyubovskiy, Vyacheslav	Russian Academy of Sciences, Russia
Zeile, Peter	Technische Universitat Kaiserslautern, Germany
Žemlička, Michal	Charles University, Czech Republic
Zivkovic, Ljiljana	Republic Agency for Spatial Planning, Belgrade
Zunino, Alejandro	Universidad Nacional del Centro, Argentina

Sponsoring Organizations

ICCSA 2016 would not have been possible without the tremendous support of many organizations and institutions, for which all organizers and participants of ICCSA 2016 express their sincere gratitude:

Springer International Publishing AG, Switzerland
(http://www.springer.com)

NVidia Co., USA
(http://www.nvidia.com)

Beijing University of Post and Telecommunication, China
(http://english.bupt.edu.cn/)

State Key Laboratory of Networking and Switching Technology, China

University of Perugia, Italy
(http://www.unipg.it)

University of Basilicata, Italy
(http://www.unibas.it)

Monash University, Australia
(http://monash.edu)

Kyushu Sangyo University, Japan
(www.kyusan-u.ac.jp)

Universidade do Minho, Portugal
(http://www.uminho.pt)

Contents – Part IV

Where the Streets Have Known Names

Paulo Dias Almeida[1(✉)], Jorge Gustavo Rocha[1],
Andrea Ballatore[2], and Alexander Zipf[3]

[1] Minho University, Braga, Portugal
{b6301,jgr}@di.uminho.pt
[2] Birkbeck College, University of London, London, UK
a.ballatore@bbk.ac.uk
[3] Universität Heidelberg, Heidelberg, Germany
alexander.zipf@geog.uni-heidelberg.de

Abstract. Street names provide important insights into the local culture, history, and politics of places. Linked open data provide a wealth of knowledge that can be associated with street names, enabling novel ways to explore cultural geographies. This paper presents a three-fold contribution. We present (1) a technique to establish a correspondence between street names and the entities that they refer to. The method is based on Wikidata, a knowledge base derived from Wikipedia. The accuracy of this mapping is evaluated on a sample of streets in Rome. As this approach reaches limited coverage, we propose to tap local knowledge with (2) a simple web platform. Users can select the best correspondence from the calculated ones or add another entity not discovered by the automated process. As a result, we design (3) an enriched OpenStreetMap web map where each street name can be explored in terms of the properties of its associated entity. Through several filters, this tool is a first step towards the interactive exploration of toponymy, showing how open data can reveal facets of the cultural texture that pervades places.

Keywords: Digital humanities · Toponymy · OpenStreetMap · Wikidata · Linked open data · Volunteered geographic information

1 Introduction

All web maps show street names, supporting us in wayfinding. What is overlooked is that, behind each street name, that there is a rich and complex story. Street names are dedicated to notable people, places or events. They are frequently used to honor notable citizens or celebrate events and revolutions. Therefore, they often provide important insights into the culture, politics, and history of a locale.

In this pilot project we aim at creating an interactive web application where users can trace the stories behind street names, relying on OpenStreetMap[1] and

---

[1] http://www.openstreetmap.org.

© Springer International Publishing Switzerland 2016
O. Gervasi et al. (Eds.): ICCSA 2016, Part IV, LNCS 9789, pp. 1–12, 2016.
DOI: 10.1007/978-3-319-42089-9_1

other open data sources. As a first step, users can explore streets named after individuals, filtering them by gender, date of birth, and profession. Wikipedia is used as an information source. More specifically, we use Wikidata[2] and DBpedia[3], two knowledge bases designed to extract structured information from Wikipedia, to link the street name with the corresponding resource described in the knowledge bases. To show the potential of linked open data, the process will be as automated as possible.

This paper describes the automatic mapping of street names with resources from these knowledge bases and rank those resources according to their relevance. The preliminary results, obtained on a sample of streets in Rome, show that there are many missing relations. To increase the coverage, we propose a web tool to that knowledge from human contributors.

The remainder of this paper is organized as follows. We start by presenting related work in Sect. 2. We then elaborate on our approach in Sect. 3. Section 4 evaluates our automated solution, and the preliminary results are discussed in Sect. 5. The design proposal for the web platform that expands and complements the automated solution is presented in Sect. 6. Finally, we present our conclusions in Sect. 7.

2   Related Work

To link street names to the relevant entities, we adopt concepts and techniques from a variety of research areas, including toponymy, geographic information science (GISc), and Semantic Web and Linked Open Data research.

2.1   Toponymy and Street Names

Toponymy is the study of place names (toponyms), with respect to their origins, meanings, use and typology. Place names provide an extremely useful geographical reference system in the world. Consistency and accuracy are essential in referring to a place to prevent confusion in everyday activities. Toponymy is crucial to establish officially recognized geographical names, and relies local written and oral histories to study and record how place names evolve and why.

Many geographers, historians, and linguists have found that toponyms provide valuable insight into the historical geography of a particular region. They play a symbolic role in the expression of local culture, being used many times to promote values related to political and religious beliefs [7]. Unsurprisingly, place names are then given an important role in territorial conflicts and landscape transformation [9]. Place names are so important that, even outside of armed conflicts, altering place names in official maps to reflect a different context and culture is regarded as a possible act of cultural aggression [9]. Consequently, place names represent an extremely important data source for analysing cultural changes across different locations over time.

---

[2] http://www.wikidata.org.
[3] http://wiki.dbpedia.org.

2.2   Linked Open Data

Linked data aims to provide knowledge in a structured and simple manner, allowing it to be understood by humans and machines [2]. This is done through the adoption of design principles and standards in order to express data in the simplest form. Data is organized in a set of triples; subject, object, and a named relation connecting them (predicate). This design goal of establishing named and directed links between typed data enables the creation of useful semantic queries and facilitates the integration of heterogeneous data sources.

The linked data paradigm has also emerged as a promising approach to structuring and sharing geospatial information. The simplification in the way data is expressed has a dramatic impact in spatially and temporally referenced data, usually modeled as complex relational schemata [10]. User-generated content, of major importance for many current applications, can also benefit from the linked data approach. Triples are seen as statements made by a know author, with great potential applications in collaborative geographic information production.

The process of linking a new dataset to existing ones is called 'bootstrapping', and is usually performed on semantic hubs, such as Wikidata and DBpedia [4,12]. These are important community efforts to extract structured information from Wikipedia according with linked data principles. These projects allow you to ask sophisticated queries against datasets derived from Wikipedia and to link other datasets on the Web to Wikipedia data.

The process of inter-linking data is complex and faces several technical and cultural challenges. Knowledge bases are built on heterogeneous and incompatible vocabularies and ontologies [3]. Several efforts have been made to ease and automate the linking process, using semantic similarity and relatedness measures [1].

3   Street Name Matching Method

Our main goal is to establish links between street names and the entities that the street names refer to. This mapping will then be used to enable a different visualization of streets and neighbourhoods, putting these entities and their historical and cultural context. We start by retrieving the street names from OpenStreetMap. Each toponym is then used in a semantic query to a knowledge bases such as Wikidata and DBpedia to retrieve relevant entities. As a case study, we selected street names in Rome, expressed in Italian.

3.1   Toponym Retrieval

In order to identify the entity represented by a street name, we start by isolating the part of the name that corresponds to said entity. Street names are normally composed with prefixes or suffixes that describe the feature type (examples in English include *avenue*, *street*, and *boulevard*). These linguistic tokens have to be filtered out before querying the knowledge base. For this purpose, we use a set of stop words. This process is language-specific and entails the a priori definition of the stop words used as prefixes and suffixes for each target language–Italian in this case study.

3.2   Entity Retrieval

After having filtered the street name, the resulting string is used in a query to the knowledge bases. In order to be incorporated in the process, a knowledge base needs to support a text-based query for relevant entities, and their properties and relations. In order to make the process as automated as possible, all the query endpoints and parameters, as well as result format, can be defined in a configuration file. This way, the application can interact with any knowledge base that supports either SPARQL or HTTP queries. The pilot tool works seamlessly with Wikidata and DBpedia, two major open knowledge bases.

3.3   Entity Ranking

After receiving the results from a knowledge base, we have to estimate their relevance. Each time we query the knowledge base for a keyword, several entities can be retrieved. Some street names in Rome, like *Via Mazzarino* will match multiple resources in Wikipedia. Mazzarino can be either a place name[4] or the surname of several notable people.[5] In order to establish which entity is referred to by a street name, a ranking of the relevance of the results needs to be calculated. Working with semantically rich data results in more adequate rankings than simple keyword-based search. Properties like the entity's location provide additional context to the query to determine the relevance of a result.

Our ranking algorithm takes into account the location of the input street and of the entities. This approach has already been proven to generate more appropriate rankings in related projects [6,11], and is crucial to this application, where the local context plays a major role. Based on the process defined by Shuyao et al. [11], we define a formula to incorporate location into a custom ranking algorithm. The relevance r of an entity e is defined as follows:

$$r_e = \beta \, i_e + (1 - \beta)(1 - dist(\lambda_q, \lambda_e)) \tag{1}$$

where i_e represents the informativeness associated with the entity, λ_q and λ_e represent the locations of the query and of the entity respectively, and β is a factor between 0 and 1 used to balance the importance of the informativeness calculated for the entity in relation to the distance between the query and an entity.

In order to calculate the informativeness of an entity (i_e), we first estimate its global relevance. In order to calculate this relevance we chose to consider the number of different language entries Wikipedia has for said entity. However,

---

[4] https://it.wikipedia.org/wiki/Mazzarino_(Italia).
[5] https://it.wikipedia.org/wiki/Mazzarino.

this approach is biased in favor of very general entities, such as countries, cities, and family names. To solve this problem, we weight the informativeness of an entity with its inverse entity frequency (ief), similar to the method by Zaragoza et al. [13]:

$$ief = \log(N/n_e) \qquad (2)$$

where N represents the totality of query results and n_e those that contain entity e. In the next section, the results obtained by this entity ranking algorithm are explored.

4 Evaluation

To validate the methodology presented and assess the suitability of the custom ranking algorithm, we proceed to evaluate the performance of our approach. For this evaluation, a central area of Rome, including the neighborhoods of Trastevere and Testaccio, was chosen as the target geographic area. The set of named streets used was obtained with the following query, on the OpenStreetMap API called Overpass:[6]

```
[out:json];
(
  way
    ["highway" = "residential"] ["name"]
    (41.8642593, 12.4612841, 41.9030756, 12.4945021)
);
out body;
>;
out skel qt;
```

The query originated a set of 1,709 named streets. When applied to this set, our street name matching algorithm found an entity for 66 % of the streets. To evaluate the quality of the matching, 20 queries with more than one entity were randomly selected. 79 entities were mapped to these queries, with the number of entities associated per query varying from two to seven. Then, the relevant entities for each query were manually selected and ranked from most to least relevant. The set of relevant documents per query is thus limited, in this evaluation, to those returned by the knowledge base for the selected sample. Table 1 illustrates an example of the process executed on Wikidata.

For this sample, the proposed solution was able to map a relevant entity for 85 % of the queries, however, only about 35 % of the retrieved entities were considered relevant results. This fact accentuates the importance of ranking the relevance of these results. Hence, we take as base line the default ranking returned by the knowledge base, and proceed to analyse how it compares with the custom ranking. The performance of the ranking algorithms is based on the mean average precision (mAP), mean reciprocal rank (MRR), and normalized discounted

---

[6] http://overpass-turbo.eu.

Table 1. Sample query used in the evaluation. *Sommergibile* means submarine.

Street name	"Via Galileo Ferraris"
Query	"Galileo Ferraris"
Wikidata entities (default ranking)	1. Galileo Ferraris (sommergibile 1914)
	2. Galileo Ferraris (sommergibile 1935)
	3. Galileo Ferraris (Italian physicist)
Wikidata entities (custom ranking)	1. Galileo Ferraris (Italian physicist)
	2. Galileo Ferraris (sommergibile 1935)
	3. Galileo Ferraris (sommergibile 1914)
Top entity	Galileo Ferraris (Italian physicist)

Table 2. Performance of the ranking methods, using mean average precision (mAP), mean reciprocal rank (MRR), and normalized discounted cumulative gain (nDCG). Best results in bold.

Algorithm	mAP	MRR	nDCG
Default	0.70	0.74	0.74
Custom	**0.75**	**0.76**	**0.80**

cumulative gain (nDCG), three common information retrieval measures. Table 2 summarizes the comparison of these two ranking approaches.

Our custom ranking outperforms the baseline in all three measures, thanks to the inclusion of location awareness and entity informativeness, calculated from the number of different Wikipedia language entries. The custom ranking algorithm proposed outperforms the default ranking, demonstrating a 7 % relative increase in mAP and a 8 % relative increase in nDCG.

Despite demonstrating encouraging results, this evaluation also brings to light a significant level of noise, in the form of incorrect entities being mapped to the queries. This can be a significant problem, regardless of the quality of the ranking algorithm, suggesting a coverage issue in the entities in the knowledge base. The limited scope of the evaluation sample is also a point of concern, highlighting the necessity to involve local communities in the matching and validation process.

Through a crowdsourced solution, users will be able to either select the best correspondence from the suggested ones, or add new entities when the automated process fails. The custom ranking algorithm developed for the automated part of the solution can also benefit from this user interaction, learning better disambiguation strategy. Cases where the user solves a conflict between entities of different classes, for example, a street is matched to an entity representing a person and also to a number of entities representing places named after that person, can be learned by the ranking algorithm to apply in the subsequent queries.

5   Results and Discussion

As stated in the previous section, our solution was tested on an area of Rome, and for the 1,709 retrieved roads with names, of which 1,121 were matched to Wikidata entities. This association between street names and known entities in a knowledge base allows for analysis over the properties and characteristics of the entities referred to by street names.

First, we consider the streets that are named after people. In this case, we determined that 630 of the 1,121 matched streets are named after individuals (approximately 56 %). Subsequently, we observe the gender of the referred people. As is possible to notice in Fig. 1, only approximately 6 % (36 instances) are named after females, while a staggering 94 % of names refer to males.

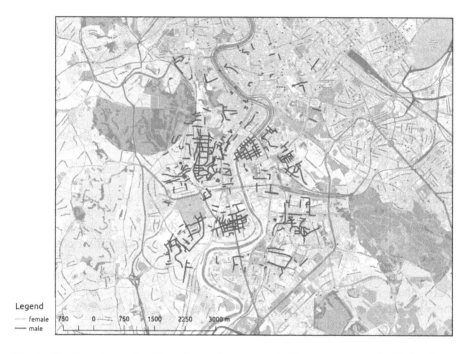

Fig. 1. Gender of people after whom the streets in Rome are named. (Color figure online)

Based on the information contained in Wikidata, we can also analyze the thematic areas in which these individuals became notable. Figure 2 shows that, for this region of Rome, the top three thematic areas of activity are science (109 instances), religion (90 instances), and politics (84 instances). The map also attracts attention to the existence of spatial clusters of streets named after writers, scientists, and politicians, reflecting clear planning choices of the Roman administration.

Fig. 2. Professions of people after whom the streets in Rome are named. (Color figure online)

This analysis exemplifies the richness and depth of the cultural data contained in Wikidata and similar knowledge bases, and how, following the linked data paradigm, cultural exploration of the geographic space can be supported. However, as mentioned in Sect. 4, there are some important limitations. First, there is a significant number of irrelevant entities being mapped to street names, introducing some noise in the data. Second, the coverage of the knowledge base is particularly limited for entities that are notable only in local contexts, such as people whose existence never became known beyond their hometown. The next section proposes a solution to these issues.

6   Crowd Sourcing Local Knowledge

The previous section discussed preliminary results obtained with our street name matching approach. Both the OpenStreetMap streets and the Wikidata knowledge base used in this study were built through crowd sourcing. It is therefore reasonable to get the communities involved, taking advantage of the local knowledge of citizens [8]. The locals, in fact, are the ones that are most likely to interpret the meaning of their local street names. Hence, these volunteer communities can contribute to the linking process as follows:

- From the lists of related entities, they can pick the best match based on their locally-situated knowledge.
- If only irrelevant entities are listed, they can add a link to Wikipedia or other web resources that refer to the relevant entities.

To achieve this goal, we outline a dedicated web platform. Designing a successful crowd sourcing platform is not easy. In another words, the success of such platforms does not depend only on technical aspects, such as design principles and the usability of tools [5], but are deeply rooted in the motivation of contributors. To advance the design of this platform, we outline its requirements, present a story board documenting a typical use case, and describe its current software architecture.

6.1   Requirements

The requirements for this VGI platform are derived directly from the aforementioned goals:

- The user will be able to navigate a map and pick any street;
- The entities associated to a selected street are presented in a list;
- The user has the power to remove/add entities from this list;
- The user can select any entity from the list as the best match for that street;
- The user can contribute with properties/relations missing from an entity;
- The user can access a page with revision history for the match and propose a change;
- The user can access a history of his contributions;
- Users can see a listing of overall top contributors to the platform;
- The user will be able to contribute with a Wikipedia login.

6.2   Storyboard

According to the requirements, the web platform will have as its main screen a map where the user can select any street and visualize its associated entities. The user also has access to a discussion page were the review history can be accessed. The sketch in Fig. 3 represents this scenario. In the case where a suitable match still has not been selected for the street, the user will be able to see the list of associated entities and edit as desired. The user can also select any entity in the list as the best match for the street. This part of the interface is represented by the sketch in Fig. 4.

6.3   Software Architecture

In terms of its overall architecture, this application will try to integrate and take advantage of other collaborative projects. The raster tiles for the map and the vector layer with street information is provided by OpenStreetMap. Tiles can

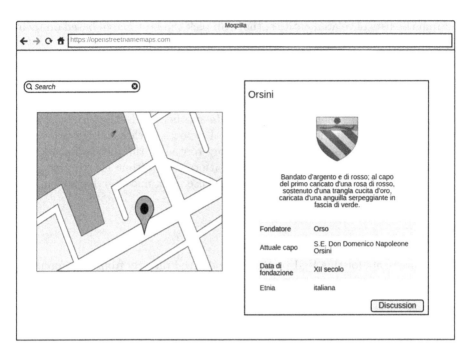

Fig. 3. Selecting a street on the web platform.

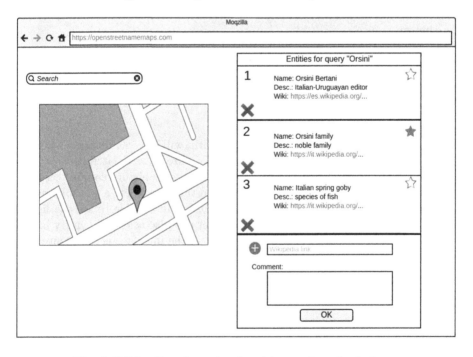

Fig. 4. Editing list of associated entities on the web platform.

be retrieved from several OSM tile servers and vector data can be efficiently obtained using the Overpass API.

Wikipedia will serve as the source for the information associated with each entity, complemented by its structured knowledge bases (Wikidata and DBpedia). The mapping between streets and entities, along with associated revision history, will be stored in a PostgreSQL/PostGIS database.

The authentication can be provided by Wikipedia OAuth service API. Users can use their existing credentials. The platform must follow the best practices that has been proven to reliably lead to community engagement and participation. For this reason, the application must keep track of who did what and when, providing the full history of changes, allowing quick undo for error and vandalism correction. All these software components will be arranged in an open source tool, freely usable and editable.

7   Conclusion and Future Work

In this paper we presented an approach to create and improve the mapping between OpenStreetMap street names and entities represented in structured knowledge bases. The ultimate goal is to generate maps where each street name can be explored in terms of the properties of its related entity, enabling a deeper analysis that provides insights into the geographic culture, history, and politics of a place. The automated mapping of entities with street names was complemented by the implementation of a custom ranking algorithm, which improved upon the default ranking obtained from the knowledge base. After identifying the limitations of an automated solution, we designed a web platform to better fulfill our goal.

This approach will enrich the semantics of OpenStreetMap and Wikipedia, creating a new, machine-readable information layer that connects toponyms with known entities. This knowledge can be returned to the community to be freely used for general exploration, tourism, and research. In the short term, we will make this mapping available online to be easily processed with other tools and linked to other knowledge bases. In parallel, we will continue the development of the web platform and its algorithms, extending it from the Italian case to other languages. Increasing access to this kind of cultural geo-information will trigger new and engaging ways to navigate and understand the intricacies of places.

References

1. Ballatore, A., Bertolotto, M., Wilson, D.: Linking geographic vocabularies through WordNet. Ann. GIS **20**(2), 73–84 (2014)
2. Ballatore, A., Wilson, D., Bertolotto, M.: A survey of volunteered open Geo-Knowledge bases in the semantic web. In: Pasi, G., Bordogna, G., Jain, L. (eds.) Quality Issues in the Management of Web Information. ISRL, vol. 50, pp. 93–120. Springer, Heidelberg (2013)

3. Ballatore, A., Mooney, P.: Conceptualising the geographic world: the dimensions of negotiation in crowdsourced cartography. Int. J. Geogr. Inf. Sci. **29**(12), 2310–2327 (2015)
4. Bizer, C., Lehmann, J., Kobilarov, G., Auer, S., Becker, C., Cyganiak, R., Hellmann, S.: DBpedia-A crystallization point for the Web of Data. Web Semant. Sci. Serv. Agents World Wide Web **7**(3), 154–165 (2009)
5. Brown, M., Sharples, S., Harding, J., Parker, C., Bearman, N., Maguire, M., Forrest, D., Haklay, M., Jackson, M.: Usability of geographic information: current challenges and future directions. Appl. Ergon. **44**(6), 855–865 (2013)
6. Buscaldi, D., Magnini, B.: Grounding toponyms in an Italian local news corpus. In: Proceedings of the 6th Workshop on Geographic Information Retrieval, GIR 2010, pp. 15:1–15:5. ACM, New York (2010)
7. Cohen, S.B., Kliot, N.: Place-names in Israel's ideological struggle over the administered territories. Ann. Assoc. Am. Geogr. **82**(4), 653–680 (1992)
8. Goodchild, M.F.: Citizens as sensors: the world of volunteered geography. Geo-Journal **69**(4), 211–221 (2007)
9. Kadmon, N.: Toponymy and geopolitics: the political use-and misuse-of geographical names. Cartographic J. **41**(2), 85–87 (2004)
10. Kuhn, W., Kauppinen, T., Janowicz, K.: Linked data - a paradigm shift for geographic information science. In: Duckham, M., Pebesma, E., Stewart, K., Frank, A.U. (eds.) GIScience 2014. LNCS, vol. 8728, pp. 173–186. Springer, Heidelberg (2014)
11. Qi, S., Wu, D., Mamoulis, N.: Location aware keyword query suggestion based on document proximity. IEEE Trans. Knowl. Data Eng. **28**(1), 82–97 (2016)
12. Vrandečić, D., Krötzsch, M.: Wikidata: a free collaborative knowledge base. Commun. ACM **57**(10), 78–85 (2014)
13. Zaragoza, H., Rode, H., Mika, P., Atserias, J., Ciaramita, M., Attardi, G.: Ranking very many typed entities on wikipedia. In: Proceedings of the Sixteenth ACM Conference on Conference on Information and Knowledge Management, CIKM 2007, pp. 1015–1018. ACM, New York (2007)

Functions and Perspectives of Public Real Estate in the Urban Policies: The Sustainable Development Plan of Syracuse

Laura Gabrielli[1(✉)], Salvatore Giuffrida[2], and Maria Rosa Trovato[2]

[1] Department of Architecture, University of Ferrara, Ferrara, Italy
gbrlra@unife.it
[2] Department of Civil Engineering and Architecture,
University of Catania, Catania, Italy
{sgiuffrida,mrtrovato}@dica.unict.it

Abstract. This study deals with the problem of the negotiation between public and private actors in the urban planning, in the case study of the Plan for the Sustainable Development of Syracuse (Italy). The contribution focuses on the modalities of execution of the Plan that envisages the tool of the Public-Private Partnership (PPP). The study intends to verify the equity of the negotiation mechanism and the advantage gained by the public actor from conferring two large buildings to a Real Estate Fund. The contribution is structured in three parts. The first part provides the general programmatic and valuation frame referring to the features of the area. It describes the overall development perspectives and therefore the whole process of real estate development that would be supported by means of the contribution of the fund. The second part describes the implementation of a cash flow analysis based on a hypothesis of use of the buildings previously outlined. The third part provides the elements of the analysis of the investment that are retroactive on the design hypotheses converging on the value assessed to determine the quota of participation of the Municipality in the Real Estate Fund.

Keywords: Strategic urban planning · PPP · Real estate finance · Sustainable planning · Property funds

1 Introduction

Urban regeneration constitutes the field of urban planning where the progressive approach between collective value and individual interest has been more significantly realized. The necessity to achieve an equal integration between private and public functions has led to the diffusion of negotiation forms where the fulfilment of the economic goals on both parts constitutes the premise for the success of the operation.

The advancement of the real estate finance, and the consequent vertical integration of the functions and the activities linked to the production of the urban space, have originated the idea that the densification of the income functions is the safest tool for the financing of public works and that the implementation of "cold works" is possible

© Springer International Publishing Switzerland 2016
O. Gervasi et al. (Eds.): ICCSA 2016, Part IV, LNCS 9789, pp. 13–28, 2016.
DOI: 10.1007/978-3-319-42089-9_2

only through the envisagement of "warm works"[1]. The science of valuation includes some real estate finance approaches, but goes beyond its intentions, giving a "further" sense to the feasibility calculation. In particular, in the field of investment planning at the urban scale in fragile contexts from the point of view of the balance between functional and landscape features, the consideration of wide spans and the availability to appreciate their values is required, instead of the capability of foreseeing the prices.

More specifically: the calculation of the feasibility of real estate investments, mainly based on indexes such as IRR and PBP, privileges high transformation intensity interventions and high exploitation rate activities. In this way, it leads urban programming to relax the restraints on the most qualified real estate stock, which is characterized instead by low capitalization rates and is selected by operators more available to carry out virtuous forms of real hoarding.

In the spirit of the *economic-valuative logic* embodied in the perspective of the "beautiful city", it is possible to assume the opposite point of view: first, the typical value terms of the urban, landscape and socioeconomic context from which the assets acquire their real estate identity are identified. Second, design hypotheses coherent with this identity are elaborated; the conditions for the convenience and financial feasibility are verified, establishing the invariants and generating the possible and most significant alternatives; the profile of the most suitable economic subject to carry out this type of investment is outlined. In general, the combination of restraints and opportunities in the historic city helps to select economic operators motivated by a perspective typical of the owners instead of the entrepreneurs, a perspective that participates in the urban dynamics instead of the dynamics of capital assets.

2 Case Study

Nowadays the urban regeneration process of Syracuse almost completely developed the potentials of the historic center, Ortigia, consolidated as the core of the architectural-landscape values and the brand of the city and its province. Such a fast achievement has spread towards the adjacent areas similar development perspectives, showing their territorial and economic-real estate complementarity. In particular, the coexistence of two clearly distinct harbor areas – the Porto Grande harbor west, and the Porto Piccolo East – close to Ortigia's privileged waterfront, has fostered different ventures linked to the real estate finance tools and the possibility to obtain European financing.

Starting from the direct grant financing model for the redevelopment of the building stock (Special Law for Ortigia, 1976), more dynamic approaches based on foreign financing and consultation have been subsequently carried out. In this field, the objectives of the architectural-urban quality, the social welfare, and the economic growth have been integrated into a strategic vision concerning the use of financing and the success of the public investment.

---

[1] "Cold works" are usually infrastructures with no appreciable cash flows, while "warm works" have a quite high cash flow capable to pay back the investment costs.

Among the tools that may be framed inside this approach, the following are worth mentioning: The Community Venture Program (PIC) Urban Syracuse (1996 and 1998–2002), PIT n. 9 Ecomuseum of the Mediterranean inside the POR Sicily 2000–2006, Ortigia's Master Plan and the 2003 Sustainable Development Plan.

In the latter, the project for the redevelopment of the waterfront includes the realization of the touristic harbor and some residential buildings in the area of the S. Antonio Pier, the realization of the Urban Center in an historic building to be restored, the extension of the Pier itself to improve the mooring of big cruise ships, and the contextual realization of the Port Terminal, a mixed use and public-private management building.

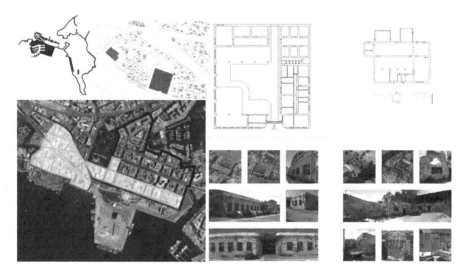

Fig. 1. The area, the sample and the two properties. (Color figure online)

The implementing program envisages the use of the Public-Private Partnership (PPP) tool and the institution of the real estate fund in which the Municipality participates by conferring two public buildings: one of them, the smaller and more ancient, is currently in ruin; the other one, more recent, is currently used as a garage and must be renovated. The percentage of participation in the fund is given by the current value of the two buildings, valuated as the 16 % of the total value.

The proposed contribution intends to verify the coherence of this valuation considering the potentials of the context in the perspective of the regeneration process currently on its course. From point of view of the urban quality and the architectural value, this process has been simulated inside two areas that are different part of the Umbertino district, and part of the Borgo S. Antonio district. It is possible to calculate the surplus of value corresponding to each of the generated strategies in the hypothesis that this process may be implemented by activating the equalization and compensation mechanisms. This expected surplus contributes to the dimensioning of the financial and monetary variables included in the calculation of the value of public buildings to be conferred to a real estate fund.

This experimentation is aimed at integrating two distinct models of real estate finance: the first one is generalized, that means referred to the conditions of access to the equalization mechanisms with which the owners are stimulated to realize the works of the redevelopment of the urban fabric (Fig. 1 left). The second one is specific and refers to the two public buildings (Fig. 1 right) the Municipality confers to the real estate fund obtaining in return the right to share the profits of real estate investment, proportionally to the value of the two buildings themselves.

3 Methods and Procedures

3.1 Urban Perspective: The Redevelopment of the Borgo S. Antonio Quarter

The urban-finance pattern concerns the convenience for the householders to join the redevelopment program, thus contributing to its partial self-financing. The proposed pattern is carried out through the analysis, valuation and programming stages.

Analysis. The detailed analysis of the urban context was carried out identifying a sample of 21 Blocks, comprising 89 Architectural Units (AU) and 287 Functional Units (FU) [11].

Valuation. The valuation stage included the standard rehabilitation costs and revenues in terms of consequent real estate market value increase, according to a set of 9 different strategies, ranging from conservative to transformative (the first one regards the status quo). The resulting real estate value surplus is considered the base for the internalization of the externalities due to the general enhancement program, according to each of the supposed strategies. Given the natural trade-off between urban landscape layout and economic feasibility (financial sustainability, self-financing ratio), the pattern allows the decision makers to arrange the strategy that best matches the two conflicting performances.

Programming. The strategies were arranged by progressively relaxing the constraints concerning:

- the usage of the ground floors – from the current to the most profitable layout;
- the intervention categories – from maintenance to renovation;
- the cubage increase of the AUs – from the least to the most valuable one, once established some invariants.

It is possible to set the physical mechanisms able to make permission fees and incentives fit the real estate value surplus for each AU, basing on this knowledge and valuation system. Permission fees or incentives S can be calculated as follows: $S = [V^* - v - k(1 + wacc)^m (1 + r')^n] / [(1 + wacc)^m (1 + r')^n]$, where: V^* is the final real estate value (after transformation), v is the current real estate value, k is the building cost, $wacc$ is the average weighted cost of capital, r' is the business risk premium rate, m is the short loan life and n the transformation period.

3.2 Public Properties Appraisal

In the perspective of the Public-Private Partnership (PPP) adopted by the Municipality to integrate into its urban policy the private venture opportunities, the appraisal of the public properties needs to fit the fair value for the public and the feasibility for the company. The PPP provides for the creation of a joint venture in which the Municipality participates by conferring two important properties to a real estate fund. Moreover, the Municipality shares the capital and profits in proportion to the value of the transferred assets. The company assumes the obligation to realize the full amount of public and private works in exchange for the planning permits for the construction of new buildings and the restoration of the existing ones.

The appraisal of the two properties has been carried out by means of the residual value method as explained in the previous paragraph: $S = [V^* - k^*(1 + wacc)^m(1 + r')^n]/ [(1 + wacc)^m(1 + r')^n]$, where the symbols have the same meaning; k^* is the building cost including the planning permission fees S, which in this case does not vary. New functions have been envisaged for both properties according to the development perspectives as outlined in the nine strategic layouts; the building costs have been calculated in detail, considering all costs for the renovation works as listed in the preliminary design drafts that have been specifically drawn up.

3.3 PPP – Public Private Partnership

The Green paper published by the European Commission in 2004 [6] affirms that the term PPP refers to forms of cooperation between public authorities and the world of business, which aim to ensure the funding, construction, renovation, management or maintenance of an infrastructure or the provision of a service.

The main characteristics of PPP are the long duration of the relationship between public and private partners; the funding of the investment, which in part is from the private sector, but it can be partially funded with public funds; multiple tasks assigned to private players; the distribution of risk between private and public sectors.

The same document distinguishes between contractual PPP and institutionalized PPP. The first involves the creation of a mixed equity company, held jointly by the public and the private partner while the second concerns to a partnership, which is based only on contractual connections between different actors.

PPP and project finance have developed around Europe, mainly in Spain, Portugal, France, Italy, UK, and Germany, which account for 95 % of all European PPPs by number. Italian examples concerns turnpikes, car parking, underground railway lines, new hospitals replacing the old ones, kindergartens, sports facilities, cemeteries and gas distribution networks [5]. The turmoil of the market has slowed down the number of projects and initiatives under the umbrella of PPPs.

In the last period, practitioners and academics began paying more attention to PPPs as viable solutions to urban challenges and urban regeneration, stressing the benefit of those approaches. The major attraction is to bring the private players and their finance into the project. The partnership can also improve the understanding of the different approaches adopted by the two partners and their various perspective, combining responsibility for its community and shareholders for the investments. The PPP can also improve the management of the process, which underpins the success of regeneration projects. The private sector brings expertise and capital to the redevelopment projects so that actions can be delivered quickly, with a long-term sustainability.

3.4 Italian Real Estate Investment Funds

Italian Real Estate Investment Funds (REIFs) were introduced in 1994. The REIFs are a pool of assets represented by units relating to a plurality of investors and managed on their behalf and in their interest by a Saving Managing Company (Società di Gestione del Risparmio – SGR).

Independent valuers assess the fair value of the assets included in the funds twice a year. Italian REIFs mainly invest in [10] real estate assets, real estate rights and shares in real estate companies; listed and un –listed financial instruments; bank deposits; receivables and securities embedding receivable and any other asset that is traded and that has a value determinable on a biannual basis.

Italian REIFs can be categorized according type of investors, which includes public funds ("retail funds"), namely when every investor could participate in the fund and private funds ("institutional funds"), in which only "qualified" investors (as banks, property investors, etc.) have shares of the funds and speculative funds. The speculative funds ("hedge funds") are highly risky funds with a few investors, greater decision-making freedom and flexibility.

Based on how the funds were established, REIFs are divided into ordinary funds (or "blind pool funds"), which collect money through subscription first, and then invest in properties; contribution funds (or "seeded funds") buy properties or interests in real estate company and then sell shares in the funds; mixed if shares are subscribed both with properties and money.

Finally, funds can be created with contribution of private property/money or through public properties. The latter is formed by transferring public properties in the ownership of public administration and other territorial public bodies. Italian REIFs can be closed-end funds, and the entire amount of the capital is determined during subscription and cannot be modified; semi closed-end funds are allowed to increase or change their value by issuing new shares.

The growth since 1999 has been rapid and at the end of June 2015, there were 262 REIFs and the overall investment value was €31.45 M with a cut of 0.4 % in comparison to the previous year. After a constant growth during the last decade, the market now seems quite stable. The Italian REIFs have targeted both private and institutional investors. 24 funds are open to all investors, which represent the 9 % of all market, while 238 are reserved for qualified investors, and take 91 % of the market. This last type of investment is the largest component of the Italian funds, both in terms of the number of funds and value. Funds with contributions represent now the 71 % of the total market, showing a decrease in numbers in the last four years.

The majority of funds invests in offices (48,8 %), followed by retail (13 %) and residential property (13,4 %), showing a poor diversification. Investments in the industrial sector have been reduced in the last few years (3,4 %) while RSA is 1,5 %. The logistic sector represents 2,9 % and other sectors amount 12,7 %.

The geographical area of Italy that plays the most important role in REIFs [1] is the North West (44 %), followed by the Centre (34 %), the North East (12 %) and South with the Islands (8 %). Foreign investments are only 2 % and have fallen (in relative terms) with the increase in new domestic investments.

The need to diversify more fully is the main issue for Italian funds [9], in terms of sector and regional allocation. Currently, the funds seem to be poorly diversified and prone to risk with investments concentrated mainly in offices in Rome and Milan and trophy buildings, located in the best locations in these cities or other main centers.

3.5   Jessica

JESSICA is the acronym of *Joint European Support for Sustainable Investment in City Areas*. This initiative of European Commission works in collaboration with the European Bank for Investment Bank (EIB) and the Council of Europe Development Bank (CEB), and it aims at promoting sustainable urban development and regeneration. European countries can choose to use part or some of their EU structural fund allocation in the revolving fund to channel the financial resources and the investments into projects concerning urban areas. JESSICA supports projects in areas as urban infrastructure, heritage or cultural sites, redevelopment of brownfield sites, creation of the new commercial floor space for SMEs, IT and/or R&D sectors, university buildings, energy efficiency actions. Contributions of the European Regional Development Fund (ERDF) are assigned to Urban Development Funds (UDFs), which invest them in public-private partnerships or other projects included in an integrated plan for sustainable urban development. These investments may be granted in the form of equity, loans, and/or guarantees.

Alternatively, managing authorities can decide to allocate funds to UDFs using Holding Funds (HFs) designed to invest in several UDFs. Owing to the revolving nature of the instruments, investment returns are reinvested in new urban development projects, reusing this way public funds and promoting the sustainability and impact of EU and national funds. Forms of investment used by the FSU can be of three types, namely loans, guarantees, equity.

JESSICA has many advantages for the development of strategic action in urban areas. For example, as a financial mechanism, it offers the opportunity of repayable assistance from the structural fund for investments, which should generate returns and so pay back the investors. This could also be improved with the mix of structural funds with other sources of funding and the ability to involve other public and private investors, resulting in a better support to a great number of projects. The instrument is characterized by flexibility, in terms of the use of funds (possible use of resources in the form of loans, guarantees or equity investments). JESSICA also implements the expertise of the private sector, banks, structural fund managing authorities, cities and towns in project implementation and management. These instruments are also a catalyst for the establishment of a partnership between different countries, cities, EIB, CEB, banks, investors in order to face the problems, which affect the urban areas.

JESSICA helps to increase efficiency and effectiveness of financial instruments, creating stronger incentives for the implementation of an urban development project, thanks to additional financial resources for PPP with a strong emphasis on sustainability.

3.6 Risk Approach

The perception of risk on the part of real estate investors plays an important role in the investment decision. In the economic literature, the ways by which the investors choose in risk condition has been extensively analyzed. In particular, this literature shows as the investors choose based on the tradeoff between the possible expectations of profit that the investment is capable of generating and the perception of risk associated with this investment [17].

These studies show that investors take a different behavior in relation to the decision time for the investment. In fact, normally in the short time, the behavior of investors is what in literature is called "mixed feelings" driven by irrationality. In the long run, the investor has the ability to contain the irrationality, find comfort in a complex information and then choose according to the principles of economic rationality [17]. The investors usually make a long-term choice, in fact, if on one hand, it is due to a substantial financial commitment and the other has long waiting times to achieve revenue objectives. It is obvious as that in this case the risk grows in proportion to the commitment of capital resources and time. The general investor, in any case, regardless of its particular attitude has the task of minimizing the uncertainty related to the activity to be undertaken, and if you can turn the uncertainty into risk. The types of risk of real estate are as follows [14]:

1. *Market risk*: the risk that the operating cash flows do not develop as expected, due to changes in demand and supply; structural changes in the population; Macroeconomic factors such as interest rates, growth and inflation rate.
2. *Environmental risk*: the environment that will have a negative impact on property value (public health or a factor that reduces the attractiveness of the property).
3. *Risk related to construction*: problems in a building that go beyond the level of maintenance, or the wrong design of a building, which can be reduced at the purchase.
4. *Legislative risk*: the impact that many laws determine on real estate investments, in particularly in the field of taxation, transaction costs, regional planning.
5. *Liquidity risk*: the liquidity of a property is related to marketing period and the number of the transactions.
6. *Management risk*: it concerns several elements such as marketing issues, rent collection, maintenance planning.
7. *Financial risk*: it is related to the use and the amount of debt and the interest rates.

The traditional methods for the treatment of risk instead of objectively quantifying the perception of risk, involve the addition of a premium to the return required under "normal" conditions of risk. This additional premium results in recovery time shortened, in demands for higher returns and downsizing of cash flows.

Among the traditional methods for the treatment of risk there are: Risk Adjusted Discount Rate (which consists of an "adjustment" in the discount rate in order to bring back the recovery time in the limit of the maximum acceptable and chosen on a subjective basis); certain equivalents (the certain equivalent is the amount to be cashed, on a given date t, which makes indifferent the choice between that same amount to be cashed that can be considered reliable and the one expected from an investment) [17].

The methods are based on the analysis of probability. In this case, to the static formulation of the NPV, which assumes a certain structure and certain cash flow, and parameters for which it is impossible to construct a probability distribution, it is possible replacing a spectrum of forecasts based on specific terms. The risk control through a probabilistic approach needs to identify the risk factors that mostly affect the outcomes that can be conducted using the partitioning of the NPV technique.

This method is based on the division of cash flows, net of tax, because of its fundamental components, which are expressed as a percentage of the total present value, thus defining the relative importance of each of them.

An extension of this methodology, which limits any errors in the forecasts from the original approach, is the sensitivity analysis. The technique is based on modifying one at a time the inputs and on studying the resulting impact on the output.

In the case of probabilistic approaches, often it can be difficult to estimate directly all possible combinations of values that the various parameters can take on the base of the respective probability distributions. To solve this problem, the approach of the Monte Carlo simulation is used. Introduced in the '60 [12], it simulates a statistically high number of possible combinations of key parameter resulting from the assumption of probability distributions.

Associating each NPV its frequency or the number of times has seen the same value repeated in the simulation, it is possible to build a certain probability distribution of values of the NPV, and so determine the net present value expected from the investment. Furthermore, the Monte Carlo simulation allows overcoming a weakness of the analysis of the sensitivity, i.e. the lack of correlation between the variables under investigation. In fact, it is not plausible to assume for variables such as prices or costs, as their swing downwards or upwards can occur without affecting other parameters.

Another approach is the *Build - up approach*, which is generally used for the determination of the discount rate (where the rate is given as a sum of various risk factors). It can also be applied to determine the cap rate (cap rate = risk-free interest specific rate + risk premium r_{pBuA}), since also this rate expresses the risk resulting from the sum of several elements: $r_{pBuA} = \alpha_r + c_r + p_r + l_r + i_r + f_r + s_r + b_r + t_r + m_r + ecp_r$, inflation, context, property, lessee, insurable, financial, system, building, test, management, exchange-country-political, respectively. The risk-free interest rate can be identified by the value assumed by a bond with the highest rating, normally a debt instrument issued by a State with a strong economy. The identification of the second component of the rate presents critical issues related to the identification and quantification of the factors that influence the specific risk premium. The factors affecting the specific risk premium can be clustered into three groups [14]: the economic and financial factors, the type of property and the specific risk of properties (which can be grouped into six categories: location, physical characteristics, legislation and taxation, the rental and contractual situation). The *RER (Real Estate Risk) pattern* determines an overall risk coefficient of a real estate in relation to the risk that can be attributed to the characteristics of the building tenant, to the context characteristics and endogenous factors of a building unit [4]. The RER aggregates the risk components into three main factors in the hypothesis that these have equal influence on the size of the risk: *Tenant specific risk, Context specific risk, Property specific risk*: $r_{pRER} = (t_r + c_r + p_r)/3$. For measuring *tenant specific risk* t_r: arises from the assignment of a rating to four different factors: the cyclical nature of the industry belongs the tenant; commodity characteristics of the activity; the solvency degree can be attributed to the company; the numbers of tenants.

For measuring *context-specific risk* (exogenous characteristics of the goods) c_r must be assessed: the rank of the city in which the property is located; the position of the same in the urban context; the intrinsic characteristics of the local housing market. The measurement *property-specific risk* p_r includes the size of the building, the typological and qualitative characteristics of the building, and the replaceable of the building.

The RER model lends itself not only to the valuation of the risk of a single real estate investment but with the appropriate methodological cautions, also to the appreciation of the risk of a real asset portfolio.

4 Results and Discussions

4.1 Generating Strategies: Urban Equalization and Regeneration

Basing on the architectural, technological, materials and urban features of the buildings, the restraints and the attitudes to transformation, and therefore, the intervention categories ranked according to eight strategies – from the most conservative to the most transformative one – have been defined for each functional unit (UF) (Fig. 2). Figure 3 summarizes the results and the relations between the main variables, which provide useful indications for the definition of the risk model.

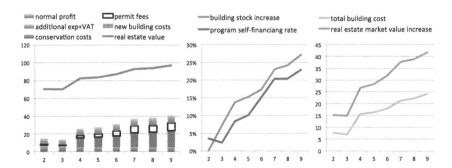

Fig. 2. Comparison between the eight strategies in terms of: cost/revenue components; building stock and program self-financing rate increase; building costs and real estate value increase. (Color figure online)

4.2   Real Estate Development of the Two Properties: Appraisals

The basic elements for the valuation of the real estate investment in the transformation of the two buildings concern the layouts of the buildings themselves defined on the grounds of an outline scheme, and the corresponding costs and revenues. The costs are equal to the sum of the prices of the elementary works b_k defined basing on the Official Price List for Public Works of the Sicilian Region (2013) applied to the quantities q_k, and its unit prices p_k: $c^* = \sum_k q_k p_k$ (Fig. 3).

areas	masonries	horiz struct	finishigs	doors/windows	equipments	dump	total costs
entrance	€ 932	€ 4.649	€ 11.427	€ 2.933	€ 1.559	€ 43	€ 21.543
hallway	€ 1.165	€ 18.450	€ 11.099	€ 2.123	€ 6.406	€ 89	€ 39.331
warehouse	€ 1.459	€ 9.847	€ 17.258	€ 4.215	€ 3.985	€ 55	€ 36.819
toilets	€ 1.201	€ 8.173	€ 15.042	€ 5.370	€ 31.272	€ 46	€ 61.104
plaza	€ 8.927	€ 44.604	€ 57.488	€ 19.713	€ 16.412	€ 688	€ 147.832
restaurants	€ 11.669	€ 57.058	€ 50.722	€ 39.924	€ 16.182	€ 808	€ 176.362
offices	€ 5.572	€ 28.030	€ 40.665	€ 18.576	€ 10.371	€ 267	€ 103.481
public area	€ 5.589	€ 101.987	€ 68.104	€ 31.200	€ 11.820	€ 743	€ 219.442
total costs	€ 36.513	€ 272.798	€ 271.804	€ 124.053	€ 98.006	€ 2.738	€ 805.913

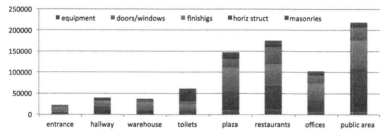

Fig. 3.   Costs by functions and work type. (Color figure online)

The revenues r^* are given by the sum of the market values V_i of the several parts of the buildings depending on the dimension a_i, and the unit value v_i corresponding to the use f_i; $v_i = n_i/r$ is the capitalization value at cape rate r of the Net Operating Income ($n_i = g_i(f_i) - e_i$) envisaged basing on the unit current market annual rents g_i for the new buildings, net of the owner's expenses (e_i), prudentially considered as the 50 %: $r^* = \sum_i v_i$: r is given by the medium rate r' corrected basing on the plus/minus-valuation coefficient, which takes into account the risk factors h indicated in paragraph 3.6: $r = r'h$ (Fig. 4). The residual value of the largest property – calculated at a cap rate of 3 % for the capital asset value, and at a *WACC* of 8,5 % supposing a construction period of two years – is 1,7 million, which is quite greater than the one defined by the independent valuer of the property funds. The same difference occurs in the case of the smaller property. A further in-depth analysis regarding the larger property is illustrated in the following paragraph.

4.3   Real Estate Finance: Scenario Analyses

The valuation of the investment is developed from a REIF point of view using a Financial Approach based on Discounted Cash Flow Analysis (DCF). The model aims to consider, from the perspective of a real estate fund, the opportunity of a participation

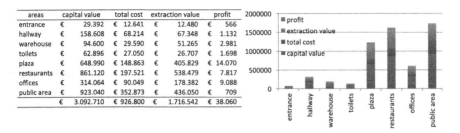

areas	capital value	total cost	extraction value	profit
entrance	€ 29.392	€ 12.641	€ 12.480	€ 566
hallway	€ 158.608	€ 68.214	€ 67.348	€ 1.132
warehouse	€ 94.600	€ 29.590	€ 51.265	€ 2.981
toilets	€ 62.896	€ 27.050	€ 26.707	€ 1.698
plaza	€ 648.990	€ 148.863	€ 405.829	€ 14.070
restaurants	€ 861.120	€ 197.521	€ 538.479	€ 7.817
offices	€ 314.064	€ 90.049	€ 178.382	€ 9.088
public area	€ 923.040	€ 352.873	€ 436.050	€ 709
	€ 3.092.710	€ 926.800	€ 1.716.542	€ 38.060

Fig. 4. Total costs, revenues, residual value and profit by functions and work type. (Color figure online)

in the process of transformation and exploitation of the area under study. The DCF, normally used for valuation purposes in real estate development operations [16], already analyzed over a span of life the trend of revenues (lease or sale) and costs (constructions and enhancement costs, as well as operational costs for the maintenance of the efficiency of the structures) discounting the different cash flows according to their respective discount factors. This more accurate valuation takes into account the developer/investor likely cash flow, so that the capital outstanding at any point in time is identified, and a precise estimation of finance charges can be made. The use of DCF analysis allows the valuer to reflect the different moment of refurbishment and selling period, as well as allowing for any change in value or cost over the period.

The model is based on certain required assumptions about the periods of time for the refurbishment and the sale of the properties; the investment costs; the calculation of the discount rates using the Weighted Average Cost of Capital (WACC), and using assumptions about equity and debt and their costs; property prices.

The DCF has allowed the fund to identify the fair value of the properties for the property funds, which is the most likely HBU (Highest and Best Use, assuming all technically possible uses, legally permissible and economically feasible). In this output (the fair value) it was conducted a sensitivity and risk analysis, because of the variation of this value caused by the independent variables. The risk analysis was developed regarding the capitalization rate, the rent, the profit of the developer and the WACC.

The capitalization rate and the rent were extrapolated through an analysis of the local property market, which produced average prices of 8 Euro/m^2/month for services, 12 Euro/m^2/month for office and 15 Euro/m^2/month for the commercial sector. A similar analysis of the property market was undertaken to ascertain the cap rate, which is set at 2,25 %. Using an income approach, the cap rate, and the rent has been used in order to identify the sale price of the completed refurbishment and this input was used in the DCF model. The expected profit from the development has been identified with a market analysis, which shows that the expected the profit on a net present value (NPV) basis is around 15 % of the GDV, the gross development value.

A multi-factorial approach, well established and widely adopted, was used for determining the WACC using different component of risk. Specifically, the "liquidity risk" is a function of all those factors (level of asset marketability, size, etc.) that potentially hinder the selling of the assets. The "sector or investment risk" expresses the specific investment risk, thus, it is a function of both the characteristics of the area (location, local market...) and the real estate and economic characteristics of the buildings (type, functional mix, conditions of lease and maintenance). Finally, the "urbanistic risk" is an expression of the urban framework (presence of planning permission, planning process, uses that do not comply with the urban law). All these components have been added to the risk-free rate (Government Bond). Thus, the cost of equity to be used in the WACC construction, with the ratio Debt/Equity 50/50, is 8,15 %, considering a Government Bond rate of return of 5 % and an urbanistic risk of 3 %. The cost of debt is 3,70 % (considering a Euribor, six months adding up a spread of the bank).

The following steps of the model look at how uncertainty can be accounted for in the DCF model. This can be achieved by recognizing that the variables used in the DCF are not single values, but they are a possible range of figures that can be represented, statistically, by a probability distribution. In order to consider uncertainty within the development process [2, 3, 8, 15].

The idea that a valuer can exactly estimate price is sophistry [7]. Conventional development appraisal models that give single point outcomes fail to direct the user to understand the importance of the inputs, and the uncertainty of each, in the overall process. In the Monte Carlo simulation, the same underlying analysis is undertaken, but adding the identification of the characteristics of the uncertainty that applies to each of the inputs. The end effect of such an analysis is a better understanding of the process and the interaction during the development process, of the developer's and investor's preferences and move towards a more informed choice.

There are advantages and disadvantages of adopting the Monte Carlo Simulation, even if it helps optimizing design specification, identifies the proper project, and helps the developer and/or investor to understand and compare the risk involved in carrying out a specific project [18]. Among the criticisms, most of the issues arise from the fact that information is not always appropriate and sometimes subjectivity plays a great role in the definition of probability distribution between variables. In our model, the ascribed inputs were ascertained with a market analysis but during the real estate market crisis, which is still ongoing, the amount of direct comparable information falls in number and quality, so that the input are less certain, and it is possible to ascribe a degree of uncertainty to the model. The sensitivity and the simulation analysis tests the robustness of the single point estimates and produces a range of possible results, the mean of which can be considered the expected value of the investments.

All the variables have been described by a normal distribution, using a minimum and maximum value, which have been defined by the market analysis. The distribution of the rent has been set as a triangular distribution, providing three absolute figures: the most likely, the minimum and the maximum. Although the normal distribution is statistically more robust, French and Gabrielli [7] argued that the triangular distribution was a more appropriate approximation of the thought process of an appraiser. A negative and positive correlation have been introduced in the simulation, defying the interrelationships between the variables. All those parameters are shown in the Table 1.

The program Crystal Ball was used to run the DCF 10.000 times, using the Monte Carlo simulation. This simulation produced the outcomes as illustrated in Fig. 5.

The expected value of 1.592.879 € is not that different from the 1.562.065 € that we get using the discrete DCF but the figures provide additional information about the

Table 1. Inputs, distribution and statistics.

Input	Distribution	Mean	SD	Minimum	Maximum	Correlation
Cap rate	Normal	2,75 %	0,30 %	2,25 %	3,35 %	Rent
Rent	Triangular	92.781	–	83.503	102.059	Cap rate, Profit, WACC
Profit	Normal	17 %	1,50 %	10 %	25 %	Rent
WACC	Normal	8,15 %	0,82 %	5 %	10 %	Rent

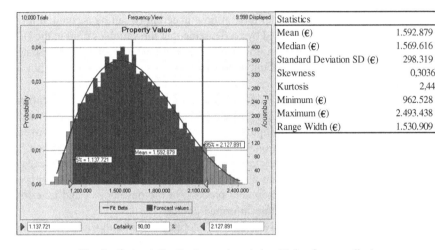

Fig. 5. Output distribution and statistics (Color figure online)

uncertainty of the results. The skewness (0,30) represents the degree of asymmetry (positive): in this case, the distribution (and so the output) is moderately skewed right, as its right tail is the longest and most of the distribution is on the left. Mean and median are different, showing that there is a higher probability that the expected value could be less than 1.562.065 € and the standard deviation (298.319 €) is a representation of the uncertainty.

As expected the outcome of this simulation is to provide the real estate fund with a display output range from € 962.528 to € 2.493.438. Statistically, the majority of outcomes (95,45 % percent) lies within two standard deviations of the mean. In this case, the important range is from € 972.978 to € 2.166.254 and it is a distribution that helps the fund to assess the uncertainty.

In terms of the sensibility of the inputs, a Tornado analysis has been performed in Fig. 6 and it shows a Tornado representation. The variables with the widest ranges are listed at the top of the diagram. These variables will likely cause the greatest variability in the possible property value (represented by the scale at the top). The property value is most sensitive to changes in the top two: capitalization rate and rent. Therefore, these variables can be defined as assumptions in Monte Carlo.

The other variables have less impact on the analysis, and they will not significantly affect the results.

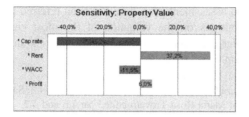

Fig. 6. Sensitivity analysis.

5 Conclusions

The proposed study involved the case of the application of a PPP process for the implementation of some addresses for Sustainable Development Plan of Syracuse. The study is particularly interested in the comparison of two different approaches to the valuation of public assets that can be involved in urban development policies through the implementation of the typical tools of real estate finance.

The context studied gets through a gentrification process started from the previous one and quite successful that affected Ortigia. As a result, the set of risk factors implemented in the robust financial tests assessing the investment's risk/return profile in the shadow of the logic of the discount rate, can be relaxed with reference to the future perspectives in the light of the spirit of the enhancement rate taken into account by the capitalization rate.

The dialectic between the discount rate and capitalization rate has been well known in the economic literature and the theory of capital since the Keynesian approach. The law of the propensity for investment focuses on the difference between interest ratio (the cost of capital) and marginal efficiency of an investment (more like to cap rate as a general category). Such an approach is addressed by Hicks [13], assuming the difference between the standard stream and the expected stream interest ratios as "crescendo" if positive, as "diminuendo" as negative.

In our case study regarding the redevelopment of two public properties, it is typically "crescendo" because of the positive difference between the value of the financial analysis (DCF) and that attributable by a more immediate property valuation (residual value). In fact, the residual value is reported to be 1.7 million euros. This value, in the DCF analysis and in its integration with a sensitivity analysis amounts to 1.57 million, then lower than that previously estimated. In addition no doubt that the developer is influenced by the best information available on the real estate market and the ability to analyze such information to define his strategic behavior in the future.

The valuation process was divided into two basic steps: one is the valuation, focusing on the costs-revenues surplus given a set of eight scenarios, ranging from the conservative to transformative, reported in a sample consisting of 89 buildings. This valuation provided the relation between the volume increase in the sample area and the self-financing ratio of the program. On this basis, the valuation of the two public properties has been carried out basing on the residual value, which provided early information about the value of the properties, in a general optimistic perspective.

The entire proposed assessment procedure has been referred to the theoretical and methodological contributions on the valuation of the most current and significant investments. In particular, the contribution focused on the risk and its modeling of the entire financial valuation.

Acknowledgements. Laura Gabrielli edited par. 3.3, 3.4, 3.5 and 4.3; Salvatore Giuffrida edited par. 2, 3.1, 3.2, 4.1 and 5; Maria Rosa Trovato edited par.3.6 and 4.2.

References

1. Assogestioni: Fondi immobiliari. Rapporto semestrale, June 2015
2. Byrne, P., Cadman, D.: Risk, Uncertainty and Decision-Making in Property Development, 2nd edn. E & FN Spon, London (1996)
3. Byrne, P.: Fuzzy analysis, a vague way of dealing with uncertainty in real estate analysis? J. Property Valuat. Invest. **13**(3), 22–41 (2015)
4. Cacciamani, C.: Real Estate. Economia, diritto, marketing e finanza immobiliare, III edizione. Egea, Milano (2012)
5. Cori, R., Giorgiantonio, C., Paradisi, I.: Risk Allocation and Incentives for Private Contractors: An Analysis of Italian Project Financing Contracts. Bank of Italy, Occasional Paper 82 (2010)
6. European Commission: Green Paper on public-private partnerships and Community law on public contracts and concessions (2004)
7. French, N., Gabrielli, L.: Discounted cash-flow: accounting for uncertainty. J. Property Invest. Finance **23**(1), 75–89 (2005)
8. French, N., Gabrielli, L.: Uncertainty & feasibility studies: an Italian case study. J. Property Invest. Finance **24**(1), 49–67 (2006)
9. Gabrielli, L.: Italian property funds: opportunities for investors. J. Real Estate Lit. **15**(3), 399–427 (2007)
10. Gabrielli, L.: Fifteen years of Italian real estate investment funds across different market cycles. In: McGreal, S., Sotelo, R. (eds.) Real Estate Investment Trust in Europe: Evolution, Regulation, and Opportunities for Growth, pp. 149–168. Springer, Heidelberg (2013)
11. Giuffrida, S., Ferluga, G.: Renewal and conservation of the historic waterfront. Analysis, evaluation and project in the grand harbor area of Syracure. BDC Università di Napoli, vol. 12, pp. 735–754 (2012)
12. Hertz, D.B.: Risk analysis in capital investment. Harv. Bus. Rev. 42(1), 95–106 (1979)
13. Hicks, J.R.: Value and Capital. Oxford University Press, London (1939)
14. Hoesli, M., Morri, G.: Investimento immobiliare. Hoepli, Milano (2010)
15. Kelliher, C.F., Mahoney, L.S.: Using Monte Carlo simulation to improve long-term investments decisions. Appraisal J. **68**, 44–56 (2000)
16. Kishore, R.: Discounted cash flow analysis in property investment valuations. J. Property Valuat. Invest. **14**(3), 63–70 (1996)
17. Manganelli, B.: La valutazione degli investimenti immobiliare. FrancoAngeli, Milano (2013)
18. Royal Institution of Chartered Surveyors (RICS): RICS Appraisal and Valuation Standards, RICS, London (2013)

Soil Loss, Productivity and Cropland Values GIS-Based Analysis and Trends in the Basilicata Region (Southern Italy) from 1980 to 2013

Antonella Dimotta[1,2], Mario Cozzi[1], Serverino Romano[1],
and Maurizio Lazzari[2(✉)]

[1] SAFE - School of Agricultural, Forestry, Food and Environmental Sciences,
University of Basilicata, 85100 Potenza, Italy
antonella.dimotta@unibas.it
[2] CNR IBAM, C/da S. Loja Zona Industriale, 85050 Tito Scalo (PZ), Italy
m.lazzari@ibam.cnr.it

Abstract. This paper concerns the trends assessment of the productivity values and croplands values of specific crops (*cereals (arable cereals land), vineyards, olive-growing lands*) in the Basilicata region at regional scale, from 1980 to 2013, in relation to the soil loss evaluated through the USPED method. The comparative analysis shows the interrelations between the soil loss by erosion and the economic value deriving from the erosive phenomenon affecting the croplands considered.

Keywords: Soil consumption · USPED · Land value · Erosion · GIS · Basilicata · Southern Italy

1 Introduction

Soil erosion is one of the major environmental and agricultural problem worldwide. Although erosion has occurred throughout the history of agriculture, it has intensified in recent years, also due to climate changes and extreme climatic events [1–3]. Each year, 75 billion metric tons of soil are removed from the land by wind and water erosion, with most coming from agricultural land [4]. The loss of soil degrades arable land and eventually renders it unproductive. Worldwide, about 12×10^6 ha of arable land are destroyed and abandoned annually because of non sustainable farming practices, and only about 1.5×10^9 ha of land are being cultivated [4]. The use of large amounts of fertilizers, pesticides, and irrigation help offset deleterious effects of erosion but have the potential to create pollution and health problems, destroy natural habitats, and contribute to high energy consumption and unsustainable agricultural systems. The eroded soil is mobilized to other places even outside from the cultivated areas, producing often widespread damages. In fact, erosion, not only damages the immediate agricultural area where it occurs, but it also negatively affects the surrounding environment. Off-site problems include roadway, sewer, and basement siltation, drainage disruption, undermining of foundations and pavements, gullying of roads, earth dam

© Springer International Publishing Switzerland 2016
O. Gervasi et al. (Eds.): ICCSA 2016, Part IV, LNCS 9789, pp. 29–45, 2016.
DOI: 10.1007/978-3-319-42089-9_3

failures, eutrophication of waterways, siltation of harbors and channels, loss of reservoir storage, loss of wildlife habitat and disruption of stream ecology, flooding, damage to public health, plus increased water treatment costs. The most serious of off-site damages are caused by soil particles entering the water systems and being deposited in streams and rivers [4].

In relation to these possible damage scenarios, in this study, we have considered the potential soil erosion assessed by empirical methods well-known in literature (Unit Stream Power – Erosion Deposition or USPED) and applicable in the contexts of study examined [5]. So, in this paper, we propose a methodology to assess the economic costs of soil erosion applied to arable cereals lands, olive growing lands and vineyards considering the study sample area of the Basilicata region (southern Italy). These crops have been selected because the territory of the Basilicata region is characterized by a historical agricultural vocationality based primarily on these three types of crops.

It is proposed also to verify the possible economic impact of soil erosion under a general regional trend, from 1980 to 2013, of decrease of agricultural areas, with particular reference to the arable areas in cereals and productivity. In fact, the estimation of costs of soil erosion is an issue of fundamental importance in view of the current worldwide discussions on sustainability.

2   Materials and Methods

In order to perform the croplands and productivity values trends assessment, datasets analysis was carried out by utilizing the time frame containing the smallest data gap within each database used in this work. Using the different datasets - related to the databases realized in order to correlate the historical data and the recent ones – an analysis of the trends and the valued soil loss in the regional territory located in the Mediterranean area was performed. The datasets used have been divided in relation to the different sources and contents of the themes addressed by making them concern into specific following databases:

- **Database (1): Land Values and Croplands quotation price** and available period:
  - *Land Values Trends (Basilicata)* – dataset coming from INEA – *National Institute of Agricultural Economy* – available period: 1960 – 2014;
  - *Croplands quotation price* (**Basilicata**) – dataset coming from *Italian Yearbook of Agriculture, Vol. XLIV - L, 1990 – 1996.*
- **Database (2): Utilised Agricultural Area (UAA)** and available period:
  - *Utilised Agricultural Area (UAA)* (**Basilicata**) – dataset coming from ISTAT (National Institute of Statistics) - *Italian Agricultural Censuses* – available period: 1970 – 2014;
  - *Utilised Agricultural Area (UAA) – Arable cereal lands, vineyards, olive-growing lands* (**Basilicata**) – dataset coming from ISTAT – National Institute of Statistics - from *3rd Italian Agriculture Census* to *6th Italian Agriculture Census* – 1980, 1990, 2000, 2010.

- **Database (3): Croplands - Agricultural Surface and Production** – utilised in order to evaluate the Productivity value at regional scale. The resulting datasets are the following:
  - *Utilised Agricultural Area* – *Arable cereal lands (1999 – 2012), vineyards (1990; 1999 – 2012), olive-growing lands* (1990; 1999 – 2011) (Basilicata) – dataset coming from ISTAT (National Institute of Statistics) – *Agriculture section: "Surface and Production", 1999 – 2011.*
  - *Utilised Agricultural Area (UAA)* **(Basilicata)** – *Arable cereal lands* (2013), *vineyards (1990; 2013), olive-growing lands (2012 – 2013)* – data coming from INEA – National Institute of Agricultural Economy;
  - **Productivity values - *Arable cereal lands*:**
    time frame: 1997 – 1998: data coming from INEA dataset;
    time frame: 1999 – 2011: data resulting from evaluation deriving from ISTAT *dataset;*
    time frame: 2012 – 2013: data coming from AGRIT - Statistic Program by Mipaaf (Ministry of Agriculture, Food and Forestry Policies), Italy.
  - **Productivity values – *Vineyards*:**
    time frame: 1990; 1999 – 2011: data resulting from evaluation deriving from ISTAT dataset;
    time frame: 2012 – 2013: data coming from INEA dataset.
  - **Productivity values – *Olive growing-lands*:**
    time frame: 1999 – 2012: data resulting from evaluation deriving from ISTAT dataset;
    time frame: 2012 – 2013: data coming from INEA dataset.
- **Database (4): Average Agricultural Values** for the following croplands:
  - *arable cereal lands* (intensive and extensive crops), *olive growing-lands* and *vineyards* – dataset coming from the Basilicata Region – Department of Agriculture, Rural Development, Mountain Economy – *Provincial Commission for determining compensation for expropriation and the average agricultural land values* – Available period: 2008 – 2013.
- **Database (5): GIS geodatabase: Corine Land Cover 2012 – 4th level -** shapefile coming from ISPRA – *Superior Institute of Environmental Protection and Research,* Italy.
  - Layer derived: layer of the Basilicata region used to extrapolate the different croplands areas/polygons of arable cereal lands, olive growing-lands and vineyards.

In order to carry out a spatial analysis of the amount of land subjected to erosion phenomenon, a quantitative analysis has been based on either the spatial and economic aspect. Indeed, the Corine Land Cover 2012 – 4th hierarchical level – has been used to define the areas interested by the different crops considered in this study, in which the USPED method has been applied to show the amount of land subjected to the potential erosion phenomenon in 2012. From this analysis, a comparative economic analysis has been carried out in order to evaluate the variation of the crop-productivity and then, consequently, of the land value by comparing the trends land values assessment from 1980 to 2013 with the scenario represented by the analysis of 2012.

2.1 Methodology - USPED

The Unit Stream Power Erosion and Deposition model (USPED) has the basic model structure of the USLE and RUSLE [6, 7].

$$A = R \times K \times (LS) \times C \times P \tag{1}$$

where, A is the computed site soil loss (Mg ha^{-1} year^{-1}), R is the rainfall erosivity factor (MJ mm ha^{-1} h-1 year^{-1}), K is the soil erodibility factor (Mg h MJ^{-1} mm^{-1}), L is the slope length factor that is often combined with S, a slope steepness factor, to yield a dimensionless terrain factor (LS), C is the dimensionless vegetation cover factor and P is the dimensionless erosion control practice factor.

Equations for the computation of the LS factor are based on upslope contributing area and have been developed by [8]:

$$LS = A^{m}(\sin \beta) \tag{2}$$

where A is upslope contributing area, β is slope angle, and m and n are constants depending on the type of flow and soil properties; where rill erosion dominates these parameters are usually set to m = 1.6 and n = 1.3, whereas where sheet erosion prevails, they are set to m = n = 1.0 [9]. This LS factor is equivalent to the traditional LS factor on planar surface but has the added benefit of being applicable to complex slope geometries [9, 10]. Then, for the estimation of the erosion and deposition (ED), which are both computed as a change in sediment transport capacity across a GIS grid cell, the following equation has to be computed:

$$ED = d(A \cos \alpha)/dx + d(A \sin \alpha)/dy \tag{3}$$

where α is the slope aspect of the terrain in degrees in the direction of the steepest slope. Net erosion areas coincided with areas of profile convexity and tangential concavity (flow acceleration and convergence), and net deposition areas coincided with areas of profile concavity (decreasing flow velocity).

The USPED model was critically applied by using an integrated GIS approach in a raster environment to obtain maps for each factor.

3 Data Analysis and Results

In relation to the Database 1, an analysis has been carried out aimed at observing the temporal trend of the agrarian and economic development in the Basilicata region from 1980 to 2013 by considering the dataset coming from INEA – *Land Values Database*. From data analysis it was possible to verify that, during the period considered, the trend of land values has steadily grown with values ranging from a minimum of 3200 €/ha in 1980 to 7000€/ha in 2013 (Fig. 1).

The Database 2 analysis has been carried out by surveying the *Total Utilised Agricultural Area* [ha] (TUAA) – at the regional scale - in relation to a large time frame – from 1980 to 2013 – in order to assess the related trend shown in Fig. 1. The same

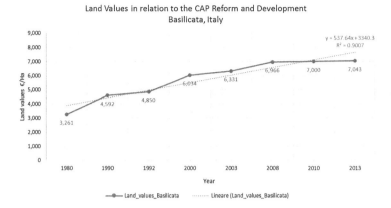

Fig. 1. Land values trend in the Basilicata region with particular attention on the time frames from 1980 to 2013. Source: INEA, Italy.

dataset analysis has been carried out by surveying the following time frame: 1980, 1990, 1992, 2000, 2003, 2008, 2010, 2013 - for evaluating particular changes in relation to the CAP Reform and Development from 1980 to 2013:

- From 1980, well-known as the *3rd phase of the CAP Structural Policy*, to 1992 – known as *McSharry's Reform* - it is possible to evidence an increase of the land values due to the central element deriving from the 'ratio' related to the compensation strongly related to the 'land' considered as a *production factor* and so this compensation of the farmers with a payments system related to the number of cultivated or non-cultivated hectares has favored an increase of the land value because of the *'production factor'*.
- In 1993 – 2nd *Structural Policy Reform* (after the first one happened in 1988) – the vision of the Agricultural enterprise was renewed as Agricultural-Multifunctional Company. This new vision provides for the safeguard of the environment and landscape, economic and social development and promotion of culture.
- In 2000 – well-known as *Agenda 2000* – was a reform conceived in anticipation of a Mid-Term Review (MTR).
- In 2003 – *Fischler's Reform* – new goals have come into a play into the CAP's scenario, such as improving the European agricultural competitiveness, re-orienting the production to the market, promoting a sustainable agriculture socially acceptable and reinforcing the rural development. In this period, conditionality represented one of the principal tools used by European Union for including the environmental thematic about the CAP – Reg. Ce. 1782/2003. In relation to the annexed document attached to this Reg. No. 1782/2003 (Annex 4), it is important to consider a significative aim, such as keeping agricultural land in good agricultural and environmental conditions.
- In 2008, the well-know *Health Check* CAP – *Fisher Boel's Reform* -, refers the proposals of the CAP changes: result of a detailed analysis on health check of CAP. Health Check based on three principle rules, such as:

1. Reg. No. 72/2009 – *Market Tools*
2. Reg. No. 73/2009 – *Direct Payments*
3. Reg. No. 74/2009 – *Rural Development Policy.*

New environmental and climatic challenges and risk management belong to the elements coming from Health Check.

An additional correlated analysis has been carried out – on the same parameter, *Utilised Agricultural Area* [ha] (UAA) – by considering the three croplands types, such as *arable cereal lands, vineyards* and *olive-growing lands* at regional scale.

At the temporal scale, this analysis has been carried out for the 1980, 1990, 2000, 2010 years, related to the Italian Agricultural Censuses – *from 3rd to 6th Italian Agricultural Census* - analyzed over time until 2013 (Fig. 2).

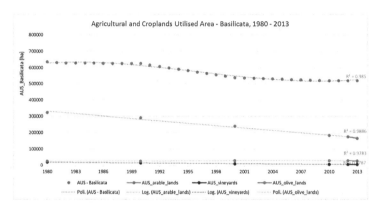

Fig. 2. Agricultural and croplands utilised surface (Agricultural Utilised Surface - AUS) trends [ha/year] in the Basilicata region from 1980 to 2013. (Color figure online)

The next step of the data analysis and the related trends assessments concerned the calculation of a new parameter, such as the *Productivity* [quintals/hectares] of the three crops considered in this work: *cereals, vineyards* and *olive trees*. This evaluation has been carried out by considering two data coming from the production scope, such as the *productive surface* (cropland) [ha] and the related *crop production* [quintals] *(ISTAT – Agriculture section: Surface and Production)*. By calculating the division ratio between both of them – *Production* [q]/*Productive Surface* [ha] – it has been calculated the *Productivity value* for each crop. Thus, in order to observe the productivity trend, it has been necessary to work on spreadsheets for developing the graph of the series of data evaluated in the time frame considered. Figure 3 shows a complete framework of the three linear trends.

In relation to the data gap related to the surface and the production from 2012 to 2013 coming from ISTAT datasets, it has been considered the data coming from INEA database.

In order to analyze and evaluate the economic-agrarian aspect related to the *Agricultural Land Values*, has been carried out an analysis of the *Agricultural Average Values* – well-known as VAM – primarily used in the context of expropriation

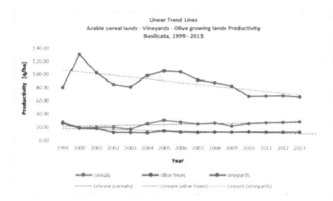

Fig. 3. Trend Line application to the croplands productivity - arable cereal lands, vineyards, olive-growing lands trends – in the Basilicata region – from 1999 to 2013. (Color figure online)

procedures for public use of non-building areas of pursuant to the Italian Presidential Decree No. 327/2001.

The first step of the following analysis has been to calculate the average agricultural value by using the database of the 21 *Agrarian Zones* of the Basilicata region, evaluating an unique Average Agricultural Value in relation to each cropland for every year of the time frame considered – from 2008 to 2013 (Fig. 4).

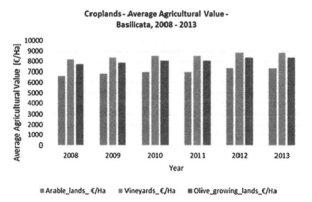

Fig. 4. Average Agricultural Values - arable cereal lands, vineyards, olive-growing lands trends – in the Basilicata region – from 2008 to 2013. (Color figure online)

3.1 USPED Method Application

Estimates of the factors related to the USPED model relative to the Basilicata region have been drawn as follows. For the support practice factor (P), which is the soil loss ratio with a specific support practice to the corresponding soil loss with up–and–down slope tillage, we adopted a unitary value, as is usually the case for natural slopes with no conservation practices.

3.1.1   Rainfall Erosivity

The R factor has been calculated by applying the equation developed by Capolongo et al. [11] for Basilicata, derived from 20-min and hourly precipitation data. It takes into account daily precipitation ≥ 10.0 mm:

$$EI_{30} = 0.0009 \cdot \left[P_k \cdot \left(9.0 + 3.0 \cdot \cos \left(2\pi \left(\frac{J - 3\sqrt{P_k}}{365} \right) + 3.0 \right) \right) \right]^2 \qquad (4)$$

where EI_{30} is the rainfall erosivity in MJ ha^{-1} mm h^{-1}, Pk the daily total rainfall in mm and J the Julian day number. To extract the spatial distribution of erosivity in the whole region, we have considered the average erosivity values of 53 pluviometric stations calculated for 2012. Thus we have drawn an erosivity map by adopting the kriging geostatistical method, as suggested by [12] (Fig. 5d).

3.1.2   Soil Erodibility

The soil erodibility factor (K) map (Fig. 5a) has been calculated by using the USLE equation [6]:

$$K = [2.1 \times 10 - 4\,(12 - M)[(Si + fS)(100 - C)]1.14 + 3.25(A - 2) + 2.5(P - 3))]/100 \quad (5)$$

where M is the organic matter content (%), Si is the silt content (%), 2 to 50 µm, fS is the fine sand content (%), 50 µ to100 µm, C is the clay content (%), less than 2 µm, S is the sand content (%), 50 µm to 2 mm, A is the structure and P is the permeability class (within top 0.60 m). Soil data were derived from different sources. Texture, structure and permeability data derived from soil map of Basilicata [13]. This soil map, based on more than 1750 direct observations from 2002 to 2004 or on observations drawn from previous studies, laboratory analyses, drill holes and profiles, allowed researchers to obtain information on the physical characteristics of each soil type in the region. Organic matter data were derived from the European Soil Database [14].

3.1.3   Slope-Length Factor

The LS factor measures the impact of slope-length on soil erosion under steep slope conditions. In the traditional USLE and RUSLE methods it is evaluated as the horizontal distance from the overland flow origin to the point where either the slope gradient decreases to a point where deposition begins, or runoff becomes concentrated in a defined channel. As clearly explained by Garcia Rodriguez and Gimenez Suarez (2012) [15], in a real two-dimensional landscape, overland flow and the resulting soil loss do not actually depend upon the distance from the divide or upslope border on the field, but rather on the area per unit of contour length contributing runoff to that point, which is strongly affected by flow convergence and/or divergence [16].

Thus the slope-length unit (L) is replaced by the unit – contributing area. Digital Elevation Models of 1955 (20 × 20 m raster cell resolution) from the Istituto Geografico Militare Italiano has been used. Here the upslope contributing area is calculated by using the SINMAP extension in ArcViewGIS [17], which employs the D-∞ algorithms proposed by Tarboton (1997) [18].

3.1.4 Cover and Management Factor

The cover and management factor (C), which reflects the effects of cropping and management practices on soil erosion rates, has been calculated by using data from the fourth level of the land use map Corine Land Cover 2012 (Fig. 5c), which is available from the *Network of the National Environmental System of the Superior Institute for Environmental Protection and Research* (ISPRA, SINANET).

3.1.5 Model Application and Validation

The USPED Model was applied to the Basilicata region for the year 2012 (Fig. 6). Results from each raster cell were separated into seven classes of erosion and deposition as listed below:

1. High erosion (< -20 Mg ha^{-1} y^{-1})
2. Moderate erosion (-12 to -20 Mg ha^{-1} y^{-1})
3. Low erosion (-1 to -12 Mg ha^{-1} y^{-1})
4. Very low erosion (-1 to 0 Mg ha^{-1} y^{-1})
5. Very low deposition (0 to 1 Mg ha^{-1} y^{-1})
6. Low–Moderate deposition (1 to 20 Mg ha^{-1} y^{-1})
7. High deposition (> 20 Mg ha^{-1} y^{-1}).

This classification was derived by adopting (i) the erosion risk classes used for Italy by van der Knijff et al. (1999) [19], (ii) the concepts formulated for Mediterranean environments by Morgan (1995) [20] and (iii) the criteria of tolerable erosion that established the soil losses at 1 Mg ha^{-1} y^{-1} [21].

Starting from the USPED raster image (Fig. 6), the values of the erosion classes, included in the Corine Land Cover polygons relative to the most economically significant for the Basilicata region (arable cereals lands, olive growing lands and vineyards) have been considered. For each cultivation has been evaluated the *Potential eroded surface* [ha] obtained through USPED method by considering only the first three classes of erosion in relation to the following range: < -20 Mg ha^{-1} y^{-1} $\div$ < -1 Mg ha^{-1} y^{-1} – from high erosion to low erosion. This parameter has been related to soil loss average erosion to obtain/assess the potential economic loss for each cropland considered, so that the economic result – defined as "Potential Economic Loss" (PEL [€/ha]) – could be put in relation with the outputs coming from the USPED method application (Fig. 7).

Proceeding, it has been necessary to consider the soil density (C_Mg/m^3) parameter, in order to obtain the weight of eroded soil (P_Mg) and, consequently, the relative volume (V_m^3). Obtained these results, it has been possible, after having calculated the relative surface (Area m^2), imposing the average value of the topsoil width of 0.1 m [22], to assess two economic parameters reported in Table 1 (PEL – *Potential Economic Loss* [€] – and PEL/B €/ha – *Potential Economic loss*), put in relation to the potential eroded surface (B) related to the three selected croplands. This last result provides the economic value, evaluated in Euros per hectare, potentially lost in condition of eroded surface, in this case, in relation to the three croplands surface evaluated

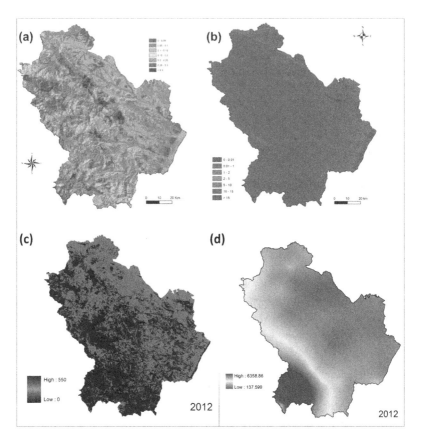

Fig. 5. Legend: 5(a) Soil erodibility factor map in the Basilicata region in 2012 (K factor); 5(b) Topographic Index Map, 2012; 5(c) Land cover and management factor map (C factor), 2012; 5 (d) Erosivity map (R factor), 2012. (Color figure online)

by considering the Average Agricultural Values (D €/ha) of the year 2012, coming from the database of the Basilicata Region Department (Table 1)

The Table 1 permits to define:

- *Potential Economic Loss* (PEL) €: parameter calculated by correlating the Area value [ha] with the Average Agricultural Values of the year 2012 for the three croplands selected. It has been defined a *potential value* because the soil loss values (average erosion [Mg/ha/y]) are considered as 'potential': indeed, the surface obtained through the USPED method application is a *potential eroded surface* [ha].
- *Economic erosion value* (PEL/B) €/ha: it is possible to notice these results in potential erosion conditions (economic value in relation to the Agricultural Average Values of the year 2012):
  • the arable lands would lose 1.50 Euros per hectare;
  • the olive growing lands would lose 0.80 Euros per hectare;
  • the vineyards would lose 1.88 Euros per hectare.

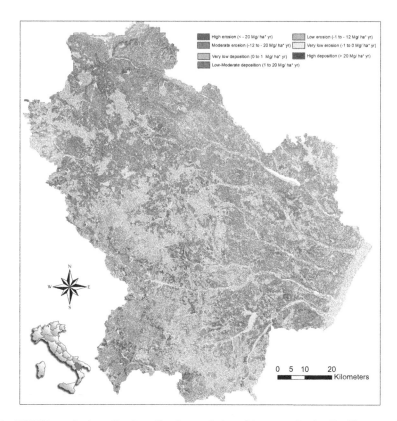

High erosion (< - 20 Mg/ ha* yr)   Low erosion (-1 to - 12 Mg/ ha* yr)
Moderate erosion (-12 to - 20 Mg/ ha* yr)   Very low erosion (-1 to 0 Mg/ ha* yr)
Very low deposition (0 to 1 Mg/ ha* yr)   High deposition (> 20 Mg/ ha* yr)
Low-Moderate deposition (1 to 20 Mg/ ha* yr)

0 5 10   20
Kilometers

Fig. 6. USPED method application. Erosion and deposition map in the Basilicata region in 2012. (Color figure online)

Proceeding to the characterization of the environments occupied by cultivated areas of three croplands, in order to determinate the surfaces related to the total and the non-potential eroded ones, respectively, the estimation of the mean erosion has been carried out through the USPED method application – as reported into the following table (Table 2).

The results concerning *not potential eroded surface* [ha] has been obtained by calculating the difference between the *total surface* [ha] and the *potential eroded surface* [ha] (reported into the Table 1).

The percentage of potential eroded surface for every cropland considered, obtained by the ratio *Potential Eroded Surface (B) [ha]/Total Surface [ha],* show a strong potential eroded surface, around 99 % of the total (Table 2).

Table 1. Calculation of soil loss caused by erosion – USPED method application outputs. Legend: (A) *Soil Loss – average erosion* [Mg/ha/y]; (B) *Potential eroded surface* [ha] obtained through USPED method by considering only the first three classes of erosion in relation to the following range: < -20 Mg ha^{-1} y$^{-1} \div < -1$ Mg ha^{-1} y^{-1} – from high erosion to low erosion); (C) *Soil density* [Mg/m³] = 1.9; (D) *Average Agricultural Values* [€/ha]; (P) *Weight of eroded soil* [Mg]; (V) *Volume* [m³] = P/C; Area [m²] = V/H (h = 0.1 m); PEL: *Potential Economic Loss* [€] = Area [ha] * D [€/ha]; PEL/B: *Economic erosion value - soil loss value* [€/ha].

Croplands	(A) [Mg/ha/y]	(B) [Ha]	(C) [Mg/m³]	(D) [€/Ha]	P (Mg)	V P/C [m³]	Area [m²]	PEL [€]	PEL/B [€/Ha]
arable_lands	2.56	372385.62	1.9	7334	145463.13	76559.54	765595.44	561487.69	1.51
olive growing_lands	5.34	27945.29	1.9	8365	5233.20	2754.32	27543.16	23039.85	0.82
vineyards	2.48	1119.73	1.9	8841	451.50	237.63	2376.34	2100.92	1.88

Table 2. Characterization of the environments related to the croplands in relation to the USPED method application.

CROPLANDS	ENVIRONMENTS	Surface [ha]	Mean erosion USPED method	Total surface [ha]	Not potential eroded surface [ha]	Potential eroded surface
Arable lands	Mountain	65998.12	4.06	376444.78	**4059.16**	**0.99**
	Hill	176537.70	2.72			
	Plain	133908.96	1.59			
Olive growing lands	Mountain	726.24	8.96	28122.45	**177.16**	**0.99**
	Hill	17892.98	6.16			
	Plain	9503.23	3.67			
Vineyards	Mountain	0.00	0.00	1151.50	**31.77**	**0.97**
	Hill	694.09	3.28			
	Plain	457.41	1.14			

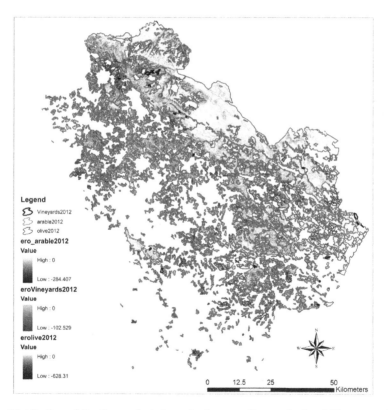

Fig. 7. Distribution of the three selected croplands, according to the Land Use map - Corine Land Cover 2012 – 4th hierarchical level, in combination with the USPED method application related to the erosion assessment (Color figure online)

4 Discussion and Final Remarks

In relation to the *Average Agricultural Values* (VAM - 2012) – coming from the Region of Basilicata - Department of Agriculture, Rural Development, Mountain Economy – *Provincial Commission for determining compensation for expropriation and the average agricultural land values* (available period: 2008 – 2013), it is necessary to premise that the value used in this study has been obtained by calculating the Average Value of each cropland considered, since the values, coming from the Table deriving from the abovementioned source, are divided into agrarian areas of the Basilicata region. This choice has been made because of the different spatial scale, as this study doesn't deal with the data at agrarian areas scale but at regional scale.

According to Bakker et al. (2004) [22], in order to obtain a clear framework of the structural analysis of the variables playing an important role into the erosion-productivity scenario, it is necessary to consider the following cases:

(a) topsoil removal may often result in a nutrient deficit;
(b) erosion may also lead to physical hindrance to root growth.

Regarding erosion-productivity relationship, it is important to consider that erosion reduces productivity so slowly that the reduction might not be recognized until crop production is no longer economically viable [22].

Moreover, putting in relation the geomorphological role with the economic one, it is interesting to mention the importance of considering the erosion and productivity, because - according Pimentel et al. [4]; El-Swaify et al. [23] and Troeh et al. [24] erosion by water and wind negatively affects soil quality and productivity by reducing infiltration rates, water-holding capacity, nutrients, organic matter, soil biota and soil depth.

The economic loss values calculated in Table 1 are comparable with those calculated for other sites synthesized by Telles et al. (2011) [25], considering that, in most of those papers, the methods used to assess the erosion were the USLE and RUSLE, which tend to amplify and emphasize the measured data and those generated by the application of USPED method [5].

The soil exposure to the erosion equal to 99 % of their surface, as is apparent by the USPED map, produces an economic loss, preliminarily estimated in €/ha (Table 1), which does not result in terms of physical removal of the soil particles by water, wind or anthropogenic land use for urban sprawl, but in a gradual destruction of the soil properties and pauperization of nutrients, organic matter and productivity.

The action of linear and areal erosion exerted by the water produces, however, a re-deposition and accumulation of the soil in other areas, but penalizing those of origin.

In this context, in this paper the estimation of costs due to soil erosion was made on the basis of on-site and not of those off-site, linked to sedimentation, flooding, flash floods, landslides and so on. By analyzing the on-site effects' framework related to erosion on crop yield, according to Lal (1998) [26], these effects might be induced by several interacting factors, such as reduction of soil organic carbon (SOC), loss of plant nutrients, decline in soil structure, loss of effective rooting depth and decrease in available water capacity (AWC). These factors, obviously, play an important role in causing a decline in soil quality and productivity. Also, reasons of decline in soil quality include edaphological factors affecting crop growth, decrease in soil structure and reduction in effective rooting depth [26].

Soil loss by erosion tends to increase production costs in the medium and long term, with an increasing demand for liming and fertilizer applications and reduce operational efficiency of machines, incurring costs to control the situation [27].

Soil erosion has, therefore, evident effects both on and off production sites, which have economic consequences on agriculture and on the land use policy. This paper provides a scenario that draws attention to the urgent need to prevent and control soil degradation processes. Considerable applied research on erosion control techniques has been done, for example, in the U.S. and elsewhere and can be grouped into two broad categories: (1) soil management techniques that improve infiltration rate, and (2) runoff management techniques that permit safe management/removal of surplus runoff. Conservation tillage [28], use of cover crops (e.g., crimson clover) to restore productivity of degraded soils [29], and summer fallowing [30] are common examples of soil management techniques. Constructing terracing, installing waterways, and other engineering devices are some examples of runoff management techniques.

For this propose, the data on erosion cost are of fundamental importance, especially in a region like Basilicata, whose economy depends by farming.

Database and Archive Sources

Land Values Trends (Basilicata), 1960 – 2014, INEA – National Institute of Agricultural.

Italian Yearbook of Agriculture – Vol. XLIV - L, 1990 – 1996 - INEA - National Institute of Agricultural Economy – Società Editrice Il Mulino, Bologna, Italy.

ISTAT – National Institute of Statistics – from 3rd to 6th Italian Agriculture Census, 1980;1990;2000;2010.

ISTAT - Agriculture section: "Surface and Production", 1999 – 2011.

Utilised Agricultural Area (UAA) (Basilicata), 1970 – 2014, ISTAT - Italian Agricultural Censuses. Utilised Agricultural Area (UAA) – Arable cereal lands (1999 – 2012), vineyards (1990; 1999 – 2012), olive-growing lands (1990; 1999 – 2011) (Basilicata) – ISTAT dataset – Agriculture section: "Surface and Production", 1999 – 2011. Utilised Agricultural Area (UAA) (Basilicata) – Arable cereal lands (2013), vineyards (1990; 2013), olive-growing lands (2012 – 2013), INEA Productivity values - Arable cereal lands, 1997 – 1998: INEA dataset; 1999 – 2011: evaluation deriving from ISTAT dataset; 2012 – 2013: AGRIT dataset, Statistic Program – Mipaaf, Ministry of Agriculture Food and Forestry Policies (Italy). Productivity values – Vineyards, 1990; 1999 – 2011: evaluation deriving from ISTAT dataset; 2012 – 2013: INEA dataset. Productivity values – Olive growing-lands: 1999 – 2012: evaluation deriving from ISTAT dataset; 2012 – 2013: INEA dataset.

Mipaaf – Ministry of Agriculture Food and Forestry Policies (Italy) - *La Politica Agricola Comune dalle origini ad oggi* – Direzione Generale delle Politiche Internazionali e dell'UE.

References

1. Piccarreta, M., Capolongo, D., Miccoli, M.N., Bentivenga, M.: Global change and long-term gully sediment production dynamics in Basilicata, southern Italy. Environ. Earth Sci. **67**(6), 1619–1630 (2012)
2. Piccarreta, M., Lazzari, M., Pasini, A.: Trends in daily temperature extremes over the Basilicata region (southern Italy) from 1951 to 2010 in a Mediterranean climatic context. Int. J. Climatol. **35**, 1964–1975 (2015)
3. Piccarreta, M., Pasini, A., Capolongo, D., Lazzari, M.: Changes in daily precipitation extremes in the Mediterranean from 1951 to 2010: the Basilicata region, southern Italy. Int. J. Climatol. **33**(15), 3229–3248 (2013)
4. Pimentel, D., Harvey, C., Resosudarmo, P., Sinclair, K., Kurz, D., McNair, M., Crist, S., Shpritz, L., Fitton, L., Saffouri, R., Blair, R.: Environmental and economic costs of soil erosion and conservation benefits. Science **267**(5201), 1117–1123 (2004). New Series, Office of technology Assessment, 1982, Impacts of Technology on U.S. Cropland and Rangeland Productivity, Washington, DC
5. Lazzari, M., Gioia, D., Piccarreta, M., Danese, M., Lanorte, A.: Sediment yield and erosion rate estimation in the mountain catchments of the Camastra artificial reservoir (Southern Italy): a comparison between different empirical methods. Catena **127**, 323–339 (2015)
6. Wischmeier, W.H., Smith, D.D.: Predicting Rainfall Erosion Losses: a Guide to Conservation Planning. Science and Education Administration, US Department of Agriculture, Washington, DC (1978). Agricultural Handbook 587
7. Renard, K.G., Foster, G.R., Weesies, G.A., McCool, D.K., Toder, D.C.: Predicting Soil Erosion by Water: A Guide to Conservation Planning with the Revised Universal Soil Loss Equation (RUSLE). US Government Printing Office, Washington DC (1997). 384 pages
8. Mitasova, H., Mitas, L., Brown, W.: Modeling topographic potential for erosion and deposition using GIS. Int. J. Geogr. Inf. Syst. **10**, 629–641 (1996)
9. Moore, I.D., Wilson, J.P.: Length–slope factors of the revised Universal soil loss equation: simplified method of estimation. J. Soil Water Conserv. **47**, 423–428 (1992). doi:10.1016/S0341-8162(03)00088-2
10. Moore, I.D., Burch, G.J.: Physical basis of the length–slope factor in the Universal soil loss equation. Soil Sci. Soc. Am. J. **50**, 1294–1298 (1986)
11. Capolongo, D., Diodato, N., Mannaerts, C.M., Piccarreta, M., Strobl, R.O.: Analyzing temporal changes in climate erosivity using a simplified rainfall erosivity model in Basilicata, southern Italy. J. Hydrol. **356**, 119–130 (2008)
12. Goovaerts, P.: Geostatistics in soil science: state-of-the-art and perspectives. Geoderma **89**, 1–45 (1999). Department of Civil and Environmental Engineering, University Of Michigan, Ann Arbor, USA
13. Cassi, F., Viviano, L., et al.: I Suoli della Basilicata - Carta pedologica della Regione Basilicata in scala 1:250.000. Regione Basilicata - Dip. Agricoltura e Sviluppo Rurale. Direzione Generale (2006)
14. Jones, R.J.A., Hiederer, R., Rusco, E., Loveland, P.J., Montanarella, L.: Estimating organic carbon in the soils of Europe for policy support. Eur. J. Soil Sci. **56**, 655–671 (2005)
15. Garcia Rodriguez, J.L., Gimenez Suarez, M.C.: Methodology for estimating the topographic factor LS of RUSLE3D and USPED using GIS. Geomorphology **175–176**, 98–106 (2012). doi:10.1016/j.geomorph.2012.07.001

16. Desmet, P.J.J., Govers, G.: GIS-based simulation of erosion and deposition patterns in an agricultural landscape: a comparison of model results with soil map information. Catena **25**, 389–401 (1995). Laboratory of Experimental Geomorphology, Redingenstraat 16B, 3000 Lenven, Belgium
17. Pack, R.T., Tarboton, D.G., Goodwin, C.N.: Terrain Stability Mapping with SINMAP, Technical Description and Users Guide for Version 1.00, Report 4114-0. Terratech, Salmon Arm, B.C, Canada (1998)
18. Tarboton, D.G.: A new method for the determination of flow directions and upslope areas in grid digital elevation models. Water Resour. Res. **33**(2), 309–319 (1997). Utah Water Research Laboratory, Utah State University, Logan
19. Van der Knijff, J.M., Jones, R.J.A., Montanarella, L.: Soil erosion risk assessment in Italy. Office for Official Publications of the European Communities Luxembourg, Luxemburg (1999). EUR 99044 EN, European Soil Bureau, p. 52
20. Morgan, R.P.C.: Soil Erosion and Conservation, 2nd edn. Longman, Essex (1995). http://eusoils.jrc.it/ESDB_Archive/pesera/pesera_cd/pdf/er_it_new.pdf
21. EEA.: Europe's Environment. The Second Assessment. Office for Official Publications of the European Communities, Luxemburg (1998). http://themes.eea.eu.int/
22. Bakker, M., Govers, G., Rounsevell, M.D.A.: The crop productivity-erosion relationship: an analysis based on experimental work. Catena **57**, 55–76 (2004)
23. El-Swaify, S.A., Moldenhauer, W.C., Lo, A.: Soil Erosion and Conservation. Soil Conservation Society of America, Ankeny (1985)
24. Troeh, F.R., Hobbs, J.A., Donahues, R.L.: Soil and Water Conservation. Prentice-Hall, Englewood (1991)
25. Telles, T.S., Guimaraes, M.F., Dechen, S.C.F.: The costs of soil erosion. Revista Brasileira de Ciência do Solo **35**, 287–298 (2011)
26. Lal, R.: Soil erosion impact on agronomic productivity and environment quality. Crit. Rev. Plant Sci. **17**(4), 319–464 (1998). The Ohio State University, School of Natural Resources, OH
27. Uri, N.D.: The environmental implications of the soil erosion in the United States. Environ. Monit. Assess. **66**, 293–312 (2001)
28. Richardson, C.W., King, K.W.: Erosion and nutrient losses from zero tillage on a clay soil. J. Agric. Eng. Res. **61**, 81–86 (1995)
29. Bruce, R.R., Langdale, G.W., West, L.T., Miller, W.P.: Surface soil degradation and soil productivity restoration and maintenance. Symposium: erosion impact on soil productivity. Soil Sci. Soc. Am. J. 59, 654–660 (1995)
30. Peterson, G.A., Westfall, D.G., Cole, C.V.: Agroecosystem approach to soil and crop management research. Soil Sci. Soc. Am. J. **57**, 1354–1360 (1993)

Fair Planning and Affordability Housing in Urban Policy. The Case of Syracuse (Italy)

Grazia Napoli[1(✉)], Salvatore Giuffrida[2], and Maria Rosa Trovato[2]

[1] Department of Architecture, Palermo, Italy
grazia.napoli@unipa.it
[2] Department of Civil Engineering and Architecture, Catania, Italy
{sgiuffrida,mrtrovato}@dica.unict.it

Abstract. Equalization can be implemented in the planning process by means of several tools. The Syracuse's Master Plan has used "urban negotiation" to obtain land for facilities and public infrastructure in different urban areas basing on the rule of the transfer of a portion of land in return for the building permission for the remaining part of each property to be developed. The Master Plan also aimed at providing social housing because the economic crisis has amplified the gap between housing market prices and household income. This study proposes an equalization and compensation model to support the urban negotiation for providing the indexes of a fair and convenient development of several interstitial urban areas. Some different scenarios, based on an equalization pattern, are prefigured to provide affordable housing for low-income households.

Keywords: Equalization · Transformation value · Affordability housing · Land use policy

1 Introduction

The Master Plan of Syracuse (Italy), approved in 2007 [1], may be considered a general and unitary planning framework based on some clear and recognizable principles: urban sprawl control, immaterial values empowerment (Dionigi's Walls archeological urban park), enhancement of mobility infrastructures, reorganization of the water front, environmental and urban coastal rehabilitation, additional areas for crafts and factories, redevelopment of the old town centres of Ortigia and Santa Lucia, identification of pilot-projects, urban redevelopment and landscaping of the former railway line, definition of a plan for facilities. The Master Plan integrates within the traditional unitary approach the feasible perspective of actualization of the renovation process by independent stages, using negotiation and basing on criteria inspired by effectiveness, efficiency and fairness. In particular, the equalization pattern has been envisaged for the completion of the urban fabric and the redevelopment of interstitial areas where new facilities and infrastructures can be settled in order to fill the current gap of 1,823,150 m^2. The implementation of the equalization program supposes 208 sectors – sometimes split into sub-sectors – characterized by shared targets and perspectives to solve or overcome the traditional "gainers *vs* losers" opposition.

© Springer International Publishing Switzerland 2016
O. Gervasi et al. (Eds.): ICCSA 2016, Part IV, LNCS 9789, pp. 46–62, 2016.
DOI: 10.1007/978-3-319-42089-9_4

This contribution proposes an equalization-compensation pattern – involving 150 urban sectors – aimed at assuming social housing as the main issue of fairness of the Master Plan (Fig. 1). In such a perspective, the typical purposes of social housing replace the current ones, mainly oriented to develop real estate profitable investments. Moreover, a well-organized building program, directed and granted by the Municipality, can reduce the market risk and, as a consequence, the expected profit rate.

Nonetheless, the proposed pattern is based on a Discounted Cash Flow Analysis (DCFA) according to the Residual Value-based approach, which allows us to make the relevant elements for choices and the references for measuring and comparing the results more objective and robust – although partial and relative–.

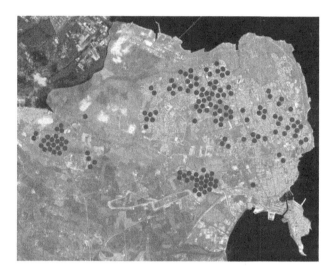

Fig. 1. Location of the 150 analyzed sectors within the Municipality of Syracuse (Italy).

In these perspectives, the Master Plan rules two main real estate regimes: the one of the "fabric-areas" and the one of the "transformation areas". The latter typically overcomes the obstacles interposed by the effects of the positional rent: the sectors are identified as the areas in which consistent public/private investment programs allow profits to be shared equally among the land owners – proportionally to the current land market value of each parcel that is involved by the transformation process – and the externalities of the whole development process to be internalized. Therefore, the increase in value process doesn't take into account either the different intrinsic features of the parcels, or the location of the profitable and unprofitable developments. As a consequence all areas are equally involved within the same coordinated valorization process. As the consistency of the process is limited to the small size of the sector, the internal equalization might not be sufficient to establish a real external equity. Therefore a general pattern in which the results of all sectors are taken into account simultaneously should be constructed in order to identify the more and the less advantaged sectors and actualize the necessary compensations.

The exchange of building areas within the "transformation areas" for building permissions envisaged by the Master Plan still actualizes a kind of "objective approach" (based on an exchange of objects). In fact, in return for each built-up area planning permission, the Master Plan identifies an area of the same size to be transferred to the Municipality for public facilities. This rule differs from the typical compensatory approach, which instead turns in value the land to be given to Municipality and the planning permissions as well.

2   Materials

As its general purpose, the Master Plan of Syracuse aims at satisfying the needs of the settled citizens, according to some past demographic projections. Since the negative migratory balance in the '90 s' (1,500 inhabitants migrated from other municipalities, and over 2,000 inhabitants emigrated) overcomes the natural positive one (1,100–1,200 births vs 900–950 deaths), the global demographic balance is negative, mostly due to both the fluxes of non-EU migrants, and the affordability of the hinterland areas dwellings. Nonetheless, at the end of the Master Plan recording period (2001, last National Statistics Institute survey) the number of registered families (47,171) had grown (+1,600) due to the reduction in the number of components (from 2,76 to 2,66). Three demographic scenarios and the consequent housing needs have been envisages by considering three different data periods, as summarized in Table 1.

The total amount of buildings to be provided and the total area to be occupied have been calculated for each scenario taking into account the average surface and volume of a dwelling, and the three different typologies: detached/semi-detached houses (1–2 dwellings houses), 3–6 dwellings blocks, 10–15 dwellings buildings with shops and offices, basing on two estimates, as displayed in Table 2.

Table 3 provides the range for total cubage, the quota of social housing, and the occupied land in comparison to the other usages. The Master Plan supposes the housing cubage to be implemented within the dense urban area and in the residual areas in which the previous bonds decayed. Social housing and mixed-use development operations are supposed to be implemented by using equalization [2]. The process is carried out by: identifying the different intervention areas for developments; concentrating the new buildings taking into account the urban quality of the different areas to be transferred to the municipality for primary and secondary developments in exchange for planning permissions; diversifying the cubage ratio; dividing the areas to be developed in sectors and sub-sectors. The areas to be developed are: the "Borgata di Santa Lucia" (150,000 m^3); the new urban area (1,817,500 m^3); the building programs areas (560,000 m^3); the hamlet of Cassibile (143,000 m^3). The areas needed for secondary developments (S1 – Education; S2 – Public facilities; S3 – Green areas; S4 – Parking) are; 940,125 m^2 in urban areas and "Borgata"; 645,777 m^2 in the southern water front; 77,234 m^2 in Cassibile. The total amount is 1,663,136 m^2.

Table 1. Master Plan housing need based on three different demographic scenarios

Senarios	Period	Population dynamic	Families annual increase	by 2013	main	secondary	previous needs	total need
A	1990-2000	stable	180	2400	2158	216	788	3162
B	1995-2000	decreasing	90	1200	1192	119	788	2099
C	1990-1995	increasing	235	3137	3116	312	788	4216

Table 2. Master Plan housing need based on typology

Scenarios	Hypo theses	Dwellings	1-2 dwellings houses %	m²	m³	3-6 dwellings houses %	m²	m³	10-15 dwellings houses %	m²	m³
A	1	3162	50%	221.326	649.868	30%	132.796	389.921	20%	88.530	265.591
	2		40%	177.061	519.894	30%	132.796	389.921	30%	132.796	398.387
B	1	2099	50%	146.944	431.464	30%	88.166	258.878	20%	58.778	176.333
	2		40%	117.555	345.171	30%	88.166	258.878	30%	88.166	264.499
C	1	4216	50%	295.092	866.463	30%	177.055	519.878	20%	118.037	354.110
	2		40%	236.074	693.170	30%	177.055	519.878	30%	177.055	531.166

Table 3. Master Plan housing need based on typology

Scenarios	Dwellings m²	m³	Shops and offices m²	m³	Social n. dwellings from	to	Total amounts m²	m³	occupied land m²
A	442.652	1.327.956	124.015	372.044	1.200	2.200	566.667	1.700.000	1.292.500
B	m²	881.664	92.779	278.336	840	1.500	386.667	1.160.000	820.000
C	590.184	1.770.552	169.816	509.448	1.600	2.900	760.000	2.280.000	1.800.000

3 Methods

3.1 Equalization and Compensation Pattern

The proposed valuation process aims at investigating the effectiveness – convenience for owners and fairness for public – of the equalization process as supposed for each sector. A specific application of the "residual value" [3] has been carried out starting from the basic formula (1) where the extraordinary permission fees are expressed:

$$v_t = v_f - k - f^* - \pi \tag{1}$$

the normal profit can be assumed as a quota of the total investment including the extra-ordinary permission fees (2):

$$\pi = r(v_t + c + f^*) \tag{2}$$

assuming the normal global profit rate as the sum of the weighed average cost of the capital paid in advance and the premium for risk and organization, and considering the supposed loan term n for each of the single development processes in each sector (3)

$$r = \left[(w+r')(1+w+r')^n - 1\right]/(w+r') \tag{3}$$

the permission fees can be easily calculated (4):

$$f^* = \left\{v_f - \left[(v_t+k)(1+r)\right]\right\}/(1+r) \tag{4}$$

the expected real estate market value and the building cost can be expressed in terms of unit market prices and unit building costs, making explicit the relevant quantities, namely the permitted area, the weighted average height, the unit price and the unit costs and taking into account all the envisaged uses (residential, offices, shops and so on) (5):

$$v_f = \frac{s \cdot i}{\bar{h}}\bar{p}; \; k = \frac{s \cdot i}{\bar{h}}\bar{k} \tag{5}$$

moreover, as required by the equalization process, basing on the area to be given to the municipality for each sector, the appropriate cubage ratio is calculated (6):

$$i = \frac{\left[v_t\bar{h}(1+r)\right]}{s\left[\bar{p} - \bar{k}(1+r)\right]} \tag{6}$$

similarly, given the cubage ratio, the percentage of the area to be given to the Municipality can be calculated (7):

$$a = \frac{v_t s(1+r)}{v_t s(1+r) + v_f \frac{s \cdot i}{h} - k \frac{s \cdot i}{h}(1+r)} \tag{7}$$

at last, the quota of the area dedicated to social housing can be calculated in case of positive revenue-cost surplus in each sector (8):

$$\begin{cases} f^* = \frac{\frac{s \cdot i}{h}\bar{p} - \left(v_t + \frac{s \cdot i}{h}\bar{k}\right)(1+r)}{(1+r)} \\ f^* > 0 \rightarrow i^* \le i \end{cases} \tag{8}$$

where:

v_t is the current real estate value of the total private developable area comprised in the sector;

v_f is the value of the private property at the end of the development process;

k is the building cost including ordinary permission fees;

r is the global profit rate in each loan term;

w is the weighed average cost of capital;

r' is the annual profit rate for the promoter's risk and organization;

f^* is the extraordinary permission fees;

n is the loan term (years);

i is the cubage ratio;

i^* is the social housing cubage ratio;

$\bar{h}$ is the weighed average height of the buildings included in the sector;

s is the permitted area;

$\bar{p}$ is the weighed average market price of the buildings included in the sector;

$\bar{k}$ is the weighed average unit building cost of the buildings included in the sector;

a is the percentage of the area s to be given to the Municipality in exchange for the building permission.

The proposed model actualizes an "axiological approach" through the well known "equalization of values" which aims at integrating, and someway overcoming, the pattern supposed by the Municipality, that instead actualizes an "object/performances approach" through the more usual "equalization of objects" based on urban qualities and building quantities (cubage ratios) within each sector [4].

Basically, an equalization of values pattern (1) deals with substitutable values and (2) represents the urban objects as values. As a consequence, (1) it is able to capture environmental, landscape and social values as well and (2) it allows planners and decision makers to implement equalization processes regardless of the actual fragmentation of properties. Thus it makes planning more flexible and consistent with the general land-social issues [5, 6].

Due to the different urban-environmental characteristics of the examined sectors, the comparison between revenues and costs may result unfair independently form the area transferred to the municipality [7]. In some cases, these inequalities need to be compensated by increasing the cubage ratios. In these cases some sectors could be grouped in over-sectors within which the development rights can be balanced without distorting the urban landscape [8, 9]. In the proposed pattern, as the convenience for owners (normal profit rate) is a constraint that can be negotiated at the beginning: (a) given the land market value, the larger the transferred area, the higher the cubage ratio; (b) given the transferred area, the higher the land market value, the higher the cubage ratio; as a consequence, (c) given the transferred area and the land market value, a fairness ratio can be calculated for each sector by comparing the amount of the permission fees and the value of the transferred area.

Furthermore, if a part of the development process is supposed to be funded by private profit surplus (extra-normal profit), the extra-ordinary permission fees should increase by arising the cubage ratio. Then, given the transferred area, the self-financial ratio (fairness) is calculated and, vice versa, the area to be transferred is calculated by setting a minimum self-financing ratio. If the self-financing ratio is zero, the value of the transferred area should be at least equal to the value of the extraordinary permission fees. As a result, the proposed pattern allows municipality to actualize a "polar equalization by values" aiming at balancing the different convenience profiles for each sector.

3.2 Housing Affordability

This study also proposes a methodology to measure the degree of housing affordability at urban level, in order to support the public decision making process concerning the housing policy. This measurement is particularly useful in this period of economic crisis, when the reductions in GDP, added value and demand have generated job cuts and high levels of unemployment, and have contributed to wide inequalities and social exclusion amplifying the difficulties of many people, including the middle class, in finding homes.

In Europe there are two models of social housing related to the debate on welfare or neo-liberal State vision and planning theories [10, 11]: a "targeted model", when the public administration has to provide housing to target groups that cannot enter the market; a "universalistic model", founded on the principle of providing affordable housing to anybody [12, 13]. Social housing is not uniquely stated, but instead it can take different meanings in different countries, such as "housing at moderate rent" in France, "rental dwellings, affordable home ownership, and shared ownership" in Great Britain; "not for profit housing for rent" in Denmark. Furthermore social housing can be provided for immediate or postponed purchase; temporary or long-term rental; co-housing, etc. [14].

Several types of financial tools can support affordable housing: national funds for grants of subsidization of loans, funds not subject to tax, tax relief, real estate fund, favorable interest rate, individual housing allowances, discounted land, development contributions, increase of cubage index, extraordinary permission fees etc. [15]. In any case, it is essential to integrate the social housing projects into the planning process for achieving urban equity.

The financial ability to afford the housing can be measured by means of different approaches and indicators, some of which are relied on the standard rule that households should not spend more than 30 percent of their income on housing expenditures. The simplest approach is referred to the "rent-to-income ratio" or "house-price-to-income ratio". Other measures of housing affordability are: "purchase affordability", that is the ability to borrow funds to purchase a house; "purchase repayments", that is the ability to afford housing finance repayments [16]; the HAI, National Association of Realtors' Housing Affordability Index [17]. Different approaches are based on: the opportunity cost defining the affordability in terms of "residual income", that means there is an affordability problem when a household has to limit the non-housing expenditures below a basic level after paying for housing [18]; or the amenity-based housing affordability index that tries to incorporate important aspects affecting the well-being of the households, such as public goods and amenities) [19].

This study proposes to measure the housing affordability through the income-threshold approach [20], which can be defined as the minimum yearly income to acquire housing at market prices and to qualify for a loan and for a given priced home. The income-threshold IT is calculated applying the formula 9 (based on the HAI's formula):

$$IT = \frac{instalment[i, T, (P \cdot LTV\%)]}{R\%} \tag{9}$$

where: *instalment* = yearly instalment loan; i = mortgage rate; T = loan term; P = house price; *LTV% (Loan To Value)*= percentage of housing price covering by loan; $R\%$ = rent-to-income ratio threshold.

According to the NAR's parameters, T is fixed in 20 years, *LTV* is 80 % (assuming a 20 % down payment of the home price), while according to the OMI's parameters, $R\%$ is 30 % [21]. The mortgage rate i and the house market prices P are variable and depend on values in the corresponding financial and real estate markets.

The methodology proceeds by analyzing the distribution pattern of income and the housing market at urban level. These data are combined and, subsequently, some income-thresholds are calculated to measure the access to the housing market per income group of households and per different location of housing. The identification of the local demand and detailed social needs is aimed to delineate local development policies and social housing practices on which the goal of increasing the social equity could be based. Moreover different financial tools and rules could be implemented to achieve the best social housing project option.

The methodology is applied, as a case study, to the above mentioned sectors where a quota of social housing can be achieved. The main steps of the methodology are:

- Analysis of the local income (per households' income level);
- Analysis of the real estate prices (per urban area and building type);
- Appraisal of the income-thresholds (per income level, urban area and building type);
- Analysis of the results to support social housing policy and projects.

Analysis of the Local Income. Many indicators can describe the adverse socioeconomic condition of the Province of Syracuse. In 2014 there were more than 36,000 unemployed, the unemployment rate was 25.5 % of the workforce (much higher than the national average which is 12.7 %) [22]. The construction sector was severely affected by the economic crisis with a significant reduction in the number of companies, active workers and wage mass (−4,432 millions of euro in 2013) caused by the contraction of the private sector and public works that decrease both in number of works and amount of investments (respectively −23.33 % and −81.25 %). Conversely, the tourism sector registered in 2013 an increase in tourist flow (+4.0 % of arrivals) as a positive outcome of policies promoting the local historical, archaeological, architectural and cultural heritage. Regarding the housing problem, it can be noticed that the number of requests for eviction execution increased significantly, +47 % in 2013–14, and 88 % of the total evictions were due to the unpaid rents.

A detailed analysis on demographic and income data is carried out in Fig. 2 and Table 4 showing that: 38.2 % of the total district income is concentrated in Syracuse; the percentage of taxpayers in Syracuse corresponds to 45.1 % of its population and it is much higher than the average percentage on a district base, as well as the declared yearly average income per taxpayer and pro capita yearly income.

The declared incomes in Syracuse show that the earning capacity of the citizens varies widely. By matching the levels of income and the corresponding percentages of household declaring those income levels, the households are segmented into three groups (Fig. 3):

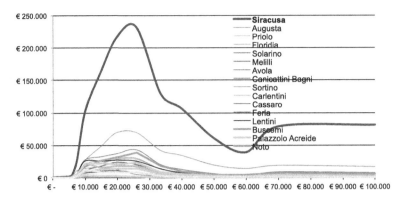

Fig. 2. Total households' income per income level in each town of Syracuse District (2013)

Table 4. Demographic and income data in Syracuse District (2013)

	Household	Population	% household	Total Income	Income/household	Income/pop.
Syracuse	53.440	118.442	45,1%	€ 1.253.827.011	€ 23.462	€ 10.586
District	157.415	399.892	39,4%	€ 3.279.369.423	€ 20.833	€ 8.201
% Syracuse/distr.	33,9%	29,6%		38,2%		

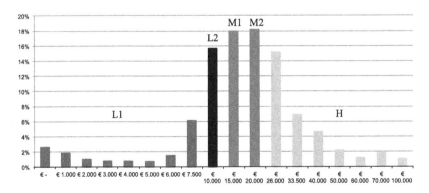

Fig. 3. Percentage of households per income level in Syracuse (2013)

- *group L*, low-income households (31 %) of whom 15 % declared an annual income of 0–10,000 euro (group *L1*), and around 16 % of them earned 10,000–15,000 euro a year (group *L2*);
- *group M*, moderate-income households (36 %) of whom 17,9 % declared an annual income of 15,000–20,000 euro (group *M1*), and around 18,2 % of them earned 20,000–26,000 euro a year (group *M2*);
- *group H*, high-income households (33 %) declaring an annual income higher than 26,000 euro.

Analysis of the Real Estate Prices. The households' decision concerning their location in a specific urban area depends on many factors, such as economies of agglomeration, accessibility to public facilities, infrastructures, potentiality of interaction and transport cost, but it has to respect in particular the threshold of the market price affordability [23, 24].

The real estate market of the city of Syracuse has been analyzed and the collected minimum and maximum housing prices are shown in Fig. 4. The prices are referred to the OMI's "civil housing" typology, located in several urban zones as defined by OMI - Osservatorio del Mercato Immobiliare (*B1, B2, B3, B4, C1, C2, C3, D1, D2, D3* and *D5*). The highest price (1,800 €/m²) has been collected in the zone *B1*, called Ortigia, where the buildings have great historical and architectural values, whereas in the other areas the maximum prices range from 1,050 to 1,400 €/m², conversely, the minimum prices range from 830 to 1,050 €/m² (2nd semester 2015) [25]. The distribution of housing rental prices between all areas is similar to the one defined above (Fig. 5). The maximum (minimum) price is not an absolute maximum (minimum), but it is rather the average of the highest (lowest) prices.

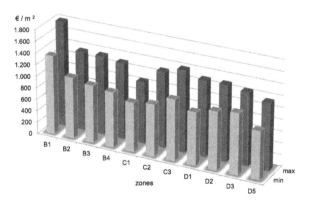

Fig. 4. Market prices (€/m²) of "civil housing" typology per OMI zone (2nd semester 2015)

Appraisal of the Income-Thresholds. The income-threshold is calculated (9) for the maximum market prices of each zone hypothesizing the renting or purchase of an 80 sq.m dwelling (a mortgage rate of 2.60 %, which is a fixed rate loan for 20 years-term, is adopted). The resulting income-thresholds (*T_B1, T_B2, T_B3, T_B4, T_C1, T_C2, T_C3, T_D1, T_D2, T_D3, T_D5*) are shown in Fig. 6, where they are combined with the income groups, so that the degree of housing affordability, the percentage of households in each income group, and the correspondent "income gap" for the housing purchase or renting in a given zone may be known (Fig. 6).

The graphs show that the low-income households (*group L*) cannot have access to the real estate market for purchasing or renting a dwelling, and the *sub-group L2*'s income gap is wide and varies between 5,000–10,000 euros per year. The *sub-group*

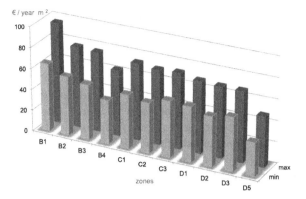

Fig. 5. Yearly rents (€/m²) of "civil housing" typology per OMI zone (2nd semester 2015)

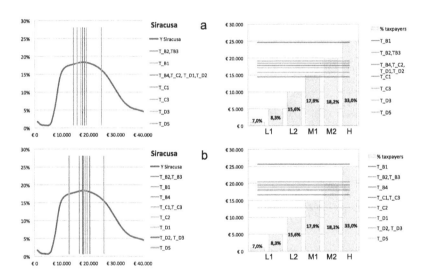

Fig. 6. Purchase (a) and rent (b). Income groups of households and threshold-income per zone in Syracuse

M1 can afford to purchase or rent a dwelling only in suburban zones (zones *C1* and *D5*). The *sub-group M2* can afford most of the dwellings except for the best ones located in Ortigia (zone *B1*), that are affordable just for the *group H*.

Since the right to rent public housing depends on certain income ceilings established by law, and the stock of public housing is insufficient for all the needs, the *subgroups L2* and *M1* could be forced to demand social housing or could need some subsidies to pay the rent or the housing price.

4 Application of the Equalization Pattern

The investigation we carried out handles a sample of 150 sectors taking into account: the existing functions and their opportunity-costs; the supposed uses (dwellings, offices, shops etc.), the development costs and the real estate expected prices; the secondary development process, their building costs and any revenue if a private/public management is supposed (the database is sampled in Fig. 7). Each different mix of time/financial variables (loan term which is the economic cycle duration, interest rate, opportunity cost of equity, leverage, profit rate) corresponds to a different yield/risk profile, and provides different economic-financial layouts.

Figure 8 shows the sectors (rows) in surplus (right bar) or deficit (left bar) from the two points of view of the internal balance (extraordinary permission fees) and the external balance (self financing ratio).

sector id	private area	1. housing	2. accomodations	3. tertiary	4. shops-offices	5. shops	1. housing	2. accomodations	3. tertiary	4. shops-offices	5. shops	6. education	7. community facilities	8. parks-sport facilities	9. parkings	10. local streets	11. public park	12. main streets	13. other facilityes	trasferred area	max height	max covered area	permeability
30	5075	100					5075					4943				977	2531			8451	16,8	0,25	0,10
31	2974	100					2974					3089				296	1567			4952	16,8	0,25	0,10
32	9698	100					9698					14808					1286			16094	16,8	0,25	0,10
33	2022	100					2022					3381								3381	16,8	0,35	0,10
34	1065	100					1065					1763								1763	16,8	0,35	0,10
35	5875	100					5875					3290	2292			738	3921			10241	16,8	0,25	0,10
36	442	100					442														16,8	0,25	0,10
37	63000			100					63000			9950	15944	26076	2343	70941	30478			155732	19	0,25	
38	781	100					781								1300					1300	16,8	0,25	0,10
39	25043	40		60			10017		15026							10373	2578		39158	52109	13,5	0,25	0,10
40	5612	100					5612														10,8	0,25	0,10
41	1494	100					1494														10,8	0,35	0,10
42	807	100					807														7	0,35	0,10
43	1173	100					1173					3675				605	198			4478	7	0,35	0,10
44	251	100					251									907	210			1117	7	0,35	0,10
45	9058	100					9058					15082								15082	16,8	0,25	0,10
46	2066	100					2066					2357				739	344			3440	16,8	0,35	0,10
47	1956	100					1956					2299				958				3257	16,8	0,35	0,10
48	1858	100					1858							2166			927			3093	16,8	0,35	0,10
49	1531	100					1531								2264		288			2552	16,8	0,35	0,10
50	3982	100					3982					5933					703			6636	16,8	0,35	0,10
51	13341	30			70		4002			9339		4713	5325			936	3835			14809	16,8	0,25	0,10
52	4087	30			70		1226			2861		4130					407			4537	16,8	0,35	0,10
53	20319	30			70		6096			14223		15600				2740	4215			22555	16,8	0,25	0,10
54	2998		100					2998				9492					2989			12481	7,5	0,25	0,10
55	2683		100					2683				5117					6051			11168	7,5	0,25	0,10

Fig. 7. A portion of the general database (sectors 30-55 only).

The results of the convenience calculation for each sector lead to a global assessment of the current land policy by equalization in the perspective of the social housing. The equilibrium between private interests and public targets has been pursued in most cases; in 20 cases, mostly large areas, both the internal and the external consistencies are widely disregarded basically due to the excessive size of the envisaged public works, comparing them to the private ones. Regarding the ratio between the extraordinary permission fees and the amount of the promoter's investment ($R1$), Fig. 9 shows the distribution of the 150 sectors between the 15 of $R1$ classes in which the range from -100 % to $+100$ % is divided. The same way, regarding the self-financing ratio ($R2$), Fig. 9 shows the related distribution of the sampled sectors. In both cases, the larger the range, the more distant from 0 % the mode of $R1$ and $R2$.

| | | promoter (tousand euros) | | | | | | | | municipality (tousand euros) | | | | | |
| | | costs | | | | | | revenues | costs | | | | | revenues | |
sector id	current land value	building cost	technical expenses	taxes	ordinary permission fees	extraordinary permission fees	promoter normal profit	real estate market value	building cost	technical expenses	taxes	ordinary permission fees	extraordinary permission fees	management	self-finance ratio
30	180	13145	1114	3137	7	679	972	10696	474	33	104	7	-7679	21	100%
31	106	7940	669	1894	4	809	570	6268	264	18	58	4	-4809	6	100%
32	344	31725	2591	7550	13	298	1858	20439	193	14	42	13	-23298	0	100%
33	72	3907	383	944	4	186	542	5966	6413	449	1411	4	186	0	2%
34	38	2057	202	497	2	98	286	3142	235	16	52	2	98	48	49%
35	215	12227	1080	2927	8	985	1126	12382	3765	264	828	8	-4985	16	100%
36	6	605	60	146	1	36	85	932	0	0	0	1	36	0	
37	2916	79866	7695	19264	428	5111	12436	136800	28193	1973	6202	428	17111	999	51%
38	28	1086	106	262	1	42	150	1646	173	12	38	1	42	35	35%
39	1029	28840	2781	6957	189	888	4965	54611	37937	2656	8346	189	10880	225	23%
40	75	4871	480	1177	5	251	678	7463	0	0	0	5	251	0	
41	20	1808	179	437	2	103	253	2781	0	0	0	2	103	0	
42	8	488	48	118	0	24	68	746	0	0	0	0	24	0	
43	75	3644	278	863	1	474	131	1442	88	6	19	1	-3474	13	100%
44	18	217	20	52	0	-8	28	309	119	8	26	0	-8	20	8%
45	322	31174	2528	7414	12	774	1735	19090	0	0	0	12	-23774	0	
46	73	3991	392	964	4	190	554	6096	4594	322	1011	4	190	16	4%
47	70	7744	649	1846	4	997	525	5771	93	6	20	4	-4997	21	100%
48	66	3590	352	867	4	171	498	5482	378	26	83	4	171	79	52%
49	54	2958	290	715	3	141	411	4517	262	18	58	3	141	49	57%
50	142	17927	1472	4268	8	993	1068	11749	105	7	23	8	-12993	0	100%
51	375	24133	2202	5794	128	164	3342	36762	7880	552	1734	128	1164	20	13%
52	115	7814	770	1888	55	806	1433	15767	7895	553	1737	55	3806	0	38%
53	572	47133	4080	11267	195	773	5090	55991	897	63	197	195	-11773	0	100%
54	206	7358	559	1742	5	845	272	2990	448	31	99	5	-6945	0	100%
55	185	4727	370	1121	5	790	243	2676	908	64	200	5	-3790	0	100%

Fig. 8. Sample of the general convenience calculation for each sector (30–55 only).

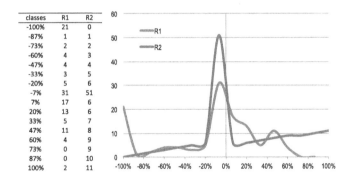

classes	R1	R2
-100%	21	0
-87%	1	1
-73%	2	2
-60%	4	3
-47%	4	4
-33%	3	5
-20%	5	6
-7%	31	51
7%	17	6
20%	13	6
33%	5	7
47%	11	8
60%	4	9
73%	0	9
87%	0	10
100%	2	11

Fig. 9. Sample of the general convenience calculation for each sector (just 30–55).

5 Discussion. Compensation Scenarios for Social Housing

About the fairness of the plan with regard to the social housing issues, and according to the general perspectives of the Master Plan, the pattern allows us to select the sectors in which a reasonable quota of the supposed residential areas can be dedicated to social housing. The sectors are selected if a favorable ratio R2 occurs; once calculated the area to use for social housing by dividing the extraordinary permission fees by the unit social housing building cost, a threshold for the minimum area to be assigned to this function is established; the sectors are selected according to the threshold; finally, the total capacity of the whole area for social housing is calculated, and a quota of it can be assumed as relevant in order to satisfy the Master Plan needs. The overall economic-financial and evaluation profile is synthesized in Table 5, in which all sectors are included.

In order to take into account just those sectors that are significant from an economic point of view, a progressive reduction of the sample has been done in order to identify the sectors in which a reasonable surplus allows the decision makers to finance a reasonable quota of social housing. Thus, the sectors have been filtered on the basis of the $R2$ coefficient. The $R2$ threshold (minimum value) is set equal to 5 %: 52 sectors have been selected and the result is shown in Table 6.

At this stage some trade-offs can be outlined by connecting the following elements: the $R2$ coefficient; the sectors whose value of $R2$ overcomes the $R2$ threshold; the amount of extra-ordinary permission fees; the total amount of gross letting area developable in the selected sectors; the quota of the residential area to be dedicated to social housing as envisaged by the Master Plan; the housing market price in each zone where the sectors are located; the affordable housing price for the low-income house-holds; the gap between market and affordable price; the quota of the extra-ordinary permission fees to be funded by the social housing program.

Table 5. Overall private and public balance

ECONOMIC-FINANCIAL ELEMENTS		
costs	private	public
land value	€ 18.282.187	
building cost	€ 504.608.830	€ 199.490.780
additional expenses	€ 44.611.793	€ 13.964.355
Vat	€ 120.825.413	€ 43.887.972
ordinary permission fees	€ 1.246.206	
extra-ord permission fees	-€ 119.591.551	
cost of capital	€ 18.282.187	
total cost	€ 569.982.878	€ 257.343.106
revenues	private	public
real estate value	€ 606.870.761	
ordinary permission fees		€ 1.246.206
extra-ord permission fees		-€ 119.591.551
management		€ 13.899.687
total revenues	€ 606.870.761	-€ 104.445.658

EVALUATION PROFILE	
convenience	private
building cost and land value	688.328.223
ordinary permission fees	€ 1.246.206
extra-ord permission fees	-€ 119.591.551
cost of capital	€ 18.282.187
profit rate	3,21%
convenience	public
total building costs	€ 257.343.106
management	€ 13.899.687
ordinary permission fees	€ 1.246.206
extra-ord permission fees	-€ 119.591.551
Net social capital value	-€ 361.788.764
self-financing rate	-40,59%

Table 7 shows the data resulting from the implementation of the model in the scenario in which affordable housing is provided for the groups of households *L2* and *M1* and the sectors are located in the OMI zones *C3* e *D3*.

The most relevant trade-offs are between:

- the population to take up residence in the social housing settlements and the housing market prices of the different areas: the higher the housing prices – because of the central location of the sector – the fewer people can afford social housing;
- social housing and other public facilities: the more the quota of extra-ordinary permission fees is dedicated to social housing, the less other types of public facilities can be constructed;

- income gap and population to take up residence in the social housing settlements: given a fixed quota of extra-ordinary permission fees, the more the household group's income gap is wide – because the household's income is low – the fewer people can set themselves up in social housing settlements.

Table 6. Private and public balance of the 52 sectors with R2>5 %

ECONOMIC-FINANCIAL ELEMENTS		
costs	private	public
land value	€ 7.411.120	
building cost	€ 217.547.806	€ 108.414.995
additional expenses	€ 20.806.762	€ 7.589.050
Vat	€ 52.434.881	€ 23.851.299
ordinary permission fees	€ 906.930	
extra-ord permission fees	€ 35.840.280	
cost of capital	€ 7.411.120	
total cost	€ 334.947.779	€ 139.855.344
revenues	private	public
real estate value	€ 360.290.325	
ordinary permission fees		€ 906.930
extra-ord permission fees		€ 35.840.280
management		€ 4.935.221
total revenues	€ 360.290.325	€ 41.682.431

EVALUATION PROFILE	
convenience	private
building cost and land value	298.200.568
ordinary permission fees	€ 906.930
extra-ord permission fees	€ 35.840.280
cost of capital	€ 7.411.120
profit rate	2,21%
convenience	public
total building costs	€ 139.855.344
management	€ 4.935.221
ordinary permission fees	€ 906.930
extra-ord permission fees	€ 35.840.280
Net social capital value	-€ 98.172.913
self-financing rate	29,80%

Table 7. Balance for providing social housing in some sectors with R2>5 %

R2	sector N.	extra-ordinary permission fees €	residential gross letting area m²	SH %	SH gross letting area m²	SH inhab	OMI zones	housing price €/m²	house hold group	affordable housing price €/m²	income gap €/m²	permission fees compensing gap affordable/market price €	fees devoted to SH %
5%	52	35.840.280	71.448	30	21.434	714	C3	1.400	L2	734	666	14.267.177	40%
									M1	1.094	306	6.564.241	18%
10%	42	36.142.264	49.408	30	14.822	494	C3	1.400	L2	734	666	9.866.113	27%
									M1	1.094	306	4.539.338	13%
20%	35	34.959.957	38.466	30	11.540	385	C3	1.400	L2	734	666	7.681.184	22%
									M1	1.094	306	3.534.066	10%
40%	17	19.686.215	24.678	30	7.403	247	D3	1.250	L2	734	516	3.817.312	19%
									M1	1.094	156	1.156.761	6%

6 Conclusions

This study has pointed out that the issue of fair planning and affordable housing cannot be dealt with by basing only on the exchange of quantities (areas or volumes), but on the exchange of monetary and social values as well. The private e public balances (costs and revenues) can provide the indexes (thresholds) for verifying the profitability of urban developments and calculating the surplus that can be assigned to public facilities and in particular to social housing. Moreover the exchange of monetary values implies the exchange of social values both in terms of urban quality and social equity, which should be implemented in the political decision process.

Acknowledgements. Within the paper G. Napoli edited par. 3.2, 5 and 6; S. Giuffrida edited par. 1, 2, 3.1, 4 and 6; M.R. Trovato edited par. 1, 2 and 3.1.

References

1. GURS: Parte prima supplemento ordinario, n. 46, Repubblica Italiana (2007)
2. Cheshire, P., Nathan, M., Overman, H.: Urban Economics and Urban Policy. Challenging Conventional Policy Wisdom. Edward Elgar Publishing, Cheltenham (2014)
3. Giuffrida, S., Gagliano, F.: Sketching smart and fair cities WebGIS and spread sheets in a code. In: Murgante, B., Misra, S., Rocha, A.M.A., Torre, C., Rocha, J.G., Falcão, M.I., Taniar, D., Apduhan, B.O., Gervasi, O. (eds.) ICCSA 2014, Part III. LNCS, vol. 8581, pp. 284–299. Springer, Heidelberg (2014)
4. Cadell, C., Falk, N., King, F.: Regeneration in European cities: Making connections. Joseph Rowntree Foundation, York (2010)
5. Atkinson, A.B., Stiglitz, J.E.: Lectures on Public Economics. Princeton University Press, Princeton and Oxford (2015)
6. Dempsey, N., Bramley, G., Power, S., Brown, C.: The social dimension of sustainable development: Defining urban social sustainability. Sustain. Dev. **19**(5), 289–300 (2011)
7. Trovato, M.R., Giuffrida, S.: The choice problem of the urban performances to support the Pachino's redevelopment plan. Int. J. Bus. Intell. Data Min. **9**, 330–355 (2014)
8. Calabrò, F., Della Spina, L.: The public-private partnerships in buildings regeneration: a model appraisal of the benefits and for land value capture. In: 5nd International Engineering Conference 2014 (KKU-IENC 2014). Advanced Materials Research, vols. 931–932, pp. 555–559. Trans Tech Publications, Switzerland (2014)
9. Napoli, G.: Financial sustainability and morphogenesis of urban transformation project. In: Gervasi, O., Murgante, B., Misra, S., Gavrilova, M.L., Rocha, A.M.A.C., Torre, C., Taniar, D., Apduhan, B.O. (eds.) ICCSA 2015. LNCS, vol. 9157, pp. 178–193. Springer, Heidelberg (2015)
10. Geddes, M., Le Galés, P.: Local partnerships, welfare regimes and local governance. In: Geddes, M., Benington, J. (eds.) Local Partnerships and Social Exclusion in the European Union. New Form of Local Social Governance? Routledge, London (2001)
11. Gyourko, J.E., Kahn, M.E., Tracy, J.: Handbook of regional and urban economics: applied urban economics. In: Handbooks in Economics, pp. 1413–1454. North-Holland, Amsterdam (1999)
12. Lang, R.E., Anacker, K.B., Hornburg, S.: The new politics of affordable housing. Hous. Policy Debate **19**(2), 231–248 (2008)
13. Pittini, A., Ghekiere, L., Dijol, J., Kiss, I.: The state of housing in EU. Housing Europe, Brussels (2015). http://www.housingeurope.eu/resource-468/the-state-of-housing-in-the-eu-2015
14. CDP Cassa Depositi e Prestiti: Report monografico 03 – Social Housing. (2014). http://www.cdpisgr.it/static/upload/cdp/cdp-report-monografico_social-housing.pdf. Accessed Feb 2016
15. Nesticò, A., Galante, M.: An estimate model for the equalisation of real estate tax: A case study. Int. J. Bus. Intell. Data Min. **10**(1), 19–32 (2015)
16. Gan, Q., Hill, R.J.: Measuring housing affordability: looking beyond the median. J. Hous. Econ. **18**(2), 115–125 (2009)
17. National Association of Realtors: Housing Affordability Index (2005). http://www.realtor.org. Accessed Feb 2016

18. Stone, M.E., Burke, T., Ralston, L: The Residual Income Approach to Housing Affordability: The Theory and the Practise (2011). http://works.bepress.com/michael_stone/7
19. Fisher, L., Pollakowski, H.O., Zabel, J.E.: Amenity-Based Housing Affordability Indexes. Real Estate Economics **37**(4), 705–746 (2009)
20. Napoli, G.: The Economic Sustainability of Residential Location and Social Housing. An Application in Palermo City. Aestimum, Jan 2016, pp. 257–277 (2016). http://www.fupress. net/index.php/ceset/article/view/17896/16729
21. OMI: Rapporto immobiliare 2015. Il settore residenziale (2015). http://www.agenziaentrate. gov.it/wps/content/Nsilib/Nsi/Documentazione/omi/Pubblicazioni/Rapporti+immobiliari +residenziali/. Accessed Feb 2016
22. Regione Siciliana: Annuario Statistico Regionale. Sicilia 2014 (2014). http://pti.regione. sicilia.it/. Accessed Feb 2016
23. Camagni, R.: Economia urbana: principi e modelli teorici. NIS, Roma (1992)
24. Alonso, W.: Location and Land Use: Towards a General Theory oh Land Rent. Harvard University Press, Cambridge (1964). (Mass)
25. OMI: Banche dati. http://www.agenziaentrate.gov.it/wps/content/Nsilib/Nsi/Documentazione/ omi/. Accessed Apr 2016

Cap Rate and the Historic City. Past and Future of the Real Estate of Noto (Italy)

Salvatore Giuffrida[✉], Salvatore Di Mauro, and Alberto Valenti

Department of Civil Engineering and Architecture, University of Catania, Catania, Italy
sgiuffrida@dica.unict.it, dimaurosalvatore@live.com,
albvlnt79@gmail.com

Abstract. A real estate market survey has been carried out within the old town of Noto, in order to describe the progressive increase of the property market prices performed in the last decade. A sample of 56 properties was analyzed by considering a large amount of characteristics grouped in a frame of five main features progressively detailed. Due to the architectural, historical and landscape performances of this urban-land context, many expectations influence prices and the kind of and exploitation of these properties, so that a further study concerning the costs of renovation, the rents and cap rates of the surveyed properties has been carried out in order to identify the economic, financial and monetary profile of each of them and some recursive rules for investments' layout. A multi-layer clustering analysis was carried out to integrate two approaches, the empirical one sorting by characteristics, the analytic one based on the scored attributes.

Keywords: Old towns · Real estate market · Imperfect markets · Cluster analysis

1 Introduction

The Evaluation Science teaches the attribution of value judgments to goods, projects and investments in queries where the territorial-real estate and therefore the economic category of rent are prevalent. In urban units with a high historic-cultural gradient, as in the case of the town of Noto, the "position rent" is connected to the "information rent" [1]. Economic rent is considered "a not earned revenue" [2] that originates unfavourable and not easily arguable ethical options, among which the issues of the redistribution of the social product and the access to central places are prevalent.

The most controversial aspect of rent is its influence on the real estate market price, depending both on its dimension and its quality [3]; a real estate is worth (a) as much as it yields it, and is worth (b) as much as it is capable of making it.

The development of the real estate capitalism has inverted the relation between production and wealth hoarding, that means that most part of the hoarded capital constitutes the condition for the production of income, and not vice versa, due to the succession of the real estate cycles as well [4, 5].

The real estate capital is one of the most resistant forms of the process of transformation of the produced wealth in capital value, and in the historic city it assumes a particular characterization due to the uniqueness of the strongly characterized old town

© Springer International Publishing Switzerland 2016
O. Gervasi et al. (Eds.): ICCSA 2016, Part IV, LNCS 9789, pp. 63–78, 2016.
DOI: 10.1007/978-3-319-42089-9_5

contexts. Real estate is durable carries out three complementary economic functions, as it has utilitarian, productive [6] and speculative value. The convergence, the complementarity and the substitutability of these three economic functions select economic subjects that own capital reserves compensating the recessive cycles, and the knowledge of the administrative system and the rental market [7, 8].

Real estate is differently characterized by liquidity, in whose respect the real estate field is very heterogeneous. In the particularly active markets and in contexts where the offered real estate assets are articulated, different expectations that differently motivate the actors to sell and purchase – due to the evident income, cultural and entrepreneurial differences characterizing the profiles of offer and demand – are prevalent. The capitalization rate is one of the most dense information variables concerning these characteristics.

Referring to the case of the Sicilian historical town of Noto, this paper addresses the phenomenon of the capitalized – explicit and implicit – urban rent within a context characterized by many unexpressed potentialities and getting through a gentrification process led by outside entrepreneurs and foreign investors. A sample of 56 properties was analysed by considering a relevant amount of characteristics grouped in a frame of five main features. A multi-layer clustering analysis was carried out aiming at defining consistent housing market segments. In order to represent the matters that can be relevant for investors, a parametric cost model was drew up and a further survey about ordinary and extraordinary rental was carried out. At last, a comparison between the expected Net Operating Income (from rent) and the stock value (ask prices + costs) of each property was carried out in order to define the relation between capitalization rate and urban-architectural quality in the perspective of the two typically different real estate investment objectives – productive and/or speculative.

2 Materials. Noto Real Estate

Noto is one of the eight towns of "Val Di Noto" declared Unesco World Heritage in 2002. After the earthquake in 1693 it was rebuilt in two areas at different altitudes. The more recent area extends on the Meti hillside, and its town centre is located inside the Unesco area (Fig. 1).

Noto's urban structure and its building heritage have largely maintained their original character due to the prevalence of an agricultural, pre-Industrial and tertiary Economy which has not led to a significant growth of the construction industry.

The recent income increase, recording approximately a 4 % growth rate in the last decade, has led to a new phase in which the old town centre has been definitively considered as one of the main resources of the territory, not only within the local area, but also at provincial level. As a consequence several industrial and foreign real estate investors focused on it just before the sharp price rising.

Despite the magnificence of the squares, the religious and civil monuments and the beauty of the aristocratic buildings, Noto's real estate continued to have a mainly utilitarian function until 2000, when the real estate market got over the lethargic phase in which the market prices were even slightly lower than the construction costs.

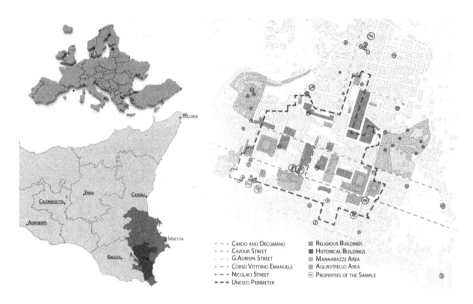

Fig. 1. Geographical and topographical localization of Noto and the survey area. (Color figure online)

Landscape and architectural elements and amazing views helped foster both mass and cultural tourism and make the Noto's housing market boom. As a consequence new activities such as catering services and business accommodation spread through the town centre. It was possible to record an imbalance between the town centre, in which the extra-ordinary profitability expectation prevailed, variously capitalized by the demanded prices, and the less desired area whose demand is for residential and not investment purpose.

A sample of 56 properties around Via dei Mille and Via Roma, Via Napoli and Via Tommaso Fazzello was analysed. The data were obtained by considering a high number of characteristics grouped in a framework of five progressively detailed main features. The sample includes all properties of this market area. The in-depth analysis has been carried out basing on 43 attributes aggregated in 5 groups of characteristics: location k_e, intrinsic k_i, technological k_t, economic k_p, architectural k_a. The attributes are expressed in a standard scale from 1 (lowest) to 5 (highest quality condition) and then standardized in z-scores. Each property of the sample can be described by means of progressively more aggregated regressors: from 44 to 5, up to one, k^* is calculated by aggregating all scores through a system of weights. Table 1 shows a synthesis of the large and detailed analysis carried out.

Table 1 shows for 28 of the 56 properties: in the first block, identification, dimensions (total and unit), ask prices and scores of the aggregated attributes; in the second block, the calculation of the renovation costs of the properties according to their maintenance status; in the third group, the calculation of the expectable annual rents; the last column reports the capitalization rates; the latter three blocks will be afterwords explained. The complexity of the sample can be considered well represented by the different distributions of prices and characteristics corresponding to the different number of classes (Fig. 2): the greater the number of classes (higher resolution of the observation), the greater the number of waves of the frequency functions, both for prices and for characteristics.

The distribution of the attributes scores shows a more marked heterogeneity in comparison to the total score k tot and the unit prices; the latter indicate the prevalence of lower-category prices. As the image resolution increases (from the upper paragraph to the bottom) rising prices become evident (right).

Table 1. Synthesis of the market survey with renovation costs, rents and capitalization rates.

id	floor	rooms	rentable area	ask price (000€)	unit prices (000€) /room	/sqm	location ke	intrinsic ki	technological kt	economi kp	architectural ka	overall score k*	standard overall score k*st	total cost	unit costs (000€) /room	/sqm	total ordinary (000€)	daily unit /room	extraordinary days	capitalization rate
1	0/1	7,6	188,2	315	41,4	1,7	4,6	2,9	4,6	3,4	2,7	3,9	0,76	21,3	0,3	0,01	5,8	24	96	1,85%
2	0	5,5	120,0	150	27,3	1,3	4,3	1,9	4,3	2,6	2,4	3,4	0,12	50,1	0,9	0,04	3,7	23	92	2,45%
3	0	2,4	35,0	50	21,3	1,4	2,6	1,3	4,5	3,4	3,3	3,1	-0,23	12,3	0,5	0,04	1,2	23	82	2,44%
4	0/1	4,3	78,8	130	30,2	1,7	2,6	2,5	4,2	1,8	2,1	3,3	-0,08	21,1	0,5	0,03	2,6	23	87	1,98%
5	1	5,5	104,5	110	20,0	1,1	2,6	2,5	4,3	3,4	3,1	3,4	0,09	38,3	0,7	0,04	3,5	23	89	3,21%
6	1	3,0	48,0	50	16,7	1,0	1,7	1,9	4,6	3,4	3,4	3,2	-0,09	10,9	0,4	0,02	1,7	23	86	3,48%
7	0/1	4,2	85,0	115	27,7	1,4	1,2	1,5	3,9	2,6	1,8	2,6	-0,95	112,4	2,7	0,13	2,3	22	92	2,02%
8	-1	2,3	35,0	35	15,2	1,0	1,2	1,4	3,1	1,8	1,4	2,1	-1,56	131,4	5,7	0,44	1,0	21	82	2,73%
9	0/1	2,8	35,5	60	21,4	1,7	3,0	2,1	4,6	3,4	3,7	3,5	0,23	4,0	0,1	0,01	1,5	23	81	2,57%
10	0/1	4,0	65,0	130	32,5	2,0	1,0	2,1	4,2	2,6	3,2	3,0	-0,43	32,3	0,8	0,05	2,2	22	86	1,71%
11	0/1	5,2	86,0	55	10,6	0,6	1,6	1,3	3,5	2,6	1,7	2,4	-1,26	108,0	2,1	0,13	2,5	21	86	4,51%
12	0/1	3,6	60,8	120	33,3	2,0	2,7	3,4	4,6	3,4	2,7	3,8	0,59	13,6	0,4	0,02	2,3	24	85	1,89%
13	0	2,5	50,0	50	20,0	1,0	2,7	1,4	3,2	3,4	2,0	2,5	-1,13	184,2	5,3	0,37	1,6	21	82	3,28%
14	1	3,3	50,7	50	15,2	1,0	2,7	3,3	2,9	4,2	2,3	3,0	-0,47	251,6	7,6	0,50	1,7	22	82	3,49%
15	0	6,8	143,0	90	13,2	0,6	1,8	3,1	2,7	3,4	2,0	2,6	-0,91	775,9	11,4	0,54	3,8	22	89	4,19%
16	1	3,5	55,0	120	34,3	2,2	4,9	2,2	4,1	5,0	2,9	3,6	0,32	48,7	1,4	0,09	2,1	24	84	1,76%
17	0	4,3	85,0	75	17,4	0,9	3,5	2,0	4,1	3,4	2,2	3,2	-0,12	55,3	1,3	0,07	2,7	23	90	3,60%
18	0	2,2	30,0	70	32,6	2,3	3,3	1,1	4,5	3,4	3,7	3,3	-0,08	6,8	0,3	0,02	1,2	23	83	1,69%
19	0	1,8	30,0	30	16,7	1,0	2,8	1,7	4,2	3,4	2,7	3,1	-0,26	21,4	1,2	0,07	1,0	23	86	3,47%
20	0	3,8	50,0	25	6,7	0,5	2,7	1,3	1,0	2,6	1,0	1,3	-2,60	488,9	13,0	0,98	1,1	19	79	4,48%
21	0	6,8	130,0	250	36,8	1,9	5,0	3,0	4,7	1,8	4,1	4,2	1,12	52,5	0,8	0,04	5,0	25	89	1,99%
22	0/1	5,6	120,0	175	31,3	1,5	4,5	3,3	4,7	1,8	3,5	4,1	1,02	41,2	0,7	0,03	4,3	25	92	2,45%
23	0	5,1	101,7	120	23,5	1,2	2,6	2,6	2,6	3,4	1,6	2,5	-1,08	513,7	10,1	0,51	2,7	21	88	2,22%
24	0	3,6	72,5	80	22,2	1,1	1,1	1,3	3,2	2,6	2,0	2,2	-1,46	228,0	6,3	0,36	1,7	21	87	2,10%
25	0/1	6,6	166,0	350	53,0	2,1	4,5	3,3	4,7	3,4	4,6	4,2	1,22	22,0	0,3	0,01	5,5	25	98	1,57%
26	0/1	6,3	83,5	105	16,8	1,3	1,6	1,9	2,3	2,6	1,6	2,0	-1,76	394,4	6,3	0,47	2,4	20	80	2,30%
27	1	6,8	133,8	220	32,4	1,6	3,2	3,9	4,5	3,4	3,5	4,0	0,94	66,3	1,0	0,05	4,9	25	89	2,25%
28	0	15,2	333,0	1100	72,4	3,3	4,5	4,5	5,0	3,4	4,8	4,7	1,87	125,4	0,8	0,04	12,7	26	93	1,16%

Furthermore the relation between unit prices and aggregate quality index shows a heterogeneity that cannot be reduced but by accepting coefficients almost zero or negatives in the multiple linear regression function about the very significant regression, such as the location and the architectural quality; more reasonable results are obtained with multiple logarithmic regression (Table 2) but with a great difference between the two R^2 with equivalent regressors coefficients.

The pattern shows that the bigger and high-priced properties cost less than the smaller and low-priced ones. The real estate recording the highest price and the best quality is however overestimated if compared to the generality of these sample relations. We are referring to a long time unsold property whose owner does not accept to lower prices (Fig. 3).

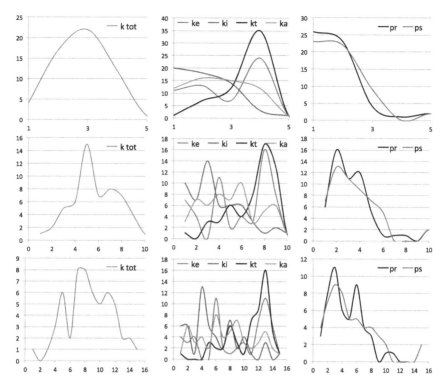

Fig. 2. Scores (left) and unit prices (right) distribution in classes of ranges (9 up, 15 down). (Color figure online)

Table 2. R^2 and coefficients of the sample for both linear and logarithmic multiple regressions.

Linear regression					
y		x_1	x_2	x_3	x_4
	R^2	ke	ki	kt	ka
p/room	0,47	0,05	0,48	0,50	−0,03
p/sq.m	0,34	0,01	0,40	0,65	−0,06
Logarithmic regression					
y		x_1	x_2	x_3	x_4
	R^2	ke	ki	kt	ka
p/room	0,51	0,23	0,26	0,29	0,22
p/sq.m	0,36	0,23	0,26	0,29	0,22

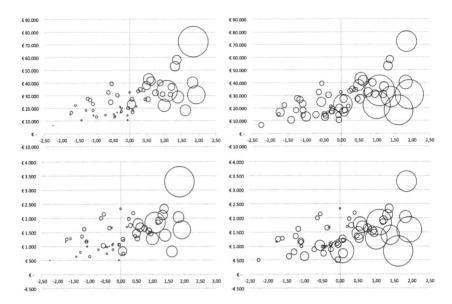

Fig. 3. Top: (left) relation between prices per room (ordinate), number of rooms (bubbles diameters) and aggregated value (abscissa). Down: (left) relation between prices per sq m (ordinate), surface (bubbles diameters) and aggregated value (abscissa); (right) relation between price per sq m (ordinate), total price (bubbles diameters) and aggregated value (abscissa).

3   Methods

3.1   General Issues

On the basis of the sample analysis our contribute aims at providing an organized set of knowledge to infer the economic-real estate profile of Noto properties in the perspective of the Capital Theory [1, 4, 9] exposed elsewhere [10] and summarized below:

1. The value of a capital asset, depends on:
   a. its temporal shape and size, that is extent, distribution and duration of income that can be connected to it;
   b. its monetary shape and size, that is ability to induce business deals over (increasing) or under (decreasing) current prices.
2. This ability derives from the ratio between implicit and explicit liquidity – that should be considered as a rate of "propensity to hoarding":
   a. the explicit liquidity is the regular income or Net Operating Income (NOI);
   b. the implicit liquidity, is a kind of psychological income that replaces NOI when not earned, voluntarily or not, by the owner. The latter: in the first case, due to the management risk is motivated to keep the asset unprofitable in order to preserve its income and exchange potentialities; in the second case, he keeps the source of income, despite – due to the decrease of the yield – the payback period extends; the capital value of the implicit liquidity stream is the expected capital

gain; this expectation keeps the ask prices higher than the bid prices and, as a consequence, the viscosity of the market grows; sometimes this viscosity turns into paralysis; in this sense, real estate property develops an anti-cyclical function regarding to economic fluctuations collecting liquidity lost by the other assets or money.

3. The temporal shape and size is represented by capitalization coefficient corresponding to the "average period" of the expected stream, regarding the specific asset: if it is greater than the normal stream "average period", than property has a *crescendo*, otherwise it has a *diminuendo*; in case of *crescendo* or *diminuendo* the asset capitalizes a greater or smaller number of incomes than those capitalized by the normal stream; an increasing asset is a "money-good", it represents a value reserve and contributes to the "wealth effect"; otherwise a decreasing asset does not;

4. The *crescendo/diminuendo* trend may regard value, income and liquidity; in this sense the capitalization rate is different from the profitability one [11] since it capitalizes not earned, just represented, incomes.

3.2   Procedure

Our goal to define the monetary profile of Noto's real estate considering our market analysis, is achieved by calculating the capitalization rate of each unit taken into account in our study. The procedure has been articulated in four stages.

The first stage, connected to the need to distinguish the structural aspects of this market, concerns the delimitation of segments that might include sample's real estate depending on their characteristics: for this purpose two procedures have been carried out. The first is a classification about some significant features of the real estate considering the identity of the Urban context: the real estate are classified into three or four groups, from the most to the least valuable [12–14].

The second classification has been carried out according to the k-means method [15] aggregating the scores of the four main features. Productive characteristics have been excluded instead. Furthermore it has been elaborated an evaluation model of the parametric costs for each of the required machining from the state of maintenance: real estate unit; the building within it is inserted, calculated as a function of the relevant portion (according to competence rate). This in-deep analysis has allowed to calibrate the score assigned to the real estate from the technological characteristics point of view and, in detail considering the preservation status, the score has been calculated by normalizing the calculated unit cost.

The third stage concerned the calculation of the expected profitability of each real estate with reference to the ordinary income (NOI) and extra-ordinary more specifically connected to the seasonal rent potentialities, really common in a city like Noto, characterized by great tourist potential. Two investigations have been developed. The first deals with income from ordinary rent, the second one concerns the real estate segment referring to real estate not intended for habitation as (B&B; Holiday-houses and rooms for rent). Gross Annual incomes have been calculated according to multiple logarithmic evaluating function considering: for ordinary incomes, the annual rent per sq.m of gross letting area; for extra-ordinary incomes, the annual rent according to the number of really usable rooms, calculated on the daily room rates that may vary depending on the type and size of each property.

The fourth stage concerns the calculation of the cap-rates of each property for sale; according to Rizzo's theoretical approach (1999) and Forte's procedure [16], the cap-rate is the ratio of NOI and the sum of: 1. the ask price; 2. The rehabilitation cost; 3. the expected capital gain/loss due to the potential extra-income. Two variants were implemented: in the former, the ask price is replaced by the value calculated by a multiple logarithmic regression trend; in the latter, assuming a "highest-and-best-use-oriented behaviour", only capital gains – not losses – were taken into account. Capital gains and losses were calculated by capitalizing the extra-income coming from the seasonal exploitation for tourist accommodation. In the calculations done by multiple logarithmic regression the components of k_t score reporting the maintenance status degree are turned to the highest level.

Combining C. Forte procedure and F. Rizzo's theoretical model, each real estate sample has been calculated by adding to the type sample, obtained by dividing ordinary income for asked price, current asset linked to the extraordinary expected profitability. The latter has been calculated as it follows: two gain expectations, the ordinary and extra-ordinary ones, have been compared and it has been calculated net income. For each real estate has been calculated the number of days per year of extra-ordinary rent that balances two net incomes: it has been capitalized extra income differential, both positive and negative, compared to the trend income, by calculating the expected capital gain (loss); furthermore it has been defined the two perspective earnings-costs profile allowing us to obtain net ordinary income, ant the extra-ordinary component. The latter has been calculated according to the value increase that the property might have due to extra income.

4   Results and Discussions

4.1   Clustering Analysis

Clustering by Typical Characteristics. Clustering by *typology* (T1 historic building; T2 small historic building; T3 ordinary building) provides a wide price range for the T3 group and gradually decreasing price ranges for the T2 and T1 groups. The average prices decrease less than the overall quality standard score k^*, so that it can be supposed that the high prices of the more valued properties drag up the prices of the less valued ones.

Clustering by *location* (L1 pedestrian area; L2 adjacent areas; L3 outside areas; L4 high town areas) a wide fluctuation of the prices of the central properties may be observed. The prices generally decrease to a high degree from the first to the second cluster, and to a much lesser degree towards the other ones. This highlights the relevance of a central location independently from the typology. The aggregated value decreases as well, but as far as the third cluster, that means that the external areas have not yet expressed potentials, due to the favourable relation between value k^* and price (Fig. 4).

Clustering by *maintenance status* (M1 no work required; M2 ordinary maintenance works; M3 extra-ordinary maintenance works; M4 renovation or restoration works) it is possible to observe a wide fluctuation of the prices of the properties in good condition,

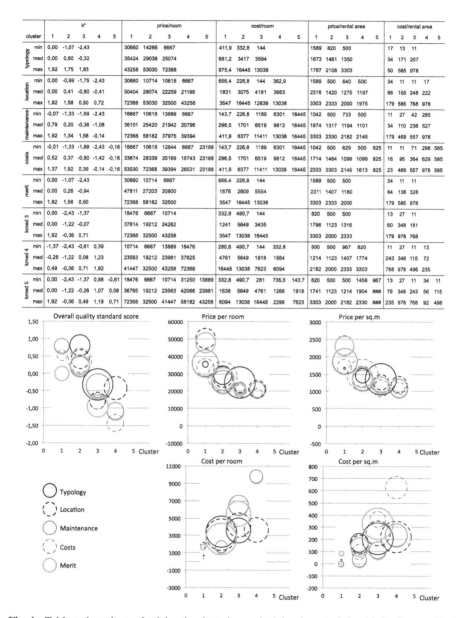

Fig. 4. Table and graphs synthesizing the clustering analysis by characteristics. (Color figure online)

which are lower than the ones of the buildings requiring significant renovation interventions; the importance of the maintenance status may be observed to a much higher degree in the characterization, due to the more marked variation of k^*.

Clustering by *merit* it is possible to observe wider price fluctuations in the two higher clusters. The most significant feature is the price range in the first cluster in comparison to the other ones, and in marginal terms as well; this highlights the role of guide segments performed by the high value properties.

Clustering by Scores: The Method of k-means. The aggregation in three clusters provides a precise segmentation of the sample concerning the correspondence between values and prices. In the first cluster a wide fluctuation of prices corresponds to a modest fluctuation of values. The spatial distribution reproducing differently clustering by localization, and more markedly the aggregation of the best properties around the central area, is significant too.

The aggregation in four clusters divides the central cluster in two not easily identifiable parts. Basing on the data in the table it is easy to verify how these further clusters differ in centrality and architectural quality that compensate each other originating overall rather similar values and prices.

The aggregation in five clusters divides the first cluster in two subgroups which are not distinguishable in this case as well. Basing on the data in the table it is possible to observe that the subgroups differ from the points of view of the productivity and archi-tectural-environmental characteristics; as the first ones don't influence price particularly, the modest variation of the latter is ascribable to the difference in architectural quality (Figs. 5 and 6).

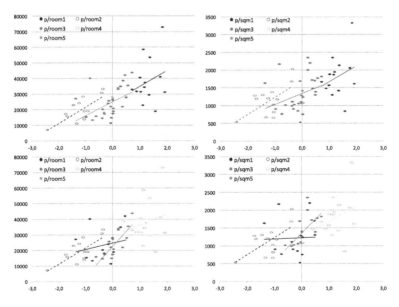

Fig. 5. Synthesis of the clustering analysis by scores 3 and 4 clusters hypotheses (5 clusters hypothesis omitted). (Color figure online)

Fig. 6. Location of the *k-means* four clusters within the sample area.

4.2 Costs Calculation

The concern about the effects of the rapid transformation of the real estate prices at the evolving of the architectural values, suggests that we should extend the property market analysis towards the issue of the real estate investments convenience. In this regard a research about the parametric costs for the technological performance improvement of the surveyed properties was carried out in order to identify the most profitable ones, in terms of capitalization rate as well. Therefore it was necessary to build a model combining the detailed and specified degree of technological quality with the redevelopment cost. To this end the sample was divided according to the intervention categories defined at art. 3 of D.P.R. (Italian Law Decree) 6 June 2001 n° 380 as: No works; Ordinary maintenance; Extraordinary maintenance; Renovation and Restoration (comprising a quota of the works for the communal parts of the building).

The parametric costs per room and per sq m of all the above mentioned intervention categories have been calculated envisaging a progressively increasing quota of the single works at the changing of the intervention category from the lightest and most conservative one to the most consistent and transformative one. Consequently, an intervention category has been associated to each property of the sample and – basing on the knowledge of the dimensions and the effects of the components of each building (facades, walls, roofs, ceilings) and the quota of participation in the expenses for the communal parts – the redevelopment cost of each property has been calculated. A synthesis of the results is shown in Fig. 7 where costs and unit prices are correlated to the variation of the conservation status and the dimension of the census buildings.

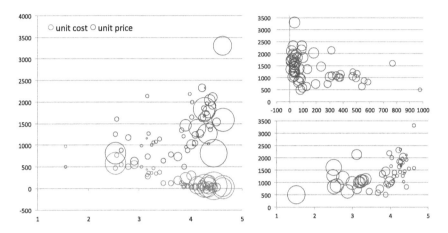

Fig. 7. Relationships between unit costs, unit prices and technological quality. (Color figure online)

The comparisons show: (left graph) the robust inverse relation between unit cost and technological quality (k_t) $(R^2 = 0,97)$; the direct connection between properties size (bubble size) and k_t; the direct connection between properties size and k_t; the weak and non-linear relation between unit house-market price and k_t $(R^2 = 0,31)$; the increase of the unit price fluctuation at the k_t increasing (top-right): the non-linear price/cost trade-off; (down-right); the direct relation between price and k_t, and the inverse relation between cost (bubble size) and k_t.

4.3 Ordinary and Extra-Ordinary Incomes

Ordinary Incomes. Basing on a market survey on the rented properties, the relation between the variation of the aggregated quality index and rents was defined, allowing determining the ordinary income of the properties of the sample.

Extra-Ordinary Incomes. The census of the complementary lodging structures, carried out basing on data obtained from the portal Booking.com and the Noto's Municipality, concerns a sample of 75 b&b, 4 landlords, 61 holiday homes and 33 apartments, and has allowed to calculate the minimum, medium and maximum prices per person for each category and in the two different seasons. The values used for the calculation of the extra-ordinary profitability don't take into account the top extremes, but the interval comprised between 60 € and 110 € (Figs. 8 and 9).

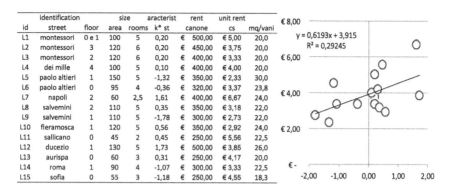

Fig. 8. Results of the survey of the residential segment for rent.

Fig. 9. Results of the survey of the tourist accommodation segment.

4.4 Capitalization Rates

In this phase the two perspectives of ordinary and extra-ordinary rental profitability have been compared. Basing on the qualitative profiles of the properties for sales the following rents have been calculated: the ordinary unit monthly rents per sq.m and the total ordinary rents; the extra-ordinary daily rents, and, basing on the latter, the total extra-ordinary rents (by multiplying the daily rents per room by the number of rooms reduced at the 70 % to keep into account the functional stiffness of the properties to be converted in lodging structures) for the envisaged annual rental days.

The expenses on the owner's part have been deducted from the ordinary income in a percentage variable from 30 % to 55 %. The following operative costs have been deducted from the extra-ordinary gross income: used materials and services; furniture depreciation; ordinary maintenance; management.

Once obtained the annual gross income, the following expenses on the owner's part have been deducted from it: extra-ordinary maintenance from 3 % to 6 %, according to the technological features and the maintenance status; administration from 1 % to 2 %, according to the ratio between the dimensions of the real estate unit and the building where it is located; insurance from 1 % to 2 %, according to the commercial surface; vacancy and uncollectable credit risk from 3 % to 6 %, inversely proportional to the overall quality; taxes, on average 33 %.

Once obtained the extra-ordinary net income, it was possible to calculate the lodging days necessary to equalize the net income obtained from long term rentals, that vary from 75 to 100 with a median of 88 days. According to the theory of capital by Hicks (1959) and Rizzo (1999), the properties that require a number of days higher than the median have been considered "in decreasing" (of income); vice versa, the properties requiring a lower number of days have been considered "in increasing" (of income). In a similar way it is possible to detect the properties "in increasing" of liquidity, as they have a capitalization rate lower than the average, inside the sample and inside each cluster individuated by means of the k-means method.

Figure 10 compares the results of the two ways for the cap rates calculation. The former (left), shows greater cap rates, especially in the lowest-k clusters. Cap rate decreases significantly if k increases. The latter (right) takes into account of the renovation cost in addition to the ask price. In this case, according to the Keynesian propensity to investments, the expected income stream needs to extend over the ordinary payback period and the range of the cap rate significantly reduces, especially the ones associated to the less performing properties needing renovations.

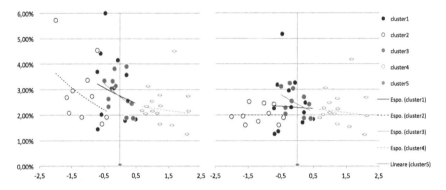

Fig. 10. Comparison between current and expected cap rates. (Color figure online)

As a consequence, the difference between the cap rates calculated by the two ways, can be considered an index of the implicit liquidity that characterizes such an imperfect and challenging local real estate market [17] (Fig. 11).

The articulation of the matters influencing the decisions concerning the real estate investment in such an articulated housing market, needs to be made explicit in order to provide helpful reports and tools to the municipalities aiming at supporting renovation and preservation urban policies. The wide sets of data we processed allowed us to achieve significant results about the two main correlations, the one concerning ask price and the one concerning cap rates. Because articulation of the variables considered in addition, and the different ways in which they fit together, the latter should be considered the most significant.

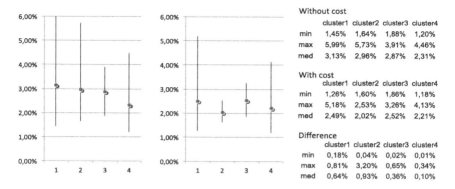

Fig. 11. Implicit liquidity index calculation.

5 Conclusions

The analysis of the sample from the point of view of the capitalization rates has allowed combining and comparing incomes and values, streams and stocks, highlighting the typical characteristics of a real estate market characterized by significant perspectives of growth tied to wealth, quality and variety of the local context.

The case of Noto is a specimen of the attention that the advanced capitalism, characterized by financial surplus, shows on all assets with unexpressed high potentials.

The analysis carried out in this context has highlighted the typical characters of these phenomena: the complexity of the market and its articulation by segments; the weak relation between economic expectancies and symbolic variables; the extension of such economic expectancies to less worth properties as well; the higher explanatory capability of the capitalization rates in comparison to prices; the strong relation between "monetary form" and "architectural form".

In the spirit of the logic of the capitalization rate [4] in the context of the investments in the historic city, it is necessary to depurate the territorial-urban-architectural approach to real estate evaluation from the most typically business like – and therefore accounting and financial – tendencies: the two approaches differ concerning the "real estate ethics" profile they put in place, and consequently the type of subject and context they address: the business approach refers to the cash flow and valorizes risk [18]; the territorial approach refers to value and valorizes the expected capital income.

The historic city is the most suitable context for the application of the territorial approach: it is a general form in whose identity each architecture unit participates according to its type-morphological profile; consequently the real estate values of the architectural units influence each other realizing the process of abstraction of value from the specific characteristics, that differentiates the motivations for the income variations from the motivations for the value change.

Acknowledgements. S. Giuffrida edited par. 1, 3, 4.1, 4.4 and 5; S. Di Mauro edited par. 2, 4.2 and 4.3, and carried out the house market survey; Alberto Valenti carried out the calculation for the application of k-means procedure.

References

1. Rizzo, F.: Valore e Valutazioni. La scienza dell'economia o l'economia della scienza, FrancoAngeli, Milano (1999)
2. Camagni, R.: Economia urbana. Principi e modelli teorici, Nuova Italia Scientifica, Roma (1992)
3. Brown, G.R., Matysiak, G.A.: Real Estate Investment. A Capital Market Approach. Prentice Hall-Pearson Education, Harlow UK (2000)
4. Rizzo, F.: Dalla rivoluzione keynesiana alla nuova economia. Dis-equilibrio, trasinformazione e coefficiente di capitalizzazione, FrancoAngeli, Milano (2002)
5. Gabrielli, L., Copiello, S.: Marginal costs and benefits in building energy retrofitting transaction. In: Conference Proceedings - SBE16 Hamburg International Conference on Sustainable Built Environment Strategies – Stakeholders – Success factors, pp. 836–845. Ed. ZEBAU – Centre for Energy, Construction, Architecture and the Environment GmbH (2015). ISBN 978-3-00-052213-0
6. Manganelli, B., Morano, P.: Sull'analisi finanziaria di Ellwood, un modello che razionalizza la stima del saggio di capitalizzazione, Aestimum, n. 46 (2005)
7. Trovato, M.R.: A fuzzy measure of the ability of a real estate capital to increase in value. The real estate decision problem for Ortigia, in Appraisals. Evolving proceedings in global change **2**, 697–720, Firenze University Press, Firenze (2012)
8. Trovato, M.R.: The real estate decision problem. A model to support the real estate market analysis. In: Rèsumés/Abstracts, 78 Meeting of the European Working Group "Multiple Criteria Decision Aiding" Catania 2013, pp. 65–66. University of Catania (2013)
9. Rizzo, F.: Nuova economia. Aracne, Roma (2013)
10. Giuffrida, S., Ferluga, G., Valenti, A.: Capitalisation rates and 'real estate semantic chains' an application of clustering analysis. Int. J. Bus. Intell. Data Min. **10**(2), 174–198 (2015). doi: 10.1504/IJBIDM.2015.069271. ISSN: 17438187, Inderscience Enterprises Ltd., Genève, Switzerland
11. Simonotti, M.: Ricerca del saggio di capitalizzazione nel mercato immobiliare. Aestimum **59**(1), 171–180 (2011)
12. Everitt, B., Landau, S., Morvene, L., Stahl, D.: Cluster Analysis, 5th edn. Wiley, Chichester (2011)
13. Ball, G., Hall, D.: ISODATA, a novel method of data analysis and pattern classification. Stanford Research Institute, Menlo Park, Calif (1965)
14. Halkidi, M., Batistakis, Y., Vazirgiannis, M.: On clustering validation techniques. J. Int. Inf. Syst. **17**(2/3), 107–145 (2001)
15. Anderberg, M.R.: Cluster analysis for application. Academic Press, New York (1973)
16. Forte, C.: Elementi di estimo urbano. Etas Kompass, Milano (1968)
17. Hepşen, A., Vatansever, M.: Using hierarchical clustering algorithms for turkish residential market. Int. J. Econ. Finan. **4**, 138–150 (2012). doi:10.5539/ijef.v4n1p138
18. Taltavull, P., Gabrielli, L.: Housing supply and price reactions: a comparison approach to Spanish and Italian Markets. Hous. Stud. **30**(7), 1036–1063 (2015)

Industrial Areas and the City. Equalization and Compensation in a Value-Oriented Allocation Pattern

Salvatore Giuffrida[1(✉)], Grazia Napoli[2], and Maria Rosa Trovato[1]

[1] Department of Civil Engineering and Architecture, Catania, Italy
{sgiuffrida,mrtrovato}@dica.unict.it
[2] Department of Architecture, Palermo, Italy
grazia.napoli@unipa.it

Abstract. This study deals with the allocation of the firms in a large industrial area of Quarto, a town in the Naples' district subject to a "Piano di Insediamenti Industriali" – PIP (Industrial Settlement Masterplan). The main concern of the Municipality is the fair integration between environmental issues, economic development and urban identity. Therefore a structured evaluation process, based on a survey about the company profiles and their geographical location, has been carried out in order to make the plan meet the needs of the firms, and to select the best companies to settle in the planned area. A "generative" MAVT pattern has been designed to outline several layout options and to select the best ones. The model also includes equalization and compensation elements by whose means it is possible to determine the extraordinary planning permission fees for the different areas where the firms are located.

Keywords: Industrial areas · Sustainable land planning · MAVT · Equalization · Compensation

1 Introduction

Over the last years many land development projects concerning the location of production activities were made without taking into any account the local social and territorial features. This often prevented the full achievement of the objectives such proposals were aiming at, largely crippling the work of the public and private stakeholders involved in the design and realization of these projects. In Southern Italy this is confirmed by the numerous IDAs (Industrial Development Areas) that are still empty or largely incomplete because of the failure or lacking of a preventive assessment of the local needs and opportunities of integration with the territory where they have been placed [1]. This study deals with the "Piano di Insediamenti Industriali" PIP of Quarto, a town located in the Flegreo area, Naples District. PIP, which is an industrial area settlement plan, has identified in its rural territory the areas to allocate as pole of development for the enterprises of the whole district. In this area, the original agricultural use is still prevalent although it has been gradually replaced by residential and industrial uses as the result of a slow but inexorable process of urbanization. This urbanization has been a spontaneous and illegal process and, recently, has been accompanied by the construction of

© Springer International Publishing Switzerland 2016
O. Gervasi et al. (Eds.): ICCSA 2016, Part IV, LNCS 9789, pp. 79–94, 2016.
DOI: 10.1007/978-3-319-42089-9_6

road infrastructure planned by an "extraordinary" commissioner, thanks to laws and programmes enacted after the earthquake that damaged the area in 1980. The paper proposes an innovative intervention approach centred on the features and the real needs of the connective fabric of the enterprises – that are local and from the whole Flegreo area as well – and capable of achieving an efficient process of settlement for the enterprises themselves. This approach uses a model which integrates the results of preliminary analysis, evaluation, choice and design, for selecting those firms which are more qualified in kick-start local economy while respecting cultural, environmental, and urban values.

Quarto's PIP is a tool for managing the realization of an integrated centre of research and development, according to the following purposes:

- To promote social-economic development (high level and stability of employment, quality of work) giving preference to firms that are linked to local economic networks, cooperatives owned by young people, etc.;
- To protect environmental resources by locating those innovative firms that produce low-level of pollution and use high-level of information;
- To select compatible enterprises able to activate positive synergies;
- To achieve a functional integration of industrial areas into the urban structure.

The change of land use and accessibility through planning produces the unavoidable effect of modifying the land market values, at urban level between PIP's areas and other urban areas, and at micro-territorial level among areas on which the construction of new buildings or urban infrastructures is planned. Equalization between landowners is been widely discussed and two models of equalization are been proposed: "areal" equalization, when the development rights are equals for all lots in an urban area; "extended" equalization, when the development rights are the same across the entire city. The latter model can be implemented by means of the Transferable Development Rights (TDRs) based on the idea that the development right can be separated from a specific parcel and transferred to another one. The two families of TDRs (zoning-integrative and zoning-alternative TDRs) have been in use for several decades in U.S. and Europe despite some criticism [2–5].

The PIP's drafting constitutes, therefore, the opportunity to reconfigure the flow of benefits and the equalization of wealth (rent) produced by planning. The flow of benefits is generated by natural landscapes, low cubage index, etc., and comes from the stock of local territorial resources and the cost opportunity which must be sustained by the citizens in the status quo (referring to the lost development of underused areas). The rent is a plus-value also called betterment, unearned increment or windfall profit [6]. There are two types of plus-value capture mechanism for financing public works and facilities [7]: direct by means of taxation and indirect by means of many tools such as developer obligations, infrastructure levy, impact fee or linkage, developer agreements, incentive zoning, transfer of development rights [6]. In Italy, many plans have adopted plus-value capture indirect mechanisms such as compensation, negotiation, or private-public partnership [8–11].

A multicriteria model has been elaborated to help the planning decision process realize PIP by selecting and locating the enterprises in the area; this model also includes a model of financial analysis used to base PIP's sizing on principles of equalization and compensation, and consequently to decide which areas should be given to enterprises and which planning permission regulations should be applied. The fair integration of subjective (land owners, entrepreneurs, public) and objective (land, identity, environment) targets is the general aim of the proposed model.

2 Materials

The PIP's final draft may be considered as the conclusive phase of the design process that began many years ago with the preliminary plan, which was an intermediate confirmation of some design solutions and proposed rules, and a tool for the consultation of the potential users of the plan itself, that identify the motivation for their investment and entrepreneurial risk in the project quality and in the procedural credibility. In particular, the preliminary plan highlighted some key issues:

- The relationship with the existing Master Plan and the objective difficulty in interpreting its choices and provisions;
- The demand analysis and the pre-qualification of firms to be located;
- The implementation procedure that, proposing a tool alternative to expropriation, introduced some basic choices subsequently adopted in the final draft.

PIP comprises a mostly flat area of 50 hectares at the Southwest base of the hills surrounding the Quarto's valley (Fig. 1). In this local context, the main infrastructures, such as the historical railway and the road system, are instrumental for defining the geographical references and the most important boundaries of the PIP's area. The North side of the planned area is delimited by the Circumflegreo railway and the Quarto's channel (the latter covered by infrastructures built after the earthquake); the East side is delimited by the Railway Naples-Rome; the South side is delimited by via Grotta del Sole (which defines the border with the Pozzuoli municipality); the West side is delimited by Via Spinelli.

Fig. 1. Modelled view of Quarto's PIP

The approved Master Plan allocates most part of the territory comprised in the triangle delimited by the former Quarto's channel, the railway Napoli-Roma and Via Grotta del Sole, to the industrial-handicraft use. The plan also identifies several areas with different functions: Da (existing industrial area), Db (industrial area to be planned),

Dc (areas for research, programming, testing, and industrial facilities), and Dd (areas for small firms and handicraft).

PIP can be also considered as an urban redevelopment project that uses the envisaged new constructions and public spaces to give order to the unregulated spontaneity of the buildings realized in that area. PIP pursues a set of objectives that can be grouped into five classes of needs (that are themselves divided into some classes of requirements) on which the Implementing Technical Standards (ITS) are based (Fig. 2). The classes of needs are: security, usability, environmental quality, morphological quality and management. The evaluation process that supports the executive project of PIP comprises the following steps:

- Selection model of the firms basing on the specific features of the companies currently settled or to be settled, and the requirements of environmental, social and economic qualification;
- Assessment model of private and public costs and revenues of the investment property for new buildings;
- Equalization and compensation model that guarantees a transparent participation and redistribution of revenues and costs between public and private stakeholders involved in the transformation process of the areas.

Functions	Development			Building area			Green void area			Covered area			Built-up area		
	Sect 1	Sect 2	Sect 3	Sect 1	Sect 2	Sect 3	Sect 1	Sect 2	Sect 3	Sect 1	Sect 2	Sect 3	Sect 1	Sect 2	Sect 3
Existing buildings															
Crop fields		34.000	14.000	34.000	14.000						5.100			5.100	
Agric. outbuildings												2.100			2.100
Factories/crafts	47.130	104.619	22.542	47.130	104.619		9.017	18.852	41.847	9.017	18.852	41.847	10.820	22.622	50.217
Advanced tertiary	47.130	104.619	22.542	47.130	104.619		451	943	2.092	1.127	2.356	5.231	2.254	2.356	5.231
Tertiary service	47.130	104.619	22.542	47.130	104.619		451	943	2.092	1.127	2.356	5.231	2.254	2.356	5.231
Public facilities	2.409	10.553	3.045	2.409	10.553		1.066	1.204	3.693	913	723	3.166	913	1.445	6.332
Public green	1.579	15.955	17.254	1.579	15.955										
Sport equipment			8.658			8.658									
Publi voids															
Driveways	520														
Pedestrian/cycle	115														
Squares		228	4.434	409	228	4.434									
Parking lots		3.981	8.275	2.503	3.981	8.275									

Fig. 2. Synthesis of the main quantities of PIP (by cubage ratio equal to 2 cm/sqm; cover ratio equal to 0,3–0,4 sqm/sqm).

3 Methods and Procedures

3.1 Objectives, Data Sources, Information

In the perspective of the actual feasibility of PIP and its adequacy to the stakeholders' expectations, an innovative solution is proposed, and it is based on the integration between knowledge, assessment and design through a continuous and constant feedback between all the different skills involved.

The innovations introduced in the process can be summarized as follows:

- Analysis: study of the PIP's end-user, by means of the call for the preliminary qualification of the enterprises aimed at the knowledge of the real demand of localization.

- Assessment: identification of the economic (private and public conveniences) and financial (ability to sustain current and future costs) conditions suitable for establishing a consortium of firms and a consortium of landowners as well. These consortiums may become unitary interlocutors able to dialog with the municipality to create the conditions to have access to the lots and to apply equalization and compensatory processes to make development and urbanization harmonious.
- Design: dimensioning of lots and public works on the basis of the demand for industrial land use.

In particular, the analysis of the demand for localization in suitable areas by the firms of the Flegreo territory is divided into two phases: the first one is developed through a sample field survey; the second one is carried out by launching a call for the preliminary qualification of some firms that could be potentially located. The references for this qualification process are related to the production layout and the used spaces, and the management and organization profile as well, in order to know the demand for space, the best organization of space itself, and the demand for facilities for local firms. Information from various sources (such as ISTAT, Chamber of Commerce, trade associations, consortia) and from a specific questionnaire is collected in a sheet for each of the 94 companies that have been consulted. The collected data concern: Company name; Localization; Type of entrepreneur; Type of activity; Company size; Availability to form consortia; Availability to relocate; Typology of the production process; Standard working hours; Settlement model (lot geometry and aggregation scheme); spatial organization (surfaces, volumes and functions); Agglomeration economies; environmental monitoring capability; Location with respect to the city centre and to the residential fabric; Compliance with environmental standards; Job security; Geography of activities, processing and equipment; Innovation; Development programs; Liquidity (assets and liabilities in the last three years); Financial autonomy (net assets, fixed assets). For new enterprises it is also specified: Project validity; Type of initiative; Potential markets; Employment impact; Environmental impact; Cover expenditure.

The detailed data are translated into scores and made available for the development in the multidimensional model for analysis and assessment. The model, that could be also linked to a Gis-Spreadsheets model [12], incorporates economic aspects in order to activate equalization and compensatory mechanisms by means of which the externalities due to the development process may be internalized. The most important aspect of this integration is the correspondence between the extraordinary planning permission fees and the profile of those enterprises considered the most suitable ones for being located in a certain area. The more an area is suitable for receiving activities that have the greatest impacts and the highest profits, the higher the fees (and vice versa). The need for efficiency, transparency and simplicity of the process of allocation of the areas to firms, and for integration of quantitative-monetary and qualitative assessments, requires the implementation of a multi-criteria evaluation model of MAVT type.

3.2   The MAVT Approach to Select and Allocate the Enterprises in a Sustainable Way

MCDA is a field of research which is concerned with the structured evaluation and support of decision problems with multiple criteria and uncertainty [13]. They are considered as explicitly subjective decision analytic technique in which value judgements are obtained and modelled through multi-attribute value and utility functions [14]. The distinction between value and utility functions is that the former incorporate no notion of risk attitude and thus apply in conditions in which there is no or, more likely, negligible uncertainty. The latter explicitly acknowledge risk and are suited to decision making under uncertainty, but require more information from the decision makers. The multi-attribute value theory (MAVT) (MultiAttribute Value Theory) [13–16] provides methods to structure and analyse decision problems by means of attribute trees (also called value trees) and to elicit the relative importance of criteria in this setting. In an attribute tree, the overall goal or objective is divided hierarchically into lower level objectives (also called criteria) and measurable attributes (also called lowest level or leaf criteria). A decision alternative x is evaluated on each attribute, i, by means of a value function $v_i(x)$. Under the assumption of mutual preferential independence of attribute, the standard additive aggregation rule can be used [14]. Them the overall value of an alternative x is evaluated as:

$$V(x) = \sum_{i=1}^{n} w_i v_i(x) \qquad (1)$$

where n is the number of attribute, w_i is the weight of attribute i and $v_i(x)$ is the rating of an alternative x with respect to attribute i. The sum of the weights is normalized to one and the component value functions $v_i(x)$ have values between 0 and 1. The weights w_i indicate the relative importance of attribute i changing from its worst level to its best level, compared with the changes in the other attributes [14, 15].

It should be stressed, however, that for a justified implementation of the additive model some requirements of MAVT concerning the problem under investigation should be held, especially the preferential independence requirements [14, 15]. MAVT relies on the assumption that the decision-maker is rational, preferring more value to less value, and that the decision-maker has perfect knowledge, and is consistent in his judgments.

MAVT can be used to address problems that involve a finite and discrete set of alternative options that have to be evaluated on the basis of conflicting objectives. For any given objective, one or more different attributes or criteria, which typically have different measurement scales, are used to measure the performance in relation to that objective [14]. MAVT can handle quantitative as well as qualitative data. If quantitative data are not available, expert judgments can be used to estimate the impacts on a qualitative scale. The intention of MAVT is to construct a means of associating a real number with each alternative, in order to produce a preference order on the alternatives consistent with the decision maker value judgments. This function is used to transform the evaluation of each alternative option on considered attributes into one single value. The alternative with the best value is then pointed out as the best [17]. Following this reasoning, it becomes clear that MAVT aggregates the options' performance across all the criteria to form an overall assessment and is thus a compensatory technique [18].

The process to be followed to build a MAVT model consists of the following five fundamental steps [19]:

1. defining and structuring the fundamental objectives and related attributes;
2. identifying alternative options;
3. assessing the scores for each alternative in terms of each criterion;
4. modeling preferences and value trade-offs: elicitation of value functions associated with objectives and attributes and assessment of their weights;
5. ranking of the alternatives: a total score is calculated for each alternative by applying a value function to all criteria scores.

In the development of a MAVT model is importance to express the perceived values on the impact that the options under consideration can have. The perceived values can be used to measure the relative worthiness of each impact of options. The preferences can be modeling by constructing a value function, that is a mathematical representation of human judgments on the considered options. The value functions are an analytical description of the value system of the individuals involved in the decision and aim at capturing the human judgments involved in the evaluation of alternatives. In particular, a value function translates the performances of the alternatives into a value score, which represents the degree to which a decision objective (or multiple decision objectives) is achieved. The value is a dimensionless score: a value of 1 indicates the best available performance and a high objective achievement, while a value of 0 indicates the worst performance and a low objective achievement. The decision makers don't express their preferences and values by using a similar dimensionless score from 1 at 0, and then the value functions have to be estimated through a specially designed interviewing process for them. A focused interviewing process to extract the preferences and values of decision makers can be used to identify the relevant judgments for the decision. The generated relevant judgments can be organized and represented analytically in to support the MAVT model. In this regard, the generating process of the values functions can be considered as an approximate representation of human judgments that have been identified in the interviewing process.

In literature many approaches define the value functions [20]: direct rating technique [15]; curve fitting; bisection; standard differences; parameter estimation; semantic judgment. About the weighting criteria, the literature proposes many different methods. It has been generally agreed that the meaning and the validity of these criteria weights are crucial in order to avoid improper use of MCDM models and the procedures for deriving criteria weights should not be independent of the manner they are used [21–23].

In the MAVT approach, the weights in the additive model are scaling constants, which allow marginal value functions to take on values in the same interval. Weights or scaling constants can be estimated using several techniques. There are two broad categories of weights, namely numerical estimation and indifference weights [15]. Ranking, direct rating, ratio estimation and swing weights belong to the numerical assessment category while the trade-off method comes under the indifference judgment category. This study uses the direct technique rating for the determination of the value functions and the swing-weights technique for the weights.

3.3 Equalization and Compensation

PIP's profitability and financial sustainability are appraised from the point of view of: equalization of advantages gained by the PIP's lots - it is achieved establishing that the increasing differential rent should be fairly redistributed among all landowners -; compensation of territorial and environmental costs sustained by the citizens due to the development projects, transferring to the municipality a part of the gained over-profit converted in extraordinary permission fees. Consequently the quantitative and qualitative PIP's profile is based on the possibility of producing wealth to be transferred and redistributed on the base of contribution from involved stakeholders that are:

- The Municipality, with regard to its capacity to provide a framework;
- Landowners, with regard to willingness to make land available at the price fixed by Consortium;
- Enterprises, with regard to their capacity to create real wealth by their participation to a unitary district.

To undertake such an action, it is assumed that PIP could be implemented by a Consortium of landowners (supported and facilitated by the Municipality) for the unitary management of the areas that will be partly transformed and partly transferring to the Municipality for the construction of public works. Each landowner participates in the Consortium's profits according to the value (quantity and quality) of its land involved in the development project.

The Consortium has the faculty to: bargain over the price for the use of the areas by enterprises to be located, with respect to the objectives of maximization of rent and protection and development of property; dispose of a significant share of lands, also by hire purchase with or without option to buy; contribute to the protection of environmental values by means of a more direct control of the land use.

As the economic surplus over the total cost depends on both the size of the transformed area and the permitted cubage ratio, these two quantities can be assumed as the core of the negotiation, and as a tool for supporting the development of the area from both a private and a public perspective. Extraordinary permission fees are therefore calculated by balancing technical feasibility and private profitability through a fair return of the resources provided by the entrepreneurs. The permission fees are supposed to be paid in advance – in cash or by the realization of public facilities – when the use value of the areas becomes effective and the state of affairs is actually modified, or are quantified in terms of future management of public facilities by discounting the cash flow. Consequently, they constitute one of the anticipated revenues from financial capital that are taken into account in the model. The size of the extraordinary permission fees results from the integration of feasibility and fairness. Feasibility concerns the appropriate income for each asset; fairness depends on the self-financing ratio of the plan, which is the quota of public cost covered by the extraordinary permission fees.

The PIP's area is divided in three sectors, and for each of them a public (Municipality) and private (the landowners' consortium) financial feasibility analysis has been carried out comparing total costs, revenues, ordinary and extraordinary permission fees, and profits [24]. Given the demand for PIP areas as a result of the demand analysis, a specific application of the "transformation value" (extraction method) has been carried out starting from the basic formula (2) where the extraordinary permission fees are expressed. The normal profit can be assumed as a quota of the total investment (3) and the normal global profit rate as the sum of the weighted average cost of the capital paid in advance and the premium for risk and organization (4), so that the permission fees can be easily calculated (5). Furthermore, as required by the equalization process, for each sector the appropriate cubage rate is calculated (6).

$$v_t = v_f - k - f^* - \pi \tag{2}$$

$$\pi = r(v_t + k + f^*) \tag{3}$$

$$r = \left[(w + r')(1 + w + r')^n - 1 \right] / (w + r') \tag{4}$$

$$f^* = \left\{ v_f - \left[(v_t + k)(1 + r) \right] \right\} / (1 + r) \tag{5}$$

$$i_f = \left[v_t \bar{h}(1 + r) \right] / s_f \left[\bar{p} - \bar{k}(1 + r) \right] \tag{6}$$

Where: v_t is the current real estate value of the total private developable area comprised in the sector; v_f is the value of the private property at the end of the development process; k is the building cost including ordinary permission fees; f^* indicates the extraordinary permission fees; r is the global profit rate for each loan term; w is the weighted average cost of capital; r' is the annual profit rate for the promoter's risk and organization; n is the loan term (years); i_f is the cubage ratio; $\bar{h}$ is the weighed average height of the buildings included in the sector; s_f is the area where the development is permitted; $\bar{p}$ is the weighted average market price of the buildings included in the sector; $\bar{k}$ is the weighted average unit cost of the buildings included in the sector.

4   Applications and Results

4.1   The Allocation of the Areas to the Firms

The decision maker's overall aim is to achieve the best PIP layout from the perspectives of economic development, environmental sustainability, land and urban identity. As a consequence, the firms have been selected according to economic, environmental and urban features, while the areas have been allocated to them by applying the evaluation model shown in Fig. 3.

Criteria	Sub-criteria	Indicators
Economy	Market	Market strength
		Innovation attitude
	Capital	Liquidity
		Financial independence
Environment	Natural environment	Adaptability to environmental regulation
		Accessibility to the urban scale
	Artificial environment	Compatibility of location 1
		Compatibility of location 2
	Human environment	Impact on employment
		Health and safety at work
Social system	Participation	Permission fees
	Agglomeration	Size of the requested area
	Landscape	Volume of buildings

Fig. 3. Evaluation model for allocating areas to firms

Figure 4 presents the results of the assessment and allocation process, and the comparison between two different modes of allocating the areas according to: profile of each sector; characteristics of the selected firms; criterion for the selection of the firms (a score-threshold that is explained below).

The features of the three sectors are outlined by three different weight lists; the 94 firms can be selected by means of their physical size (left) or overall quality based on their attributes (right); the sum of the requested settlement areas must be smaller than the permitted area within each sector (first three columns of the two parts). The coloured cells indicate where the firm is located (sector 1, 2 or 3) and the two blocks of the table show the comparison between the two different selection modes.

The other three groups of columns report the quality scores from the three perspectives (economy, environment, social system) and for each sector. The shade of the cells allows us to compare the values in the cells of the same column (the darker, the higher), so that the table provides an overall diagram of the position and qualification of the firms.

The selection by size allocates the smaller firms in Sector 1, the medium ones in Sector 2 and the larger ones in Sector 3, according to the sizes of the sectors themselves. The selection by attributes is carried out establishing, for each group of attributes, a score-threshold above which firms are allowed, and below which they are rejected; this type of selection allows the decision makers to assign the firms to the sectors to which they are best suited.

The pattern allows many in-depth analyses useful to assess the overall profile of the strategies that can be generated by changing the weights recursively, in order to identify the best overall layout, fitting tactic (valuation of a single firm) and strategic (valuation of the overall layout of each sector) approaches as well.

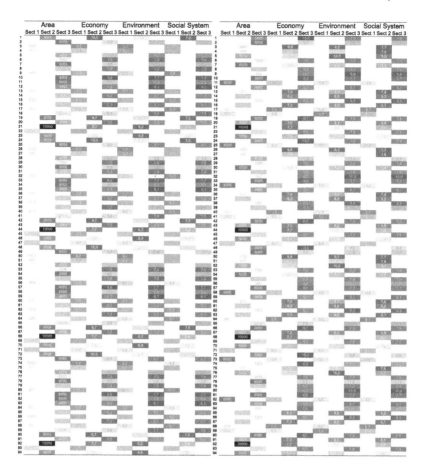

Fig. 4. Allocation of the areas of each sector by volumes (left) and attributes (right)

4.2 Results of the Equalization and Compensation Process

The calculation of the extraordinary fees requested for planning permissions in the different PIP's areas is carried out counterbalancing revenues and costs, both private and public, as shown in the summary tables (Fig. 5).

The costs included in the private budget are: cost of construction; cost for the landscaping of the external areas, general expenses, taxes (VAT), cost opportunity of the existing areas and buildings (expressed by their current market price); ordinary permission fees (Article 3 of Law 10/1977); extraordinary permission fees that are equal to potential extra-profits. The private revenues consist of the increases of the real estate value involved in the PIP's transformation process, and the capital value of flows of income coming from the possible private management of public works. The entrepreneur's profit (Consortium of landowners) constitutes the condition for the PIP's feasibility, comprises the remuneration for the anticipations, for the organization of the process, and for business risk, and is diversified relating to the profit rate of equity

		Private Balance Sheet				Public Balance Sheet		
Sector 1	Costs	Land and existing buildings	€	747.830	Costs	Construction	€	2.795.060
		Construction	€	11.291.810		Other costs	€	279.400
		Other costs	€	1.128.970		total costs	€	3.074.460
		Ordinary permission fee	€	240.150				
		Extraordinary permission fee	€	406.970				
		total costs	€	13.815.730				
	Revenues	Real estate market values	€	17.127.780	Revenues	Ordinary permission fee	€	240.150
		Management of public works	€	-		Extraordinary permission fee	€	406.970
	Profit	24%	€	3.312.040		total revenues	€	647.120
	Indexes	Net value	€	3.312.040	Indexes	Net value	-€	2.427.340
		Extraordinary permission fee (equal to extra-profit)	€	406.970		% public costs financed		21,0%
Sector 2	Costs	Land and existing buildings	€	1.062.870	Costs	Construction	€	3.140.570
		Construction	€	21.588.930		Other costs	€	314.010
		Other costs	€	2.158.790		total costs	€	3.454.580
		Ordinary permission fee	€	385.280				
		Extraordinary permission fee	€	1.117.100				
		total costs	€	26.312.970				
	Revenues	Real estate market values	€	32.689.140	Revenues	Ordinary permission fee	€	385.280
		Management of public works	€	-		Extraordinary permission fee	€	1.117.100
	Profit	24%	€	6.376.180		total revenues	€	1.502.380
	Indexes	Net value	€	6.376.180	Indexes	Net value	-€	1.952.200
		Extraordinary permission fee (equal to extra-profit)	€	1.117.100		% public costs financed		43,5%
Sector 3	Costs	Land and existing buildings	€	2.371.050	Costs	Construction	€	12.250.870
		Construction	€	41.804.600		Other costs	€	1.225.040
		Other costs	€	4.180.720		total costs	€	13.475.910
		Ordinary permission fee	€	855.770				
		Extraordinary permission fee	€	2.045.690				
		total costs	€	51.257.830				
	Revenues	Real estate market values	€	63.621.810	Revenues	Ordinary permission fee	€	855.770
		Management of public works	€	-		Extraordinary permission fee	€	2.045.690
	Profit	24%	€	12.363.980		total revenues	€	2.901.460
	Indexes	Net value	€	12.363.980	Indexes	Net value	-€	10.574.450
		Extraordinary permission fee (equal to extra-profit)	€	2.045.690		% public costs financed		21,5%
Total	Indexes	Net value	€	22.052.200	Indexes	Net value	-€	14.953.990
		Extraordinary permission fee (equal to extra-profit)	€	3.569.760		% public costs financed		25,2%

Fig. 5. Summary tables of private and public balance by sector

(including areas and buildings conferred to the Consortium) and capital of debt; the profit rate is the WACC.

In the public budget the costs comprise: potential indemnity of expropriation; costs for the landscaping of the areas; costs of construction; general expenses and taxes. The public revenues are ordinary and extraordinary permission fees.

Public and private budgets are made for each area for the status quo and for a basic project, which has invariant and variable (morphogenetic) features [25], assuming a mix of functions and uses.

Once a profit rate is determined, the budgets allow the calculation of the possible entity of the extra-profits. If the budget has a deficit, the morphogenetic features of the project are changed modifying cubature and land use according to cubature limits,

	Green areas		Parking lots		Public facilities		Urban infrastructures	
	Total costs	% financed	Total costs	% financed	Total costs	% financed	Total costs	% financed
Sector 1	€ 785.010	83%	€ 495.800	0%	€ 754.030	0%	€ 1.038.080	0%
Sector 2	€ 108.460	100%	€ 790.180	100%	€ 1.322.130	46%	€ 1.234.330	0%
Sector 3	€ 2.448.010	100%	€ 1.647.500	28%	€ 5.804.980	0%	€ 3.579.050	0%
Total	€ 3.341.480	96%	€ 2.933.480	42%	€ 7.881.140	8%	€ 5.851.460	0%

Fig. 6. Public works financed by means of extraordinary permission fees

planning standards, morphological quality, and historical or environmental constrains. The process of adjustment of the features of a project is iterative and stops when the private budget provides those extra-profits that, turned into extraordinary permission fees, can finance significant quotas of public works (Figs. 6 and 7).

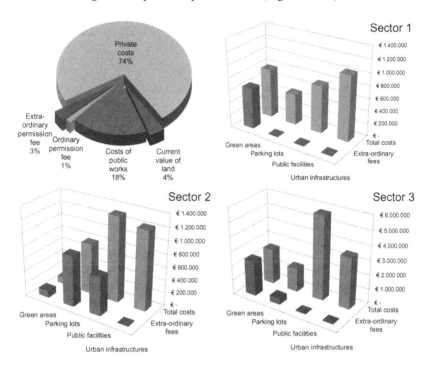

Fig. 7. Public works financed by means of extraordinary permission fees

5   Discussions and Conclusions

Evaluation of the Strategies. Combining the procedures of selection (by volume or attributes) and the preference systems by criteria it is possible to verify the best allocation of the areas to the enterprises, which is the best location of the enterprises in the three sectors as well. The exemplification in Fig. 8 synthesizes the evaluation of the two formerly explained layouts, and shows the weights of the criteria by each sector (bar histograms) and the general evaluation of each sector from the point of view of the three fundamental criteria (economy, environment, and social system).

When the selection is made by volumes (on top), the features of the enterprises that are located in the three sectors provide a global value lower than the one obtained by selecting the enterprises by attributes (at the bottom), where the global value for the sectors 2 and 3 is significantly higher, while we observe a slight lowering of the value of the sector 1. This difference of evaluation depends on the system of weights that, as shown in the second procedure of selection (diagrams at the bottom), allows to fully

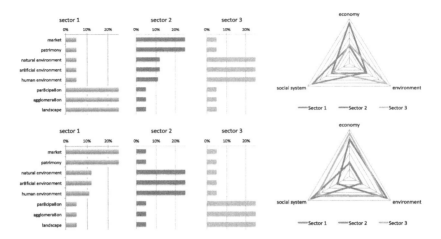

Fig. 8. Weight of criteria and global evaluation by sector and by layout (Color figure online)

appreciate the environmental performances of the enterprises located in the sector 2 and the social-systemic performances of the ones located in the sector 3.

Sensitivity Analysis. The process can be reiterated until the solution that maximises the total value is achieved. Figures 9 shows 10 iterations that allow the individuation of the best arrangement modifying the weights and the thresholds for selecting the firms.

strategy	weights									valuations									settled firms			total value	treshold 1	treshold 2
	economy			environment			social-system			economy			environment			social-system								
	S1	S2	S3	S1	S2	S3	S1	S2	S3	S1	S2	S3	S1	S2	S3	S1	S2	S3	S1	S2	S3			
1	25%	5%	5%	12%	25%	5%	5%	5%	25%	1,0	6,6	5,9	0,2	8,7	5,7	7,1	5,8	5,7	30	40	118	5,2	1,5	4,5
2	25%	12%	10%	5%	20%	5%	12%	5%	22%	0,4	6,6	5,7	0,3	8,7	6,1	6,2	6,1	6,0	22	40	126	5,1	1,2	4,5
3	5%	25%	5%	5%	12%	25%	25%	5%	5%	0,1	8,0	4,5	0,0	6,7	6,0	5,5	5,9	6,0	12	54	122	4,8	1,5	4,5
4	7%	19%	8%	12%	14%	16%	18%	8%	12%	0,1	7,4	5,6	0,4	7,9	5,9	6,3	5,3	5,9	22	42	124	5,0	1,5	4,2
5	19%	8%	7%	14%	16%	12%	8%	12%	18%	0,7	7,3	6,0	0,2	8,5	5,6	7,1	5,4	5,8	30	40	118	5,2	1,5	4,0
6	8%	7%	19%	16%	12%	14%	12%	18%	8%	0,4	7,6	5,7	0,2	8,5	5,7	7,1	6,4	5,6	30	36	122	5,2	1,5	3,5
7	10%	21%	8%	8%	12%	16%	19%	8%	12%	0,2	7,1	5,3	0,0	7,7	6,4	5,9	6,7	5,7	16	50	122	5,0	1,5	4,2
8	21%	8%	10%	12%	16%	8%	8%	12%	19%	0,8	7,7	5,7	0,1	8,6	5,6	7,4	5,4	5,9	30	40	118	5,2	1,5	4,2
9	8%	10%	21%	16%	8%	12%	12%	19%	8%	0,4	6,8	5,2	0,2	7,2	5,0	7,1	7,2	5,4	30	68	90	4,9	1,5	3,0
10	13%	13%	13%	13%	13%	13%	13%	13%	13%	0,2	6,7	5,5	0,3	7,8	6,0	6,2	5,9	6,0	22	56	110	4,9	1,5	3,5

Fig. 9. Valuation of different strategies

The iterations indicate a set of strategies (1, 5, 6 and 8) that maximize the total value. The procedure can be repeated until the configuration that maximizes the total value is achieved. The graphs in Fig. 10 show the partial values and, more generally, the strong prevalence of the values of the sectors 2 and 3 in each iteration. In this regard, the different dimension of the areas affects the selection model, because the sector 1 is much smaller than the other ones and, therefore, it is less likely that firms with an elevated score can be selected. This constitutes a limit of the proposed model and represents a new starting point for further study.

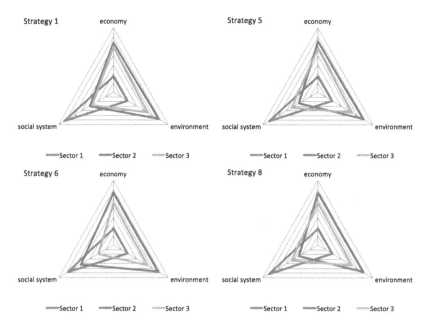

Fig. 10. Strategies that best maximize the total value (Color figure online)

Acknowledgements. The paper is based on documents of Quarto's PIP drawn up by prof. C. Gasparrini (scientific advisor and project leader). The data have been provided or elaborated by: dott. A Corapi and R. Staffa (economic-productive data of the firms); the working group and the Municipality (data of the firms); F. Forte and. M. D'Amato (real estate market data). S. Giuffrida elaborated the economic and qualitative evaluation model. In the paper S. Giuffrida edited par. 3.1, 3.3, 4.1 and 5; G. Napoli edited par. 1, 4.1 and 4.2; M.R. Trovato edited par. 2, 3.2 and 4.1.

References

1. Martinico, F.: Il Territorio dell'industria. Nuove strategie di pianificazione delle aree industriali in Europa. Gangemi, Roma (2001)
2. Chiodelli, F., Moroni, S.: Zoning-integrative and zoning-alternative transferable development rights: compensation, equity, efficiency. Land Use Policy **52**, 422–429 (2016)
3. Woodbury, S.: Transfer of development rights: a new tool for planner. J. Am. Plan. Assoc. **41**(1), 3–14 (1973)
4. Henger, R., Bizer, K.: Tradable planning permits for land-use control in Germany. Land Use Policy **27**, 843–852 (2010)
5. Camagni, R.: Perequazione urbanistica "estesa", rendita e finanziarizzazione immobiliare: un conflitto con l'equità e la qualità territoriale. Scienze Regionali **13**(2), 29–44 (2014)
6. Alterman, R.: Can the 'unearned increment' in land values be harnessed to supply affordable housing? In: Financing Affordable Housing and Infrastructure in Cities: Towards Innovative Land and Property Taxation, pp. 17–18. UN Habitat, Varsaw (2010)
7. Camagni, R.: Il finanziamento della città pubblica: la cattura dei plusvalori fondiari e il modello perequativo.: In: Curti, F. (ed.) Urbanistica e fiscalità locale, pp. 321–342. Maggioli, Ravenna (1999)

8. Micelli, E.: Development rights markets to manage Urban plans in Italy. Urban Stud. **39**(1), 141–154 (2002)
9. Stanghellini, S. (ed.): Il negoziato pubblico-privato nei progetti urbani. DEI, Roma (2012)
10. Calabrò, F., Della Spina, L.: The public-private partnerships in buildings regeneration: a model appraisal of the benefits and for land value capture. In: 5nd International Engineering Conference 2014 (KKU-IENC 2014). Advanced Materials Research, vols. 931–932, pp. 555–559. Trans Tech Publications, Switzerland (2014)
11. Nesticò, A., Galante, M.: An estimate model for the equalisation of real estate tax: a case study. Int. J. Bus. Intell. Data Min. **10**(1), 19–32 (2015)
12. Giuffrida, S., Gagliano, F.: Sketching smart and fair cities WebGIS and Spread sheets in a code. In: Murgante, B., et al. (eds.) ICCSA 2014, Part III. LNCS, vol. 8581, pp. 284–299. Springer, Heidelberg (2014)
13. Belton, V., Stewart, T.J.: Multiple Criteria Decision Analysis: An Integrated Approach. Kluwer Academic Press, Boston (2002)
14. Keeney, R., Raiffa, H.: Decisions with Multiple Objectives: Preferences and Value Trade-offs. Wiley, New York (1976)
15. Von Winterfeldt, D.W., Edwards, W.: Decision Analysis and Behavioral Research. Cambridge University Press, Cambridge (1986)
16. Greco, S., Ehrgott, M., Figueira, J.: Multiple Criteria Decision Analysis: State of the Art Surveys, vol. 78. Springer, New York (2005)
17. Herwijnen, M.V.: Spatial Decision Support for Environmental Management. Vrije Universiteit, Amsterdam (1999)
18. Trovato, M.R., Giuffrida, S.: The choice problem of the urban performances to support the Pachino's redevelopment plan. Int. J. Bus. Intell. Data Min. **9**(4), 330–355 (2014)
19. Montibeller, G., Yoshizaki, H.: A framework for locating logistic facilities with multi-criteria decision analysis. In: Takahashi, R.H., Deb, K., Wanner, E.F., Greco, S. (eds.) EMO 2011. LNCS, vol. 6576, pp. 505–519. Springer, Heidelberg (2011)
20. Beinat, E.: Value Functions for Environmental Management. Kluwer Academic Publishers, Dordrecht (1997)
21. Roy, B., Mousseau, V.: A theoretical framework for analysing the notion of relative importance of criteria. J. Multi-Criteria Decis. Anal. **5**, 145–159 (1996)
22. Choo, E.U., Bertram, S., Wedley, W.: Interpretation of criteria weights in multicriteria decision making. Comput. Ind. Eng. **37**, 527–541 (1999)
23. Poyhonen, M., Hamalainen, R.P.: On the convergence of multi-attribute weighting methods. Eur. J. Oper. Res. **129**, 569–585 (2001)
24. Trovato, M.R., Giuffrida, S.: The choice problem of the urban performances to support the Pachino's redevelopment plan. Int. J. Bus. Intell. Data Min. **9**(4), 330–355 (2014)
25. Napoli, G.: Financial sustainability and morphogenesis of Urban transformation project. In: Gervasi, O., et al. (eds.) ICCSA 2015. LNCS, vol. 9157, pp. 178–193. Springer, Heidelberg (2015)

Environmental Noise Sensing Approach Based on Volunteered Geographic Information and Spatio-Temporal Analysis with Machine Learning

Miguel Torres-Ruiz[1]([⊠]), Juan H. Juárez-Hipólito[1], Miltiadis Demetrios Lytras[2], and Marco Moreno-Ibarra[1]

[1] CIC, UPALM-Zacatenco, Instituto Politécnico Nacional, 05320 Mexico City, Mexico
{mtorres,marcomoreno}@cic.ipn.mx, a140097@sagitario.cic.ipn.mx
[2] The American College of Greece, 15310 Athens, Greece
mlytras@acg.edu

Abstract. In this paper a methodology for analyzing the behavior of the environmental noise pollution is proposed. It consists of a mobile application called 'NoiseMonitor', which senses the environmental noise with the microphone sensor available in the mobile device. The georeferenced noise data constitute Volunteered Geographic Information that compose a large geospatial database of urban information of the Mexico City. In addition, a Web-GIS is proposed in order to make spatio-temporal analysis based on a prediction model, applying Machine Learning techniques to generate acoustic noise mapping with contextual information. According to the obtained results, a comparison between support vector machines and artificial neural networks were performed in order to evaluate the model and the behavior of the sensed data.

Keywords: Volunteered geographic information · Support vector machine · Artificial neural network · Mobile application · Noise sensing

1 Introduction

Environmental noise represents a high level of pollution in big cities. People who lives in urban areas are suffering the effects of the environmental noise. It affects the life quality of people, and it may cause stress, high blood pressure, sleep loss, and heart diseases [1, 9].

With the emerging of new technologies and mobile devices, it is predicted that there will be as much data created than the entire history of the Earth planet [2]. During the past decade, machine-learning systems have been widely adopted in many areas such as astronomy, biology, climatology, medicine, finance, and economy [3].

Nowadays, sensors embedded in mobile devices are increasingly available for multiple sensing tasks. Such sensors are in charged of measuring and registering sensed data that represent the dynamic of the cities, even including the spatial and temporal dimensions. This is environmental information, which is gathered by sensors via its sensing mobile. Thus, these sensors generate huge quantity of data per year and decisions

© Springer International Publishing Switzerland 2016
O. Gervasi et al. (Eds.): ICCSA 2016, Part IV, LNCS 9789, pp. 95–110, 2016.
DOI: 10.1007/978-3-319-42089-9_7

need to be made in real-time, like how much data to analyze, and how much to transmit for further analysis, among others.

The rapid growth in the population density in megalopolis demands tolerable provision of services and infrastructure, in order to know the needs of the city inhabitants [4]. Thus, the raise in the request for embedded devices, such as sensors, actuators, and smartphones, is an important potential business towards the new era of Internet of Things (IoT); in which all the devices are capable of interconnecting and communicating with other over Internet. Therefore, Internet technologies provide a way towards integrating and sharing a common communication medium for collaborative urban systems [4].

The mobile sensing revolution is coming, and we are seeing these systems in everyday use. This progress is accelerated by the development of smartphones as a viable sensing platform. Today, most mobile phones include various sensors, such as GPS, accelerometers, microphones, and cameras. Classification models can exploit such data to allow, for instance, a mobile phone can understand our actions and environment. These models are driving key mobile application domains including heath, traffic congestion, air pollution, and green energy awareness, to name a few. However, significant challenges exist in the real-world with respect to human activity modeling. For example, key obstacles are the difference in contextual conditions and user characteristics encountered in large-scale mobile sensing systems. This leads to the discriminative features in sensor data, used by classifiers to recognize different human activities, varying from user to user [5]. The capacity for sensing the state of different factors such as specific particles, flow of people, environmental conditions, etc., have contributed to mitigate natural disasters like earthquakes, fires, control and prevention of epidemics, vehicular congestion, and urban planning [6].

According to [7], urban computing is a process of acquisition, integration, and analysis of big and heterogeneous data, generated by a diversity of sources in urban spaces, such as sensors, devices, vehicles, buildings, and human, to tackle the major issues that cities face, *e.g.*, air and noise pollution, energy consumption and traffic congestion.

The present work proposes a general framework for generating and sensing volunteered geographic information (VGI) related to environmental noise. In this approach, citizens play important roles, as users and map producers. Additionally, a mobile application in order to sense noise, using the microphone sensor embedded in the mobile devices is proposed in order to generate collaborative information. Later, machine learning-based methods are used to characterize the phenomenon and analyze with an acoustic noise prediction model the spatio-temporal sensed data. The information visualization is presented as heat maps that reflect the concentration and distribution of the acoustic noise information related to particular areas of the Mexico City in a web-mapping application.

The remainder of the paper is organized as follows. Section 2 presents a brief discussion of the related work. Section 3 describes the general framework for analyzing and monitoring environmental noise with a mobile application. Section 4 presents a case study in which the proposed methodology is applied. Section 5 outlines the experimental results as well as the comparison with other machine learning approaches. Section 6 highlights the conclusion and future work.

2   Related Work

Many works have addressed to VGI, its representation and analysis, and its relevance for urban planning, as well as the sustainability of smart cities.

Other kind of pollution is related to the noise, which affects the population of big cities, and causes physical and mental illnesses. A first step towards the understanding of urban noise is to measure real noise levels. In [8] two categories of methods for measuring these levels were introduced.

Moreover, in [10] a noise map of New York City, based on four ubiquitous data sources: 311 complaint data, social media, road networks, and Point of Interests (POIs) was presented. The noise situation of any location in the city, consisting of a noise pollution indicator and a noise composition, which is derived through a context-aware tensor decomposition approach. According to authors, this application highlights two components: (1) ranking locations based on inferred noise indicators in various settings, *e.g.*, on the weekdays (or weekends), in a time slot (or overall time), and in a noise category (or all categories); (2) revealing the distribution of noises over different noise categories in a specific location.

The case study presented in [11] consisted of an aethelometer, for continuous elemental carbon measurement, which was located with a continuous noise monitor near a major urban highway in New York City for six days. Hourly elemental carbon measurements and hourly data on overall noise levels and low, medium and high frequency noise levels were collected. Hourly average concentrations of fine particles and nitrogen oxides, wind speed and direction and car, truck and bus traffic were obtained from nearby regulatory monitors. Overall temporal patterns, as well as day–night and weekday–weekend patterns, were characterized and compared for all variables. The noise levels were correlated with car, truck, and bus traffic and with air pollutants. Summing up, the noise levels were temporally correlated with traffic and combustion pollutants and the correlations were modified by the time of day, noise frequency and wind.

On the other hand, a mobile application to sense the noise pollution with citizen participatory was proposed in [12]. The application was designed following gamification techniques to encourage users to participate using their personal smartphones. Thus, users are involved in taking and sharing noise pollution measurements in their cities that other stakeholders can use in their analysis and decision making processes.

According to [13], the prediction of environmental noise in urban environments requires the solution of a complex and non-linear problem. The inclusion of the spatial heterogeneity characteristic of urban environments could be essential in order to achieve an accurate environmental-noise prediction in cities. This work proposed to use a procedure based on feature-selection techniques and machine-learning regression methods. Thus, [14] presented a study of temporal and spatial variability of traffic noise in the Toronto City, with two cycles of intensive field measurement campaign to collect real-time measurements of traffic noise at 554 locations across Toronto.

Finally, [15] proposed a set of participatory techniques to achieve the same accuracy such as the standard noise mapping techniques. The 'NoiseTube' application collected sensed data of participative grassroots noise mapping campaigns, from the user perspective. Measurement-based noise maps of the target area with error margins comparable to those of official simulation-based noise maps were implemented.

3   The Proposed Methodology

The methodology has been conceptualized with the VGI foundations. This volunteered noise information is generated by the participation of people, using a mobile application to sense noise measures in particular places (see Fig. 1). Later, Support Vector Machine (SVM) technique is applied to the noise sensed data, in order to generate a prediction model that provides noise levels on predefined places, which are used to create noise maps with the measures and predicted values.

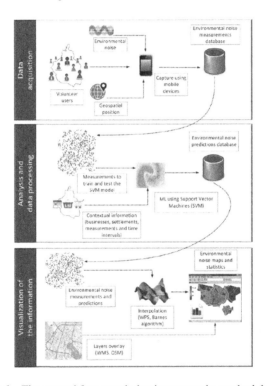

Fig. 1. The general framework that integrates the methodology.

This methodology is composed of three stages: (1) *Data acquisition*. Users act voluntarily as sensors, capture environmental noises and their geospatial position with a specific mobile application. The streaming of data measures is sent to the server, in which these measures are stored into a geospatial database. (2) *Analysis and data processing*. The sensed measures and contextual information (data about time intervals, nearby businesses, settlements, etc.) are used to train and test a machine learning model. The SVM approach describes the noise behavior and predicts noise levels, which will be stored into a geospatial database. (3) *Information visualization.* The sensed measures and the prediction values are interpolated by using a Web Processing Service (WPS); in this case, the Barnes algorithm with a standard color code was used. So, in a web-mapping application, the obtained layer is overlaid as a Web Map Service (WMS), other layers from OpenStreetMap (OSM) were selected in order to obtain an environmental noise map, which is presented to user with relevant noise statistics.

3.1 Prediction Model at Geographic Neighborhood-Level

Some factors related to noise measurements were taken into account as relevant for the prediction model. According to the geographic neighborhood of each measure, the following features were considered:

- *Business density.* It represents a higher concentration of businesses with a lot of people in them around a geographic point. It is equivalent to a greater movement of people, and higher levels of noise.
- *Previous noise measures.* They are the noise measures made previously in the same geographic area at similar time intervals, and they are somewhat similar to the current measure value.
- *Type of settlements.* It establishes the number and the type of settlements around a geographic area, in order to determine partially the existing noise level in that region.
- *Measure time.* It defines the date and time of the measure, determining the noise level recorded in the geographic area for that measure.

These features were defined to implement the prediction model, which are presented in Fig. 2.

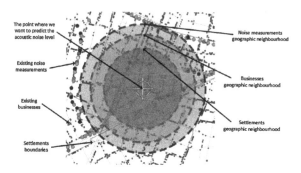

Fig. 2. Different features involved in the geographic neighborhood of the measurements for predicting the noise level.

According to Fig. 2, the radius of the concentric circles represents the influence area around the point, where the noise level located within the geographic neighborhood of the measure will be predicted. Data about the businesses were obtained from the DENUE, which is a public geospatial database created by the Mexican government.

It contains the identification and geographic location of 5 million active businesses that are registered in the economic census until 2015.

Thus, in a free space (without obstacles), a noise emitted from a point source describes the movement of a spherical surface. When the distance r increases, the sound energy radiated from the point source r covers a larger area. So, the sound pressure p decreases, it means that $p \sim 1/r$ is known as the law $1/r$. It indicates that the sound pressure p is inversely proportional to the distance r from the source. Then, the noise intensity falls with the distance, assuming a space with these characteristics. All these factors are computed considering the distance from the geographic point of the measure to the point that represents each factor. These are computed around each point and all the obtained values are used by the SVM to predict noise levels at any other geographic point.

3.2   Time Intervals

According to studies, the human body cannot be adapted to high and constant noise levels. Although people have different sensitivity to noise, the effects change with factors like time or making noise more annoying at night than during the day [8]. So, it is important to analyze noise using certain time intervals.

The analysis was carried out considering time intervals with respect to the most general daily activities of population in the Mexico City. Thus, different noise maps were obtained from each time interval. The intervals defined, according to the daily activities are the follows:

- *HL-00-time interval.* In the wee hours, from 00:00 to 06:00 h, when most people sleep and there is usually less traffic.
- *HL-06-time interval.* In the morning, from 06:00 to 12:00 h, when most people go to work or school in the morning shift.
- *HL-12-time interval.* In the afternoon, from 12:00 to 18:00 h, when most people eat out or go to school in the afternoon shift.
- *HL-18-time interval.* At night, from 18:00 to 24:00 h, when most people come back from work and go home to rest.

3.3   Generalities of the Prediction Model for Environmental Noise

The learning model consists of two sets of points: the learning set (C_A) and the prediction set (C_P). It was used to train and evaluate the task based on SVM. The union of these sets define the universe set (C_U) (see Fig. 3).

The points of georeferenced measurements represent places in which users made noise measures, by using their mobile devices. So, they compose the sets C_A and C_P; in additionally, C_P contains the centroids of the settlements, in order to predict the noise level in each settlement, regardless of whether there is or not noise measures inside it. A lattice of points within the area of interest is proposed to make the prediction in areas where there are not settlements or few of them exist. The distance that separates the arrangement of points in the lattice is arbitrary, and it can be configured to increase or decrease the values. The C_A is divided randomly into two subsets: the training set (C_{AI})

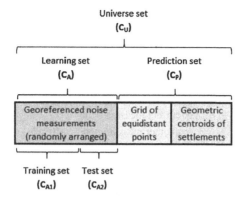

Fig. 3. Sets of geographic points used to predict environmental noise.

and the test set (C_{A2}). By using *k-fold cross validation* with $k = 10$, it is possible to deny the ratio of C_{A1} as the 90 % of C_A, and C_{A2} as the remaining 10 %. With C_A the SVM is trained and evaluated.

Summing up, Table 1 presents the general equations, parameters, and restrictions that were defined for the prediction model. In addition, Table 2 describes the variables, units, as well as the range of each established factor.

Table 1. Description of factors with equations and restrictions for the prediction model.

Factor	Description	Equation	Restrictions
f_1	DENUE density	$f_1 = \sum_i \dfrac{employeesAverage(p_i)}{distance(p_i, q)^2}$	$distance(p_i, q) \leq r_1$ $r_1 = 200$ m
f_2	Surrounding settlements	$f_2 = \sum_i \dfrac{1}{distance(centroid(p_i), q)^2}$	$distance(border(p_i), q) \leq r_2$ $r_2 = 1000$ m
f_3	Previous noise measures	$f_3 = \sum_i \dfrac{noiseAverage(p_i)}{distance(p_i, q)^2}$	$distance(p_i, q) \leq r_3$ $distance(p_i, q) \leq r_3$
f_4	Number of measures	$f_4 = \sum_i 1$	$distance(p_i, q) \leq r_3$ $distance(p_i, q) \leq r_3$
f_5	Measure time	$f_5 = time(q)$	None

4 The Case Study

In order to verify the effectiveness of the proposed methodology, a case study for environmental noise sensing in a specific district of the Mexico City was implemented. The case study was divided into two analysis levels: *macro* and *micro* partitions. The first one considers the entire district, and the second only uses the streets around the Historic Center.

In the *macro analysis level*, HL-12-time interval was used to sense the noise in the Cuauhtémoc District. This area and interval were chosen by the following characteristics:

Table 2. Description of units, variables, and ranges for the prediction model.

Factor	Units	Variables	Range	Normalized range
f_1	$\dfrac{number\_of\_people}{squared\_meters}$	p: DENUE geographic point q: Current measure point	$[0, \infty)$	$[0, 1]$
f_2	$\dfrac{units}{square\_meters}$	p: surrounding settlement q: current measure point	$[0, \infty)$	$[0, 1]$
f_3	$\dfrac{dB(SPL)}{squared\_meters}$	p: another measure point q: current measure point	$[0, \infty)$	$[0, 1]$
f_4	$units$	p: another measure point q: current measure point	$[0, \infty)$	$[0, 1]$
f_5	$hours$	q: current measure point	$[0, \infty)$	$[0, 1]$

- *Location.* The selected area includes the Mexico City downtown.
- *Movement.* It is one of the most active areas in terms of human mobility.
- *Business.* It contains a high concentration of businesses, so it is expected that the measurements are most useful for the learning model.
- *Safety.* This place contains several settlements with greater security and better infra-structure, so it represents a lower risk during the measurement time intervals.

In the *micro analysis level,* the HL-06, HL-12, HNL-06, and HNL-12 time intervals to monitor the noise in the streets of the Historic Center of Mexico City were defined. The reasons for choosing this place and its surroundings are the follows:

- *Noise.* It is one of the regions with the highest noise generation, because it is a place of broad commercial activity and people transit.
- *Movement.* It is one of the most active areas in terms of human mobility.
- *Safety.* It is the most important tourist destination in the city. This place offers constant surveillance, so it also represents the lowest risk during the noise measures by walking.

5   Experimental Results

According to the framework presented in Fig. 1, the methodology is composed of three stages. With respect to the *Data acquisition stage*, a mobile application called 'Noise-Monitor' was implemented. It collects noise measures, by using the mobile device. It is based on Android, with support for version 2.3.3 (API level 10) and higher. The mobile application consists of a main screen, from which user can move between three sections: the screen to capture the noise levels, the screen to visualize the noise map, and the screen to configure data about user and device (see Fig. 4).

In the *Information visualization stage*, a web-mapping application was implemented. It is based on four sections: the first tab shows the noise map, the second one allows user to search geographic places, the third tab depicts some statistics about noise levels in different regions, and the fourth one allows user download other resources. This application was developed using PHP, HTML5, CSS, JSON, JavaScript, and AJAX.

Fig. 4. The mobile application for monitoring environmental noise.

To visualize the noise maps, the JavaScript framework, OpenLayers for reading data from the GeoServer, via WMS with the PostGIS extension were used (see Fig. 5).

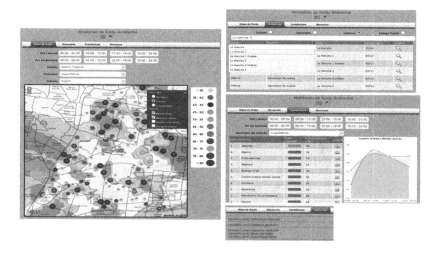

Fig. 5. The web-mapping application.

On the other hand, the noise sampling within the interest area was split in two groups. The first group of samples was produced between July 6 and July 10, 2015, on routes that attempted to cover at least a part of each settlement in the Cuauhtémoc District, during the mentioned previously time intervals. Figure 6 shows the maps for the routes (right) and the recorded measures (left). The second group of samples was produced in the area of the Historic Center by the mobile application, following the path shown in Fig. 7, with the time intervals mentioned previously. The total of samples that were collected from the mobile device was 2,355 georeferenced environmental noise measures, covering around 83 km.

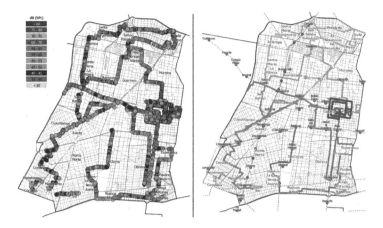

Fig. 6. Maps of the routes and the noise measurements generated in the micro analysis level. (Color figure online)

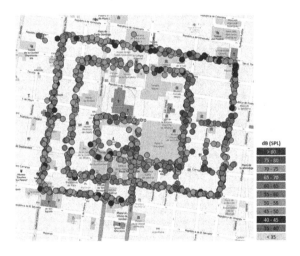

Fig. 7. Noise measurements sensed in the streets of the Historic Center of Mexico City. (Color figure online)

5.1 Prediction Model Based on Support Vector Machine

In order to calibrate the prediction model, appropriate values were searched, using a program implemented in the R language to evaluate the SVM performance. This model was compared against Artificial Neural Networks (ANN). From the set with 2,355 noise measures, 1,871 (about 80 %) were randomly taken to establish the training set, and the remaining 484 (about 20 %) to compose the test set. Both sets were evaluated with SVM and ANN techniques. When the training process was finished in each model, it was possible to compare the noise measures (previously known) against the new noise levels (recently predicted by the model). Thus, the differences between values were used to compute the mean absolute error (MAE) and the correlation (CORR). With these results, the accuracy of each model was determined. In this case, the model with more accurate is presented when its MAE value is the lowest and its CORR value is the highest. For geographic neighborhoods, different radius lengths were evaluated, a radius of 200 m for DENUE points, 1,000 m for the radius of the nearby settlements, and 1,000 m for the radius of the nearby noise measures.

5.1.1 Performance of the Artificial Neural Networks

In order to evaluate the ANN performance, the *neuralnet* function was considered. Different amount of nodes was used in the hidden layers, and the default algorithm, *rprop+* (resilient back propagation with weight back tracking) were considered. We built a ANN, with 0 hidden nodes (ANN *nod* = 0), and other ANN with a hidden layer of 1, 2, 3, 4, 5 and 10 nodes. An ANN with two hidden layers of 5 nodes in each (ANN *nod* = 5-5), other with two hidden layers of 10 nodes in each (ANN *nod* = 10-10), and finally a ANN with three hidden layers of 5 hidden nodes in each (ANN *nod* = 5-5-5) were also implemented. By using a logarithmic scale, Fig. 8 depicts the performance of those different ANN configurations. The CORR is relatively incremented to other ANN with the same number of hidden layers, while increasing the number of nodes in the hidden layers. The MAE and the sum of squared errors (SSE) had similar behaviors, because the SSE only amplifies the effect of the MAE. The training time for the model (in seconds) depends directly on the number of steps performed by the algorithm, and generally it grows exponentially as nodes in the hidden layers' increase. The best results of the prediction accuracy were obtained with an ANN *nod* = 10-10, The training time was approximately 3.3 min, which is considerably high in comparison with an ANN *nod* = 3, which took about 1.2 min for training process. Regarding the training time, the worst results were obtained with an ANN *nod* = 5-5 with about 7.7 min, although it was one of the best ANN with respect to the prediction accuracy.

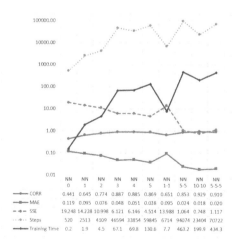

Fig. 8. Performance of different ANN configurations.

5.1.2   Performance of the Support Vector Machine

In order to evaluate the SVM performance, different kernels, parameters, and the *ksvm* function from the *kernlab* library of R language were considered. The evaluated kernels were the radial basis or Gaussian kernel (*rbfdot*), the Laplacian kernel (*laplacedot*), the Bessel kernel (*besseldot*), the RBF Anova kernel (*anovadot*), the polynomial kernel (*polydot*), the linear kernel (*vanilladot*), and the hyperbolic tangent kernel (*tanhdot*).

For each kernel, the parameter *C* was used to compute the cost of restrictions violation. It was also evaluated in a range of 0.1 to 3.1 at intervals of 0.5. For each kernel and value of *C*, the ξ parameter (defined in the regression insensitivity loss function) was evaluated in a range of 0.1 to 2.6 at intervals of 0.5. Thus, 294 combinations were obtained. The parameter $k = 10$ was established to perform *k-fold cross validation* with the training set in order to improve the model quality.

A comparison of the performance of the different kernels, considering the minimum MAE, the maximum CORR, and the minimum training time, is shown in Fig. 9. With respect to the MAE and CORR, the best results were obtained with the Gaussian kernel, followed by the Laplacian kernel, while the worst performance corresponded to the polynomial kernel of second degree, the hyperbolic tangent and kernel and the linear kernels respectively. Regarding the minimum training time achieved, the best performance also corresponded to the Gaussian kernel, with 6 s, followed by the linear kernel and Laplacian kernels, both with 8 s; the polynomial kernel obtained the worst performance, with 3.8 min, even it was configured with grade 2.

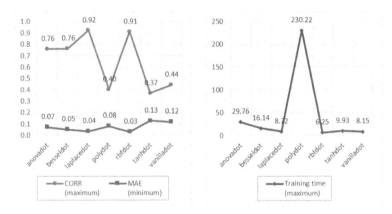

Fig. 9. Performance of different SVM kernels.

5.1.3   Performance Comparison Between ANN and SVM

Once the machine learning models with ANN and SVM were analyzed, a comparison to evaluate which model is more suitable for this dataset was carried out. By using a logarithmic scale, Fig. 10 depicts a summary about the performance obtained with different models, SVM with different kernels against ANN with different amount of nodes in the hidden layers. In this case, the performance was determined in terms of CORR (maximum), MAE (minimum), and the training time (minimum) by each model.

The columns were sorted by the type of model (SVM or ANN) and the obtained MAE. For example, the MAE and CORR obtained with SVM *ker = vanilladot* proved to be comparable to a ANN *nod* = 0, although this ANN took much less time in training than that the SVM. The better models are SVM *ker = rbfdot* and ANN *nod* = 10-10, because they reported the greater prediction accuracy in terms of the lowest MAE and the highest CORR for SVM and ANN, respectively. With the ANN *nod* = 10-10, we obtained slightly better MAE and CORR results than the obtained results with the SVM *ker = rbfdot*. Regarding the training time, the SVM took about 6 s, while the ANN required approximately 3.3 min (30 times more than the SVM).

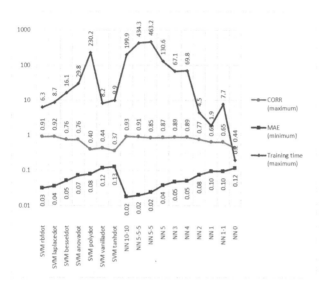

Fig. 10. Performance comparison between SVM and ANN.

5.2   Visualization of Environmental Noise Maps

An environmental noise maps with different time intervals are presented in Figs. 11 and 12. These maps were obtained by applying the interpolation method IDW with the parameter $C = 2$, and the color code corresponds to the standard ISO 1996-2:1987. The highest noise levels recorded during the tours, were located in the zone around the Historic Center. With respect to the Cuauhtémoc district, the highest noise concentration was detected in the Centro settlement (almost entirely), and in the Morelos settlement, which is located near the Merced market zone (see Fig. 11).

Moreover, in the Centro settlement, Fig. 12 shows that the highest noise concentration was found in the streets around the National Palace (streets with high vendor's density). In general, the highest noise levels were registered in the *HNL-12-time interval*, while the lowest noise levels were registered in the *HL-06-time interval*. Thus, we found the highest average of noise levels were caused by the constant noise of people (for example, the shout of street vendors and the continuous talk of crowds), followed by the noise caused by cars.

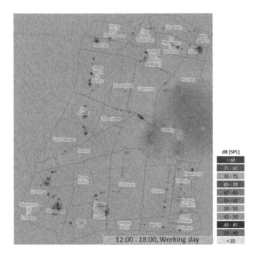

Fig. 11. Environmental noise map for settlements in the Cuautémoc district. (Color figure online)

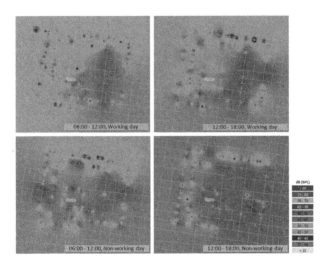

Fig. 12. Environmental noise maps for the streets around the Historic Center. (Color figure online)

6   Conclusion and Future Work

In this paper, a methodology to sense, analyze and characterize the environment noise was described. We proposed a general framework, which consists of three stages from the data acquisition to information visualization. In this approach, citizens play an important role, which can change from simple users to data producers (human sensors), representing the benefits of generating volunteered geographic information.

This approach is focused on obtaining environmental noise measures by means of a mobile application. The sensed data were processed and used as parameters in the prediction model based on two machine learning techniques (SVM and ANN). The result is the characterization of the phenomenon and a spatio-temporal analysis, which was focused on predicting the status of the noise levels.

With respect to the experiments, the comparison of the results of the SVM prediction model against the ANN model showed that ANN has a higher correlation and a lower mean absolute error. The training time is much greater than the SVM model. However, we preferred to use a SVM more than an ANN, since the correlation and mean absolute error of SVM are acceptable, while its training time is much less than the ANN, and it allows making faster predictions.

On the other hand, recent advances in wireless communications, mobile devices, and social computing applications are enabling new urban sensing and management opportunities. Summing up, urban sensing is an interdisciplinary issue, which explores how real-time technologies can help us better understand the dynamics of our cities, as well as conceiving possibilities how these technologies are sustainable for smart cities.

Future works are oriented towards finding a good balance between system scalability and sensing accuracy for city-wide deployment environments. In addition, the integration of heterogeneous data sources could improve the accuracy and enrich the statistics and the spatio-temporal analysis.

Acknowledgments. This work was partially sponsored by the Instituto Politécnico Nacional (IPN), the Consejo Nacional de Ciencia y Tecnología (CONACYT), and the Secretaría de Investigación y Posgrado (SIP) under grant 20162006.

References

1. Leuenberger, M., Kanevski, M.: Extreme learning machines for spatial environmental data. Comput. Geosci. **85**, 64–73 (2015)
2. Nil, J.: Managing data ood is industry challenge (2005)
3. Al-Jarrah, O.Y., Yoo, P.D., Muhaidat, S., Karagiannidis, G.K., Taha, K.: Efficient machine learning for big data: a review. Big Data Res. **2**(3), 87–93 (2015)
4. Rathore, M.M., Ahmad, A., Paul, A., Rho, S.: Urban planning and building smart cities based on the internet of things using big data analytics. Comput. Netw. **101**, 63–80 (2016)
5. Torres-Ruiz, M., Lytras, M.D.: Urban computing and smart cities applications for the knowledge society. Int. J. Know. Soc. Res. **7**(1), 113–119 (2016)
6. Sekimoto, Y., Shibasaki, R., Kanasugi, H., Usui, T., Shimazaki, Y.: Pflow: reconstructing people flow recycling large-scale social survey data. IEEE Pervasive Comput. **4**, 27–35 (2011)
7. Zheng, Y., Capra, L., Wolfson, O., Yang, H.: Urban computing: concepts, methodologies, and applications. ACM Trans. Intell. Syst. Technol. **5**(3), 38–96 (2014)
8. Liu, T., Zheng, Y., Liu, L., Liu, Y., Zhu, Y.: Methods for sensing urban noises. Technical report. MSR-TR-2014–66 (2014)
9. Bulter, D.: Noise management: sound and vision. Nature **5**, 280–481 (2004)

10. Wang, Y., Zheng, Y., Liu, T.: A noise map of New York City. In: Proceedings of the ACM International Joint Conference on Pervasive and Ubiquitous Computing: Adjunct Publication, pp. 275–278. ACM (2014)

11. Ross, Z., Kheirbek, I., Clougherty, J.E., Ito, K., Matte, T., Markowitz, S., Eisl, H.: Noise, air pollutants and traffic: continuous measurement and correlation at a high-traffic location in New York City. Environ. Res. **111**, 1054–1063 (2011)

12. Martí, I.G., Rodríguez, L.E., Benedito, M., Trilles, S., Beltrán, A., Díaz, L., Huerta, J.: Mobile application for noise pollution monitoring through gamification techniques. In: Herrlich, M., Malaka, R., Masuch, M. (eds.) ICEC 2012. LNCS, vol. 7522, pp. 562–571. Springer, Heidelberg (2012)

13. Torija, A.J., Ruiz, D.P.: A general procedure to generate models for urban environmental-noise pollution using feature selection and machine learning methods. Sci. Total Environ. **505**, 680–693 (2015)

14. Zuo, F., Li, Y., Johnson, S., Johnson, J., Varughese, S., Copes, R., Liu, F., Wu, H.J., Hou, R., Chen, H.: Temporal and spatial variability of traffic-related noise in the City of Toronto. Can. Sci. Total Environ. **472**, 1100–1107 (2014)

15. D'Hondt, E., Stevens, M., Jacobs, A.: Participatory noise mapping works! an evaluation of participatory sensing as an alternative to standard techniques for environmental monitoring. Pervasive Mob. Comput. **9**(5), 681–694 (2013)

A Knowledge-Based Approach
for the Implementation of a SDSS
in the Partenio Regional Park (Italy)

Maria Cerreta[✉], Simona Panaro[✉], and Giuliano Poli[✉]

Department of Architecture (DiARC),
University of Naples Federico II, Naples, Italy
{cerreta,simona.panaro,giuliano.poli}@unina.it

Abstract. The paper recommends a methodology for data gathering and processing through the spatial analysis techniques and the combinatorial
multi-criteria procedure of Weighted Linear Combination (WLC). The purpose
concerns the spatial problem structuring in a complex decisional context lacking
in the geographical dataset. The processing of data and information provided by
VGIs and Open Systems is crucial for the enrichment of spatial datasets in these
circumstances, but it is advisable to make attention about the data reliability and
the known problems of the geographic dataset, i.e. Modifiable Areal Unit
Problem (MAUP). The method was tested with the case study of 27 Municipalities around the Partenio Regional Park, in the South of Italy. Within the
SDSS, the multidimensional landscape's indicators were combined with data
gathering on the field, in order to build an evolving informative system.
A multidimensional approach, focused on the recognition of environmental,
social, economic and cultural resources, was chosen providing some strategies
of enhancement for the overviewed landscape of the Park. The evaluation of the
policy and actions for the examined regions generated scenario-maps through
multi-criteria procedures and GIS tools.

Keywords: Landscape · Spatial Decision Support System (SDSS) ·
Volunteered Geographic Information (VGIs) · Weighted Linear Combination
(WLC) · Spatial multi-criteria analysis

1 Introduction

The paper introduces an application of a Spatial Decision Support System (SDSS) for
the landscape evaluation, focused on the development of tourism and recreation services in a region around a natural park, the Partenio Regional Park, in the South of
Italy. The decision problem examines many issues concerning the relationships and
trade-off among economic, social, environmental and cultural values. In order to
investigate the different components, the recent literature recommends gathering hard
and soft data about the region, understanding the spatial effects of a decision on the
landscape (Cerreta et al. 2014; Fusco Girard et al. 2014) and combining institutional
data with open source. During the last twenty years, the progress in remote-sensing and
power computing extended the spatial component evaluation to the decision-making

© Springer International Publishing Switzerland 2016
O. Gervasi et al. (Eds.): ICCSA 2016, Part IV, LNCS 9789, pp. 111–124, 2016.
DOI: 10.1007/978-3-319-42089-9_8

process. Moreover the spatial analysis tools provided by Geographic Information System (GIS) aid the decision-maker (DM) in the data management and analysis of spatial features, in the solution of ill-structured problems in an iterative way, in the scenario evaluation, in report generation and visualization of spatial indicators (Sugumaran and de Groote 2010). Including the multidimensional landscape features within the SDSS procedures is a practice that can be continuously improved and empowered (Cerreta and Fusco Girard 2016). An open issue concerns the spatial problem structuring in a complex decisional context where institutional geographic dataset can lack. Nowadays new open-source software for the production of digital geographic information are widely available and everybody can create his own maps through Volunteered Geographic Information (Goodchild and Li 2012). This skill makes inhabitants more aware of their place and increases geographic data for that territory. One of the main issues regards how to relate institutional data with those produced by open-source's users. This stage is critical for the decision-making process since far more information can be made explicit and available to public debate, increasing, as well, the evaluation transparency (Golub 1997; van der Sluijs 2002). Always more frequently, traditional and new dataset are being used in the SDSS, trying to overcome the lack of geographic data. OpenStreetMap is the best-known platform to create an alternative to the products of official agencies and exported data can be used to provide geographic information on non-spatial indicators too. With its normative, spatial, temporal, environmental, cultural, social, and cognitive features, the landscape becomes the framework where planning and project responses can be shaped.

The first part of the paper (Sect. 2) defines the literature review; the second one (Sect. 3) explains the methodological framework of the SDSS; the third (Sect. 4) shows the study case and the outcomes to test the methodological framework; while the fourth (Sect. 5) concludes about the VGI usefulness in policy-making and landscape planning.

2   A Spatial Decision-Making Process for the Landscape Evaluation. A Literature Review

The increasing complexity of landscape planning and policy-making is related to the impacts of the urbanization processes, the irregular development in spatial planning and the growth of the big data. In this context, many authors report the lack of coordination and adoption of advanced technologies to share information (Li et al. 2013).

On the other hand, the landscape knowledge in a decisional context lacking in geographical data is a critical phase of the spatial decision-making processes since it is necessary to guarantee openness and sharing also in the evaluation processes (Golub 1997; van der Sluijs 2002). It is possible to consider the geographical dataset as a segment of a knowledge-based system that aids the DM for sharing strategies of development and transformation/conservation of the landscape characters. Indeed, geographical data implementation within the landscape evaluation aims the community at identifying own landscape; analysing characteristics, dynamics of transformation and pressures; monitoring environmental and anthropic systems; identifying the values that the people assign to the landscape. Moreover, the representation of the territorial system and the processes simulation are two critical models for the landscape evaluation according to Steinitz's Geodesign framework (Steinitz 2012; Cocco et al. 2015).

The spatial feature add-on within the Decision Support System (DSS), making explicit the relationships between the socio-economic and geo-morphological characteristics of the landscape, aids to understanding the transformation processes of the territory and to identify actions, tactics, and strategies of development (Murgante et al. 2011; Attardi et al. 2014). Moreover, the PGIS tools and the multi-criteria methods integrated to GIS software simplify the spatial evaluations and they aim to convert the qualitative judgments to measurable functions through the landscape metrics approach (Brown and Weber 2011). There are many utilities using spatial data, i.e. visualization and data mapping, proactive communication of critical issues, decision-making simplification, etc. However some care is indispensable about the choice of a consistent spatial reference frame and fixed scale of analysis. Indeed the misunderstanding of the Modifiable Areal Unit Problem (MAUP) can compromise the spatial statistics and the final results of the analysis (Openshaw 1983). In presence of census data, moreover, the unit of aggregation for sampling must be evaluated such as household, neighbourhood or country scale. It is really important to understand that the choice of a different scale can lead to completely different outcomes because of different patterns and relationships within the spatial features (O'Sullivan and Unwin 2010). According to Malczewski (2006), the spatial evaluation criteria can be classified into two macro-category: explicitly and implicitly spatial criteria. The first criteria are composed by inherently spatial data, i.e. geomorphology, natural areas, etc., while the latter use the geographic features in order to transfer a spatial representation of themselves, i.e. the ecological integrity index, the number of employers in tourism per census zone, etc. In this way, both the criteria aid the experts to achieve spatial representation of no spatial explicitly indicators to broaden and improve the knowledge of the landscape.

3 The Methodological Steps of the Knowledge-Based Approach

The purpose of the research aims at forecasting new scenario of the suitability for the touristic development and safeguarding the local potentials and environmental assets. The management of the development and the identification of new strategies require a multidimensional approach, in order to merge different components aiming at supporting the identification of innovative place-based actions (Cerreta 2010).

The transformations characterizing the landscape and local systems depend on multiple factors, such as demographic, social and professional changes within the population; the outplacement in new houses and workplaces; the changes of the specialization; the shift of the transportation and communications both in infrastructural and functional way (Istat 2015). The modelling phase of the SDSS for the landscape of "Partenio" generated the spatial indicators using both spatial explicitly and implicitly criteria. Both the raster and vector-based approach was performed in order to produce spatially referenced data and indicators able to describe real-world features in a virtual environment. Moreover, the geographical dataset architecture allowed to manage numerous and heterogeneous data and to produce useful changing scenario (Cerreta and Poli 2013). Specifically, the variety of information picked was classified into six main domains that characterize the Smart Cities (Economy, Environment, People, Living, Mobility and Governance). The six domains identified the physical, economic, intellectual and social capitals for the development of a territory (Giffinger and Haindl 2007) in order to reach the sustainable use of resources. Lastly, any spatial geo-statistics were performed in order to produce new indicators bridging the gap due to the lack of the geographical data and updates. Specifically, the SDSS was structured in the following steps (Fig. 1):

1. "Data gathering" concerns the selection of the data for the study area through various sources, i.e. the field research, the Web, the surveys.
2. "Spatial data representation" aims at building a representation model in GIS environment.
3. "Spatial indicators" aims at the data processing and classification of the indicators in six domains according to Smart Cities approach.
4. "Normalization" of the indicators have been done in order to make homogeneous the values for the next evaluation phase.
5. "Data processing and clustering" step uses the conversion tool "shape to raster" to elaborate suitable data.
6. "Reclassify" of data was performed in order to give semantic judgments to the values according to a scale from low to very high.
7. Multi-criteria method "WLC" was applied to obtain the overlay maps.
8. "Smart maps" shows the weakness and potential of the study area.
9. "Weighting" step was performed through "swing weight method" (Bodily 1985).
10. "Scenario" simulation was run and two final maps were performed.

In this way, the complex problems can be analysed simultaneously in a "what-if" perspective because of the power computing, knowledge domains and organizational skills of ICT.

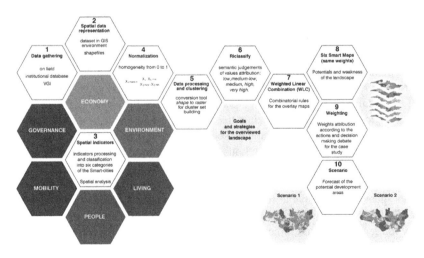

Fig. 1. Methodological framework: steps and contents

4 Case Study

The Spatial Decision Support System for the 27 municipalities around the Partenio Regional Park in Italy (Fig. 2) aims at simplifying the knowledge process of the landscape's tangible and intangible assets in order to recognize the relationships among these assets and to provide guidelines for the DM about the enhancement and the local network strategy. A multidimensional approach focused on the recognition of environmental, social, economic and cultural assets, was chosen and it has made possible to compare some strategies of enhancement with each smart domain for the overviewed landscape.

The above-mentioned steps of the data gathering and indicators processing (steps from 1 to 3) led to identifying the representation model and the process model. The classification of the indicators according to smart domains, indeed, brought to identify an evaluation model composed of six composite maps (steps from 4 to 8) and two final scenario maps (steps 9 and 10).

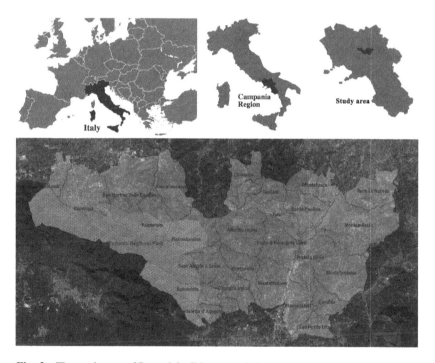

Fig. 2. The study area: 27 municipalities around the Partenio Regional Park in Italy

4.1 The Representation and the Process Models

In the following subsections, the six smart domains (Sect. 3) were synthetically described according to the meaning of each domain and the indicators developing the representation model, and the process model. In Table 1, the structure of the indicators

Table 1. The spatial indicators set categorized into smart domains

Domain	Indicator	U.M.	Year	Source	ID
Economy	Number of beds/Km^2 (Tourism index)	num.	2015	field research	ECO_1
	Mean price of accommodations	€	2015	field research	ECO_2
	Density of accommodations in 5 km	num.	2015	OSM/Web	ECO_3
	Density of food services in 5 km	num.	2015	OSM/Web	ECO_4
	Number of employers in tourism	num.	2009	dps.gov.it	ECO_5
	Number of wine firms	num.	2015	galpartenio.it	ECO_6
	Mean value of agricultural soils	€/ha	2015	CLC/Agenzia delle Entrate	ECO_7
	Number of people with income	num.	2011	ISTAT	ECO_8
Environment	Safeguard surface	ha	2015	Natura 2000	ENV_1
	Density of interest sites in 5 km	num.	2015	OSM	ENV_2
	Ecological integrity index	num.	2012	CLC	ENV_3
	Uninhabited houses	num.	2011	ISTAT	ENV_4
	Number of families in renting house	num.	2011	ISTAT	ENV_5
	Number of house owners	num.	2011	ISTAT	ENV_6
People	Number of people with master degree	%	2011	ISTAT	PEO_1
	Number of residents	num.	2011	ISTAT	PEO_2
	Housing density	Inh/Km^2	2015	ISTAT	PEO_3
	Youth index (20–35 age)	%	2011	ISTAT	PEO_4
	Old age index (over 65)	%	2011	ISTAT	PEO_5
	Employment rate	%	2011	ISTAT	PEO_6
	Unemployment rate	%	2011	ISTAT	PEO_7
Living	Number of months per municipality with cultural events	num.	2015	galpartenio.it	LIV_1
	Variety of cultural events	num.	2015	galpartenio.it	LIV_2
	Number of cultural events	num.	2015	galpartenio.it	LIV_3

(*Continued*)

Table 1. (*Continued*)

Domain	Indicator	U.M.	Year	Source	ID
	Number of municipalities with naturalistic path and itinerary	num.		OSM	LIV_4
Mobility	Class of accessibility per municipality	class	2009	dps.gov.it	MOB_1
	Presence or absence of a station per municipality	binary	2009	dps.gov.it	MOB_2
	Accessibility network	Km	2015	OSM	MOB_3
	People moving outside of their municipality	num.	2015	field research	MOB_4
Governance	Number of stakeholders	num.	2015	field research	GOV_1
	Number of projects	num.	2015	field research	GOV_2

is shown and the following fields are highlighted: domain, indicator's name, unit of measure (U.M.), year, source and ID (Table 1).

Economy. The "economy" domain aims at identifying the zones where the density of the economics and production for tourism development establish a spatial correlation. The selected local resources are classified in the following thematic areas and indicators: the accommodation (number, location, type, average prices); the quality of the agricultural production (number of wineries per municipality and market value of the agricultural soil) and the catering facilities. Specifically the indicator "value of the agricultural soils" (ECO_9) was built through the combination of the specific classes of CLC and the mean value of the land provided by the institutional dataset of the Italian "Agenzia delle Entrate". This processing has made spatially explicit the approximated quality of the agricultural production. Furthermore, some indices as the density of accommodation and catering facilities were processed through the Kernel Density Estimation method in order to build the process model. The geostatistics tool assesses the number of point events per unit of surface within a point-pattern (O'Sullivan and Unwin 2010). Therefore, the processing of point data produces new indicators on the areal surface of the landscape to identify the areas with the greatest concentration of tourism services (Fig. 3).

Environment. The "environment" domain includes both the potential and the weak components for the improvement of the touristic fruition. The natural and cultural landscape indicators were processed by the manipulation of the selected row data, i.e. the safeguard level of the natural surfaces and the ecological integrity index (van Berkel and Verburg 2014); furthermore the institutional dataset of the census zones provided information about the state of the housing abandon and the number of the family living in the analysis area. Same as above, the kernel density estimation identified the highest concentration of the touristic facilities points (Fig. 4).

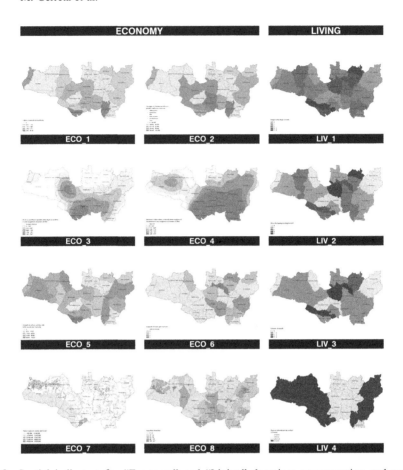

Fig. 3. Spatial indicators for "Economy" and "Living" domains: representation and process models

People. The "people" domain contains information about the population more exposed to the impact of the decisions. The indicators in this domain measure mainly the structure of the population per age and education. The major weakness of these indicators set is the MAUP. For this reason, while the "people" class is useful to understand the social and economic systems in order to build a broader and bright representation model, it is necessary to make a new indicators selection before proceeding to the multicriteria method application and evaluation (Fig. 5).

Living. The "living" domain aims at the identification of the cultural vitality of the examined region. The number of the cultural events and their type/frequency, the naturalistic path, the geographic itineraries were selected in order to identify the local resources improving the touristic network. The selected indicators in this category are samples of no spatially explicit criteria.

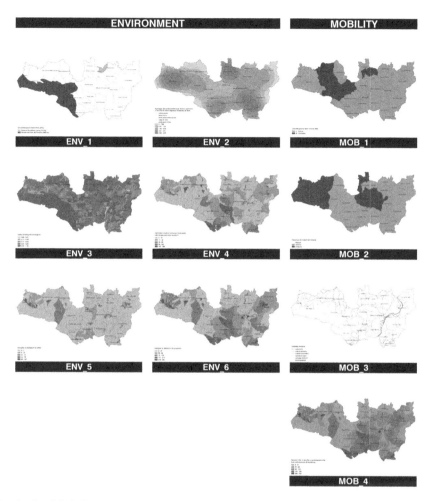

Fig. 4. Spatial indicators for "Environment" and "Mobility" domains: representation and process models

Mobility. The "mobility" domain shows the outside/inside accessibility of the municipalities. The indicators summarizing these issues was chosen within an institutional dataset that makes a classification of the municipalities according to the presence or absence of a railway station. In this regard, the number of railway stations and the other infrastructures of the study area were identified through different sources (Istat, DPS, OpenStreetMap). Furthermore, the number of people moving outside the municipalities was selected. Also in this category, the MAUP can be crucial.

Governance. Lastly, the "governance" domain contains indicators that measure the network of the stakeholders and the financing projects on the landscape.

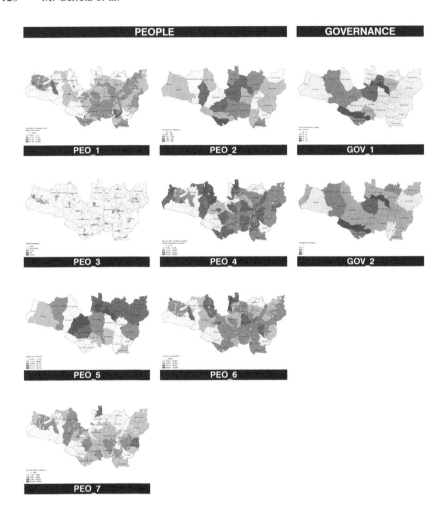

Fig. 5. Spatial indicators for "People" and "Governance" domains: representation and process models

4.2   Outcome: Two Evaluation Scenario for the Tourism Development Through WLC Method

The spatial indicators were worked out through the multi-criteria WLC method and six composite maps show the state of the local assets and the processes in the territory (Fig. 6).

According to the main purpose of the research, the weighting phase was implemented with the "swing weights method" (Bodily 1985). This method can be preferable when geographical data are available, since it simplifies the attributes outranking and weighting according to stakeholders preferences (Malczewski 1999).

The gradual scale of colours, from red to green, was chosen in order to identify areas with a different degree of suitability for tourism services development,

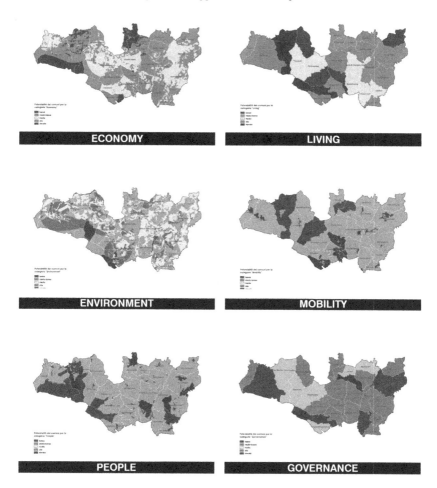

Fig. 6. Smart domains maps: weakness and potential of the study area through WLC method (Color figure online)

considering the red zones as negative and the green ones as positive, while the others mean intermediate values. The weighting phase provides two scenarios showing the suitable zones for tourism development according to the provided policy and planning strategies. The scenario 1, defined "Sulfur-Line" (Fig. 7), aims at improving the local resources through the wine and food paths strategy. This strategy is able to make explicit the history of the old mining quarries of the landscape, visiting naturalistic places and tasting local products.

The purpose aims at improving a network of municipalities to guarantee the touristic flows. The Table 2 shows the weights assigned to the scenario 1.

The scenario 2, defined "Welfare-Line" (Fig. 8), aims at fostering the religious and naturalistic tourism, implementing the quality of life through the preference of a slow-mobility, the enhancement of the amenities and the use of the touristic path and guides in the Partenio Regional Park.

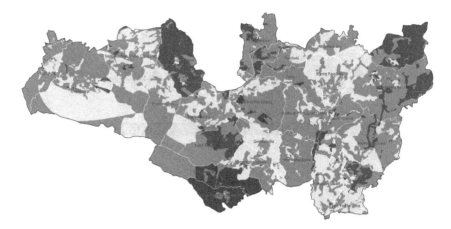

Fig. 7. Scenario 1 "Sulfur-Line". The gradual scale of colours, from red to green, allows to identify areas with major functioning of the tourism development strategy, considering the red as low and green as high, while the others mean intermediate judgments (Color figure online)

Table 2. Scenario 1 "Sulfur-Line". Weights of the domains

Ranking	Domain	Weight	
1	Environment	0,245	0,49
	Living	0,245	
2	Economy	0,165	0,33
	Mobility	0,165	
3	People	0,09	0,18
	Governance	0,09	

Fig. 8. Scenario 2 "Welfare-Line". The gradual scale of colours, from red to green, allows to identify areas with major functioning of the quality of life enhancement strategy, considering the red as low and green as high, while the others mean intermediate judgments (Color figure online)

The Table 3 shows the weights assigned to the scenario 2.

Table 3. Scenario 2 "Welfare-Line". Weights of the domains

Ranking	Domain	Weight	
1	Economy	0,245	0,49
	Living	0,245	
2	Governance	0,165	0,33
	Mobility	0,165	
3	People	0,09	0,18
	Environment	0,09	

5   Conclusions

In the paper, it has been tested the SDSS for landscape evaluation. Its methodological approach wants to improve the acknowledgement of the complex values of the landscape, by defining a model of representing and processing data. These models have made possible to draw up appropriate spatial indicators for both geographical explicit data and implicit ones. This has improved the understanding of the landscape resources and transformation ways ongoing in the municipalities around the Partenio Regional Park.

By integrating, thus, data coming from public sources with VGIs', it has been given a picture of the information on the context and the touristic enhancement goals. In details, the arranging of the information by following the domains of smart grammar aims to supervise available resources and the vigour of the place, highly regarding both environmental preservation and the needs of inhabitants and tourists. The issues dealt so far are about the geographical acknowledgement of the information and the chance to improve it through open source data.

The fact-finding survey developed here can be, therefore, improved and refined by users' contribution. The tested SDSS opens the path to public debate on future scenario. Specifically, those simulated in step 9 and 10 allow a preliminary evaluation of the policies and the planning strategies currently in action. Thus, the scenario maps show new geography of complex values, where the green colour areas have the major opportunity about the functioning of the tourism development strategy, while the intermediate colour areas can be understood as a bridge among the strong zones and the weak one.

References

Attardi, R., Pastore, E., Torre, C.M.: Scrapping of quarters and Urban renewal: a geostatistic-based evaluation. In: Murgante, B., et al. (eds.) ICCSA 2014, Part III. LNCS, vol. 8581, pp. 430–445. Springer, Heidelberg (2014)

Bodily, S.E.: Modern Decision Making: A Guide to Modelling with Decision Support Systems. Mcgraw-Hill College, New York (1985)

Brown, G., Weber, D.: Public participation GIS: a new method for National Park planning. Landscape Urban Plan. **102**(1), 1–15 (2011)

Cerreta, M.: Thinking through complex values. In: Cerreta, M., Concilio, G., Monno, V. (eds.) Making Strategies in Spatial Planning. Knowledge and Values, vol. 9, pp. 381–404. Springer, Dordrecht (2010)

Cerreta, M., Fusco Girard, L.: Human smart landscape: an adaptive and synergistic approach for the National Park of Cilento, Vallo di Diano and Alburni. Agric. Agric. Sci. Procedia **8**(5), 489–493 (2016)

Cerreta, M., Inglese, P., Malangone, V., Panaro, S.: Complex values-based approach for multidimensional evaluation of landscape. In: Murgante, B., et al. (eds.) ICCSA 2014, Part III. LNCS, vol. 8581, pp. 382–397. Springer, Heidelberg (2014)

Cerreta, M., Poli, G.: A complex values map of marginal Urban landscapes: an experiment in Naples (Italy). Int. J. Agric. Environ. Inf. Syst. **4**(3), 41–62 (2013)

Cocco, C., Fonseca, M.B., Campagna, M.: Applying geodesign in Urban planning case study of Pampulha, Belo Horizonte. Braz. J. Cartography **67**(5), 929–940 (2015)

Fusco Girard, L., Cerreta, M., De Toro, P.: Integrated assessment for sustainable choices. Scienze Regionali **13**(1), 111–141 (2014)

Giffinger, R., Haindl, G.: Smart cities: ranking of European medium-sized cities. Centre of Regional Science (SRF), Vienna University of Technology, Vienna, Austria (2007)

Golub, A.L.: Decision Analysis: An Integrated Approach. Wiley, New York (1997)

Goodchild, M.F., Li, L.: Assuring the quality of volunteered geographic information. Spat. Stat. **1**, 110–120 (2012)

Istat: La nuova geografia dei sistemi locali, Istituto Nazionale di Statistica, Rome, Italy (2015)

Li, W., Li, L., Goodchild, M.F., Anselin, L.: A geospatial cyberinfrastructure for Urban economic analysis and spatial decision-making. ISPRS Int. J. Geo-Inf. **2**, 413–431 (2013)

Malczewski, J.: GIS and Multi-criteria Decision Analysis. Wiley, New York (1999)

Malczewski, J.: GIS-based multicriteria decision analysis: a survey of the literature. Int. J. Geogr. Inf. Sci. **20**(7), 703–726 (2006)

Murgante, B., Tilio, L., Lanza, V., Scorza, F.: Using participative GIS and e-tools for involving citizens of Marmo Platano - Melandro area in European programming activities. J. Balkans Near East. Stud. **13**(1), 97–115 (2011)

Openshaw, S.: The modifiable areal unit problem. CATMOG - Concepts Tech. Mod. Geogr. **38**, 3–41 (1983)

O'Sullivan, D., Unwin, D.J.: Geographic Information Analysis. Wiley, Hoboken (2010)

Steinitz, C.: A Framework for Geodesign: Changing Geography by Design. Esri Press, Redlands (2012)

Sugumaran, R., de Groote, J.: Spatial Decision Support Systems: Principles and Practices. CRC Press Taylor & Francis Group, Boca Raton (2010)

van Berkel, D.B., Verburg, P.H.: Spatial quantification and valuation of cultural ecosystem services in an agricultural landscape. Ecol. Ind. **37**, 163–164 (2014)

van der Sluijs, J.P.: A way out of the credibility crisis of models used in integrated environmental assessment. Futures **34**, 133–146 (2002)

Factors of Perceived Walkability: A Pilot Empirical Study

Ivan Blečić[1(✉)], Dario Canu[2], Arnaldo Cecchini[2], Tanja Congiu[2],
and Giovanna Fancello[3]

[1] Department of Civil, Environmental and Architectural Engineering
(DICAAR), University of Cagliari, Cagliari, Italy
ivanblecic@unica.it
[2] Department of Architecture, Design and Urban Planning (DADU),
University of Sassari, Alghero, Italy
{dacanu, cecchini, tanjacongiu}@uniss.it
[3] CNRS-Lamsade, Université Paris Dauphine, Paris, France
giovanna.fancello@dauphine.fr

Abstract. We present preliminary results of a pilot empirical study designed to examine factors associated with pedestrians' perception of walkability, i.e. the perception of the quality, comfort and pleasantness of streets, and their conductivity to walk. Through a contingent field survey we collected 18 observable street attributes (independent variables), and a synthetic subjective perception of walkability (dependent variable), for the entire street network (408 street segments) of the city of Alghero in Italy. Regression analysis yields high goodness of fit (R-squared = 0.60 using all 18 variables), and points at 9 out of 18 as the most significant factors of perceived walkability ("useful sidewalk width"; "architectural, urban and environmental attractions"; "density of shops, bars, services, economic activities"; "vehicles-pedestrians separation"; "cyclability"; "opportunities to sit"; "shelters and shades"; "car roadway width"; "street lighting"; R-squared = 0.59). Among those, the first five factors in particular show as jointly most important as predictors of perceived walkability.

Keywords: Walkability · Regression analysis · Walkability perception · Urban design · Walkability audit

1 Introduction

This empirical study contributes to the ongoing multidisciplinary effort to pin down factors, their relative importance and their interactions, relevant for pedestrians' *perception* of walkability, that is to say, of the quality, comfort and pleasantness of streets, and their conductivity to walk.

In attempt to describe and explain people's propensity and decision to walk, their choices of pedestrian route and the qualitative perception thereof, scholars have examined a series of factors, related to individual characteristics (e.g. age, gender, income, etc.), mobility opportunities (e.g. availability of public transportation), trip types (purpose, frequency, available time, etc.), and features of the walking environment [1].

© Springer International Publishing Switzerland 2016
O. Gervasi et al. (Eds.): ICCSA 2016, Part IV, LNCS 9789, pp. 125–137, 2016.
DOI: 10.1007/978-3-319-42089-9_9

Our study focuses on this latter family of factors, related to the physical urban environment, and attempts to determine their correlation with the subjective, qualitative perception of the walking environment. Ultimately, the purpose is to provide useful indications both for modelling and evaluating urban walkability [2, 3], as well as for suggesting the most effective levers urban design and planning may be able pull to encourage walking behaviour by improving the pedestrian friendliness of cities [4].

Our study is based on a survey of the entire street network of the city of Alghero, a coastal town of approximately 40.000 inhabitants in the North-West Sardinia in Italy. Every street segment was audited for 18 *analytic* descriptive attributes, and was independently scored for its perceived overall walkability. This data allowed us to perform a regression analysis and to estimate the relationship between the analytic attributes and the synthetic perception of walkability.

In the next section we provide a brief review of approaches, methods and findings of similar studies reported in literature. Following we present the experimental design and settings of the study, and discuss its main findings.

2 Background

The question we attempt to address in our study is how the physical features of urban space influence the (qualitative) perception of its walkability. Scholars have employed a range of experimental designs, sets of dependent and independent variables, survey methods, and analytical tools to tackle this issue.

As dependent variables, one can find measurements of degree of satisfaction with urban environment [5, 6], the perception of its quality [7, 8], the perceived pedestrian-centred *Level of Service* (LOS) of street segments [9] and of street crossings [10, 11], the willingness to pay for improvements [12], the easiness of crossing [13], pedestrian accessibility [14], perceived safety and comfort [15], children safety and its perception by the parents [16, 17], and the relationship between children's and parents' choices of routes [18]. When attempting to acquire "objective" measures and observed behaviours, rather than declared qualitative perceptions and evaluative judgements, scholars have (with variable success) employed data on physical activity (and inactivity) [19, 20], use of public transportation [21], fraction of trips by foot [22], route choices [23], their feasibility [4] and their relation to personal traits [24].

Among independent variables, assumed as "predictors" of walkability, in literature we encounter three types of variables, along objective-subjective axis: (1) physical, functional and urban design features of space; (2) practices of use of space (frequencies, densities, flows, rates of use, etc.); and finally (3) individual perceptions or reactions to space [e.g. 25].

The first type of variables cover physical features (such as walkway width, number of car lanes, presence of green areas, landmarks and other "attractions", as well as the degree of maintenance). These measures may be strictly quantitative on cardinal scales (e.g. width in meters, car speed in km/h), or more qualitative usually evaluated on ordinal scale (e.g. degrees of maintenance).

The second type of variables describe phenomena related to how the space is being used, such as land uses, economic activities (bars, restaurants, shops, services, etc.), population densities, traffic flows, pedestrian flows, and so on.

Finally, the third type of variables are those more related to perception and reaction to space, such as sense of security, perceived urban quality, "sense of place", and so on.

As for the data collection and survey methods, scholars have been undertaking different routes. Direct, on-street survey methods can be classified [10] into:

- *observational method*, evaluating the LOS based on *in-situ* observation of pedestrian behaviour (pedestrian density, pedestrian flow rate, walking speed, etc.);
- *intercept survey*, interviewing pedestrians after they have traversed a crosswalk at intersection or a street segment and asking them to grade the crossing or the segment;
- *contingent field survey* (CFS), involves subjects walking along routes and instructed to grade each crosswalk or street segments immediately after they have traversed the intersection or the street;
- *controlled field valuation* (CFV) involves taking subjects to different intersections and letting surveyors observe and then grade the crosswalk without actually undertake the crossing; usually used for intersections, this method can also be adapted for the street segments;
- *laboratory/simulation studies* (LSS) involve subjects observing and evaluating a representation of the pedestrian environment; simulations may comprise various techniques to describe, represent and visualize the walking environment, from 3D renderings [e.g. 18], to photographs and photomontages [e.g. 12], to video clips [e.g. 6].

Among those mentioned, there are a few studies drawing methodological resemblances with ours.

Koh and Wong [23] conducted a survey to examine the influence of "infrastructural compatibility factors" on pedestrians' and cyclists' choices of commuting route. They combined interviews for stated preferences with walkability and bikeability audits, and compared the commuters' chosen routes with shortest available routes based on 11 infrastructural compatibility factors (for pedestrians: weather protection, distance, comfort, security, traffic accident risk, crowdedness, detour, number of road crossings/delay, stairs/slope, directional signs, good scenery and shops along route). For pedestrians, the study revealed a preference for routes that are comfortable, with shops and good scenery and preferably with the presence of other people (crowdedness).

Lamíquiz and López-Domínguez [22] use bivariate correlation and multivariate regression modelling to examine the association between features of the built environment (independent variables) and the proportion of pedestrians on all trips (dependent variable) in different parts of a city. The independent variables were organized in three groups: (1) street network (line length, segment length, etc.) and its configurational accessibility (connectivity, integration, etc.); (2) land use (including density and mix; and (3) non-built-environment variables (socio-economic characteristics, such as age and car ownership). Bivariate correlation showed a relatively high level of relation between pedestrian trips and about ten variables: their multivariate regression model yielded R-squared (adjusted) of 70.63 % using variables: mean line length, mean line density, percentage of culs-de-sac, "radius 5 integration", intelligibility, "resid. + jobs + stud./Ha", retail food units/Ha, retail units/Ha, jobs/residents, retail units/residents, retail food units/Res., distance to city center, percentage of residents > 65, and percentage of residents between 45–65 years.

In a study carried out by Evers *et al.* [16], parents volunteered to audit streets and intersections leading to seven elementary schools in a suburban school district. The parents were asked to report their agreement with the statement "I would feel comfortable letting an unsupervised 8-year-old child travel along/across this street/intersection", on a 5-point Likert scale centered on neutral. Logistic regression models were created for street segments and intersections with the variables: traffic lines, turning lane, paved/planted median strip, trees presence, cul-de-sac end, likely place to walk, wheelchair accessibility, walking path wide, tripping hazards, obstructions, driveway hazard for the street segments and traffic control, size of intersection, "bumped-out" corner, curb cuts lack, crossing medians and crossing marked for the interactions. The final model predicting perceived lack of safety for street segments encompassed five predictors and performed well (R-squared = 0.632). Significant association were found with variables: street trees presence, most likely place to walk, traversable by wheelchair, free of tripping hazards, path obstructed intersection and size of intersecting roads.

Ling *et al.* [10] estimated the perceived pedestrian level of service (LOS), using correlation analysis and stepwise regression analysis. In this study, the LOS reflected pedestrians' perception of crossing in safety and comfort. A *contingent field survey* was used to ask pedestrians to score crosswalks from 1 to 5. The relevant variables estimated by the stepwise regression (R-squared = 0.65) were found to be: entering right-turning motorized vehicles, leaving left-turning non-motorized vehicles, pedestrians volume at the beginning of green time, mixed cyclists volume, pedestrian delay (s), presence of refuge island, presence of two-step crossing.

Muraleetharan *et al.* [13] attempted to identify factors affecting pedestrian LOS at intersections. Factors examined were space at corner, visible cross markings, separate bicycle path, pathfinder tiles, curb ramps, number of lanes, refuge islands, turning vehicles, delay and pedestrian-bicycle interaction. A stepwise multivariable regression analysis was used to model the pedestrian LOS. The study revealed that the factor "turning vehicle" has greater influence on pedestrian LOS than any other. When the number of turning vehicles increases, the result shows a corresponding decrease in the perceived safety to the pedestrian. Furthermore, the factors "delays at signals" and "pedestrian-bicycle interaction" were also found to be significant factors in determining pedestrian LOS at intersections.

Jensen [6] attempted to determine the key variables influencing pedestrian satisfaction (stated preferences). The final regression model of LOS yielded the R-squared value of 0.55 for walkability, using as dependent variables: motor vehicles flow rate, average speed of motor vehicles, type of pedestrian facility (sidewalk - no sidewalk), type of bicycle facility (one-way bicycle track - bicycle lane - drive lane) and type of land use/buildings (shopping - residential - mixed use).

3   Experimental Design

Building on aforementioned studies, we conducted a *contingent field survey* (CFS) of the entire street network of the city of Alghero (Italy). The purpose of the survey was to collect two separate measures for each street segment of the city: (1) an *analytic* description of the street segment, through 18 observable street attributes; and (2) a *synthetic*, subjective perception of its quality of walkability.

The streets were divided into 408 homogeneous street segments, and the city was subdivides into 10 sectors (see Fig. 1). The survey was carried out in January 2016 by 24 graduate students split into 12 pairs, each pair assigned (1) to undertake a walkability audit of one urban sector (collecting 18 attributes for each street segment), and (2) to provide their subjective *synthetic* evaluation of the street segments in another sector. The two sectors assigned to each pair were different not to have their previous analytic knowledge influence their synthetic evaluation of streets.

A total of 483 records were collected, and then reduced to 408, one for each street segment, by averaging the values of multiple records referring to the same segment.

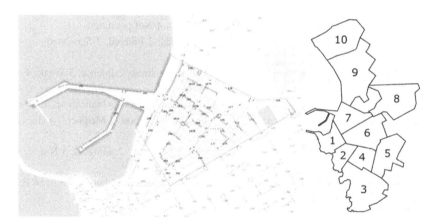

Fig. 1. 10 city sectors (right), and homogeneous street segments (example for Sector 7).

The list of the 18 analytic attributes are reported in Table 1. While some attributes do use qualitative levels, we provided auditors with detailed definitions and exemplifications to limit their interpretative ambiguity. So for example, for the five levels of the attribute "X1 Useful sidewalk width", the levels and their interpretations we provided as instructions to the auditors were: "1. Wide: allows comfortable passage for at least 4 people without obstacles; 2. Comfortable: allows passage for 3 people, even if with few minor nuisances; 3. Minimal: allows passage for 2 people, with obstacles that occasionally force to divert; 4. Inadequate: allows passage for only one person, with numerous obstacles along the route and detours; 5. Missing: no sidewalk, or impossible to use".

As for the *synthetic* subjective perception of the quality of walk, the auditors were asked to express their overall evaluative judgment about street segments by answering to the following question: "Express a *synthetic* evaluation of your perception of the

Table 1. List of street segment attributes.

Attributes (variables)	Scale levels
X1 Useful sidewalk width	1 Wide, 2 Comfortable, 3 Minimal, 4 Inadequate, 5 Missing
X2 Objects of architectural, urban and environmental attractions	1 Many, 2 Some, 3 Absent, 4 Some disturbing, 5 Many disturbing
X3 Density of shops, bars, services, economic activities	1 Plenty, 2 Some, 3 Few, 4 Absent
X4 Opportunity to sit (benches, etc.)	1 Extended, 2 Sparse, 3 Absent
X5 Shelters and shades	1 Strong, 2 Weak, 3 Absent
X6 Car traffic direction	1 Pedestrian street, 2 One way, 3 Two way
X7 Car roadway width	1 Pedestrian street, 2 One lane, 3 Two lanes, 4 Three lanes, 5 Four (or more) lanes
X8 Speed limit	1 Pedestrian street, 2 ≤20 km/h, 3 30 km/h, 4 50 km/h, 5 ≥70 km/h.
X9 Bicycle track (cyclability)	1 Off-road exclusive lane, 2 On-road exclusive lane, 3 On-road shared with vehicles, 4 Not permitted
X10 Degree of integration with surrounding space	1 Integrated, 2 Filtered, 3 Separated
X11 Vehicles-pedestrians separation	1 Pedestrian street, 2 Intense, 3 Weak, 4 Absent
X12 Street lighting	1 Excellent, 2 Good, 3 Inadequate, 4 Absent
X13 Sidewalk degree of maintenance	1 Excellent, 2 Good, 3 Mediocre, 4 Uneven, 5 Absent
X14 Street-level parking	1 Pedestrian street, 2 Allowed, 3 Not Allowed, 4 Illegal
X15 Physical car-speed reducers (hump, raised crossings, traffic islands, mini roundabouts)	1 Pedestrian street, 2 Many, 3 Some, 4 Few, 5 Absent
X16 Non-physical car speed reducers (traffic lights density, *enclosure*)	1 Pedestrian street, 2 High, 3 Medium, 4 Low, 5 Absent
X17 Crossings density (crossing opportunity)	1 Pedestrian street, 2 High, 3 Low, 4 Absent
X18 Road type	1 Pedestrian street, 2 Car street

quality and walkability of the street, from your point of view as pedestrian. The evaluation must be expressed on a qualitative scale from 1 ("insufficient") to 5 ("excellent"), taking into account the physical features of the pedestrian walkway, the overall qualitative characteristics of the urban space, and in general how you perceive the street to be safe, comfortable, pleasant, attractive and usable.

In order to express your evaluation, you *must not* take into account the distance from the city centre, nor the temporary sources of disturbance (such as public works, construction in progress, etc.)".

Following are the meanings of each evaluation level:

5. Excellent: maximum pedestrian comfort; the street is very pleasant to walk, with particularly attractive and valuable surrounding urban space or landscape where it is interesting to walk, sit and hangout.
4. Very good: the street is comfortable to walk, the pedestrian transit is pleasant and free of obstacles.
3. Good: the street can be walked, the pedestrian transit is without obstacles, but the surrounding urban space or landscape is not attractive.
2. Sufficient: the street is difficult to walk, there are obstacles to the pedestrian transit or the quality of the urban space and landscape is low, unpleasant or very disturbing.
1. Insufficient: the street is impossible to walk or feels very unsafe, the quality of the surrounding space or landscape is very disturbing.

4 Results

After uniformly re-scaling all the evaluations and grades on a scale from 0 to 1, we ran several multivariate linear regressions to explore models of correlation between the synthetic evaluation of walkability (dependent variable) and the street attributes (independent variables).

Table 2. Multivariate linear regression models

	Model A (R-squared = 0.60)			Model B (R-squared = 0.59)			Model C (R-squared = 0.56)		
	Est.	St. err.	p-val.	Est.	St. err.	p-val.	Est.	St. err.	p-val.
(Incpt.)	−0.209	0.049	3.e-05***	−0.210	0.046	8.e-06***	−0.049	0.032	0.126
X1	0.150	0.054	0.006**	0.202	0.048	3.e-05***	0.277	0.046	3.E-09***
X2	0.229	0.059	1.e-04***	0.236	0.056	3.e-05***	0.287	0.056	4.E-07***
X3	0.085	0.033	0.011*	0.101	0.032	0.002**	0.130	0.032	6.E-05***
X4	0.096	0.047	0.042*	0.107	0.046	0.020*			
X5	0.066	0.037	0.077	0.067	0.034	0.048*			
X6	−0.028	0.048	0.555						
X7	0.162	0.081	0.046*	0.119	0.067	0.077			
X8	0.116	0.092	0.209						
X9	0.169	0.067	0.012*	0.136	0.061	0.026*	0.197	0.0572	6.E-04***
X10	0.001	0.034	0.968						
X11	0.141	0.069	0.042*	0.129	0.055	0.019*	0.174	0.048	4.E-04***
X12	0.169	0.055	0.002**	0.212	0.050	3.e-05***			
X13	0.052	0.057	0.364						
X14	0.097	0.059	0.102						
X15	−0.048	0.071	0.498						
X16	−0.059	0.042	0.159						
X17	0.075	0.049	0.131						
X18	−0.094	0.066	0.160						

The first model, using all the available independent variables, yields R-squared = 0.60 (see Model A in Table 2). A subsequent model, using the 9 most significant variables from model A (and excluding "Street level parking" for low variability in the data) yields R-squared = 0.59 (see Model B in Table 2). These results point at the following nine variables as jointly most strongly associated with the overall synthetic perception of walkability: "Useful sidewalk width" (X1), "Objects of architectural, urban and environmental attractions" (X2), "Density of shops, bars, services, economic activities" (X3), "Opportunity to sit" (X4), "Shelters and shades" (X5), "Car roadway width" (X7), "Bicycle track (cyclability)" (X9), "Vehicles-pedestrians separation" (X11), and "Street lighting" (X12).

A graphical representation of two-way contingency tables between each of the nine variables and the dependent variable are shown in Fig. 2.

For a comparison of relative importance of independent variables, we ran separate monovariate linear regressions for those nine variables. From these results, reported in Table 3, we further note that the strongest individual effect may be observed on the variables X1, X2, X3, X9 and X11, each *individually* yielding R-squared > 0.20. A multivariate linear regression model using only those five variables yields R-squared = 0.56 (See Model C in Table 2.)

Using the model with nine variables for prediction (Model B), in Fig. 3 we compare the actual and the predicted values of synthetic evaluation of walkability.

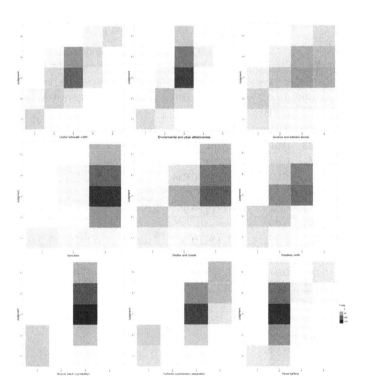

Fig. 2. Tile plot of judgments and street characteristics distribution of frequency.

Table 3. Monovariate linear regressions for the most significant independent variables

	X1	X2	X3	X4	X5	X7	X9	X11	X12
Coef. est.	0.6474	0.7637	0.3861	0.4598	0.3535	0.5540	0.6797	0.5804	0.5845
Std. error	0.0399	0.0559	0.0374	0.0520	0.0399	0.0676	0.0489	0.0379	0.0592
p-value	<2e-16	<2e-16	<2e-16	<2e-16	<2e-16	3.4e-15	<2e-16	<2e-16	<2e-16
(Intercept)	0.2381	0.2017	0.3500	0.1979	0.3323	0.3509	0.1626	0.1939	0.2919
(Std. error)	0.0249	0.0317	0.0276	0.0483	0.0336	0.0340	0.0339	0.0290	0.0343
(p-value)	<2e-16	5.3e-10	<2e-16	5.1e-05	<2e-16	<2e-16	2.3e-06	7.3e-11	3.5e-16
R-squared	0.3931	0.3152	0.2082	0.1617	0.1620	0.1418	0.3220	0.3659	0.1935

Fig. 3. Precision of street walkability predictions, using model B.

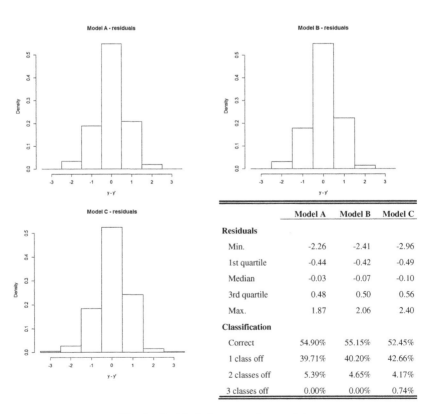

	Model A	Model B	Model C
Residuals			
Min.	-2.26	-2.41	-2.96
1st quartile	-0.44	-0.42	-0.49
Median	-0.03	-0.07	-0.10
3rd quartile	0.48	0.50	0.56
Max.	1.87	2.06	2.40
Classification			
Correct	54.90%	55.15%	52.45%
1 class off	39.71%	40.20%	42.66%
2 classes off	5.39%	4.65%	4.17%
3 classes off	0.00%	0.00%	0.74%

Fig. 4. Residuals for regression models

The distribution of residuals of this model are shown and reported in Fig. 4. From the summary data in the figure, one can note that the Model B predicts approximately 55 % of street segments in the correct class, and classifies over 95 % of street segments correctly or at most one class off from the actual synthetic evaluation assigned by the auditors.

5 Conclusion

The purpose of our study was to determine which urban features and design characteristics of the streets are most strongly correlated with a qualitative synthetic perception of the quality and walkability of streets. With respect to other similar studies, we have undertaken walkability audits to collect comparatively more detailed descriptions of the streets, both in terms of the number of descriptive attributes and in terms of modalities for some of the attributes. Furthermore, we used both qualitative and quantitative descriptors in a way to reduce equivocation and misunderstandings of the meaning of their respective scales of measurement.

In our *contingent field survey*, we were able to estimate the importance of street attributes in relation to the declared *synthetic* evaluation, and thus to avoid possible errors of direct declared valuation of relative importance of attributes by the interviewees. As noted by Guo *et al.* [4] "contingent rating based on stated preference may overestimate the importance of more tangible attributes, such as distance and safety, because pedestrians were often unable to articulate intangible amenities, such as streetscapes and façade designs". Judgments and walkability audits were also collected *in-situ* to capture as much as possible the real perceptions of the space, which could get lost in standard survey methods.

The results of regression analysis in particular show the following nine attributes to be highly significant and jointly yield a relatively high R-squared of 0.59: "Useful sidewalk width", "Objects of architectural, urban and environmental attractions", "Density of shops, bars, services, economic activities", "Vehicles-pedestrians separation", "Bicycle track (cyclability)", "Opportunity to sit", "Shelters and shades", "Car roadway width", and "Street lighting". These attribute are related to the pleasantness, comfort and safety, and are thus in accordance with Alfonzo's *et al.* [26] hierarchy of walking needs. Out of the nine attributes listed before, the first five in particular are revealed to be jointly most strongly associated with the perceived *synthetic* walkability (R-squared = 0.56).

As a prediction tool, the regression model using the above mentioned nine most significant attributes shows a fairly high precision of predictions (55 % streets classified correctly, 95 % classified correctly or at most ±1 class off).

In future, we will intend to widen the sample and explore different statistical approaches, such as ordinal model, conjoint analysis, and part-worth function models. Also further investigation is needed to explore interactions between variables, which may me undertaken through choice modeling approaches.

References

1. Mateo-Babiano, I.: Pedestrian's needs matter: examining Manila's walking environment. Transp. Policy **45**, 107–115 (2016)
2. Blečić, I., Cecchini, A., Congiu, T., Fancello, G., Trunfio, G.A.: Evaluating walkability: a capability-wise planning and design support system. Int. J. Geog. Inf. Sci. **29**(8), 1350–1374 (2015)
3. Blečić, I., Cecchini, A., Congiu, T., Fancello, F., Fancello, G., Trunfio, G.A.: Walkability explorer: application to a case-study. In: Gervasi, O., Murgante, B., Misra, S., Gavrilova, M. L., Rocha, A.M.A.C., Torre, C., Taniar, D., Apduhan, B.O. (eds.) ICCSA 2015. LNCS, vol. 9157, pp. 758–770. Springer, Heidelberg (2015)
4. Guo, Z., Loo, B.P.Y.: Pedestrian environment and route choice: evidence from New York City and Hong Kong. J. Transp. Geography. **28**, 124–136 (2013)
5. Van Dyck, D., Cardon, G., Bedorche, B., De Bourdeaudhuij, I.: Do adults like living in high-walkable neighborhoods? Associations of walkability parameters with neighborhood satisfaction and possible mediators. Health Place **17**, 971–977 (2011)
6. Jensen, S.: Pedestrian and bicyclist level of service on roadway segments. Transp. Res. Rec. J. Transp. Res. Board **2031**, 43–51 (2007)
7. Hanák, T., Marović, I., Aigel, P.: Perception of residential environment in cities: a comparative study. Procedia Eng. **117**, 495–501 (2015)
8. Ewing, R., Handy, S., Crownson, R.C., Clemente, O., Winston, E.: Identifying and measuring urban design qualities related to walkability. J. Phys. Act. Health **3**(1), 223–240 (2006)
9. Kang, L., Xiong, Y., Mannering, F.L.: Statistical analysis of pedestrian perceptions of sidewalk level of service in the presence of bicycles. Transp. Res. Part A **53**, 10–21 (2013)
10. Ling, Z., Ni, Y., Cherry, C.R., Li, K.: Pedestrian level of service at signalized intersections in China using contingent field survey and pedestrian crossing video simulation. Transp. Res. Board Annu. Meet. **14**, 41–52 (2014)
11. Muraleetharan, T., Adachi, T., Uchida, K., Hagiwara, T., Kagaya, S.: A study on evaluation of pedestrian level of service along sidewalks and at crosswalks using conjoint analysis. Infrastruct. Planning Rev. **21**, 727–735 (2004)
12. Ng, W.Y., Chau, C.K., Powell, G., Leung, T.M.: Preferences for street configuration and street tree planting in urban Hong Kong. Urban For. Urban Greening **14**, 30–38 (2015)
13. Muraleetharan, T., Adachi, T., Hagiwara, T., Kagaya, S.: Method to determine pedestrian level-of-service for crosswalks at urban intersections. J. East. Asia Soc. Transp. Stud. **6**, 127–136 (2005)
14. Nakamura, K.: The spatial relationship between pedestrian flows and street characteristics around multiple destinations. Int. Assoc. Traffic Safety Sci. **39**(2), 156–163 (2016). http://dx.doi.org/10.1016/j.iatssr.2015.08.001
15. Landis, B.W., Vattikuti, V.R., Ottenberg, R.M., McLeod, D.S., Guttenplan, M.: Modeling the roadside walking environment: a pedestrian level of service. Transp. Res. Rec. **1773**, 82–88 (2001)
16. Evers, C., Boles, S., Johnson-Shelton, D., Schlossberg, M.: Parent safety perceptions of child walking routes. J. Transp. Health **1**, 108–115 (2014)
17. Gallimore, J.M., Brown, B.B., Wener, C.M.: Walking routes to school in new urban and suburban neighborhoods: an environmental walkability analysis of blocks and routes. J. Environ. Psychol. **31**, 184–191 (2011)
18. Nasar, J.L., Holloman, C., Abdulkarim, D.: Street characteristics to encourage children to walk. Transp. Res. Part A **72**, 62–70 (2015)

19. Cerin, E., Macfarlane, D.J., Ko, H.H., Chan, K.C.A.: Measuring perceived neighbourhood walkability in Hong Kong. Cities **24**(3), 209–217 (2007)
20. Gauvin, L., Richard, L., Craig, C.L., Spivock, M., Riva, M., Forster, M., Laforest, S., Laberge, S., Fournel, M.C., Gagnon, H., Gagné, S., Potvin, L.: From walkability to active living potential, an "ecometric" validation study. Am. J. Prev. Med. **28**(2), 126–133 (2005)
21. Spears, S., Houston, D., Boarnet, M.G.: Illuminating the unseen in transit use: a framework for examining the effect of attitudes and perceptions on travel behavior. Transp. Res. Part A **58**, 40–53 (2013)
22. Lamíquiz, P.J., López-Domínguez, J.: Effects of built environment on walking at the neighbourhood scale. A new role for street networks by modelling their configurational accessibility? Transp. Res. Part A **74**, 148–163 (2015)
23. Koh, P.P., Wong, Y.D.: Influence of infrastructural compatibility factors on walking and cycling route choices. J. Environ. Psychol. **36**, 202–213 (2013)
24. Ramezani, S., Pizzo, B., Deakin, E.: Built environment versus personal traits: an application of integrated choice and latent variable model (ICLV) in understanding modal choice in Rome, Italy. In: 14th International Conference on Computers in Urban Planning and Urban Management (2015)
25. Noriza, R., Ariffin, R., Zahari, R.K.: Perceptions of the urban walking environments. Soc. Behav. Sci. **105**, 589–597 (2013)
26. Alfonzo, M., Boarnet, M.G., Day, K., McMillan, T., Anderson, C.L.: The relationship of neighbourhood built environment features and adult parents' walking. J. Urban Des. **13**, 29–51 (2008)

Evaluating the Effect of Urban Intersections on Walkability

Ivan Blečić[1(✉)], Arnaldo Cecchini[2], Dario Canu[2], Andrea Cappai[2], Tanja Congiu[2], and Giovanna Fancello[3]

[1] Department of Civil, Environmental and Architectural Engineering,
University of Cagliari, Cagliari, Italy
ivanblecic@unica.it
[2] Department of Architecture, Design and Urban Planning,
University of Sassari, Alghero, Italy
{cecchini,dacanu,andrea2.cappai.studenti,
tancon}@uniss.it
[3] CNRS-Lamsade, Université Paris Dauphine, Paris, France
giovanna.fancello@dauphine.fr

Abstract. This study proposes an analytical and evaluative method of the performances of urban intersections from the perspective of pedestrians. We further present a case study assessment of walkability of crossings and their conduciveness to walk. Implications in integrated urban and transport planning practice are emphasized as the method is suited to support decision makers involved in urban roads management to identify major spatial and operational problems and to prioritize improvement interventions.

Keywords: Urban intersections · Pedestrian accessibility · Walkability · Evaluation · Multicriteria decision aiding

1 Introduction

This study deals with the limiting effect of street intersections on walking in urban areas. We are interested in understanding to what extent spatial and operational configuration of crossings (geometry, design, facilities, road signs, crossing control devices, etc.) affect the convenience, pleasantness and quality of walk across the city.

For this purpose, we have developed a method for analyzing and evaluating urban intersections, hopefully useful as a decision support for the development of transportation policies and for prioritizing interventions aimed at enhancing pedestrian mobility. We have tested the evaluation method on nine intersections with different characteristics in the town of Alghero (Italy). The selected intersections have been rated with an ELECTRE TRI rating/classification procedure, based on their geometric and operational characteristics and then classified with respect to their level of impedance to walking. Since the proposed method is reasonably simple to implement and communicate we believe it can be practically used by decision makers interested in making pedestrian crossings more comfortable and safe, promoting thus walking in cities. Furthermore, the proposed evaluation procedure may be incorporated into other

© Springer International Publishing Switzerland 2016
O. Gervasi et al. (Eds.): ICCSA 2016, Part IV, LNCS 9789, pp. 138–149, 2016.
DOI: 10.1007/978-3-319-42089-9_10

evaluation models of urban walkability, which frequently put only minor attention on (when not entirely ignoring) the effects of junctions on the overall walkability of places.

2 Background and Related Research

In urban and transportation planning there is a growing attention on how built environment correlates to walking. However, models of pedestrian behavior, accounting for the characteristics of urban space, are mostly concentrated on street segments, with less attention given to intersections. When taken into account as punctual elements of the transport network characterized by a high potential of conflict between vehicular and pedestrian flows, intersections are often studied in relation to concerns of safety and comfort. However, depending on their physical and operational characteristics (geometry, volume of vehicular traffic, control system, etc.), urban intersections represent a major interruption of pedestrian trajectory, and can in general significantly affect the perception of and the attitude towards urban space.

Scholars largely agree on the fact that the distance between origin and destination is one of the determinants of the decision to walk, along with directness and continuity of paths, i.e. the absence of interruptions along pedestrian routes. According to some authors, the convenience of walking is further influenced by street connectivity, which in turn depends on urban grid and land-use patterns. For this reason, in measures of non-motorized accessibility, such as walkability indices, intersections are accounted for in the form of *density* (number of intersections per unit area), *typology* (number of approaches for each node, i.e. number of 3/4-way intersections), *connected node ratio*, etc. [1–4]. However, connectivity indicators often show not to be adequate for measuring the effects of intersections on walking.

In literature, the level of observation remains often general, neglecting information on spatial and operational characteristics of crossing areas. In this regard, one prolific field of studies gathers pedestrian level of service (PLOS) measures, aimed at evaluating road and crossing performances from the point of view of pedestrians. PLOS ranks roads or intersections into classes, each describing how the design of a given one is conducive to walk. Most common factors considered in those measures are traffic volumes, car speed, visibility, presence of physical barriers and facilities for impaired [5].

There are examples of comparative procedures of evaluating intersections [e.g. 6] which oppose the real crossing area to an ideal one defined through a combination of spatial configurations supposed to ensure high levels of safety and comfort standards for pedestrian. Other studies combine physical and operational variables with behavioral features: "Walking Security Index" (WSI) [7] and Muraleetharan's *et al.* index [8] are two examples of this approach. Both attempt to rank intersections according to their performances expressed in terms of level of safety, comfort and convenience expected and experienced by pedestrians.

With time, measures of PLOS became more detailed including environmental factors considered integral part of pedestrian experience.

Abley and Tuner [9] designed two separate predictive regression models, one for path lengths and one for road crossings, aiming at detecting the elements of the walking environment that make a place more attractive to be traversed by foot. The authors correlated objective physical and operational characteristics of urban space with subjects' perception of the walking environment.

Following the literature, several physical and operational factors of urban intersections correlate to walking. Safety conditions of pedestrian seems to be the main issue of concern when analyzing intersections and their effect on walkability. Comfort and convenience are other recurring qualities.

The first and perhaps most obvious factor is the effect of traffic volume and speed on pedestrian safety. It has been largely demonstrated that high speed traffic impedes pedestrian crossing: narrow streets with lower traffic speeds and volumes are easier to cross, wider streets with higher traffic volumes and speeds cause discomfort by increasing safety risks and delays [10].

According to Litman [11], the "barrier effect" of vehicles represents an indirect cost people bear when changing route or travel mode to avoid crash risk and disturbance. Traffic entails delays or extensions of the itinerary, steering the choice of travel mode towards motorized conveyances. It is thus a constraint factor which reduces the viability of non-motorized travel, especially undermining the freedom of choice of citizens with limited accessibility abilities such as children, elderly people, people with disability, household with no car, etc. Langlois *et al.* [12] argued that difficulties of older adults to cross the street relate mainly to insufficient time to cross the road because of large distance from curb to curb, slower walking speed and visual impairments. Moreover, excessive noise, inadequate lighting and heavy traffic are main environmental factors affecting functional ability and independence of elderly [13]. Similarly, traffic around home, crash risk, long distances to travel and general sense of insecurity produce a sense of danger among parents who decide not to allow their children to walk alone [14, 15].

The physical distance to cross is another important variable. It can be estimated as a geometric measure (length of road) or in the form of number of lanes or types of roads (arterial, collector, local).

The number and extension of incurred stops, their geometry, facilities and equipments further influence pedestrian behavior [16]. Reported among relevant variables are: presence and characteristics of the space at corners composed by hold area and circulation area, crossing facilities (elevated crossing, zebra and other pavement markings including detectable warnings for alert impaired pedestrians before entering a vehicular way), medians or refuge islands, controlling measures such as traffic lights.

With regard to the comfort, curb ramps at opposite corners of the intersection make the elevation transition from sidewalk to street level and vice-versa more comfortable. The same role is played by medians and refuge islands, better if raised and wide enough to handle pedestrian during two-stage crossing.

Other reported factors are the quality of the connection between opposite sides and the "atmosphere" (sense of place) of surrounding environment created by good lighting, mix of functions, "enclosure", seats, trees, shade and signs [17]. Road furniture (street lamps, directional and informational signs, trees, transit shelters) are at once both useful and hazardous as they guide pedestrian movements but, depending on location, can obscure visual contact and compromise pedestrian's ability to safely cross the road. The limitative effect to visibility occurs even in presence of on street parking, both regular or irregular, which can hinder the vision of the crossing [18, 19].

Pedestrian delay at crossings is also considered in many studies [5, 20] as a factor conducive to risky behaviors. Moreover, as time saving represents an important benefit for those who decide to travel by foot, pedestrian time delays are perceived as a real disutility.

3 The Evaluation Model

The aim of this study is to identify and measure critical factors of intersection design and management which limit pedestrian accessibility. For this purpose, we propose an evaluation method of urban intersections. The procedure resorts to an audit field analysis with a rating evaluation model to combine both qualitative and quantitative characteristics that affect the performance of intersections. The main outcome of the method is thus a ranking of intersections, from most to less hindering for pedestrians.

As we already mentioned, the accomplishment of pedestrian safety, comfort, convenience and attractiveness should be among the basic requirements for promoting urban walkability. Therefore, it can be useful to have evaluation methods to support: (1) detecting critical factors, such as the adverse state of certain intersection characteristics; (2) assessing their effect on people's capacity to experience and use the city by foot; and (3) decision making and prioritizing of improvements. From a practical point of view, the model provides local decision makers with a formal method for prioritizing interventions on intersections that will yield benefits in terms of improved pedestrian safety and increased number of people who decide to walk.

In order to evaluate the relationships between built environment and non-motorized travel we employ a multicriteria evaluation method ELECTRE TRI [21–23]. ELECTRE TRI is an outranking rating/classification method endowed with properties useful for the present study: (*i*) it allows an exhaustive classification of elements in ordered prioritizing categories; (*ii*) criteria aggregation is flexible and allows to express the relative weights, clusters of coalitions (majority rule) and eventual vetoes; (*iii*) it allows a prudential non-compensatory aggregation of information with limited loss of information during consecutive stages of evaluation; (*iv*) it is a procedure that resembles individual models of thinking. These features make the ELECTRE TRI procedure appropriate for our assessment purposes.

When choosing the criteria and designing the model, we privileged those features of the built environment which can be clearly conceptualized and objectively or unambiguously measured. Furthermore, we have incorporated attributes used in the CAWS walkability evaluation model (implemented in Walkability Explorer) [24] in order to make the two interoperable and to lay foundations for further integrations. Table 1 summarizes the attributes of intersections we collected in our study.

Table 1. Attributes of geometric and operational characteristics of intersections

Variables		Type/Unit
Number of approaches at each intersection		number
Number of lanes of each road		
Dedicated bike lanes (B_{lanes})		Present (1)
Traffic lights (T_{light})		Absent (0)
Cross markings (Z)		
Couples of curb cut (C_{cutx2})		
Sidewalk extensions (S_{ext})		
Elevated sidewalk (S_{elev})		
Physical obstacles to visibility (V_{ObEl})	Parked Cars (C_{parked})	Present (1)
	Sticking out Buildings(B_{protr})	Absent (0)
	Signs (S)	
Physical elements directing pedestrian movements (D_{PEI})	Fencings (F)	Present (1)
	Expedients to Direct Pedestrians(D_{exp})	Absent (0)
Corner area ($A_P = H_P + W_P$)	Holding space (H_P)	m^2
	Space for Walking (W_P)	
Carriageway area (A_v)		m^2
Road width from curb to curb (R_{width})		m
Number of pedestrians in all directions (P_T)		n_{ped}/h
Numbers of motorized vehicles (in Vehicle Equivalency Units) (M_T)		VEU*/h
Speed limit (V_S)		km/h

4   A Case Study of Application of the Evaluation Method

4.1   Study Area

We experimented the evaluation model in the city of Alghero (Italy) on nine intersections in the "Pietraia" neighborhood (Fig. 1). The urban fabric stands out for medium urban density and mixed use with residential, retail activities and services. The population (of about 10.000 inhabitants, 1/4 of Alghero) consists for the majority of households of low to medium income, with a notable incidence of elderly people. A range of urban attractors make Pietraia a vibrant area with roads interested by medium and high volumes of motorized and non-motorized traffic depending on the time of day and on seasons. The Via Don Minzoni, one of the most important urban collectors passes through the area connecting important facilities and services of the city: the hospital, one hypermarket and a few supermarkets, public offices, a police station, primary and high schools, and the railway station. The Viale Europa, another important urban road providing both distributional and connecting function, marks the subdivision of the study area into two parts: on the east side the former residential borough with permanent households and services prevailing; on the west the seaside sector that holds many tourists facilities and accommodations (summer residences, hotels, B&B, restaurants). The beach side of the neighborhood shows a great variability on the number of inhabitants during the year due to the season of summer tourism, and consequently presents different intensity of flows (motorized and non motorized) across the area with related problems of accessibility.

Furthermore, Pietraia is involved in an important project of urban regeneration promoted by the municipality aiming at improving the livability of the district by reducing physical, functional and perceptive distances from the town centre. One of the fundamental strategies of the project is to restore and activate urban relationships starting from physical re-connections. According to this design, the continuity of pedestrian routes becomes a basic requirement in order to allow people to move across destinations making them to perceive as walking over a unified urban space. This objective appears attainable due to the relative proximity of local facilities and because of the fairly short distances between Pietraia and other areas of Alghero such as the town centre (3.5/4 km) or Lido (< 1 km).

4.2   Data and Evaluation Criteria

The nine intersections considered in the study are representative of Pietraia everyday life and of most common problems with regard to pedestrian's experience of urban space. They have a strategic location for daily urban activities at both neighbourhood and urban level. Different combinations of roads (collectors and local streets) converge in the selected nodes with crossing distances ranged from 6.0 m to 20.0 m and with variable numbers of lanes. This entails differences in geometry, operational organizations of crossing and in land-use structure.

A field survey was conducted by students at each selected site. Physical features of crossings were collected through direct site inspections and measurements. Figure 2 shows examples of two surveyed intersections.

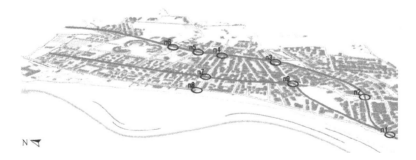

Fig. 1. Geographic distribution of the nine surveyed intersections in Pietraia, Alghero

Video graphic method was used to collect data on mobility. Traffic volume (motorized, bikes and pedestrians) for each direction of intersections was recorded in peak hours of a normal day (from 10:00 a.m. to 12:00 a.m. and from 5:30 to 19:30 p.m.) and under fair weather conditions.

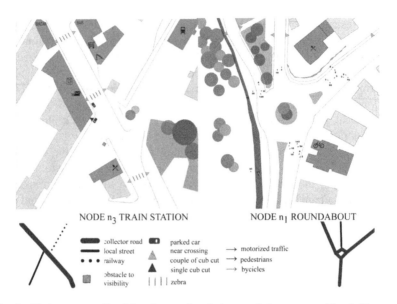

Fig. 2. Nodes n_1, n_3 - Spatial and operational characteristics measured by field survey

4.3 The Evaluation Procedure

In order to consider all the routing alternatives, the first step in evaluating an intersection is to subdivide it into individual lane crossings a pedestrian can walk.

Each individual crossing is then evaluated on five criteria $H = \{H_1, ..., H_m\}$, whose performances are derived from the respective variables of geometric and operational qualities of the crossing. Then, based on classification rules and class thresholds on those criteria/factors, each individual crossing is rated in one of the three possible classes: C_1 "Conducive to walk", C_2 "Sufficient to support walk", C_3 "Obstacle to walk". The criteria, their underlying variables, and the rating rules and thresholds are reported in Table 2.

Finally, by adopting equal weight of the five criteria ($w = 1/5$) we assign through ELECTRE TRI two classifications to each individual crossing, one more conservative (majority threshold $\gamma = 75$ %) and one less conservative ($\gamma = 50$ %).

Once single crossings belonging to an intersection have been evaluated, we compute the intersection's *overall rating* as a weighted average of individual crossing evaluations, by using crossings' relative importance (calculated as the share of all pedestrians at the intersection using each crossing) as weights.

As an example in Fig. 3 and Table 3 we represent the evaluations on five criteria for the four crossings (a_{31}, a_{32}, a_{33}, a_{34}) composing intersection n_3 "Train Station".

According to this classification, the crossing a_{31} is conservatively ($\gamma = 75$ %) rated as C_3 "obstacle" and less conservatively ($\gamma = 50$ %) as C_2 "sufficient". In the same, way a_{32}, a_{33} and a_{34} are rated as C_3 for both majority thresholds.

In Table 4 we report the classification of all the nine intersections.

Table 2. Assessment criteria and definition of categories (cfr. Tab. 1 for variables)

Criterion	Variables	Rating rules /thresholds		
		C_1 Conducive	C_2 Sufficient	C_3 Obstacle
H_1– Crossing control	T_{light}	T_{light} and Z	T_{light} xor Z	Neither
	Z			
H_2–Continuity & Ease of movement	C_{cutx2}	(C_{Cut} and S_{ext})	C_{Cut} xor S_{ext}	Neither
	S_{ext}	or S_{elev}	if no S_{elev}	
	S_{elev}			
H_3 – Comfort	$P_{ratio} = A_P/A_v$	$P_{ratio} \gg 1$	$P_{ratio} \cong 1$	$0 \ll P_{ratio} < <1$
H_4 – Safety	V_{ObEl}	No V_{ObEl}	no V_{ObEl}	V_{ObEl} and
	D_{PEl}	and D_{PEl}	xor D_{PEl}	no D_{PEl}
H5–Barrier effect	$(R_{width} + M_T + V_S)/3$	$0 \le B_{Eff} \le 1.4$	$1,5 \le B_{Eff} \le 2.4$	$2.5 \le B_{Eff} \le 3$

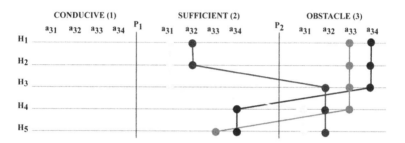

Fig. 3. Evaluation of intersection n_3 with respect to the five criteria

Table 3. Assessment results of node n_3 - "Train Station"

Intersection	Crossing	Weight (Relative importance)	Class		Score	
			$\gamma = 75\%$	$\gamma = 50\%$	$\gamma = 75\%$	$\gamma = 50\%$
n_3	a_{31}	0,07	C_3	C_2	3	2
	a_{32}	0,03	C_3	C_3	3	3
	a_{33}	0,23	C_3	C_3	3	3
	a_{34}	0,67	C_3	C_3	3	3
	Final class (weighted average)		C_3	C_3	3	2.93

Table 4. Assessment results of all nine surveyed intersections

Intersection	n_1		n_2		n_3		n_4		n_5		n_6		n_7		n_8		n_9	
γ	(a)	(b)	(a)	(b)	(a)	(b)	(a)	(b)	(a)	(b)	(a)	(b)	(a)	(b)	(a)	(b)	(a)	(b)
Scores	2	2.2	3	2.5	3	2.9	2.7	2	3	2	2.6	2.5	3	2.9	2	2	3	3
Class	C_3	C_2	C_3	C_3	C_3	C_3	C_3	C_2	C_3	C_2	C_3	C_3	C_3	C_3	C2	C_2	C_3	C_3

4.4   Prioritizing Intersections

The above classification of intersections highlights the physical and operational features which are obstacles for walking, and is suggestive of possible corrective interventions which may be implemented.

For that purpose, we further attempt to establish an order of priority among intersections, in terms of their need for improvements. The ordering is carried out using the following two priority criteria:

- *Walkability class*: based on the previous rating classification, each intersection is assigned to a walkability class according to the following rules:
  - class W_1, if the intersection is classified as C_3 for both majority thresholds ($\gamma = 75 \%$ and $\gamma = 50 \%$),;
  - class W_2, if the intersection is classified C_3 under the majority threshold $\gamma = 50 \%$, but not under $\gamma = 75 \%$,
  - class W_3, if the intersection is classified C_2 for both majority thresholds ($\gamma = 75 \%$ and $\gamma = 50 \%$),
  - class W_4, otherwise.
- *Intersection importance*: four classes of importance (I_1, …, I_4), based on the share of pedestrians on each intersection, with respect to the total number of pedestrians on all the nine intersections, observed in the same time period during the audit.

The evaluations of intersections on the two criteria are reported in Table 5.

Table 5. Walkability class and importance of intersections

Walkability class	Importance			
	I_1	I_2	I_3	I_4
W_1	n_3	n_2, n_7	n_6	n_9
W_2	n_1	n_4		n_5
W_3	n_8			
W_4				

By assigning numeric values (Borda rule) to classes in Table 5 equivalent to their scale order, and calculating the weighted average (with weights 60 % for "walkability class" and 40 % for "importance") we obtain the final order of priority among intersections: n_3 in the highest class of priority; n_1, n_2, n_6 and n_7 in the second class of priority; n_4, n_8 and n_9 in the third class of priority; and finally n_5 classified in the lowest class of priority.

5   Discussion of Results

According to the employed assessment method, one intersection resulted of the highest priority, while four out of nine were assigned to the second level of intervention priority. Among the latter, the nodes n_2 and n_7, which, based on their spatial and functional characteristics were less conducive for walking (class C_3), were in terms of priority classified in the lower category when taking into account their pedestrian level of use. A similar circumstance affected the intersections n_6 and n_9, whose relatively limited importance determined their prioritisation into second and third class of priority.

By working backwards through the assessment procedure, it is possible to better identify which geometric and operational factors most affect the pedestrian accessibility at intersections. The strongest factors of impedance to walking in the nine Pietraia's intersections were related to the barrier effect (criterion H_5), produced by the distance to cross and the intensity of the motorised traffic.

Crossings often result not to be conducive to walk also with regard to the criterion H_4, expressing the ratio between the space available for pedestrians and that for the vehicles: on that account, 77 % of crossings were classified as "obstacles" for walking, while only 20 % resulted as favourable to pedestrians.

With respect to the criteria H_1, H_2 and H_3, the majority of crossings present an intermediate degree of pedestrian friendliness (respectively 63 %, 60 % and 74 % of the crossings belong to the category C_2 "sufficient"). In case of the criterion H_2 concerning safety condition, 40 % of crossings are assessed as "obstacle", suggesting thus a need for interventions aimed at improving the continuity of pedestrian paths and the ease of movement for all users, including the disabled.

Considering intersections as a whole (comprising multiple crossings), our assessments report a generalised need of improvement in order to promote walking. Under the most conservative procedure of evaluation (corresponding to majority threshold $\gamma = 75$ %) almost all intersections in their current state are evaluated as fairly unsuitable to comfortably accommodate pedestrians (eight out of nine intersections were classified as C_3 "obstacle").

The scenario among Pietraia's intersections somewhat improves when using the majority threshold $\gamma = 50$ %. In this case five out of nine intersections keep category C_3 and the remaining four are classified as "sufficient" for walking.

6   Conclusions and Future Work

Most of the research on urban walkability reported in literature is generally focused on spatial features on uninterrupted sidewalks. Much less attention has been devoted to intersections, although they are often the predominant factor of the quality of walk and of pedestrians' perception of safety, comfort and pleasantness of a walking route. In this paper we presented an application of an evaluation model, based on ELECTRE TRI, for rating of urban intersections according to their effect in limiting or facilitating walking, and a subsequent procedure for prioritising interventions of improvement. The approach we propose provides meaningful information on the spatial conditions of intersections related to features which mainly matter to people on foot. Therefore, we believe it represents a useful tool for planners and decision makers to address policies and interventions aimed at improving the liveability of the city by enhancing the safety, comfort, usefulness and pleasantness of pedestrian space.

Further investigation will comprise a field survey on pedestrian's perception of intersections in order to verify whether the main assumptions and parameters used in the evaluation procedure hereby presented are substantiated by empirical evidence. Moreover this study represents another step in our ongoing research effort on urban walkability [25], aiming at improving the original model of walkability evaluation by incorporating considerations related to intersections.

References

1. Cervero, R., Kockelman, K.: Travel demand and the 3Ds: density, diversity, and design. Trans. Res. D **2**, 199–219 (1997)
2. Dill, J.: Measuring network connectivity for bicycling and walking. In: 83rd Annual Meeting of the Transportation Research Board, pp. 11–15, Washington DC (2004)
3. Forsyth, A., Hearst, M., Oakes, J.M., Schmitz, K.H.: Design and destinations: factors influencing walking and total physical activity. Urban Stud. **45**(9), 1973–1996 (2008)
4. Schlossberg, M., et al.: Refining the grain: using resident-based walkability audits to better understand walkable urban form. J. Urbanism **8**(3), 260–278 (2015)
5. Transportation Research Board: Highway Capacity Manual, TRB National Research Council, Washington D.C (2000)
6. Canale S., Distefano N., Leonardi S.: Analisi comparativa del rischio di incidentalità pedonale in corrispondenza delle intersezioni stradali urbane. In: XVII Convegno Nazionale S.I.I.V. Università Kore di Enna (2008)
7. Wellar, B.: The walking security index and pedestrians' security in urban areas. In: Jannelle, D., Warf, B., Hansen, K. (eds.) WorldMinds: Geographical Perspectives on 100 Problems, pp. 183–189. Kleuwer Academic Publishers, Boston (2004)
8. Muraleetharan, T., Adachi, T., Hagiwara, T., Kagaya, S.: Method to determine pedestrian level-of-service for crosswalks at urban intersections. J. Eastern Asia Soc. Transp. Stud. **6**, 127–136 (2005)
9. Abley, S. Tuner, S.: Predicting walkability. Research Report 452, New Zealand Transport Agency, Wellington (2011)
10. Jacobsen, P.L., Racioppi, F., Rutter, H.: Who owns the roads? How motorised traffic discourages walking and bicycling. Inj. Prev. **15**(6), 369–373 (2009)
11. Litman, T.: Transportation Cost Analysis; Techniques, Estimates and Implications. Victoria Transport Policy Institute (2000). www.vtpi.org
12. Langlois, J.A., Keyl, P.M., Guralnik, J.M., et al.: Characteristics of older pedestrians who have difficulty crossing the street. Am. J. Public Health **87**, 393–397 (1997)
13. Balfour, J.L., Kaplan, G.A.: Neighborhood environment and loss of physical function in older adults: evidence from the Alameda county study. Am. J. Epidemiol. **155**(6), 507–515 (2002)
14. Kadali, B.R., Vedagiri, P.: Evaluation of pedestrian crosswalk level of service (LOS) in perspective of type of land-use. Transp. Res. A **73**, 113–124 (2015)
15. Nasar, J.L., Holloman, C., Abdulkarim, D.: Street characteristics to encourage children to walk. Transp. Res. A **72**, 62–70 (2015)
16. Bian, Y., Lu, J., Zhao, L.: Method to determine pedestrians level of service for unsignalized intersections. Appl. Mech. Mater. **253–255**, 1936–1943 (2013)
17. Ewing, R., Handy, S.: Measuring the unmeasurable: urban design qualities related to walkability. J. Urban Des. **14**(1), 65–84 (2009)

18. Martin, A.: Factors Influencing Pedestrian Safety: A Literature Review. TRL Limited, London (2012)
19. Peprah, C., Oduro, C.Y., Afi Ocloo, K.: On-street parking and pedestrian safety in the Kumasi metropolis: issues of culture and attitude. Developing Country Stud. **4**(20), 85–94 (2014)
20. Li, B.: A model of pedestrians' intended waiting times for street crossings at signalized intersections. Transp. Res. B **51**, 17–28 (2013)
21. Bouyssou, D., Marchant, T., Pirlot, M., Tsoukiàs, A., Vincke, P.: Evaluation and Decision Models with Multiple Criteria: Stepping Stones for the Analyst. Springer Science & Business Media Inc., New York (2006)
22. Roy, B.: Decision Aid and Decision Making. Eur. J. Oper. Res. **45**, 324–331 (1990)
23. Roy, B., Bouyssou, D.: Aide Multicritère à la Décision: Méthodes et Cas. Economica, Paris (1993)
24. Blečić, I., Cecchini, A., Congiu, T., Fancello, G., Trunfio, G.A.: Evaluating walkability: a capability-wise planning and design support system. Int. J. Geogr. Inf. Sci. **29**(8), 1350–1374 (2015)
25. Blečić, I., Cecchini, A., Congiu, T., Fancello, G., Trunfio, G.A.: Walkability explorer: an evaluation and design support tool for walkability. In: Murgante, B., Misra, S., Rocha, A.M. A., Torre, C., Rocha, J.G., Falcão, M.I., Taniar, D., Apduhan, B.O., Gervasi, O. (eds.) ICCSA 2014, Part IV. LNCS, vol. 8582, pp. 511–521. Springer, Heidelberg (2014)

Coupling Surveys with GPS Tracking to Explore Tourists' Spatio-Temporal Behaviour

Ivan Blečić[1(✉)], Dario Canu[2], Arnaldo Cecchini[2], Tanja Congiu[2],
Giovanna Fancello[3], Stefania Mauro[4], Sara Levi Sacerdotti[4],
and Giuseppe A. Trunfio[2]

[1] Department of Civil and Environmental Engineering and Architecture,
University of Cagliari, Cagliari, Italy
ivanblecic@unica.it

[2] Department of Architecture, Design and Urban Planning, University of Sassari,
Alghero, Italy
{dacanu,cecchini,tancon,trunfio}@uniss.it

[3] CNRS-Lamsade, Université Paris Dauphine, Paris, France
giovanna.fancello@dauphine.fr

[4] SiTI – Higher Institute for Territorial Innovation, Turin, Italy
{stefania.mauro,levi}@siti.polito.it

Abstract. Position tracking technologies developed in the last decade are a valuable addition to the traditional toolbox for data collection, as they offer the opportunity to gather a great amount of unprecedented information on tourists behaviour. In particular, they allow to collect detailed information on spatial and temporal behaviour with respect to different categories/profiles of tourists. We present the results of a survey of tourists' spatial behaviour coupling GPS movement tracking and questionnaires, and furthermore discuss how this kind of studies may prove useful in providing guidelines for territorial, tourist and transportation policies.

Keywords: Tourist behaviour · GPS tracking · Tourism policy

1 Introduction

"Know thy tourists" is a useful maxim for any smart local government official or destination manager responsible for tourism policies. For that, position tracking technologies developed in the last decade are a valuable addition to the traditional toolbox for data collection, as they offer the opportunity to gather a great amount of unprecedented information on tourists behaviour is space and time [1, 2].

Tourism policies are foremost territorial policies, so such information, if wisely used, may become of particular relevance when investigating the relationship between tourists' behaviour, their individual characteristics and interests [3], and spatial distribution and accessibility of urban and territorial attractors and activities [4]. It may, for instance, serve to better understand preferences and choices of different populations of tourists, in order to put in place policies for tourism deseasoning; or to more

© Springer International Publishing Switzerland 2016
O. Gervasi et al. (Eds.): ICCSA 2016, Part IV, LNCS 9789, pp. 150–160, 2016.
DOI: 10.1007/978-3-319-42089-9_11

efficiently coordinate activities, attractions and transportation services; or to build tourist fidelity. In general, to better understand tourists' behaviours in space and time is to have useful information for public policies aiming at the development and governing of tourism.

In what follows we first present a brief review of some experiences of collection and analysis of spatio-temporal data, reported in literature, and carried out specifically for the study of tourists' behaviour and related to tourism policies. Subsequently we present and discuss the results of a survey and GPS tracking of tourists' spatio-temporal behaviour we conducted in October 2014 in the city of Alghero (Italy).

2   Background

Until recently, the most common way to study tourists' behaviours in time and space were various methods of diary (re)construction [2]. New possibilities have arisen by the development of automated tracking, above all the satellite-based (e.g. GPS) technologies for automatic position tracking with high time and spatial resolution.

Even if methods for surveying and analysing spatio-temporal behaviour are, of course, becoming highly developed in transportation research and in social sciences in general, comparatively little attention was being paid to the spatial and temporal behaviour of tourists, and systematic studies taking advantage of the technological developments offered by the high precision position tracking are still relatively few [1, 2].

In recent years scholars have started to break ground in this specific domain of applied research, experimenting and exploring advantages and disadvantages of satellite-based positioning technologies for the study of tourists' spatio-temporal behaviours [5].

Different methods have been reported for tracking tourist spatio-temporal behaviour [3]:

- direct observation of tourists' activities, with interviews or remote observations;
- time-space budget techniques [6] which analyse tourists' activities within destinations by using diaries, questionnaires and interviews;
- video-based tracking analysis, used to track tourists movements through video footages;
- smartphones with apps using positioning systems;
- specialized GPS tracking devices;
- land-based tracking systems that collect data thanks to radio technology.

Many studies report observations on a small scale, focusing on particular urban areas or activities that have a clearly defined entry and exit point, such as natural parks or historical centres [2]. Also many urban contexts have been analysed, among which Rome [7], Lago del Garda [8], Lago d'Orta [9], Canberra and Sydney [10], Salzburg [3]. On the other side, there are studies that attempt to analyse tourist behaviours at very large scale (e.g. on the national scale, by using mobile phones data [11]).

Different statistical and visualisation techniques may be employed to analyse and represent such spatial data. Frequently spatial temporal data are used only to describe the tourists track in maps. Sometimes these spatial maps are combined further information collected through questionnaires. This allows to link spatial temporal behaviour to a specific tourist category of population. Furthermore, of interest are studies trying to predict visitors whereabouts, how long they will stay in a place, or carry out a specific activity [12].

3   The Case Study

We hereby present the results of a study we conducted in October-November 2014 in the city of Alghero (Italy), a tourist destination of approximately 40.000 inhabitants in the North-West Sardinia in Italy. The purpose of the study was to explore tourists' movements, expectations and degree of satisfaction with the destination during a period of "low-season", considering that Alghero's peek period is the summer season, with highest tourist concentration between July and September. The collected data was thus meant to provide useful information and hints on possible policies to attract tourists outside the summer season.

Participants were recruited in five venues in Alghero, both hotels and bed & breakfasts, selected taking into account their geographic distribution. The study combined interviews with GPS movement tracking. Each volunteering tourist was interviewed twice: in the morning for the outgoing interview to collect information on his/her socioeconomic characteristics, preferences, and expectations; in the evening for the return interview to gather information on his/her whereabouts and personal evaluation of the day. During the morning interview, tourists were provided with a small GPS data logger which registered their movements during the entire day.

A total of 75 questionnaires and tracks were collected, referred to 225 tourists involved in the survey (205 adults, 17 teenagers and 3 children).

3.1   Exploratory Data Analysis of Interviews and Tourists Profiling

In Table 1 we report some basic features of the sample. The tourists were mainly Italian (28 %), Swedish (22.67 %) and German (16 %) (due to the availability at the time of low-cost flights from Stocholm and Göteborg), with a relatively high level of education (12 % post-degree level; 64 % degree level).

Around half of all interviewed tourists travel in couple (46,67 %). Frequent are also tourist travelling alone (17,33 %), while groups of families, couples with children and groups of friends cover each about 9 % share of the sample.

Almost all tourists have arrived by plane (90 %), confirming the importance of the local low-cost airport for the tourist development of the city and of the whole territory. Only 10 % of tourists have chosen to travel by ship or by other means of transportation.

The median vacation length is four days, with peaks of eight days (Fig. 1). The distribution estimate shows bimodal shape with two peaks in correspondence of about fourth and seventh days of vacation.

Among the sampled tourists, almost two third are in Alghero for the first time (65 %), while one third have come twice or more often (24.67 %) (Table 1-c).

Table 1. Features of the tourists' sample

Country of Origin (a)	
Italy	28.00%
Sweden	22.67%
Germany	16.00%
UK	8.00%
Benelux	8.00%
Other	17.31%

Education level (b)	
Post-graduate degree	12.00%
Graduate degree	64.00%
High school	22.67%
Diploma	1.33%

How often in Alghero? (c)	
"First time"	65.33%
"Second time"	22.67%
I" come often"	12.00%

Travelling with... (d)	
couple	54.67%
(with children)	(8.00%)
alone	17.33%
family	12.00%
friends	8.00%
Organized trip	4.00%
colleague	4.00%

Employment (e)	
employee	28.00%
teacher	22.66%
Self-employed	18.66%
manager	9.33%
retired	6.67%
housewife	4.00%
other	11.00%

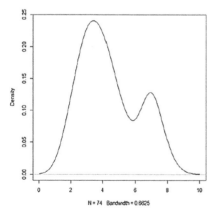

Length of vacation in days	
Min.	2.00
1° quartile	3.00
Median	4.00
Average	4.36
3r° quartile	5.75
Max.	8.00

N = 74   Bandwidth = 0.6825

Fig. 1. Length of vacation in Alghero: kernel density estimation (left), summary (right)

When asked to declare the importance (on a qualitative scale from 1 to 5) of the motives they have chosen Alghero as destination for this vacation, the interviewed tourists most frequently grade highly favourable climates, the possibility to relax, the quality of the environment, and food & wine (see Fig. 2). Despite the low cost of the trip the interviewees did not put this among the principal reasons to choose Alghero. These grading were confirmed by their travel plans. Tourists in Alghero would like, above all, to explore environmental resources and to discover the territory and the villages near Alghero city. The analysis of expectations with respect to actual activities carried out that tourists did reveal interesting information: of the 20 % of tourists interested in tourist services and leisure activities, only 1 % had effectively enjoyed these during the day; but, above all, 61 % of tourists declared they would like to enjoy food & wine, but only the 4.3 % declared to have done that.

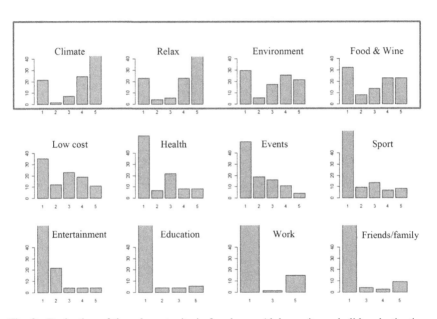

Fig. 2. Evaluation of the relevant criteria for choose Alghero city as holiday destination

Comparing expectations (outgoing interview) to the declared (return interview) use of means of transportation during the day (Table 2), we observe that 20 % of tourists intended to move in the territory by public transports, but only 6.67 % actually used it. This points at (and confirms previous concerns about) the inadequacy of the public transportation system to serve tourists well.

With regard to expenditures (Table 3), the comparison shows that tourists frequently expected and planned to spend more than they actually did, which raises the question whether there is a mismatch between the supply and tourists' demand for goods and services (another matter deserving further examination).

Table 2. Means of transportation, expected and declared

	"Which means of transportation do you plan to use during the day?" (outgoing interview)	"Which means of transportation did you use during the day?" (return interview)
Bike	18.67 %	12.00 %
Car rental	37.33 %	41.33 %
Public transportation	20.00 %	6.67 %

Table 3. Spending: expected (a) vs. declared (b)

	Expected **total** expenditures[a] (outgoing interview)	Expected **daily** expenditures[a] (outgoing interview)	Effective (declared) **daily** expenditures[a] (return interview)
Min.			**€3.00**
1st quartile	€200.00	€50.00	€29.25
Median	€300.00	€80.00	€49.00
Mean	€503.30	€89.97	€64.95
3rd quartile	€500.00	€100.00	€80.25
Max.			€295.00

[a] Accommodation and incoming/outgoing travel excluded

3.2   Tourists' Spatial Behaviour

In this subsection we present summary analysis of the GPS tracking of tourists. Their daily spatio-temporal schedules may be subdivided into three types: "Alghero – town", "Alghero - surrounding territory", and "outside Alghero". The shares of GPS tracks of each type of schedule is shown in Table 4. Having the sea swimming season ended, the

Table 4. Shares (time) of daily spatio-temporal schedules among different areas/zones

Macro-area		Area		Zone	
Alghero and its territory	80.29 %	Town	64.01 %	Historical centre	37.74 %
				Garibaldi and Lido waterfront	30.34 %
				Dante/Valencia waterfront	7.82 %
				Other	24.10 %
		Surrounding territory	35.99 %	Park and marine area	43.26 %
				Bombarde-Lazzaretto	25.59 %
				Maria Pia beaches	22.26 %
				Fertilia	8.89 %
Outside Alghero	19.71 %				

tracks within the town area rarely include beaches, as tourists show to be more focused on the historical centre, the harbour and the waterfront.

In addition to classifying the tracks, we analysed the totality of points marked by the GPS data loggers to represent the daily distribution of tourists in space. The GPS data loggers register individual position in space each 20 s during entire day. This permits to observe the spatial distribution of tourists in space in relation to the geography of activities, resources and services that are known touristic attractors in the Alghero's territory.

Figure 3-A shows the distribution of tourists within the town of Alghero. The point representation and the kernel density estimations emphasise main places of interest, concentrated between the historical centre and the waterfront, while the route representation shows the flows and the main ways of communications used by tourists.

Figure 3-B is indicative of different points of interests internal to the historical centre. Even if almost all parts of the historical centre are visited by tourists, pedestrian zones, city walls and main squares result the busiest. Many tracks (35.99 % of all of Alghero's) refer to tourists visiting Alghero's surrounding territory. Main points of

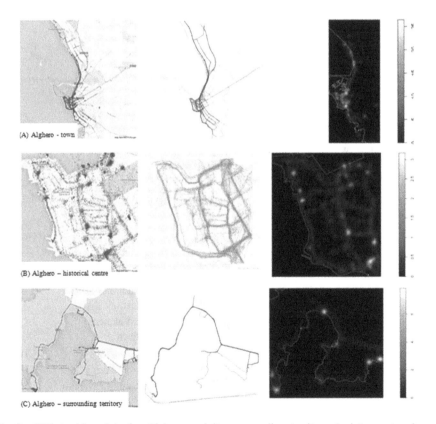

Fig. 3. GPS tracking data for Alghero and its surrounding territory (points, routes, kernel density estimation): (A) Alghero – town; (B) Historical centre; (C) Alghero surrounding territory. (Color figure online)

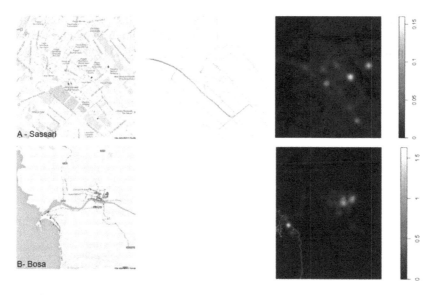

Fig. 4. GPS tracking data for Sassari (A) and Bosa (B). (Color figure online)

interest are the Neptune's Cave, the Porto Conte Natural Park and the protected marine area, the Bombarde and Lazzaretto beaches (see Table 4).

Figure 3-C shows the distribution of tourists around the area of Porto Conte Natural Park. The kernel density estimation places in evidence several points of interest: the beaches, the caves and the panoramic viewpoints.

For "outside Alghero" day-schedule type (about 20 % of GPS-recorder time share), Alghero served as a foothold for visiting other cities and places in Sardinia, among which the towns of Bosa, Castelsardo and Stintino are the most visited.

In Fig. 4 we offer a focus on two such destinations outside Alghero: Sassari (the provincial administrative centre) and Bosa, one of the main tourist attraction in the wider area. Only four of the registered GPS tracks contained a visit to the city of Sassari, with visit duration of only a couple of hours. The points visited in the city were the historical centre and the public garden, with no visits to city monuments and others points of interest in the city. This fact should be taken into account by policy makers when defining urban and territorial strategies aimed at tourist development.

On the other hand, the city of Bosa collected numerous visits. The map highlights frequent visits of the Malaspina's castle, the historical centre, the riverfront and the sea hamlet. This destination has of lately became a point of interest destination on regional scale.

3.3 Time Use by Tourist Profiles

A further possible analysis is to explore behaviour by different profiles of tourists. We propose a preliminary study at the city scale with respect to tourists segmented by company, age and expenditures. A spatialisation of this information is represented in Figs. 5, 6 and 7.

The comparison between couples and groups of people does not show significant differences (Fig. 5), while there are some interesting variability by age (Fig. 6): young people spend relatively less time in the city centre, and appear to be more interested in

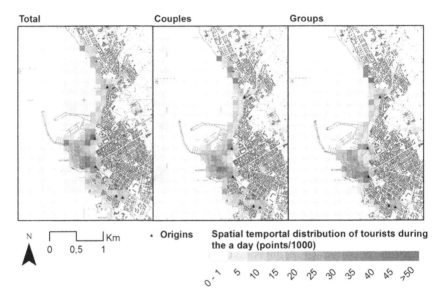

Fig. 5. Spatial temporal distribution for the total sample, for couples and for groups (Color figure online)

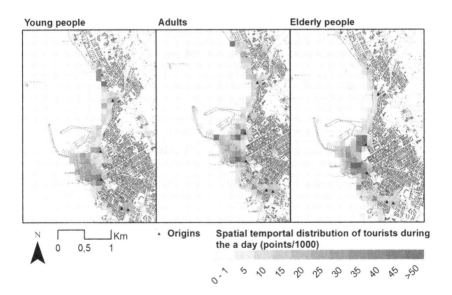

Fig. 6. Spatial distribution of the sample in respect to the age category (Color figure online)

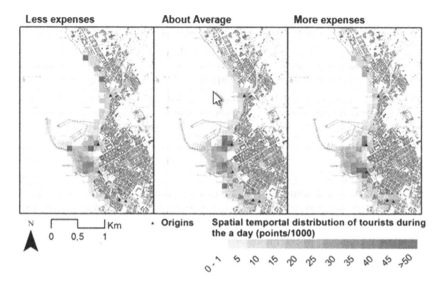

Fig. 7. Spatial distribution of the sample in respect to the amount of spending (Color figure online)

beaches and the surrounding territory; adults are more frequent in three points of interest of the historical centre and in the beach; while elderly people concentrate their time in two point of interest of the historical centre and appear to be less interested by the beaches (which may be due to the season).

Finally the comparison in terms of expenditures (Fig. 7) reveals that those who spend more also spend more time in the city centre (and are more concentrated in fewer places), while those who spend less are also relatively less concentrated in the space and spend more of their time in different parts of the city.

4   Conclusions

In this paper we have presented and explored tourists' spatio-temporal behaviour in urban and territorial context of the city of Alghero. The analysis of tourist populations in general pointed at some weaknesses and lack of "urban opportunities" which appear to limit the "urban capabilities" of tourists.

Among these, two types of urban opportunities seem to stand out among those in need of greater improvements: leisure-time activities/services and public transportation. Furthermore, a relatively low degree of loyalty of tourists requires specific attention by marketing and destination management policies.

Among the findings we presented, it is interesting to stress that the comparison of preference with the actual activities carried out by tourists points at a possible mismatch between supply and demand and a lack of leisure activities and those related to the wine and food traditions; indicative of this is the greater declared willingness to spend with respect to the actual daily expenditure (a gap of 30 € per day per person).

Equally remarkable are the limits to the development of the movement capabilities in the territory, due to the lack of a quality public transportation system: a disadvantage for tourists but also for the residents. Furthermore, we experienced difficulties with including in the survey accommodation facilities located in rural areas because they were at the time largely lacking tourists. This is indicative of a reduced attendance in these areas in this period of the year (which is probably related to the lack of public transportation, inadequate information/marketing, and absence of tourist services), even if the interviewed tourists frequently declared individual preference for relax, enjoyment of the environment, wine & food traditions, all opportunities possible to develop in rural areas.

In conclusion, this study offers an example of how GPS tracking technologies may be coupled with more traditional survey methods to produce meaningful information and profiling information related to spatio-temporal behaviour of tourists, and in this way to construct a better and more purposeful knowledge for tourism and territorial policies and development programmes.

References

1. Shoval, N., Isaacson, M.: Tracking tourists in the digital age. Ann. Tourism Res. **34**(1), 141–159 (2007)
2. Shoval, N., Isaacson, M., Chhetri, P.: GPS, smartphones, and the future of tourism research. In: The Wiley Blackwell Companion to Tourism, pp. 251–261 (2014)
3. Kellner, L., Egger, R.: Tracking tourist spatial-temporal behavior in urban places, a methodological overview and GPS case study. In: Inversini, A., Schegg, R. (eds.) Information and Communication Technologies in Tourism 2016, pp. 481–494. Springer International Publishing, Heidelberg (2016)
4. Blečić, I., Cecchini, A., Congiu, T., Fancello, G., Trunfio, G.A.: Evaluating walkability: a capability-wise planning and design support system. Int. J. Geogr. Inform. Sci. IJGIS **29**, 1350–1374 (2015). doi:10.1080/13658816.2015.1026824
5. Shoval, N., Isaacson, M.: Tourist mobility and advanced tracking technologies. Routledge, London (2010)
6. Pearce, D.G.: Tourist time-budget. Ann. Tourism Res. **15**(1), 106–121 (1988)
7. Calabrese, F., Ratti, C.: Real time rome. Netw. Commun. Stud. **20**(3–4), 247–258 (2006)
8. Bruno, A., Gasca, E., Mauro, S., Pollichino, G., Sacerdotti, S.L., Stupino, F.: Understanding tourist behaviour in wide áreas using GPS technologies. E-Review of Tourism Research (2010)
9. Levi Sacerdotti, S., Mauro, S., Gasca E.: Visitor Manager. Turismo, territorio, innovazione, CELID, Torino (2011)
10. Edwards, D.: Using GPS to Track Tourists Spatial Behaviour in Urban Destinations, 1 June 2009. SSRN: http://ssrn.com/abstract=1905286
11. Ahas, R., Aasa, A., Mark, Ü., Pae, T., Kull, A.: Seasonal tourism spaces in estonia: case study with mobile positioning data. Tourism Manage. **28**(3), 898–910 (2007)
12. Chhetri, P., Corcoran, J., Arrowsmith, C.: Investigating the temporal dynamics of tourist movement: an application of circular statistics. Tourism Anal. **15**(1), 71–88 (2010)

Countryside vs City: A User-Centered Approach to Open Spatial Indicators of Urban Sprawl

Alessandro Bonifazi[1](✉), Valentina Sannicandro[2], Raffaele Attardi[1],
Gianluca Di Cugno[1], and Carmelo Maria Torre[1]

[1] Department of Civil Engineering and Architecture, Polytechnic of Bari, Bari, Italy
{alessandro.bonifazi,carmelomaria.torre}@poliba.it,
raffaeleattardi@gmail.com, gianlucadicugno@gmail.com
[2] Department of Architecture, University of Naples Federico II, Naples, Italy
sanni.vale@gmail.com

Abstract. The interplay between land take and climate change is reviving the debate on the environmental impacts of urbanization. Monitoring and evaluation of land-cover and land-use changes have secured political commitment worldwide, and in the European Union in particular – following the agreement on a "no net land take by 2050" target. This paper addresses the ensuing challenges by investigating how open data services and spatial indicators may help manage urban sprawl more effectively. Experts, scholars, students and local government officials were engaged in a *living lab* exercise centered around the uptake of geospatial data in planning, policy making and design processes. Main findings point to a great potential, and pressing need, for open spatial data services in mainstreaming sustainable land use practices. However, urban sprawl's elusiveness calls for interactive approaches, since the actual usability of proposed tools needs to be carefully investigated and planned for.

Keywords: Urban sprawl · Open data · Land take · Soil sealing · Spatial indicators · e-governance · Italy

1 Introduction: Actionable Knowledge About Urban Sprawl

Urban sprawl is an elusive concept that has come to conjure up images of the in-between spaces where the city and the countryside mix in manifold combinations of land covers and uses. Two different understandings of urban sprawl, and of the related notion of peri-urbanization, seem to be dominant. The first, which might be labelled as "geographical/descriptive", assumes peri-urban landscape is the buffer area separating urban and rural contexts [1, 2], and showing intermediate traits (in terms of density, fragmentation, environmental quality, *etc.*). The second one follows a "strategic/evaluative" approach and it stresses the socio-spatial processes that on the one hand are giving rise to highly dissipative and often exclusive urban development trajectories, and on the other hand are home to a fuzzy cloud of more sustainable practices [3, 4]. These include local commoning, ecosystem-based design and management, and economic innovations

© Springer International Publishing Switzerland 2016
O. Gervasi et al. (Eds.): ICCSA 2016, Part IV, LNCS 9789, pp. 161–176, 2016.
DOI: 10.1007/978-3-319-42089-9_12

(among them urban farming and peer-to-peer platforms), which are altogether challenging the specialization paradigm that has long affected urban planning.

In order to be properly addressed, urban sprawl need be framed in a more complex, and constantly evolving, set of connected processes and related lines of inquiry. Among the many relevant issues that have a bearing on urban sprawl theory and practice, it is our opinion that the links between local land-taking processes and global environmental change [6], on the one hand, and the expanding role of geospatial information in contemporary digital revolution [6], on the other hand, are of overriding importance. Hence, our contribution targets the growing field of open spatial data, mapping and indicator platforms, to test their relevance and usability in understanding urban sprawl and contrasting low density, fragmented or dispersed urbanization patterns.

The paper continues with a conceptual background section: we focus on the European context, since the empirical work addressed below concerns the Italian region of Puglia. Section 3 covers research design, in terms of both methodology and a brief introduction to the study settings. In the following section, results, observations and discussion are merged in a stepwise account of the research work, before concluding remarks are provided.

2   Conceptual Background: Open Spatial Indicators and Urban Sprawl in a European Perspective

While trying to make sense of the ever more pervasive expansion of low-density urban land covers into rural landscapes, in an age of globally skyrocketing population, scholars are prone to depict sprawl as a worldwide epidemic [7], which broke out in the US nearly a century ago, later spread over Europe, and nowadays affects even those continents that had long been home to the most crowded cities – making them "giant, *sprawling* and dominated by cars" [8].

Urban sprawl unanimously signals physical patterns of low-density expansion, although it always takes further specifications to complete its conceptual boundaries. The European Environment Agency [9] stresses that sprawl takes place in agricultural areas under market conditions, it implies little planning control, and it tends to dig a wedge between urbanization and population growth – indeed, the former has increased by around 80 % in the EU since the mid-1950s, whilst the latter had grown by 33 % in the same period. Similar trends have been observed, among other areas, in the US [10]. Beyond sheer land take in terms of the loss of agricultural, forest and other semi-natural and natural land to human developments, sprawl is therefore marked by a decline in urban density, a decentralization of urban functions and the transformation of a compact urban form to a discontinuous and dispersed pattern [11].

In the European Union (EU), experts and policy makers have become increasingly aware of the environmental challenges posed by land take, as well as of its toll on the degradation of soil functions and ecosystem services. Accordingly, the 7th EU Environment Action Programme commits EU member states to implement targeted policies by 2020 so as to achieve *no net land take by 2050* [12]. Although the European Commission eventually withdrew the proposal to set up a comprehensive land policy from its 2015

Work Programme, a thorough study on the state of the art in sustainable land use management has been issued [13]. Five priority areas are identified: land take, land recycling (focusing on brownfields and infill development), land degradation (in terms of water erosion and loss of organic matter), land multi-functionality, and the global impact of EU land demand (including indirect effects in the rest of the world). However, even with full political support, the scientific and administrative challenges that need be tackled to implement the objectives, targets and indicators outlined in the study should not be underestimated.

It has indeed been noted that the sheer combination of alternative definitions (land take, soil sealing, land free for development) and analytical methods (automatic classification of high resolution satellite imagery, interpretation of orto-photographs, statistical surveys, and land registries) is bringing about such a great ambiguity that coherent monitoring and evaluation processes at EU level might be jeopardized [14].

The beginning of the third millennium in the Common Era, however, has also seen unprecedented developments in information and communication technologies – with big data management and visualization, social networking and software-as-a-service applications at the cutting edge of the process. Although few countries have already secured full Freedom of Information legislation, scholars and social innovators are reaching beyond this passive approach to engage in open data practices and policies that potentially harness public sector information for strengthening both the common good and profit-oriented business models [15].

It is against the background of this switchover in communication, learning, and policy making landscapes that innovative land take data and indicators should be framed. The spread of open data, taken to mean "*information that can be freely used, re-used and redistributed by anyone – either free or at marginal cost*" [16], meets with different barriers, and spatial data are no exception. To overcome operational and technical hindrances, a specific body of literature and experience on spatial data infrastructures has been developed, which in the EU revolves around the implementation of the 2007 INSPIRE Directive. Aiming at ensuring spatial data services' accessibility and compatibility – not least, in a trans-boundary context – these implementing rules need to be complemented by further action, if the ready and transparent access to geographic information is also to be granted to those who have no familiarity with data handling [17].

In line with its spatial data infrastructure policy, the EU develops and disseminate several data services, the most relevant of which may be considered the Copernicus Land Monitoring Services[1], the European Space Agency's Earth Observations Missions[2], and the monitoring systems maintained by the European Environment Agency[3] and Eurostat – of the like of the *Productivity of artificial land*, a dashboard indicator for the assessment of the Europe 2020 Strategy[4].

However, for open data to become valuable, networks of various actors are needed (involving data providers, developers and users) in order to treat raw data, analyze it,

---

[1] http://land.copernicus.eu/.

[2] https://earth.esa.int/web/guest/missions/esa-operational-eo-missions.

[3] http://www.eea.europa.eu/data-and-maps/indicators#c5=&c0=10&b_start=0.

[4] http://ec.europa.eu/eurostat/en/web/products-datasets/-/T2020_RD100.

combine it[5], make it available and interpret it as information [18]. Hence, usability becomes key in delivering open data services, all the more in e-government services, including those regarding urban governance and management [19]. Unfortunately, usability is too often overlooked, and only when breakdowns in the flow of action between humans and technological artifacts occur, do policy makers and service providers become aware of the need to place the embodied experience of intended users – and their focus on goals rather than on tasks – at the center of policy and service design right form the onset [20]. In other words, as the civic, commercial or professional routines become ever more imbued with technology (including open data and services), artifacts are not only used, but lived with [21].

Back to urban sprawl, our research aims to address the inherently multidimensional nature of land-taking and soil-sealing processes by looking at the situated needs of users of open spatial data and indicators services, as they engage in planning, policy making and design interaction. Given spatial data infrastructures are still is to be found in their foundational stage, we maintain that getting better insights into innovative approaches to user-centered design practices may help improve the usability – and hence, the effectiveness [22, 23] – of urban sprawl monitoring and evaluation tools. This in turn is supposed to foster the co-creation of integrated technological and social innovations in the field of efficient use of land and soil resources.

We are cognizant of alternative takes on the elusiveness of this research issue, which are rather tailored to novel analytical methods [24–26], and we deem it promising that in certain respects our findings resonate with the reflections offered in these related bodies of literature [14].

3   Research Design

This Section frames the empirical work, presented in the following one, in terms of the chosen research approach, methods, data, and geographical context.

3.1   Research Approach and Process

This exploratory work has been carried out under a national funding program aimed to strengthen research capacity in less developed EU regions. Within the framework of the Multimedia Information for Territorial Objects (MITO) project[6], a research group based at the Polytechnic of Bari (MITO-Lab[7]), focused on the uptake of geospatial data in planning, policy making and design processes. Among other activities, a *living lab* format was implemented [22, 27, 28] by:

- modelling how prospective users interact with open spatial data and indicator services in their daily civic, administrative or professional practice;

---

[5] In the context of this paper, into indicators and maps.

[6] Further information may be found at http://www.ponrec.it/open-data/progetti/scheda-progetto?ProgettoID=7431.

[7] www.mitolab.poliba.it.

- setting up a research infrastructure, in terms of both a physical (the MITO-Lab facility) and virtual space for continuous interaction among participants;
- gathering a mixed group of participants (designers/providers of services and prospected users), including six experts (in geographic information systems, multi-criteria decision aid, information engineering, urban economics and valuation, environmental planning), undergraduate students, professional planners and designers, and a limited number of civil servants and town councilors;
- implementing an iterative research process, through subsequent steps of prototyping, evaluation, co-creation, and validation of services; and
- fostering knowledge dissemination and mutual learning processes, by establishing a network with other expert centers and groups[8].

3.2 Data and Analytical Methods

The research work has been based exclusively on secondary data, briefly described in the present Sub-Section. We have however developed or fine-tuned further indicators or applied original data analysis methods, as explained below, and implemented (either public or restricted access) services for both data and indicators.

In 2015, the Italian National Environment Protection Agency (ISPRA) released an open access package comprising, among others[9]:

- a very high resolution imperviousness map (raster) of Italy, based on semi-automatic classification of RapidEye satellite imagery (temporal coverage 2012), with a minimum mapping unit (mmu) of 25 m^2;
- a classification of Italian municipalities' urban forms and patterns into five classes (Compact, Compact with full artificial cover of the municipal territory, Compact but sprawling, Sprawled, Polycentric), based on the combination of three landscape metrics applied in a stepwise approach (*Largest Class Patch Index*, *Edge Density*, and *Residual Mean Patch Size*).

As for the regional spatial data infrastructure[10], we made primarily use of the Regional Thematic Land Use Maps (RTLUM), covering respectively 2006 and 2011, a vector map with an mmu of 2.500 m^2 and a classification that is consistent with the Corine Land Cover nomenclature, extended for most classes to the 4th level.

---

[8] These include the national research center on land take (consumosuolo.org), a research group based at the University of Basilicata (lisut.org/it), the in-house company of the Puglia regional administration that is in charge of managing the regional geoportal (sit.puglia.it), the national association of urban and regional planners (inu.it), and some members of the regional open data movement (openpuglia.org).

[9] All data are available at: www.consumosuolo.isprambiente.it. See ISPRA [29] for detailed methodological information.

[10] Browse at www.sit.puglia.it (in Italian) to view and download data.

·

The research center on land take and soil sealing[11] (CRCS) drafted a study on Puglia, at landscape area and municipal level, which also included a quantitative assessment of land-use changes (temporal coverage 2006–11) and the calculation of the Urban Shape Index, a measure of compactness.

It would be impractical to provide details for all data services managed by the national statistics office (ISTAT) we made use of during the project: readers are therefore invited to browse through the data warehouse[12]. Suffice it to report here that the minimum survey units, named census tracts, may fall into one of four types of localities: three residential (urban areas, small inhabited areas, and non-urban areas) and one industrial and commercial; non-urban areas host less than five households, or dwellings that are further away than 30 m.

Among MITO-Lab's methodological contributions, the following are to be mentioned:

- a *Land Use Efficiency Trend Indicator*, measuring the variation in per capita artificial land cover area compared to population change (m^2/inhabitants) at municipal level, according to the formula:

  $100 * A (B-C)$, where

  A = artificial land cover area (a.l.c.a.) 2011/population 2011

  B = (a.l.c.a. 2011 − a.l.c.a. 2006)/a.l.c.a. 2006

  C = (population 2011 − population 2006)/population 2006;

- a *Land Use Intensity Seasonality Indicator*, measuring the seasonal variation in artificial land use intensity (under a summer peak assumption), by using monthly waste production (kg) as a proxy, according to the formula:

  $100 * $ (August Waste Production 2011 − Average Monthly Waste Production 2011)/Average Monthly Waste Production 2011.

3.3 Research Context

Puglia is the south-easternmost region in Italy, stretching over almost 20.000 km^2 and hosting around 4 million inhabitants. It is mostly a flat area, except in the North – where the Monti Dauni range and the Gargano headland face each other across the main plain (Tavoliere). Once dominated by extensive agriculture and few prominent trading ports, Puglia evolved into a mixed-economy with a growing service sector and few large industrial poles. Its 258 municipalities are mainly medium to small towns, with only about 15 cities having a population greater than 50.000 inhabitants.

Urban dispersion has long been very limited, as population used to be highly concentrated in towns rather than in small villages or isolated settlements. Beginning from the second half of the 20th century, however, the trend has reversed, and Puglia now ranks very high in urban sprawl, with estimates ranging from 6 through 9 %, depending on the focus being, respectively, on soil sealing or land take [29, 30]. Sprawl is especially

---

[11] A national partnership established by the Polytechnic University of Milan, the National Planning Association (INU), and an environmental NGO (Legambiente), to act as a think-tank in addressing land take and soil sealing.

[12] Available at: http://en.istat.it.

evident along the coast and in most rural areas – where very small patches, often made of single buildings, correspond to what has been labelled as *sprinkling* [31].

4 Discussion of Preliminary Results

We wrapped up results in 4 Sub-Sections, each ending with a brief discussion of the usefulness and main shortcomings of specific data and indicator services. Section contents reflect interaction among participants, although the stepwise arrangement of activities and arguments has been adapted to meet the readers' need for clarity.

4.1 Classifying Urban Forms Based on Landscape Metrics

As a trigger, participants were confronted with the sprawl-detection oriented interpretation of urban form and patterns, developed by ISPRA [29] and based on very high resolution imperviousness map data (Fig. 1). Most participants were puzzled by the inclusion in class 3 (Sprawled) of cities that were perceived as very different landscapes. We therefore focused on this class, to address the acceptability and potential utilization of ISPRA's sprawl classification by planners, citizens and policy-makers.

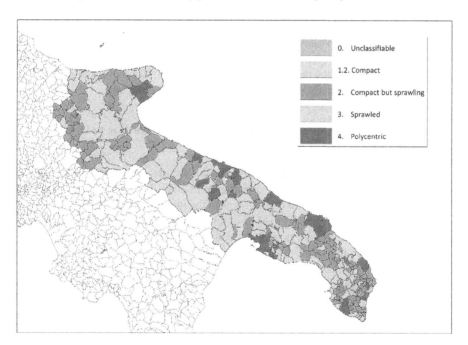

Fig. 1. Classification of the municipalities of Puglia, according to urban forms and patterns. Based on ISPRA [29]. No class 1.1 case is represented (compact cities where artificial land cover stretches beyond municipal borders in all directions). (Color figure online)

Firstly, an underestimation of the relationship between sprawl and the magnitude of land take was put into question, since impervious land-cover ratio to overall municipal area ranged from a minimum of 1.5 %, and remained below 3 % for most municipalities in the Monti Dauni area, up to a maximum bracket of 15 to 22 % - which mostly coastal cities fell within, and only in the Salento area. The sprawl classification was then contrasted with the Urban Shape Index (USI) developed by CRCS based on the RTLUM, and therefore factoring in larger artificial land areas rather than impervious soils alone. The two analyses seemed to be consistent, since the value distributions of the USI in ISPRA's Sprawled and Compact but sprawling classes only slightly overlap (arithmetic mean and standard deviation of, respectively: 6,10/2,97 and 9.56/3,85).

Secondly, participants argued that the sprawl classification did not reflect the landscape diversity they had direct experience of (Fig. 2). We have restrained from showing extremely different cases (for instance, provincial capitals vs very small towns), and focused on the 10–50.000 bracket, the most numerous in the region.

It should be stressed that, although some patterns were easily interpreted by most participants, e.g. the sealed coastal strip in Trani (top-right) or the sprinkled [31] residential developments in Martina Franca (mid-left), others were more elusive – such as the apparently less densely built-up San Severo (top-left) or the agglomeration of small towns in Salento, with few to no gaps in the urban fabric (bottom-left, there are three towns around Racale). Ostuni (bottom-right) was considered an ideal case to illustrate (almost) all sprawl patterns to be found in the region, since it shares with all other municipalities in the Itria Valley the very-low density rural settlements extending southward, whilst moving northward one comes across a satellite, high-density, urban area, before meeting the leap-frog developments along the coast. Coasts looked remarkably sealed in most areas, even more than what participants had expected.

4.2 Land Use Matters: Sorting Out Sprawl Patterns According to Urban Functions

To unravel divergent patterns based on less evident factors, we turned to the share of dwellings in non-urban areas (in total dwellings). In 2011, less than 50 municipalities exceeded the 10 % value, but in all major towns in the Itria valley (Locorotondo, Ostuni, Martina Franca, and Cisternino) 35 to 46 % of total dwellings were located in non-urban areas. The share decreases with no evident pattern, until the very opposite end of the range, where almost only small towns in Salento are to be found. This is in line with well documented historical trends in the region at landscape scale [30].

To investigate the relative importance of predominantly residential urban in driving sprawl, we rearranged the 3rd or 4th level classes of RTLUM artificial land covers to highlight the contribution of specific industrial and commercial units that had caught the attention of both researchers and participants.

The outcome of this analysis is illustrated in Fig. 3. We looked again at the same municipalities we have shown in Fig. 2 (except Ostuni), to learn that specialization is evident in many cases, be that energy facilities in Mesagne (mid-right), quarries in Trani (top-right), or industrial and transport units in San Severo (mid-right).

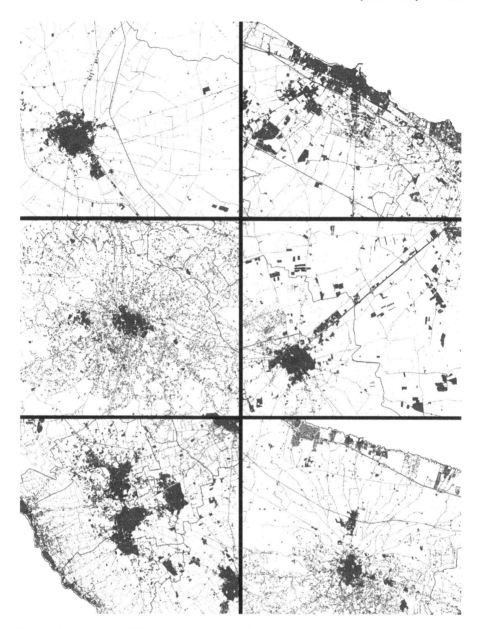

Fig. 2. An overview of different urban forms and patterns falling within the "Sprawled" class – as in ISPRA [29]. Map data: very high resolution imperviousness, mmu 25 m², temporal coverage 2012. Maps are centered on the following municipalities: San Severo (top-left), Trani (top-right), Martina Franca (mid-left), Mesagne (mid-right), Racale (bottom-left), and Ostuni (bottom-right).

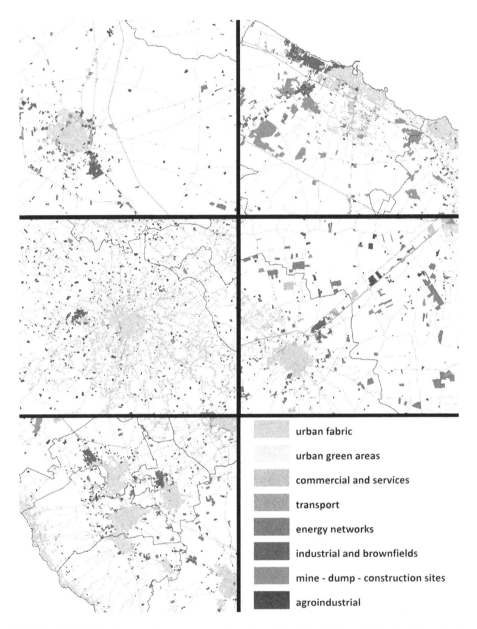

Fig. 3. Artificial land-use patterns in a sample of municipalities falling within the "Sprawled" class – as in ISPRA [29]. Map data: Regional Thematic Land Use Map, mmu 2.500 m², temporal coverage 2011. Maps are centered on the following municipalities: San Severo (top-left), Trani (top-right), Martina Franca (mid-left), Mesagne (mid-right), and Racale (bottom-left). (Color figure online)

By disaggregating land uses, participants were prompted to articulate their views on urban expansion. This in turn triggered interesting mutual learning processes, while they envisaged the underlying socio-economic factors that could explain the observed patterns, or tried to match their own experiential knowledge with the spatial statistics they were confronted with.

The position and dispersion measures for all 258 municipalities is given through a box and whisker plot in Fig. 4. Besides *urban fabric*, accounting for 50 % or more of artificial land use in half of the municipalities, the combined figure of non-residential land uses raised concern. *Transport units*, and to a much lesser extent other *industrial & brownfields, mines, dump and construction sites* and *agroindustrial* installations show a remarkable weight in many cases, and the latter in particular had never been considered a main factor by any participant. *Energy networks*, though less relevant on the whole, included a series of 12 outliers whose values ranged from 10 through 25 %, and which could be traced to a recent boom in ground-mounted solar power plants.

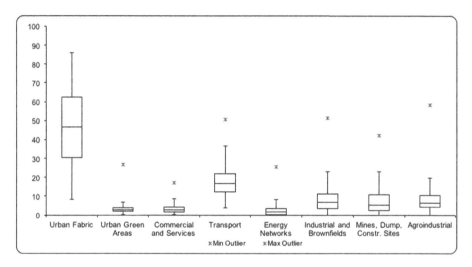

Fig. 4. Measures of position and dispersion for artificial land use classes in municipalities, based on shares (%) in the total artificial land. Lower and upper edge and internal line of each box correspond to, respectively, the first and the third quartile, and the median; ends of whiskers are set at 1,5* Interquartile Range, above and below edges – unless min and max values happen to be less extreme.

At this stage, technical and operational complications started to hinder the living lab's activities, as most processing had to be performed offline by researchers, and not all results could be easily grasped by participants. As far as value positions are concerned [33], those who shared planning skills or responsibilities seemed to become more cautious when evaluating the magnitude of land take at municipal level, as though – once disaggregated – each land use was more likely to command an acceptable share.

4.3 Putting Socio-Economic Drivers into the Picture: Land Use Efficiency

Rather than providing definite answers, the foray into land-use patterns enticed participants and researchers into investigating the socio-economic drivers behind so diverse a range of combinations. On the one hand, it was suggested to look once again to ISTAT's population statistics to screen rural dwellings for their actual use. The resulting evidence was mixed: only some of the highly appealing (to tourists and second-homers) rural municipalities in the Itria Valley had (in 2011) a share of non-urban population about equal to that of rural dwellings (Locorotondo, Cisternino), whereas in others the ratio was less than 1/2 (Ostuni, Martina Franca)[13]. The small towns in Monti Dauni revealed contrasting trends: whereas some showed a dramatic drop in population with respect to dwellings (Faeto), others had a much greater share of population as compared to dwellings (Alberona, Celle di San Vito) – although they all experienced a declining population over the last six decades.

On the other hand, we aimed at a systematic assessment of land-use efficiency, and we assumed participants would have found it easier to relate to a measure of how much land (in m^2) is taken to cater for each inhabitant or job, rather than to alternative land utilization density indicators [13, 24]. The resulting map is shown in Fig. 5. The relatively low efficiency of residential land uses in rural areas or in tourist hotspots lining along the coast mirrored previous observations. Likewise, the relatively high efficiency of larger cities and/or active economic centers in the central part of the region called into play well-established socio-spatial processes. Conversely, the least efficient municipalities were either small towns with declining population or a few towns with an almost exclusively tourist economy. To benchmark the values of the land use efficiency indicator, we might recall that a maximum per capita urbanization rate of 400 m^2 was established by the Swiss Federal Council in 2002 [24], and the average for 13 European countries was (in 2006) about 350 m^2/inh., with Italy stopping well below (at 250 m^2/inh.) [32].

Interestingly enough, some argued that low per capita urbanization rates, that is, high densities, relate to urban quality in ambivalent terms, because they were achieved to the detriment of an adequate supply of open and green areas – for instance, in the regional capital of Bari.

4.4 Sprawling in Time: Tracking Urban Change to Understand Land Taking Processes

The living lab exercise came to an end by relating urban sprawl to dynamic variations in time. Firstly, we refined the previous approach to land-use efficiency, by looking at short term trends, based on changes in per inhabitant/land take between 2006 and 2011 (Fig. 6). Following the *Land Use Efficiency Trend Indicator* formula (provided in Sub-Sect. 3.2), municipalities with a negative value had increased their land-use efficiency in the 2006–11 period, as compared to the reference year (2006).

---

[13] It should be stressed, however, that these are census data and there are different reasons why declared residents' figures might depart from actual ones.

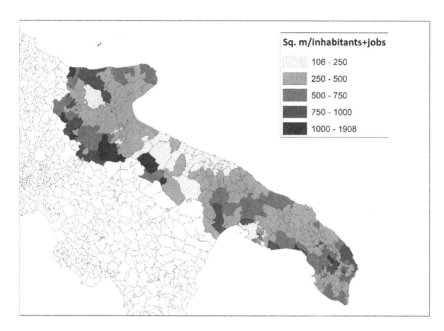

Fig. 5. A Land Use Efficiency map of Puglia, showing per capita artificial land cover (m²/user, both inhabitants and jobs). Map data: artificial land covers from RTLUM; inhabitants and jobs at municipal level from ISTAT. Temporal coverage is 2011 for all data sources. (Color figure online)

In fact, the indicator relates the two rates (population and land-use changes), so absolute land take might still be increasing with negative values being measured. However, it is a useful early warning that, should the trend continue in the longer term, the city would be either sprawling (positive values) or densifying (negative values) – for instance, as a consequence of infill developments or land-recycling policies [24]. While working on more sophisticated tools, some participants questioned the open data approach, which they considered an unfair business model in that it does not secure adequate returns on required time investments.

Secondly, and lastly, we took advantage of the availability of an open data service provided by the regional government, to track *seasonality* in land-use intensity. Hence, we developed a proxy indicator to measure monthly waste production under a summer peak assumption. August production exceeded the monthly average (in 2011) in almost all cities and towns (acting therefore as sinks of urban users in summer), save for some 10 % municipalities (notably including the regional capital and two provincial capitals) that were rather sources of summer city users. When compared to the land use efficiency indicator or the rate of officially vacant rural houses, this indicator was better at sorting seaside resorts from small inland towns with declining population.

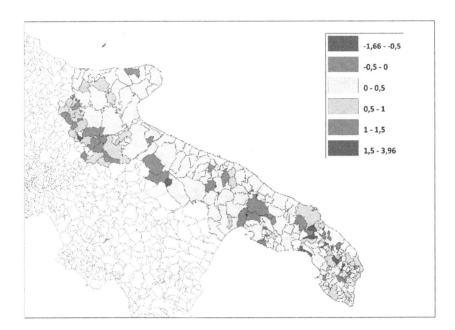

Fig. 6. A Land Use Efficiency Trend map of Puglia, showing per capita artificial land cover (m^2/ inh.). Map data: artificial land covers from RTLUM; inhabitants and jobs at municipal level from ISTAT. Temporal coverage is 2006 and 2011 for all data sources. (Color figure online)

5   Concluding Remarks and Future Developments

Progress towards no net land take targets in the near future requires far reaching innovations in dominant socio-technological practices worldwide [4, 6]. Spatial indicators and open data may play a key role in this perspective, by facilitating the uptake of geospatial information in civic engagement, planning and design processes [15, 17].

In this paper, we reported about a living lab exercise to test the usability of spatial indicators of urban sprawl. Planners, designers, experts in data analysis, monitoring and evaluation, students and civil servants joined in iterative prototyping, co-creation, and validation of services [22, 27]. In line with the relevant literature [11, 13, 14], findings point to the multidimensional nature of urban sprawl and call for either conceptually complex evaluation tools or flexible and context-specific adaptation pathways, or both. User-centered open data and indicator services may help bridge the gap between a rapidly growing body of environmental knowledge and its poor uptake in decision-making processes. In that respect, however, this work highlights the difficulties in reproducing fully realistic planning and design settings while implementing innovative open data and spatial indicator services.

Insights in users' value positions on open spatial data and urban sprawl were the most interesting results [33], together with a better understanding of what aspects of indicators are likely to render them more or less usable. Future developments might include a better consideration of both the ecological implications of soil sealing, and the role of market and public finance as drivers for land take and urban sprawl.

Acknowledgments. The research illustrated in this paper was carried out within the framework of the Multimedia Information for Territorial Objects (MITO) project, funded under the European Regional Development Fund. The authors thank all participants in the living lab experience at MITO-Lab, Department for Civil Engineering and Architecture, Polytechnic of Bari.

References

1. Meeus, S., Gulinck, H.: Semi-urban areas in landscape research: a review. Living Rev. Landscape Res. **2**, 5–45 (2008)
2. Attardi, R., Cerreta, M., Sannicandro, V., Torre, C.M.: The multidimensional assessment of land take and soil sealing. In: Gervasi, O., Murgante, B., Misra, S., Gavrilova, M.L., Rocha, A.M.A.C., Torre, C., Taniar, D., Apduhan, B.O. (eds.) ICCSA 2015. LNCS, vol. 9157, pp. 301–316. Springer, Heidelberg (2015)
3. Donadieu, P.: Landscape urbanism in europe: from brownfields to sustainable urban development. J. Landscape Archit. **1**(2), 36–45 (2006)
4. Friedmann, J.: Planning theory revisited. Eur. Plan. Stud. **6**, 245–253 (1998)
5. World Resources Institute (WRI): Ecosystems and human well-being – a framework for assessment. Island Press, Washington (DC) (2003)
6. Carpenter, J., Snell, J.: Future Trends in Geospatial Information Management: The Five to Ten Year Vision. United Nations Committee of Experts on Global Geospatial Information Management, New York (2013)
7. Nilsson, K., Pauleit, S., Bell, S., Aalbers, C., Nielsen, T.S.: Introduction. In: Nilsson, K., Bell, S., Nielsen, T.S. (eds.) Peri-urban Futures: Scenarios and Models for Land Use Change in Europe, pp. 1–9. Springer, Heidelberg (2013)
8. The Economist. http://www.economist.com/news/china/21640396-how-fix-chinese-cities-great-sprawl-china
9. European Environment Agency: Urban sprawl in Europe: The ignored challenge. European Environmental Agengy, Copenhagen (2006)
10. Fulton, W., Pendall, R., Nguyen, M., Harrison, A.: Who Sprawls Most? How Growth Patterns Differ Across the U.S. The Brookings Institution, Washington DC (2001)
11. Siedentop, S., Fina, S.: Monitoring urban sprawl in Germany: towards a GIS-based measurement and assessment approach. J. Land Use Sci. **5**(2), 73–104 (2010)
12. European Union: Decision 1386/2013/EU of the European Parliament and of the Council of 20 November 2013 on a General Union Environment Action Programme to 2020 'Living well, within the limits of our planet' (2013)
13. BIO by Deloitte: Study supporting potential land and soil targets under the 2015 Land Communication, Report prepared for the European Commission, DG Environment in collaboration with AMEC, IVM and WU; Publications Office of the European Union, Luxembourg (2014)
14. Decoville, A., Schneider, M.: Can the 2050 zero land take objective of the EU be reliably monitored? a comparative study. J. Land Use Sci. 1–19 (2015)
15. Zuiderwijk, A., Helbig, N., Gil-Garcia, J.R., Janssen, M.: Special issue on innovation through open data - a review of the state-of-the-art and an emerging research agenda: guest editors' introduction. J. Theoret. Appl. Electron. Commer. Res. **9**, 1–2 (2014)

16. European Commission: Turning Government Data into Gold, Press release from the speech of the Commissioner Neelie Kroes. http://europa.eu/rapid/pressReleasesAction.do?reference =IP/11/1524&format=HTML&aged=0&language=EN&guiLanguage=en

17. Borzacchiello, M.T., Craglia, M.: The impact on innovation of open access to spatial environmental information: a research strategy. Int. J. Technol. Manag. **60**, 114–129 (2012)

18. Lindman, J., Rossi, M., Tuunainen, V.K.: Introduction to open data services minitrack. In: 2013 46th Hawaii International Conference on System Sciences, p. 1238. IEEE Computer Society (2013)

19. Balena, P., Bonifazi, A., Mangialardi, G.: Smart communities meet urban management: harnessing the potential of open data and public/private partnerships through innovative e-governance applications. In: Murgante, B., Misra, S., Carlini, M., Torre, C.M., Nguyen, H.-Q., Taniar, D., Apduhan, B.O., Gervasi, O. (eds.) ICCSA 2013, Part IV. LNCS, vol. 7974, pp. 528–540. Springer, Heidelberg (2013)

20. Inglesant, P., Sasse, M.A.: Usability is the best policy: public policy and the lived experience of transport systems in London. In: Proceedings of HCI 2007, The 21st British HCI Group Annual Conference, vol. 1 (no page numbers). British Computer Society, Swindon (2007)

21. McCarthy, J., Wright, P.: Technology as Experience. MIT Press, Cambridge (2004)

22. Liedtke, C., Welfens, M.J., Rohn, H., Nordmann, J.: Living Lab: user-driven innovation for sustainability. Int. J. Sustain. High. Educ. **13**, 106–118 (2012)

23. Pollino, M., Modica, G.: Free web mapping tools to characterise landscape dynamics and to favour e-Participation. In: Murgante, B., Misra, S., Carlini, M., Torre, C.M., Nguyen, H.-Q., Taniar, D., Apduhan, B.O., Gervasi, O. (eds.) ICCSA 2013, Part III. LNCS, vol. 7973, pp. 566–581. Springer, Heidelberg (2013)

24. Jaeger, J.A.G., Schwick, C.: Improving the measurement of urban sprawl: Weighted Urban Proliferation (WUP) and its application to Switzerland. Ecol. Ind. **38**, 294–308 (2014)

25. Manganelli, B., Murgante, B.: Analyzing periurban fringe with rough set. world academy of science. Eng. Technol. **71**, 111–117 (2012)

26. Modica, G., Praticò, S., Pollino, M., Di Fazio, S.: Geomatics in analysing the evolution of agricultural terraced landscapes. In: Murgante, B., Misra, S., Rocha, A.M.A., Torre, C., Rocha, J.G., Falcão, M.I., Taniar, D., Apduhan, B.O., Gervasi, O. (eds.) ICCSA 2014, Part IV. LNCS, vol. 8582, pp. 479–494. Springer, Heidelberg (2014)

27. Dell'Era, C., Landoni, P.: Living Lab: a methodology between user-centred design and participatory design. Creativity Innov. Manag. **23**, 137–154 (2014)

28. Cerreta, M., Inglese, P., Malangone, V., Panaro, S.: Complex values-based approach for multidimensional evaluation of landscape. In: Murgante, B., Misra, S., Rocha, A.M.A., Torre, C., Rocha, J.G., Falcão, M.I., Taniar, D., Apduhan, B.O., Gervasi, O. (eds.) ICCSA 2014, Part III. LNCS, vol. 8581, pp. 382–397. Springer, Heidelberg (2014)

29. ISPRA (Istituto Superiore per la Protezione e la Ricerca Ambientale). Rapporto 218/2015: Il consumo di suolo in Italia, Edizione 2015. ISPRA, Rome (2015)

30. Bonifazi, A., Balena, P., Sannicandro, V.: I suoli di Puglia fra consumo e politiche per il risparmio. In: Arcidiacono, A., Di Simine, D., Oliva, F., Ronchi, S., Salata, S. (eds.) Nuove sfide per il suolo – Centro di Ricerca sui Consumi di Suolo, Rapporto 2016, pp. 99–104. INU Edizioni, Rome (2015)

31. Romano, B., Zullo, F.: Half a century of urbanization in southern European lowlands: a study on the Po Valley (Northern Italy). Urban Res. Pract. 1–22 (2015)

32. Romano, B., Zullo, F.: Models of urban land use in Europe: assessment tools and criticalities. Int. J. Agric. Environ. Inf. Syst. **4**(3), 80–97 (2013)

33. Rose, J., Persson, J.S., Heeager, L.T., Irani, Z.: Managing e-Government: value positions and relationships. Inf. Syst. J. **25**, 531–571 (2015)

Integrating Financial Analysis and Decision Theory for the Evaluation of Alternative Reuse Scenarios of Historical Buildings

Carmelo M. Torre$^{(\boxtimes)}$, Raffaele Attardi, and Valentina Sannicandro

MITO Lab-Department of Civil Engineering and Architecture,
Polytechnic of Bari, Bari, Italy
`carmelomaria.torre@poliba.it`,
`raffaeleattardi@gmail.com`, `sannivale@gmail.com`

Abstract. The expected utility theory assumes that the advantage of an agent under conditions of uncertainty can be calculated as a weighted average of the utilities in each state as possible, by using as weights the likelihood of the occurrence of individual states.

The expected utility is thus an expected value (according to the terminology of the theory of probability). In order to determine the utility according to this method, it is supposed that the decision maker is be able to order their preferences with regard to the consequences of different decisions.

The experiment described in the paper shows that different actors of a decision process tend "to move" the Centre of gravity of the decision to their preference. Arrow's theorem taught that there is no unanimity.

The starting point of the case of study is an analysis of the probability for different way of use of a Heritage property, related to tourism and leisure activities: Catering, Conferences and hosteling. The different actors have different preferences on each one of the three activities. Their vision partially contrasts with the likelihood of generating income through the activities. Each activity can create income as a function of use of the Fabric (for one use or for the other). The coexistence of the three forces of the combined use has a limitation, so some use combinations generate more income than others, according to a probability curve.

Each actor will attempt to shift the business mix toward equilibrium that appreciates most. Whereas, as an element of interpretation of the behavior of multi-actor, Kannehman's approach consider that each subject will see the advantage linked to a different combination from that with most likely utility; the different combination is affected by the expectations of actors, and described by the "perspective theory".

Keywords: Expected utility · Perspective theory · Reuse of buildings · Business plan · AHP

1 Introduction

The expected utility theory states that the convenience of a given event depends on its probability of occurrence, under the assumptions of completeness, transitivity, independence and continuity [1, 2].

© Springer International Publishing Switzerland 2016
O. Gervasi et al. (Eds.): ICCSA 2016, Part IV, LNCS 9789, pp. 177–190, 2016.
DOI: 10.1007/978-3-319-42089-9_13

The expected utility is thus an expected value according to the terminology of the probability theory. In order to determine the utility according to this method, the decision maker must be able to order his preferences with regard to the impacts of different decisions.

However, in a multi-actor decision-making context the perceived utility of a given event and its realistic chance of occurrence are not always related since different actors can show different expectations and visions about the event itself.

In such decision context, [3, 4] rationality suggest that the perceived utility of all the actors playing a role in the decision process should be assessed. The prospect theory [5] describes different behaviors when social actors choose between probabilistic alternatives that involve a risk under uncertain conditions. The prospect theory states that people make decisions on the basis of individual losses and gains, rather than on the probabilistic outcome. Since each social actor has different expectations and values, his own perceived utility differs from those of other actors.

In this sense, each actor assumes a specific reference point when defining the value function of a set of events. This differs from expected utility theory, in which a hypothetical completely rational agent is indifferent to the reference point. The application of the prospect theory returns a correction of the expected value through subjective valuations of the relative importance of alternative events. When dealing with evaluation of possible reuse scenarios for cultural heritage and historical buildings [6–8], a number of actors can show different expectations with regard to reuse alternatives [9]. In this paper we propose an integrated approach, combining financial analysis and multi-criteria with Analytic Hierarchy process [10, 11] and a multi-group analysis based on coalition formation as decision tools in order to compare the Net Present Value (NPV) deriving from an activity of refurbishment and reuse of an historic building, when the NPV is firstly, calculated by the use of a probabilistic approach, and secondly by the reference of the expected utility of the outcomes for the individual decision makers.

2 Case of Study

2.1 The Context

We apply the methodology to the analysis of reuse scenario for an historic monastery in a village in Southern Italy.

Instead of associating to each variable of the NPV the probability that it will occur in maximum, medium or minimum extent, we will join the significant weight of the acceptability assigned by each decision maker to them. This phase represents the shift from the likelihood of utility to the perspective of utility.

At the end we will represent the Decision framework showing the "distances" among the different expectation of the actors.

Serino is a village about 10 km from Avellino and 30 km from the Salerno in Campania Region, Southern Italy. The St. Francis Monastery in Serino is a monument with significant architectural and artistic value: its cloister is decorated with frescoes by Ricciardi in 1700. Today the monastery is in a state of abandonment. In Serino other architectural and cultural assets are existing: the seventeenth-century Hermitage of San Gaetano, the Medieval Castle, the sixteenth-century Monastery of St. Lucia and some Roman archaeological sites. The monastery should host new functions and activities:

- hosteling for visitors of the area of Mount Terminio, pilgrims of the Montevergine Shrine and person belonging to the staff of the University of Fisciano, as well as casual visitors;
- rent for religious ceremonies and conferences with banquet opportunity (Fig. 1).

Fig. 1. The convent of san francis in serino

2.2   Calculation of Likelihood Function of NPV

The most significant uncertainties relate to the occupancy forecasts of the building facilities and the likelihood to find funding for the restoration work.

After the phase of fund raising a Cost-Revenue Analysis has been carried out.

The financial analysis of the reuse alternatives for the St. Francis Monastery is needed in order to verify the economic sustainability of the restoration.

The framework of these uncertainties is made more evident through the financial analysis that describes the variability of the Net Present Value (NPV) of the monastery after the restoration as a function of some key variables, namely:

- the amount of public funding for the restoration work, in order to reduce restoration costs (variable X1);
- the average annual occupancy rate of accommodation facilities (Y1), the number of conferences per year (Y2) and the number of religious ceremonies (Y3).

The NPV is calculated as the sum of each yearly difference between revenues and costs actualized at the present time:

$$NPV = \sum_{i=1}^{10} \frac{R_i(Y1, Y2, Y3) - C}{(1+r)^i}$$

The average occupancy of accommodation facilities in the area varies between 50 % and 65 % of the total annual number of beds per night offered.

For the monastery, it is that the average level of occupancy is achieved in the next 10 years from the restoration.

Therefore in the formula of Fig. 2, the difference R-C is depending by Y1, Y2, Y3 (if the cofounding is considered also X1 becomes constant).

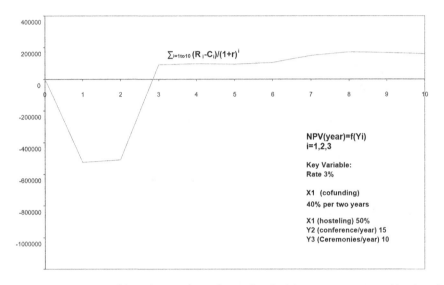

Fig. 2. The average condition of convenience for reusing the Monastery, represented by the rule of variation of the Net Present value in the first ten Year.

The financial analysis shows that the economic sustainability of the restoration is based on a public funding equal to 40 % of the total costs, it is achievable in 10 years (considering an Rate r of 3 %) of activity of the newly introduced facilities and it is based on some fixed values of the variables Y1, Y2, Y3, namely 50 %, 15 and 10. The cost of renewal is assumed as constant.

3   Probabilistic Analysis of the Re-Use Revenue for Determining the Expected Value

The expected value of a prediction is the probability that an event will occur in the mode identified by the same prediction.

In order to determine the probability that an event will occur, it must be known or built a law of frequency distribution (discrete or continuous).

The probability will result from the frequency data related to variable values. If the NPV of the example is then treated according to the key variables X1,Y1,Y2,Y3,, identified the chances PX1*,PY1*,PY2*, PY3* that the variables assume the values X1*, Y1*,Y2*, Y3*, the consequent expected NPV is function of PX1*,PY1*,PY2*, PY3*.

They can then build a set of discontinuous distributions based on permutation, such as those in the Table 1, relative to NPV obtainable in the different hypotheses of reuse, by varying between 50 % and 62 % the intensity of hosteling (variable Y1), between 10 and 20 the number of annual congresses (variable Y2), and between 6 and 14 the number of annual ceremony events (variable Y3).

Permutations generally refer to the three values (intermediate and the two extremes lower or higher). Having the ability to determine the frequency of minimum usage situations, maximum and intermediate structures for the activities (reception, events, ceremonies) in the reference group, they consider the discrete probability distribution corresponding to the detected frequency.

Table 1. The relationship between probability of extremes and average conditions of variables generating the occurrence of profitable activities after the refurbishment

	Hypothesis 1		Hypothesis 2		Hypothesis	
Constant	X1=40% Y2=15, Y3=10		X1=40% Y1=50%, Y3=10		X1=40%, Y1=50%, Y2=15	
Variable	Y1	$\partial \dfrac{VAN}{\partial Y_1} Y_1$	Y2	$\partial \dfrac{VAN}{\partial Y_2} Y_2$	Y3	$\partial \dfrac{VAN}{\partial Y_3} Y_3$
MAX	62,5%	VAN>0	20 conference	VAN>0	15 ceremonies	VAN>0
MED	57,5%	VAN>0	15 conference	VAN>0	10 ceremonies	VAN>0
MIN	50%	VAN>0	10 conference	VAN<0	6 ceremonies	VAN<0

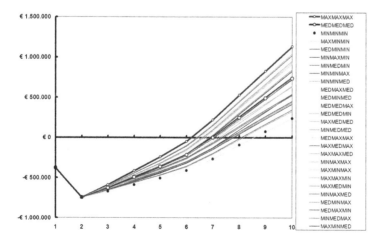

Fig. 3. The range of variation of the NPV in the time interval of 10 years according to the permutations of the revenue (Min Med Max) of each activity (Y1, Y2, Y3) (Color figure online)

In order to improve the financial analysis, we select three different values for each of the variables Y1, Y2, Y3, in order to consider an optimistic (MAX), a pessimistic (MIN) and an intermediate (MED) alternative (see Table 1). We then calculate the NPV for all permutations of the set values of Y1, Y2, Y3, as shown in Fig. 3.

If we associate the composed likelihood of each permutation of the vector (Y1, Y2, Y3) we can calculate the probability function of NPV (Fig. 4).

The graph below shows how the construction of the cumulative probability corresponds to an approximation of the points representing each combination (probability of NPV) due to the permutations of singular probability (represented in the discrete

Permutation of Y₁, Y₂, Y₃	NPV (rate 3%)	Composed probability P(Y₁) x P (Y₂) x P(Y₃)	P (Y₁,)	P (Y₂)	P(Y₃)
MAX-MAX-MAX	1132608	2,37%	28,2%	28,7	29,3%
MAX-MAX-AVER	1043282	3,72%	28,2%	28,7	46,0%
MAX-AVER-MAX	1025417	3,72%	28,2%	45,0%	29,3%
MAX-MAX-MIN	953956	2,00%	28,2%	28,7	24,7%
MAX-AVER-AVER	936091	5,84%	28,2%	45,0%	46,0%
AVER-MAX-MAX	932312	4,08%	48,5%	28,7	29,3%
MAX-MIN-MAX	918226	2,17%	28,2%	26,3%	29,3%
MAX-AVER-MIN	846765	3,64%	48,5%	45,0%	24,7%
MAX-MIN-AVER	828900	3,41%	28,2%	26,3%	46,0%
MED-AVER-MAX	825121	6,39%	48,5%	45,0%	29,3%
AVER-MAX-MIN	753660	3,44%	48,5%	28,7	24,7%
	NPV > 750000	**Prob.=40,78%**			
MAX-MIN-MIN	739574	1,83%	28,2%	26,3%	24,7%
AVER-AVER-AVER	735795	10,04%	48,5%	45,0%	46,0%
AVER-MAX-AVER	735795	6,40%	48,5%	28,7	46,0%
AVER-MIN-MAX	717930	5,87%	48,5%	26,3%	29,3%
MED-AVER-MIN	646469	6,26%	48,5%	45,0%	24,7%
MIN-MAX-MAX	631868	1,96%	23,3%	28,7	29,3%
MED-MIN-AVER	628604	5,87%	48,5%	26,3%	46,0%
MIN-MAX-AVER	542543	3,08%	23,3%	28,7	46,0%
AVER-MIN-MIN	539278	3,15%	48,5%	26,3%	24,7%
MIN-AVER-MAX	524677	3,07%	23,3%	45,0%	29,3%
MIN-MAX-MIN	453217	1,65%	23,3%	28,7	24,7%
MIN-MED-AVER	435351	4,82%	23,3%	45,0%	46,0%
MIN-MIN-MAX	417486	1,80%	23,3%	26,3%	29,3%
MIN-AVER-MIN	346026	2,59%	23,3%	45,0%	24,7%
MIN-MIN-AVER	328160	2,82%	23,3%	26,3%	46,0%
MIN-MIN-MIN	238835	1,51%	23,3%	26,3%	24,7%
	NPV < 750000	**Prob.=59,22%**			

Fig. 4. The likelihood of variation of the NPV according to the permutations of the revenue (Min Aver Max) of each activity (Y1, Y2, Y3)

probability distribution in the chart above) to an area under a limit function (represented in the graph below).

The cumulate probability is more less equal to 40 %, to overpass the average NPV and equal to the 60 % to stay under the average NPV.

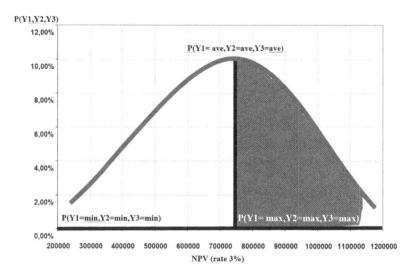

Fig. 5. The composite cumulate likelihood of the average NPV according to the permutations of the revenue (Min Aver Max) of each activity (Y1, Y2, Y3)

4   The Gaming Among DM

When we look at a future event, we can assume that: the probability of manifestation of a given same event is due by a combination of constrains that we can assume "external" respect to the decision problem and respect to the behavior of the actors involved in the decision process.

We will call these actors Decision Makers (DM).

The following DM are involved in the decision process, discriminating their point of view on the reuse of the Convent on the size of the different activities envisaged:

– the Friars, that is the Owners of the Convent,
– the Municipality,
– the County,
– he Ministry's local Delegate for Architectural Heritage and Landscape.

The constrain are the element conditioning the probability of occurring of a given event. Outside the decision problem, constrains will encounter the favor of one or another decision maker in a different way.

The most the constrain occurs in favor of one decision maker, the most this decision maker will see profitable the event. Despite to the expectation of actors, the relationship between constrains and probability is independent by their expectation.

Arrows' Theorem of Impossibility [12] states that there is no unanimous decision in multi actor decision processes. We intend to measure the degree of conflict between the different DM, as a potential cost/obstacle to reach the agreement [13].

It is not true, for example, that all the DM tend to prefer the maximization of net present value, per se, through the maximization of all key variables size. According to the different expectations and different targets some DM may find desirable to strike a balance between economic functions represented by the key variables not pushed to the extreme use of the building [14–16]. In our case the expectation of each actor is described as follows.

The Friars intend to maximize returns (an extreme expectancy for all three key variables).

The Municipal government (assuming a co-management of assets) expresses a greater preference for more stable assets (the key variable to be maximized is then Y1: hosteling business).

The provincial government believes prioritize the activities linked to congresses and cultural events, according to a local promotion policy that aims to reach potential users, mainly from the cultural and academic world. (the key variables to maximize Y2 are: conferences and events).

The Ministry Delegate intends to limit the massive use of the Convent, preferring activities diluted over time that does not involve excessive pressure, favoring the hosteling (the key variable to be maximized is Y1: hosteling business).

For three key variables (business accommodation, activities for conferences and ceremonies), and four decision makers, they will have design assumptions with respect to the minimum, average, and maximum of activities twelve sets of weights. We could consider the relationship between the expectation of DM and the probability of occurring a NPV as a measure of the "distortion" of the expected value given by the study of probability respect to the "desiderata" of DM.

The use of Analytic Hierarchy Process is useful to define the expected value for each actor, but giving a weight to the importance that each dimension of the three activities (hosteling, conferences, ceremonies) assumes for each DM (see example in Fig. 6).

With the use of the Saaty' Method, each decision matrix expresses a qualitative preference defined a combination of the Degree of Acceptability of Option (DAO) referring to a given event compared to another, translated into weights.

You then will have a distribution of weights generating the DAO of each DM for each key variable. For the ieth decision maker (DM) you will have the following Set of Option (SO):

$$
SO_i \quad
\begin{matrix}
UiA\,(Y1max) & UiA\,(Y1med) & UiA\,(Y1min) \\
UiA\,(Y2max) & UiA\,(Y2med) & UiA\,(Y2min) \\
UiA\,(Y3max) & UiA\,(Y3med) & UiA\,(Y3min)
\end{matrix}
\tag{1}
$$

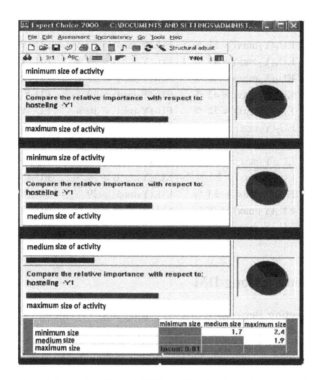

Fig. 6. An example of the application of AHP to identify the ratio among the expectation of Friars in case of maximum, average, and minimum intensity of Hosteling.

The results of the identification, which gives a short description hereinafter, are reported in the expectations as follows in Table 2.

The similitude between the point of view of DM4 (the County) and DM1 (the Friars) in the table, is evident where the high Expected Utility is in the same tern (the first) for both Decision Makers.

As already reminded, it is visible in Table 2 that the DM "Friars" want to maximize each function related to intensity of use. Therefore, the more profitable combination according to the care of Friars is the permutation corresponding to the NPV obtained by the tern (max EU1-maxY2-maxY3), as expressed in the first three row of the Table 2. As already done, is possible to represent the cumulative curve of Expected Utility for Friars, in the same way of the cumulative likelihood. As in Fig. 7 is represented the variation of Expected Utility od DM1 (Friars), and in Fig. 8 the EU function of Friars and in Fig. 9 a comparison between probability and expected utility for the Friars.

Note that we can observe only one table of variation of Probable NPV (such as shown in Fig. 4), while the Expected NPV can be represented for each DM.

Table 2. The DAO (EU) for the SO of each DMi (i = 1,2,34)

Friars (DM$_1$)	**EU$_1$(Y$_1$max) = 51 %**	EU$_1$(Y$_1$med) = 30 %	EU$_1$(Y$_1$min) = 19 %
	EU$_1$(Y$_2$max) = 49 %	EU$_1$(Y$_2$med) = 29 %	EU$_1$(Y$_2$min) = 22 %
	EU$_1$(Y$_3$max) = 46 %	EU$_1$(Y$_3$med) = 29 %	EU$_1$(Y$_3$min) = 24 %
County (DM$_2$)	**EU$_2$(Y$_1$max) = 42 %**	EU$_2$(Y$_1$med) = 30 %	EU$_2$Y$_1$min) = 28 %
	EU$_2$(Y$_2$max) = 52 %	EU$_2$(Y$_2$med) = 30 %	EU $_2$Y$_2$min) = 18 %
	EU$_2$(Y$_3$max) = 28 %	EU$_2$(Y$_3$med) = 29 %	**EU$_2$(Y$_3$min) = 43 %**
Ministry	**EU$_3$(Y$_1$max) = 58 %**	EU$_3$(Y$_1$med) = 25 %	EU$_3$(Y$_1$min) = 17 %
Delegate	EU$_3$(Y$_2$max) = 21 %	**EU$_3$(Y$_2$med) = 30 %**	**EU$_3$(Y$_2$min) = 49 %**
(DM$_3$)	EU$_3$(Y$_3$max) = 20 %	EU$_3$(Y$_3$med) = 29 %	EU$_3$(Y$_3$min) = 51 %
County (DM$_4$)	**EU$_4$(Y$_1$max) = 53 %**	EU$_4$(Y$_1$med) = 29 %	EU$_4$(Y$_1$min) = 18 %
	EU$_4$(Y$_2$max) = 40 %	EU$_4$(Y$_2$med) = 31 %	EU$_4$(Y$_2$min) = 29 %
	EU$_4$(Y$_3$max) = 42 %	EU$_4$(Y$_3$med) = 31 %	EU$_4$(Y$_3$min) = 27 %

5   The Coalition Among DM

5.1   The Intersection Between Expected Utility and Likelihood

Then is the time to create a coalition diagram. A diagram of coalition needs to assess the similarity of points of view, and the probability of solving conflicts among decision makers.

In the case of study the construction of the coalition diagram starts from the following considerations:

- respect to the same financial revenue (represented by a given NPV) each subject feeds a different expectation;
- the average Expected Utility of each DM (corresponding to a wait of 50 % utility) corresponds to different financial advantages (and therefore in net present values) for each DM.

The procedure at this point considers the following steps:

(a) the identification of the "average acceptable NPV", corresponding to the expected 50 % (degree of expectation) on a given decision maker (in our case the promoter of the intervention, that is, DM1 "Friars").
(b) the identification of the "average acceptable" NPV, corresponding to an expected utility of 50 % of the other decision makers
(c) identification of expected utilities by the other DMs corresponding the "average acceptable" NPV by the DM1.
(d) Comparison of the results with the construction of a "coalition diagram", in which the determinations of NPV shows

The coalition diagram represented in a Cartesian Plane, the distance between the point of view of different DM; on the vertical axes is represented the EU of each DM, and on the Horizontal axis is represented the NPV having the average of EU.

Value of partial EU Y₁, Y₂, Y₃	rate 3%	Expected Utility U_A(Y₁) x U_A (Y₂) x U_A (Y₃)	EU₁ (Y₁,)	EU₁ (Y₂)	EU₁ (Y₃)
MAXMAXMAX	1132608	11,50%	51%	49%	46%
MAXMAXMED	1043282	7,32%	51%	49%	29,3%
MAXMEDMAX	1025417	6,80%	51%	29%	46%
MAXMAXMIN	953956	6,17%	51%	49%	24,7%
MAXMEDMED	936091	4,33%	51%	29%	29,3%
MEDMAXMAX	932312	6,76%	30%	49%	46%
MAXMINMAX	918226	5,16%	51%	22%	46%
MAXMEDMIN	846765	7,25%	30%	29%	24,7%
MAXMINMED	828900	3,29%	51%	22%	29,3%
MEDMEDMAX	825121	4,00%	30%	29%	46%
MEDMAXMIN	753660	3,63%	30%	49%	24,7%
	VAN > 750000	EU=66,21%			
MAXMINMIN	739574	2,77%	51%	22%	24,7%
MEDMEDMED	735795	2,55%	30%	29%	29,3%
MEDMAXMED	735795	4,31%	30%	49%	29,3%
MEDMINMAX	717930	1,93%	30%	22%	46%
MEDMEDMIN	646469	4,26%	30%	29%	24,7%
MINMAXMAX	631868	4,28%	19%	49%	46%
MEDMINMED	628604	1,93%	30%	22%	29,3%
MINMAXMED	542543	2,73%	19%	49%	29,3%
MEDMINMIN	539278	1,63%	30%	22%	24,7%
MINMEDMAX	524677	2,53%	19%	29%	46%
MINMAXMIN	453217	2,30%	19%	49%	24,7%
MINMEDMED	435351	1,61%	19%	29%	29,3%
MINMINMAX	417486	1,92%	19%	22%	46%
MINMEDMIN	346026	1,36%	19%	29%	24,7%
MINMINMED	328160	1,22%	19%	22%	29,3%
MINMINMIN	238835	1,03%	19%	22%	24,7%
	VAN < 750000	EU=33,78%			

Fig. 7. The likelihood of the expected utility for Friars due to NPV according to the permutations of the expected revenue (Min Med Max) of each activity (Y1, Y2, Y3)

In the diagram in this example (Fig. 10), the points corresponding to the positions of the Ministry Delegate (DM3) and of the County (DM4) are very close, as well as the points corresponding to the positions of the Friars (DM1) and the Municipal Administration (DM2). At a first glance it is visually evident that the diagram shows the presence of two coalitions (DM3 and DM4; DM1 and DM2).

A first rough interpretation could attribute a more similarity to local actors on one hand, and super-territorial actors on the other side.

It is clear that in the presence of a greater number of actors this representation can be even more significant in presence of an effective "cloud" of point.

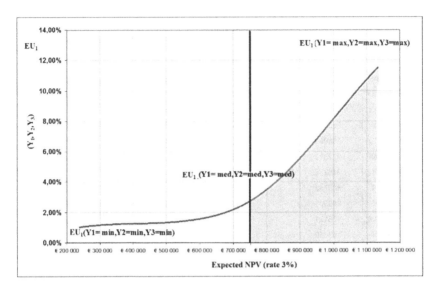

Fig. 8. The composite cumulate Expected utility according to the permutations of the expected revenue (Min Aver Max) of each activity (Y1, Y2, Y3) for DM1 Friars.

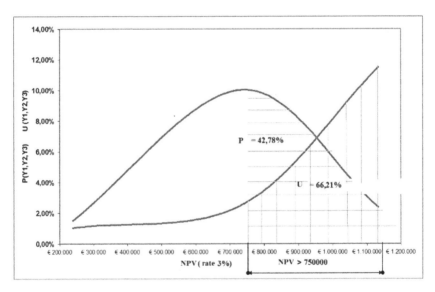

Fig. 9. The intersecton of the composite cumulate Expected Utility for DM1 Friars (in Fig. 8), according to the permutations of the expected revenue for DM1 Friars and the Likelihood curve of NPV (in Fig. 5)

On the one hand it is obvious that the expectations of each subject described here, through the use of the Saaty method, are determined by elements of a different nature, in an index that synthesizes, but implicitly, various elements themselves, such as objectives, conveniences, ethical positions, etc. [16, 17].

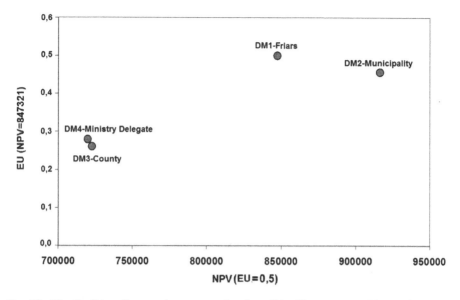

Fig. 10. The Coalition diagram shows a couple of possible alliance among DM1 and DM2 on one side, and among DM3 and DM4 on the other side

It appears that local actors (DM1 and DM2) have a more ambitious point of view, in terms of expected NPV, and in terms of expected utility of the reuse. This is evident when looking at their position on the more far corner of the Diagram

6   Conclusion

The integration of the two indicators, Likelihood and Expected Utility, considered as magnifying glass of NPV analysis in a multiactor process, allows a step forward a strictly financial analysis, by providing a social key of interpretation; furthermore, show the need for which it will be useful to argue with new methodological research and future experiments [14–16].

The expanded financial analysis thus becomes a supporting element to the negotiation stage, passing through a process of integrate assessments, taking into account the exercised influence on the final decision by different actors and the fragmentary nature of positions of DM, showing possible convergences or conflicts of interest on a same result.

The experiment can be ascribed in the family of Decision Making test and Theory, starting from seminal and basic assessment of perception and expectation.

Acknowledgements.   This study has been supported by the MITO Lab, at the Department Dicar of Bari Polytechnic.

The work was the result of a collaboration among authors. R. Attardi wrote the first and the second paragraph, C. M. Torre wrote the third and the fourth paragraph, V. Sannicandro wrote the fifth paragraph.

References

1. Von Neumann, J., Morgenstein, O.: Theory of Games and Economic Behavior. Princeton University Press, Princeton (1944)
2. Turner, R.K., Pearce, D.W., Bateman, I.: Environmental Economics: An Elementary Introduction. Johns Hopkins University Press, Baltimore (1993)
3. Munda, G.: Social multi-criteria evaluation: Methodological foundations and operational consequences. Eur. J. Oper. Res. **158**(3), 662–677 (2004). Elsevier
4. Kahneman, D., Tversky, A.: Prospect theory: an analysis of decision under risk. Econometrica **47**(2), 263–292 (1979)
5. Arrow, K.: Alternative approaches to the theory of choice in risk-taking situations. Econometrica **19**(4), 404–437 (1951)
6. Tajani, F., Morano, P.: An evaluation model of the financial feasibility of social housing in urban redevelopment. Property Manag. **33**(2), 133–151 (2015)
7. Girard, L.F., Torre, C.M.: The use of ahp in a multiactor evaluation for urban development programs: a case study. In: Murgante, B., Gervasi, O., Misra, S., Nedjah, N., Rocha, A.M. A., Taniar, D., Apduhan, B.O. (eds.) ICCSA 2012, Part II. LNCS, vol. 7334, pp. 157–167. Springer, Heidelberg (2012)
8. Vizzari, M., Modica, G.: Environmental effectiveness of swine sewage management: a multicriteria AHP-based model for a reliable quick assessment. Environ. Manage. **52**(4), 1023–1039 (2013)
9. Torre, C.M.: Socio-economic dimension in managing the renewal of ancient historic centers. In: Rotondo, F., Selicato, F., Marin, V., Galdeano, J.L. (eds.) Cultural Territorial Systems, pp. 97–106. Springer, Heidelberg (2016)
10. Torre, C.M., Morano, P., Taiani, F.: Social balance and economic effectiveness in historic centers rehabilitation. In: Gervasi, O., Murgante, B., Misra, S., Gavrilova, M.L., Rocha, A. M.A.C., Torre, C., Taniar, D., Apduhan, B.O. (eds.) ICCSA 2015. LNCS, vol. 9157, pp. 317–329. Springer, Heidelberg (2015)
11. Saaty, T.L.: A note on the AHP and expected value theory. Soc. Econ. Plann. Sci. **20**(6), 397–398 (1986)
12. Arrow, K.: Uncertainty and the welfare economics of medical care. Am. Econ. Rev. **53**(5), 941–973 (1963)
13. Sen, A.: The impossibility of a paretian liberal. J. Polit. Econ. **78**(1), 152–157 (1970)
14. Pollino, M., Modica, G.: Free web mapping tools to characterise landscape dynamics and to favour e-participation. In: Murgante, B., Misra, S., Carlini, M., Torre, C.M., Nguyen, H.-Q., Taniar, D., Apduhan, B.O., Gervasi, O. (eds.) ICCSA 2013, Part III. LNCS, vol. 7973, pp. 566–581. Springer, Heidelberg (2013)
15. Morano, P., Tajani, F., Locurcio, M.: Land use, economic welfare and property values: an analysis of the interdependencies of the real estate market with zonal and macro-economic variables in the municipalities of Apulia Region (Italy) International Journal of Agricultural and Environmental. Inf. Syst. **6**(4), 16–39 (2015)
16. Attardi, R., Cerreta, M., Franciosa, A., Gravagnuolo, A.: Valuing cultural landscape services: a multidimensional and multi-group sdss for scenario simulations. In: Murgante, B., Misra, S., Rocha, A.M.A., Torre, C., Rocha, J.G., Falcão, M.I., Taniar, D., Apduhan, B.O., Gervasi, O. (eds.) ICCSA 2014, Part III. LNCS, vol. 8581, pp. 398–413. Springer, Heidelberg (2014)
17. Attardi, R., Cerreta, M., Poli, G.: A collaborative multi-criteria spatial decision support system for multifunctional landscape evaluation. In: Gervasi, O., Murgante, B., Misra, S., Gavrilova, M.L., Rocha, A.M.A.C., Torre, C., Taniar, D., Apduhan, B.O. (eds.) ICCSA 2015. LNCS, vol. 9157, pp. 782–797. Springer, Heidelberg (2015)

Spatial Analysis for the Study of Environmental Settlement Patterns: The Archaeological Sites of the Santa Cruz Province

Maria Danese[1], Gisela Cassiodoro[2], Francisco Guichón[2],
Rafael Goñi[3], Nicola Masini[1], Gabriele Nolè[4],
and Rosa Lasaponara[4(✉)]

[1] CNR-IBAM, C.da S. Loja, 85050 Tito Scalo, PZ, Italy
[2] CONICET, INAPL, Buenos Aires, Argentina
[3] INAPL, University of Buenos Aires, Buenos Aires, Argentina
[4] CNR-IMAA, C.da S. Loja, 85050 Tito Scalo, PZ, Italy
rosa.lasaponara@imaa.cnr.it

Abstract. The aim of this work is to use spatial analysis to study relationships between environmental parameters and archaeological sites in the Santa Cruz province (Patagonia, Argentina) and to develop an analysis protocol that could reveal including and excluding factors, useful for the research and the discovery of new archaeological sites on the territory. Consequently, a model that considers interactions between sites and some parameters such as the elevation, the slope, the aspect, the landforms, the land use and the distances from water was constructed. Moreover a final sensibility map was produced, that wants to help the archaeologist to know, in every point in the space, how many and which are parameters that could increase the probability to find new archaeological sites and that are survey priorities on the study region.

Keywords: Spatial analysis · Patagonia · GIS · Prediction model

1 Introduction

Spatial analysis is very important in the study of settlement patterns in archaeological sites, because often it helps to discover hidden or however not clear settlements dynamics. With spatial analysis it is possible to develop useful archaeological predictive models [1] as Support Decision Systems.

The aim of this work is to use spatial analysis to study relationships between environmental parameters and archaeological sites in the Santa Cruz province (Patagonia, Argentina) and to develop an analysis protocol that could reveal including and excluding factors, useful for the research and the discovery of new archaeological sites on the territory. Consequently, a model that considers interactions between sites and some parameters such as the elevation, the slope, the aspect, the landforms, the land use and the distances from water was constructed. Moreover a final sensibility map was produced, that wants to help the archaeologist to know, in every point in the

© Springer International Publishing Switzerland 2016
O. Gervasi et al. (Eds.): ICCSA 2016, Part IV, LNCS 9789, pp. 191–203, 2016.
DOI: 10.1007/978-3-319-42089-9_14

space, how many and which are parameters that could increase the probability to find new archaeological sites and that are survey priorities on the study region.

The analysis and the constructed model were applied on the case study of the archaeological sites at central-western Santa Cruz province (Fig. 1). The aim of the archaeological investigations is to establish the relationship between the fluctuating Holocene environmental conditions and hunter-gatherer populations [2]. Investigations currently under way are approached from a wide spatial scale and from multiple lines of evidence (technological evidence, archaeofaunal remains, rock art and human remains). Different ecological areas are considered; low basins, high basins and plateaus. There are different types of archaeological sites: hunting blinds, rock shelters, human burials and superficial concentrations of lithic and archaeofaunistical materials [3].

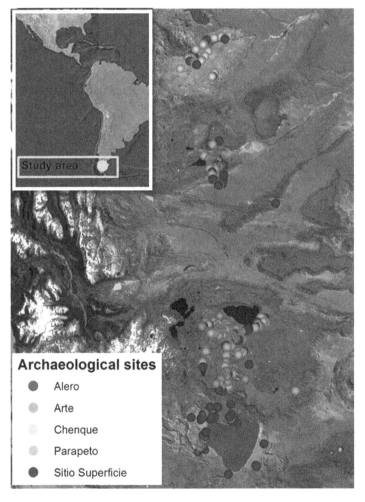

Fig. 1. The study area and archaeological sites in Santa Cruz Province (Patagonia). (Color figure online)

2   Methods

2.1   Map Algebra

The term "map algebra" relates to the use of elementary operators in sequence to be able to solve complex spatial problems. It is analogous to an algebraic expression in which the combination of values provides a result. It involves the use of logical and mathematical expressions applied to spatial data. The terminology has been formalized by [4] which defined the operators and the ways in which you could create a structure that could be called map algebra.

The definition of map algebra was carried out and implemented in a raster environment, where the potential of the model calculation provide the best results. The applications in map algebra as a scheme is also possible in the vector field, where, however, is not optimized.

A fundamental condition the use map algebra is that the data sets are exactly the same size, both in terms of the domain and in terms of spatial resolution. To properly operate the map algebra operations imply a perfect "overlapping" of the individual layers. Important to note that the result of the map will have the same size of the input maps. In order to perform the processing of raster you can use different operators: local, zonal, focal, global and utility.

Local operators allow you to assign to each element of the output raster a value, which is a function of the values of the matching of the input raster. A typical operator is the "sum Local" thanks to which the values attributed to individual cells expressing the sum performed between the corresponding maps of the input cells. The speech is similar to any other operation, not only mathematical: operators can indeed be mathematical (arithmetic, trigonometric, statistical, logarithmic, and so on), logical (true or false), comparison (higher or lower). The uses of local players are many, can be used for a simple conversion of units of measurement (multiplication), for the assignment of values obtained by algebraic operations or statistics on number of values, or reclassification.

The zonal operators allow you to get a cell value of the output map as a function of the cell values of an input layer, grouped according to the areas defined in another input layer. For example, considering again the sum, in the first input layer they are added together only the cells belonging to a same area (defined by the second layer).

In addition to the sum, are typical zonal operations, the calculation of the average (in which all the cells of a zone is assigned the value of the average of the values of the single cells), and the calculation of the minimum and maximum value or a range of values.

The focal operators allow to obtain a value of the raster cell in output as a function of the values of cells belonging to a neighbourhood considered. In zonal operators the operation result value is assigned to an entire area, while in the case of focal operators the calculated value between multiple cells is attributed only to an element already chosen. It is necessary to choose the type of operator (such as mean, standard deviation, range, sum) and around of interest, in terms of size and shape. The filters used in the measurement values of processes, in order to eliminate the peaks or background noise, can provide an example of a focal operator.

A global operation is a function that is performed on each output cell using all of the cells of the input raster. To perform this calculation, it is obviously necessary to know the value of all cells in the input layer, and that is why these global operators are defined.

In this paper map algebra is used to develop an analysis protocol and to construct a predictive model for the analysis of the archaeological sites in the Santa Cruz County.

2.2 Predictive Models

Predictive models in archaeology constitute an opportunity to support the archaeological research from the scientific but also from the practical point of view. In fact, to design an archaeological sensibility map, allows to find areas where there is a high priority to do more investigation, but also, all the modelling process support the archaeological thinking by highlighting relationships between the settlements and the environment [5].

This is done, in literature, with different methods and by considering different parameters.

The methods most use statistics [6–8], but also map algebra is often used [9–11].

In the models developed in literature, the parameters that are used vary consistently, depending also from the study case. The environmental parameters most used are the elevation, the land use and the proximity to the water bodies [9], while less frequently are considered the view shed, the sun exposure, distances from morphological elements [11–13]. Also social parameters, such as distances between sites and their density over the space [1, 14] are rarely considered.

3   The Study Case

3.1   The Study Area

The region under study is located in the Central-Western area of Santa Cruz province. It is bounded by the Andes on the West and by National Route 40 on the East. Within it we have established two distinct sectors based on altitudinal and ecological terms: lowlands (below 400 masl) and highlands (over 900 masl).

Lowlands, such as Cardiel and Salitroso Lakes basins, are characterized by an arid/semi-arid environment with rainfall ranging from 100 to 270 mm annually. They correspond to a grass and shrub steppe [15] and exhibit low winter snow load.

On the other hand, highlands have a cold-temperate climate with annual rainfalls between 200 and 400 mm. These sectors are characterized by an environment of shrub steppe [16]. One of the peculiarities of them is that there is a marked seasonality, as during the winter months snowfalls turn these areas uninhabitable. These spaces include basaltic plateaus (Strobel and Pampa del Asador/Guitarra) and lakes (Belgrano, Nansen, Burmeister and Azara Lakes).

3.2 The Archaeological Settings and the Sites of Santa Cruz

The ongoing archaeological investigations in central-western Santa Cruz province are framed within the systematic research that started in the 80's in the Rio Belgrano-Lago Posadas area [17]. Within this peculiar landscape and under adverse climatic conditions, numerous evidences from hunter gatherers populations that inhabited this area were found. These evidences are present in caves, rock shelters and open air sites.

For the purpose of this work, archaeological sites were classified as follows:

- Alero (Rockshelter): Shelter in which lithic artifacts, animal remains and rock art are found (mainly negative painted hands) (Fig. 2a).
- Arte (Rock art): Rocky basaltic walls on which rock art is found. Mainly engravings footprint and abstract motives (Fig. 2b).
- Chenque (Burial): Human burials structures made of collected stones in an oval or circular form (Fig. 2c).
- Parapeto (Hunting blind): Semicircular stone structure that would have been used for hunting guanaco (*lama guanicoe*) (Fig. 2d).
- Sitio superficie (Surface site): Surface concentration of lithic and faunal archaeological materials (Fig. 2e).

Fig. 2. (a) An Alero, (b) an Arte site, (c) a Chenque, (d) a Parapeto and (e) a Sitio superficie.

3.3 The Archaeological Sensibility Evaluation (ASE) Model

For the construction of the ASE model, the analysis protocol schedule the following five steps.

First Step. In this step it is important to understand if, for each site type, there are environmental preferences in the positioning of each site. So it is needed to look for classes of parameters (for example, a slope range) where there is a greater frequency of sites. Parameters considered in the analysis are:

- Elevation (DEM; pixel size = 5 m)
- Slope
- Aspect

Consequently pure statistics were calculated, for each of the 4 parameters considered and for each site type.

Second Step. The results found in the previous step were summarized. Each class derived represents a sort of "including factor" for the environment, to be used to characterize settlement choice for that particular kind of site.

It was considered, for each parameter:

- The full-range values: the range of values in a single parameter, where the chosen site type was found.
- The more-frequent-range values: the range of values in a single parameter, where it is more frequent (probability > 90 %) to find a site of that type.

On this basis, the following classes were derived (Table 1):

- pixels belonging to the more frequent range as areas with high probability to find new sites (assigned value = 2);
- (pixels in the full range)-(pixels in the more frequent range) as areas with medium probability to find new sites (assigned value = 1);
- the remaining pixels as areas with low probability to find new sites (assigned value = 0).

Third Step. Classes found in the second step and validated in the third step were used to calculate single prevision maps for each parameter. Each map tells to the viewer which are areas sensible from a 0 to 2 score to find a type of archaeological site for the considered parameter (Fig. 3) With this aim Map Algebra was used.

Fourth Step. Starting from the maps obtained in the previous step, with the help of Spatial analysis, in particular Map Algebra (see the box beside), a resuming archaeological sensibility map (here called ASE, as Archaeological Sensibility Evaluation, but also as the order of parameters used for its calculation) for each site type was calculated in the following way:

$$ASE = Aspect \cdot 102 + Slope \cdot 101 + Elevation \cdot 100 \qquad (1)$$

where aspect, slope and elevation can assume a value from 0 to 2 according to what expressed in the second and fourth step, by determining the three probability (or sensibility) classes.

Consequently, the resulting sensibility raster (ASE) is probably not very readable if we try to colour and show all the combination of the many unique values that we obtain from the combination, but it is very significant as information system, better as Support

Table 1. Including factors. Classes extracted.

Site type	Parameter	Class	Values
Alero	Elevation	Low (0)	[0, 321) (581, 2697]
		Medium (1)	(374, 581]
		High (2)	[321, 374]
	Slope	Low (0)	[0, 3.2) (23, 90]
		Medium (1)	(18.9, 23]
		High (2)	[3.2, 18.9]
	Aspect	Low (0)	S-SW-W
		Medium (1)	NE-SE-NW
		High (2)	N-E
Arte	Elevation	Low (0)	[0, 336) (510, 670) (1183, 2697]
		Medium (1)	[336, 510] [670, 840)
		High (2)	[840, 1183]
	Slope	Low (0)	[0, 0.4) (27, 90]
		Medium (1)	(16.3, 27]
		High (2)	[0.4, 16.3]
	Aspect	Low (0)	W
		Medium (1)	S-SW-NW
		High (2)	N-NE-E-SE
Chenque	Elevation	Low (0)	[0, 139) (236, 382) (479, 576) (624, 2697]
		Medium (1)	[382, 479] [576, 624)
		High (2)	[139, 236]
	Slope	Low (0)	[0, 0.6) (18, 90]
		Medium (1)	(11.5, 18]
		High (2)	[0.6, 11.5]
	Aspect	Low (0)	S
		Medium (1)	N-NE-SW
		High (2)	E-SE-W-NW
Parapeto	Elevation	Low (0)	[0, 640) (1251, 2697]
		Medium (1)	[640, 880)
		High (2)	[880, 1251]
	Slope	Low (0)	(20.5, 90]
		Medium (1)	(8.25, 20.5]
		High (2)	[0, 8.25]
	Aspect	Low (0)	-
		Medium (1)	SW-W-NW
		High (2)	N-NE-E-SE-S
Sitio superficie	Elevation	Low (0)	[0, 136) (1193, 2697]
		Medium (1)	(560, 770)
		High (2)	[136, 560] [770, 1193]
	Slope	Low (0)	[20.7, 90]
		Medium (1)	(13, 20.7]
		High (2)	[0, 13]
	Aspect	Low (0)	-
		Medium (1)	SW-NW
		High (2)	N-NE-E-SE-S-W

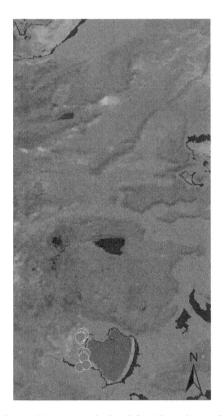

Fig. 3. Example of single prevision map calculated for Alero sites and for the DEM parameter. Light gray areas have medium sensibility to find Alero sites. Black areas have high sensibility to find Alero sites. The remaining areas have low probability to find Alero sites. (Color figure online)

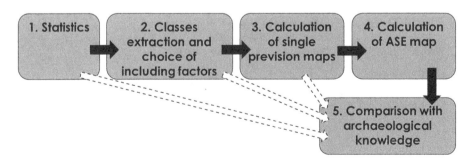

Fig. 4. Flow chart resuming the ASE models.

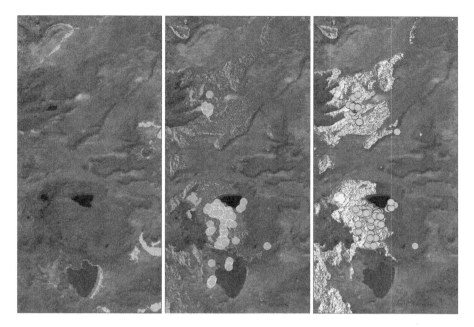

Fig. 5. Sensibility maps for Alero (a), Arte (b), Parapeto (c), Green: areas with high probability for the DEM and with medium or low probability for the other parameters (002, 012, 102, 112); yellow: areas with high probability for the DEM and with one another high-value parameter between aspect and slope (022, 122, 202, 212); red: areas with high value of all the three parameters (222). (Color figure online)

Decision System. In fact, we can query each pixel and to know how much and which parameters are propitious to the presence of that site category and also if its influence is low, medium or high. So we can classify according to the visualization needs.

Fifth Step. It is very important to compare classes, maps and results found so far with the historical site characteristics and knowledge. Therefore, this was done in all the phases of the process and in particular in this final step to understand if previous steps are consistent or not from the archaeological point of view.

The flow chart in Fig. 4 resume the analysis protocol.

4 Results

Different results were obtained for the different type of site, according to the archaeological interpretation:

- Alero (Fig. 5a): the sensibility map generates bounded zones where Aleros can be found.
- Arte (Fig. 5b): unlike what was observed in the other cases, the sensibility map does not generate very bounded future survey areas where Arte can be found. Since art is

Fig. 6. Sensibility maps for Chenque. Green: areas with high probability for the DEM and with medium or low probability for the other parameters (002, 012, 102, 112); yellow: areas with high probability for the DEM and with one another high-value parameter between aspect and slope (022, 122, 202, 212); red: areas with high value of all the three parameters (222); orange areas with medium probability for the DEM (001, 011,021, 101, 111, 121, 201, 211, 221). (Color figure online)

found in basaltic walls, geology can be a good parameter to consider to add in the future in the ACE model.

– Chenque (Fig. 6): the sensibility map generates bounded zones where concentrations of human burials can be found. In this case, to the legend used for Alero, Arte and Parapeto sensibility maps another set of classes was higlighted: areas with medium probability for the DEM. It is important to have a more additional class because, according to the knowledge of the territory, in the high-values DEM classes there's a more chance to find clustered chenque sites, while in the medium-values DEM class there's a more chance to find isolated chenque sites.

– Parapeto (Fig. 5c): the sensibility map generate bounded zones where Parapetos could be found.

– Sitio superficie (Fig. 7): the ubiquity of surface archaeological sites hinders the generation of a clear sensibility map. Provide more information on each archaeological site (for example the presence/absence of ceramic) could clarify the map.

5   Final Discussion

This work revealed that not all of the environmental parameters considered have the same sensitivity to help in the discovery of different types of archaeological sites. The elevation parameter is more sensitive than others to predict the location of

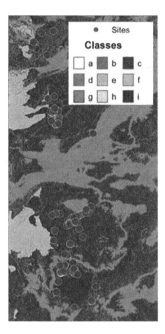

Fig. 7. Legend created just considering the statistical distribution. Classes defined: (a) all low including factors (000); (b) one medium inc. factor (100, 001, 010); (c) one high inc. factor (200, 002, 020); (d) two medium inc. factors (101, 011, 110); (e) two high inc. factors (202, 022, 220); (f) two mixed inc. factors (120, 201, 210, 102, 021, 012); (g) three mixed inc. factors (112, 121, 211, 122, 212, 221); (h) all medium inc. factors (111); (i) all high inc. factors (222). (Color figure online)

archaeological sites, mainly chenques, parapetos, arte and aleros. This is important considering that the height above sea level is a significant variable that defines the seasonality of the different areas in Patagonia. For example, the highlands tend to be inaccessible during winter because of the large amount of snow they have.

The ASE model allows us to rank the sectors not yet disclosed. This may guide researchers in planning and refining of future fieldwork for the discovery of new archaeological sites.

As a future perspective, it is necessary to link the information of new parameters with the ones generated here. For example along with elevation, vegetation and distance to water are environmental parameters that are clearly related to settlement patterns of hunter-gatherer societies.

Another future work may consist in provide more information corresponding to each of the archeological sites (for example, size of the archaeological site, associated technology, etc.). Thus, increasing the variability of the archaeological record we could evaluate parameters' sensibility in new spatial scale.

Moreover the analysis could be improved also from the modelling point of view. For example the use of the right cell size is not an easy choice, because a smaller cell size returns a result too disaggregated and probably unuseful, while a bigger cell size is not good to map environmental phenomena or particular existing structure that could be useful to determine settlement pattern and choices and so to determine better the sensibility areas.

Another example is that the model could reveal in real time which parameters are not useful for the analysis. For example, in this study, at the start four parameters were used, while, during the process it appeared clear that the landform one was not indicative of settlement choices, so it was not used to derive the sensibility maps.

Acknowledgments. The authors thank the Ministero degli Affari Esteri e Cooperazione Internazionale (MAECI) for supporting this activity in the framework of the Project "Smart management of cultural heritage sites in Italy and Argentina: Earth Observation and pilot projects" PGR00189 2014-2016.

References

1. Danese, M., Masini, N., Biscione, M., Lasaponara, R.: Predictive modeling for preventive archaeology: overview and case study. Central Eur. J. Geosci. **6**(1), 42–55 (2014). doi:10.2478/s13533-012-0160-5
2. Goñi, R.: Arqueología de momentos históricos fuera de los centros de conquista y colonización: un análisis de caso en el sur de la Patagonia. In: Belardi, J.P., Marina, F.C., Espinosa, S. (editado) En Desde el País de los Gigantes. Perspectivas Arqueológicas en Patagonia, pp. 283–296. Universidad Nacional de la Patagonia Austral, Río Gallegos (2000)
3. Cassiodoro, G., Rindel, D., Goñi, R., Re, A., Tessone, A., Guraieb, S.G., Belardi, J., Espinosa, S., Delaunay, A.N., Dellepiane, J., Coni, J.F., Guichón, F., Martínez, C., Pasqualini, S.: Arqueología del Holoceno medio y tardío en Patagonia meridional poblamiento humano y fluctuaciones climáticas. Revista Dialogo Andino **41**(5), 5–23 (2013)
4. Tomlin, D.: Geographic Information Systems and Cartographic Modeling. Prentice-Hall, Englewood Cliffs (1990)
5. Vaughn, S., Crawford, T.: A predictive model of archaeological potential: an example from northwestern Belize. App. Geog. **29**(4), 542–555 (2009)
6. Niknami, K.A.: A stochastic model to simulate and predict archaeological landscape taphonomy: monitoring cultural landscapes values based on an Iranian survey project. Archeologia e Calcolatori **18**, 101–120 (2007)
7. Canning, S.: BELIEF in the past: dempster-shafer theory, GIS and archaeological predictive modeling. Aust. Archaeol. **60**, 6–15 (2005)
8. Zhongxuan, L., Gang, H., Cheng, Z.: Multi-fractal analysis on neolithic sites around the Sanxia reservoir area. In: RSETE, pp. 491–494 (2011)
9. Podobnikar, T., Veljanovski, T., Stanèiè, Z., Oštir, K.: Archaeological predictive modelling in cultural resource management. In: Konečný M. (Ed.) GI in EUROPE: Integrative – Interoperable – Interactive, Proceedings of 4th Agile Conference on Geographic Information Science, pp. 535–544 (2001)
10. Brandt, R., Groenewoudt, B.J., Kvamme, K.L.: An experiment in archaeological site location: modeling in the Netherlands using GIS techniques. World Archaeol. **24**(2), 268–282 (1992)
11. Stanèiè, Z., Kvamme, K.: Settlement pattern modelling through Boolean overlays of social and environmental variables. In: Barcelo J.A., Briz I., Vila, A. (eds.), New Techniques for Old Times, (CAA 1998), BAR International Series, vol. 757, pp. 231–237 (1999)

12. Verhagen, P.: Testing archaeological predictive models: a rough guide. In: Posluschny, A., Lambers, K., Herzog, I. (eds.) Layers of Perception, Proceedings of the 35th International Conference on Computer Applications and Quantitative Methods in Archaeology (CAA), Berlin, Germany, vol. 10, pp. 285–291 (2007)

13. Garcia, A.: GIS-based methodology for Palaeolithic site location preferences analysis: a case study from Late Palaeolithic Cantabria (Northern Iberian Peninsula). J. Archaeol. Sci. **40**, 217–226 (2013)

14. Alexakis, D., Sarris, A., Astaras, T., Albanakis, K.: Integrated GIS, remote sensing and geomorphologic approaches for the reconstruction of the landscape habitation of thessaly during the neolithic period. J. Archaeol. Sci. **38**, 89–100 (2011)

15. Oliva, G., González, L., Rial, P., Livraghi, E.: El ambiente en la Patagonia Austral. In: Borreli, P., Oliva, G. (eds.), Ganadería Ovina Sustentable en la Patagonia Austral. Tecnologías de Manejo Extensivo, pp. 19–82. Ediciones INTA, Buenos Aires (2001)

16. Cabrera, A., Willink, A.: Biogeografía de America Latina. Monografía, no. 13. Secretaría General de la OEA, Washington (1980)

17. Aschero, C., Bellelli, C., Goñi, R.: Avances en las investigaciones arqueológicas del Parque Nacional Perito Moreno (Provincia de Santa Cruz, Patagonia Argentina). Cuadernos del Instituto Nacional de Antropología y Pensamiento Latinoamericano, vol. 14, pp. 143–170 (1992–1993)

Self-renovation in Rome: Ex Ante, in Itinere and Ex Post Evaluation

Maria Rosaria Guarini[✉]

Department of Architecture and Design (DIAP), Faculty of Architecture,
Sapienza University of Rome, Rome, Italy
mariarosaria.guarini@uniroma1.it

Abstract. In Europe, self-construction/self-renovation are innovative and additional tools to meet the needs of a part of "disadvantaged" social groups that can not buy or rent dwelling at market prices. At the end of the 90 s of the twentieth century, the Municipality of Rome has set the first trial at the national level (still not completed and remained almost unique) related to disused building self-renovation (especially school buildings). The text shows the results of a research, still ongoing, aimed at ex post evaluation of items that have prevented to conclude timely and as provided such interventions and, consequently, doesn't meet the housing needs for which had been started.

This allows to highlight as the assessment tools, in the different phases of the development process of these initiatives, ex ante, ongoing and ex post may help to reduce the risks of "failure" of self-construction/self-renovation initiatives.

Keywords: Ex ante, in itinere, Ex post evaluation · Multi-criteria decision analysis · Self-construction · Self-renovation · Social housing

1 Introduction

In Europe, within the sector of public policy on social housing, initiatives of self-construction/self-renovation are held to be an innovative, supplementary tool when it comes to meeting the need for housing of the new (and old) "disadvantaged" segments of society, or those that find themselves unable to purchase or rent a dwelling at market prices (Housing Europe 2015, CECODHAS 2012, BSHF 2011; Czischke and Berthon 2008; AAVV 2000). In Italy, which presents a variety of significant manifestations of such unmet housing needs (by now "chronic" in nature, especially - though not exclusively - in large urban agglomerations) (Guarini and Battisti 2014a) projects of self-construction/self-renovation constitute an opportunity limited to "weak" family units that present specific characteristics: a low capacity for earning income, but suitable conditions - as well as a willingness - to join together with other family units (as members of a cooperative) and work in their free time on the construction of a residential building suited to their needs and to the number of participating families, all of which share in meeting the risks tied to a similar operation. From the outlook of government bodies and authorities, be they regional, provincial or municipal administrations or agencies responsible for managing public housing, efforts of self-construction/self-renovation

© Springer International Publishing Switzerland 2016
O. Gervasi et al. (Eds.): ICCSA 2016, Part IV, LNCS 9789, pp. 204–218, 2016.
DOI: 10.1007/978-3-319-42089-9_15

should be viewed solely as a specific approach that provides an alternative to other forms of intervention, meaning that such initiatives should always be considered and undertaken within the context of more extensive, fully effective housing policies, though, at the same time, examples of self-construction/self-renovation already implemented in Italy and Europe show that the approach can also serve as a way of furthering policies of social support and integration. In fact, the resulting efforts can be seen as contributing to: a lessening of instances of social exclusion and spatial segregation; the implementation of strategies to fight poverty; a heightening of the sense of responsibility on the part of the subjects to whom the housing units are assigned, in terms of the care and management of the dwellings during the operational phase (Housing Europe 2015; Bertoni and Cantini 2008; Cittalia 2011; BSHF 2015; Mullins and Sacranie 2014).

The self-restoration projects promoted by the municipal administration of Rome within the framework of the policies undertaken, starting from the late 1990's, to deal with the sharp peaks in the housing crisis that occurred in those years within the municipal territory provide a specific set of case studies involving the types of self-renovation/ self-construction projects that were carried out in Italy and Europe as a whole. Identification and evaluation of positives and negative factors that have marked the start and development of the mentioned experience after almost two decades still not completed, may be useful in reducing the risk of failure of any similar initiatives that will be undertaken in Italy and in Europe; in this way will be possible to reach the objectives and the expected results related to process, cost and schedule.

2 Aims of the Work and Assessment Approach

2.1 General Aims and Particular Focus

The text provides an overview of the results of the portion of a research (still underway)[1] aimed at identifying and evaluating, in terms of the planned procedures and those actually implemented, the operative management of the projects of self-renovation carried out by Rome Municipality. A particular focus of the research is to formulate an analysis and assessment, of this specific approach to policies providing housing support to disadvantaged segments of society promoted by the municipal administration of Rome from the moment of their start-up (1998) though June of 2015, so as to be able to:

---

[1] The research was developed, in part under a working relationship established with the Municipal Department for the Development of Outlying Areas, Infrastructures and Urban Maintenance of the City of Rome, as per a convention signed in December of 2014 with the Department of Architecture and Design. Some of the results of the research were illustrated in the course of an international workshop entitled, "Faced with the problem of real-estate speculation, what responses are provided by housing cooperatives in the Latin sector of Europe, meaning Italy and Spain?" (Rome. 10-11 June 2015), organised jointly by CHAIRECOOP & the LAC (Laboratoire d'Art Civique, the University of Rome, Campus III) and the Rome Tenants Union. Claudia Buccarini (during doctoral course) and Serena Sbaffoni (during the preparation of the fist level degree thesis), participated in the research collecting and processing some of the data related to self-renovation cases presented in this text.

- highlight (after the fact/while the efforts are underway) the factors and causes that have led to failure to achieve the expected results - within the forecast time periods and cost parameters - for problems involving: the housing crisis; the environmental and social deterioration of certain urban zones; social cohesion and integration; the consumption of land (Nesticò and Pipolo 2015; Tajani and Morano 2015);
- formulate possible lines and modes of action to be implemented (in advance) for the future development of new initiatives of self-renovation in Italy and Europe.

In the research it was examined and organized:

- in general and concise way, information and related data: (i) a survey of documentation on of self construction/self recovery initiatives ongoing (or proposed) at the international, European, national, and (ii) the legislation adopted in Italy at regional level and local;
- in specific and detailed way, as auto construction was regulated and implemented in the Lazio Region, particularly in the City of Rome.

The goals pursued have been identified, highlighted and compared, as well as the different subjects involved and their roles, together with the criteria followed in procuring and assigning the funds allocated for the development of these initiatives, plus the budgets and timing of the projects, both as forecast and as recorded in actual fact (European Commission 2014).

To develop initiatives relating to the deepening of self-recovery carried out in the municipality of Rome, data and information contained in the resolutions of the administration Capitoline Council (*Determinazione di Giunta Capitolina* -DGC) and a part of the Managerial Determinations (*Determinazione Dirigenziale* - DD) - taken by AC offices responsible for managing these interventions - has been collected, analyzed and processed.

2.2 Assessment Approach and Perimeter of Investigation

In this research an evaluation approach has adopted to assess the level of correspondence between objectives and expected results expected and actually achieved, with a perinductive comparison process between than assumed ex ante, what happened in itinere and what is encountered ex post.

This is in order to identify: (i) aspects, moments, decision nodes which provide the quality and success of an intervention self construction/self renovation (satisfaction the objectives undertaken in the planning stage); (ii) the main risk factors that can hinder the success of this type of initiatives.

In this way it has been possible to build a useful framework for defining strategies for actions that have to be implemented in the planning, design, implementation of similar future initiatives, not only in Italy but also in other European countries.

For this purpose, of the 11 efforts (197 housing units) being drawn up in 2008 Fig. 1) (http://www.comune.roma.it/pcr/it/dip_pol_riq_per_aut.page), there was detailed examination of 7 projects that involved school buildings (on the Via Saredo; the Via Marica; the Via Colomberti; the Via dei Lauri; the Via di Grotta Perfetta; the Via F. De Grenet; and the

Via Alzavole) and that, at the time the research began (2013), were still, in varying stages, underway. These initiatives, financed with funds from the Ministry of Infrastructures, the City of Rome and the Lazio Region were all carried out by the "Inventare l'abitare" ("Inventive Housing") Cooperative.

Fig. 1. The 11 self-renovation intervention being drawn up in Rome in 2008 (source: https://www.comune.roma.it/pcr/it/dip_pol_riq_per_aut.page)

The detailed analysis did not cover a project in the Piazza Sonnino involving an historic structure (on the part of the cooperative "Vivere 2000" Srl) or three efforts that had been suspended (at the Via Appiani, the Via Monte Meta and Largo Monte San Giusto). The data provided by the documentation to be to be found on the website of the city of Rome, as well as the documentation made available (in part) by the Department of Policies for the Renewal of Outlying Urban Areas of the City of Rome and by the "Inventare l'abitare" Cooperative were processed to determine what factors prevented completion, within the time periods and according to the procedures stipulated, of the project undertaken, leading to failure to satisfy the need for housing that they were initiated to fulfil.

The results recorded to date, though based on data that are not complete, nonetheless make it possible to draw up a fairly precise of the critical problems underlying the failure to achieve (as of June 2015) the objectives that the municipal government intended to pursue with the projects, and namely:

- satisfaction of at least a portion of the demand for social housing over periods of time that are shorter than those needed for the construction of new residential housing, and without causing further consumption of land;
- experimentation with an "alternative" response to the housing crisis;
- solving of certain manifestations of environmental and social deterioration found in a number of the city's urban settings, potentially with the involvement of occupants of buildings (mainly school facilities) to be retrofitted as residential structures (European Commission 2014).

The paragraphs that follow succinctly illustrate how, in the course of the research, the various critical problems identified were found to be closely interrelated, all being attributable, in general terms, to the manner in which the procedure followed had been structured and implemented, though specific failings regarded:

- the ways in which the buildings to be reclaimed through the efforts of self-renovation were identified;
- the role and the procedures for involvement of the future holders of the residential units during the phase in which the financing was procured and the internal restoration of the units was performed;
- the instructions and the decisions on questions of timing, financing, planning solutions and investment costs needed to carry out the work required to transform the school buildings into residential housing;
- the business and financial capabilities of the enterprises selected to perform the work.

At first descriptions shall be given of the decisions made, together with their repercussions, with regard to the procedures for identifying the structures and the role and manner of involvement of their future holders (Sect. 3.1), as well as those addressing aspects of the planning, design and execution of the initiatives (Sect. 3.2). Then, prior to the presentation of the conclusions (Sect. 5), a number of assessments are drawn up, together with operating guidelines to be implemented in advance of any future initiatives of self-construction (Sect. 4).

3 Projects of Self-renovation of School Buildings in the City of Rome for Residential Purposes

3.1 Decisions and Repercussions Regarding the Procedures for Identifying the Structures and Determining the Role of the Future Holders of the Housing Units

The city government undertook the procedures of self-construction in accordance with statutory and regulatory measures of the following types pertinent to self-renovation/self-construction: (i) general, stipulated under regional statutes (Law no. 55 of the Lazio Region, issued on 11 December 1998); (ii) specific, promulgated by the municipal administration for the enactment of the projects, starting with Resolution no. 248 of 20 November 1998 of the Municipal Council.

Examination of the documentation considered shows that the city government identified the buildings to be retrofitted for residential purposes by considering the housing crisis that, since the late 1990's, had manifested itself, in part, through the occupation, with the support of movements for more available housing, of unused public structures (school buildings in particular) on the part of family units in search of dwellings. Resolution no. 753/2002 of the Municipal Council – on the approval of the first "Draft Agreement. Draft of the public procedure for assignment and authorisation of the establishment of statutory first mortgages covering the work involved in the construction transformation through self-renovation" of the school buildings on: the Via Saredo; the Via Marica; the Via Colomberti; the Via dei Lauri; the Via di Grotta Perfetta; and the Via F. De Grenet – describes the "climate" under which decisions were made regarding both the buildings involved in the initiatives of self-renovation and the positions and roles assigned to the parties occupying the structures as the process of self-construction moved forward. The document in question states that: "certain school buildings owned by the municipal government are improperly utilised as family dwellings following the illicit occupation of the structures; this practice arose due to the inadequate and insufficient supply of housing offered by the rental market, as well as the impossibility of finding alternative arrangements, combined with the modest economic means of the majority of the illicit occupants; the housing situation inside the occupied school buildings – apart from the significant concerns it raises in terms of public security – must be restored to a situation of legitimacy; to that end, and as part of a more extensive program for the regularisation of the large number of different types of illicit occupations, consideration has bene given to the possibility, among other options, of maintaining the residential use of certain school structures". Furthermore, regional and municipal measure specified that, for each building involved in an initiative of self-construction, the financing, planning and performance of the works was to be:

- managed directly by the municipal government with regard to the jointly held portions of the structure, both internal and external (primary renovation). The municipal government was responsible for procuring the financial coverage needed for the preliminary and final planning of the entire effort, as well as for the formulation of the working plans and the performance of the work of primary renovation, doing so both by drawing from funds in its own budget and through financing from the central state (the Ministry of Infrastructures – Committee on Residential Housing, from the allocations of Law 457/78) and the regional government. Assignment of the planning and performance of the primary renovation work must be made through a procedure open to public scrutiny;
- the work inside the apartments (secondary renovation) was to be delegated to the cooperative that won the competitive procedure for assignment of the building. The cooperative was to be selected in accordance with the parameters of quality indicated in the regional directive and specified in the call to tender drawn up for the purpose by the municipal government. In responding to the call to tender of the municipal government, the competing cooperatives were to: (i) present a working plan for the activities of secondary renovation to be carried out; (ii) agree to: advance the sums

need to carry out the work by taking out a loan (guaranteed by the municipal government) through registration of a mortgage on the assigned structure and to arrange for the performance of the work, under the supervision of the municipal government, by a film selected through direct assignment. The cooperative that won the project tender was to repay the sum advanced by the bank within the stipulated time period, doing so through a mechanism that took ongoing deductions of the amounts due from the monthly rental payments owed by the families to which the housing units had been assigned.

If one of the purposes of self-renovation efforts was to achieve objectives tied to the active involvement of the inhabitants in the process of modifying the structures, then the procedure, as structured and implemented in Rome, constitutes only a "partial" form of self-renovation, in light of both the role give to the future residents in the financing process and the procedures for carrying out the work needed to retrofit the school buildings for residential use. Indeed, the calls to tender did not require that there be, with regard to the activities of secondary renovation, an active contribution on the part of the occupants of the structure (who, having formed cooperatives, were to be assigned the right to live in the buildings). The documentation shows that the establishment of the cooperatives by the occupants (who, in this way, gave up their places on the waiting lists for public housing) was essentially geared towards:

- obtaining access to mortgage loans under favourable conditions (guaranteed by the municipal government as the owner of the building), in order to procure the sums needed to carry out the work of secondary renovation;
- satisfying the housing needs of the individuals occupying the school buildings by regularising their irregular positions, assuming they meet the requirements for obtaining public housing.

The fact is that the presence in a cooperative submitting a proposal of individuals "considered to be occupants of the buildings to be restored, or otherwise falling within the context of the buildings covered by the program agreement to resolve the housing crisis endorsed, of common accord, by the Ministry of Infrastructures and the Lazio Region", as well as the municipal government, was one of the three criteria of priority stipulated in the calls for tender for the assignment of the buildings to the cooperatives. However this priority, though meant to provide the occupying family units in need of housing with reasonable safeguards, resulted, in practice, in each family unit automatically being assigned the space that it already occupied inside the building. This outcome constitutes an "anomaly", compared to how initiatives of self-construction/self-renovation normally unfold, both in Italy and abroad. In the majority of such initiatives, in fact, the residential units are not assigned to the families until all the work has been completed, with this being done, in part, to obtain the active cooperation, as well as the integration, of the building's future inhabitants, thanks to the bond of their joint effort to make their dream of adequate housing become reality. Further critical problems arising from the anomalous role given to the family units involved in these initiatives have to do with the delays in the various phases of the development (planning, design, execution) of the initiatives (illustrated in greater detail further on), which also led to discrepancies in the treated afforded to the different family units. In fact, in terms of timing, there were differences in: (i) the points in time when the occupying families were able to leave the occupied buildings and move to "temporary" accommodations so that the work needed to upgrade the school buildings for residential purposes could be carried out; (ii) the periods of time during which the family units stayed in these "temporary" accommodations, where the majority still live (2015).

On account of unforeseen complications and accumulated delays, the cooperative "Inventare l' abitare" was able to take possession of only some of the buildings, and at different points in time, to carry out the secondary restoration work. This work was performed when, in actual fact, the activities for which the city administration was responsible had not yet bene completed (see further on). Only a few of the families of cooperative members to which residential units had been assigned continued to live in precarious and disadvantaged conditions in a number of the structures (which still lacked documentation attesting to the final performance testing of the work done, as well as certificates of fitness for use and utility hook-ups), in order to guard the premises and prevent acts of vandalism.

Given this state of things, further repercussions to be taken into consideration include the fact that: (i) many family units have begun paying back their quotas of the loan provided on eased terms (taken out by the cooperative to finance the secondary renovation work), even though they do not yet live in the assigned housing; (ii) in the ensuing years, the city government has been obliged to sustain elevated costs to carry out evictions (following subsequent occupations of the buildings by other families), including the expense of moving and housing the occupants in residential hotels or through rental vouchers at market prices.

3.2 Decisions and Repercussions Tied to the Procedures for the Planning, Design and Execution of the Initiatives

The documentation show that, in the case of the majority of the projects, the phases of planning and design occupied a timeframe that ran from 1998 to 2004 (Fig. 2). Specifically, it took:

- approximately 10 months to arrive at the signing (on 22 September 1999) of the Memorandum of Understanding between the city government of Rome, the Ministry of Public Works (Committee on Residential Housing) and the Lazio Region (Department of Urban Planning and Housing), a document "meant to implement a complex program of residential construction, in order to move displaced inhabitants from residential hotels and absorb the peak manifestations of the housing crisis, eventually by means of self-renovation";
- more than a year and a half for (outside) assignment of the formulation of the final plans (3 August 2000) and for the holding of the Services Conference (12 September 2000) to obtain the preliminary opinions of the departments responsible for the efforts, both inside and outside of the municipal administration;
- approximately 2 years for the city government to approve: (i) modification of the zoning designation of use (from educational to residential) of the buildings, under a waiver of the regulations for the enactment of the General Regulatory Plan (16

November 2000); (ii) the Residential Housing Program covering the renovation of the buildings earmarked for residential retrofitting, together with authorisation (25 January 2001) for the Mayor to sign the resulting Program Agreement;

- approximately 3 years to arrive at the signing (3 October 2001), ratification (18 October 2001) and publication of the Program Agreement between the Ministry of Infrastructures and Transportation, the Lazio Region and the City of Rome (Official Bulletin of the Lazio Region, issue no. 32 of 20 November 2001, ordinary supplement no. 2), the document under which the localisations and amounts of the financing for the initiatives of self-renovation involving the housing stock of the city administration were approved.
- approximately 4 years for approval (17 December 2002) of the first draft of the Convention and Call for Tender for the assignment and authorisation of the stipulation of statutory first mortgages for performance of the work of residential retrofitting, under a self-renovation approach, on the interiors of 6 school buildings. Another five years were to go by before approval (20 February 2008) of the draft for the other 5 buildings for which projects were underway since 1998 (making for a total of 9 years);
- more than 4 and a half years for the municipal government to receive, towards the middle of 2003, the regional allocations disbursed under Law 457/1978 for the performance of the initiatives of self-renovation.
- approximately 6 years for approval of the final plans assigned to outside planners.

In the case of all the projects, the approval of the working plans, the commitment of the funding and the holding of the tender to assign the work of (primary) renovation took place at practically the same point in time. Given the impossibility of examining all the documentation regarding these acts, the timeframe that led to the opening of the worksites can be calculated only by taking the last date common to all the efforts, as corroborated in the documentation (approval of the final plans). Only in the case of two of the efforts (the buildings on the Via Marica and the Via dei Lauri) did it take less than a year to go from the approval of the final plans to the drafting and confirmation of the working plans, followed by enactment of the procedures of the call to tender for the execution of the primary restoration work and, finally, consignment of the structures to the contractors. In the other cases, the additional time needed was traceable to delays in the presentation of the plans and/or in the holding of the contract tender and/or in the procurement of certification of proper payment of taxes on the part of the contracting enterprises. As a result of all the above, the performance of the primary restoration work began anywhere from 6 to 9 years following approval of the first regional and municipal regulatory measures (Fig. 2). Naturally the winners of the tenders were selected, based on the starting levels for bids, at amounts, time periods and volumes of work, as indicated in the working plans, that were subject to noteworthy reductions (an average of approximately 37 %).

The documentation shows that the cost estimates for the projects, as illustrated in the economic overviews for the final plans and the working plans, remained essentially unchanged, in terms of the overall amounts, from those indicated in the preliminary plans (drawn up by the municipal government). The estimated time for the presumed duration of the performance of the work was also held to a maximum of 360 consecutive calendar days, starting from the official notification of the presentation of the projects

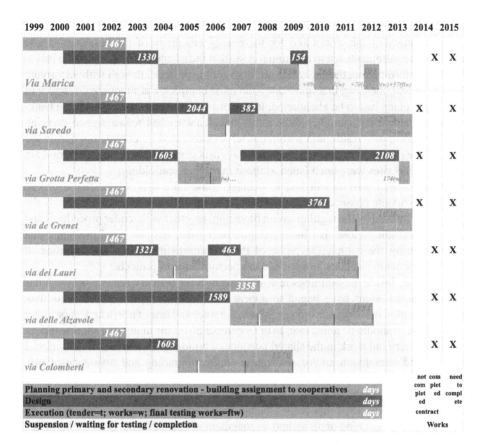

Fig. 2. Real time programming and design and execution of the primary recovery work

to the tender winners, in accordance with what was stipulated in the Program for Public Residential Housing approved by the municipal government. As in the case of all such initiatives, specific solutions were called for to increase energy efficiency, in keeping with the provisions of Regional Law 55/1998 and with the criterion in the calls to tender under which the buildings were assigned to the cooperatives (use of biocompatible materials and technologies).

The documentation shows that, right from the start of the demolition work (primary recovery work), problems arose with regard to: (i) the conditions and the construction quality of the buildings; (ii) the structural layouts of the buildings and the load-bearing elements; (iii) the noteworthy deterioration of the structures, both in terms of the original parts (as a result, to give just one of the possible examples, of the infiltration of moisture from the ground) and as regards the numerous works added at later dates (first and foremost, partitions) by the occupants (following the occupations), to be able to use the facilities as residences.

In many cases the delays in the timeframe made it necessary to completely revise the existing plans, or to draft new ones, with the attendant need to take into account new requirements promulgated under legislative or regulatory measures that had been passed in the meantime regarding technical standards for construction in seismic zones, for the structural safety of the works and for the energy efficiency of the buildings. These unforeseen requirements led to additional costs that the municipal administration had to absorb by reformulating the economic outlooks for the projects, though without varying the overall cost, meaning that quite often the most complex and innovative features of the original plans had to be abandoned. It also proved necessary to deal with the bankruptcy and breaches of a number of the forms to which work had been assigned, carrying out new tenders as a result, and with the reserves expressed by contractors during the performance of the work, as well as additional costs generated by extended periods in which the worksites were open (such as those involving scaffolding).

The municipal government had to manage some critical problems that emerged during the worksite phase, coming up with solutions that took into account not only the "constraints" posed by the regulations on financing, together with tender procedures and contracts, including those that have changed over time, but also vital problems that emerged during the completion phase of the procedures involving: the awarding of contracts; the taking of consignment and the performance of projects.

Over time, inside the buildings in which much of the primary renovation work has been completed, signs have begun to appear of deterioration traceable to: less than perfect execution of works; acts of vandalism; the wear and tear of use; a lack of ordinary maintenance. It should be noted that, as of December 2014, the municipal administration still had to carry out work in the shared portions of buildings, consisting of activities of finishing and completion and/or renovation and/or upgrading and assurance of safe conditions (for an estimated amount, based on the figures bid, of 300,000 euros), involving the hook-ups to sewage and electricity networks, infiltrations and upward seeping of water, accessibility for the handicapped and miscellaneous finishing work (fencing, railings, pruning of plants and arrangement of outside areas) in all the buildings being considered, except for the one on the Via Grotta Perfetta, where solutions must still be found for critical structural problems detected in the building immediately after the start of the work, despite the numerous assessments and structural designs that had been drawn up.

In the final analysis, the value of the primary recovery work completed, or still to be carried out, did not, in actual fact, correspond to the estimated figures, as the works themselves did not match, from an economic standpoint, the activities performed or the resulting functional features, as these had been defined during the phases of preliminary, final and working planning and design, and not only in terms of dimensions, forms and quantities, but as regards the quality of materials and components as well (Table 1).

4   Considerations and Strategies of Action to Be Followed (in Advance) for the Development of Future Initiatives of Self-renovation

Based on the considerations presented above, it appears safe to say that the procedures, both those planned and those actually implemented, were not well suited to guaranteeing achievement of the results expected in terms of timing (brief periods of completion) and

Table 1. Snapshot of the processed data relating to the progress of self-recovery operations

Phase			Self-renovation intervention considered						
			Via Marica	Via Saredo	Via di Grotta Perfetta	Via de Grenet	Via dei Lauri	Via delle Alzavole	Via Colomberti
Planning	Homes		27	11	20	8	23	8	10
	Resident		84	35	55	26	65	21	23
	Primary rennovation	Cost (planned) €	774,685.35	396,122.00	697,216.81	370,816.27	515,739.54	516,461.21	329,558.89
	Secondary rennovation	Cost (planned) €	580,123.08	235.504,35	391,328.67	198,515.18	373,429.92	224,820.62	233,376.52
Design	Detailed proposals	Assignment (a)	03/08/2000	03/08/2000	03/08/2000	03/08/2000	03/08/2000	27/06/2002	03/08/2000
		Validation (b)	10/03/2004	22/07/2005	12/05/2004	18/02/2004	18/02/2004	10/11/2004	26/05/2004
		Δ (b − a)	≈ 4 years	≈ 5 years	≈ 4 years	≈ 4 years	≈ 4 years	≈ 2,5 years	≈ 4 years
	Final proposals	Assignment (c)							
		Validation (d)							
Execution	Contract tender	Contract Award (e)	10/11/2004	18/10/2004	25/01/2006	21/03/2005	13/10/2004	29/01/2008	26/09/2005
		Δ (e − b)	8 months	39 months	20 months	13 months	8 months	38 months	16 months
	Primary rennovation	Start (planned)	2005	2003	2006	2003	2003	2003	2007 200?
		End (planned)	2005	2003	2006	2003	2003	2003	2007 200?
		State (2015/06)			in progress	relieved	relivied	In progress	≈ end 2011
		Cost (2014)	1,072,875.00	396,122.00	1,106,320.00	370,816.27	898,051.00	516,461.21	329,558.89
		Finishing work required (2014)	X	X		X	X	X	X
	Secondary rennovation	Assignment (f)	23/06/2003	17/01/2006	19/12/2003	23/06/2003	23/06/2004	18/01/2008	17/06/2005
		Consign (g)	2008/03	2009/10	No	2009/04	No	2012/11	2008/03
		Δ (g − f)	5 years	3 years		6 years		16 months	3 years
		End (h)	2010/10	2010/11		2010/05		2013/05	2008/10
		Δ (h-g)	16 months	12 months		12 months		7 months	7 months
Operational	Building Home Loan	Inhabitate	(safeguard)	(safeguard)		No	No	(safeguard)	No
		Pre-allocation	Yes	Yes		Yes		Yes	Yes
		Payment (from)	2008/02	2009/11		2009/06		2012/11	2009/08

operational performance (the quality and innovative nature of the initiatives to be carried out: experimentation with bio-climatic building or bio-construction techniques). There can be no question that the undertaking of a procedure structured with the primary renovation work placed under the responsibility of one and the secondary renovation work under the responsibility of another, though useful for the purposes of procuring financing, with one portion covered by the public sector and the other by the beneficiaries, would not appear to be the best approach for achieving the expected results. But more to the point, the observations made lead to the reasonable supposition that the preliminary, working and final plans for both the primary and secondary works, together with the resulting cost estimates, were formulated in the absence of an accurate preliminary assessment of the features, as well as the state of construction and the structural condition, of the buildings, which, furthermore, deteriorated over time, on account of the periods during which they were occupied, as well as the persistent delays in the phases preceding the start of the work to be carried out, plus the prolonged suspensions of the working activities.

The "political" decision to meet the housing needs of the occupants of the school buildings by retrofitting these structures for residential purposes did not lead to the hoped-for results and benefits within the time periods, and in the manner, counted upon to respond to the compelling needs of the occupants and the body public, for the simple reason that it was not supported by assessments expressed according to objective criteria, following a preliminary evaluation of the possibilities, in terms of urban planning, categories of construction, static conditions and funding, for transforming and upgrading the occupied buildings for use as residential units (Morano and Tajani 2013), in keeping with their current state, all leading to a final decision on the advisability of proceeding with a renovation project rather than, for example, demolishing and rebuilding the structure as part of a self-construction effort.

During the planning phase (prior to the decision on what type of initiative to undertake), a "material" knowledge of the structures being phased out of operation, and slated for retrofitting through self-renovation projects, should be obtained. Before initiating a self-renovation effort, the following key questions must be answered: which/how many buildings no longer used as active municipal assets could undergo self-renovation initiatives? Do the features of these buildings make them well suited to retrofitting for residential purposes? Are the costs and timing required for a restoration initiative (to be accurately estimated) more sustainable than those for carrying out a project of demolition/reconstruction? (Guarini 2014).

Tools of multi-criteria evaluation make it possible to come up with objective responses to these questions, in support of decisions leading to well-informed, sustainable initiatives (European Commission 2006). Given that the decision-making problems to be addressed include spatial factors as well, it might prove advisable to resort to the joint, integrated application of a multi-criteria analysis (MCA) (Guarini and Battisti 2014b) and a territorial evaluation employing geographic information systems (GIS).

5   Conclusions

With reference to the examined case study, the adopted evaluation approach allowed to assess the level of correspondence between objectives and expected results expected and actually achieved, with a per-inductive comparison process between than assumed ex ante, what happened in itinere and what is encountered ex post.

The research work carried out made it possible: (i) to formulate an overview of what has effectively been done and by what means; (ii) to identify the technical and regulatory problems that have hindered proper execution; (iii) to illustrate the results obtained in terms of costs and benefits. More specifically, it highlighted the fact that, in order to carry out an initiative of self-renovation/self-construction suited to the underlying objectives of such efforts, care must be taken: (i) to review the procedure for the enactment of the initiatives of self-renovation/self-construction; (ii) to use tools of evaluation that prove adequate when it comes to examining all the elements potentially able to contribute to: determining the criteria of feasibility for initiatives of self-renovation involving buildings that are no longer active municipal, as well as selecting the buildings on which to undertake self-construction projects, with the ultimate goal of reaching the stipulated objective within the in accordance with the expected time periods and methods.

Compared to a more general framework, the analyzes of this research have identified: (i) aspects, moments, decision nodes which provide the quality and success of an intervention self construction/self renovation (satisfaction the objectives undertaken in the planning stage); (ii) the main risk factors that can hinder the success of this type of initiatives.

In this way it has been possible to build a useful framework for defining strategies for actions that have to be implemented in the planning, design, implementation of similar future initiatives, not only in Italy but also in other European countries.

References

AAVV. "Autocostruzione e autorecupero una pratica sociale della casa". La Nuova Città, 7/2000, Angelo Pontecorboli Editore, Firenze (2000)

Bertoni, M., Cantini, A.: Autocostruzione associate ed assistita in Italia. Progettazione e processo edilizio di un modello di housing sociale, Editrice Dedalo, Roma (2008)

BSHF - Building and Social Housing Foundation. Supporting Self-Help Housing (2011). https://bshf-wpengine.netdna-ssl.com/wp-content/uploads/2011/06/Supporting-Self-Help-Housing-Empty-Homes-Programme.pdf

BSHF - Building and social housing foundation. Self-Help Housing in the North of England (2015). https://www.bshf.org/world-habitat-awards/winners-and-finalists/self-help-housing-in-the-north-of-england/

CECODHAS Housing Europe's Observatory. Housing Europe Review 2012. The nuts and bolts of European social housing systems (2011). http://www.housingeurope.eu/resource-105/the-housing-europe-review-2012

Cittalia - Fondazione Anci Ricerche. Progetto L'abitare sociale. Strategie locali di lotta alla povertà: città a confronto, La povertà e le famiglie (2011). http://www.cittalia.com/images/file/POVERTA_FAMIGLIE.pdf

Czischke D., Berthon, B.: Casa e accessibilità nell'Unione Europea. In: CECODHAS Housing Europe's Observatory, Research briefing, 1(1) (2008). http://www.federcasa.it/news/osservatorio_casa/01_Research_Briefing_UE_Casa_e_accessibilità.pdf

European Commission. Evaluation methods for the european union's external assistance. Evaluation tool, 4 (2006). http://ec.europa.eu/europeaid/sites/devco/files/evaluation-methods-guidance-vol4_en.pdf

European Commission. The urban dimension of EU policies – key features of an EU urban agenda, COM 490 (2014). http://ec.europa.eu/regional_policy/sources/consultation/urb_agenda/pdf/comm_act_urb_agenda_en.pdf

Guarini, M.R.: Costi finanziari ed economici nell'autocostruzione. In: Ferretti, L.V., Mariano, C. (eds.) La città dimenticata una proposta per l'emergenza abitativa, pp. 96–101. Prospettive, Roma (2014)

Guarini, M.R., Battisti, F.: Social housing and redevelopment of building complexes on brownfield sites: the financial sustainability of residential projects for vulnerable social groups. In: Xu, Q., Li, H., Li, Q. (eds.) Sustainable development of industry and economy, 3rd International Conference on Energy, Environment and Sustainable Development (EESD 2013), vol. 869–870, pp. 3–13 (2014a). doi:10.4028/www.scientific.net/AMR.869-870.03. http://www.scientific.net/AMR.869-870.03. WOS: 000339125800001

Guarini, M.R., Battisti, F.: Benchmarking multi-criteria evaluation: a proposed method for the definition of benchmarks in negotiation public-private partnerships. In: Murgante, B., Misra, S., Rocha, A.M.A., Torre, C., Rocha, J.G., Falcão, M.I., Taniar, D., Apduhan, B.O., Gervasi, O. (eds.) Computational Science and Its Applications – ICCSA 2014, Part III, LNCS, vol. 8581, pp. 208–223. Springer, Heidelberg (2014b). doi:10.1007/978-3-319-09150-1_16. http://link.springer.com/chapter/10.1007%2F978-3-319-09150-1_16. WOS: 000349442800016

Housing Europe - European Federation of Public, Cooperative Social Housing. The State of Housing in the EU 2015. A Housing Europe Review (2015). http://www.housingeurope.eu/resource-468/the-state-of-housing-in-the-eu-2015

Malczewski, J.: GIS and Multicriteria Decision Analysis. Wiley, New York (1999)

Morano, P., Tajani, F.: Estimative analysis of a segment of the bare ownership market of residential property. In: Murgante, B., Misra, S., Carlini, M., Torre, C.M., Nguyen, H.-Q., Taniar, D., Apduhan, B.O., Gervasi, O. (eds.) Computational Science and Its Applications – ICCSA 2013, Part IV. LNCS, vol. 7974, pp. 433–443. Springer, Heidelberg (2013). doi:10.1007/978-3-642-39649-6_31. http://link.springer.com/chapter/10.1007/978-3-642-39649-6_31#page-1

Mullins, D., Sacranie, H.: Evaluation of the Empty Homes Community Grants Programme (EHCGP) - Midlands region Baseline Case Studies Report; Housing and Communities Research Group, University of Birmingham (2014). https://bshf-wpengine.netdna-ssl.com/wp-content/uploads/2016/03/EHCGP-Midlands-FIN-APRIL-7-2014.pdf

Nesticò, A., Pipolo, O.: A protocol for sustainable building interventions: financial analysis and environmental effects. Int. J. Bus. Intell. Data Min. **10**(3), 199–212 (2015). doi:10.1504/IJBIDM.2015.071325. http://www.inderscience.com/offer.php?id=71325

Tajani, F., Morano, P.: An evaluation model of the financial feasibility of social housing in urban redevelopment. Property Manage. **2**, 133–151 (2015). doi:10.1108/PM-02-2014-0007

Enhancing an IaaS Ontology Clustering Scheme for Resiliency Support in Hybrid Cloud

Toshihiro Uchibayashi[1], Bernady Apduhan[2(✉)], Kazutoshi Niiho[2],
Takuo Suganuma[1], and Norio Shiratori[3]

[1] Research Institute of Electrical Communication, Cyberscience Center,
Tohoku University, 2-1-1 Katahira, Aoba-ku, Sendai, Japan
uchibayashi@ci.cc.tohoku.ac.jp, suganuma@tohoku.ac.jp
[2] Kyushu Sangyo University, 3-1 Matsukadai 2-Chome,
Higashi-ku, Fukuoka, Japan
bob@is.kyusan-u.ac.jp, kl2jk097@st.kyusan-u.ac.jp
[3] Global Information and Telecommunication Institute, Waseda University,
3-4-1 Okubo, Shinjuku-ku, Tokyo, Japan
norio@shiratori.riec.tohoku.ac.jp

Abstract. When an undesirable situation occurs in a hybrid cloud computing environment, vital issues arise when searching for IaaS cloud services that best-match to the user's requirements. This includes the different descriptions/ naming of IaaS cloud services, i.e., CPU, memory, and others, adapted by different companies making it difficult and ambiguous to select the best-match cloud services. Initially, we considered utilizing ontology technology and typical clustering methods to narrow down the selection process. In this paper, we proposed an improved ontology clustering scheme and describe the methodology. Preliminary experiments shows promising results showing a fair gathering of related elements in the cluster and the speedup of processing depicting a viable resiliency support for hybrid cloud.

Keywords: Hybrid cloud · Ontology · Clustering · IaaS cloud service

1 Introduction

With the advent of cloud computing, the use and management of software or data has changed dramatically. The typical way of costly acquiring and the hassle of constructing one's own server has changed since the availability of "pay as you use" server infrastructure services offered by cloud service providers. These server services and other available infrastructure cloud services (IaaS) are remarkably gaining acceptance in the academic, research and industry communities due to its easy deployment which spur the widespread use of cloud computing services.

As a result, the number of institutions and companies are venturing into the development of different cloud services to meet the increasing demand which have burgeoned in recent years. However, the standardization of cloud services, e.g., cloud service names, has not coped up with the development pace. For example, the auxiliary

© Springer International Publishing Switzerland 2016
O. Gervasi et al. (Eds.): ICCSA 2016, Part IV, LNCS 9789, pp. 219–231, 2016.
DOI: 10.1007/978-3-319-42089-9_16

storage device of a virtual machine maybe called as HDD by a certain provider, while another provider may call it Storage.

Furthermore, ensuring security when using cloud services is a problem issue. It would be easy to link a number of virtual machines into one cloud to build and provide cloud services. However, aggregating these virtual machines into a single cloud is accompanied by risks, such as the occurrence of a natural disaster, system failures, information leakage, etc., that would hamper the provisioning of the services. This situation can be avoided by distributing some services to different public cloud service providers.

However, by distributing the services in order to overcome the risks will on the other hand raise a new difficult problem, i.e., providing continuity of the service when a failure occurs in any one of the member clouds. To this, the system must be able to quickly find an alternative cloud service provider to provide service continuity whenever failures occur.

To solve these problems, we proposed a method using ontology [1] to perform a unified representations of cloud service names to quickly discover and select the same or the closest available cloud service which will assume the continuity of provisioning the service. Ontology is a hierarchical technique of performing a semantic search, as in Semantic Web [2], by describing the meaning of the relationship between hierarchy and concepts.

By unifying the cloud service representations using ontology, it will be possible to discover the cloud service equivalent to the failed service. As the number of cloud service providers offering different cloud services increases, cloud service search and selection processing becomes slow. To address this issue, we proposed to adapt ontology clustering to speed up the search process.

In the following section, we discuss some related works, and introduce the problem issues in using cloud services in Sect. 3. In Sect. 4, we introduce our initial approach and describe the IaaS cloud ontology. Section 5, we describe the ontology similarity computation and conduct experiments to some existing clustering schemes. In Sect. 6, we introduce an enhanced ontology clustering scheme. The verification and performance tests results are describe in Sect. 6.3. Our conclusion and future work are describe in Sect. 7.

2 Related Work

Ontology defines the meaning of an element and is often used in conjunction with RDF [3] which model the object description. RDFS (RDF Schema) is one of the description methods in writing ontology. Therefore, RDF and ontology are very similar in structure.

Yonglin Leng [4] proposed an RDF partitioning algorithm based on hybrid hierarchical clustering (BRDPHHC) which combines Affinity Propagation and K-means clustering algorithm. Their study shows and proved the structure similarity of ontology and RDF. However, they use objects as elements to cluster; whereas in our case, we cluster the ontology.

Furthermore, Soraya Setti Ahmed [5] proposed a revision and enhancement of K-means clustering algorithm based on a new semantic similarity measure for partitioning a given ontology into high quality modules. Their methodology is aimed to cover a broader scope of ontology, whereas in our case we want to focus on cloud ontology in IaaS cloud services.

3 Problem Issues

With the availability of cloud services, it is now easy to use computing resources in which the user could easily increase or reduce the number of resources as the user may deem it necessary. When using public cloud IaaS services, the provider takes care of the service management and so the user can just concentrate on its maintenance. Typically, a large number of users are using the same services in the same Data Center of a public cloud service provider.

When using a public cloud, the user stores the program and data on the cloud service whose management of which will be left to the provider. Since program and data are sources of corporate competitiveness, companies are required to strictly safeguard and protect their secrets.

However, for some reasons, data or services of a user may have the danger of being leak to unwanted entities which may prove detrimental to the user. This may be due to the carelessness of the provider or outside malicious acts exposing the unauthorized use of the program and data. As a countermeasure, it is proposed to distribute the services to multiple public clouds. However, when a disaster occurs and a member cloud is affected, the whole system will fail or in-operational. To this, there is a need to quickly discover a substitute or alternate the same type of cloud service to restore and continue service provisioning.

4 The IaaS Service Discovery

4.1 Initial Approach

In order to maintain the service provisioning from different cloud providers, we proposed a mechanism to discover a service which is equivalent to the failed service, and facilitate the transition.

In the absence of cloud service notation standards, some companies may call it HDD cloud service while another company may call it Storage. In paper [6], we proposed the implementation of a matching algorithm utilizing IaaS cloud service ontology in order to discover the best-fit cloud service. With this mechanism, the best-fit cloud service can be found and allows the system to quickly restore its operation.

4.2 The IaaS Cloud Service Ontology

In [7], they proposed IaaS cloud services based on IaaS cloud ontology. In this IaaS cloud ontology, we added the SLA (System Level Agreement) element [8, 9]. Figure 1 shows the IaaS cloud ontology software, services, and resources.

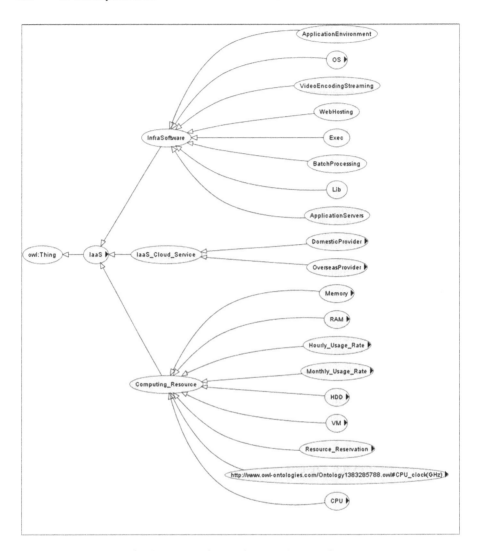

Fig. 1. A part of IaaS cloud services ontology.

The lower layer of IaaS is composed of Infrasoftware, IaaS Cloud Service, and Com-puting Resources. Under Infrasoftware are the Application Environment, OS, Video-EncodingStreaming, WebHosting, Exec, BatchProcessing, Lib, and Application Serv-ers. Whereas, below the IaaS Cloud Service are Domestic Provider and Overseas Provider. Furthermore, below the Computing Resource are the Memory, RAM, Hourly Usage Rate, Monthly Usage Rate, HDD, VM, Resource Reservation, CPU clock (GHz), and CPU.

Figure 2 shows 15 domestic IaaS cloud service providers, namely; FUJITSU, Sakura Internet, ITOCHU Techno-Solutions, at + link, GMO Cloud, KDDI, IDC

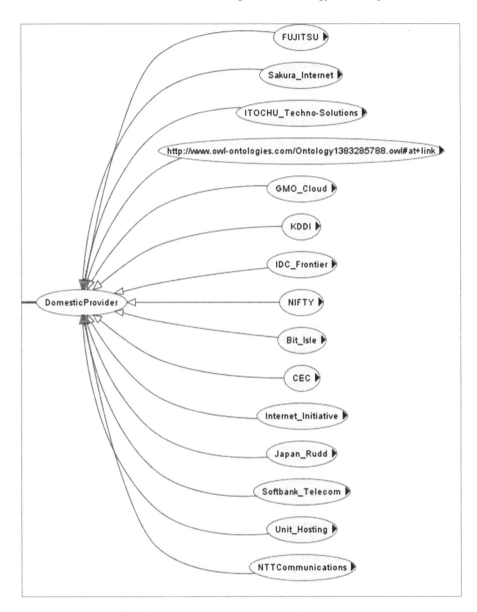

Fig. 2. Domestic cloud service providers elements

Frontier, NIFTY, Bit Isle, CEC, Internet Initiative, Japan rudd, Softbank Tele-com, Unit Hosting, and NTT Communications.

Figure 3 shows the FUJITSU provider and its service FUJITSUCloudInitiative with different types such as Double High, Advanced Type, High Performance Type, Economy Type, Quad High, Standard Type, and FUJITSUCloudSLA.

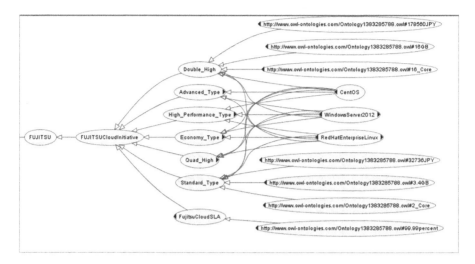

Fig. 3. A part of FUJITSU cloud provider elements

By utilizing ontology, we were able to discover an alternative cloud to continuously provide the service. However, as the amount of ontology data increases, the amount of information to handle also increases resulting in the slowing down of ontology processing. So, to address this issue and speed up the processing, we adapted ontology clustering. In paper [10], we proposed an ontology clustering mechanism to enhance IaaS service discovery and selection. With the proposed method, the ontology can be well divided into clusters. However, as the ontology grows larger the processing cost increase. So, we proposed a new clustering method and compare it with some known clustering methods.

5   Clustering Using Some Existing Methods

To conduct clustering in IaaS cloud ontology, we used some generic clustering schemes; namely, the nearest neighbor method, furthest neighbor method, group average method, centroid method, and Ward's method. We measured the distance between the elements using ontology semantic similarity.

5.1   Similarity Computation

In [11], Andreasen proposed an algorithm to determine the semantic similarity using the depth between the ontology elements. Equation 1 is the equation for obtaining the similarity of elements x and y. Equation sim(x, y) shows the similarity of elements x and y, where $\alpha(x)$ and $\alpha(y)$ is the depth of each element, and $\alpha(x) \cap \alpha(y)$ is the common parent of the depth of $\alpha(x)$ and $\alpha(y)$.

$$\text{sim}(x, y) = \frac{1}{2}\frac{|\alpha(x) \cap \alpha(y)|}{|\alpha(x)|} + \frac{1}{2}\frac{|\alpha(x) \cap \alpha(y)|}{|\alpha(y)|} \tag{1}$$

We calculate Fig. 4 using Eq. 2. When measuring the similarity between D and E, x = D, and y = E. The common parent of x = D and y = E is $\alpha(x) \cap \alpha(y)$ = B". The depth of A is 1, B is 2, C is 3, D is 4, and E is 5. Therefore, similarity D and E becomes 7/12.

$$\text{sim}(D, E) = \frac{1}{2}\frac{|2|}{|4|} + \frac{1}{2}\frac{|2|}{|3|} = \frac{7}{12} \tag{2}$$

Equation 2 implies that the pair of ontology in the cluster having the least similarity value has the shallowest depth of relationships.

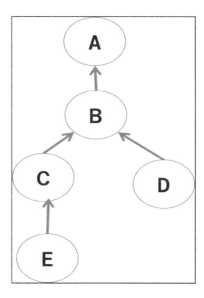

Fig. 4. Elements of ontology

5.2 Clustering Results

Table 1 show the clustering results of IaaS cloud ontology with the number of clusters set at 20. Using Eq. 1, the Ward method has the smallest deviation and elements were more or less equally divided. With the number of clusters set to 20, part of the screenshot of the clustering results using the nearest neighbor method, furthest neighbor method, group average method, centroid method, and Ward's method are shown in Fig. 5. The numbers below the cluster number shows the number of elements that gathered in that particular cluster of a certain clustering algorithm.

Cluster 1(labeled as Group 1) contains 60 elements, i.e., Copper_4, Gold_2,14.2-JPY, SUSE Linux Enterprise Server, such as Silver_1. The cluster 2 (or Group 2) elements includes Sakura_Internet, 96000JPY, 5200JPY, Type_6,3 GB, Type_5,

Type_19, which contains 56 elements, such as Type_18. While cluster 3 (or Group 3) contains Type_S, IDC_Frontier, 90300JPY, 2.4 GHz, Type_L, Type_M2,7.9JPY, 30 pieces of elements such as 84000JPY.

Table 1. Clustering results

Clustering Algorithm	Cluster Number																				
	#1	#2	#3	#4	#5	#6	#7	#8	#9	#10	#11	#12	#13	#14	#15	#16	#17	#18	#19	#20	sum
The Nearest Neighbor Method	628	1	1	1	1	1	1	1	1	1	1	1	1	1	1	1	1	1	1	1	647
Furthest Neighbor Method	59	50	34	47	24	38	32	125	47	32	28	17	14	9	15	20	11	19	22	4	647
Median Method	628	1	1	1	1	1	1	1	1	1	1	1	1	1	1	1	1	1	1	1	647
Centroid Method	628	1	1	1	1	1	1	1	1	1	1	1	1	1	1	1	1	1	1	1	647
Group Average Method	189	79	228	26	10	7	17	17	9	9	15	8	8	5	4	8	4	1	2	1	647
Ward Method	60	56	30	26	38	27	42	20	44	35	43	17	46	30	14	15	44	22	16	22	647

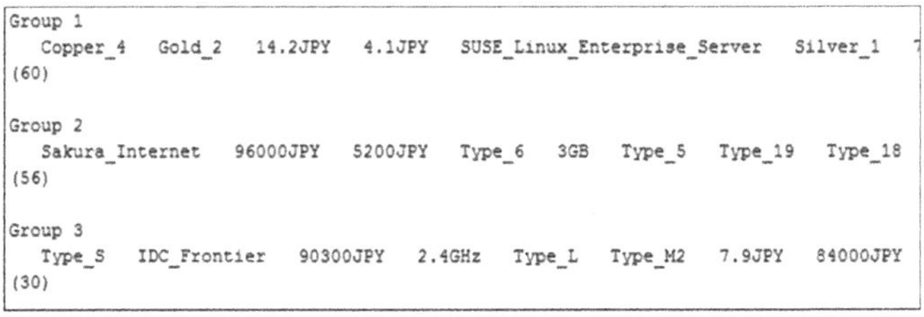

```
Group 1
    Copper_4    Gold_2    14.2JPY    4.1JPY    SUSE_Linux_Enterprise_Server    Silver_1
(60)

Group 2
    Sakura_Internet    96000JPY    5200JPY    Type_6    3GB    Type_5    Type_19    Type_18
(56)

Group 3
    Type_S    IDC_Frontier    90300JPY    2.4GHz    Type_L    Type_M2    7.9JPY    84000JPY
(30)
```

Fig. 5. Clustering results using Ward's method

6   Enhancing the Clustering Scheme

6.1   Problem Issues

In Ward's method, the elements were sparsely distributed to the clusters. The problem is that it exhibits no relationships between the elements. For example, Group3 in Fig. 5, there is no relationship between IDC Frontier and Type_S, Type_L, and Type_M2.

6.2   The Clustering Procedure

At first, all the elements are contained in a single cluster. Next, create an empty cluster. Find the elements set with the lowest degree of similarity and move one element of the set to the empty cluster, but keep the other element in the original cluster. Repeat the same procedures, i.e. finding the succeeding elements set with the lowest degree of similarity among the remaining elements, until the number of elements remaining in the original cluster is half or one greater than the new cluster. Do the same procedures to the two clusters (separately, to the original and new cluster) and to the following ones until the total number of clusters is equal to the specified number of clusters.

6.3 Verification and Performance Test

We conducted experiments to compare the clustering results using Ward's method and our proposed scheme. We also compare the speed in searching the required services using our proposed clustering scheme and without using any clustering method.

6.4 Test Environment

Table 2 shows the computer specifications we used in the experiments.

Table 2. Computer specifications

OS	Windows 10
Processer	Intel (R) Core i (TM)7 4770 CPU@2.40 GHz
Memory	8.0 GB

6.5 Experiment Results

Figure 6 shows part of the screenshot showing results of clustering using the Ward's method. As shown, the elements is biased to one of the cluster (labeled as Group).

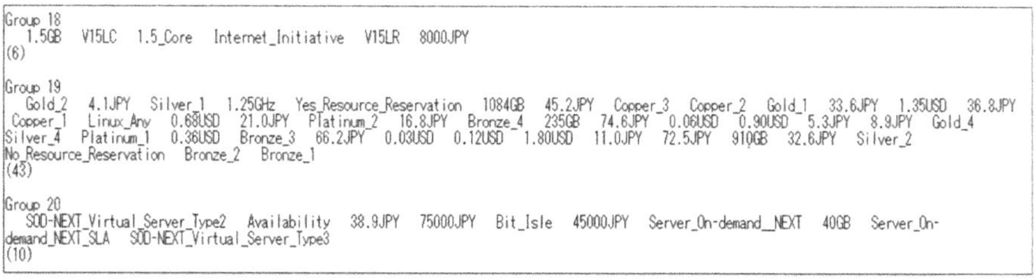

Fig. 6. Part of the results showing Ward's method (Sample 1).

Figure 7 shows part of the screenshot showing the results of clustering using our proposed scheme depicting no bias in all of the clusters.

Figure 8 shows another part of the screenshot showing the results of clustering using the Ward's method. Only few elements gathered in some clusters and shows no relationship between elements.

```
Cluster 18: High_Memory_2XL Silver_4 Gold_3 Platinum_1 Bronze_3 Type_1 Standard 0.5GB 64500JPY 19600JPY
11

Cluster 19: IBM_SLA Medium Type_20 Spec_Type13 Type_12 Type_M IIJ_GIO_Component_Nation_Service 18600JPY 132000JPY 5300JPY
11

Cluster 20: IBMPowerSystemsOS MaxOS BatchProcessing Exec ApplicationServers Option_8 Option_7 5250JPY 290000JPY 25000JPY
11
```

Fig. 7. Part of the results showing our proposed scheme (Sample 1).

```
Group 62
    x86Option    VM
(2)

Group 63
    Type_XS    0.8GHz    3000JPY
(3)

Group 64
    Industria    High_Performance_Medium    Industria_SLA    16GB    17000JPY    Japan_Rudd
(6)
```

Fig. 8. A part of the results using Ward's method (Sample 2).

Figure 9 shows part of the screenshot showing the results of clustering using our proposed scheme depicting the relationships of elements in the cluster.

```
Cluster 31: WindowsServer2003 Android WindowsMe WindowsServer2008 Windows2000 Windows05 Windows98 WindowsXP Windows OS: OS
10
Cluster 32: MobileLinux EmbeddedLinux PalmOS IbmAIX6.1 Linux WindowsCE iPhoneOS Unix Embedded IBMPowerSystemsOS: OS
10
Cluster 33: 60GB 64bit S AmazonEC2SLA 1200GB 788.40USD Ubuntu 209.0JPY 28.4JPY XXX-Large; Hourly_Usage_Rate OS HDD
```

Fig. 9. Part of the results of our proposed scheme (Sample 2).

We conducted further experiments to verify the proposed method with 64 number of clusters. Using Ward's method, the number of elements in the cluster varies, exhibiting a bias. In our proposed method, it shows a very slight variation on the number of elements in the cluster, depicting no bias (Tables 3, 4 and 5).

In order to further evaluate the results of the algorithm, we consider the typical search without clustering as the standard reference and compare it with the search using our clustering scheme. Figure 10 shows the results when clustering was not used, and Fig. 11 shows the same search results with our proposed clustering scheme. These justifies the validity of our algorithm.

Next, we perform comparison on the search speed. The search conditions are: "CPU is 2_Core", "OS is CentOS", "Memory is 512 MB", and "HDD is 50 GB." The graph in Fig. 12 depicts the average search speed conducted 50 times. The average search speed without clustering is 88.69 ms, while with clustering is 47.56 ms. The search speed using our clustering scheme is about 2 times faster.

Table 3. Clustering results with 64 number of clusters (1-of-3)

Clustering Algorithm	Cluster Number																			
	#1	#2	#3	#4	#5	#6	#7	#8	#9	#10	#11	#12	#13	#14	#15	#16	#17	#18	#19	#20
Ward's Method	9	15	4	16	2	10	9	18	14	5	20	32	4	7	43	13	15	8	43	10
Proposed Method	10	10	10	10	10	10	11	10	10	10	10	10	10	11	10	10	10	10	11	10

Table 4. Clustering result with 64 number of clusters (2-of-3)

Clustering Algorithm	Cluster Number																			
	#21	#22	#23	#24	#25	#26	#27	#28	#29	#30	#31	#32	#33	#34	#35	#36	#37	#38	#39	#40
Ward's Method	7	8	9	10	11	15	8	12	25	9	22	8	4	8	10	11	4	10	5	8
Proposed Method	10	10	10	10	10	10	10	10	10	10	10	10	10	10	10	10	10	10	10	11

Table 5. Clustering result with 64 number of clusters (3-of-3)

Clusterin Algorithm	Cluster Number																							
	#41	#42	#43	#44	#45	#46	#47	#48	#49	#50	#51	#52	#53	#54	#55	#56	#57	#58	#59	#60	#61	#62	#63	#64
Ward's Method	12	4	4	19	12	3	22	5	5	4	5	4	4	3	4	5	4	12	5	3	4	2	3	6
Proposed Method	10	10	10	10	10	10	10	11	10	10	10	10	10	10	10	11	10	10	10	10	10	10	10	11

\<Ponit>	provider	Service
3	CEC	Option_6
3	CEC	Option_5
3	CEC	Option_4
3	CEC	Option_1
3	ITOCHU_TechnoSolutions	Specif
2	Microsoft	M
2	GOGRID	Small
2	GOGRID	Medium
2	at+link	at+link_Case4
2	at+link	at+link_Case3

Fig. 10. Results of search without using clustering (Top10)

<Ponit>	provider	Service
3	CEC	Option_6
3	CEC	Option_5
3	CEC	Option_4
3	CEC	Option_1
3	ITOCHU_TechnoSolutions	Specif
2	Microsoft	M
2	GOGRID	Small
2	GOGRID	Medium
2	at+link	at+link_Case4
2	at+link	at+link_Case3

Fig. 11. Results of search using clustering (Top10)

Compared to the above generic clustering approaches, our proposed scheme makes relevant elements to gather in the same cluster. However, a few unrelated elements were detected in some clusters. This can be considered as one limitation with hard clustering wherein an element can only be a member in one specific cluster. In order to include the relevant elements in a cluster, soft clustering can be used wherein an element can be a member in one or more clusters.

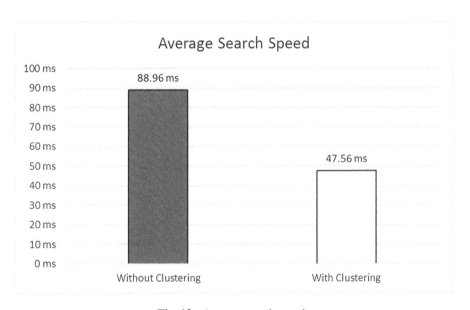

Fig. 12. Average search speeds

7 Conclusion and Future Work

In our present study, we proposed an ontology clustering scheme for high-speed search and to discover the best-fit IaaS cloud service substitute based on user's requirements. Our scheme differs from some existing clustering methods in the sense that it gathers the elements of the cluster considering the relationship between elements. Furthermore, with our proposed method, the clustering of elements forms a bias-free cluster, or the elements are fairly distributed to the preset number of clusters.

As this study is still at preliminary stage, we plan to carry out more validation and verification experiments to our proposed algorithm and conduct performance evaluation.

Acknowledgment. This research was supported in part by the Japan Society for the Promotion of Science Grants-in-Aid for Scientific Research 24500100.

References

1. W3C: OWL Web Ontology Language Overview. http://www.w3.org/TR/oel-features/
2. Berners-Lee, T., Hendler, J., Lassila, O.: The semantic web. Sci. Am. **284**(5), 34–43 (2001)
3. RDF Scheme (online). http://www.w3.org/TR/rdf-schema/
4. Leng, Y., Chen, Z., Zhong, F., Zhong, H.: BRDPHHC: a balance RDF data partitioning algorithm based on hybrid hierarchical clustering. In: The 2015 IEEE 17th International Conference on High Performance Computing and Communications, 2015 IEEE 7th International Symposium on Cyberspace Safety and Security, and 2015 IEEE 12th International Conference on Embedded Software and Systems (Multi-Conference), pp. 1755–1760 (2015)
5. Ahmed, S.S., Malki, M., Benslimane, S.M.: Ontology partitioning: Clustering based approach. Int. J. Inf. Technol. Comput. Sci. **7**(6), 1–11 (2015)
6. Uchibayashi, T., Apduhan, B.O., Shiratori, N.: Towards a resilient hybrid IaaS cloud with ontology and agents. In: The 14th International Conference on Computational Science and Its Applications (ICCSA), The 2014 International Conference on Computational Science and Its Applications (ICCSA 2014), pp. 70–73 (2014)
7. Han, T., Mong Sim, K.: An ontology-enhanced cloud service discovery system. In: International Multi-Conference of Engineers and Computer Scientists (IMEC 2010), Hong Kong, pp. 644–649 (2010)
8. Casalicchio, E., Silvestri, L.: Mechanisms for SLA provisioning in cloud-based service providers. Comput. Netw. **57**, 795–810 (2013)
9. Kona, S., Bansal, A., Blake, M.B., Bleul, S., Weise, T.: WSC-2009: a quality of service-oriented web services challenge. In: IEEE Conference on Commerce and Enterprise Computing, 2009. CEC 2009, pp. 487–490 (2009)
10. Uchibayashi, T., Apduhan, B., Shiratori, N.: Towards a cloud ontology clustering mechanism to enhance IaaS service discovery and selection. In: Gervasi, O., Murgante, B., Misra, S., Gavrilova, M.L., Rocha, A.M.A.C., Torre, C., Taniar, D., Apduhan, B.O. (eds.) ICCSA 2015. LNCS, vol. 9155, pp. 545–556. Springer, Heidelberg (2015)
11. Andreasen, T., Bulskov, H., Knappe, R.: From ontology over similarity to query evaluation. In: Bernardi, R., Moortgat, M. (eds.) The 2nd CoLogNET-ElsNET Symposium—Questions and Answers: Theoretical and Applied Perspectives, pp. 39–50 (2003)

A Simple Stochastic Gradient Variational Bayes for Latent Dirichlet Allocation

Tomonari Masada[1(✉)] and Atsuhiro Takasu[2]

[1] Nagasaki University, 1-14 Bunkyo-machi, Nagasaki, Japan
masada@nagasaki-u.ac.jp
[2] National Institute of Informatics, 2-1-2 Hitotsubashi, Chiyoda-ku, Tokyo, Japan
takasu@nii.ac.jp

Abstract. This paper proposes a new inference for the latent Dirichlet allocation (LDA) [4]. Our proposal is an instance of the stochastic gradient variational Bayes (SGVB) [9,13]. SGVB is a general framework for devising posterior inferences for Bayesian probabilistic models. Our aim is to show the effectiveness of SGVB by presenting an example of SGVB-type inference for LDA, the best-known Bayesian model in text mining. The inference proposed in this paper is easy to implement from scratch. A special feature of the proposed inference is that the logistic normal distribution is used to approximate the true posterior. This is counterintuitive, because we obtain the Dirichlet distribution by taking the functional derivative when we lower bound the log evidence of LDA after applying a mean field approximation. However, our experiment showed that the proposed inference gave a better predictive performance in terms of test set perplexity than the inference using the Dirichlet distribution for posterior approximation. While the logistic normal is more complicated than the Dirichlet, SGVB makes the manipulation of the expectations with respect to the posterior relatively easy. The proposed inference was better even than the collapsed Gibbs sampling [6] for not all but many settings consulted in our experiment. It must be worthwhile future work to devise a new inference based on SGVB also for other Bayesian models.

Keywords: Text mining · Topic models · variational Bayesian inference

1 Introduction

When we use Bayesian probabilistic models for data mining applications, we need to infer the posterior distribution. While the Markov Chain Monte Carlo (MCMC) is undoubtedly important [5,14], this paper focuses on the variational Bayesian inference (VB). Therefore, we first present an outline of VB.

Let $\boldsymbol{x}$ be a set of the random variables whose values are observed. A probabilistic model for analyzing the observed data $\boldsymbol{x}$ can be specified unambiguously by its full joint distribution $p(\boldsymbol{x}, \boldsymbol{z}, \boldsymbol{\Theta})$, where $\boldsymbol{z}$ denotes the discrete latent variables and $\boldsymbol{\Theta}$ the continuous ones. In VB, we maximize the log of the evidence $p(\boldsymbol{x})$, which is obtained from the full joint distribution by marginalizing $\boldsymbol{z}$ and

© Springer International Publishing Switzerland 2016
O. Gervasi et al. (Eds.): ICCSA 2016, Part IV, LNCS 9789, pp. 232–245, 2016.
DOI: 10.1007/978-3-319-42089-9_17

Θ out, i.e., $p(\boldsymbol{x}) = \int \sum_{\boldsymbol{z}} p(\boldsymbol{x}, \boldsymbol{z}, \Theta) d\Theta$. However, the maximization of $\log p(\boldsymbol{x})$ is generally intractable. Therefore, we instead maximize its lower bound:

$$\log p(\boldsymbol{x}) = \log \int \sum_{\boldsymbol{z}} q(\boldsymbol{z}, \Theta) \frac{p(\boldsymbol{x}, \boldsymbol{z}, \Theta)}{q(\boldsymbol{z}, \Theta)} d\Theta \geq \int \sum_{\boldsymbol{z}} q(\boldsymbol{z}, \Theta) \log \frac{p(\boldsymbol{x}, \boldsymbol{z}, \Theta)}{q(\boldsymbol{z}, \Theta)} d\Theta. \quad (1)$$

We have introduced an approximate posterior $q(\boldsymbol{z}, \Theta)$ in Eq. (1) to apply Jensen's inequality. If we put the true posterior $p(\boldsymbol{z}, \Theta|\boldsymbol{x})$ in place of the approximate posterior, Jensen's inequality holds with equality. However, the true posterior is typically intractable. Therefore, in VB, the inference of the approximate posterior is the main task.

In this paper, we consider the latent Dirichlet allocation (LDA) [4], the best-known Bayesian model in text mining, as our target. LDA and its extensions have been applied to solve a wide variety of text mining problems [7,10,12,15,17]. It is known that the performance of LDA measured in terms of test set perplexity heavily depends on how the inference is conducted [2]. Therefore, to provide a new proposal relating to the posterior inference for LDA is highly relevant to text mining research.

The main contribution of this paper is to propose a new VB-type inference for LDA. The proposed inference was better than the VB presented in [4] in terms of test set perplexity in all situations consulted by our experiment. For brevity, we call the VB presented in [4] standard.[1]

In the standard VB, the true posterior distribution is approximated by the Dirichlet distribution. This is because the Dirichlet is obtained analytically by taking the functional derivative after applying a mean field approximation. It has been shown experimentally that the standard VB works as well as other inference methods [2]. In our proposed inference, we apply the same mean field approximation. However, we do not use the Dirichlet for approximating the true posterior. Nevertheless, the proposed inference could achieve a better perplexity than the standard VB in our experiment. Interestingly, our method was better even than the collapsed Gibbs sampling (CGS) [6] in not all but many situations.

The proposed inference for LDA is based on the stochastic gradient variational Bayes (SGVB), which has been proposed by the two papers [9,13] almost simultaneously. SGVB can be regarded as a general framework for obtaining VB-type inferences for a wide range of Bayesian probabilistic models. Precisely, SGVB provides a general framework for obtaining a Monte Carlo estimate of the log-evidence lower bound, i.e., the lower bound of $\log p(\boldsymbol{x})$ in Eq. (1). In this paper, we utilize SGVB to devise an inference easy to implement for LDA. We use the logistic normal distribution [1] for approximating the true posterior. While the logistic normal is more complicated than the Dirichlet, we can obtain a simple VB-type inference owing to SGVB.

---

[1] Precisely speaking, the VB presented in [4] performs a point estimation for the per-topic word multinomial distributions. In the VB we call standard here, a Bayesian inference is performed also for the per-topic word multinomial distributions, not only for the per-document topic multinomial distributions.

In the next section, we describe SGVB based on [9], which gives an explanation easy to understand for those familiar with LDA. Our description does not cover the full generality of SGVB, partly because we focus only on LDA as our target. We then provide the details of our proposal in Sect. 3. Section 4 presents the results of an evaluation experiment, where we compared our proposal with other methods including the standard VB and CGS. Section 5 concludes the paper with discussion on worthwhile future work.

2 Stochastic Gradient Variational Bayes

The log-evidence lower bound in Eq. (1) can be rewritten as follows:

$$\mathcal{L}(\boldsymbol{\Lambda}) = \mathbb{E}_{q(\boldsymbol{z},\boldsymbol{\Theta}|\boldsymbol{\Lambda})}[\log p(\boldsymbol{x},\boldsymbol{z},\boldsymbol{\Theta})] - \mathbb{E}_{q(\boldsymbol{z},\boldsymbol{\Theta}|\boldsymbol{\Lambda})}[\log q(\boldsymbol{z},\boldsymbol{\Theta}|\boldsymbol{\Lambda})], \qquad (2)$$

where $\boldsymbol{\Lambda}$ denotes the parameters of the approximate posterior $q(\boldsymbol{z},\boldsymbol{\Theta}|\boldsymbol{\Lambda})$, and $\mathbb{E}_{q(\boldsymbol{z},\boldsymbol{\Theta}|\boldsymbol{\Lambda})}[\cdot]$ denotes the expectation with respect to $q(\boldsymbol{z},\boldsymbol{\Theta}|\boldsymbol{\Lambda})$. We assume that $q(\boldsymbol{z},\boldsymbol{\Theta}|\boldsymbol{\Lambda})$ factorizes as $q(\boldsymbol{z}|\boldsymbol{\Lambda_z})q(\boldsymbol{\Theta}|\boldsymbol{\Lambda_\Theta})$. Then we can write $\mathcal{L}(\boldsymbol{\Lambda})$ as

$$\begin{aligned}\mathcal{L}(\boldsymbol{\Lambda}) =& \mathbb{E}_{q(\boldsymbol{z},\boldsymbol{\Theta}|\boldsymbol{\Lambda})}[\log p(\boldsymbol{x},\boldsymbol{z},\boldsymbol{\Theta})] \\ &- \mathbb{E}_{q(\boldsymbol{z}|\boldsymbol{\Lambda_z})}[\log q(\boldsymbol{z}|\boldsymbol{\Lambda_z})] - \mathbb{E}_{q(\boldsymbol{\Theta}|\boldsymbol{\Lambda_\Theta})}[\log q(\boldsymbol{\Theta}|\boldsymbol{\Lambda_\Theta})].\end{aligned} \qquad (3)$$

Our task is to estimate the expectations on the right hand side of Eq. (3).

In this paper, we estimate the log-evidence lower bound of the latent Dirichlet allocation (LDA) by using the stochastic gradient variational Bayes (SGVB) [9,13]. SGVB is a general framework for obtaining a Monte Carlo estimate of the log-evidence lower bound for a wide variety of Bayesian probabilistic models. Note that SGVB cannot provide an estimate of the expectation with respect to the distribution for the discrete random variables [11]. However, we can perform an estimation as in the standard VB for LDA [4] with resect to $\boldsymbol{z}$.

In SGVB, we can assume that the approximate posterior $q(\boldsymbol{\Theta}|\boldsymbol{\Lambda_\Theta})$ depends on the observed data $\boldsymbol{x}$. We do not consider this option here and thus do not explore the full generality of SGVB. However, by making the approximate posterior not dependent on $\boldsymbol{x}$, we can make the proposed inference simple.

When we apply SGVB, the approximate posterior should meet at least two requirements. SGVB estimates the expectations for the continuous variables $\boldsymbol{\Theta}$ by the Monte Carlo method. Therefore, the approximate posterior $q(\boldsymbol{\Theta}|\boldsymbol{\Lambda_\Theta})$ should be a distribution from which we can draw samples. This is the first requirement. Let $\boldsymbol{\Theta}^{(l)}$, $l = 1,\ldots,L$ be the samples drawn from $q(\boldsymbol{\Theta}|\boldsymbol{\Lambda_\Theta})$. Then $\mathcal{L}(\boldsymbol{\Lambda})$ in Eq. (3) is estimated as

$$\begin{aligned}\hat{\mathcal{L}}(\boldsymbol{\Lambda}) =& \frac{1}{L}\sum_{l=1}^{L}\Big\{\mathbb{E}_{q(\boldsymbol{z}|\boldsymbol{\Lambda_z})}[\log p(\boldsymbol{x},\boldsymbol{z},\boldsymbol{\Theta}^{(l)})] - \log q(\boldsymbol{\Theta}^{(l)}|\boldsymbol{\Lambda_\Theta})\Big\} \\ &- \mathbb{E}_{q(\boldsymbol{z}|\boldsymbol{\Lambda_z})}[\log q(\boldsymbol{z}|\boldsymbol{\Lambda_z})].\end{aligned} \qquad (4)$$

In SGVB, we maximize $\hat{\mathcal{L}}(\boldsymbol{\Lambda})$ in Eq. (4) in place of $\mathcal{L}(\boldsymbol{\Lambda})$ in Eq. (3). To maximize $\hat{\mathcal{L}}(\boldsymbol{\Lambda})$, we need to obtain its derivatives with respect to the relevant variables.

Therefore, $q(\boldsymbol{\Theta}|\boldsymbol{\Lambda_\Theta})$ should be a distribution that makes $\hat{\mathcal{L}}(\boldsymbol{\Lambda})$ differentiable. This is the second requirement. As described below, the inference proposed in this paper for LDA can be regarded as an example of SGVB.

3 Our Proposal

3.1 Lower Bound Estimation

We first describe LDA. Let D, K, and V denote the numbers of documents, latent topics, and vocabulary words, respectively. The parameters of the per-document topic multinomial distributions and the parameters of the per-topic word multinomial distributions are represented as $\boldsymbol{\theta}_d = (\theta_{d1}, \ldots, \theta_{dK})$ for $d = 1, \ldots, D$ and $\boldsymbol{\phi}_k = (\phi_{k1}, \ldots, \phi_{kV})$ for $k = 1, \ldots, K$, respectively. Then the full joint distribution of LDA is written as follows:

$$p(\boldsymbol{x}, \boldsymbol{z}, \boldsymbol{\theta}, \boldsymbol{\phi}|\alpha, \beta) = \prod_{d=1}^{D} p(\boldsymbol{x}_d|\boldsymbol{z}_d, \boldsymbol{\phi})p(\boldsymbol{z}_d|\boldsymbol{\theta}_d) \cdot \prod_{d=1}^{D} p(\boldsymbol{\theta}_d|\alpha) \cdot \prod_{k=1}^{K} p(\boldsymbol{\phi}_k|\beta)$$

$$= \prod_{d=1}^{D} \prod_{i=1}^{N_d} \phi_{z_{di} x_{di}} \theta_{dz_{di}} \cdot \prod_{d=1}^{D} \frac{\Gamma(K\alpha)}{\Gamma(\alpha)^K} \prod_{k=1}^{K} \theta_{dk}^{\alpha-1} \cdot \prod_{k=1}^{K} \frac{\Gamma(V\beta)}{\Gamma(\beta)^V} \prod_{v=1}^{V} \phi_{kv}^{\beta-1}, \qquad (5)$$

where N_d is the length of the dth document. x_{di} is an observed variable whose value is the vocabulary word appearing as the ith token of the dth document. z_{di} is a latent variable whose value is the topic to which the ith word token of the dth document is assigned. The notation $\phi_{z_{di} x_{di}}$ is equivalent to ϕ_{kv} when $x_{di} = v$ and $z_{di} = k$. α and β are the hyperparameters of the symmetric Dirichlet priors for $\boldsymbol{\theta}_d$ and $\boldsymbol{\phi}_k$, respectively.

We propose a new inference method for LDA based on SGVB explained in Sect. 2. However, SGVB is applicable only to the continuous latent variables. Therefore, in LDA, SGVB works only for the $\boldsymbol{\theta}_d$s and the $\boldsymbol{\phi}_k$s.

With the mean field approximation $q(\boldsymbol{z}, \boldsymbol{\theta}, \boldsymbol{\phi}) \approx \prod_{d=1}^{D} \prod_{i=1}^{N_d} q(z_{di}) \cdot \prod_{d=1}^{D} q(\boldsymbol{\theta}_d) \cdot \prod_{k=1}^{K} q(\boldsymbol{\phi}_k)$, the lower bound of $\log p(\boldsymbol{x})$ (cf. Eq. (3)) is obtained as follows:

$$\mathcal{L}(\boldsymbol{\Lambda})$$

$$= \sum_{d=1}^{D} \sum_{i=1}^{N_d} \mathbb{E}_{q(z_{di})q(\boldsymbol{\phi}_{z_{di}})}\big[\log p(x_{di}|z_{di}, \boldsymbol{\phi}_{z_{di}})\big] + \sum_{d=1}^{D} \sum_{i=1}^{N_d} \mathbb{E}_{q(z_{di})q(\boldsymbol{\theta}_d)}\big[\log p(z_{di}|\boldsymbol{\theta}_d)\big]$$

$$+ \sum_{d=1}^{D} \mathbb{E}_{q(\boldsymbol{\theta}_d)}\big[\log p(\boldsymbol{\theta}_d|\alpha)\big] + \sum_{k=1}^{K} \mathbb{E}_{q(\boldsymbol{\phi}_k)}\big[\log p(\boldsymbol{\phi}_k|\beta)\big]$$

$$- \sum_{d=1}^{D} \mathbb{E}_{q(\boldsymbol{\theta}_d)}\big[\log q(\boldsymbol{\theta}_d)\big] - \sum_{k=1}^{K} \mathbb{E}_{q(\boldsymbol{\phi}_k)}\big[\log q(\boldsymbol{\phi}_k)\big] - \sum_{d=1}^{D} \sum_{i=1}^{N_d} \mathbb{E}_{q(z_{di})}\big[\log q(z_{di})\big]. \quad (6)$$

In the standard VB [4], we obtain the approximate posteriors by a functional derivative method after using the mean field approximation given above. The result is that the posterior $q(\boldsymbol{\theta}_d)$ for each d and the posterior $q(\boldsymbol{\phi}_k)$ for each k are a Dirichlet distribution. However, it is one thing that approximate posteriors can be found analytically by a functional derivative method, and it is a different thing that such approximate posteriors lead to a good evaluation result in terms of test set perplexity. Therefore, we can choose a distribution other than the Dirichlet for approximating the true posterior.

In this paper, we propose to use the logistic normal distribution [1] for approximating the true posterior. We define θ_{dk} and ϕ_{kv} with the samples $\epsilon_{\theta,dk}$ and $\epsilon_{\phi,kv}$ from the standard normal distribution $\mathcal{N}(0,1)$ as follows:

$$\theta_{dk} \equiv \frac{\exp(\epsilon_{\theta,dk}\sigma_{\theta,dk} + \mu_{\theta,dk})}{\sum_{k'=1}^{K} \exp(\epsilon_{\theta,dk'}\sigma_{\theta,dk'} + \mu_{\theta,dk'})} \text{ and}$$

$$\phi_{kv} \equiv \frac{\exp(\epsilon_{\phi,kv}\sigma_{\phi,kv} + \mu_{\phi,kv})}{\sum_{v'=1}^{V} \exp(\epsilon_{\phi,kv'}\sigma_{\phi,kv'} + \mu_{\phi,kv'})}. \tag{7}$$

Note that $\epsilon\sigma + \mu \sim \mathcal{N}(\mu,\sigma)$ when $\epsilon \sim \mathcal{N}(0,1)$. μ and σ in Eq. (7) are the mean and standard deviation parameters of the logistic normal. Equation (7) gives the reparameterization trick [9] in our case, where we assume that the covariance matrix of the logistic normal is diagonal to make the inference simple.

We can draw $\boldsymbol{\theta}_d \sim \text{LogitNorm}(\boldsymbol{\mu}_{\theta,d}, \boldsymbol{\sigma}_{\theta,d})$ and $\boldsymbol{\phi}_k \sim \text{LogitNorm}(\boldsymbol{\mu}_{\phi,k}, \boldsymbol{\sigma}_{\phi,k})$ efficiently based on Eq. (7). Therefore, the first requirement given in Sect. 2 is met. For the approximate posterior $q(\boldsymbol{z})$, we assume as in the standard VB that we have a different discrete distribution $\text{Discrete}(\boldsymbol{\gamma}_{di})$ for each word token x_{di}, where γ_{dik} is the probability that $z_{di} = k$ holds, i.e., the probability that the ith token of the dth document is assigned to the kth topic.

However, the algebraic manipulation of the expectation with respect to the logistic normal distribution is highly complicated. Here SGVB has an advantage, because it estimates the expectations with respect to approximate posteriors by the Monte Carlo method. $\mathcal{L}(\boldsymbol{\Lambda})$ in Eq. (6) is estimated with L samples $\boldsymbol{\theta}_d^{(l)} \sim \text{LogitNorm}(\boldsymbol{\mu}_{\theta,d}, \boldsymbol{\sigma}_{\theta,d})$ and $\boldsymbol{\phi}_k^{(l)} \sim \text{LogitNorm}(\boldsymbol{\mu}_{\phi,k}, \boldsymbol{\sigma}_{\phi,k})$ for $l = 1,\dots,L$ as

$$\hat{\mathcal{L}}(\boldsymbol{\Lambda}) = \frac{1}{L}\sum_{l=1}^{L}\sum_{d=1}^{D}\sum_{i=1}^{N_d} \mathbb{E}_{q(z_{di})}\big[\log p(x_{di}|z_{di}, \boldsymbol{\phi}_{z_{di}}^{(l)})\big]$$

$$+ \frac{1}{L}\sum_{l=1}^{L}\sum_{d=1}^{D}\sum_{i=1}^{N_d} \mathbb{E}_{q(z_{di})}\big[\log p(z_{di}|\boldsymbol{\theta}_d^{(l)})\big] - \sum_{d=1}^{D}\sum_{i=1}^{N_d} \mathbb{E}_{q(z_{di})}\big[\log q(z_{di})\big]$$

$$+ \frac{1}{L}\sum_{l=1}^{L}\sum_{d=1}^{D} \mathbb{E}_{q(\boldsymbol{\theta}_d)}\big[\log p(\boldsymbol{\theta}_d^{(l)}|\alpha) - \log q(\boldsymbol{\theta}_d^{(l)})\big]$$

$$+ \frac{1}{L}\sum_{l=1}^{L}\sum_{k=1}^{K} \mathbb{E}_{q(\boldsymbol{\phi}_k)}\big[\log p(\boldsymbol{\phi}_k^{(l)}|\beta) - \log q(\boldsymbol{\phi}_k^{(l)})\big]. \tag{8}$$

We maximize the above estimate, denoted as $\hat{\mathcal{L}}(\boldsymbol{\Lambda})$, in place of $\mathcal{L}(\boldsymbol{\Lambda})$ in Eq. (6).

Due to the limit of space, we only discuss the first two expectation terms of the right hand side of Eq. (8). The first term can be rewritten as follows:

$$\frac{1}{L}\sum_{l=1}^{L}\sum_{d=1}^{D}\sum_{i=1}^{N_d}\mathbb{E}_{q(z_{di})}\left[\log p(x_{di}|z_{di},\phi_{z_{di}}^{(l)})\right] = \frac{1}{L}\sum_{l=1}^{L}\sum_{d=1}^{D}\sum_{i=1}^{N_d}\sum_{k=1}^{K}\gamma_{dik}\log\phi_{kx_{di}}^{(l)}$$

$$= \frac{1}{L}\sum_{l=1}^{L}\sum_{d=1}^{D}\sum_{i=1}^{N_d}\sum_{k=1}^{K}\gamma_{dik}\log\left\{\frac{\exp(\epsilon_{\phi,kx_{di}}^{(l)}\sigma_{\phi,kx_{di}}+\mu_{\phi,kx_{di}})}{\sum_{v'=1}^{V}\exp(\epsilon_{\phi,kv'}^{(l)}\sigma_{\phi,kv'}+\mu_{\phi,kv'})}\right\}\quad(\text{cf. Eq. 7})$$

$$= \frac{1}{L}\sum_{l=1}^{L}\sum_{d=1}^{D}\sum_{i=1}^{N_d}\sum_{k=1}^{K}\gamma_{dik}(\epsilon_{\phi,kx_{di}}^{(l)}\sigma_{\phi,kx_{di}}+\mu_{\phi,kx_{di}})$$

$$-\frac{1}{L}\sum_{l=1}^{L}\sum_{d=1}^{D}\sum_{i=1}^{N_d}\sum_{k=1}^{K}\gamma_{dik}\log\left\{\sum_{v=1}^{V}\exp\left(\epsilon_{\phi,kv}^{(l)}\sigma_{\phi,kv}+\mu_{\phi,kv}\right)\right\},\tag{9}$$

where we use the definition of ϕ_{kv} in Eq. (7). The summation term on the last line in Eq. (9) can be upper bounded by using the Taylor expansion [3]:

$$\log\sum_{v}\exp(\epsilon_{\phi,kv}^{(l)}\sigma_{\phi,kv}+\mu_{\phi,kv}) \leq \frac{\sum_{v}\exp(\epsilon_{\phi,kv}^{(l)}\sigma_{\phi,kv}+\mu_{\phi,kv})}{\eta_{\phi,k}^{(l)}}-1+\log\eta_{\phi,k}^{(l)},\tag{10}$$

where we have introduced a new variational parameter $\eta_{\phi,k}^{(l)}$. Consequently, we can lower bound Eq. (9) as follows:

$$\frac{1}{L}\sum_{l=1}^{L}\mathbb{E}_{q(z|\gamma)}\left[\log p(x|z,\phi^{(l)})\right] \geq \frac{1}{L}\sum_{l=1}^{L}\sum_{d=1}^{D}\sum_{i=1}^{N_d}\sum_{k=1}^{K}\gamma_{dik}(\epsilon_{\phi,kx_{di}}^{(l)}\sigma_{\phi,kx_{di}}+\mu_{\phi,kx_{di}})$$

$$-\frac{1}{L}\sum_{l=1}^{L}\sum_{d=1}^{D}\sum_{i=1}^{N_d}\sum_{k=1}^{K}\gamma_{dik}\left\{\frac{\sum_{v}\exp(\epsilon_{\phi,kv}^{(l)}\sigma_{\phi,kv}+\mu_{\phi,kv})}{\eta_{\phi,k}^{(l)}}+\log\eta_{\phi,k}^{(l)}-1\right\}.\tag{11}$$

Let us define $N_{kv} \equiv \sum_{d=1}^{D}\sum_{i=1}^{N_d}\gamma_{dik}\delta(x_{di}=v)$, where $\delta(\cdot)$ is an indicator function that evaluates to 1 if the condition in parentheses holds and to 0 otherwise. N_{kv} means how many tokens of the vth vocabulary word are assigned to the kth topic in expectation. Further, we define $N_k \equiv \sum_{v=1}^{V}N_{kv}$. Then Eq. (11) can be presented more neatly:

$$\frac{1}{L}\sum_{l=1}^{L}\sum_{d=1}^{D}\sum_{i=1}^{N_d}\mathbb{E}_{q(z_{di})}\left[\log p(x_{di}|z_{di},\phi_{z_{di}}^{(l)})\right] \geq \sum_{k=1}^{K}\sum_{v=1}^{V}N_{kv}(\sigma_{\phi,kv}\bar{\epsilon}_{\phi,kv}+\mu_{\phi,kv})$$

$$-\frac{1}{L}\sum_{l=1}^{L}\sum_{k=1}^{K}N_k\left\{\frac{\sum_{v}\exp(\epsilon_{\phi,kv}^{(l)}\sigma_{\phi,kv}+\mu_{\phi,kv})}{\eta_{\phi,k}^{(l)}}+\log\eta_{\phi,k}^{(l)}-1\right\},\tag{12}$$

where we define $\bar{\epsilon}_{\phi,kv} \equiv \frac{1}{L}\sum_{l=1}^{L}\epsilon_{\phi,kv}^{(l)}$.

In a similar manner, we can lower bound the second expectation term of the right hand side in Eq. (8) as follows:

$$\frac{1}{L}\sum_{l=1}^{L}\sum_{d=1}^{D}\sum_{i=1}^{N_d}\mathbb{E}_{q(z_{di})}\big[\log p(z_{di}|\boldsymbol{\theta}_d^{(l)})\big] \geq \sum_{d=1}^{D}\sum_{k=1}^{K}N_{dk}(\sigma_{\theta,dk}\bar{\epsilon}_{\theta,dk}+\mu_{\theta,dk})$$

$$-\frac{1}{L}\sum_{l=1}^{L}\sum_{d=1}^{D}N_d\bigg\{\frac{\sum_k\exp(\epsilon_{\theta,dk}^{(l)}\sigma_{\theta,dk}+\mu_{\theta,dk})}{\eta_{\theta,d}^{(l)}}+\log\eta_{\theta,d}^{(l)}-1\bigg\}, \qquad (13)$$

where $\eta_{\theta,d}$ is a new parameter introduced in a manner similar to Eq. (10). N_{dk} is defined as $\sum_{i=1}^{N_d}\gamma_{dik}$. N_{dk} means how many word tokens of the dth document are assigned to the kth topic in expectation.

We skip the explanation for other expectation terms in Eq. (8) and only show the final result. We can lower bound $\hat{\mathcal{L}}(\boldsymbol{\Lambda})$ as follows:

$$\hat{\mathcal{L}}(\boldsymbol{\Lambda}) \geq \sum_{d,k}(N_{dk}+\alpha)(\sigma_{\theta,dk}\bar{\epsilon}_{\theta,dk}+\mu_{\theta,dk}) + \sum_{k,v}(N_{kv}+\beta)(\sigma_{\phi,kv}\bar{\epsilon}_{\phi,kv}+\mu_{\phi,kv})$$

$$-\frac{1}{L}\sum_{l=1}^{L}\sum_{d=1}^{D}(N_d+K\alpha)\bigg\{\frac{\sum_k\exp(\epsilon_{\theta,dk}^{(l)}\sigma_{\theta,dk}+\mu_{\theta,dk})}{\eta_{\theta,d}^{(l)}}+\log\eta_{\theta,d}^{(l)}-1\bigg\}$$

$$-\frac{1}{L}\sum_{l=1}^{L}\sum_{k=1}^{K}(N_k+V\beta)\bigg\{\frac{\sum_v\exp(\epsilon_{\phi,kv}^{(l)}\sigma_{\phi,kv}+\mu_{\phi,kv})}{\eta_{\phi,k}^{(l)}}+\log\eta_{\phi,k}^{(l)}-1\bigg\}$$

$$+\sum_{k=1}^{K}\sum_{v=1}^{V}\log\sigma_{\phi,kv} + \sum_{d=1}^{D}\sum_{k=1}^{K}\log\sigma_{\theta,dk} - \sum_{d=1}^{D}\sum_{i=1}^{N_d}\sum_{k=1}^{K}\gamma_{dik}\log\gamma_{dik}$$

$$+D\log\Gamma(K\alpha)-DK\log\Gamma(\alpha)+K\log\Gamma(V\beta)-KV\log\Gamma(\beta), \qquad (14)$$

where the constant term is omitted. We refer to the right hand side of Eq. (14) by $\tilde{\mathcal{L}}(\boldsymbol{\Lambda})$, where $\boldsymbol{\Lambda}$ denotes the posterior parameters $\{\boldsymbol{\mu},\boldsymbol{\sigma},\boldsymbol{\eta},\boldsymbol{\gamma}\}$. In our version of SGVB for LDA, we maximize $\tilde{\mathcal{L}}(\boldsymbol{\Lambda})$. Note that $\tilde{\mathcal{L}}(\boldsymbol{\Lambda})$ is differentiable with respect to all relevant variables owing to the parameterization trick in Eq. (7). Therefore, the second requirement given in Sect. 2 is met.

3.2   Maximization of Lower Bound

We maximize $\tilde{\mathcal{L}}(\boldsymbol{\Lambda})$, i.e., the right hand side of Eq. (14), by differentiating it with respect to the relevant variables. With respect to $\eta_{\theta,d}^{(l)}$, we obtain the derivative:

$$\frac{\partial\hat{L}(\boldsymbol{\Lambda})}{\partial\eta_{\theta,d}^{(l)}} = \frac{1}{L}(N_d+K\alpha)\bigg\{\frac{1}{\eta_{\theta,d}^{(l)}}-\frac{1}{(\eta_{\theta,d}^{(l)})^2}\sum_{k=1}^{K}\exp\big(\epsilon_{\theta,dk}^{(l)}\sigma_{\theta,dk}+\mu_{\theta,dk}\big)\bigg\}. \qquad (15)$$

The equation $\frac{\partial\hat{L}(\boldsymbol{\Lambda})}{\partial\eta_{\theta,d}^{(l)}}=0$ gives the solution: $\eta_{\theta,d}^{(l)}=\sum_{k=1}^{K}\exp(\epsilon_{\theta,dk}^{(l)}\sigma_{\theta,dk}+\mu_{\theta,dk})$. Similarly, $\frac{\partial\hat{L}(\boldsymbol{\Lambda})}{\partial\eta_{\phi,k}^{(l)}}=0$ gives the solution: $\eta_{\phi,k}^{(l)}=\sum_{v=1}^{V}\exp(\epsilon_{\phi,kv}^{(l)}\sigma_{\phi,kv}+\mu_{\phi,kv})$. Note that each of these solutions appears as the denominator in Eq. (7).

Further, with respect to $\mu_{\theta,dk}$, we obtain the derivative:

$$\frac{\partial \hat{L}(\boldsymbol{\Lambda})}{\partial \mu_{\theta,dk}} = (N_{dk} + \alpha) - \frac{\exp(\mu_{\theta,dk})}{L}(N_d + K\alpha) \sum_{l=1}^{L} \frac{\exp(\epsilon_{\theta,dk}^{(l)}\sigma_{\theta,dk})}{\eta_{\theta,d}^{(l)}}. \tag{16}$$

Therefore, $\exp(\mu_{\theta,dk})$ is estimated as $\frac{N_{dk}+\alpha}{N_d+K\alpha} \cdot \frac{L}{\sum_l \exp(\epsilon_{\theta,dk}^{(l)}\sigma_{\theta,dk})/\eta_{\theta,d}^{(l)}}$. Based on the definition of $\theta_{dk}^{(l)}$ in Eq. (7), we obtain the following update for $\exp(\mu_{\theta,dk})$:

$$\exp(\mu_{\theta,dk}) \leftarrow \exp(\mu_{\theta,dk}) \cdot \left(\frac{N_{dk}+\alpha}{N_d+K\alpha}\right) \Big/ \left(\frac{\sum_l \theta_{dk}^{(l)}}{L}\right). \tag{17}$$

Similarly, $\exp(\mu_{\phi,kv})$ is updated as follows:

$$\exp(\mu_{\phi,kv}) \leftarrow \exp(\mu_{\phi,kv}) \cdot \left(\frac{N_{kv}+\beta}{N_k+V\beta}\right) \Big/ \left(\frac{\sum_l \phi_{kv}^{(l)}}{L}\right). \tag{18}$$

We can give an intuitive explanation to the update in Eq. (17). $\frac{N_{dk}+\alpha}{N_d+K\alpha}$ is an estimate of the per-document topic probability based on the word count expectation N_{dk}. In contrast, $\frac{\sum_l \theta_{dk}^{(l)}}{L}$ is an estimate of the same probability based on the logistic normal samples $\theta_{dk}^{(1)}, \dots, \theta_{dk}^{(L)}$. We adjust $\exp(\mu_{\theta,dk})$ based on to what extent the latter estimate deviates from the former. A similar explanation can be given to the update in Eq. (18).

For the standard deviation parameters $\sigma_{\theta,dk}$ and $\sigma_{\phi,kv}$, we cannot obtain any closed form update. Therefore, we perform a gradient-based optimization. For numerical reasons, we change variables as $\tau = \log(\sigma^2)$. Then the derivatives with respect to $\tau_{\theta,dk}$ and $\tau_{\phi,kv}$ are obtained as follows:

$$\frac{\partial \hat{L}(\boldsymbol{\Lambda})}{\partial \tau_{\theta,dk}} = \frac{1}{2} + \frac{1}{2}\exp\left(\frac{\tau_{\theta,dk}}{2}\right) \frac{\sum_{l=1}^{L} \epsilon_{\theta,dk}^{(l)}\{(N_{dk}+\alpha) - (N_d+K\alpha)\theta_{dk}^{(l)}\}}{L}, \tag{19}$$

$$\frac{\partial \hat{L}(\boldsymbol{\Lambda})}{\partial \tau_{\phi,kv}} = \frac{1}{2} + \frac{1}{2}\exp\left(\frac{\tau_{\phi,kv}}{2}\right) \frac{\sum_{l=1}^{L} \epsilon_{\phi,kv}^{(l)}\{(N_{kv}+\beta) - (N_k+V\beta)\phi_{kv}^{(l)}\}}{L}. \tag{20}$$

With respect to γ_{dik}, we obtain the following derivative:

$$\frac{\partial \hat{L}(\boldsymbol{\Lambda})}{\partial \gamma_{dik}} = \frac{1}{L}\sum_{l=1}^{L}(\epsilon_{\theta,dk}^{(l)}\sigma_{\theta,dk} + \mu_{\theta,dk}) + \frac{1}{L}\sum_{l=1}^{L}(\epsilon_{\phi,kx_{di}}^{(l)}\sigma_{\phi,kx_{di}} + \mu_{\phi,kx_{di}})$$
$$- \frac{1}{L}\sum_{l=1}^{L}\left\{\frac{\sum_v \exp(\epsilon_{\phi,kv}^{(l)}\sigma_{\phi,kv} + \mu_{\phi,kv})}{\eta_{\phi,k}^{(l)}} + \log\eta_{\phi,k}^{(l)} - 1\right\} - \log\gamma_{dn,k} - 1. \tag{21}$$

$\frac{\partial \hat{L}(\boldsymbol{\Lambda})}{\partial \gamma_{dik}} = 0$ gives the following update:

$$\gamma_{dik} \propto \left(\prod_{l=1}^{L}\theta_{dk}^{(l)}\right)^{\frac{1}{L}} \cdot \left(\prod_{l=1}^{L}\phi_{kx_{dn}}^{(l)}\right)^{\frac{1}{L}}. \tag{22}$$

Algorithm 1. An SGVB for LDA with logistic normal

1: Split the document set into small batches
2: **for** each iteration **do**
3:     **for** each small batch **do**
4:         **for** each topic k **do**
5:             Draw $\phi_k^{(l)} \sim \text{LogitNorm}(\boldsymbol{\mu}_{\phi,k}, \boldsymbol{\sigma}_{\phi,k})$ for $l = 1, \ldots, L$
6:             Update $\exp(\mu_{\phi,kv})$ based on Eq. (18)
7:             Update $\tau_{\phi,kv}$ based on the gradient in Eq. (20)
8:         **end for**
9:         **for** each document d **do**
10:             Draw $\boldsymbol{\theta}_d^{(l)} \sim \text{LogitNorm}(\boldsymbol{\mu}_{\theta,d}, \boldsymbol{\sigma}_{\theta,d})$ for $l = 1, \ldots, L$
11:             **for** $i = 1, \ldots, N_d$ **do**
12:                 Update γ_{dik} based on Eq. (22)
13:                 Update N_{dk}, N_{kv}, and N_k
14:             **end for**
15:             Update $\exp(\mu_{\theta,dk})$ based on Eq. (17)
16:             Update $\tau_{\theta,dk}$ based on the gradient in Eq. (19)
17:         **end for**
18:     **end for**
19: **end for**

The right hand side of Eq. (22) is the product of the geometric mean of the sampled per-document topic probabilities $\theta_{dk}^{(l)}$ and the geometric mean of the sampled per-topic word probabilities $\phi_{kx_{dn}}^{(l)}$. Interestingly, other inference methods for LDA also represent the per-token topic probability as the product of the per-document topic probability and the per-topic word probability [2].

Algorithm 1 gives the pseudocode. As in CVB for LDA [16], we only need to maintain one copy of γ_{dik} for each unique document/word pair. Therefore, we can reduce the number of iterations of the loop on line 11 from N_d to the number of different words in the dth document. Consequently, the time complexity of the proposed inference for each scan of the data is $\mathcal{O}(MK)$, where M is the total number of unique document/word pairs. For updating $\tau_{\theta,dk}$ and $\tau_{\phi,kv}$ based on the gradients in Eqs. (19) and (20), we use Adam [8] in this paper.

3.3   Estimation Without Sampling

By setting all standard deviation parameters $\boldsymbol{\sigma}$ to 0, we obtain an estimate of the log-evidence lower bound without sampling as a degenerated version of our proposal. In this case, we only update the parameter γ_{dik} by

$$\gamma_{dik} \propto \left(\frac{N_{dk} + \alpha}{N_d + K\alpha} \right) \cdot \left(\frac{N_{kv} + \beta}{N_k + V\beta} \right). \tag{23}$$

This is almost the same with the update of γ_{dik} in CVB0 [2] except that the contribution of γ_{dik} is not subtracted from N_{dk}, N_{kv}, and N_k. In the evaluation experiment presented in the next section, we compared the proposed method also with this degenerated version to clarify the effect of the sampling.

Table 1. Specifications of the document sets

	# documents (D)	# vocabulary words (V)	# word tokens $(\sum_d N_d)$	Average document length $(\sum_d N_d/D)$
NYT	99,932	46,263	34,458,469	344.8
MOVIE	27,859	62,408	12,788,477	459.0
NSF	128,818	21,471	14,681,181	114.0
MED	125,490	42,830	17,610,749	140.3

4 Experiment

4.1 Data Sets

In the evaluation experiment, we used the four English document sets in Table 1. NYT is a part of the NYTimes news articles in "Bag of Words Data Set" of the UCI Machine Learning Repository.[2] We reduced the number of documents to one third of its original number due to the limit of the main memory. MOVIE is the set of movie reviews known as "Movie Review Data."[3] NSF is "NSF Research Award Abstracts 1990–2003 Data Set" of the UCI Machine Learning Repository. MED is a subset of the paper abstracts of the MEDLINE®/PUBMED®, a database of the U.S. National Library of Medicine.[4] For all document sets, we applied the Porter stemming algorithm and removed highly frequent words and extremely rare words. The average document lengths, i.e., $\sum_d N_d/D$, of NYT and MOVIE are 344.6 and 459.0, respectively. In contrast, those of NSF and MED are 114.0 and 140.3, respectively. This difference comes from the fact that NSF and MED consist of abstracts.

4.2 Evaluation Method

By using the above four data sets, we compared our proposal with the following three inference methods for LDA: the standard VB [4], CGS [6], and the degenerated version described in Sect. 3.3. The evaluation measure is the test set perplexity. We ran each of the compared inference methods on the 90 % word tokens randomly selected from each document and used the estimated parameters for computing the test set perplexity on the other 10 % word tokens as follows:

$$perplexity \equiv \exp\left\{ - \frac{1}{N_{\text{test}}} \sum_{d=1}^{D} \sum_{i \in \mathcal{I}_d} \log \left(\sum_{k=1}^{K} \theta_{dk}\phi_{kx_{di}} \right) \right\}, \qquad (24)$$

where $\mathcal{I}_d$ is the set of the indices of the test word tokens in the dth document, and N_{test} is the total number of the test tokens. For K, i.e., the number of topics, we tested the following three settings: $K = 50, 100,$ and 200.

---

[2] https://archive.ics.uci.edu/ml/datasets.html.
[3] http://www.cs.cornell.edu/people/pabo/movie-review-data/polarity_html.zip.
[4] We used the XML files from medline14n0770.xml to medline14n0774.xml.

4.3   Inference Settings

The proposed inference was run on each data set in the following manner. We tuned the free parameters on a random training/test split that was prepared only for validation. On lines 7 and 16 in Algorithm 1, $\tau_{\phi,kv}$ and $\tau_{\theta,dk}$ are updated by using the gradients. For this optimization, we used Adam [8]. The stepsize parameter in Adam was chosen from $\{0.01, 0.001, 0.0001\}$, though the other parameters are used with their default settings. The common initial value of the parameters $\tau_{\theta,dk}$ and $\tau_{\phi,kv}$ was chosen from $\{-10.0, -1.0, -0.5\}$. The sample size L was one, because larger sample sizes gave comparable or worse results.

Based on the discussion in [2], we tuned the hyperparameters α and β of the symmetric Dirichlet priors by a grid search. Each of α and β was chosen from $\{0.00005, 0.0001, 0.0002, 0.0005, 0.001, 0.002, 0.005, 0.01, 0.02, 0.05, 0.1, 0.2, 0.3, 0.4, 0.5\}$. The same grid search was used for the Dirichlet hyperparameters of the compared methods. The number of small batches in Algorithm 1 was set to 20. The number of iterations, i.e., how many times we scanned the entire document set, was chosen as 500. We computed ten test set perplexities based on the ten different random splits prepared for evaluation. The test set perplexity in Eq. (24) was computed at the 500th iteration.

4.4   Evaluation Results

Figs. 1 and 2 present the evaluation results. We split the results into two figures, because the two data sets NYT and MOVIE are widely different from the other two NSF and MED in their average document lengths. This difference may be part of the reason why we obtained different evaluation results. The horizontal axis of the charts in Figs. 1 and 2 gives the three settings for the number of topics: $K = 50$, 100, and 200. The vertical axis gives the test set perplexity averaged over the ten different random splits. The standard deviation of the ten test set perplexities is presented by the error bar. LNV and DEG are the labels for our proposal and its degenerated version, respectively. VB and CGS refer to the standard VB [4] and the collapsed Gibbs sampling [6], respectively.

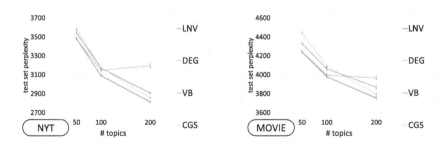

Fig. 1. Evaluation results in terms of test set perplexity for the two document sets NYT (left) and MOVIE (right), whose average document lengths are relatively long.

Figure 1 presents the results for the two data sets NYT and MOVIE, which consist of relatively long documents. Our proposal LNV led to the best perplexity for four cases among the six cases given in Fig. 1. When we set $K = 50$, DEG could give almost the same test set perplexity with LNV for both of the NYT and MOVIE data sets. However, the differences were not statistically significant, because the p-values of the two-tailed t-test were 0.463 and 0.211, respectively. For the other four cases, LNV could give the best perplexity. The differences for all these four cases were statistically significant. For example, when we set $K = 100$ for the MOVIE data set, the p-value of the two-tailed t-test where we compare LNV and DEG was 0.00134. It can be said that our proposal worked effectively for these two document sets.

Figure 2 shows the results for the two data sets NSF and MED, which consist of relatively short documents. Our proposal LNV could provide the best perplexity for the following three cases: $K = 50$ for the NSF data set, $K = 50$ for the MED data set, and $K = 100$ for the MED data set. Note that the difference between LNV and CGS when we set $K = 50$ for the NSF data set was statistically significant, because the p-value of the two-tailed t-test was 0.000121. For the other three cases, CGS gave the best perplexity. For these two data sets, our proposal worked only when the number of topics was not large.

Note that LNV was superior to VB for all settings presented in Figs. 1 and 2. This proves empirically that the Dirichlet distribution is not necessarily the best choice for approximating the true posterior even when we use the same mean field approximation with the standard VB. In sum, we can draw the following conclusion. When LNV is available, there may be no reason to use VB, DEG, or CGS for relatively long documents. Also for relatively short documents, our proposal may be adopted when the number of latent topics is small.

However, in terms of computation time, LNV has a disadvantage. For example, when we set $K = 200$ for the NYT data set, it took 43 h for finishing 500 iterations with LNV, though it took 14 h with CGS and 23 h with VB. However, inferences for LDA are generally easy to parallelize, e.g. by using GPU [18,19]. It may be an advantage for parallelization that each sample $\epsilon \sim \mathcal{N}(0,1)$ in the proposed method can be drawn independently.

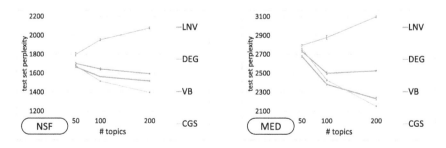

Fig. 2. Evaluation results in terms of test set perplexity for the two document sets NSF (left) and MED (right), whose average document lengths are relatively short.

4.5 Effect of Sampling

The degenerated version of our proposal gave a comparable perplexity only for the limited number of cases in our experiment. When the number of latent topics was large, or when the average document length was short, the degenerated version led to quite a poor perplexity. Especially in the two charts of Fig. 2, it seems that the degenerated version exhibits substantial overfitting. Therefore, sampling is indispensable. Recall that L, i.e., the number of the samples from the standard normal, was set to 1, because larger numbers of samples did not improve the test set perplexity. However, a single random sample could work as a kind of *perturbation* for the update of the corresponding parameter. Without this perturbation, the inference tended to get trapped in local minima as shown in Fig. 2. A single sample can change the course of inference through perturbation. This may be the reason why our proposal gave better perplexities than its degenerated version in many of the situations consulted in the experiment.

Based on our experience, it is important to optimize the standard deviation parameters $\tau_{\theta,dk}$ and $\tau_{\phi,kv}$ carefully in order to avoid overfitting. When the stepsize parameter of Adam was not tuned, the standard deviation parameters stayed around at their initial values. This made the perplexity almost similar to that of the degenerated version. In addition, the initial values of $\tau_{\theta,dk}$ and $\tau_{\phi,kv}$ also needed to be tuned carefully. However, the tuning was not that difficult, because the optimization was almost always successful when the parameters $\tau_{\theta,dk}$ and $\tau_{\phi,kv}$ took values widely different from their initial values. Further, we only needed to test several setting for the combination of the stepsize parameter in Adam and the common initial value of $\tau_{\theta,dk}$ and $\tau_{\phi,kv}$. In our experiment, the stepsize parameter was chosen from $\{0.01, 0.001, 0.0001\}$. The common initial value of the parameters $\tau_{\theta,dk}$ and $\tau_{\phi,kv}$ was chosen from $\{-10.0, -1.0, -0.5\}$. Therefore, at most nine settings were checked. However, the combination of 0.001 for the stepsize parameter and -0.5 for the initial value of $\tau_{\theta,dk}$ and $\tau_{\phi,kv}$ often worked. Only when this setting did not work, we considered other settings.

5 Conclusion

In this paper, we proposed a new VB-type inference method for LDA. Our method is based on the stochastic gradient variational Bayes [9,13] and approximates the true posterior with the logistic normal distribution. The proposed method was better than the standard VB for all situations consulted in the experiment and was better even than the collapsed Gibbs sampling for not all but many situations. Further, when deprived of sampling, the inference tended to get trapped in local minima. Therefore, sampling worked.

While we use the logistic normal distribution in the proposed inference, we can choose other distributions as long as they meet the two requirements given in Sect. 2. Further, we can propose a similar inference also for other Bayesian probabilistic models. One important merit of SGVB is that the expectation with respect to the approximate posterior for continuous variables is estimated by the Monte Carlo method. Even when the full joint distribution of the target

Bayesian model is complicated, SGVB may make the computation relating to such expectations efficient. Therefore, it is worthwhile future work to provide a new inference for the existing Bayesian models with the distribution that has not been considered due to the complication in handling the expectations.

References

1. Aitchison, J., Shen, S.-M.: Logistic-normal distributions: some properties and uses. Biometrika **67**(2), 261–272 (1980)
2. Asuncion, A., Welling, M., Smyth, P., Teh, Y.W.: On smoothing and inference for topic models. In: UAI, pp. 27–34 (2009)
3. Blei, D.M., Lafferty, J.D.: Correlated topic models. In: NIPS, pp. 147–154 (2005)
4. Blei, D.M., Ng, A.Y., Jordan, M.I.: Latent Dirichlet allocation. JMLR **3**, 993–1022 (2003)
5. Brooks, S., Gelman, A., Jones, G., Meng, X.-L.: Handbook of Markov Chain Monte Carlo. CRC Press, Boca Raton (2011)
6. Griffiths, T.L., Steyvers, M.: Finding scientific topics. PNAS **101**(Suppl 1), 5228–5235 (2004)
7. Kang, J.-H., Lerman, K., Getoor, L.: LA-LDA: a limited attention topic model for social recommendation. In: Greenberg, A.M., Kennedy, W.G., Bos, N.D. (eds.) SBP 2013. LNCS, vol. 7812, pp. 211–220. Springer, Heidelberg (2013)
8. Kingma, D.P., Ba, J.: Adam: a method for stochastic optimization. In: ICLR (2015)
9. Kingma, D.P., Welling, M.: Stochastic gradient VB and the variational auto-encoder. In: ICLR (2014)
10. Lin, T.-Y., Tian, W.-T., Mei, Q.-Z., Cheng, H.: The dual-sparse topic model: mining focused topics and focused terms in short text. In: WWW, pp. 539–550 (2014)
11. Mnih, A., Gregor, K.: Neural variational inference and learning in belief networks. In: ICML, pp. 1791–1799 (2014)
12. O'Connor, B., Stewart, B.M., Smith, N.A.: Learning to extract international relations from political context. In: ACL, pp. 1094–1104 (2013)
13. Rezende, D.J., Mohamed, S., Wierstra, D.: Stochastic backpropagation and approximate inference in deep generative models. In: ICML, pp. 1278–1286 (2014)
14. Robert, C.P., Casella, G.: Monte Carlo Statistical Methods. Springer, New York (2004)
15. Sasaki, K., Yoshikawa, T., Furuhashi, T.: Online topic model for Twitter considering dynamics of user interests and topic trends. In: EMNLP, pp. 1977–1985 (2014)
16. Teh, Y.-W., Newman, D., Welling, M.: A collapsed variational Bayesian inference algorithm for latent Dirichlet allocation. In: NIPS, pp. 1353–1360 (2007)
17. Vosecky, J., Leung, KW.-T., Ng, W.: Collaborative personalized Twitter search with topic-language models. In: SIGIR, pp. 53–62 (2014)
18. Yan, F., Xu, N.-Y., Qi, Y.: Parallel inference for latent Dirichlet allocation on graphics processing units. In: NIPS, pp. 2134–2142 (2009)
19. Zhao, H.-S., Jiang, B.-Y., Canny, J.F., Jaros, B.: SAME but different: fast and high quality Gibbs parameter estimation. In: KDD, pp. 1495–1502 (2015)

On Optimizing Partitioning Strategies for Faster Inverted Index Compression

Xingshen Song$^{(\boxtimes)}$, Kun Jiang, Yu Jiang, and Yuexiang Yang

College of Computer, National University of Defense Technology, Changsha, China
{songxingshen,jiangkun,jiangyu14,yyx}@nudt.edu.cn

Abstract. Inverted index is a key component for search engine to manage billions of documents and fast respond to users' queries. While substantial effort has been made to compromise space occupancy and decoding speed, what has been overlooked is the encoding speed when constructing the index. VSEncoding is a powerful encoder that works by optimally partitioning a list of integers into blocks which are efficiently compressed by using simple encoders, however, these partitions are found by using a dynamic programming approach which is obviously inefficient. In this paper, we introduce compression speed as one criterion to evaluate compression techniques, and thoroughly analyze performances of different partitioning strategies. A linear-time optimization is also proposed, to enhance VSEncoding with faster compression speed and more flexibility to partition an index. Experiments show that our method offers a far more better compression speed, while retaining an excellent space occupancy and decompression speed.

Keywords: Inverted index · Index compression · Optimal partition · Approximation algorithm

1 Introduction

Due to its simplicity and flexibility, inverted index gains much popularity among modern IR systems since 1950s. Especially in large scale search engines, inverted index is now adopted as their core component to maintain billions of documents and respond to enormous queries. In its most basic and popular form, an inverted index is a collection of sorted sequences of integers [10,16,18]. Growing size of data and stringent query processing efficiency requirement have appealed a large amount of research, with the aim to compress the space occupancy of the index and speed up the query processing.

While state-of-the-art encoders do obtain very good space-time trade-offs, we argue that one important evaluation criterion has been neglected is the compression speed of these methods [10,13,17]. The reason can be attributed to the fact that index is always preprocessed offline before deployment, and once being taken into effect, update can be committed in an asynchronous and parallel manner. Therefore index designers usually traverse the sequences more than

© Springer International Publishing Switzerland 2016
O. Gervasi et al. (Eds.): ICCSA 2016, Part IV, LNCS 9789, pp. 246–260, 2016.
DOI: 10.1007/978-3-319-42089-9_18

once to find the optimal parameters for space-efficiency gains, making index construction speed rather slow. However, timely update for unexpected queries is becoming more and more stringent in search engine, especially in twitter and other social network sites. Compression speed should also be an important factor to evaluate an index compression algorithm. Early techniques like Simple-9 and Simple-16 [2,3] evaluate all the possible schemes to decide the best partition, PFOR [9,17,18] splits each sequence into blocks of fixed size (say, 128 integers) and goes over the whole block to decide the exception ratio and block width. In a nutshell, compression speed is compromised to achieve better space occupancy and decoding speed.

Modern encoders are designed to compress lists of integers, that is the input posting list is split into blocks with fixed or variable lengths to be encoded. Intuitively, partitioning posting list aligning to its clustered distribution, can effectively minimize the compressed size while keeping partitions separately accessed. Works from literature [1,6,13] give another perspective on index compression. The integer sequence is considered a particular directed acyclic graph (DAG), the partitioning problem is then treated as optimal path finding problem, unfortunately the DAG is complete with $\frac{n(n+1)}{2} = \Theta\left(n^2\right)$ edges, a trivial traversal may not suffice to obtain an efficient solution for this problem. Scheme from [1] uses a *greedy* mechanism to yield a sub-optimal partition with a small amount of effectiveness exchanged for speed of compression and ease of implementation. AFOR from [6] computes the block partition by using a series of fixed-sized sliding windows over a block and determines the optimal configuration of frames and frame lengths of the current window. VSEncoding from [13] finds the optimal partition by using a *dynamic programming* approach and is said to be able to encode groups of integers beating the entropy of the gaps distribution. However, dynamic programming can be very inefficient since it needs to recalculate all the edges when a new vertex is added in the current graph, to mitigate this problem VSEncoding simply restricts the length of the longest block to h, reducing its time complexity from $O(n^2)$ to $O(nh)$, but barely satisfactory in practice.

Recently, Ottaviano overcomes this drawback by introducing a new compression scheme called Partitioned Elias-Fano Index (PEF) [12], in which a linear-time approximation algorithm is presented to find a solution at most $(1 + \epsilon)$ times larger than the optimal one, for any given $\epsilon \in (0,1)$. Its core idea is to generate a pruned graph $\mathcal{G}_\epsilon$ in linear time directly without explicitly constructing the whole graph $\mathcal{G}$ which, otherwise, would require quadratic time. Edges with a heavy weight are dropped according to a predefined pruning policy, reducing the time and space complexities to linear at the cost of finding slightly suboptimal partitions. The same idea is also adopted in [7].

In this paper we follow the compression techniques studied in [4,9,15,17]. However, while previous work has focused on improving compression ratio and speed up decoding, we consider compression speed as one criterion. In particular, we extensively study various compression schemes on their space-time trade-offs, and propose our optimization on VSEncoding to achieve a faster compression speed while keeping its space and decoding-time efficiencies, namely substituting an approximation algorithm for the dynamic programming used by VSEncoding [13] when partitioning the input sequence into blocks, an experiment is also performed on TREC GOV2 collection to validate the proposed method, results show that our method significantly improves the compression speed with a slight loss at index size overhead.

The rest of this paper is organized as follows. Section 2 provides a background on compression techniques and partitioning strategies; Sect. 3 proposes our optimization on speeding up partitioning procedure for VSEncoding; Sect. 4 shows our experimental results and analyses of the original methods and their optimizations; conclusion and future work follow in Sect. 5.

2   Background

2.1   Index Compression

Compressing the index has long been a key issue for researchers to ensure both the time- and space-efficiency of inverted index, various encoders with different properties have been put up to settle this problem, and they can be roughly divided into two classes, namely the *integer-oriented encoders* and the *list-oriented encoders*. The integer-oriented encoders assign an unique codeword to each integer of the input sequence, then the compression procedure turns into a mapping or substitution from the integer space to code space. As they compress integers without considering their neighborings, the integer-oriented encoders are also called oblivious encoders [4], such as *unary code, Elias Gamma/Delta codes* and *Golomb/Rice codes*. Most integer-oriented encoders are hard to decode since they need bitwise operations to cross computer word boundaries, so byte/word-aligned encoders, are proposed to solve this problem, like *Variable Byte* and *Group Varint*, more importantly, they can be further improved by SIMD instructions of modern CPUs [14,15].

List-oriented encoders are designed to exploit the cluster of neighboring integers, each time a fixed-sized or variable-sized group of integers is binary packed with an uniform bit width, providing equivalent compression ratio and faster decoding speed, the technique used by these encoders is called *frame-of-reference* (FOR), or *binary packing* [8]. Basically, their compression ratios are inferior to these of the first category as a batch of integers are encoded indiscriminately, and useless zeros are padded in the codeword to keep word-aligned, however, when decoded, list-oriented encoders can obtain an entire block while the formers just decode one integer at a time. More importantly, with the help of *skip pointers* or *skip list*, it is possible to step along the codewords compressed by list-oriented encoders and stop when the required number of blocks has been bypassed. Examples of these encoders are *Simple Family*, AFOR and *Patched FOR* (PFOR, OptPFOR and FastPFOR).

2.2   Directed Acyclic Graph

One thing to be noted is that list-oriented encoders may cost equivalent time to compress the input sequence as the integer-oriented encoders even they are designed to compress a list of integers at the same time. Before compressing, a partitioning strategy is needed to traverse the whole input list to search for an optimal partition, in consideration of compression ratio and decoding speed. Even after that, an uniform bit width has to be chosen to fit every element in for each block. As for Simple Family, a descriptor is decided after enumerating all the possible partitioning cases; OptPFOR needs an additional computation to choose the optimal proportion of exceptions in each block in order to achieve a better space efficiency. To speed up the compression speed, a proliferation of partitioning schemes have been seen in the last few years [1,6,12,13].

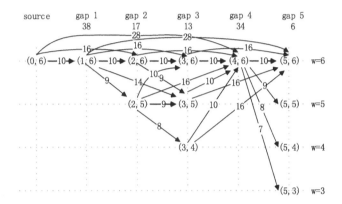

Fig. 1. Here is a DAG for sequence with 6 integers represented using gap (differences between them). Optional bit widths range from 3 to 5, each gap is fitted in the tuple (*position, available-bit-width*), if the available-bit-widths are more than one, they are placed in different rows, the number in the edge denotes the cost for this path, the fixed cost is set to 4.

The common foundation for the abovementioned methods is to recast the integer sequence $S[0, n]$ to a particular DAG $\mathcal{G}$, each integer is represented by a vertex, plus a dummy vertex marking the end of the sequence, the graph $\mathcal{G}$ is complete, which means that for any i and j with $i < j \leqslant n$, there exists an edge connecting v_i and v_j, denoted as (v_i, v_j). In fact, the edge is an exact correspondence of a partition in the sequence $S[i,j]$, the problem of fully partitioning S is converted to finding a path π in $\mathcal{G}$, for instance, $\pi = (v_0, v_{i_1})(v_{i_1}, v_{i_2}) \ldots (v_{i_{k-1}}, v_n)$ with k edges corresponds to the partition $S[0, i_1 - 1]S[i_1, i_2 - 1] \ldots S[i_{k-1}, n - 1]$ of k blocks. The weight of an edge in the graph is equal to the cost in bits consumed by the partition. Thus, the problem of optimally partitioning a sequence is reduced to the problem of Single-Source Shortest Path (SSSP) Labeling, as shown in Fig. 1. An intuitive way to solve this is to firstly set the cost of each vertex in $\mathcal{G}$ to $+\infty$, then an iteration starts from the left vertex to the rightmost, when it comes to a vertex v_j with $0 \leqslant j < n$, a subproblem of find the optimal path from v_j to v_n shows up, assuming the optimal path from v_0 to v_j has been correctly computed. Each edge (v_j, v') outgoing from v_j will be assessed and cost of vertex v' is updated if it becomes smaller. As can be seen, the time complexity of this algorithm is proportional to the number of edges in $\mathcal{G}$.

However, the $\mathcal{G}$ transformed from integer sequence $\mathcal{S}$ is complete with $\Theta\left(n^2\right)$ edges, especially some posting lists for popular terms will be quite large, finding their optimal partitions will be intolerable. Since dynamic programming is inefficient and greedy mechanism is too coarse, an elaborate approximation algorithm which reduces the time and space complexities to linear at the cost of finding slightly suboptimal solutions will be feasible.

3   Optimizing Partitioning Strategy via Pruning DAG

While PEF using Elias-Fano code gets an impressive compression performance, we are aiming at revitalizing encoders using binary packing with optimizations. Since Elias-Fano code compresses integer in its complete form which may leads to poor space efficiency, binary packing uses gapped integer instead will obtain better compression ratio, it is still an promising method with potential for faster compression speed.

3.1   VSEncoding

VSEncoding is similar to PFOR, however, it neither applies a fixed-sized block length nor appends a patch for outliers at the end of blocks. In order to maximize the compression while retaining simple and fast decompression, VSEncoding partitions each posting list into blocks of variable length, and binary packs the integers inside of each block with the number of bits, say b, required to encode the largest one. Finally the the value of b and the length of the block, k, are encoded using distinct encoders $\mathcal{M}_1$ and $\mathcal{M}_2$. The basic form of each block can be seen as $\{< k_i, b_i >: data_i\}_{block_i}$, $data_i$ is the k_i integers of $block_i$ using b_i bits each.

Given a sequence $\mathcal{S}$ and a vector of partitions, the number of bits required to encode $\mathcal{S}$ can be computed in constant time, this quantity is calculated by summing up the costs of all the blocks, as the length of each block is variable, the problem of minimizing bits used relies on the problem of finding the optimal partitions. Let P denote the vector of m vertexes indicating the boundary of each block, with $P[0] = 1$, and $P[m] = n$. The problem can be represented as follow:

$$\min_{P \in \mathcal{S}} \sum_{i=0}^{m-1} c\left(P[i] - 1, P[i+1]\right)$$

where $c\left(P[i] - 1, P[i+1]\right) = |\mathcal{M}_1\left(b_i\right)| + |\mathcal{M}_2\left(k_i\right)| + k_i b_i$, namely the cost to encode i-th block.

VSEncoding adopts a dynamic programming algorithm to obtain the optimal partitions, each time a subproblem t consists in encoding in the best way all the integers starting from 0 to t, with a memo to look up when deciding whether to merge or split current partition. The whole procedure starts by setting $t = 0$ and goes down to $t = n$, which represents the solution to the original problem. In order to implement a faster algorithm, the block lengths are restrained to some small constants between 16 and 64, say h, thus the time complexity drops from $O(n^2)$ to $O(n \log^2 h)$.

To further improve decompression speed and keep the block representations word aligned, VSEncoding reorganizes the layout of blocks, first the description parts are stored together in their order (i.e., $b_0 k_0, b_1 k_1 \ldots, b_m k_m$), then the data parts are written separately into each group: first the values that have to be represented with 1 bit, then with 2 bits, and so on. The decompression procedure is done by binary-unpacking and permuting the data parts into the correct order according to description parts.

3.2   Optimized Partitioning Strategy

Next we describe our modification to VSEncoding that achieve significant improvements over its original version in [13]. As mentioned before dynamic programming used in optimal partitioning costs too much time of index compression, to overcome this problem, we present a new partitioning scheme, which uses approximation algorithm in place of dynamic programming, thus reducing the time and space complexities to linear by finding a suboptimal partition. The partitioning problem can be reduced to SSSP over a DAG $\mathcal{G}$ with $\Theta(n^2)$ edges, and its time complexity is proportional to the number of edges, our aim is to design a pruning strategy removing edges of large costs while retaining edges which costs no more than $(1 + \epsilon)$ times what the shortest paths do, for any given $\epsilon \in (0, 1)$.

We use $c(v_i, v_j)$ to denote the cost function of the edge (v_i, v_j), which is also the cost of a partition $\mathcal{S}[i, j - 1]$, U as the upper bound of cost by representing $\mathcal{S}$ as a single partition, and F as the fixed cost of each partition(e.g. the descriptor). There is an obvious fact that, given any $0 \leqslant i < j < k \leqslant n$, it is $0 < c(v_i, v_j) \leqslant c(v_i, v_k) \leqslant \ldots \leqslant c(v_i, v_{n+1})$.

By adopting a nontrivial pruning strategy a subgraph $\mathcal{G}_\varepsilon$ of the original $\mathcal{G}$ is produced, in which the shortest path and the suboptimal path which increases a little are preserved. Any edge (v_i, v_j) in $\mathcal{G}_\varepsilon$ follows at least one of the following conditions: (1) there exits a positive integer k such that $c(v_i, v_j) \leqslant F(1 + \varepsilon)^k < c(v_i, v_{j+1})$; (2) $j = n + 1$. These edges of $\mathcal{G}_\varepsilon$ are called $\varepsilon - maximal$ edges. As we have set the upper bound to U, there exists at most $\log_{1+\varepsilon} \frac{U}{F}$ possible values for k, thus for each vertex $\mathcal{G}_\varepsilon$ has at most $\log_{1+\varepsilon} \frac{U}{F}$ outgoing $\varepsilon - maximal$ edges. [7] has proved that the shortest path distance on $\mathcal{G}_\varepsilon$, which is at most $(1 + \varepsilon)$ times larger than the one in $\mathcal{G}$, can be computed in $O(n \log_{1+\varepsilon} \frac{U}{F})$ time (Fig. 2).

The above procedure is like the trimming scheme for the subset-sum problem [5], which is to choose a representative for a compact range of its neighbor. In other words, we are sparsifying the complete graph $\mathcal{G}$ by quantizing its edge costs into classes of cost between $(1 + \epsilon)^i$ and $(1 + \epsilon)^{i+1}$, for each cost class of each node, only one $\epsilon - maximal$ edge is retained. If to prune $\mathcal{G}$ in a more coarse-grained but faster way, we can further remove edges which span too many vertex,

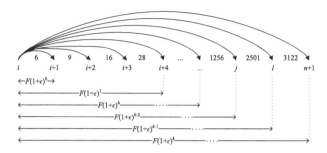

Fig. 2. Curves represent outgoing edges from vertex v_i, the number under each curve is the cost of it, postulating the fixed cost is $F = 4$. Lines below are cost classes with different k, for each class we can choose one edge as $\epsilon - maximal$ edge. This process is called sparsification.

intuitively, long edges are more vulnerable and sensitive to outliers, resulting in high cost and poor efficiency, thus are less likely to be enrolled in optimal partitioning. At the very beginning, we set the upper bound U to be the cost of representing the whole sequence S as a single block, however this is an asymptotic upper bound which might never be reached, by refining U to a more compact bound we can further lower the time complexity to linear without losing the approximation guarantees. Before giving our U we first state the following proposition to be based on:

Proposition 1. *For any $0 \leqslant i < j < k \leqslant n+1$, the weighting function satisfies* $c(v_i, v_k) + c(v_k, v_j) \leqslant c(v_i, v_j) + F + 1$.

It is easy to prove as we split one edge into two shorter edges, the correspondent interval in S is also partitioned. The worst case is the split cuts out nothing but only adds cost of one descriptor. For any edge (v_i, v_j) in $\mathcal{G}_\epsilon$, we have $c(v_i, v_j) \leqslant \overline{U}$, if $c(v_i, v_j) > \overline{U}$, then they are pruned and replaced by sub-edges in $\mathcal{G}_\epsilon$. These sub-edges can be found in a greedy way, in which the cost of each edge equals $\overline{U}$ (optimal ones are probably better), thus the number of edges cannot be larger than $\frac{c(v_i,v_j)-F}{\overline{U}-F} + 1$. Postulating all the sub-edges are the worst cases which only add costs, the overall cost follows the inequality:

$$\sum_{i \leqslant k < l \leqslant j} c(v_k, v_l) \leqslant c(v_i, v_j) + \left(\frac{c(v_i, v_j) - F}{\overline{U} - F} + 1 \right)(F + 1) \qquad (1)$$

our ultimate goal is to keep the shortest path distance in $\mathcal{G}_\epsilon$ no more than $(1 + \epsilon)$ times the optimal one in $\mathcal{G}$, if (v_i, v_j) is one edge of the optimal path, we get the following inequality:

$$\sum_{i \leqslant k < l \leqslant j} c(v_k, v_l) \leqslant (1 + \epsilon) \cdot c(v_i, v_j) \qquad (2)$$

combining these two we get an inequality for $\overline{U}$:

$$\overline{U} \geqslant F + \frac{F+1}{\epsilon} \tag{3}$$

One thing to be noted is that, even parameters ϵ and F are predefined by user, there is an implicit criterion that for any edge (v_i, v_j) the cost function must satisfy, since the correctness of pruning strategy relies on it: $\epsilon \cdot c(v_i, v_j) - F - 1 > 0$, which is not mentioned in [12]. Different lower bound of $c(v_i, v_j)$ determines the minimum of ϵ, for instance, if $c(v_i, v_j) \geqslant 2(F+1)$, then $\epsilon \in [0.5, 1)$. Finally we can set our $\overline{U}$ to $F + \frac{F+1}{\epsilon}$, reducing the time complexity from $O\left(n \log_{1+\epsilon} \frac{U}{F}\right)$ to $O\left(n \log_{1+\epsilon} \frac{1}{\epsilon}\right) = O(n)$, more importantly, the approximation guarantee is preserved.

Even providing the edge cost can be computed in constant time, we cannot check every edge of the complete graph $\mathcal{G}$ to determine whether it is $\epsilon - maximal$ or not, since it would take $\Theta(n^2)$ time. To construct the pruned graph $\mathcal{G}_\epsilon$ on the fly we need to deploy a dynamic data structure that maintains a set of $sliding$ $windows$ over $\mathcal{S}$ denoted by $\omega_0, \omega_1, \ldots, \omega_{\log_{1+\epsilon} \frac{1}{\epsilon}}$, each of them represents a cost class of $F(1+\epsilon)^k$, starting from the same vertex v_i but covering a different range of $\mathcal{S}$. For each vertex v_i, each sliding window ω_j begins to expand its size from the starting position to the position where the cost is larger than $F(1+\epsilon)^j$ for the first time, thus we generate all the $\epsilon - maximal$ edges outgoing from v_i. By performing a scan on $\mathcal{S}$ for each sliding window, we get and evaluate all the $\epsilon - maximal$ edges on-the-fly. Thus, the algorithm returns the optimal partition in $O\left(n \log_{1+\epsilon} \frac{1}{\epsilon}\right)$ time.

4 Experiments

4.1 Experimental Setup

In our experiments, we use the posting lists extracted from the TREC GOV2 collection, which consists of 25.2 million web pages and about 32.8 million terms in the vocabulary crawled from the gov Internet domain. The uncompressed size of these web pages is 426 GB. Also, all the terms have the Porter stemmer applied, and stopwords have been removed, docids are assigned in two ways: according to the lexicographic order of their URLs or to the order that they appear in the collection, thus we can see how docid reordering influence indexing performance. Then the docids and term frequencies are extracted from the collection in a non-interleaved way and applied with compression methods separately. To highlight the improvements between the original methods and our optimizations, the compression methods used in experiments are AFOR, VSEncoding via Dynamic Programming (VSE-DP) and VSEncoding via Optimal Partitioning (VSE-OP), we do not compare other methods like the Simple Family or PFOR in our benchmark as they have been throughly studied in the literature [4,9,11,17].

All the implementations are carried out on an Intel(r) Xeon(r) E5620 processor running at 2.40 GHz with 128 GB of RAM and 12,288 KB of cache. The default physical block size is 16 KB, algorithms are implemented using C++ and compiled with GCC 4.8.1 with O3 optimizations. In all our runs, the whole inverted index is completely loaded into main memory, in order to warm up the execution environment, each query set is run 4 times for each experiment, and the response times only measured for the last run. Our implementations are available at https://github.com/Sparklexs/myVS.

4.2 Indexing Performance

The performance of indexing is based on the index size and compression speed. Before comparing the spaces obtained by different methods, we first set the parameters needed by AFOR, as mentioned before, to keep byte-aligned, we set the frame length to be (8, 16, 32), these configurations give the best balance between performance and compression ratio. Offering additional frame lengths will slightly increase the compression ratio, at the cost of linearly decreasing the compression speed.

Table 1. Total Size in GB, and corresponding average bits per integer (bpi) for docid and frequency, compressed by different methods on GOV2

methods	original				reordered			
	docid	bpi	freq	bpi	docid	bpi	freq	bpi
AFOR	16.30	30.17	8.10	14.95	9.78	18.05	6.39	11.81
VSE-DP	4.13	7.64	2.08	3.84	1.97	3.65	1.89	3.50
VSE-OP	4.22	7.80	2.25	4.17	2.04	3.76	2.04	3.77

Table 1 shows the index space, it is divided into two classes, the original and the reordered, the former denotes docids are assigned in the order they appear and the latter denotes docids are assigned in the lexicographic order of their URLs. The overall size and bits per integer of docid and frequency are shown separately. To facilitate reading, we fill docid-related cells with gray, so are the other tables below. We can observe that VSE-OP gets slightly worse result than its initial version, since we are focusing on accelerating compression speed, the partitioning scheme we adopt in VSE-OP is suboptimal, but it is still competitive compared with the optimal one. We can also observe VSEncoding achieves far more better compression ratio than AFOR, no matter for docid or frequency, the size of AFOR is nearly 4 times larger than VSEncoding, which demonstrates that a partitioning strategy, which offers more optional frame lengths, can more effectively utilize the distribution of integers to compress the sequence. Also note that docid after reordering is half the size it is ordered by the sequence of appearance, while frequency stays less sensitive to reordering, this can be explained by the fact that docid is stored in ascending order and reordering by URL further narrows the gaps between consecutive docids, however the frequency is aligned with docid and stored in an unordered way, thus reordering does not make too much difference. Also the docids are quite sparse while the frequencies are fairly concentrated, rendering the docid's compression ratio twice larger than the frequency's.

Table 2. Total time elapsed in seconds and performance in million integers per second (mis) when compressing docid and frequency

methods	original				reordered			
	docid	mis	freq	mis	docid	mis	freq	mis
AFOR	78.98	58.95	72.56	64.17	75.43	61.72	66.92	69.58
VSE-DP	8303.96	0.56	8005.42	0.58	7849.07	0.59	7966.30	0.58
VSE-OP	5982.24	0.78	5819.72	0.80	5601.14	0.83	5648.14	0.82

Table 2 shows the compression speed of different methods when constructing index for docid and frequency. Combining with the index size shown in Table 1, we can find that there is a clear trade-off between compression efficiency and effectiveness, while index compressed by AFOR is quadruple the size of the ones compressed by VSEncoding, however, its compression speed is two orders of magnitude faster than VSEncoding. Due to a lack of attention, the construction time of experimented methods are rarely mentioned in the literature, we cannot find their performances on other platforms to compare. In our experiment, the optimized method outperforms its original version, the average time saved constitutes nearly 20 % of the time cost by the original. We can notice that the performance gap between docid and frequency is not very apparent, for that they contain the same number of integers. Also, the compression methods traverse the lists in a fixed way, which is irrelevant to the symbols the lists may contain. The only difference exists is that docid is sparser and larger than frequency, thus leading to the result that compressing docid is a little more time-consuming than compressing frequency. One exception is that VSE-OP partitions a list under the influence of the predefined parameter ϵ and the upper bound of representing the list in a single block, we can observe that compression speed between docid and frequency using VSE-OP is quite different. Another thing to be noted is that compressing reordered lists cost less time than the original lists both for docid and frequency.

Table 3 further details the performance of different methods. As is shown, Table 3 lists the number of partitioning schemes evaluated when traversing the integer sequences, comparing these different encoders, we can see that the partitioning step is crucial for compression speed. The optimized method sharply reduces the calculation needed to partition a posting list. By pruning edges with large cost, more than half of the calculation is saved by VSE-OP. We can also note that all methods maintain the same calculation in the four columns except VSE-OP, it is easy to explain by the fact that time complexities of the first two methods are only relevant to the length of input sequences, while VSE-OP is relevant to both the length and the upper bound of the sequences. Calculation of frequency for VSE-OP stays the same after reordering because the changed integer distribution makes no difference to the upper bound of the sequence. However calculation for docid is slightly reduced as upper bound of the sequence is narrowed.

Table 3. Number of evaluated partitions by different methods

methods	original		reordered	
	docid	freq	docid	freq
AFOR	718,920,570	718,920,570	718,920,570	718,920,570
VSE-DP	293,805,525,502	293,805,525,502	293,805,525,502	293,805,525,502
VSE-OP	128,325,159,726	126,661,641,588	111,438,988,684	126,661,641,588

4.3 Decompression Performance

In this subsection, we are going to discuss the decompression performance of different methods, before reporting our results on the decompression speed, we first display the distribution of partition lengths, which can help us in understanding the differences among these partitioning schemes, and the correlation between clusters in a collection and partitions produced on it by variant methods.

As is discussed before, AFOR and VSE-DP use a partitioning strategy with fixed-sized lengths (8, 16, 32 for AFOR, and 1, 2, 4, 6, 8, 16, 32, 64 for VSE-DP), while VSE-OP uses a more flexible partitioning strategy, its partition lengths are determined by a series of sliding windows when traversing the sequence. To show the distribution clearly, we display the first two methods in histograms and VSE-OP in scatter plots as shown in Figs. 3 and 4.

There is one important proposition we need to declare: there exists a compromise among partition length, compression ratio, compression and decompression speed. All the codewords that fall into one partition share the same bit width, thus the smaller the partition length is, the less bits will be wasted, however, smaller partition lengths also produce more fragmentations in one sequence, resulting in more CPU cycles and disk I/O needed to write and read these partitions. Vice versa, larger partition lengths are easier to access but less space-friendly. In Fig. 3, each bar indicates number of specified lengths used by one or more methods in pairs, each two adjacent blocks in one bar indicate the different number before and after reordering for one method, more exactly, the pale color represents the original and the dark color represents the reordered.

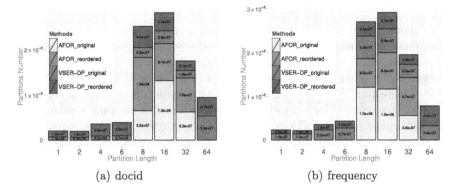

(a) docid   (b) frequency

Fig. 3. Distribution of partition lengths produced by AFOR and VSE-DP

Figure 3(a) shows that after reordering these two methods produce more blocks with small lengths than before, so the compressed size contracts sharply. However, these methods produce more large lengths on frequency as shown in Fig. 3(b), resulting an inferior compression ratio than that on docid. When it comes to decompression speed (which will be listed below in Table 4), larger partitions imply faster access speed for a list of integers. From Fig. 3, we can find that AFOR has larger partition length than VSE-DP, so the disk I/O for it decreases orderly, however this only includes the disk transfer time, their time complexities on calculation do not change. Overall, these methods do not have a large variation in terms of partition lengths, which shows that they are able to adapt to the skewness of a dataset.

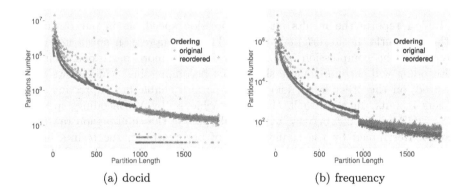

(a) docid   (b) frequency

Fig. 4. Distribution of partition lengths produced by VSE-OP

Things go quite different for VSE-OP: as shown in Fig. 4, the partition lengths vary from 0 to larger than 1500 rather than being confined to a small range. In order to improve the interpretability of the result, we apply *log transformation* to the original graph. As for the docid on the left hand, we can observe that partitions shorter than 500 compose the main proportion of the whole distribution, at the point of 1000, there exists a sharp segmentation, specially for the docid without reordering, number of such partitions drops to 1. However, reordering increases the number by almost one order of magnitude. With much more long partition allocated, VSE-OP still keeps its compression ratio close to VSE-DP, this illustrates its superiority on choosing partitions while avoiding wasting space caused by outliers. Also, different from Fig. 3, with quantity of longer partitions growing, size of VSE-OP after reordering decreases to half of the original as shown in Table 1, which again confirms that long partitions chosen by VSE-OP do not add load on space occupancy. Figure 4(b) for frequency shows similar result with that for docid, the only difference is that reordering does not affect the distribution as acutely as that in Fig. 4(a), this can be explained by the fact that reordering has no effect on the value range of frequency sequences.

Table 4. Total time elapsed in seconds and performance in million integers per second (mis) when decompressing docid and frequency

methods	original				reordered			
	docid	mis	freq	mis	docid	mis	freq	mis
AFOR	36.53	127.04	31.24	148.55	33.01	140.56	28.77	161.36
VSE-DP	69.09	67.17	71.97	64.48	52.98	87.60	70.11	66.19
VSE-OP	48.24	96.20	49.81	93.16	40.15	115.58	48.56	95.56

Table 4 reports the results on decompression speed, again our optimized method outperforms its original version. The decompression speed are much faster than the compression speed, however, with more partitions included, VSEncoding will encounter more skips when decompressing, which slows down its speed; on the other hand, long partitions also enable decompressing more integers in bulk. In a nutshell, the speed achieved by VSEncoding, specially by VSE-OP, is still competitive with AFOR, taking the compression ratio into account. Restrained by the configuration of our platform, the results in our experiment are quite different from that in [13], where VSE-DP is said to reach a speed of 450 ± 20 mis.

5   Conclusion and Future Work

In this paper we introduced and motivated the study of shortening compression time of inverted index via optimizing partitioning strategies. We first summarized a series of compression techniques which fall into the same category by treating the partitioning problem as SSSP over a DAG. Then we presented our optimization on VSEncoding with a better space-time trade-off, namely to enhance its partitioning procedure with more flexibility and faster speed, while keeping its compression ratio and decompression speed competitive. At last, an extensive experimental analysis was given, which showed that our optimization significantly improve the compression speed as well as the decompression speed, with a little loss at in space efficiency.

There are still many open problems and opportunities for future research, since our solution only focus on optimizing VSEncoding, further experiments will investigate the consequence of optimizing other compression techniques and design a compression that offers better space-time trade-offs. An interesting problem would be using SIMD instructions to further accelerate the compression and decompression speed, also it would be promising to decrease the time complexity of the approximation algorithm adopted to compute optimal partitions.

References

1. Anh, V.N., Moffat, A.: Index compression using fixed binary codewords. In: Proceedings of the 15th Australasian Database Conference, vol. 27, pp. 61–67. Australian Computer Society, Inc. (2004)
2. Anh, V.N., Moffat, A.: Inverted index compression using word-aligned binary codes. Inf. Retr. **8**(1), 151–166 (2005)
3. Anh, V.N., Moffat, A.: Index compression using 64-bit words. Softw. Pract. Exp. **40**(2), 131–147 (2010)
4. Catena, M., Macdonald, C., Ounis, I.: On inverted index compression for search engine efficiency. In: de Rijke, M., Kenter, T., de Vries, A.P., Zhai, C.X., de Jong, F., Radinsky, K., Hofmann, K. (eds.) ECIR 2014. LNCS, vol. 8416, pp. 359–371. Springer, Heidelberg (2014)
5. Cormen, T.H., Leiserson, C.E., Rivest, R.L., Stein, C.: Introduction to Algorithms, 3rd edn. MIT Press, Cambridge (2009)
6. Delbru, R., Campinas, S., Samp, K., Tummarello, G.: Adaptive frame of reference for compressing inverted lists. Technical report, DERI-Digital Enterprise Research Institute, December 2010
7. Ferragina, P., Nitto, I., Venturini, R.: On optimally partitioning a text to improve its compression. Algorithmica **61**(1), 51–74 (2011)
8. Goldstein, J., Ramakrishnan, R., Shaft, U.: Compressing relations and indexes. In: Proceedings of 14th International Conference on Data Engineering, pp. 370–379. IEEE (1998)
9. Lemire, D., Boytsov, L.: Decoding billions of integers per second through vectorization. Softw. Pract. Exp. **45**(1), 1–29 (2015)
10. Manning, C.D., Raghavan, P., Schütze, H., et al.: Introduction to Information Retrieval, vol. 1. Cambridge university press, Cambridge (2008)
11. Ottaviano, G., Tonellotto, N., Venturini, R.: Optimal space-time tradeoffs for inverted indexes. In: Proceedings of the Eighth ACM International Conference on Web Search and Data Mining, pp. 47–56. ACM (2015)
12. Ottaviano, G., Venturini, R.: Partitioned elias-fano indexes. In: Proceedingsof the 37th International ACM SIGIR Conference on Research & Development in Information Retrieval, pp. 273–282. ACM (2014)

13. Silvestri, F., Venturini, R.: Vsencoding: efficient coding and fast decoding of integer lists via dynamic programming. In: Proceedings of the 19th ACM International Conference on Information and Knowledge Management, pp. 1219–1228. ACM (2010)
14. Stepanov, A.A., Gangolli, A.R., Rose, D.E., Ernst, R.J., Oberoi, P.S.:SIMD-based decoding of posting lists. In: Proceedings of the 20th ACM International Conference on Information and Knowledge Management, pp. 317–326. ACM (2011)
15. Trotman, A.: Compression, SIMD, and postings lists. In: Proceedings of the 2014 Australasian Document Computing Symposium, p. 50. ACM (2014)
16. Witten, I.H., Moffat, A., Bell, T.C.: Managing Gigabytes: Compressing and Indexing Documents and Images. Morgan Kaufmann, San Francisco (1999)
17. Yan, H., Ding, S., Suel, T.: Inverted index compression and query processing with optimized document ordering. In: Proceedings of the 18th International Conference on World Wide Web, pp. 401–410. ACM (2009)
18. Zobel, J., Moffat, A.: Inverted files for text search engines. ACM Comput. Surv. (CSUR) 38(2), 6 (2006)

The Analysis for Ripple-Effect of Ontology Evolution Based on Graph

Qiuyao Lv[1,2], Yingping Zhang[1,2], and Jinguang Gu[1,2(✉)]

[1] School of Computer Science and Technology,
Wuhan University of Science and Technology, Wuhan 430065, China
simon@wust.edu.cn
[2] Key Laboratory of Intelligent Information Processing and Real-time Industrial
System in Hubei Province, Wuhan 430065, China

Abstract. Ontology is an important foundation for the Semantic Web and ontology-driven applications. While real network and application requirements are constantly changing, in order to adapt to these changes, the ontologies need to be updated in time. There are various types of semantic relationships, between the elements in ontology that the strength varies as well as their impacts of the process of evolution. To reflect them to the analysis of the ripple-effect of the ontology evolution, the common types of semantic relationships in ontology was extracted, and on this basis, proposed SRG (Semantic Relationship Graph) graph model in which the property-related semantic type substitutes the property and established the matrix and reachability matrix of the relationships among the elements in the ontology corresponding to this graph model. Then, the ripple-effect during the process of ontology evolution was analyzed via using matrix operations to calculate the comprehensive influence of the node in ontology. Experimental results show that the proposed method can provide accurate quantitative analysis for ripple-effect of ontology evolution under the premise of efficiency is possible.

Keywords: Ontology evolution · Volatility · Semantic relationship · Ontology graph · Reachability matrix

1 Introduction

Ontology evolution is the process of making a timely adjustment to the changes in the ontology and maintaining the consistency of the ontology in the process of evolution [1–4]. In recent years, ontology has been more and more popular, and has become the choice of building a shared knowledge model in many fields. Over time, ontology has to be updated timely in order to adapt to the situation where the need of users for application and domain knowledge have been changing. The key point of ontology evolution is how to keep the compatibility and consistency of the ontology.

Stojanovic et al. [5–7] proposed a widely recognized ontology evolution process model in 2002 and a way to manage the process of ontology evolution based on user-driven. From the literature [5] we can know that there may

© Springer International Publishing Switzerland 2016
O. Gervasi et al. (Eds.): ICCSA 2016, Part IV, LNCS 9789, pp. 261–276, 2016.
DOI: 10.1007/978-3-319-42089-9_19

be a variety of evolutionary strategies to meet a demand. It is not hard to see from a comparative analysis of the process of ontology evolution and model evolution in the literature [8] that the process of ontology evolution is very complex, in order to quantify the cost of each evolutionary strategy, so the less cost of the evolutionary strategy will be selected correctly, we need to analyze the mechanism of the ontology evolution in order to facilitate the selection of the evolution strategy. According to the analysis of volatility of ontology evolution, Jin and Liu [9] proposed a quantitative analysis method with matrix manipulation, quantization process based on Ontology semantic, However, it does not take the semantic relation intensity between ontology elements in, Thus, this method can only reflect the relevance of the elements in the ontology and can't express the strong or weak of the correlation.

Because there is a semantic relationship between ontology elements, the change of an ontology element will lead to the corresponding changes in other elements of the ontology, and different semantic relations will bring about different effects on evolution, in order to quantify the influence of the change of semantic relationships, the author proposes a SRG graph model in which the property-related semantic type replace the property. And then establish the relationship matrix and reachability matrix corresponding to the graph model. Finally, the use of matrix operation to analyze the mechanism of the evolution of the ontology. There are two key points in the article:

1. Extracting the semantic relations among the ontology terms, and constructing the ontology semantic relation graph model which contains the attribute relations.
2. Ranking and assignment according to strength of semantic relationships among different types of semantic relationships in ontology and then injecting semantic relationships into the quantitative analysis of ripple-effect in ontology evolution.

The rest of this paper is organized as follows: In Sect. 2, we mainly introduce the role of property in ontology evolution. The discovery of semantic relationships, together with building the SRG graph model would be presented in Sect. 3. In Sect. 4, we would construct reachability matrix corresponding semantic relationship graph model and then conduct quantitative analysis of ripple-effect in ontology evolution by matrix transformation.

2   The Function of Property in Ontology Evolution

An ontology is mainly composed of class, property and instance. Due to the semantic relationships among elements in ontology, a change of one element may affect other elements [10]. In this chapter, the function of the property in the process of ontology evolution will be analyzed in detail, and then the opinion that there is no need to treat property as entity to be taken into account in the analysis of ripple effect in the process of ontology evolution proposed.

The Function of Property. In general, the domain and range of a property will not be easily changed, but the possibility of change can't be denied. By the definition of the property [11], A property is composed of domain and range, which may change through the following three ways: domain change, range change, domain and range changes simultaneously.

With respect to an ontology, All changes in the evolution process can be determined by the two basic operations—add and remove, and therefore the changes of a property above can be transformed into the two basic operations—AddProperty and RemoveProperty [12]. For ease of analysis, a simple example of the ontology is given is Fig. 1:

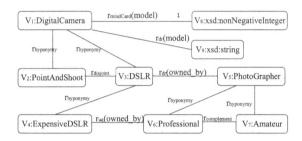

Fig. 1. A simple example of ontology

In Fig. 1, a property named owned-by who has DSLR as its domain and PhotoGrapher as range is contained. When one or all of the domain and range of owned-by have changes, we can consider it as a deletion operation: RemoveProperty (owned-by), and then the association and semantic relationships between DSLR and PhotoGrapher will no longer exist. At the same time a new owned-by property will be created according to the needs. A new connection will be made between the domain and range of the new property. By the above process, adding and deleting of a property will not cause substantial impact on the resource entity itself, only change is the semantic relationship between the two resources associated by the property. From the above analysis, we can see that the role of properties in the ontology is mainly used to construct the semantic relations among different resources. In view of this, we can transform the properties into the semantic relations between resources, and construct the ontology evolution graph model without property entities. In addition to the relationships (such as inherited, disjoint, complement, etc.) among elements in the graph model in itself, There exist a kind of semantic relations among different resources, which is based on the relationship between properties, such as domain and range, allValuesFrom, cardinality, etc. Then in the process of ontology evolution, the change of properties is used as an additional condition to update the semantic relations between different resources.

The semantic relations between the feature of attribute and the attribute constraint of the ontology terms will be refined in the following, and then it is reflected in the ontology semantic relationship graph, and finally it is reflected in the process of analyzing the evolution of ontology.

3   Extracting Semantic Relationships in Ontology

The semantic relationships among elements in ontologies may lead to conflicts between the terms defined in the ontology in the process of ontology evolution, which can result in an inconsistent state of the ontology [10]. Then, it can be inferred that the primary cause of one element's changes having the ability to affect others is the semantic relationships among them. Based on this, the influence of different semantic relationships should be taken into account on the analysis of the ripple effects of ontology evolution. This chapter will propose and perfect the different semantic relations among the ontology terms.

3.1   Semantic Relationships

In the ontology, the elements with their relationships form the semantic basis of ontology, 11 kinds of semantic relations are proposed in the process of constructing the virtual ontology by Zhang Xiang (As in the first 11 rows in Table 1). Then the 11 types of semantic relations are divided int 4 grades according to the intensity of semantic relations [13], as shown in Table 2. These 11 types of semantic relations are mainly used to represent the relationships between the classes of the ontology. However, in an ontology, there have other links which represent the relations between a class and the data type. As shown in Fig. 1, the DigitalCamera class establishes a connection with the data type xsd:nonNegativeInteger by the model property, used to indicate that a digital camera has at least one different model. These 11 semantic relations above can not cover this kind of relationship, and therefore three semantic relationships are needed to add additional: r_{card}, $r_{mincard}$, $r_{maxcard}$, The additional relationships as well as r_{uq}, r_{eq} represent the semantic relationships between two terms established by the triple structure of owl:onProperty. Among them, r_{card}, $r_{mincard}$, $r_{maxcard}$ represent constraints of owl:cardinality, owl:minCardinality and owl:maxCardinality, its meaning is shown in Table 1.

Although the semantic relationship has been proposed can express most common semantic relation types, the relations caused by other operations such as collection and reference can't be represented, such as owl:unionOf is not expressed, So the semantic relations among the ontology terms after the completion of this paper are shown in Table 1. The last column in the table represents the relational complexity of each type of semantic relations. According to the author's definition, the complexity of the semantic relationship between the terms of the class is shown by a single or multiple triplets or an implicit description, the relational complexity refers to the number of triplets contained in the structure of the triplets, the less triplets semantic relations contain, the lower the two categories of the relationship is, it is indicated that there is a more direct relationship between the two categories of the relationship [13].

In this paper, owl:minCardinality is used to describe the semantic relationships between DigitalCamera and xsd:nonNegativeInteger through (DigitalCamera, owl:subClassOf, anonymity1), (anonymity1, model, someValues) and (someValues, owl:minCardinality, xsd:nonNegativeInteger), therefore, the complexity of owl:minCardinality is 3.

3.2 Quantization for the Strength of Semantic Relationship

According to the relationship of the Table 1, the author makes a preliminary division of the relationship between the types of semantic relations, the equivalence relation is special, which represents the equivalence between the two entities, the semantic relationship is the most intense and the semantic distance is zero, so they put them separately as the highest level, and the semantic relations level is shown in Table 2 after dividing the level. In the process of ontology evolution,

Table 1. Semantic relationships

Symbol	Representation	Name	Complexity
$r_{equivalent}$	owl:equivalentClass	Equivalent relation (class)	1
$r_{hyponymy}$	rdfs:subClassOf	Hyponymy relation	1
$r_{disjoint}$	owl:disjointWith	Disjoint relation	1
$r_{complement}$	owl:complementOf	Complement relation	1
r_{uq}	owl:allValuesFrom	Universal quantification relation	3
r_{eq}	owl:someValuesFrom	Existential quantification relation	3
r_{dr}	owl:ObjectProperty	Domain-range relation	2
r_{drf}	owl:FunctionalProperty	Functional domain-range relation	2
r_{drif}	owl:InverseFunctional Property	Inverse functional domain-range relation	2
r_{drs}	owl:SymmetricProperty	Symmetric domain-range relation	2
r_{drt}	owl:TransitiveProperty	Transitive domain-range relation	2
r_{card}	owl:cardinality	Base value constraints relation	3
$r_{minCard}$	owl:minCardinality	Minimum base value constraints relation	3
$r_{maxCard}$	owl:maxCardinality	Maximum base value constraints relation	3
r_{union}	owl:unionOf	Inclusion relation	1
r_{same}	owl:sameAs	Equivalent relation (class, instance)	1

the author set up the corresponding weights according to the classification level. As shown in Table 2, "weight" is as a quantitative parameter to analyze the evolution of the ontology. It can set its value $\alpha0 = 1$ in view of the particularity of the equivalence relation. Thus, the value of the semantic relationship level has the following relations: $1 = \alpha0 > \alpha1 > \alpha2 > \alpha3 > 0$.

Table 2. Levels of semantic relationships

Relationship level	Relationship type	Weight
The first level of relationship	$r_{equivalent}, r_{same}$	$\alpha0$
The second level of relationship	$r_{hyponymy}, r_{disjoint}, r_{complement}, r_{union}$	$\alpha1$
The third level of relationship	$r_{dr}, r_{drf}, r_{drif}, r_{drs}, r_{drt}$	$\alpha2$
The fourth level of relationship	$r_{uq}, r_{eq}, r_{card}, r_{minCard}, r_{maxCard}$	$\alpha3$

3.3 Construction of SRG Graph Model

In order to make use of the method of graph theory to analyze the process of ontology evolution, we first need to construct a graph of the semantic relations among the elements in the ontology. The definition and construction method of Semantic Relationship Graph are presented below, from the above, we can know that the ontology properties are not directly involved in the evolutionary analysis in the process of evolution, but only as a link between the creation and updating of different terms. Thus, the semantic relationship graph does not contain the property entity itself, but the semantic relations between the terms which are caused by the property.

Definition 1 (SRG): The semantic relation graph model of ontology is a kind of undirected graph with labels $G = < V, E >$, where the node set $V = < v_i : entityname >$ represents the entity set in the ontology (including concepts, instances, etc.), v_i represents the node sequence of ontology elements, entityname represents the name of the entity of the node, and edge set with the form of $E = < r, v_i, v_j >$ represents that there is some semantic relationship between the class term v_i and v_j through the relation of r (semantic relationships is shown in Table 1).

For ease of understanding, a simple example of SRG Graph would like to be presented to explain as shown in Fig. 2:

In Fig. 2, $r_{hyponymy}$ represents the hyponymy relation (i.e. inheritance) between the two entities connected by the edge; $r_{minCard}$ represents the minimum base value constraints relation between V_1 and V_8, That is at least one model of DigitalCamera; $r_{disjoint}$ indicates that there is no intersection between V_2 and V_3, therefore their children and grandchildren do not have any contact too; $r_{complement}$ represents a complementary relationship between the V_6 and V_7, and then we can assume a Professional has been defined and therefore we

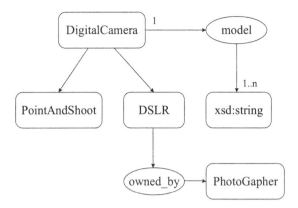

Fig. 2. A simple example for SRG model

could use the complementary relationship related to Professional to define the amateur photographer; r_{dr} (owned-by) and r_{uq} (owned-by) denote respectively the domain range relationship and universal quantification relationship.

The model consists of 9 nodes and 11 edges in the graph, the node represents the ontology elements and edge represents the semantic relationship among elements in ontology. The directions of the arcs fixed through the meanings of the labels on the edges are as follows:

$$e_1 = < r_{hyponymy}, V_1, V_2 >$$
$$e_2 = < r_{hyponymy}, V_1, V_3 >$$
$$e_3 = < r_{disjoint}, V_2, V_3 >$$
$$e_4 = < r_{hyponymy}, V_3, V_4 >$$
$$e_5 = < r_{dr}, V_5, V_3 >$$
$$e_6 = < r_{hyponymy}, V_5, V_6 >$$
$$e_7 = < r_{hyponymy}, V_5, V_7 >$$
$$e_8 = < r_{uq}, V_6, V_4 >$$
$$e_9 = < r_{complement}, V_6, V_7 >$$
$$e_{10} = < r_{minCard}, V_8, V_1 >$$
$$e_{11} = < r_{dr}, V_9, V_1 >$$

Among them, direction of the edge in $< r, v_i, v_j >$ (from v_i to v_j) represents the dependence from ending to starting node, and the label r represents the semantic relationship between the start point and the end point of the edge. By the above analysis, we can use the method of graph theory to study the ontology, the adjacency matrix corresponding to Fig. 2 is as shown in Table 3. The direction of the semantic relations in the graph is represented by rows and columns, where the row represents the first element and the column represents the second element. Zero in the Table represents the fact that there does not exist or exist indirect semantic relationship between the two elements in the ranks,

Table 3. Adjacency matrix for Fig. 2

	v1	v2	v3	v4	v5	v6	v7	v8	v9
v1	0	$\alpha1$	$\alpha1$	0	0	0	0	0	0
v2	0	0	$\alpha1$	0	0	0	0	0	0
v3	0	0	0	$\alpha1$	0	0	0	0	0
v4	0	0	0	0	0	0	0	0	0
v5	0	0	$\alpha2$	0	0	$\alpha1$	$\alpha1$	0	0
v6	0	0	0	$\alpha3$	0	0	$\alpha1$	0	0
v7	0	0	0	0	0	0	0	0	0
v8	$\alpha3$	0	0	0	0	0	0	0	0
v9	$\alpha2$	0	0	0	0	0	0	0	0

αi represents that the value of strength of the semantic relationship between the two elements in the ranks is αi.

The advantages of constructing SRG graph model:

1. The entities in the SRG graph model only contain concepts and examples, which can reflect the relationship between the various entities more clearly;
2. The edge set in the SRG graph model contains a wealth of semantic information, which can reflect the meaning of the relationship and the degree of the relationship between the various entities;
3. Different from existing Ontology graph model, SRG graph model omitting the properties of entities, it is more simple and intuitive and reduce the evolution of complexity under the premise of not losing the original ontology information.

4   Analysis on the Ripple-Effect of Ontology Matrix and Ontology Evolution

4.1   Ontology Matrix

We convert SRG graph model to corresponding adjacency matrix of semantic relationships of ontology with the form of MR = $[R_{ij}]$, R_{ij} represents the value of the strength of direct semantic relationship between ontology elements e_i and e_j, the expression is like the formula (1):

$$R_{ij} = \begin{cases} \alpha_t & \text{the intensity of direct semantic relationship between } e_i \text{ and } e_j \text{ is } \alpha_t (1 \geq \alpha_t > 0) \\ 0 & \text{there is no direct semantic relationship between } e_i \text{ and } e_j \end{cases} \quad (1)$$

In this paper, the values of α_1, α_2 and α_3 take $1/2$, $1/3$ and $1/4$ in turn, therefore, the corresponding ontology semantic relationship matrix of Fig. 2 is as follows:

$$M_R = \begin{bmatrix} 0 & 1/2 & 1/2 & 0 & 0 & 0 & 0 & 0 & 0 \\ 0 & 0 & 1/2 & 0 & 0 & 0 & 0 & 0 & 0 \\ 0 & 0 & 0 & 1/2 & 0 & 0 & 0 & 0 & 0 \\ 0 & 0 & 0 & 0 & 0 & 0 & 0 & 0 & 0 \\ 0 & 0 & 1/3 & 0 & 0 & 1/2 & 1/2 & 0 & 0 \\ 0 & 0 & 0 & 1/4 & 0 & 0 & 1/2 & 0 & 0 \\ 0 & 0 & 0 & 0 & 0 & 0 & 0 & 0 & 0 \\ 1/4 & 0 & 0 & 0 & 0 & 0 & 0 & 0 & 0 \\ 1/3 & 0 & 0 & 0 & 0 & 0 & 0 & 0 & 0 \end{bmatrix}$$

With the matrix of ontology semantic relations, the reachability of any two nodes and the semantic relationship between them can be obtained. Since the effect of the semantic relationships among elements has characteristic of propagation, the idea of multiplicative is used to compute the strength of indirect semantic relationships, for example, since $R_{13} = 1/2$ and $R_{34} = 1/2$, then $R_{14} = 1/2 * 1/2 = 1/4$. Following the reachability matrix would to be defined with the form of $M_R^+ = [R_{ij}^+]$, in which R_{ij}^+ represents the value of the strength of semantic relationship between ontology elements e_i and e_j and its expression is as the formula (2):

$$R_{ij}^+ = \begin{cases} \alpha & \text{the intensity of semantic relationship between } e_i \text{ and } e_j, (1 \geq \alpha > 0) \\ 0 & \text{there is no direct semantic relationship between } e_i \text{ and } e_j \end{cases} \tag{2}$$

According to the definition of ontology reachability matrix, the results we translate the semantic relationship matrix into the reachable matrix as follows:

$$M_R^+ = \begin{bmatrix} 0 & 0.5 & 0.5 & 0.25 & 0 & 0 & 0 & 0 & 0 \\ 0 & 0 & 0.5 & 0.25 & 0 & 0 & 0 & 0 & 0 \\ 0 & 0 & 0 & 0.5 & 0 & 0 & 0 & 0 & 0 \\ 0 & 0 & 0 & 0 & 0 & 0 & 0 & 0 & 0 \\ 0 & 0 & 0.333 & 0.166 & 0 & 0.5 & 0.5 & 0 & 0 \\ 0 & 0 & 0 & 0.25 & 0 & 0 & 0.5 & 0 & 0 \\ 0 & 0 & 0 & 0 & 0 & 0 & 0 & 0 & 0 \\ 0.25 & 0.125 & 0.125 & 0.063 & 0 & 0 & 0 & 0 & 0 \\ 0.333 & 0.167 & 0.167 & 0.083 & 0 & 0 & 0 & 0 & 0 \end{bmatrix}$$

An important benefit of the reachability matrix element is that it is convenient to conduct the quantitative analysis of the ripple-effect of ontology evolution. This will be analyzed in the next section.

4.2 Quantitative Analysis of the Ripple-Effect of Ontology Evolution

In the study of the ripple-effect of ontology evolution, the following two situations are needed:

1. What degree the other elements will be affected when one element have changes;

2. Comprehensive influence of this element on the whole ontology.

After the above analysis, the first case is easy to understood, When the element ei of ontology changes, the influence another element ej received is the value of the element in the matrix represented R_{ij}^+, That is:

$$Affected(e_j) = R_{ij}^+; \tag{3}$$

For example, when DigitalCamera changed, the intensity of ExpensiveDSLR was $R_{03}^+ = 0.25$.

A simple understanding of the second is that how many elements could be affected by the element ei and how much influence it could bring. This aspect can be achieved by summing up the elements in the row including the element ei of the reachability matrix, that is:

$$Comprehensive(ei) = \sum_{j=0}^{j=n-1} R_{ij}^+; \tag{4}$$

For example, PointAndShoot, DSLR, and ExpensiveDSLR are affected by the change of DigitalCamera. The comprehensive influence of the elements on the ontology is: $R_{01}^+ + R_{02}^+ + R_{03}^+ = 1.25$. After analysis and calculation, the result of the ontology in Fig. 1 is as shown in Table 4.

Table 4. Comprehensive influence of node

Node Vi	Number of influenced node	Comprehensive
V_1	3	1.250
V_2	2	0.750
V_3	1	0.500
V_4	0	0.000
V_5	4	1.458
V_6	2	0.750
V_7	0	0.000
V_8	4	0.562
V_9	4	0.749

5   Experiments and Evaluation

In this section, we mainly evaluate the evolution method which is mentioned above in order to demonstrate the feasibility and effectiveness of the method. In addition, we analyze the experimental results compared with results of the Literature [9].

Experiment. Experimental data in this article is from experimental ontology in the internet, whose language is OWL DL language belonging to OWL, therefore the source of data is reliable. We would get statistics about ontology by extracting and classifying entities among ontology (including concepts, properties and instances) as shown in Table 5 (♯ represents the number of unit). Which relates to the type of semantic relations and the use of the number of times of the composition in the ontology as shown in Table 6.

Table 5. Statistical information of experimental Ontology

♯ class	♯ property	♯ instance	♯ node	♯ edge
64	8	25	89	217

Table 6. Types of semantic relations and the number of references

Types of semantic relations	Number of references
$r_{hyponymy}$	69
$r_{complement}$	1
$r_{disjoint}$	38
$r_{equivalent}$	1
r_{same}	12
r_{dr}	4
$r_{minCard}$	3
r_{uq}	25
r_{eq}	60
r_{union}	4

The ordering of the nodes and edges in the ontology is ordered by the order of the nodes and edges in the ontology. In this paper, we construct the corresponding ontology relationship matrix by the ontology nodes and edge set information. The comprehensive influence of every node in ontology on the ontology is obtained by the operation of matrix, as shown attached Table 7 (part of the results). The experimental results of this paper are listed in the fourth column of the table, and the last column represents the experimental results of Literature [9]. As shown in the table, according to the experimental results, the 47 node MealCourse has the greatest comprehensive influence of the ontology is 13.569, as many as 54 nodes are affected when the entity changes. Through the analysis of the experimental results is not difficult to see that there is an nonstrict positive correlation between the influence of the node on ontology and the number of

Table 7. The experimental results of the comprehensive influence of each node

Node	Node name	Number of nodes affected	Comprehensive influence(this paper)	Comprehensive influence (Literature [8])
1	ConsumableThing	49	6.062	49
2	NonConsumableThing	47	3.013	47
3	EdibleThing	54	6.712	54
4	PotableLiquid	54	6.170	54
5	Wine	46	4.807	46
6	Juice	0	0.000	0
7	White	8	1.438	8
8	Rose	1	1.000	1
9	Red	11	1.547	11
10	Sweet	7	1.242	7
11	OffDry	2	1.062	2
12	Dry	13	1.750	13
13	Delicate	6	1.266	6
14	Moderate	9	1.469	9
15	Strong	9	1.336	9
16	Light	4	1.141	4
17	Medium	9	1.500	9
18	Full	8	1.305	8
19	Grape	1	0.500	1
20	Pasta	57	7.367	57
21	PastaWithWhiteSauce	2	1.000	2
22	PastaWithSpicyRedSauceCourse	0	0.000	0
23	PastaWithSpicyRedSauce	1	0.250	1
24	PastaWithRedSauce	0	0.000	0
25	PastaWithNonSpicyRedSauceCourse	0	0.000	0
26	PastaWithNonSpicyRedSauce	3	0.875	3
27	PastaWithLightCreamSauce	3	0.812	3
28	PastaWithLightCreamCourse	1	0.250	1
29	PastaWithHeavyCreamSauce	0	0.000	0
30	PastaWithHeavyCreamCourse	0	0.000	0
31	OysterShellfishCourse	0	0.000	0
32	OysterShellfish	1	0.250	1
33	OtherTomatoBasedFoodCourse	0	0.000	0
34	OtherTomatoBasedFood	57	7.346	57
35	NonSweetFruitCourse	0	0.000	0
36	SweetFruitCourse	0	0.000	0
37	NonSweetFruit	4	1.125	4

Table 7. (*Continued*)

Node	Node name	Number of nodes affected	Comprehensive influence(this paper)	Comprehensive influence (Literature [8])
38	NonSpicyRedMeatCourse	0	0.000	0
39	NonSpicyRedMeat	2	0.625	2
40	NonRedMeatCourse	0	0.000	0
41	NonRedMeat	6	1.195	6
42	NonOysterShellfishCourse	0	0.000	0
43	NonOysterShellfish	3	0.875	3
44	NonBlandFishCourse	0	0.000	0
45	NonBlandFish	1	0.250	1
46	Meat	56	6.762	56
47	MealCourse	54	13.569	54
48	Meal	54	7.701	54
49	LightMeatFowlCourse	0	0.000	0
50	LightMeatFowl	1	0.250	1
51	FruitCourse	0	0.000	0
52	Fruit	50	2.560	50
53	Fowl	55	4.536	55
54	FishCourse	0	0.000	0
55	DessertCourse	4	1.000	4
56	Dessert	52	3.077	52
57	SweetFruit	1	0.250	1
58	SweetDessertCourse	0	0.000	0
59	SweetDessert	3	1.000	3
60	DarkMeatFowlCourse	0	0.000	0
61	DarkMeatFowl	3	0.875	3
62	CheeseNutsDessertCourse	0	0.000	0
63	CheeseNutsDessert	5	1.250	5
64	BlandFishCourse	0	0.000	0
65	BlandFish	3	0.875	3
66	Fish	11	2.656	11
67	SpicyRedMeat	1	0.250	1
68	ShellfishCourse	0	0.000	0
69	Shellfish	5	1.438	5
70	SeafoodCourse	0	0.000	0
71	Seafood	56	4.880	56
72	RedMeatCourse	3	0.812	3
73	RedMeat	4	0.891	4
74	EatingGrape	0	0.000	0
75	SpicyRedMeatCourse	0	0.000	0

Table 7. (*Continued*)

Node	Node name	Number of nodes affected	Comprehensive influence(this paper)	Comprehensive influence (Literature [8])
76	vin;Wine	0	0.000	0
77	xsd;nonNegativeInteger	54	2.518	54
78	vin;White	0	0.000	0
79	vin;Rose	0	0.000	0
80	vin;Red	0	0.000	0
81	vin;Sweet	0	0.000	0
82	vin;OffDry	0	0.000	0
83	vin;Dry	0	0.000	0
84	vin;Delicate	0	0.000	0
85	vin;Moderate	0	0.000	0
86	vin;Strong	0	0.000	0
87	vin;Light	0	0.000	0
88	vin;Medium	0	0.000	0
89	vin;Full	0	0.000	0

nodes affected by the node. Generally speaking, the more the number of nodes affected by the node, the greater the influence of the node on ontology, but there is a small range of ups and downs, such as 2nd and 5nd node, the number of nodes affected by 2nd is slightly less than 5nd, but its comprehensive influence is greater than 5nd.

Comparison of the results of the literature [9] can be found that the number of nodes affected by nodes is just the comprehensive influence of the node to the ontology, in other words, there is a positive correlation between the influence of the node on ontology and the number of nodes affected by the node. The experimental result also can reflect the degree of importance of each node to ontology, however, compared with this paper, the quantitative granularity is obviously too extensive and the main reason is that they do not introduce the strength of semantic relations into the quantitative process. According to Table 7 can be clearly seen that when node 3 or node 4 changes, the number of nodes in the ontology are affected by 54, But in this paper considering the types of semantic relations, the comprehensive influence of node 3 and node 4 of the ontology were 6.712 and 6.170. Although the number of nodes are affected by the same, but considering the semantic relationship between nodes of different types, not directly to the number of nodes affected as the comprehensive influence of the body. In this paper, node 3 is slightly larger than node 4 through the analysis of comprehensive influence. This situation is very rare in the experiment of the literature [9] and in this article can be a very good show.

Through the comparison of the above, experimental result in this article is superior to the result of predecessors. In this paper, the algorithms is based on

Warshall algorithm [14] which is to an efficient algorithm to solve binary relation and transmission package. Its time complexity is $O(n^3)$ and space complexity is $O(n^2)$. The time and space consumed by the algorithm are mainly determined by the order of the matrix, which provides a guarantee for the feasibility and validity of this experiment.

6 Conclusion

In this paper, we start from the semantic relationship between ontology elements, extract the type of semantic relations in the ontology, analyze the function of attributes to ontology elements, and then use semantic attributes replace property type method based on this proposed SRG Figure model. The relationship matrix and the reachable matrix of the ontology elements are established according to the SRG graph model, in the end, the ripple-effect of the ontology evolution is analyzed by the operation and transformation of the matrix.

However, the paper of different types of semantic relations semantic strength and the strength of the division is also not detailed enough to quantify the size, So the analysis and quantification of the fluctuation mechanism of the ontology evolution is not accurate enough. The semantics of whether the strength of quantization coupled with the real needs of the environment in the ontology to some extent remains to be studied, these will be the next research work.

Acknowledgement. This work was partially supported by a grant from the NSF (Natural Science Foundation) of China under grant number 61272110, the Key Projects of National Social Science Foundation of China under grant number 11&ZD189, and it was partially supported by a grant from NSF of Hubei Prov. of China under grant number 2013CFB334. It was partially supported by NSF of educational agency of Hubei Prov. under grant number Q20101110, and the State Key Lab of Software Engineering Open Foundation of Wuhan University under grant number SKLSE2012-09-07.

References

1. Shang, J., Zhang, R., Lu, S., Liu, L.: Impact analysis of ontology evolution based on dependency graph model. J. Jilin Univ. **50**(1), 89–94 (2012)
2. Liu, J., Zhang, Y., Li, S., Gu, L., Zhu, C., Zhu, L.: Research progress of ontology evolution. Comput. Syst. Appl. **20**(7), 239–243 (2011)
3. Liu, L., Fan, R., Zhang, R., Lu, S., Zhang, Y.: SetPi-calculus and Modeling Method for Ontology Evolution. Sciencepaper Online (2010)
4. Liu, S.: Study on enterprise ontology evolution method based on minimal ripple-effect. Electro-Mech. Eng. **30**(2), 51–56 (2014)
5. Stojanovic, L., Maedche, A., Motik, B., Stojanovic, N.: User-driven ontology evolution management. In: Gómez-Pérez, A., Benjamins, V.R. (eds.) EKAW 2002. LNCS (LNAI), vol. 2473, pp. 285–300. Springer, Heidelberg (2002)
6. Stojanovic, L.: Methods and tools for ontology evolution. Inf. Syst. Methodol. **16**, 411–423 (2004)

7. Maedche, A., Motik, B., Stojanovic, L.: Managing multiple and distributed ontologies on the semantic web. VLDB J. **12**(4), 286–302 (2003)
8. Noy, N.F., Klein, M.: Ontology evolution: not the same as schema evolution. Knowl. Inf. Syst. **6**(4), 428–440 (2004)
9. Jin, L., Liu, L.: A ripple-effect analysis method for ontology evolution. Acta Electronica Sinica **34**(8), 1469–1474 (2006)
10. Stojanovic, L., Maedche, A., Stojanovic, N., Studer, R.: Ontology evolution as reconfiguration-design problem solving. In: Proceedings of the 2nd International Conference on Knowledge Capture K-CA 2003, pp. 162–171. ACM, New York (2003)
11. Li, Y.: A developer's Guide to Semantic Web, pp. 161–239. Springer, Heidelberg (2011)
12. Liu, C., Han, Y., Chen, W., Wang, J.: Mini: an ontology evolution algorithm for reducing impact ranges. Chin. J. Comput. **31**(5), 711–720 (2008)
13. Zhang, X., Li, X., Wen, Y., Shen, K., Hao, J.: Building virtual ontologies in semantic web. J. SE Univ. **45**(4), 652–656 (2015)
14. Long, L.: Warshall's algorithm for transitive closure of fuzzy relation matrices. Fuzzy Syst. Math. (1), 59–61 (2003)

Linearizability Proof of Stack Data

Jun-Yan Qian, Guo-Qing Yao, Guang-Xi Chen, and Ling-Zhong Zhao[✉]

Guangxi Key Laboratory of Trusted Software,
Guilin University of Electronic Technology, Guilin 541004, China
qjy2000@gmail.com, zhaolingzhong@126.com

Abstract. In order to guarantee the linearizability of stack data, linearization points are necessary. However, searching linearization points is very difficult. We propose a simpler method which restrains stack operations to guarantee the linearizability of stack data, instead of using the linearization points. First of all, based on the description of the problem, it is shown that if stack operations violate three basic properties, then data inconsistency will occur during push/pop operation. After that, it is proved that stack data is linearizable if and only if stack operations satisfy three basic properties under the situation where a complete history is available. Lastly, extending the method to general condition, it is shown that stack data is linearizable if purely-blocking history can be completed and then meets the three properties.

Keywords: Linearizability · Linearizability points · Three basic properties · Complete history · General history

1 Introduction

Concurrent programs have some uncertain interactions when they are executing, which may easily lead to mistakes and bugs. Linearizability is widely accepted as the standard correctness criterion for concurrent programss data. The theory of linearizability and linearization points were first raised by Herlihy and Wing in 1990, they proved linearizability of HW queue via invariants and abstract functions [1]. At the same time, bibliography [1] points out that linearizability cannot be simply proved by a mapping that is defined from the concrete HW queue states to the queue abstract states, and constructing the proof which is very complicated is also needed.

Linearization point typically refer to a specific operation step that the result of a method invocation is visible to other method invocations. Because linearization point may depend on the operations of a thread at present or in the future, it is very difficult to assign the linearizability points in concurrent programs. Amit used static analysis to verify linearizability of concurrent unbounded linked data structures, and singly-linked lists were analyzed as an example [2]; Vafeiadis verified linearizability of fine-grained blocking and non-blocking concurrent algorithms [3]; Vechev and Liang respectively described methods of checking linearizability for algorithms with non-fixed linearizability points [4,5]. Derrick

© Springer International Publishing Switzerland 2016
O. Gervasi et al. (Eds.): ICCSA 2016, Part IV, LNCS 9789, pp. 277–288, 2016.
DOI: 10.1007/978-3-319-42089-9_20

employed a theorem prover to mechanize linearizability proof, and applied it to a lock-free stack and a set with lock-coupling [6]. Colvin and Derrick verified linearizability of concurrent programs through forward simulation, which is easily implemented by brute-force search for some simple concurrent algorithms, for example Treibers stack [7–9]; On this basis, Dragoi extended this approach by rewriting implementations [10]; Abdulla verified linearizability of stack and queue by automaton which can detect violations [11]. Friggens showed that shape predicates allow linearizability of concurrent data structures to be verified by using canonical abstraction alone [12]. However, all these aforementioned methods need linearization points, so searching linearization point is necessary.

Henzinger proved that HW queue is linearizable by detecting properties of the queue [13]. We extend this method to stack, it is proved that stack data is linearizable if and only if stack operations satisfy three basic properties. By comparison, this method does not involve linearization points and it is easy to verify linearizability.

The rest parts of the paper are described as follows: Sect. 2 shows a brief introduction of background study. In Sect. 3, we describe the reasons of data inconsistency. In Sect. 4, we analyze and prove linearizability of stack data. Section 5 is our conclusion and future work.

2   Background Study

A stack S is a tuple $S = (D, \sum_s)$, where D is the data domain and $\sum_s = \{push, pop\}$ contains push and pop operations. An event of S is a tuple (uid, m, d_i, d_o), where uid is a unique event identifier, $m \in \sum_s$ is the stack method, and $d_i, d_o \in D$, d_i represents an input parameter and d_o represents an output parameter. Initially, an event $e = (uid, m, d_i, d_o)$ has two actions: (1) the invocation of e, denoted by $m_i^{uid}(d_i)$; (2) the response of e, denoted by $m_r^{uid}(d_o)$.

$push(x)$ represents a push operation event, PUSH represents the set of all push operation events; $pop(x)$ represents a pop operation event, POP represents the set of all pop operation events. A non redundant event sequence of S is called a behavior. Given a behavior b, PUSH(b) denotes the set of all push operation events in b, POP(b) denotes the set of all pop operation events in b, $Val_b(e)$ represents input parameter of push operation event e or output parameter of pop operation event e in b.

Now, we introduce a labeled transition system LTS to define stack semantics. A route of LTS is an alternating sequence $t_0 e_1 t_1 \cdots e_i t_i \cdots e_n t_n$ of states t_i and stack events e_i, such that for $1 \leq i \leq n$, we have $t_{i-1} \xrightarrow{e_i} t_i$. A stack behavior $b = e_1 \cdots e_n$ if and only if there is a route $t_0 e_1 t_1 \cdots e_i t_i \cdots e_n t_n$ in LTS. A behavior b is sequential, if every pop operation event has a corresponding push operation event, or pop operation event returns NULL then it does not have a corresponding push operation event, and stack operations must follow FILO (First In Last Out).

A history c of S is a sequence of invocation and response actions of event e. $Val_c(e)$ denotes the input parameters of the push operation event e or output parameters of pop operation event e in c. Given two histories c_1 and c_2, let $c_1 \bullet c_2$ denote their concatenation, and let $c_1 \sim_{perm} c_2$ hold if c_2 is a permutation of c_1. The set of history c is denoted by C.

If the invocation and response actions of an event both occur in c, then the event is complete, if only its invocation action occurs in c, then the event will be pended. $remPending(c)$ is defined as the sub-sequence of c where all pending events have been removed. A history c is complete if there are no pending events. For an incomplete history c, its completion is denoted by $\hat{c} = remPending(c \bullet c')$, where c' contains response actions of pending events. And $Compl(c)$ represents the set of all completion of c.

Given history c and behavior b, the response action of e_1 precedes the invocation action of e_2 in c or b, written $e_1 <_c e_2$ or $e_1 <_b e_2$ respectively. And $Before(e, c)$ represents the set of all events which precede e in c. The formalized definition is:

$$Before(e, c) \triangleq \{e_1 \mid e_1 <_c e\}$$

Given history c, it is serializable if invocation actions of all events are followed by their corresponding response actions after the adjustment, and the last action is permitted to be an invocation of a pending event. Given a history c, a serializable history s is a linearization of c, if there is a completion $\hat{c} \in Compl(c)$ such that $\hat{c} \sim_{perm} s$ and whenever $e_1 <_{\hat{c}} e_2$, $e_1 <_s e_2$ holds.

Definition 1 (Legal Behavior). *Given a stack behavior b, pop operation events e_1, e_2, e_3 and push operation event e_1'. b is legal if and only if there is a mapping $\mu_{seq} : POP(b) \rightarrow PUSH(b) \cup \bot$ satisfies:*

(1) If $\mu_{seq}(e_1) = e_1'$, then $Val_b(e_1) = Val_b(e_1')$. It denotes that the output parameter of pop operation event is equal to the input parameter of its corresponding push operation event.

(2) If $\mu_{seq}(e_3) = \bot$, then $Val_b(e_3) = NULL$. It denotes that if pop operation event has no corresponding push operation event, then the output parameter of pop operation event is NULL.

(3) If $\mu_{seq}(e_1) = \mu_{seq}(e_2) \neq \bot$, then $e_1 = e_2$. It denotes that if two pop operation events have the same corresponding push operation event, then the two pop operation are same.

(4) If $\mu_{seq}(e_1) = e_1'$, then $e_1' <_b e_1$. It denotes that push operation event executes before its corresponding pop operation event.

(5) If $e_1' <_b \mu_{seq}(e_2)$, then $e_2 <_b \mu_{seq}^{-1}(e_1')$. It denotes that the order of two push operation events and the order of their corresponding pop operation events are opposite.

(6) If $\mu_{seq}(e_3) = \bot$, then $\mid \{e_i \in PUSH(b) \mid e_i <_b e_3\} \mid = \mid \{e_j \in POP(b) \mid e_j <_b e_3 \wedge \mu_{seq}(e_j) \neq \bot\} \mid$. It denotes that if a pop operation event has no corresponding push operation event, namely the stack is empty, then the number of push operation events before e_3 is equal to the one of pop operation events before e_3.

The formal definition of legal behavior will be used in Sect. 4 to prove linearizability of stack data.

Definition 2 (Linearizability). *Given a stack S, if a history c of S is serializable, and its corresponding behavior is legal, then c is linearizable with respect to S. C is linearizable with respect to S if every $c \in C$ is linearizable with respect to S.*

In the following example, we will describe how to transform concurrent stack operations to a history c, and linearize c.

The push&pop operations program is described by Algorithm 1. Assuming that the stack is unbounded and its data is stored in an array $S.items$, pointer $S.top$ points to the top of the stack, pointer $S.bottom$ points to the bottom of the stack. The push operation is done in two steps, P_1: the pushing data x is stored in the position where $S.top$ points; then P_2: $S.top$ is incremented. When pop operation executes, one should first judge whether the stack is empty or not. If the stack is empty, then P_6: NULL is returned; If the stack is not empty, it executes in three steps, P_3: $S.top$ is decremented, P_4: the data that $S.top$ points to is assigned to x, then P_5: x is returned.

Algorithm 1. The push&pop operations

var $S.top$:int	procedure $pop()$: val
var $S.bottom$:int	if($S.top! = S.bottom$){
var $S.items$:array of val	$P_3 : S.top \leftarrow S.top - 1$
procedure $push(x{:}val)$	$P_4 : x \leftarrow S.items[S.top]$
$P_1 : S.items[S.top] \leftarrow x$	$P_5 : $ return x}
$P_2 : S.top \leftarrow S.top + 1$	$P_6 : $else return NULL

An execution trace is a sequence of thread identifiers coupled with program statement labels. For example, $(t : P)$ represents the thread t executes the program statement with label P. If a label is the entry of method m, then it is written as $enter(m)$. Likewise, if an label is the exit of m, then it is written as $exit(m)$. Every execution trace τ could induce a history $h(\tau)$, where $(t : enter(m))$ is replaced by $m_i^{uid}(d_i)$, $(t : exit(m))$ is replaced by $m_r^{uid}(d_o)$ and remaining symbols are removed for conciseness and easy understanding.

We assume that t, u, v, z represent threads, $\bullet$ denotes context switches between concurrent threads. In the stack with label i, z is ready to pop the data pushed by t. In the stack with label j, v is ready to pop the data pushed by u. The multi-threaded concurrent operations are described in Fig. 1.

According to Fig. 1, we can achieve the execution trace:

$$(t : P_1) \bullet (u : P_1, P_2) \bullet (t : P_2) \bullet (v : P_3, P_4) \bullet (z : P_3, P_4) \bullet (v : P_5) \bullet (z : P_5),$$

its corresponding history is:

$$push_i^t(x) \bullet push_i^u(y) \bullet pop_i^v \bullet pop_i^z \bullet pop_r^v(y) \bullet pop_r^z(x),$$

after completing the history with response actions $push_r^t$ and $push_r^u$, the history can be linearized as follows:

$$push_i^t(x) \bullet push_r^t \bullet push_i^u(y) \bullet push_r^u \bullet pop_i^v \bullet pop_i^z \bullet pop_r^v(y) \bullet pop_r^z(x),$$

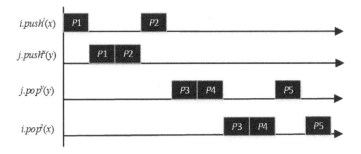

Fig. 1. Multi-threaded concurrent operations

its corresponding serializable history is:

$$push_i^t(x) \bullet push_r^t \bullet push_i^u(y) \bullet push_r^u \bullet pop_i^v \bullet pop_r^v(y) \bullet pop_i^z \bullet pop_r^z(x),$$

it can be transformed to a behavior:

$$push^t(x) \bullet push^u(y) \bullet pop^v(y) \bullet pop^z(x),$$

which is legal, so the history is linearizable.

3 Problem Proposing

There may be four kinds of data inconsistency when stacks perform push&pop operations.

Case 1 is described in Fig. 2, where thread u is ready to pop 10 pushed by thread t in the stack with label i. Thread t performs push operation, and 10 is stored in the position where $S.top$ points after P_1 executes. Then u performs pop operation and P_3 executes, $S.top$ is decremented. u will pop the datum which is next to the top of the stack not 10. Because pop operation executes before the push operation is completed, data inconsistency will occur.

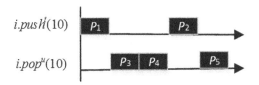

Fig. 2. Case 1: pop operation executes before push operation finishes

Case 2 is described in Fig. 3, where thread u and w both are ready to pop 9 pushed by thread u in the stack with label i. Thread u finishes push operation, then thread v performs pop operation and P_3 executes, $S.top$ is decremented.

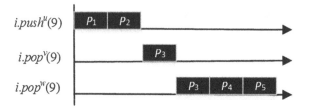

Fig. 3. Case 2: the same data performs pop operation twice

After that, w also performs pop operation and P_3 executes, $S.top$ is decremented, at last P_4 and P_5 execute. w will pop the datum which is next to the top of the stack not 9. Because the same datum is popped twice, it will lead to data inconsistency.

Case 3 is described in Fig. 4, where thread v is ready to pop 7 pushed by thread u, and thread w is ready to pop 3 pushed by thread t in the stack with label i. Thread t and u perform push operation successively, and 3 is stored in the stack preceding 7. After that, v performs pop operation and P_3 executes, $S.top$ is decremented, then w finishes pop operation, 3 is popped before 7. Because pop operations do not follow FILO, data inconsistency will occur.

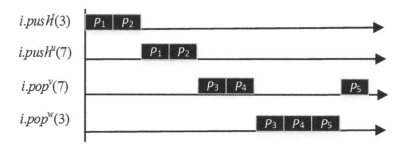

Fig. 4. Case 3: pop operations do not follow FILO

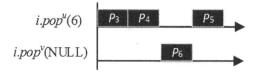

Fig. 5. Case 4: there is data returning after the stack becomes empty

Case 4 is described in Fig. 5, where thread u is ready to pop the last datum 6. Thread u performs P_3P_4, the stack is empty. Then, v performs P_6 and it returns

NULL. After that, u performs P_5 and 6 is returned. Because there is datum returned after the stack becomes empty, data inconsistency will occur.

In conclusion, to guarantee stack data linearizable, the stack operations should satisfy three properties following:

Property 1: never permit popping data which is not pushed;

Property 2: never permit the same data to be popped twice;

Property 3: the data must follow FILO and empty stacks must not perform pop operations.

In the next section, we will prove that stack data is linearizable if and only if stack operations satisfy the three properties.

4   Analysis for Stack Data Linearizability

Firstly, we will prove the linearizability of stack data on complete history. After that we extend the method to purely-blocking history, then prove the stack data is linearizable if and only if the purely-blocking history can be completed and satisfies the three properties.

4.1   Analysis on Complete History

Generally speaking, every pop operation event matches the push operation event whose value the pop operation event returns or matches nothing if the pop operation event returns NULL. Given a mapping $Match$, it is safe if it pairs every pop operation event with a corresponding push operation event. That is to say, every datum is inserted only once and removed exactly once. If a safe mapping follows FILO, then it is ordered. On this basis, an ordered mapping is linearizable if it can correctly judge whether the stack is empty or not.

Definition 3 (Safe Mapping). *Given a complete history c, a mapping $Match : POP(c) \rightarrow PUSH(c) \cup \{\bot\}$ is safe if*

- *for all $e_1 \in POP(c)$, if $Match(e_1) \neq \bot$, then $Val_c(e_1) = Val_c(Match(e_1))$.*
- *for all $e_1 \in POP(c)$, $Match(e_1) = \bot$ if and only if $Val_c(e_1) = NULL$.*
- *for all $e_1, e_2 \in POP(c)$, if $Match(e_1) = Match(e_2)$, then $e_1 = e_2$.*

Definition 4 (Ordered Mapping). *Given a complete history c and a safe mapping $Match$, $Match$ is ordered if*

- *for all $e_1 \in POP(c)$, $e_1 \not<_c Match(e_1)$ holds.*
- *for all $e_1' \in PUSH(c)$ and $e_1, e_2 \in POP(c)$, $e_1' = Match(e_1)$, if $e_1' <_c Match(e_2)$, then $e_1 \not< e_2$.*

Given a complete history c, $Match$ is an ordered mapping over c, push operation event e satisfies $e = Match(e')$, pop operation event $e_\bot \in POP(c)$ and $Val_c(e_\bot) = NULL$. The set $Bad(e_\bot, c)$ contains all push operation events e such that either e execute after $e_\bot$, or e execute before $e_\bot$ and e' execute after $e_\bot$, namely $e_\perp <_c e$ or $e <_c e_\perp <_c e'$ [14]. $Bad(e_\perp, c)$ does not contain push operation events e such that e precede $e_\perp$ and e' also precede $e_\perp$, namely $e <_c e_\perp$ and $e' <_c e_\perp$. Assuming that $e_1, e_2 \in \text{PUSH}(c)$, $Match(e_1') = e_1$, $Match(e_2') = e_2$, if $e_1, e_2 \in Bad(e_\perp, c)$, then $e_1 <_c e_2 <_c e_1'$ or $e_2 <_c e_1 <_c e_2'$ exists. The formalized definition of $Bad(e_\perp, c)$ is given inductively as followed:

- $Bad_0(e_\perp, c) = \{e_0 \in \text{PUSH}(c) | e_\perp <_c e_0 \lor \forall e_0' \in \text{POP}(c).Match(e_0') = e_0 \Rightarrow e_0 <_c e_\perp <_c e_0'\}$.
- $Bad_{i+1}(e_\perp, c) = \{e_{i+1} \in \text{PUSH}(c) \mid \exists e_i \in Bad_i(e_\perp, c), e_i' \in \text{POP}(c). Match(e_i') = e_i \land e_i <_c e_{i+1} <_c e_i' \lor \exists e_{i+1}' \in POP(c).Match(e_{i+1}') = e_{i+1} \land e_{i+1} <_c e_i <_c e_{i+1}'\}$
- $Bad(e_\perp, c) = \cup_{i \in N} Bad_i(e_\perp, c)$

If $e \in Bad(e_\perp, c)$ and e completes before $e_\perp$ does, then the stack is guaranteed to be non-empty.

Definition 5 (Linearizable Mapping). *Given a complete history c and an ordered mapping $Match$, $Match$ is linearizable if for any $e_1 \in POP(c)$ with $Val_c(e_1) = NULL$, we have $Bad(e_1, c) \cap Before(e_1, c) = \varnothing$.*

Proposition 1. *Given a stack S, complete history c is linearizable with respect to S if and only if mapping $Match$ of c satisfies the linearizable conditions.*

Proof. Given $e_1, e_2, e_3 \in \text{POP}(c)$, $Match(e_1) = e_1', Match(e_2) = e_2'$, a mapping μ_{seq}: $\text{POP}(b) \to \text{PUSH}(b) \cup \{\perp\}$, a mapping $Match$: $\text{POP}(c) \to \text{PUSH}(c) \cup \{\perp\}$.

($\Rightarrow$)c is linearizable with respect to S, then by Definiton 2, c is serializable and its corresponding behavior is legal. c is serializable, so $\mu_{seq}(e_1) = e_1' \Rightarrow Match(e_1) = e_1'$.

By Definiton 1,

(1) $\mu_{seq}(e_1) = e_1', Val_b(e_1) = Val_b(e_1') \Rightarrow Val_c(e_1) = Val_c(e_1') \Rightarrow Match(e_1) \neq \perp, Val_c(e_1) = Val_c(Match(e_1))$.
(2) $\mu_{seq}(e_3) = \perp, Val_b(e_3) = \text{NULL} \Rightarrow Match(e_3) = \perp$ and $Val_c(e_3) = \text{NULL}$.
(3) $\mu_{seq}(e_1) = \mu_{seq}(e_2) \neq \perp, e_1 = e_2 \Rightarrow Match(e_1) = Match(e_2) \neq \perp, e_1 = e_2 \Rightarrow Match$ is a safe mapping.
(4) $\mu_{seq}(e_1) = e_1', e_1' <_b e_1 \Rightarrow Match(e_1) <_c e_1 \Rightarrow e_1 \not<_c Match(e_1)$.
(5) $\mu_{seq}e_1' <_b \mu_{seq}(e_2), e_2 <_b \mu_{seq}^{-1}(e_1') \Rightarrow e_1' <_c e_2', e_2 <_c Match^{-1}(e_1') \Rightarrow e_1' <_c Match(e_2), e_2 <_c e_1 \Rightarrow e_1' <_c Match(e_2), e_1 \not<_c e_2 \Rightarrow Match$ is an ordered mapping.
(6) $\mu_{seq}(e_3) = \perp, | \{e_i \in \text{PUSH}(b) \mid e_i <_b e_3\} | = | \{e_j \in \text{POP}(b) \mid e_j <_b e_3 \land \mu_{seq}(e_j) \neq \perp\} | \Rightarrow Val_c(e_3) = \text{NULL}, e_3 = e_\perp; \forall e_k \in \text{POP}(c), e_k' \in \text{PUSH}(c), e_k' = Match(e_k)$ and $e_k <_c e_3$ hold, then there exist $e_k' <_c e_3$ and $e_k \in Before(e_3, c)$, however $e_k <_c e_3 <_c e_k'$ is not true, namely $e_k \in Bad(e_3, c) \Rightarrow Val_c(e_3) = \text{NULL}$ and $Bad(e_3, c) \cap Before(e_3, c) = \varnothing \Rightarrow Match$ is a linearizable mapping.

So, if a complete history c is linearizable with respect to stack S, then mapping $Match$ of c satisfies the linearizable conditions.

($\Leftarrow$)Mapping $Match$ is linearizable, as well as safe and ordered. If $Match(e_1) = e_1'$, then $\mu_{seq}(e_1) = e_1'$.

(1) $Match(e_1) \neq \bot, Val_c(e_1) = Val_c(Match(e_1)) \Rightarrow Match(e_1) = e_1'$,
   $Val_c(e_1) = Val_c(e_1') \Rightarrow \mu_{seq}(e_1) = e_1', Val_b(e_1) = Val_b(e_1')$.
(2) $Match(e_3) = \bot$ and $Val_c(e_3) = NULL \Rightarrow \mu_{seq}(e_3) = \bot, Val_b(e_3) = NULL$.
(3) $Match(e_1) = Match(e_2) \neq \bot, e_1 = e_2 \Rightarrow \mu_{seq}(e_1) = \mu_{seq}(e_2), e_1 = e_2$.
(4) $e_1 \nless_c Match(e_1) \Rightarrow Match(e_1) = e_1', e_1 \nless_c e_1' \Rightarrow \mu_{seq}(e_1) = e_1', e_1' <_b e_1$.
(5) $e_1' <_c Match(e_2)$ and $e_1 \nless_c e_2 \Rightarrow e_1' <_c e_2'$ and $e_1 \nless_c e_2 \Rightarrow e_1' <_c e_2', e_2 <_c$
   $e_1 \Rightarrow e_1' <_b \mu_{seq}(e_2), e_2 <_b \mu_{seq}^{-1}(e_1')$.
(6) $Val_c(e_3) = NULL$ and $Bad(e_3,c) \cap Before(e_3,c) = \varnothing \Rightarrow \forall e_k \in POP(c)$,
   $e_k' \in PUSH(c)$, $e_k' = Match(e_k)$ holds, however $e_k <_c e_3 <_c e_k'$ is not true,
   $e_k <_c e_3$ and $e_k' <_c e_3$ exist $\Rightarrow \mu_{seq}(e_3) = \bot, |\{e_i \in PUSH(b) \mid e_i <_b e_3\}| = |$
   $\{e_j \in POP(b) \mid e_j <_b e_3 \wedge \mu_{seq}(e_j) \neq \bot\}|$.

So, if mapping $Match$ of complete history c satisfies the linearizable conditions, then c is linearizable with respect to stack S.

Proposition 2. *Given a stack S, complete history c is linearizable with respect to S if and only if c satisfies the three properties listed in Sect. 3.*

Proof. Given $e_1, e_2, e_3 \in POP(c)$, $e_1', e_2' \in PUSH(c)$, there exist $Match(e_1) = e_1'$ and $Match(e_2) = e_2'$.

($\Rightarrow$)If complete history c is linearizable with respect to stack S, then by Proposition 1, mapping $Match$ over c satisfies the linearizable conditions. So, $Match$ is safe as well as ordered.

By Definition 3, $Match(e_1) \neq \bot, Val_c(e_1) = Val_c(Match(e_1)) \Rightarrow e_1' \neq \bot$, $Val_c(e_1) = Val_c(e_1')$, it indicates that if pop operation events do not return NULL, then they have corresponding push operation events; $Match(e_1) = \bot$ and $Val_c(e_1) = NULL \Rightarrow e_1' = \bot, Val_c(e_1) = NULL$, it indicates that if pop operation events return NULL, then they have no corresponding push operation events $\Rightarrow$ Property 1: never permit popping data which is not pushed.

By Definition 3, $Match(e_1) = Match(e_2) \neq \bot, e_1 = e_2$, it indicates that if two pop operation events match the same push operation event, then the two pop operation events must be the same $\Rightarrow$ Property 2: never permit the same data to be popped twice.

By Definition 4, $e_1' <_c Match(e_2)$ and $e_1 \nless_c e_2 \Rightarrow e_1' <_c e_2', e_2 <_c e_1$, it indicates that the operation events in c must follow FILO; By Definition 5, $Val_c(e_3) = NULL$ and $Bad(e_3,c) \cap Before(e_3,c) = \varnothing \Rightarrow \forall e_k \in POP(c)$, $e_k' \in PUSH$ (c), $e_k' = Match(e_k)$ holds, there exist $e_k <_c e_3$ and $e_k' <_c e_3$, it indicates that when pop operation event e_3 returns NULL, the data pushed before e_3 is already popped, and there exist no push operation events preceding e_3 and their corresponding pop operation events behind e_3, then the stack is empty $\Rightarrow$ Property 3: the data must follow FILO and empty stacks must not perform pop operations.

($\Leftarrow$)If complete history c satisfies the three properties, then Properties 1 and 2 guarantee that mapping $Match$ in c is safe; Property 3 guarantees that safe mapping $Match$ is linearizable. According to Proposition 1, complete history c is linearizable.

The linearizability of stack data on complete history can also be proved through detecting linearizability violations. The linearizability violations can be defined as follows:

- VFresh: violating Property 1 that never permit popping data which is not pushed. Namely, pop operation events return incorrect pushing data. Given $pop(x) \in \text{POP}(c)$, $x \neq \text{NULL}$, if there exists VFresh, then $push(x) \notin \text{PUSH}(c)$ or $pop(x) <_c push(x)$.
- VRepeat: violating Property 2 that never permit the same data to be popped twice. Namely, two pop operation events return the data which are inserted by the same push operation event. Given two pop operation events $e_1, e_2 \in \text{POP}(c)$, $e_1 \neq e_2$, if there exists VRepet, then $Match(e_1) = Match(e_2) \neq \bot$.
- VOrd: violating Property 3 that the data must follow FILO and empty stacks must not perform pop operations. Namely, two ordered pop operation events return data which are inserted by push operation events in the same order, or pop operation events return non-NULL after the stack is logically empty. Given data x, y and $c = c_0 \bullet pop(\text{NULL}) \bullet c_1$, where c_0 and c_1 are sub-sequence of c, $push(x) <_c push(y)$ and $push(y)$ executes in c_0, if there exists VOrd, then (1) $pop(x) <_c pop(y)$ or (2) $pop(y) <_c pop(x)$, where $pop(x)$ does not execute in c_0.

Proposition 3. *Complete history c is linearizable, if and only if it has none of the three violations VFresh, VRepeat or VOrd.*

Proof. Given a complete history c, $e_1 = pop(x) \in \text{POP}(c)$, $e_1' = Match(e_1) = push(x) \in \text{PUSH}(c)$, $e_2 = pop(y) \in \text{POP}(c)$, $e_2' = Match(e_2) = push(y) \in \text{PUSH}(c)$, $e_3 \in \text{POP}(c)$.

($\Rightarrow$)If c is linearizable, according to Proposition 1, then mapping $Match$ of c satisfies the linearizable conditions. So, $Match$ is safe as well as ordered.

(1) $Match(e_1) \neq \bot, Val_c(e_1) = Val_c(Match(e_1))$ and $e_1 \not<_c Match(e_1) \Rightarrow x = Val_c(e_1) \neq \text{NULL}$, $e_1' <_c e_1 \Rightarrow push(x) <_c pop(x) \Rightarrow$ VFresh does not exist.
(2) $Match(e_1) = Match(e_2) \neq \bot, e_1 = e_2 \Rightarrow e_1 \neq e_2$, then $Match(e_1) \neq Match(e_2) \Rightarrow$ VRepet does not exist.
(3) $e_1' <_c Match(e_2)$ and $e_1 \not<_c e_2 \Rightarrow e_1' <_c e_2'$ and $e_2 <_c e_1 \Rightarrow push(x) <_c push(y), pop(y) <_c pop(x); Val_c(e_3) = \text{NULL}$ and $Bad(e_3, c) \cap Before(e_3, c) = \varnothing \Rightarrow c = c_0 \bullet pop(\text{NULL}) \bullet c_1$, where c_0, c_1 are subsequence of c, $e_3 = pop(\text{NULL})$, $\forall e_k \in \text{POP}(c)$, $e_k' = Match(e_k)$, there exists $e_k' <_c e_3(e_k <_c e_3$ also exists) or $e_3 <_c e_k <_c e_k'$, however $e_k <_c e_3 <_c e_k'$ does not exist $\Rightarrow \forall z \neq \text{NULL}, push(z)$ is finished in c_0, and so does $pop(z) \Rightarrow$ VOrd does not exist.

($\Leftarrow$) If there are no violations in c, then the three properties are established, according to Proposition 2, c is linearizable.

We notice that there is a possible situation that there are some push operation events that have no corresponding pop operation events in complete history c. Under this situation, because there are no pop operation events returning data, VFresh, VRepet or VOrd does not exist, we can also prove whether c is linearizable or not by detecting violations.

In the following, we will prove the linearizability of stack data on purely-blocking history.

4.2   Analysis on Purely-Blocking History

If the operation events in a history c cannot execute successfully, then there are pending events in c, and c is called purely-blocking history. Because purely-blocking history is not complete, we cannot check linearizability of a stack by regarding only executed states. Next, we consider that whether $\hat{c}$, the completion of general history c, satisfies linearizability or not.

Proposition 4. C *is linearizable if and only if every* $c \in C$ *has a completion* $\hat{c}$ *whose mapping* $Match$ *satisfies the linearizable conditions.*

Proof. ($\Rightarrow$)If every $c \in C$ is linearizable, then its completion $\hat{c}$ has a serializable form s, namely $\hat{c} \sim_{perm} s$, and s can be transformed to a legal behavior of stack. By Definition 2, $\hat{c}$ is linearizable. According to Proposition 1, mapping $Match$ in $\hat{c}$ satisfies the linearizable conditions.

($\Leftarrow$)Given history $c \in C$, $Compl(c)$ is the completion of c. If mapping $Match$ in $\hat{c}$ satisfies the linearizable conditions, according to Proposition 1, $\hat{c}$ is linearizable. So $\hat{c}$ has a serializable form s, and s can be transformed to a legal behavior of stack. For c, s is also serializable of c, and s can be transformed to a legal behavior of stack. So c is linearizable, and C is linearizable.

By Proposition 4, for any history c, we can detect the linearizability of c by its completion. For purely-blocking history, it can be transformed to a complete history c by completing pending event e with its response operation, then we could analyze linearizability of purely-blocking history by detecting whether there are violations in c.

5   Conclusion and Future Work

We have proposed a method to guarantee the linearizability of stack, that is to say, stack data is linearizable if and only if stack operations satisfy the three properties. Then, proving linearizability of stack data by detecting whether there are linearizability violations, and we could detect whether stack operations violate any of the three properties to guarantee stack data linearizable, instead of seeking linearization points and complicated proof. This method can be applicable to other types of concurrent sharing data structures, for example, Graph Structure. In the next, we will build tools to detect the three properties automatically, so as to ensure linearizability of stack data.

Acknowledgments. This work is supported by the National Natural Science Foundation of China under grant No. 61262008, 61363030, 61562015, U1501252 and 61572146, the High Level Innovation Team of Guangxi Colleges and Universities and Outstanding Scholars Fund, Guangxi Natural Science Foundation of China under grant No. 2014GXNSFAA118365, 2015GXNSFDA139038, Guangxi Key Laboratory of Trusted Software Focus Fund, Program for Innovative Research Team of Guilin University of Electronic Technology, Innovation Project of GUET Graduate Education under grant No. YJCXS201537.

References

1. Herlihy, M., Wing, J.: Linearizability: a correctness condition for concurrent objects. ACM Trans. Program. Lang. Syst. **12**, 463–492 (1990)
2. Amit, D., Rinetzky, N., Reps, T., Sagiv, M., Yahav, E.: Comparison under abstraction for verifying linearizability. In: Damm, W., Hermanns, H. (eds.) CAV 2007. LNCS, vol. 4590, pp. 477–490. Springer, Heidelberg (2007)
3. Vafeiadis, V.: Shape-value abstraction for verifying linearizability. In: Jones, N.D., Müller-Olm, M. (eds.) VMCAI 2009. LNCS, vol. 5403, pp. 335–348. Springer, Heidelberg (2009)
4. Vechev, M., Yahav, E., Yorsh, G.: Experience with model checking linearizability. In: Păsăreanu, C.S. (ed.) Model Checking Software. LNCS, vol. 5578, pp. 261–278. Springer, Heidelberg (2009)
5. Liang, H., Feng, X.: Modular verification of linearizability with non-fixed linearization points. ACM SIGPLAN Not. **48**, 459–470 (2010)
6. Derrick, J., Schellhorn, G., Wehrheim, H.: Mechanically verified proof obligations for linearizability. ACM Trans. Program. Lang. Syst. **33**, 623–636 (2011)
7. Colvin, R., Doherty, S., Groves, L.: Verifying concurrent data structures by simulation. Electr. Notes Theor. Comput. Sci. **37**, 93–110 (2005)
8. Derrick, J., Schellhorn, G., Wehrheim, H.: Verifying linearizability with potential linearization points. In: Proceedings of the 17th International Symposium on Formal Methods, Limerick, Ireland, pp. 323–337 (2011)
9. Vafeiadis, V.: Automatically proving linearizability. In: Touili, T., Cook, B., Jackson, P. (eds.) CAV 2010. LNCS, vol. 6174, pp. 450–464. Springer, Heidelberg (2010)
10. Drăgoi, C., Gupta, A., Henzinger, T.A.: Automatic linearizability proofs of concurrent objects with cooperating updates. In: Sharygina, N., Veith, H. (eds.) CAV 2013. LNCS, vol. 8044, pp. 174–190. Springer, Heidelberg (2013)
11. Abdulla, P.A., Haziza, F., Holík, L., Jonsson, B., Rezine, A.: An integrated specification and verification technique for highly concurrent data structures. In: Piterman, N., Smolka, S.A. (eds.) TACAS 2013 (ETAPS 2013). LNCS, vol. 7795, pp. 324–338. Springer, Heidelberg (2013)
12. Friggens, D., Groves, L.: Shape predicates allow unbounded verification of linearizability using canonical abstraction. In: Proceedings of the 37th Australasian Computer Science Conference, Auckland, New Zealand, pp. 49–56 (2014)
13. Henzinger, T., Sezgin, A., Vafeiadis, V.: Aspect-oriented linearizability proofs. In: Proceedings of the 24th International Conference on Concurrency Theory, Buenos Aires, Argentina, pp. 242–256 (2013)
14. Chakraborty, S., Henzinger, T., Sezgin, A., Vafeiadis, V.: Aspect-oriented linearizability proofs. Log. Methods Comput. Sci. **11**, 1–33 (2015)

Anonymous Mutual Authentication Scheme for Secure Inter-Device Communication in Mobile Networks

Youngseok Chung[1,2], Seokjin Choi[1], and Dongho Won[2(✉)]

[1] Electronics and Telecommunications Research Institute,
Daejeon, Republic of Korea
{yschung11, choisj}@nsr.re.kr
[2] Department of Computer Engineering, Sungkyunkwan University,
Suwon, Republic of Korea
dhwon@security.re.kr

Abstract. An anonymous authentication scheme provides privacy preserving communication. Recently, many anonymous authentication schemes have been proposed, and various cryptanalysis and improvements have been followed. Some schemes use high-cost functions, such as symmetric and asymmetric cryptographic functions while the others use low-cost functions such as hash functions and simple bitwise operations. The previous schemes are based on the client-server model and a server anonymously authenticates clients. In this paper, we propose an anonymous authentication scheme for communication between client-side devices. The proposed scheme preserves secure inter-device communication by providing anonymous mutual authentication and fair key agreement. Since the proposed scheme uses only low-cost functions, it can apply to mobile networks considering lower energy consumption.

Keywords: Anonymity · Mutual authentication · Privacy · Mobile network

1 Introduction

A lot of mobile users move and encounter others frequently in mobile networks. Each time, they want to authenticate and communicate with others in a secure way for security reasons. Since it is difficult for them to trust others, they are unwilling to open their identities as well as sensitive information. In other words, the growth of users in mobile networks raises privacy and security concerns. Guaranteeing user's privacy and communicating securely are very important issues. In such situations, anonymous authentication and key agreement schemes are the proper ways to preserve user's privacy and secrecy. Moreover, due to the distinct characteristics of mobile networks, smaller consumption of the energy is also essential.

For several years, anonymous authentication schemes and related protocols as well as various proofs of weaknesses and improvements have been presented in wireless or mobile networks [1–25]. Some schemes are based on high-cost functions, such as symmetric cryptographic functions, asymmetric cryptographic functions, and modular operations [1–19]. The others use low-cost functions, such as one-way hash functions

© Springer International Publishing Switzerland 2016
O. Gervasi et al. (Eds.): ICCSA 2016, Part IV, LNCS 9789, pp. 289–301, 2016.
DOI: 10.1007/978-3-319-42089-9_21

and exclusive-OR operations [20–25]. If all of them guarantee the same level of security, the latter ones are more efficient and lightweight for battery powered devices in wireless or mobile networks. Meanwhile, the previous schemes are based on the client-server model and only a server can authenticate clients anonymously. In other words, anonymous mutual authentication between the entities concerned is not accomplished.

In this paper, we propose an anonymous authentication scheme which provides mutual authentication and fair key agreement for inter-device communications. Our scheme makes the registered devices authenticate anonymously and communicate with each other securely. Also, we only use low-cost functions to provide computational benefits in mobile networks.

The remainder of this paper is organized as follows: We review previous works in Sect. 2 and present the proposed scheme in Sect. 3. In Sect. 4, we analyze security of our scheme. Finally, we give a concluding remark in Sect. 5.

2 Previous Works

In this section, we briefly review several previous works by focusing on their intertwined relationship. All of them have their own characteristic and they analyze other works in a different prospective. The previous works can be divided into two categories: schemes based on high-cost functions and schemes based on low-cost functions.

In 2004, Zhu and Ma [1] proposed a new anonymous authentication scheme which uses high-cost functions. After that, Lee et al. [2] presented Zhu and Ma's scheme does not achieve perfect backward secrecy and mutual authentication, and they proposed the improved scheme by slightly modifying Zhu and Ma's scheme. Also, Wei et al. [3] pointed out Zhu and Ma's scheme does not provide anonymity, traceability, and security of user's password. Unfortunately, Wu et al. [4] proved both Zhu and Ma's and Lee et al.'s schemes fail to provide anonymity and backward secrecy, and they presented the improvements of Lee et al.'s scheme. However, the weaknesses of Wu et al.'s scheme were shown by Lee et al. [5] and Xu and Feng [6] respectively. After that, Kun et al. [7] improved Xu and Feng's scheme to provide identity anonymity and unlinkability. But Tsai et al. [8] discovered Kun et al.'s scheme is vulnerable to de-synchronization attack. Mun et al. [9] also showed Wu et al.'s scheme does not guarantee anonymity, perfect forward secrecy, and secrecy of legitimate user's password, and proposed an enhanced anonymous authentication scheme. However, Kim and Kwak [10] found Mun et al.'s scheme is flawed against replay attack and man-in-the-middle attack, and presented the improved secure scheme. Also, Zhao et al. [11] proved Mun et al.'s scheme is vulnerable to impersonation attack and insider attacks, and then they proposed a novel anonymous authentication scheme.

Independently, Chang et al. [20] pointed out Lee et al.'s scheme [2] cannot provide anonymity under the forgery attack, and also proposed an enhanced authentication scheme in 2009. It was the first scheme which is based on low-cost functions. Youn et al. [21] proved Chang et al.'s scheme does not provide not only anonymity against passive adversaries and malicious legal users but also security against the known session key attack and the side channel attack without any countermeasures. Also,

Zhou and Xu [14] showed Chang et al.'s scheme does not provide user anonymity and confidentiality of session keys, and they proposed an improved scheme at the same time. Unfortunately, the weaknesses of Zhou and Xu's scheme were presented and an improved version of theirs was proposed by Gope and Hwang [25]. Also, He et al. [22] proved Chang et al.'s scheme is not efficient in password authentication and proposed a lightweight and efficient authentication scheme. However, Jiang et al. [15] presented He et al.'s scheme is vulnerable to offline guessing attack with smart cards, mobile user tracking attack, session key and replay attack, spoofing attack, and privileged insider attack. They proposed an enhanced protocol based on quadratic residue assumption, but their protocol was proved not to guarantee security against stolen-verifier attack, replay attack, and denial of service attack by Wen et al. [16]. After that, Gope and Hwang [17] showed that the proposed scheme of Wen et al. has security weaknesses, such as offline guessing attack with smart cards, forgery attack, and unfair key agreement. And they proposed an enhanced and secure mutual authentication key agreement scheme preserving user anonymity. Meanwhile, He et al.'s scheme was also proved vulnerable to insider attack and does not provide session key secrecy and password update secrecy by Shin et al. [18]. And Shin et al. proposed an improved authentication scheme. Farash et al. [19] presented the vulnerabilities of both Wen et al.'s scheme and Shin et al.'s scheme. They proved Wen et al.'s scheme is insecure under session key disclosure attack and known session key attack, and Shin et al.'s scheme does not guarantee untraceability, secrecy of the sensitive parameter of home agent, secrecy against impersonation attack, and session key secrecy. Also, they proposed an improved scheme which remedies the weaknesses presented by them using symmetric cryptographic functions.

3 Proposed Scheme

To preserve anonymity, mutual authentication, security, and fair key agreement, we propose an anonymous mutual authentication scheme. Notations used in the proposed scheme are defined in Table 1.

Table 1. Notations

Notations	Descriptions
ID_X	The identity of an entity X
VID_X	The virtual identity of an entity X
SID_X	A shadow identity of an entity X
x	The secret key of R
x_X	The secret key for an entity X generated by R
n_X	A nonce generated by entity X
r_X	A random number generated by entity X
$h(\cdot)$	A collision free one-way hash function
$\|$	A concatenation
$\oplus$	An exclusive-OR operation

The proposed scheme consists of four phases: registration, authentication and session key establishment, and renewal. The mobile device M and the registration server R are involved in each phase. Firstly, it is necessary that M registers its real identity to R and get a virtual identity from R in the registration phase. The mobile device uses the virtual identity instead of the real one with authenticating others. Then, the mobile device which wants to communicate with other device in a secure manner tries to accomplish the authentication and session key establishment phases. In addition, the mobile device which suspects its virtual identity is revealed can accomplish the renewal phase to replace it. Although the virtual identity protects the real identity from revealing, the fixed virtual identity can link one mobile device and its messages. And it is possible to recognize that certain messages are originated by one same mobile device. To resolve this problem, the mobile device renews its virtual identity periodically by accomplishing the renewal phase.

3.1 Registration Phase

A mobile device M firstly registers its identity ID_M to the registration server R. Then, R computes M's virtual identity VID_M after generating the random secret number x_M as follows:

$$VID_M = h(ID_M||x_M||x). \tag{1}$$

Since x_M is only allocated to M, it acts as the secret key of M. It means M and R are the only entities which know x_M. After that, R stores $[VID_M, ID_M, x_M]$ secretly in its database and delivers $[VID_M, x_M, h(x)]$ to M through a secure channel. Figure 1 demonstrates the registration phase.

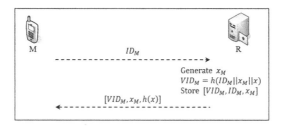

Fig. 1. Registration phase

3.2 Authentication and Session Key Establishment Phases

Let's suppose that the mobile devices M_1 and M_2 have already registered to R. Obviously, they own $[VID_{M_1}, x_{M_1}, h(x)]$ and $[VID_{M_2}, x_{M_2}, h(x)]$ respectively through the registration phase. To authenticate each other and to share a session key, they accomplish the authentication and session key establishment phases through R. Figure 2 shows these phases.

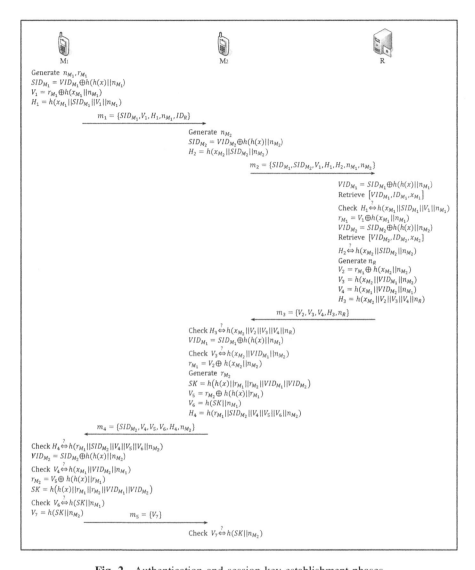

Fig. 2. Authentication and session key establishment phases

1. M_1 generates a nonce n_{M_1} and a random number r_{MN_1} firstly, and calculates a shadow identity SID_{M_1}, a verifier V_1, and a hashing value H_1 as follows:

$$SID_{M_1} = VID_{M_1} \oplus h(h(x)||n_{M_1}), \tag{2}$$

$$V_1 = r_{M_1} \oplus h(x_{M_1}||n_{M_1}), \tag{3}$$

$$H_1 = h(x_{M_1}||SID_{M_1}||V_1||n_{M_1}). \tag{4}$$

And then, M_1 sends a message $m_1 = \{SID_{M_1}, V_1, H_1, n_{M_1}, ID_R\}$ to M_2.

2. After receiving m_1, M_2 firstly checks ID_R whether R is the trusted registration server or not. If M_2 confirms ID_R successfully, M_2 generates a nonce n_{M_2} and calculates the following shadow identity SID_{M_2} and a hashing value H_2:

$$SID_{M_2} = VID_{M_2} \oplus h(h(x)||n_{M_2}), \tag{5}$$

$$H_2 = h(x_{M_2}||SID_{M_2}||n_{M_2}). \tag{6}$$

Then, M_2 sends a message $m_2 = \{SID_{M_1}, SID_{M_2}, V_1, H_1, H_2, n_{M_1}, n_{M_2}\}$ to R.

3. Upon receiving m_2, R calculates VID_{M_1} using SID_{M_1}, n_{M_1}, and h(x) as follows:

$$VID_{M_1} = SID_{M_1} \oplus h(h(x)||n_{M_1}). \tag{7}$$

To authenticate M_1, R retrieves $[VID_{M_1}, ID_{M_1}, x_{M_1}]$ from its database using VID_{M_1} as a keyword. If there is no matched information, R rejects M_2's request to terminate the connection. Since only R can compute the virtual identity using its secret key x, no matched information in R's database represents VID_{M_1} is an invalid virtual identity. Then, HA computes $h(x_{M_1}||SID_{M_1}||V_1||n_{M_1})$ and checks it equals to H_1. The equivalence implies that the owner of x_{M_1}, namely a valid mobile device M_1, calculates SID_{M_1}, V_1, and H_1 originally. After that, R computes r_{M_1} such that:

$$r_{M_1} = V_1 \oplus h(x_{M_1}||n_{M_1}). \tag{8}$$

4. In the same way, R calculates the following VID_{M_2} and checks that M_2 is a valid mobile device by searching $[VID_{M_2}, ID_{M_2}, x_{M_2}]$ successfully:

$$VID_{M_2} = SID_{M_2} \oplus h(h(x)||n_{M_2}). \tag{9}$$

And then, R checks the validity of H_2. If H_2 is equals to $h(x_{M_2}||SID_{M_2}||n_{M_2})$ computed by R, R confirms M_2 computes SID_{M_2} and H_2.

5. To make a sending message, R generates a nonce n_R and computes following verifiers V_2, V_3, and V_4, and a hashing value H_3:

$$V_2 = r_{M_1} \oplus h(x_{M_2}||n_{M_2}), \tag{10}$$

$$V_3 = h(x_{M_2}||VID_{M_1}||n_{M_2}), \tag{11}$$

$$V_4 = h(x_{M_1}||VID_{M_2}||n_{M_1}), \tag{12}$$

$$H_3 = h(x_{M_2}||V_2||V_3||V_4||n_R). \tag{13}$$

Then, R sends a message $m_3 = \{V_2, V_3, V_4, H_3, n_R\}$ to M_2.

6. After receiving m_3, M_2 checks whether H_3 is valid or not by comparing the received H_3 and the computed $h(x_{M_2}||V_2||V_3||V_4||n_R)$. If it is valid, M_2 confirms that V_2, V_3, V_4, and H_3 are computed by R which is the only entity knows x_{M_2} except M_2. After computing VID_{M_1} by following equation, M_2 checks that V_3 and $h(x_{M_2}||VID_{M_1}||n_{M_2})$ are equivalent:

$$VID_{M_1} = SID_{M_1} \oplus h(h(x)||n_{M_1}). \tag{14}$$

If they are equal, M_2 confirms the validity of VID_{MN_1} and authenticates M_1 as a valid mobile device. And then, M_2 calculates r_{M_1} such that:

$$r_{M_1} = V_2 \oplus h(x_{M_2}||n_{M_2}). \tag{15}$$

7. To compute a session key and make a sending message, M_2 generates a random number r_{M_2} and computes SK, V_5, V_6 and H_4 as shown in

$$SK = h(h(x)||r_{M_1}||r_{M_2}||VID_{M_1}||VID_{M_2}), \tag{16}$$

$$V_5 = r_{M_2} \oplus h(h(x)||r_{M_1}), \tag{17}$$

$$V_6 = h(SK||n_{M_1}), \tag{18}$$

$$H_4 = h(r_{M_1}||SID_{M_2}||V_4||V_5||V_6||n_{M_2}). \tag{19}$$

Then, M_2 sends a message $m_4 = \{SID_{M_2}, V_4, V_5, V_6, H_4, n_{M_2}\}$ to M_1.

8. Upon receiving m_4, M_1 firstly checks the equivalence of H_4 and $h(r_{M_1}||SID_{M_2}||V_4||V_5||V_6||n_{M_2})$. Since the intended target for authentication and session key establishment is M_2, it is necessary that M_1 confirms M_2 gets r_{M_1} successfully by checking H_4. To authenticate M_2, M_1 computes the following VID_{M_2} and checks its validity by comparing V_4 and $h(x_{M_1}||VID_{M_2}||n_{M_1})$:

$$VID_{M_2} = SID_{M_2} \oplus h(h(x)||n_{M_2}). \tag{20}$$

9. After that, M_1 computes r_{MN_2} from V_5 and calculates SK such that:

$$r_{M_2} = V_5 \oplus h(h(x)||r_{M_1}), \tag{21}$$

$$SK = h(h(x)||r_{M_1}||r_{M_2}||VID_{M_1}||VID_{M_2}). \tag{22}$$

After checking the equivalence of V_6 and $h(SK||n_{M_1})$, M_1 confirms SK is shared between M_1 and M_2 successfully. Then, M_1 sends a message $m_5 = \{V_7\}$ after computing V_7 expressed by:

$$V_7 = h(SK||n_{M_2}). \tag{23}$$

10. After receiving m_5, M_2 checks the equivalence of V_7 and $h(SK||n_{M_2})$. If they are equal, M_2 also confirms M_1 and M_2 share SK successfully.

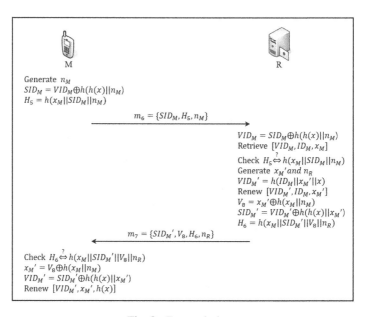

Fig. 3. Renewal phase

3.3   Renewal Phase

It is assumed that M suspects its virtual identity is revealed. To renew the virtual identity, M performs the renewal phase through R. Figure 3 presents the renewal phase.

1. M generates a nonce n_M and computes SID_M and H_5 as follows:

$$SID_M = VID_M \oplus h(h(x)||n_M), \tag{24}$$

$$H_5 = h(x_M||SID_M||n_M) \tag{25}$$

And then, M sends a message $m_6 = \{SID_{M_1}, H_5, n_M\}$ to R.

2. Upon receiving m_6, R computes VID_M and retrieve $[VID_M, ID_M, x_M]$ from its database. Since M is a valid mobile device registered to R, the matched information is founded successfully. If there is no matched information, R stops the renewal phase immediately. Then, R checks whether H_5 and $h(x_M||SID_M||n_M)$ are equal.

If they are identical, R generates the new secret key of M's x'_M and a nonce n_R. And R calculates the following VID'_M and replaces $[VID_M, ID_M, x_M]$ to $[VID'_M, ID_M, x'_M]$:

$$VID'_M = h(ID_M||x'_M||x). \qquad (26)$$

To transfer VID'_M and x'_M, R computes V_8, SID'_M, and H_6 as follows and sends a message $m_7 = \{SID'_M, V_8, H_6, n_R\}$ to M:

$$V_8 = x'_M \oplus h(x_M||n_M), \qquad (27)$$

$$SID'_M = VID'_M \oplus h(h(x)||x'_M), \qquad (28)$$

$$H_6 = h(x_M||SID'_M||V_8||n_R). \qquad (29)$$

3. After receiving m_7, M firstly checks the equivalence of H_6 and $h(x_M||SID'_M||V_8||n_R)$. If M checks the validity of H_6, M computes x'_M and VID'_M as follows:

$$x'_M = V_8 \oplus h(x_M||n_M), \qquad (30)$$

$$VID'_M = SID'_M \oplus h(h(x)||x'_M), \qquad (31)$$

By checking H_6, M confirms the fact that R computes SID'_M and V_8 legally. Since only M which knows x_M can get x'_M and VID'_M from V_8 and SID'_M, M replaces $[VID_M, x_M, h(x)]$ to $[VID'_M, x'_M, h(x)]$ obviously. As a result, renewing the virtual identity and the secret key is done successfully.

4 Security Analysis

In this section, we prove the proposed scheme guarantees anonymity, mutual authentication, security, and fair key agreement.

Anonymity Even though the adversary possesses a valid device registered for R, he or she cannot get a legitimate device's identity. Let's suppose that the adversary who has a valid device eavesdrops m_1 or m_2. Then, the adversary can get SID_{M_1} and n_{M_1} from m_1 or m_2 easily, and obtain $h(x)$ from a valid device. And it is possible to get VID_{M_1} by calculating $SID_{M_1} \oplus h(h(x)||n_{M_1})$. However, the virtual identity VID_{M_1} which is formed as $h(ID_{M_1}||x_{M_1}||x)$ does not reveal the real identity ID_{M_1} due to x_{M_1} and x. Meanwhile, an identity itself is easy to guess since it is short and has a certain format. So the adversary can guess the candidate identity like ID'_{M_1} and try to check whether VID_{M_1} equals to $h(ID'_{M_1}||x_{M_1}||x)$ or not. However, VID_{M_1} is the hashing value which contains x_{M_1} and x. It means that whoever guesses a candidate identity cannot check the equivalence between the guessed identity and the real identity of a valid device. As a result, anonymity is guaranteed in the proposed scheme.

Mutual Authentication In authentication and session key establishment phases, M_1 and M_2 authenticate each other by verifying V_4 and V_3 respectively. R computes V_4 by containing M_1's private key x_{M_1} and M_2's virtual identity VID_{M_2}. Since only R can compute valid V_4, M_1 can check the validity of VID_{M_2} through R by verifying V_4. Also, in the same manner, R computes V_3 using M_2's private key x_{M_2} and M_1's virtual identity VID_{M_1}. M_2 can convince VID_{M_1} is valid if V_3 is verified successfully through R. Finally, although both M_1 and M_2 just check the other's virtual identity, they can authenticate mutually one another with the aid of R.

Security Against Known Key Attack The session key established between M_1 and M_2 contains secret random numbers r_{M_1} and r_{M_2}. Since r_{M_1} and r_{M_2} are generated newly every time, there is no relation to each session key. Even if the session key is revealed, the adversary cannot get other session key because the eavesdropped V_6 and V_7 reveal no information about the session key. Meanwhile, the adversary who obtains a device legitimately registered for R can get VID_{M_1} and VID_{M_2} easily from SID_{M_1} and SID_{M_2}. However, it is impossible to compute SK which is formed as $h(h(x)\|r_{M_1}\|r_{M_2}\|VID_{M_1}\|VID_{M_2})$ since r_{M_1} and r_{M_2} are only known to M_1 and M_2. The adversary can eavesdrop V_2 and V_5 and try to obtain r_{M_1} and r_{M_2}. However, V_2 which is equal to $r_{M_1} \oplus h(x_{M_2}\|n_{M_2})$ contains x_{M_2} only known to M_2, and V_5 which is equals to $r_{M_2} \oplus h(h(x)\|r_{M_1})$ contains r_{M_1} only known to M_1. As a result, no one can compute the session key under any cases. It means the proposed scheme preserves backward and forward secrecy.

Fair Key Agreement When M_1 and M_2 accomplish the authentication and session key establishment phases, they compute the session key after generating random numbers r_{M_1} and r_{M_2} respectively. Since the session key contains both r_{M_1} and r_{M_2}, M_1 and M_2

Table 2. Performance comparison

	S1	S2	S3
M	$6t_{HASH} + 1t_{EXP}$	$6t_{HASH}$	$10t_{HASH}$
F(or M)	$3t_{HASH} + 1t_{SYM}$	$1t_{HASH} + 2t_{SYM}$	$11t_{HASH}$
H	$7t_{HASH} + 1t_{SYM} + 1t_{EXP}$	$5t_{HASH} + 2t_{SYM}$	$9t_{HASH}$
Total	$16t_{HASH} + 2t_{EXP} + 2t_{SYM}$	$12t_{HASH} + 4t_{SYM}$	$30t_{HASH}$
Time(sec)	1.0694	0.0408	0.015

- S1: Gope and Hwang's scheme [17]
- S2: Farash et al.'s scheme [19]
- S3: Our scheme
- M: Mobile device
- F: Foreign agent
- H: Home agent
- t_{HASH}: Execution time of a one-way hash function
  (1H = 0.0005(sec))
- t_{SYM}: Execution time of a one-way hash function (1H = 0.0005 (sec))
- t_{EXP}: Execution time of a modular exponential operation
  (1E = 0.522(sec))

have same contribution to randomness of the session key. In other words, both M_1 and M_2 contribute equally during the establishment of the session key. For this reason, key agreement is achieved fairly to both sides.

5 Performance Analysis

In this section, we show performance comparisons of our scheme with some recent schemes. Recent schemes involve two servers, namely home and foreign agents, and a mobile device while our scheme involves a home agent and two mobile devices. Despite of the different entities, we can analyze the performance of each scheme by comparing the total computational cost for all entities. We apply the experiment result of Li et al. [26] to analyze the performances. Table 2 denotes the performance comparison of each scheme.

6 Conclusions

In this paper, we propose a novel scheme which provides anonymous mutual authentication and secure communication between client-side devices in mobile networks. The trusted authority provides trustworthy information to each device, and they verify all transmitted data precisely. Also, the virtual identity keeps the real one secret from adversaries. Due to this, each device can authenticate each other anonymously and mutually, and they can share the session key securely. Moreover, the device can replace its virtual identity with a new one through the trusted authority whenever it wants. This renewal process removes possibility to link certain messages to one specific originator. Since the session key established between each device contains two random numbers generated independently, each device can contribute the randomness of the session key equally. Meanwhile, low-cost functions used in the proposed scheme increase efficiency. The proposed scheme consumes less time and energy to operate, so it is more suitable for mobile networks than the schemes based on high-cost functions.

References

1. Zhu, J., Ma, J.: A new authentication scheme with anonymity for wireless environments. IEEE Trans. Consum. Electron. **50**(1), 231–235 (2004)
2. Lee, C.C., Hwang, M.S., Lio, I.E.: Security enhancement on a new authentication scheme with anonymity for wireless environments. IEEE Trans. Ind. Electron. **53**(5), 1683–1687 (2006)
3. Wei, Y., Qiu, H., Hu, Y.: Security analysis of authentication scheme with anonymity for wireless environments. In: Proceedings of IEEE International Conference on Communication Technology, pp. 1–4 (2006)
4. Wu, C.C., Lee, W.B., Tsaur, W.J.: A secure authentication scheme with anonymity for wireless communications. IEEE Commun. Lett. **12**(10), 722–723 (2008)
5. Lee, J.S., Chang, J.H., Lee, D.H.: Security flaw of authentication scheme with anonymity for wireless communications. IEEE Commun. Lett. **13**(5), 292–293 (2009)

6. Xu, J., Feng, D.: Security flaws in authentication protocols with anonymity for wireless environments. ETRI J. **31**, 460–462 (2009)
7. Kun, L., Anna, X., Fei, H., Lee, D.H.: Anonymous authentication with unlinkability for wireless environments. IEICE Electron. Express **8**(8), 536–541 (2011)
8. Tsai, J.L., Lo, N.W., Wu, T.C.: Secure anonymous authentication protocol with unlinkability for mobile wireless environment. In: Proceedings of IEEE International Conference on Anti-Counterfeiting Security and Identification, pp. 1–5 (2012)
9. Mun, H., Han, K., Lee, Y.S., Yeun, C.Y., Choi, H.H.: Enhanced secure anonymous authentication scheme for roaming service in global mobility networks. Math. Comput. Model. **55**, 214–222 (2012)
10. Kim, J., Kwak, J.: Improved secure anonymous authentication scheme for roaming service in global mobile network. Int. J. Secur. Appl. **6**(3), 45–54 (2012)
11. Zhao, D., Peng, H., Li, L., Yang, Y.: A secure and effective anonymous authentication scheme for roaming service in global mobility networks. Wirel. Pers. Commun. **78**(1), 247–267 (2013)
12. Jeon, W., Kim, J., Nam, J., Lee, Y., Won, D.: An enhanced secure authentication scheme with anonymity for wireless environments. IEICE Trans. Commun. **95**(5), 1819–1821 (2012)
13. Nam, J., Choo, K.K., Han, S., Kim, M., Paik, J., Won, D.: Efficient and anonymous two-factor user authentication in wireless sensor networks: achieving user anonymity with lightweight sensor computation. PLoS ONE **10**(4), 1–21 (2015)
14. Zhou, T., Xu, J.: Provable secure authentication protocol with anonymity for roaming service in global mobility networks. Comput. Netw. **55**(1), 205–213 (2011)
15. Jiang, Q., Ma, J., Li, G., Yang, L.: An enhanced authentication scheme with privacy preservation for roaming service in global mobility networks. Wirel. Pers. Commun. **68**(4), 1477–1491 (2013)
16. Wen, F., Susilo, W., Yang, G.: A secure and effective anonymous user authentication scheme for roaming service in global mobility networks. Wirel. Pers. Commun. **73**(3), 993–1004 (2013)
17. Gope, P., Hwang, T.: Enhanced secure mutual authentication and key agreement scheme preserving user anonymity in global mobile networks. Wirel. Pers. Commun. **82**(4), 2231–2245 (2015)
18. Shin, S., Yeh, H., Kim, K.: An efficient secure authentication scheme with user anonymity for roaming user in ubiquitous networks. Peer-to-Peer Netw. Appl. **8**(4), 674–683 (2015)
19. Farash, M.S., Chaudhry, S.A., Heydari, M., Sadough S.M.S., Kumari, S., Khan, M.K.: A lightweight anonymous authentication scheme for consumer roaming in ubiquitous networks with provable security. Int. J. Commun. Syst. (2015). doi:10.1002/dac.3019
20. Chang, C.C., Lee, C.Y., Chiu, Y.C.: Enhanced authentication scheme with anonymity for roaming service in global mobility networks. Comput. Commun. **32**, 611–618 (2009)
21. Youn, T.Y., Park, Y.H., Lim, J.: Weaknesses in an anonymous authentication scheme for roaming service in global mobility networks. IEEE Commun. Lett. **13**(7), 471–473 (2009)
22. He, D., Chan, S., Chen, C., Bu, J., Fan, R.: Design and validation of an efficient authentication scheme with anonymity for roaming service in global mobility networks. Wirel. Pers. Commun. **61**(2), 465–476 (2011)
23. Choi, Y., Nam, J., Lee, D., Kim, J., Jung, J., Won, D.: Security enhanced anonymous multi-server authenticated key agreement scheme using smart cards and biometrics. Sci. World J. **2014**, Article 281305 (2014)
24. Kim, J., Lee, D., Jeon, W., Lee, Y., Won, D.: Security analysis and improvements of two-factor mutual authentication with key agreement in wireless sensor networks. Sensors **14**(4), 6443–6462 (2014)

25. Gope, P., Hwang, T.: Lightweight and energy-efficient mutual authentication and key agreement scheme with user anonymity for secure communication in global mobility networks. IEEE Syst. J. (2015). doi:10.1109/JSYST.2015.2416396

26. Li, C., Hwang, M., Chung, Y.: A secure and efficient communication scheme with authenticated key establishment and privacy preserving for vehicular ad hoc networks. Comput. Commun. **31**, 2803–2814 (2008)

Recommending Books for Children Based on the Collaborative and Content-Based Filtering Approaches

Yiu-Kai Ng[✉]

Computer Science Department, Brigham Young University, Provo, UT 84602, USA
ng@compsci.byu.edu

Abstract. According to a study conducted by the National Institute of Child Health and Human Development, reading is the single most important skill necessary for a happy, productive, and successful life. A child who is an excellent reader often has confident and a high level of self esteem and can easily make the transition from learning to read to reading to learn. Promoting good reading habits among children is essential, given the enormous influence of reading on students' development as learners and members of the society. Unfortunately, very few (children) websites or online applications recommend books to children, even though they can play a significant role in encouraging children to read. Popular book websites, such as goodreads.com, commonsensemedia.org, and readanybook.com, suggest books to children based on the popularity of books or rankings on books, which are not customized/personalized for each individual user and likely recommend books that users do not want or like. We have integrated the collaborative filtering (CF) approach and the content-based approach, in addition to predicting the grade levels of books, to recommend books for children. The user-based CF approaches filter books appealing to each user based on users' *ratings*, whereas the content-based filtering method analyzes the *descriptions* of books rated by a user in the past and constructs a user profile to capture the user's preferences. Recent research works have demonstrated that a hybrid approach, which combines the content-based filtering and CF approaches is more effective in making recommendations. Conducted empirical study has verified the effectiveness of our proposed children book recommender.

Keywords: Book recommendation · Content analysis · Collaborative filtering · Children

1 Introduction

Reading is an activity performed on a daily basis: from reading news articles and books to cereal boxes and street signs. We recognize that children literacy forms a foundation upon which children will gage their future reading.[1] It is

---

[1] http://www.deafed.net/publisheddocs/sub/9807kle.htm.

O. Gervasi et al. (Eds.): ICCSA 2016, Part IV, LNCS 9789, pp. 302–317, 2016.
DOI: 10.1007/978-3-319-42089-9_22

imperative to motivate young readers to read by offering them appealing books to read so that they can enjoy reading and gradually establish a reading habit during their formative years that can aid in promoting their good reading habits. As stated in [22], learning to read is a key milestone for children living in a literate society, specially given that reading provides the foundation for children's academic success. A recent study [19] highlights the fact that children who "do not read proficiently by the end of third grade are four times more likely to leave school without a diploma than proficient readers." The results of the study correlate with earlier statistics [8] which confirm that 88 % of children who are poor readers by the end of the first grade remain so by the end of the fourth grade. Moreover, young readers who successfully learn to read in the early primary years of school will more likely be prepared to read for pleasure and learning in the future [11]. The aforementioned findings constitute the essence of encouraging good reading habits early on. Identifying books appealing to children, however, can be challenging, given the amount of books made available on a regular basis that address a diversity of topics and target readers at different reading levels. It is essential to provide children with reading materials matching their preferences/interests and reading abilities, since exposing young readers to materials that are either too easy/difficult to understand or involving unappealing topics could diminish their interest in reading [1].

With the huge volume of children books available these days,[2] it is a time-consuming and tedious process for K-12[3] teachers, librarians, and parents to manually examine the topic of each book and choose the one for their students/children to read. Moreover, it is hard for children to choose books to read on their own, since they lack of experiences on choosing appropriate books to read. Online book websites, such as goodreads.com and commensemedia.org, make book recommendations to children based on the popularity and rankings. However, a child may not enjoy reading a popular book or a book with a very high ranking. For example, the book "Where the wild things are" is considered one of the most popular children books; however, some children find it unappealing to them because of the frightening scenes depicted in the book. Instead of relying on popularity or rankings on books, we have developed *CBRec*, a children book recommender, which adopts the *content-based* and *user-based collaborative filtering* approaches to make personalized book recommendations for children. The user-based and item-based collaborative filtering (CF) approaches are popular techniques for generating personalized recommendations [4]. When the data sparsity becomes a problem for certain children users, i.e., when there is not enough data to generate similar user groups or similar item groups to use

---

[2] According to a report published by the Statistics Portal (http://www.statista.com/statistics/194700/us-book-production-by-subject-since-2002-juveniles/) there are 32,624 children books published in the U.S.A. in 2012 alone.

[3] K-12, which is a term used in the educational system in the United States and Canada (among other countries), refers to the primary and secondary/high school years of public/private school grades prior to college. These grades are kindergarten (K) through 12th grades.

the CF methods, the content-based filtering approach can be adopted to make personalized recommendations for the users.

CBRec is designed for solving the *information overload* problem while minimizing the *time* and *efforts* imposed on parents/educators/young readers in discovering unknown, but suitable, books for pleasure reading or knowledge acquisition. CBRec first infers the readability level of a user U by analyzing the grade levels of books in his/her profile, which are determined using ReLAT, a robust readability level analysis tool that we have developed [17]. Hereafter, CBRec identifies a set of candidate books, among the ones archived at a website, with grade levels compatible to the inferred readability level of U. The current implementation of CBRec is tailored towards recommending books written in English and classified based on the K-12 grade level system. CBRec, however, can be easily adopted to make suggestions in languages other than English.

CBRec is a novel recommender that exclusively targets children readers, an audience who has not been catered by existing recommendation systems. CBRec is a self-reliant recommender which, unlike others, does not rely on personal tags nor access logs to make book recommendation. CBRec is unique, since it explicitly considers the ratings and content descriptions on books rated by children, in addition to the readability levels of children.

The remaining of this paper is organized as follows. In Sect. 2, we discuss existing book recommenders that have been used for suggesting books for individual readers, including children. In Sect. 3, we introduce CBRec and its overall design methodology. In Sect. 4, we present the results of the empirical study on CBRec conducted to assess its performance. In Sect. 5, we give a concluding remark and present directions for future work on CBRec.

2 Related Work

In this section, we present a number of widely-used book recommenders and compare them with CBRec.

A number of book recommenders [10, 15, 25] have been proposed in the past. Amazon's recommender [10] suggests books based on the purchase patterns of its users. Yang et al. [25] analyze users' access logs to infer their preferences and apply the traditional CF strategy, along with a ranking method, to make book suggestions. Givon and Lavrenko [6] combine the CF strategy and social tags to capture the content of books for recommendation. Similar to the recommenders in [6, 25], the book recommender in [18] adopts the standard user-based CF framework and incorporates semantic knowledge in the form of a domain ontology to determine the users' topics of interest. The recommenders in [6, 18, 25] overcome the problem that arises due to the lack of initial information to perform the recommendation task, i.e., the cold-start problem. However, the authors of [6, 25] rely on user access logs and social tags, respectively to recommend books, which may not be publicly available and are not required by CBRec. Furthermore, the recommender in [18] is based on the existence of a book ontology, which can be labor-intensive and time-consuming to construct [5].

In making recommendations, Park and Chang [15] analyze individual/group behaviors, such as clicks and shopping habits, and features describing books, such as their library classification, whereas $PReF$ [16] suggests books bookmarked by connections of a LibraryThing user. $PReF$ adopts a similarity-matching strategy that uses analogous, but not necessarily the same, words to the ones employed to capture the content of a book of interest to U. This strategy differs from the exact-matching constraint imposed in [15] and a number of content-based recommenders [7,13]. However, neither $PReF$ nor any of the aforementioned recommenders considers the readability level of their users as part of their recommendation strategies.

Vaz et al. [20] present a hybrid book recommendation system that take into account the preferences of users on the content of a book and its authors using two item-based CF approaches. Users' rankings on authors are predicted and are considered along with former book predictions for the users. Mooney and Roy [12], on the other hand, introduce a content-based book recommendation strategy which uses information about an item to make suggestions. An advantage of using the content-based approach for information filtering is that it does not rely on users' ratings on items which is useful in recommending previously unrated items. However, as shown in our empirical study (as presented in Sect. 4), the performance of a recommendation system based on either the item-based CF approach or content-based approach cannot achieve the same degree of effectiveness by incorporating the user-based and content-based filtering approaches.

As mentioned earlier, Givon and Lavrenko [6] predict user ratings on books by using tags attached to books on a social-networking sites. The authors attempt to solve the cold-start book recommendation problem by inferring the most probable tags from the text of a book. However, there are major design faults of the proposed book recommendation system. First of all, According to [2], only 7.7 % of published books in the OCLC database, a popular and worldwide library cooperative, are linked to the partial or full content of their corresponding books. For this reason, it is a severe constraint imposed on any analysis tool that relies on even an excerpt of a book due to copyright laws that often prohibit book content from being made publicly accessible. Second, tags are not widely available at children's book sites, since *personal tags* [9] assigned to books are rarely provided by children at the existing social bookmarking sites established for them.

Woodruff et al. [23] apply spreading activation over a text document (i.e., books in their case) and its citations such that nodes in the activation represent documents, whereas edges are created using the citations. The authors claim that the fused spreading activation techniques are superior compared with the traditional text-based retrieval methods. However, unlike textbooks or reference books in the book market, children books lack of references and hence the spreading activation methodology is not applicable to the design of children book recommendation systems.

Cui and Chen [3] claim that existing book recommendation systems do not offer enough information for their users to decide whether a book should be recommended to others. In solving the existing problem, the authors create recommendation pages of books which contain book information for the users to refer to. This recommendation approach, however, is not applicable to children, since the latter are interested in books recommended to them, instead of initiating the process of making recommendations on books to friends or other users on an online book websites.

3   Our Book Recommender

Content-based recommendation systems suggest books similar in *content* to the ones a given user has liked in the past, whereas recommendation systems based on the CF approaches identify a group of users S whose preferences represented by their ratings are similar to those of the given user U and suggests books to U that are likely appealing to U based on the ratings of S.

3.1   Identifying Candidate Books

We recognize that "reading for understanding cannot take place unless the words in the text are accurately and efficiently decoded" [14] and only recommends books with readability levels appropriate to its users. In order to accomplish this task, CBRec determines the readability level of a book (user, respectively) using ReLAT [17] developed by us. Due to the huge number of books written for K-12 readers, it is not feasible to analyze all the books (e.g., books posted at a social bookmarking site) to identify the ones that potentially match the interests of a site user U. Consequently, CBRec follows a common practice among existing readability analysis tools [24] and applies Eq. 1 to estimate the readability level of a user U, denoted $RL(U)$, based on the grade level of each book P_B in U's profile predicted by ReLAT, denoted ReLAT(P_B). Note that only books bookmarked in a user's profile during the most recent academic year are considered, since it is anticipated that the grade levels of books bookmarked by users gradually increase as the users enhance their reading comprehension skills over time.

$$RL(U) = \frac{\sum_{P_B \in P} ReLAT(P_B)}{|P|} \tag{1}$$

where $|P|$ denotes the number of books in U's profile and *average* is employed to capture the *central tendency* on the grade levels of books bookmarked by U.

CBRec first creates *CandBks*, the subset of books (archived at a social bookmarking site) that are compatible with the readability level of a user U which are further analyzed for making recommendations to U to ensure that recommendations made for U can be understood by and are suitable for U. *CandBks* includes a number of books considered by CBRec for recommendation, each of which is within-one-grade-level range from U's. By considering books within *one* grade level above/below U's mean readability level,[4] CBRec recommends books with

---

[4] We have experimentally determined this range to ensure the suitability of the recommended books with respect to the reading level of the corresponding user.

an appropriate level of complexity for U and grade levels approximate to the grade levels of books that have been read by U (as of the most recent academic year) and thus encourage users' reading growth which are neither too difficult nor too easy for U to understand.

Example 1. Consider a user U who has bookmarked a number of books from Dav Pilkey's "Captain Underpants" series. Based on the grade levels predicted by ReLAT for the books archived at BiblioNasium.com (see a sample of BiblioNasium books in Table 1) and U's readability level, which is 4, CBRec does not include Bk_1 nor Bk_3 in $CandBks$, since their grade levels are below/beyond the range deemed appropriate for U and thus it is not considered for recommendation by CBRec for U. □

3.2   The Content-Based Filtering Method

The content-based filtering approach recommend items to a user that are *similar* to the ones that the user prefers in the past. The approach can be adopted for identifying the common characteristics of books being liked by user u and recommend to u new books that share these characteristics. The *similarity of books* is computed by using the designated features applicable to the books to be compared. For example, if u offers very high ratings on books in the domain of adventure or a particular author, then the content-based filtering approach suggests other books to u based on the same domain, i.e., adventure, or author.

The Content-Based Filtering Approach Using the Vector Space Model. The content-based filtering method analyzes the descriptions of children books rated by a user u and construct the profile of u based on the descriptions which are used for predicting the ratings of books unknown to u. Given the attributes of a user profile that capture the preferences and interests of the user, a content-based recommender attempts to match the attributes with the ones that describe the content of another book to recommend new interesting books to the user. This method does not require the *ratings* on books given by other

Table 1. A number of BiblioNasium books

ID	Book title	Grade level
Bk_1	Mummies in the Morning	2.9
Bk_2	Captain Underpants and the Big, Bad Battle of the Bionic Booger Boy	4.7
Bk_3	The Hidden Boy	5.6
Bk_4	Dragon's Halloween	3.1
Bk_5	Junie B. Jones Smells Something Fishy	3.0
$\cdots$	$\cdots$	$\cdots$

users as in the collaborative filtering approaches to predict ratings on unknown books to u. The user profile of u is a vector representation of u's interests, which is constructed using Eq. 2.

$$X_u = \Sigma_{i \in \tau_u} r_{u,i} X_i \tag{2}$$

where τ_u is the set of books rated by user u, $r_{u,i}$ denotes the rating provided by user u on book i, and X_i is the vector representation of the description D on i with the *weight* of each keyword k in D computed by using the *term frequency (TF)* and *inverse document frequency (IDF)* of k.

The vector space model (VSM) is used to predict the *rating* of a book B unknown to user u using the profile P of u, denoted $CSim(P, B)$. The profile representation of P is computed using Eq. 2, whereas the vector representation of the description of B is determined similarly as X_i in Eq. 2, i.e., the *weight* of each keyword k in the description of B, denoted B_i, is calculated by using the TF/IDF of k.

$$CSim(P, B) = \frac{\Sigma_{i=1}^{t} (P_i \times B_i)}{\sqrt{\Sigma_{i=1}^{t} P_i^2 \times \Sigma_{i=1}^{t} B_i^2}} \tag{3}$$

where t is the dimension of the vector representation of P and B.

The Content-Based Filtering Approach Using the Naïve Bayes Model.
Besides using the vector space model for content-based filtering, machine-learning techniques have also been widely used in inducing content-based profiles. In using the machine-learning approach for text (which can be adopted for profile) classification, an inductive process automatically constructs a text classifier by learning from a set of training documents, which have already been labeled with the corresponding categories. Indeed, learning to classify user profiles can be treated as a binary text categorization problem, i.e., each item is classified as either interesting or not interesting with respect to the user preferences as specified by the attributes in a user profile. In this section, we discuss the Naïve Bayes classifier, which is a widely-used machine-learning algorithm in content-based filtering approach. Even though a constraint imposed on using the machine-learning approach for content-based filtering is that items must be labeled with their respective classes during the training process, after the classifier has been trained, it can be used to automatically infer profiles based on the trained model.

The Naïve Bayes model is a probabilistic approach to inductive learning. The probabilistic classifier developed by the Naïve Bayes model is based on the *Bayes' rule* as defined below.

$$P(C|D) = \frac{P(D|C) \times P(C)}{P(D)} \tag{4}$$

$$= \frac{P(D|C) \times P(C)}{\sum_{c \in C} P(D|C = c) P(C = c)}$$

where C (D, respectively) is a random variable corresponding to a class (document[5], respectively).

Based on the term, which is either an attribute or a keyword in an item, independence assumption, the Naïve Bayes rule yields

$$P(c|d) = \frac{P(d|c) \times P(c)}{\sum_{c \in C} P(d|c)P(c)} \tag{5}$$
$$= \frac{\prod_{i=1}^{n} P(w_i|c) \times P(c)}{\sum_{c \in C} \prod_{i=1}^{n} P(w_i|c) \times P(c)}$$

where w_i ($1 \leq i \leq n$) is a term in d, and $\sum_{c \in C} \prod_{i=1}^{n} P(w_i|c) \times P(c)$ is a *chain rule*.

The classification process is based on the following computation:

$$Class(d) = arg\ max_{c \in C} P(c|d) \tag{6}$$
$$= arg\ max_{c \in C} \frac{P(d|c) \times P(c)}{\sum_{c \in C} P(d|c) \times P(c)}$$

where $P(d|c)$ is the *probability* that d is observed, given that the class is known to be c, and $P(c)$ is the *probability* of observing class c, which is defined as

$$P(c) = \frac{N_c}{N} \tag{7}$$

where N_c is the number of training items in class c, and N is the total number of training items.

In estimating $P(d|c)$ in Eq. 6, the *Multiple-Bernoulli model* is applied, since the Multiple-Bernoulli distribution is a natural way to model the distributions over binary vectors. $P(d|c)$ is computed in the *Multiple-Bernoulli model* as

$$P(d|c) = \prod_{w \in V} P(w|c)^{\delta(w,d)}(1 - P(w|c))^{1-\delta(w,d)} \tag{8}$$

where $\delta(w,d) = 1$ if and only if term w occurs in d.

Note that $P(d|c) = 0$ if there exists a $w \in d$ that never occurs in c in the training set, which is the *data sparseness* problem and can be solved by using the *Laplacian smoothed* estimate as defined below.

$$P(w|c) = \frac{df_{w,c} + 1}{N_c + 1} \tag{9}$$

---

[5] From now on, unless stated otherwise, a document is treated as an item, such as a book.

where $df_{w,c}$ denotes the number of items in c which includes term w, and N_c is the number of items belonged to the class c.

In designing CBRec, we have decided to adopt the *vector space model* instead of the *Naïve Bayes* model, since the latter requires a trained model using a pre-defined labeled item set which imposes additional overhead. However, the *Naïve Bayes* model is an alternative model that can be considered in developing CBRec or other book recommender systems for children.

3.3   The Collaborative Filtering (CF) Approaches

The CF approaches rely on the ratings of a user and other users. The predicted rating of a user u on a book i is likely similar to the rating of user v on i if both users have rated other books similarly.

The User-Based CF Approach. The user-based CF approach determines the interest of a user u on a book i using the ratings on i by other users, i.e., the neighbors of u who have similar rating patterns [26]. We apply the Cosine similarity measure as defined in Eq. 10 to calculate the similarity between two users u and v and determine user pairs which have the lowest rating difference among all the users. Equation 10 can be applied to compute the *similarity* between a user and each one of the other users to find out the *similarity group* of each user. Two users who have the lowest difference rating value between them means that they are the *closest neighbors*.

$$USim(u, v) = \frac{\Sigma_{s \in S_{u,v}} (r_{u,s} \times r_{v,s})}{\sqrt{\Sigma_{s \in S_{u,v}} r_{u,s}^2} \times \sqrt{\Sigma_{s \in S_{u,v}} r_{v,s}^2}} \tag{10}$$

where $r_{u,s}$ ($r_{v,s}$, respectively) denotes the rating of user u (user v, respectively) on book s, and $S_{u,v}$ denotes the set of books rated by both users u and v.

Upon determining the K-nearest neighbors (i.e., KNN) of a user u using Eq. 10, we can compute the predicted rating on a book s for u using Eq. 11.

$$\hat{r}_{u,s} = \bar{r_u} + \frac{\Sigma_{v \in S_{u,v}} (r_{v,s} - \bar{r_u}) \times USim(u, v)}{\Sigma_{v \in S_{u,v}} USim(u, v)} \tag{11}$$

where $\hat{r}_{u,s}$ stands for the predicted rating on book s for user u, $\bar{r_u}$ is the average rating on books provided by user u, $S_{u,v}$ is the group of closest neighbors of u, $r_{v,s}$ is the rating of user v on book s, and $USim(u, v)$ is the similarity measure between users u and v as computed in Eq. 10.

Instead of using the user-based predicted ratings as defined in Eq. 11, another commonly-used user-based rating prediction approach is given in Eq. 12 in which the rating prediction is computed for each user u on a new book i without considering *different levels of similarity* among users.

$$\hat{r}_{u,i} = \frac{1}{|N_i(u)|} \sum_{v \in N_i(u)} r_{v,i} \tag{12}$$

Table 2. Children's books and the ratings (in the range of 1-5) by the children

	Runny Babbit	Harry Potter	Kid Athletes	Funny Bones	Finding Winnie
Noah		4	2	3	3
Alex	5	4	3	1	4
Emma	3	4	?	1	5
Ava	4	3	4	2	5
Jacob	2			5	2

where $N_i(u)$ is the group of nearest neighbors of u who have rated book i and $r_{v,i}$ is the rating of i provided by user v who is one of the nearest neighbors of u.

If the nearest neighbors of u come with different levels of similarity with respect to u, denoted $w_{u,v}$, the predicted user-based rating using the different levels of user similarity is defined as

$$\hat{r}_{u,i} = \frac{\sum_{v \in N_i(u)} w_{u,v} \times r_{v,i}}{\sum_{v \in N_i(u)} w_{u,v}} \tag{13}$$

Example 2. Consider Table 2 which includes a number of children and their ratings on different books.

Two children, Emma and Ava, have very similar ratings on the five books listed in Table 2, whereas Emma and Jacob have very dissimilar ratings on corresponding books. Both Emma and Ava enjoyed the book *Finding Winnie* and disliked the book *Funny Bones*. However, Jacob really liked the book *Funny Bones*, whereas Emma did not like it at all.

Assume that we are supposed to predict the rating of the book *Kid Athletes* for Emma using the ratings provided by Ava and Alex, the two nearest neighbors of Emma. Further assume that the similarity values between Emma and Ava and between Emma and Alex are 0.8 and 0.5. Applying Eq. 13, the predicted rating on the book *Kid Athletes* for Emma would be

$$\hat{r}_{\text{``Emma''},\text{``KidAthletes''}} = \frac{0.8 \times 4 + 0.5 \times 3}{0.8 + 0.5} \cong 3.62 \quad \Box$$

We adopt Eq. 11 for the rating predictions using the user-based CF approach, instead of Eq. 13, which requires different levels of similarity among different users to be generated in advance. However, if the similarity weights are available among different users, Eq. 13 could be adopted in place of Eq. 11 in the user-based CF rating prediction.

The Item-Based CF Approach. Contrast to the user-based CF approach which relies on similar user groups to recommend books, the item-based CF approach computes the similarity values among different books and determines sets of books with similar ratings provided by different users. The item-based CF approach predicts the rating of a book i for a user u based on the ratings of u on books similar to i. The *adjusted cosine similarity matrix* [21] is applied to compute the similarity values among different books and assign books with similar ratings into the same similar-item group as defined in Eq. 14.

$$ISim(i,j) = \frac{\Sigma_{u \in U}(r_{u,i} - \bar{r_u}) \times (r_{u,j} - \bar{r_u})}{\sqrt{\Sigma_{u \in U}(r_{u,i} - \bar{r_u})^2} \times \sqrt{\Sigma_{u \in U}(r_{u,j} - \bar{r_u})^2}} \tag{14}$$

where $ISim(i,j)$ denotes the similarity value between books i and j, $r_{u,i}$ ($r_{u,j}$, respectively) denotes the rating of user u on book i (j, respectively), $\bar{r_u}$ is the average rating for user u on all books u has rated, and U is the set of books rated by u.

Equation 14 computes the similarity between two books, whereas Eq. 15 predicts the rating for user u on book i.

$$S_{u,i} = \bar{r_u} + \frac{\Sigma_u(r_{u,i} - \bar{r_u})}{u(i) + r} + \frac{\Sigma_{j \in I(u)}(r_{u,j} - \frac{\Sigma_u(r_{u,j} - \bar{r_u})}{u(j)})}{I(u) + r} \tag{15}$$

where $S_{u,i}$ denotes the predicted rating on book i for user u, $\bar{r_u}$ denotes the *average rating* on all books u has rated, $r_{u,i}$ ($r_{u,j}$, respectively) is the rating on book i (j, respectively) provided by u, $I(u)$ is the set of books u has rated, $u(j)$ is the number of users who has rated i, and r is the book rating which is used to decrease the extremeness when there are not enough ratings available, which is determined experimentally.

The first component on the right-hand side of Eq. 15 is called the *global mean* which is the average rating on all the books u has rated. The second component is called the *item offset* which is the score for user u on book i, whereas the third component is called the *user offset* which is the user prediction on book i.

An alternative item-based CF approach is presented in [4] which considers similar items (i.e., books in our case) with similarity weights between two items, which are predefined. The predicted rating on item i for user u is computed as follows.

$$\hat{r}_{u,i} = \frac{\Sigma_{j \in N_u(i)} w_{i,j} \times r_{u,j}}{\Sigma_{j \in N_u(i)} |w_{i,j}|} \tag{16}$$

where $N_u(i)$ is the set of items rated by user u that are *most similar* to item i, and $w_{i,j}$ is the *similarity weight* between items i and j.

Example 3. Consider Example 2 again. Instead of consulting Emma's peers, CBRec considers the ratings on the books Emma and others have read in the past. Based on the ratings provided by children as shown in Table 2, the two books that are the closest neighbors, i.e., most similar in terms of ratings, of the book *Kid Athletes* are *Harry Potter* and *Finding Winnie*. Assume that the similarity values between the books *Kid Athletes* and *Harry Potter* and between *Kid Athletes* and *Finding Winnie* are 0.55 and 0.35. As shown in Table 2, the ratings given by Emma on *Harry Potter* and *Finding Winnie* are 4 and 5, respectively, the predicted rating on *Kid Athletes* for Emma is

$$\hat{r}_{\text{"Emma"},\text{"KidAthletes"}} = \frac{0.55 \times 4 + 0.35 \times 3}{0.55 + 0.35} \cong 4.3 \quad \square$$

Once again we do not adopt Eq. 16 for our item-based CF approach, since the similarity weights of two items must be predefined.

4   Experimental Results

In this section, we first introduce the datasets used for the empirical study conducted to assess the performance CBRec (in Sect. 4.1). Hereafter, we present the results of the empirical study on CBRec in Sect. 4.2.

4.1   Datasets

We have chosen a number of children book records included in the Book-Crossing dataset to conduct our performance evaluation of CBRec.[6] The book-crossing dataset was collected by Cai-Nicolas Ziegler from the book-crossing community. It contains 278,858 users who provide 1,149,780 ratings on 271,379 books. Since not all of books in the book-crossing database are children books, we pre-processed the dataset to extract only children book records. Each record includes a user_ID, the ISBN of a book, and the rating provided by the user (identified by the user_id) on the book. We used Amazon.com AWS advisement API to verify that the ISBNs from the book-crossing dataset are valid and they are children books. Out of the 271,379 books in the Book-Crossing dataset, approximately 29,000 books are children books, which is denoted as CBC_DS.

Besides using the children books and their ratings in the Book-Crossing dataset, we extracted the book description of 30 % of the children books in CBC_DS from Amazon.com, since they were missing in the children books and needed for the content-based filtering approach. The Amazon dataset yields the additional dataset used for evaluating the performance of CBRec. Figure 1 shows the differences in terms of prediction errors using only 70 % versus 100 % of book descriptions generated by the content-based filtering approach.

4.2   Performance Evaluation of CBRec

In our empirical study, we considered the similar user group size of *ten* users in the user-based CF approach, since LensKit,[7] which has implemented the user-based CF method and has been cited in a number of published papers, has demonstrated that ten is an ideal group size in predicting user ratings. We have also chosen *ten* to be the group size of books used in the content-based and item-based CF approaches, since the prediction error rate using this group side is the most ideal, in terms of size and accuracy, as demonstrated in our empirical study and reported in Fig. 1.

To evaluate the performance of CBRec, we computed the prediction error of CBRec for each user U in CBC_DS by taking the absolute value of the difference

---

[6] Other datasets can be considered as long as they contain user_IDs, book ISBNs, and rating information.

[7] http://lenskit.org/.

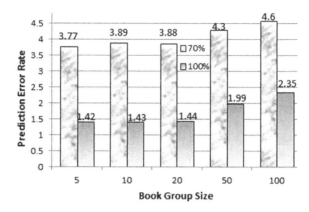

Fig. 1. Prediction errors of the content-based approach on the children books in the Book-Crossing dataset

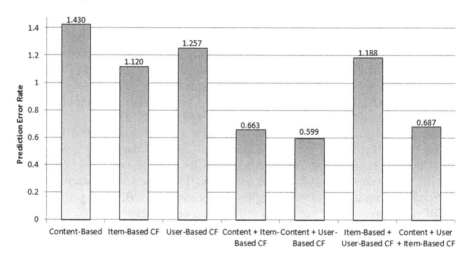

Fig. 2. Prediction error rates of the various filtering approaches and their combinations

between the real and predicted ratings on each book U has rated in the dataset. These prediction errors were added up and divided it by the total number of predictions. Figure 2 shows the prediction error rate of each filtering approaches and their combinations.

As shown in Fig. 2, the combined content-based and user-based CF approach, denoted CUB, outperforms individual and other combined prediction models in terms of obtaining the lowest prediction error rates among all the models. CUB achieves the highest prediction accuracy, which is only *half* a rating (out of 10) away from the actual rating, since the content-based filtering approach compensates the user-based CF approach when user ratings are *sparse* and vice versa. The prediction error rate, i.e., accuracy ratio, achieved by UCB is statistically significant ($p < 0.05$) over the ones based on the combined content-based and item-based CF approach and the combination of all the three filtering approaches, the next two models with prediction error rate lower than *one*, which are determined using the Wilcoxon test signed-ranked test. The experimental results have verified that UCB is the most accurate recommendation tool in predicting children's ratings on books, which is the most suitable choice for making book recommendations for children based on the rating prediction.

5   Conclusions and Future Work

Existing book recommenders either are (i) not personalized enough, since they make the *same* recommendations to all users on a given book, (ii) based on the availability of users' historical data in the form of social tags, which may not be publicly available on children, or (iii) developed for a *general audience*, instead of taking into account the reading levels of their users. To address these issues, we have developed CBRec, a book recommender tailored to children, which simultaneously considers the reading levels and interests of its users in making personalized suggestions. CBRec adopts the widely-used *content-based filtering* approach and the *user-based collaborative filtering* approach and integrates the two filtering approaches in predicting ratings on children books to make book recommendation. Predicted ratings on children books, in addition to the readability levels of the (candidate) books to be considered for recommendation, provide CBRec the wealth of useful information to suggest books with appropriate levels of complexity and topics of interest that are appealing to children.

CBRec is unique, since it makes *personalized* suggestions on books that satisfy both the *preferences* and *reading abilities* of its users. Unlike current state-of-the-art recommenders that rely on the existence of user access logs or social tags, CBRec simply considers *brief descriptions* on children's books, their *ratings*, and their *grade levels* computed using our grade-level prediction tool, ReLAT, which is different from popular readability formulas that focus solely on analyzing lexicographical and syntactical structures of the texts in books. Information, such as metadata and ratings on books, are readily available on (children) social bookmarking websites, such as goodreads.com., whereas ReLAT can determine the grade level of any book (even if a sample of the text of a book is unavailable) by analyzing the Subject Headings of books, US Curriculum subject areas identified in books, and information about the authors of books. As children continue to read more books if they can *choose* what to read [1], a significant contribution of CBRec is to provide children a selection of suitable books to choose from that are not only appealing to them, but can be comprehended by them. The conducted experiments demonstrate the effectiveness of CBRec in suggesting books for children.

As a by-product of this research work we have created a benchmark dataset consisting of users and books, in addition to metadata and readability levels of the books, which can be used to assess the performance of recommenders that provide books suggestions to K-12 readers.

As part of our future work, we intend to extend CBRec so that it can suggest reading materials for struggling readers, especially the ones with learning disabilities and those who learn English as a second language. For these readers, their readability levels can be different from ordinary students. Book recommenders can aid these users by finding books potentially of interest to them.

References

1. Allington, R., Gabriel, E.: Every child, every day. Read. Core Skill **69**(6), 10–15 (2012)
2. Chen, X.: Google books and worldcat: a comparison of their content. Online Inf. Rev. **36**(4), 507–516 (2012)
3. Cui, B., Chen, X.: An online book recommendation system based on web service. In: Proceedings of the Sixth International Conference on Fuzzy Systems and Knowledge Discovery (FSKD 2009), pp. 520–524 (2009)
4. Desrosiers, C., Karypis, G.: A comprehensive survey of neighborhood-based recommendation methods. In: Ricci, F., Rokach, L., Shapira, B., Kantor, P.B. (eds.) Recommender Systems Handbook, pp. 107–144. Springer, Heidelberg (2011)
5. Ding, Z.: The Development of ontology information system based on bayesian network and learning. In: Jin, D., Lin, S. (eds.) Advances in Multimedia, Software Engineering and Computing Vol. 2. Advances in Intelligent and Soft Computing, vol. 129, pp. 401–406. Springer, Heidelberg (2012)
6. Givon, S., Lavrenko, V.: Predicting social-tags for cold start book recommendations. In: Proceedings of the 3rd ACM Conference on Recommender Systems (ACM RecSys 2009), pp. 333–336 (2009)
7. Guan, Z., Wang, C., Bu, J., Chen, C., Yang, K., Cai, D., He, X.: Document recommendation in social tagging services. In: Proceedings of the 19th International World Wide Web Conference (WWW 2010), pp. 391–400 (2010)
8. Juel, C.: Learning to read and write: a longitudinal study of fifty-four children from first through fourth grade. Educ. Psychol. **80**, 437–447 (1988)
9. Li, H., Gu, Y., Koul, S.: Review of digital library book recommendation models. SSRN (2009). http://dx.doi.org/10.2139/ssrn.1513415
10. Linden, G., Smith, B., York, J.: Amazon.com recommendations: item-to-item collaborative filtering. IEEE Internet Comput. **7**(1), 76–80 (2003)
11. Ministry of Education of Ontario. A Guide to Effective Instruction in Reading, Kindergarten to Grade3 (2005). http://goo.gl/UCo5e3
12. Mooney, R., Roy, L.: Content-based bookrecommending using learning for text categorization. In: Proceedings of the fifth ACM Conference on Digital Libraries (DL 2000), pp. 195–204 (2000)
13. Nascimento, C., Laender, A., da Silva, A., Goncalves, M.: A source independent framework for research paper recommendation. In: Proceedings of ACM/IEEE Joint Conference on Digital Libraries (JCDL 2011), pp. 297–306 (2011)
14. Oakhill, J., Cain, K.: The precursors of reading ability in young readers: evidence from a four-year longitudinal study. Sci. Stud. Read. (SSR) **16**(2), 91–121 (2012)
15. Park, Y., Chang, K.: Individual and group behavior-based customer profile model for personalized product recommendation. Expert Syst. Appl. **36**(2), 1932–1939 (2009)

16. Pera, M.S., Ng, Y.-K.: With a little help from my friends: generating personalized book recommendations using data extracted from a social website. In: Proceedings of the 2011 IEEE/WIC/ACM Joint Conference on Web Intelligent (WI 2011), pp. 96–99 (2011)
17. Pera, M.S., Ng, Y.-K.:What to read next?: making personalized book recommendations for K-12 users. In: Proceedings of the 7th ACM Conference on Recommender Systems (ACM RecSys 2013), pp. 113–120 (2013)
18. Sieg, A., Mobasher, B., Burke, R.: Improving the effectiveness of collaborative recommendation with ontology-based user profiles. In: Proceedings of the 1st International Workshop on Information Heterogeneity and Fusion in Recommender Systems (ACM HetRec), pp. 39–46 (2010)
19. The Annie E. Casey Foundation.Early Warning Confirmed: A Research Update on Third-Grade Reading (2013). http://goo.gl/HQrPOA
20. Vaz, P., de Matos, D., Martins, B., Calado, P.: Improving a hybrid literary book recommendation system through author ranking. In: Proceedings of the 12th ACM/IEEE-CS Joint Conference on Digital Libraries (JCDL2012), pp. 387–388 (2012)
21. Wang, J., de Vries, A., Reinders, M.: Unifying user-based and item-based collaborative filtering approaches by similarity fusion. In: Proceedings of the 29th Annual International ACM SIGIR Conference on Research and Development in Information Retrieval (SIGIR 2006), pp. 501–508 (2006)
22. Whitehurst, G., Lonigan, C.: Emergent literacy: development from prereaders to readers. In: Handbook of Early Literacy Research, vol. 1. The Guilford Press (2003)
23. Woodruff, A., Gossweiler, R., Pitkow, J., Chi, E., Card, S.: Enhancing a digital book with a reading recommender. In: Proceedings of the SIGCHI Conference on Human Factors in Computing Systems (CHI 2000), pp. 153–160 (2000)
24. Wright, B., Stenner, A.: Readability and reading ability. ERIC Document Reproduction Service No. ED435979 (1998)
25. Yang, C., Wei, B., Wu, J., Zhang, Y., Zhang, L.: CARES: A ranking-oriented CADAL recommender system. In: Proceedings of ACM/IEEE Joint Conference on Digital Libraries (JCDL 2009), pp. 203–212 (2009)
26. Zhao, Z., Shang, M.: User-based collaborative-filtering recommendation algorithms on hadoop. In: Proceedings of the 16th ACM SIGKDD International Conference on Knowledge Discovery and Data Mining (ACM KDD), pp. 478–481 (2010)

Ontology Evaluation Approaches: A Case Study from Agriculture Domain

Anusha Indika Walisadeera[1,3(✉)], Athula Ginige[2],
and Gihan Nilendra Wikramanayake[3]

[1] Department of Computer Science, University of Ruhuna, Matara, Sri Lanka
waindika@cc.ruh.ac.lk
[2] School of Computing, Engineering and Mathematics,
Western Sydney University, Penrith, NSW 2751, Australia
A.Ginige@westernsydney.edu.au
[3] University of Colombo School of Computing, Colombo 00700, Sri Lanka
gnw@ucsc.cmb.ac.lk

Abstract. The quality of an ontology very much depends on its validity. Therefore, ontology validation and evaluation is very important task. However, according to the current literature, there is no agreed method or approach to evaluate an ontology. The choice of a suitable approach very much depends on the purpose of validation or evaluation, the application in which the ontology is to be used, and on what aspect of the ontology we are trying to validate or evaluate. We have developed large user centered ontology to represent agricultural information and relevant knowledge in user context for Sri Lankan farmers. In this paper, we described the validation and evaluation procedures we applied to verify the content and examine the applicability of the developed ontology. We obtained expert suggestions and assessments for the criteria used to develop the ontology as well as to obtain user feedback especially from the farmers to measure the ontological commitment. Delphi Method, Modified Delphi Method and OOPS! Web-based tool were used to validate the ontology in terms of accuracy and quality. The implemented ontology is evaluated internally and externally to identify the deficiencies of the artifact in use. An online knowledge base with a SPARQL endpoint was created to share and reuse the domain knowledge. It was also made use of for the evaluation process. A mobile-based application is developed to check user satisfaction on the knowledge provided by the ontology. Since there is no single best or preferred method for ontology evaluation we reviewed various approaches used to evaluate the ontology and finally identified classification for ontology evaluation approaches based on our work.

Keywords: Ontology · Evaluation · Validation · Delphi method · Mobile-based application · SPARQL endpoint · Agriculture

1 Introduction

Quality of an ontology very much depends on its validity. Thus by establishing the validity we can ascertain the quality. After designing the ontology, the contents of the ontology need to be validated; otherwise, defects in the ontology will spread to

© Springer International Publishing Switzerland 2016
O. Gervasi et al. (Eds.): ICCSA 2016, Part IV, LNCS 9789, pp. 318–333, 2016.
DOI: 10.1007/978-3-319-42089-9_23

subsequent design and implementation activities. Therefore an ontology validation process is needed. Such a process will prevent application from using inconsistent, incorrect, redundant information, enhancing the quality of information captured in the ontology. However, according to the current literature, there is no single best or preferred approach to ontology validation and evaluation. The literature on ontology validation and evaluation is fragmentary [1]. Most approaches address more or less specific evaluation issues but often do it unsystematically [1]. The choice of a suitable approach must depend on a purpose of validation or evaluation, an application in which ontology is to be used, and on what aspect of ontology we are trying to validate or evaluate. Ontology evaluation methods group around two concepts; verification methods that ensure the structure of the ontology, and validation methods that examine their applicability in the real world [2].

Ontology evaluation is an important task that is needed in many situations, for example, during the process of building an ontology, ontology evaluation is important to guarantee that what is built meets the application requirements [3]. The ontological commitment is also important. It is an agreement by multiple parties (e.g. persons and software systems) to adopt a particular ontology when communicating about the domain of interest, even though they do not have same experiences, theories or prescriptions about that domain [4]. In other words, it is an agreement to use the shared vocabulary in a coherent and consistent manner within a specific context [5]. The design of the ontology needs to promote the ontological commitment. The ontological commitment should be verified by validating and evaluating the ontology in proper way.

We have developed user centered ontology for Sri Lankan farmers to represent the necessary agricultural information and relevant knowledge that can be queried based on the farmer context [6]. We tried to obtain expert suggestions and assessments for the criteria used to develop the ontology as well as to get user feedback especially from the farmers to measure the ontological commitment. In this paper, we have briefly discussed our attempt. This study was carried out as a part of the Social Life Networks for the Middle of the Pyramid (SLN4MOP) (www.sln4mop.org) project. SLN4MOP is an International Collaborative research project aiming to develop a mobile-based information system to support livelihood activities of people in developing countries [7].

The remainder of the paper is organized as follows. Section 2 presents related research in this field. In this paper, the validation and evaluation procedures are discussed separately. Validation procedure to verify the contents of the ontology during the ontology development is described in Sect. 3. Applicability of the ontology in the domain of agriculture is examined in Sect. 4. Section 5 summarizes the approaches used for the evaluation as a classification and concludes the paper.

2 Related Work

According to the current literature, there is no single best or preferred method or approach for ontology validation and evaluation. However, several methods, techniques, guidelines, and tools are available for this purpose. Here we have cited some of the studies in this field that helped us to clearly identify what aspect we have to consider when validating and evaluating an ontology; and what approaches are available for this.

Brank et al. [8] have done a survey of ontology evaluation techniques. They have summarized the different levels defined by different authors that help to evaluate the complex structure of an ontology. These various levels are: *Lexical, vocabulary, or data layer* (includes concepts, instances, facts, etc.); *Hierarchy or taxonomy* (includes hierarchical is-a relations); *Other semantic relations* (contains other relations except is-a); *Context or application level* (the context of the application where the ontology is to be used); *Syntactic level* (this level may be of particular interest for ontology mostly constructed by manually. Generally the ontology is described in a particular formal language and then it must match the syntactic requirements of the language); *Structure, architecture or design* (includes certain pre-defined design principles or criteria, etc.).

Brank et al. [8] further discussed some approaches to evaluate the ontology based on each level mentioned above. Such approaches are golden standard (comparing the ontology with an existing ontology that serves as a reference), data-driven evaluation, application-based evaluation, and assessment by humans against a set of criteria, etc. In data-driven evaluation, basically an ontology may be evaluated by comparing it with existing data (collection of textual documents) about the problem domain which the ontology refers. Practically, the ontology can be used for some kind of application or task. In application-based evaluation, the outputs of the application or its performance on the given task can be evaluated. Thus the ontology can be evaluated simply by plugging them into an application. However, several drawbacks can be seen in the application-based evaluation such as it is difficult to generalize the observations; the ontology could be only a small component of the application, then its effects on the outcome may be relatively small and indirect; comparison between ontologies is possible only if they can all plugged into the same application [8].

Obrst et al. [9] have mentioned many criteria that can be used to evaluate an ontology. They are: its coverage of a particular domain and the richness, complexity and granularity of that coverage; the specific use cases, scenarios, requirements, applications and data sources it was developed to address; and formal properties such as the consistency and completeness of the ontology and the representation language in which it is modeled.

Guarino and Welty [10] have discussed the OntoClean methodology that can be used to validate taxonomies by exposing inappropriate and inconsistent modeling choices. This method is used to clean concept taxonomies (remove wrong *subclass* of relations in taxonomies) based on formal notions such as *rigidity* (a property is rigid if it is essential to all its instances), *identity* (refers to the problems of being able to recognize individual entities in the world as being the same or different), and *unity* (refers to being able to recognize all the parts that form an individual entity) [10].

Gómez-Pérez has reviewed some of the previous works on ontology evaluation and mentioned that the criteria used for evaluating and assessing the ontologies [11]. These criteria are: *consistency, completeness, conciseness* (no unnecessary or useless definitions, no redundancies), *expandability* and *sensitiveness* (refers how small changes in a definition alter the well-defined properties). Gómez-Pérez has also addressed some possible types of errors caused when structuring domain knowledge in taxonomies in an ontology. These errors are: circularity errors, exhaustive and non-exhaustive class partition errors, redundancy errors, grammatical errors, semantic errors, and incompleteness errors [11].

We have considered the criteria and types of errors mentioned in the literature during the development process of our ontology. In next two sections, we discuss ontology evaluation approaches which we applied for evaluating our ontology.

3 Ontology Validation

It is very important to check the validity of the ontology. In this study, two aspects need to be validated; the correctness or appropriateness of the contents of the ontology and the correctness of the construction of the ontology. After designing the ontology, the contents of the ontology need to be validated by domain experts against the users' requirements. Validation is the process of checking whether or not a certain design is appropriate for its purpose; meets all the constraints; and will perform as expected. The validation process for content validation and correctness of the construction is discussed in the following sub sections (Sects. 3.1, 3.2 and 3.3).

3.1 Delphi Method

The content correctness depends on definitions of concepts, relationships between concepts, hierarchical structures, concept properties, and information constraints of the ontology. The Delphi method is a research technique that is used to obtain responses to a problem from a group of domain experts [12]. Since experts' opinions are a source of information available for this purpose to clarify the complex real situations in the domain of agriculture, the Delphi method is selected to obtain expert advice and responses to validate the content of the ontology. This technique is used by the experts to suggest and confirm the criteria. The end product of this technique is a consensus among the experts by use of statistical information and includes their commentaries on each of the questionnaire items, organized as a written document by the Delphi investigator [4].

The validation process is done by agricultural experts by examining the correctness, relevancy and consistency of the ontology components (i.e. concepts, relationships and constraints). The structured paper-based questionnaires were used for validation of the content correctness according to this method. The contents were refined based on domain experts' feedbacks and comments.

First of all, Delphi investigator (one of the authors of this paper) needs to identify the issue or question that need to be addressed. Especially, this will be a complex issue or question that requiring experts' advice. Next, the investigator selects the appropriate participants who are the experts with the matter at hand. Then first set of questionnaire was created and distributed to panelists. The results produced by the panel were analyzed and second set of questionnaire was prepared based on the responses. Second questionnaire is distributed to the panel. The responses to the second questionnaire were again analyzed. The steps were taken to repeat many times as necessary or as desired by the researchers. This usually only takes two iterations, but can sometimes takes as many as six rounds before a consensus is reached. Finally, Delphi investigator prepares a report based on the information found through analysis of panel responses. This technique gives a systematic way for gathering perspectives and critiques on a problem as a basis for iterative improvement.

A number of advantages of this method were noticed. They are: respondents can take time to deem before providing responses to the questionnaire, participants can live anywhere in the world, avoids group-think, flexible with no set meeting times, information can be very powerful in predicting future trends and events, gain consensus on a large scale, cost effective, more details than typical survey responses, etc. [13]. However, there were several limitations and weakness in this method. These were: the quality of responses depends on the panel, researchers must choose participants selectively, time consuming, panel may lose interest in the topic if consensus was not reached within a reasonable time, panel may lose cohesiveness over time, interaction between participants and researchers are not face-to-face, depending on the subject matter this fact can hinder or assist in information gathering, and low response rates, etc. [13].

The aim of this study was to validate the definitions of concepts, relationships between concepts, hierarchical structures, concept properties, and information constraints of the ontology. The Delphi method was selected to obtain expert advice and responses to check the definitions of concepts, relationships, and data properties; and hierarchical structures. A questionnaire was prepared and sent to the domain experts to get the feedback and comments. The domain experts' perceptions were captured using Yes/No, True/False and written responses. The validation process was mainly done by agricultural experts from different agricultural institutes using questionnaires based on the Delphi method. However, the Delphi investigator had experienced that the response rate was very low to the questionnaire using this method and this method was not very effective to validate the constraints (criteria).

3.2   Modified Delphi Method

The predefined criteria, design models as well as the assumptions made during the design process need to be validated properly. When designing the ontology, the ontology developers have to decide ways to represent complex real situations of the domain. Therefore a consensus among the group members (domain experts) to each of the decision made by developers to represent complex situation needs to be achieved (to validate assumptions and/or judgments). However, the Delphi method had some limitations for this purpose. To make more dialogues and active collaboration among the participants in the Delphi group, we arranged a discussion based on the Modified Delphi method to validate the assumptions and/or judgments.

The Delphi Method was adapted for use in face-to-face group meeting, allowing group discussions. This technique is very effective in generating a large quantity of creative new ideas. It was designed to allow every member of the group to express their ideas and minimizes the influence of other participants [14]. If you need to generate a lot of ideas and want to assure all members participate freely without influence from other participants, to identify priorities or select a few alternatives for further examination then the modified Delphi method can be effectively used [14]. The use of the consensus is common to both techniques.

For the discussion, eleven (11) Agricultural Instructors (AIs) gathered at Lunama Govi Jana Seva Center, Ambalanthota, Sri Lanka. The main aim of this discussion was to check the criteria relevant to fertilizer application, growing problems and control methods, etc. The Delphi investigator explained the problems in detail to obtain experts' knowledge. The investigator also allowed them to discuss the problems and possible solutions.

Based on their responses, comments, and suggestions we made judgments about the validity of the design criteria and assumptions made during the design process. The feedback and comments to the questionnaire were analyzed. Table 1 shows some responses received from the experts. The contents of the ontology have been refined based on these feedback and comments.

Table 1. A summary of expert response for some design criteria

Criteria	Description of the criteria	Correctness (Agree/Disagree or Yes/No)	Expert comments
A fertilizer quantity depends on many factors	When applying a fertilizer to a specific crop or soil it involves a fertilizer quantity. A fertilizer quantity depends on many factors: • fertilizer types (e.g. Chemical, Organic, or Biological fertilizers) • its specific sources (e.g. Nitrogen, Phosphorus, Potassium, etc.) and their ratio • location, water source (water supply), soil Ph range, time of application, and application method Suggest additional depending factors (if any).	Agree	Additional depending factors: • Variety (local variety or hybrid variety) • Type: annual crop or semi-annual crop • The nature of the water source (e.g. well, pond, etc.) • Spaces (based on the number of plants per hectare) • Crop stage
A control method depends on many factors	When selecting control methods, it depends on many factors: • types of the control methods (e.g. Cultural, Chemical, or Bio) • soil type • location, time of application, and application stage Suggest additional depending factors (if any).	Agree	Additional depending factors: • Gardening farmer • Commercial farmer • Economical status of the farmer • Matured time

The Modified Delphi procedure was more effective as this approach took less time, and hence was able to maintain participant's enthusiasm throughout the process, and also make dialogues and collaboration between participants that would encourage making new ideas. However, this method also has few limitations such as it requires extended advance preparation, tends to be limited to a single-purpose (single-topic meeting), it is difficult to change topics in the middle of the meeting, needs agreement from all participants to use the same structured method.

The correctness and relevancy of the ontology components as well as validation of a set of predefined criteria were verified using the above methods. The validation procedure for the correctness of the construction is discussed in the next sub section.

3.3   Correctness of the Construction of an Ontology

One approach for checking the correctness of the construction is to analyze whether the ontology contain anomalies or pitfalls [15]. First, the common pitfalls was identified before the implementation, for examples, defining synonyms as concepts, defining wrong inverse relationships, recursive definition, misuse of *part-of* and *subclass* relationships, etc. Next, the types of Ontology Design Patterns (ODPs) were identified to avoid the pitfalls by means of adapting or combining existing ODPs [16]. Design patterns were shared guidelines to solve design problems. Some examples were Semantic Web Best Practices and Development under W3C [17], and NeOn Modelling Components under NeOn project [18].

The reasoner (used FaCT ++ reasoner plug-in with Protégé 4.2) attached with the ontology development tool was used to check the logical inconsistency. The implicit knowledge was derived from the ontology through inference and reasoning procedures attached to the Protégé ontology.

The web-based tool called OOPS! [15] was also used to detect potential pitfalls in the ontology. All the pitfalls were not equally important. Their impact on the ontology would depend on multiple factors. Based on that each pitfall was assigned an importance level indicating how important it was. There were three levels such as *Critical*, *Important* and *Minor*. Critical was crucial and it needs to be corrected. Otherwise, it could affect the ontology consistency, reasoning, applicability, etc. Although important level of pitfall was not critical for ontology function, it was important to correct this type of pitfall. Minor pitfall was not really a problem, but by correcting it we make the ontology nicer. The pitfalls of the ontology were detected using OOPS! and was fixed.

These evidences suggest that this ontology was constructed correctly. The quality of the ontology has been measured and maintained using the above methods.

4   Ontology Evaluation

Evaluation is the process of examining a system or a product to determine the user satisfaction. After implementing the ontology, the deficiencies of the ontology in use were examined. This evaluation section is mainly divided into two sections such as internal evaluation (during the design process) and external evaluation (after designing). Sect. 4.1 explains how the usefulness of the ontology is tested during the design process. Sect. 4.2 explains how the effectiveness of the ontology is measured externally using the application-based approaches.

4.1 Internal Evaluation

By comparing the features of the ontology development tools and representing languages, the Protégé as an ontology development environment and Web Ontology Language (OWL) as an ontology language were selected. In this study, the decidability is very important since the agricultural information in the user's context needs to be retrieved. Therefore the DL based (OWL 2 - DL) approach was selected to implement this ontology.

The implemented ontology using Protégé is available at http://www.sln4mop.org/ontologies/2014/SLN_Ontology (Turtle version: http://www.sln4mop.org/ontologies/2014/SLN_Ontology.ttl). This implementation was used to evaluate the ontological commitments internally and also used to test the consistency and inferences using reasoners (used FaCT ++ reasoners plug-in with Protégé 4.2). The competency questions were used to evaluate the ontological commitments by observing whether the ontology meets the farmers' requirements. Description Logic queries (DL expressions) and SPARQL queries (see the Sect. 4.2.2 for more details about the SPARQL) already available in Protégé environment were used to query the ontology. For instance, farmers were interested to know suitable cultural control methods to manage Bacterial wilt disease of Tomato. The related DL query is shown below.

```
ControlMethod and hasControlMethodType value "Cultur-
al"^^string and isRelatedControlMethodOf some
((GrowingProblemEvent and hasGrowingProblem value Bacte-
rial_Wilt) and (isGrowingProblemEventOf value Tomato))
```

The SPARQL query below is to find crops that grow in Up Country, applicable to Sandy Loamy soil, and the temperature ranges between 9 and 28 °C.

```
PREFIX sln:
<http://www.sln4mop.org/ontologies/2014/SLN_Ontology#>
SELECT ?crop WHERE {
{?crop sln:growsIn sln:UpCountry.}
{?crop sln:hasSoilFactor sln:Sandy_Loamy}
{?crop sln:hasMaxTemperature ?o.
FILTER (?o < 28)}
{?crop sln:hasMinTemperature ?c.
FILTER (?c > 9)}}
```

By evaluating the outputs of the queries the ontological commitments were checked. These internal evaluation procedures were used to improve and refine the ontology during the design process.

4.2   Application-Based Evaluation

The user satisfaction assessment and task assessment (what is supported by the ontology) were tested using application-based evaluation. It is discussed in detail in the next three sub sections.

4.2.1   Field Testing

A Mobile based application was developed to provide information to farmers using this ontology [19]. Using this application, level of user satisfaction (i.e. utility) with the knowledge provided by the ontology is evaluated.

The first evaluation (first field trial) was done only for the crop selection stage with a group of 32 farmers in Sri Lanka. These farmers were selected with the help of Agriculture extension officers from Matale District in Sri Lanka. In this district high percentage of people are involved in cultivating wide range of vegetables. The evaluation study comprised of a demonstration session where farmers were given a brief introduction about the research and what is expected from them. First, a training session was carried out to make farmers familiar with the touch screen technology. Next, the crop selection prototype was demonstrated while illustrating the key features incorporated into the application. Crop selection provides a list of crops that grow in the region based on farm location and farmer preferences. The farmers used the application to select a crop that they want to grow, and they were asked the question "Is all information for the crop selection stage provided". They recorded the answer on a 1 to 5 Likert scale; strongly agree, agree, moderately agree, disagree, and strongly disagree. The responses were 7 % strongly agree, 57 % agree and 36 % moderately agree (i.e. 100 % agreed to the question "All information for the crop selection stage is provided"). The outcomes of this field trial are summarized in Table 2. The detail information including other findings from the field trial is available in [19].

The farmers also suggested few areas where they would like to get more information. These requirements were also gathered for future refinement of the ontology. The ontology was redesigned based on the feedbacks from the first field trial. Accordingly mobile-based information system was also redesigned to incorporate changes to the ontology.

Table 2. Summary of the outcomes from first field trial

Is all information for the crop selection stage provided?	100 %
Information is sufficient to make a decision on what to grow	97 %
Provide the knowledge on different crop varieties	100 %

Second field trial was done using this mobile application (http://webe11.scem.uws.edu.au/slnfarmer/) in Dambulla and Polonnaruwa area in Sri Lanka with 30 farmers. In this trial, the user satisfaction (i.e. utility) to the knowledge provided by the ontology is evaluated again through a questionnaire. Three separate sessions were included for this evaluation namely pre-evaluation, demonstration session, and post-evaluation.

- *Pre-evaluation*: farmers' knowledge about the crops and varieties which were grown in their area; and knowledge about the characteristics of the varieties were checked before introducing the mobile-based application. They recorded their answers as a percentage between 0 % to 100 %.
- *Demonstration session*: farmers were given the mobile phones with the application (refer http://webe11.scem.uws.edu.au/slnfarmer/). All the key features with respect to the crop selection were illustrated. They have been given time to use this application and to select what crops and varieties to grow based on their region and preferences.
- *Post-evaluation*: after introducing the mobile application, the farmers' knowledge about crops and varieties grown in their area, knowledge about the characteristics of the varieties were checked again and recorded as a percentage. Farmers' feedbacks for the information on what crops to grow in the coming season, what are the factors which affect the crop selection, usage of information in deciding the crops in future, etc. were also obtained.

By analyzing the data the following conclusions were made (see Table 3). For example, the increased knowledge using this application on the crops that can be grown in farmer's area as a percentage is 63 %.

Table 3. Summary of the outcomes from second field trial

Knowledge increase on crops that can be grown in farmer's area as a percentage	63 %
Knowledge increase on crop varieties that can be grown in farmer's area as a percentage (color, shape, size, weight, average yield, etc.)	64 %
Already decided on what crop to grow in the coming season	97 %
Decision change on the selected crops made in above as a percentage	30 %
Decision change on the selected crops made in above based on the opinions of others as a percentage	28 %
Usage percentage in deciding the crops using this information (through Mobile application) in the future	69 %
Usage percentage for the cultivation using this information in the future	65 %

The ontology also provided a beneficial outcome of increasing farmers' knowledge, for example, their knowledge increased on crops, varieties, crop management practices such as fertilizers and chemicals. Some impacts of the knowledge on the society were also identified through this trial. They were: economic and social development; livelihood enhancement of farmers; wrong information can have an impact on the farming decisions; creating a new farming generation; etc. Their suggestions are also gathered for future refinements, for example, farmers need more information on crop management practices, climatic information, new farming techniques (from other farmers and/or other countries), market prices, information on floriculture, and they also stressed that this information should be in their native languages.

The Fig. 1 shows a sequence of interfaces from the mobile-based system that uses this ontology to highlight its applicability in the real world.

First, farmers need to be registered to this system and then they can login to the system. In the crop selection stage, the farmer needs to provide the information about the farm (farm location) using the map or drop down menus (see Fig. 1). Based on the farm location (regional area), a list of crops that can be grown in this area are provided. Then the farmer needs to select the crops. For example, if the farmer selects the Tomato from the list then he/she can view the suitable varieties which grow in that location. Further, farmer can view the crop properties such as color, weight, size, shape, etc. for each variety. Based on that, the farmers can select the suitable crops and varieties to grow in their farm land.

Through the mobile-based application the user satisfaction to the knowledge in the ontology especially for the crop selection stage is measured. User satisfaction to each task, for examples, fertilizer application, growing problems and its control methods can be evaluated in this manner (then can check the domain coverage).

The user feedback specially depends on the service provided by an application (e.g. mobile-based application). Sometimes it is very hard to decide if the positive or negative points are more related to the design of the application or to the knowledge provided by the ontology. To clarify this, the third field trial was carried out with 26 farmers who received the mobile phones with this application in the second field trial. They had been given more than three months after the second field trial to use this application. After that, some limitations in the design of the application (mobile application) that can affect the usefulness of system were identified.

Farmers used this application to select crops that are suitable to grow in their farm, apply suitable fertilizers, identify diseases and pest attacks, and recommended control methods, etc. In this field trial, farmers were asked the questions related to the design of

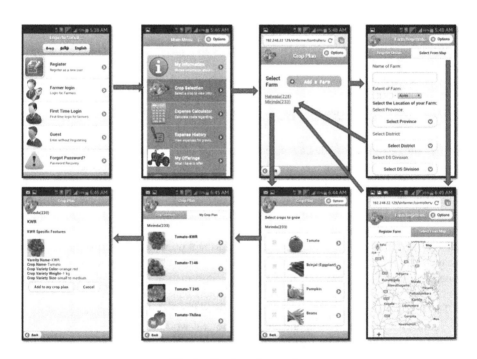

Fig. 1. Ontology in practice.

the application through a questionnaire. They recorded their answer on a 1 to 5 Likert scale (such as StronglyDisagree/MoreDifficult/MoreComplicated, Disagree/Difficult/ Complicated, ModeratelyAgree/Average, Agree/Easy/Simple, StronglyAgree/Very Easy/VerySimple). The results of this field trial are summarized in Table 4.

Table 4. Summary of the results from third field trial

Question	Mode (by Likert scale)	Conclusion
The organization of information	4	Simple
Sequence of screens	4	Simple
Reading characters on the screen	4	Easy
Use of terms throughout the system	4	Appropriate
Prompts for inputs	4	Appropriate
Error messages	4	Simple
The system is easy to use	4	Agree
The system is simple to use	4	Agree
Learning to operate the system	4	Easy
Exploring new features by trial and error	4	Easy
Remembering commands of the system	4	Easy
It is easy to learn to use it	4	Agree
It is designed for any level of users	4	Agree
I learned to use the system quickly	4	Agree
I am satisfied with the system	4	Agree

The outcomes received to the question "The organization of information" are represented in graphical format (see Fig. 2). The results in the Table 4 show that the design of the mobile-based application does not affect to the usefulness of the ontology.

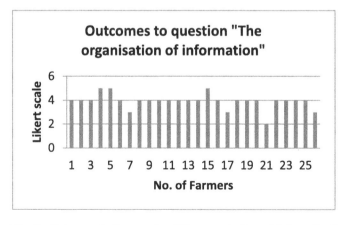

Fig. 2. Outcomes to the question "The organization of Information"

4.2.2   Online Knowledge Base

The knowledge base based on the ontology was created by populating the ontology with instances. To share and reuse the agricultural information and knowledge we need to access the knowledge base via the Web. The online knowledge base can also be used for evaluation process.

To create online knowledge base (SPARQL endpoint), first, the inferred model of the ontology has been converted into the Resource Description Framework (RDF) model. RDF is the World Wide Web Consortium (W3C) standard for representing and storing information on the Web [20]. We then created the SPARQL endpoint using ARC2 (appmosphere RDF classes) Semantic Web toolkit [21] to query the domain knowledge using SPARQL queries (refer http://webe2.scem.uws.edu.au/arc2/sep.php). The detail of creating the online knowledge base is described in [6]. The SPARQL is a standard query language developed primarily to query the RDF graphs [22].

The contextualized information on the Web can be queried via this application using SPARQL queries. For instance, the following SPARQL query lists the suitable fertilizers and in what quantities for Banana which are grown in Dry Zone with other background knowledge (i.e. application method, time of application, unit, etc.).

```
PREFIX sln:<http://www.sln4mop.org/ontologies/2014/SLN_Ontology#>
SELECT DISTINCT ?Fertilizer ?ApplicationMethod
?TimeOfApplication ?Quantity ?Unit
WHERE {{?Fertilizer sln:isFertilizerOf    ?s .}
{?s sln:hasLocationForFertilizerEvent  sln:DryZone .}
{?s sln:isFertilizerEventOf     sln:Banana  .}
{?s sln:hasFertilizerQuantity    ?Quantity .}
OPTIONAL{?s sln:hasApplicationMethodForFertilizerEvent
?ApplicationMethod .}
OPTIONAL{?s sln:hasTimeOfApplicationForFertilizerEvent
?TimeOfApplication .}
OPTIONAL{?s sln:hasWaterSourceFertilizerEvent
?WaterSource .}
OPTIONAL{?s sln:hasFertilizerUnit ?Unit .}
} LIMIT 250
```

The ontology is further evaluated by verifying outputs of the given queries. We especially checked the accuracy of the responses provided by the application. This is mostly useful for advanced users (e.g. researchers or developers) who have knowledge about SPARQL query.

4.2.3   Question Answering

An end-to-end ontology management system via web-based interface for large-scale ontology development and maintenance purposes was developed. The more details about the framework of the end-to-end system and implementation can be seen in [23]. Through this system users can query (using SPARQL endpoint) and view (using user

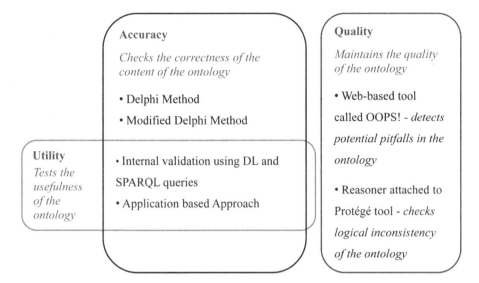

Fig. 3. A classification of ontology validation and evaluation approaches

friendly interfaces) the domain information and knowledge in user context. Especially, normal users, for example, farmers can view the domain information in their context (refer http://webe2.scem.uws.edu.au/oms/searchInformation.php - under the question answering section) using a natural language (in English, Sinhala or Tamil). These natural language questions are also used for the evaluation (see http://webe2.scem.uws.edu.au/oms/searchInformation.php).

5 Generalizing the Approaches and Conclusions

Based on the above experience we obtained from a long journey to validate and evaluate our ontology, we have classified these approaches as shown in Fig. 3. The classification we made through this research will be helpful in evaluating ontologies.

This paper discussed the validation and evaluation procedures applied to evaluate the ontology in terms of accuracy, quality, and utility.

Delphi Method, Modified Delphi Method and the OOPS! web-based tool are used to validate the ontology in terms of accuracy and quality. The Modified Delphi procedure is more effective as this approach minimizes time required, can maintain participant's enthusiasm throughout the process, and also facilitate dialogues and collaboration among the participants that will encourage making new ideas. The contents of the ontology are improved by using domain experts' feedback and their responses.

The implemented ontology is evaluated internally and externally to identify the deficiencies of the artifact in use (i.e. utility). The ontology is evaluated against the user requirements by using a mobile-based application. Very valuable feedback and comments were received from the field trials. The online knowledge base with a SPARQL endpoint was created to share and reuse the domain knowledge that can be queried based on farmer's context. This endpoint is also used for evaluation against the competency questions. Further, the ontology was evaluated using natural language question answering. The ontology has been improved and refined based on above evaluation results.

Acknowledgements. We acknowledge the financial assistance provided to carry out this research work by the HRD Program of the HETC project of the Ministry of Higher Education, Sri Lanka and the valuable assistance from other researchers working on the Social Life Network project. Assistance from the National Science Foundation to carry out the field visits is also acknowledged. Further, we would like to give our heartiest thanks to all the farmers and Agricultural Instructors (AIs) who took part in the evaluation process for their valuable insights and time. Finally, we would like to acknowledge the LK Domain Registry for funding to publish this work.

References

1. Gangemi, A., Catenacci, C., Ciaramita, M., Lehmann, J.: Modelling ontology evaluation and validation. In: Sure, Y., Domingue, J. (eds.) ESWC 2006. LNCS, vol. 4011, pp. 140–154. Springer, Heidelberg (2006)
2. Gozde, B., Dikmen, I., Birgonul, M.T.: Ontology evaluation: an example of delay analysis. Procedia Eng. **85**, 61–68 (2014)
3. Tartir, S., Arpinar, I.B., Sheth, A.P.: Ontological evaluation and validation. In: Poli, R., Healy, M., Kameas, A. (eds.) Theory and Applications of Ontology: Computer Applications, pp. 115–130. Springer, Netherlands (2010)
4. Holsapple, C.W., Joshi, K.D.: A collaborative approach to ontology design. Commun. ACM **45**(2), 42–47 (2002)
5. Ontological commitment. http://en.wikipedia.org/wiki/Ontological_commitment
6. Walisadeera, A.I., Ginige, A., Wikramanayake, G.N.: User centered ontology for Sri Lankan farmers. Ecol. Inf. **26**(2), 140–150 (2015)
7. Ginige, A.: Social life networks for the middle of the pyramid. http://www.sln4mop.org//index.php/sln/articles/index/1/3
8. Brank, J., Grobelnik, M., Mladenić, D.: A survey of ontology evaluation techniques. In: Proceedings of the Conference on Data Mining and Data Warehouses (2005)
9. Obrst, L., Ceusters, W., Mani, I., Ray, S., Smith, B.: The evaluation of ontologies. In: Baker, C.J.O., Cheung, K. (eds.) Revolutionizing Knowledge Discovery in the Life Sciences, pp. 139–158. Springer, USA (2007)
10. Guarino, N., Welty, C.: Evaluating ontological decisions with OntoClean. Commun. ACM **45**(2), 61–65 (2002)
11. Gómez-Pérez, A.: Ontology evaluation. In: Staab, S., et al. (eds.) Handbook on Ontologies, Part II, pp. 251–274. Springer, Berlin Heidelberg (2004)
12. Mattingley-Scott, M.: Delphi method. http://www.12manage.com/methods_helmer_delphi_method.html
13. Hsu, C.-C., Sandford, B.A.: The Delphi technique: making sense of consensus. Pract. Assess. Res. Eval. **12**(10), 8 (2007)
14. NOMINAL GROUP TECHNIQUE 1, HANDOUT: The Skilled Group Leader. http://www2.ca.uky.edu/agpsd/nominal.pdf

15. Poveda-Villalón, M., Suárez-Figueroa, M.C., Gómez-Pérez, A.: Validating ontologies with OOPS! In: ten Teije, A., et al. (eds.) EKAW 2012. LNCS, vol. 7603, pp. 267–281. Springer, Heidelberg (2012)

16. Poveda, M., Suárez-Figueroa, M.C., Gómez-Pérez, A.: Common pitfalls in ontology development. In: Meseguer, P., Mandow, L., Gasca, R.M. (eds.) CAEPIA 2009. LNCS, vol. 5988, pp. 91–100. Springer, Heidelberg (2010)

17. Semantic web best practices and development. W3C Recommendation. http://www.w3.org/2001/sw/BestPrctices/

18. Suárez-Figueroa, M.C., Brockmans, S., Gangemi, A.,Gómez-Pérez, A., Lehmann, J., Lewen, H., Presutti, V., Sabou, M.: Neon modeling components (2007). http://www.neon-project.org/deliverables/WP5/NeOn_2007_D5.1.1v3.pdf

19. De Silva, L.N., Goonetillake, J.S., Wikramanayake, G.N., Ginige, A.: Farmer response towards the initial agriculture information dissemination mobile prototype. In: Murgante, B., et al. (eds.) ICCSA 2013, Part I. LNCS, vol. 7971, pp. 264–278. Springer, Heidelberg (2013)

20. Klyne, G., Carroll, J.J.: Resource Description Framework (RDF): concepts and abstract syntax. world wide web consortium (2004)

21. PHP and Semantic web. http://blog.m1k.info/category/semantic-web-2/?lang=en&lang=en

22. Prud'hommeaux, E., Seaborne, A.: SPARQL query language for RDF. W3C Recommendation (2008). http://www.w3.org/TR/rdf-sparql-query

23. Walisadeera, A.I., Ginige, A., Wikramanayake, G.N., Pamuditha Madushanka, A.L., Shanika Udeshini, A.A.: A framework for end-to-end ontology management system. In: Gervasi, O., et al. (eds.) ICCSA 2015. LNCS, vol. 9155, pp. 529–544. Springer, Heidelberg (2015)

Effort Estimation for Program Modification in Object Oriented Development

Yashvardhan Sharma$^{(\boxtimes)}$

CSIS Department, Bits Pilani, Pilani, Rajasthan, India
yash@pilani.bits-pilani.ac.in

Abstract. One of the major problems faced by software developers and managers is the estimation of efforts for the development and maintenance of a programming system. In this paper, estimation of efforts needed to update programs according to a given requirement change has been discussed for the Object Oriented (OO) environment. Since the demand for these changes has to be met quickly, a method is required to estimate the efforts for making the required changes. Methods exist for estimating the efforts in OO environment but none of them cater to the needs of updating requirements. We have proposed an up-gradation to the approach for effort estimation, which makes use of certain characteristics of the OO paradigm, specifically Inheritance and Encapsulation. We found that the degree of inheritance has to be considered for effort estimation because it plays an important role for identifying which methods need to be modified and others to be reused as it is.

1 Introduction

A number of software development paradigm exists to support the software development process. Among them, the OO development is regularly gaining importance because of its capability to transform and map the complex real world structures into programming concepts [1, 4, 12, 13]. Hence OO development is widely used in industries today. Following are the advantages, when we choose OO paradigm for our software development:

- Better Mapping of the requirement to the real world problem domain: When the requirements are mapped to the real world, objects represent customers, machinery, banks, sensors or pieces of paper, they can fit naturally into human thought processes which leads to faster development and better code reusability. But such reuse does require proper planning and investment.
- Code Reuse: Object-oriented approach provides a way to write modular code, group together related component which includes objects and smaller sub-systems.
- Modular Architecture: Object-oriented approach provides a way to write modular code, group together related component which includes objects and smaller sub-systems.
- High Maintainability: OO Approach incorporates high maintainability by following S.O.L.I.D design principle.

O. Gervasi et al. (Eds.): ICCSA 2016, Part IV, LNCS 9789, pp. 334–345, 2016.
DOI: 10.1007/978-3-319-42089-9_24

- Client/Server Applications: OOD is best fit for Client/server applications which required message transmission from client to server and vice versa over network. It is for these reasons that OO development is widely used in industries today.

It can also be observed that the OO development approach is natural fit and reasonable for rapid prototyping because the Object oriented paradigms make it easy to understand the structure of system and enforce reusability of components across system.

In the OO approach, it is very important from the development and management point of view to estimate the cost of update to incorporate requirement changes. Since the requirement change requests comes frequently from customers, accurate estimation of effort can lead to improve the quality of overall development process and will help the developers to incorporate and update the changes [5, 21].

In this paper, we have proposed enhancements to the formula to estimate the cost for updating activities by using the data that can be obtained prior to coding activities. However, there are many methods and models are available for effort estimation in software development, such as COCOMO model [16] and the Function point method [15]. These methods take small amount of OO development properties into consideration. Furthermore, these existing methods require a lot of preparation in terms of customization to incorporate them in OO development environments. Also it is difficult to apply them under the restrictive cost limitations and rapid development environments.

Our objective, in this paper is to propose a simple approach to estimates efforts for modification under the OO development environment where changes to requirements are proposed frequently. In the proposed work, we have taken OO properties into consideration such as information hiding, abstraction, inheritance and updating activities at class and method level.

Here, we proposed enhancements to the formula $E(P,\sigma)$ to calculate the efforts needed to update a program, P according to a set of requirement changes σ [1]. The formula $E(P,\sigma)$ includes weighting parameters: W_{upd}, W_{type}, W_{inf-h} and W_{inht} corresponding to the OO characteristics updating tasks, updating target type, degree of information hiding and degree of inheritance, respectively. Finally, experimental evaluations are conducted by applying the formula presented to actual project data.

The detailed description of our work has been organized as follows: Sect. 2 briefly talks about related work, objective and motivation factor to state the problem statement. Section 3 is outlined about Object Oriented characteristics which we have taken into consideration while estimating efforts for updating and modification of system programs. In Sect. 4 we have presented formulas and parameter details for effort estimation. Section 5 shows experiment setup and evaluation method which we have carried out. Finally, Sect. 6 concludes our work and spoken of further improvement opportunity.

2   Related Works

During the development process, it's very obvious to have requirement and design decision changes in place, hence it becomes important to estimate total effort needed to incorporate those changes from customer and management point of view so that customer and management come to conclusion based on effort needed to incorporate changes to avoid deadline miss, schedule slippage, and project failures.

There are many algorithmic methods for estimating cost in software development, such as Function point based models [16], Putnam models, Constructive Cost Model COCOMO models, Prediction model using one hidden layer feed forward neural network with genetic algorithm [18], Optimal Neural Network model [19], some of the models which take the object oriented properties such as inheritance, encapsulation, data hiding, abstraction of the OO into the consideration while calculating estimate.

Since Constructive Cost Model (COCOMO) model is based on regression curve to accurately predict the estimates. COCOMO model is based on the fact that system and software requirement have already been defined and these requirements are stable [10, 16]. It's very difficult to adopt COCOMO model to estimate the effort accurately in practical OO environment where requirement changes and system design changes take place frequently. Moreover, the cost drivers in COCOMO do not take into account OO characteristics, directly [3]. Many other models like [6, 7, 9] simulator estimates the efforts and faults of the software project, but are targeted on the standard waterfall model.

The Function Point method is another effort estimation algorithm. This method takes complexity of software system in terms of functions that the system exposes to the user [15]. In addition to this, customization and adjustment for environmental processing complexity is needed which is not available for small development environment. However, the function point analysis based models can be used for accurate cost estimation from requirements specification or from design specification independent of the language and tools. This results in better quality and productivity which helps developers and protect manager to plan and construct and apply approaches in better way.

Some of the approaches proposed in [1, 14] calculates the effort for the updating activity in OO paradigm. The method takes into account factors like type of updating activity, degree of information hiding, type of parameters.

But we found that inheritance, which is one of the major concerns during updating activity, has to be considered in the calculation of effort. In our work, a parameter has been included for inheritance. If a class being updated has several children, then each one of them has to be handled separately. This adds to the existing effort of the updating process. Focus is more on developing a method for effort estimation which is intuitive and also easy to calculate.

Our objective is to accurately estimate the required effort for modification and updating activity in the OO development based software program. Since the requirement changes frequently throughout software development cycle, so cost estimation process should be easy to apply. Although the effective and accurate cost estimation will help to take program management decision with accuracy to incorporate changes.

Throughout this work we kept our focus to develop an effective estimation method while considering the following properties into consideration:

A. Fast calculation: The manager and developer can obtain the result immediately
B. Easy calculation: The effort estimation method should be simple to apply during the development process.
C. Simple calculation: The effort required for estimation should be simple for manager and developers.

The existing model such as COCOMO which is good solution for estimating cost for incorporate system design changes in software development. However, our purposed approach is different from the COCOMO. Since COCOMO follow regression curve based model for calculating estimates, it is difficult to understand and apply for general developer.

In addition to this, Kusumoto et al. come up with a simulator to estimates the effort and detect faults in System program [17]. However, they have followed the standard waterfall model for their simulator, so it was very hard to apply to the environment where frequent changes in the requirements take place.

Our purposed method is completely based on the OO program characteristics where we are estimating a score for each component based on the analysis of OO characteristics. In some context, it is very similar to functional point method. However, it's not easy to apply the function point method in the OO development environment directly, since the function point method depends upon number of screens or number of inputs/outputs. Therefore, we have to establish a very straightforward method to estimate the efforts for the program which is developed in OO environment.

3 Effort Estimation for Object Oriented Development

In order to estimate the efforts to update the program in the OO development environment, we chose the following four characteristics in the effort estimation: (1) the kind of updating activities, (2) the types of targets to be updated, (3) the degree of information hiding and (4) the degree of inheritance. We have identified and analyzed various factors which helped us in distinguishing them and scoring them based on the difficulty of updating them. We have briefed about OO characteristics which we are taking into consideration as follows:

3.1 Factors for Effort Estimation Calculation

Following factors, we have considered while calculating the estimated effort of modified program in OOD [1]:

A. Updating activities
B. Target to be updated
C. Object oriented paradigm
   1. Degree of information hiding.
   2. Degree of inheritance involved.

3.2 A Updating Activities

The updating activity type plays an important role in the calculation of effort. [1] Shows that there are three fundamental operations performed on the target (class/attribute/method):

- Creation: Creating new classes, methods and attributes in a classes.
- Deletion: Deleting exiting classes, attributes and methods.
- Modification: Modification to an existing class, attribute and methods (that is, changing attribute data type, changing return type of method, changing access modifier of function, changing access modifier for classes).

The effort for "adding a new class," is different from "updating an existing class". Therefore, a factor (W_{upd}) is assigned to represent the difference in the difficulty caused due to updating activities.

3.3   B Target to be Updated

The effort required also depends on the type of target. Followings are the categories that needs to be considered:

- Void Type: Void is type of Data type like int, char which specify no value. In C++ and JAVA when method doesn't need to return any value to called function then void clause is used as return type.
- Library type: The logical unit of procedure which is defined and consumed as third-party methods. For example, modifying a target with primitive type is different from that of user defined type. Separate scores (W_{type}) are assigned to these types to distinguish the effort required upon them.
- Basic type (Primitive data-type): Primitive data type is in-built data type which is included in the OO language, such as int, and char in C++ and JAVA.
- Custom type: The logical unit of procedure which is defined inside developing program.

3.4   C Object Oriented Paradigm

The characteristics of OO paradigm considered for calculation are encapsulation, inheritance. All other characteristics can be expressed in terms of these two characteristics.

- Degree of information hiding: Information hiding is a feature that controls the visibility of the target to the external classes. The following summarizes information hiding in C++:
  - Private: the attributes and methods which are defined private are only accessible within the same class where method and attribute are defined.
  - Protected: the attributes and methods which are defined as protected only accessible to inherited child classes.
  - Public: the attributes and methods which are defined as public are accessible outside of classes.

When modifying access modifier public attribute/method to other modifier or vice versa. Then more effort is required to modify the reference target classes and methods.

Thus there is a difference in the difficulty due to degree of information hiding. This is denoted by a weight W_{inf-h}.

Degree of inheritance involved: Inheritance provides a way to reuse and extend existing class's property and methods without modifying, it also produces hierarchical relation between base class and derived class [2]. In addition to this, inheritance also plays major role for calculating updating effort when requirement and design changes for a software system. Since inheritance involves parent-child relationship among classes (base class and derived classes) updating in base classes cause and force modification in derived classes. We have considered inheritance properties of OOD while calculating estimated effort and we have assigned weight for inheritance as this is required for effort calculation. As degree of inheritance increases in system, the total estimated effort also increases. To represent difficulty encountered due to the degree of inheritance, this is denoted by a weight W_{inht}.

4   Formula for Effort Estimation

Let say there are some changes in design specification, therefore program P should be updated to incorporate these changes [1]. According to design specification change R_i (Requirement changes) a class C_j is updated. In more detail, an attribute A_k in the C_j (Class) and M_l (Method) in the class C_j are updated.

The total effort required for updating is the sum of efforts required for updating individual classes.

$$E\,(P, \sigma) = \sum_{i=1}^{n} E_{requirement}(R_i)$$

$$= \sum_{i=1}^{n} \sum_{j=1}^{m} E_{class}(C_j) \tag{1}$$

$$= \sum_{i=1}^{n} \sum_{j=1}^{m} \left(\sum_{k=1}^{p} E_{attribute}(A_k) + \sum_{l=1}^{q} E_{method}(M_l) \right)$$

$$E_{req}(R) = \sum_{j=1}^{m} E_{cl}(C_j) \tag{2}$$

Where $C_1 \ldots C_m$ are the classes to be updated. The effort required to update a class is the sum of effort required to update its attributes and methods.

$$E_{cl}(C) = \sum_{k=1}^{p} E_{attr}(A_k) + \sum_{l=1}^{q} E_{meth}(M_l) \tag{3}$$

Where $E_{attr}(A_k)$ and $E'_{meth}(M_l)$ denote the effort for updating attribute and method respectively.

Effort estimation for updating the program is defined as follows [1]:

$$E_{attr}(A) = \alpha \times W_{upd} \times W_{type} \times W_{inf-h} \tag{4}$$

Here α is a basic score for updating an attribute.

Considering inheritance:

$$E_{meth}(M) = \beta \times (1 + W_{inht} \times NOC(C) + WCC(M) + WCM(M) + WPM(M))$$
$$\times W_{upd} \times W_{type} \times W_{inf-h} \tag{5}$$

β: a basic score for updating a method.

Where variables W_{upd}, W_{type}, W_{inf-h}, W_{inht} are constants to represent characteristics in previous section and WCC(M), WCM(M), WPM(M), NOC(C) are fundamental OO metrics [8]. In Eq. (5), metrics WCC(M), WCM(M) and WPM(M) represent internal properties in a method M and they are independent from their definition [8]. As the number of child class increases, the effort for handling the method also increases. Therefore, a separate factor is required to represent inheritance relationship. The sample values of these variables will be shown in Sect. 5.

4.1 A Fundamental Metrics

Metrics which we have taken into account WCC, WCM, WPM and NOC reference the complexity of a program [8].

- WCC (Weighted Coupling Classes)
  Let say a method M consist n classes that is C_1, C_2.... C_n which are referred in M. The weight W_{Ci} is represent degree of uses of classes in method M. Here WCC is defined as follows:

$$WCC\ (M) = \sum_{i=1}^{n} w_{c_i} \tag{6}$$

- WCM (Weighted Coupling Members)
  Let say a method M consist n attributes that is A_1, A_2......A_n, which are defined in the same class C and being used inside same class method M. The weight W_{Ai} is represent degree by which attribute is being used inside method M for an attribute A_i. WCM is defined as follows:

$$WCM\ (M) = \sum_{i=1}^{n} w_{A_i} \tag{7}$$

- WPM (Weighted Parameters of Method)
  Let say a method M consist n parameters P_1, P_2......P_n. The weight represent degree by which parameter P_i is being used inside method M. WPM is defined as follows:

$$WPM\ (M) = \sum_{i=1}^{n} w_{p_i} \tag{8}$$

- NOC (Number of Children)
  Represents degree of subclasses which are going to inherit the base class methods:
  NOC = the number of immediate subclasses

5 Experimental Setup

In order to demonstrate the formula which, we have used to measure the estimated effort, required to modify some of the classes in "spring-framework-0.9.0", "spring-framework-0.9.1" and "spring-framework-1.0" have been calculated [11].

The weights of parameters for the calculations are made on the following assumptions [1]:

- The effort required to create an attribute is greater than or equal to that of deletion and modification.
- Custom types attributes required more efforts to update than library types, and library types need more than basic types.
- Public attributes or methods required more efforts to update than protected ones, and protected ones need more than private ones (Table 1).

Table 1. Parameters for attributes ($\alpha = 1$).

W_{upd}	Creation	Deletion	Modification
	10	2	N/A
W_{type}	Basic	Library	Custom
	1	2	3
W_{inf-h}	Private	Protected	Public
	1	2	3

From the above Table 2, the higher the value in the effort column, greater is the effort required. The classes with value "0" indicate the absence of any change in the class (Tables 3 and 4).

Estimated effort for the classes 'in "aop" directory of "spring-framework-0.9.0" "spring-framework-0.9.1" and "spring-framework-1.0" are given in Table 5:

6 Conclusion

The focus of the work is primarily on enhancing the method for effort estimation by including degree of inheritance in an object oriented development environment. Weights are assigned to various aspects of the program like the type of updating activities, type of target to be updated, degree of encapsulation and inheritance. The degree of inheritance plays an important role in the calculation of effort.

The degree of inheritance presented in this work has been found to be useful predictors of effort. Earlier the results have been verified upon relatively small systems

Table 2. Classes in spring framework "aop" directory.

Class Name	Effort E(P,σ) (spring-framework-0.9.0)	Effort E(P,σ) (spring-framework-0.9.1)	Effort E(P,σ) (spring-framework-1.0)
WildcardAttributeRegistry	214.4	0	0
AbstractMethodPointout	155.28	0	0
AlwaysInvoked	70.8	15.98	0
AopProxy	170.4	33.84	0
AopUtils	13.44	2.48	0
DefaultProxyConfig	31.356	6.76	0
DelegatingIntroductionInterceptor	44.64	0	0
DynamicMethodPointCut	73.92	0	0
InvokerInterceptor	47.52	3.36	0
MethodInvocationImpl	228.96	26.64	0
MethodPointcut	0	2.64	0
ProxyConfig	0	0	8.64
ProxyFactory	40.32	3	22.8
ProxyFactoryBean	21.06	5.08	26.88
ProxyInterceptor	66	0	0
RegexpMethodPointcut	186.9	0	0
StaticMethodPointcut	81.84	0	0
AbstractQaInterceptor	40.32	4	0
ClassLoaderAnalyzerInterceptor	48.96	6	4.96
DebugInterceptor	40.32	6	5.28
EventPublicationInterceptor	13.44	0	0
MockObjectInterceptor	13.44	0	0
PerformanceMonitorInterceptor	48.24	6	4.96

(Continued)

Table 2. (*Continued*)

Class Name	Effort E(P,σ) (spring-framework-0.9.0)	Effort E(P,σ) (spring-framework-0.9.1)	Effort E(P,σ) (spring-framework-1.0)
Advisor	–	–	48
AfterReturningAdvice	–	–	28
BeforeAdvice	–	–	24
ClassFilter	–	–	112.4
IntroductionAdvisor	–	–	134.4
IntroductionInterceptor	–	–	73.2
MethodBeforeAdvice	–	–	85.2
MethodMatcher	–	–	53.2
Pointcut	–	–	29.6
PointcutAdvisor	–	–	201.6
TargetSource	–	–	24.4
ThrowsAdvice	–	–	73.2
Advised	–	–	873.6
AdvisedSupport	–	–	478.8
AdvisedSupportListener	–	–	177.6
AdvisorChainFactory	–	–	85.2

Table 3. Parameters for method (β = 2).

W_{upd}	Creation		Deletion	Modification
	5		1	3
W_{type}	Void	Basic	Library	Custom
	1	1	2	3
$W_{inf\text{-}h}$	Private		Protected	Public
	1		2	3
W_{inht}	0.1			

Table 4. Metrics WCC, WCM, WPM.

	Basic	Library	Custom
w_c	N/A	0.1	0.12
w_a	0.05	0.1	0.12
w_p	0.05	0.1	0.12

Table 5. Data from spring-framework-0.9.0, 0.9.1 and 1.0 "aop" directory.

Version	No of Class	LOC	E(P,σ)
$V_{0.9}$	28	2251	1651.556
$V_{0.9.1}$	28	2206	121.78
$V_{1.0}$	44	7337	2575.92

but we have tested the usefulness of the measures in large stable systems i.e. spring-framework. As a future work, more analysis needs to be done on determining the weights empirically.

References

1. Uehara, S., Mizuno, O., Kikuno, T.: A straightforward approach to effort estimation for updating programs in object-oriented prototyping development. In: Proceedings of 6th Asia-Pacific Software Engineering Conference, pp. 144–151 (1999)
2. Rumbaugh, J.: Object Oriented Modeling and Design. Prentice Hall, Upper Saddle River (1991)
3. Sommerville, I.: Software Engineering, Addison Wesley (2006)
4. Uehara, S., Mizuno, O., Itou, Y., Kikuno, T.: An MVC based analysis of object oriented system prototyping for banking related GUI applications–correlation ship between OO metrics and efforts for requirement change. In: Proceedings of 4th International Workshop on Object Oriented Real Time Dependable Systems, pp. 91–104 (1999)
5. Li, W., Henry, S.: Object oriented metrics that predict maintainability. J. Syst. Softw. **23**, 111–122 (1993)

6. Kusumoto, S., Mizuno, O., Hirayama, Y., Kikuno, T., Takagi, Y., Sakamoto, K.: A new project simulator based on generalized stochastic PetriNet. In: Proceedings of 19th International Conference on Software Engineering, pp. 293–303 (1997)
7. Kim, M.: Program Complexity Metric and Safety Verification Method for Object Oriented Software Development, Ph.D. Dissertation, Osaka University, January 1997
8. Chidamber, S.R., Kemerer, C.F.: A metrics suite for object oriented design. IEEE Trans. Softw. Eng. **20**(6), 476–493 (1994)
9. Briand, L.C., Daly, J.W., Wust, J.K.: A unified framework for coupling measurement in object oriented systems. IEEE Trans. Softw. Eng. **25**(1), 91–121 (1999)
10. Albrecht, J., Gaffney, J.E.: Software function source lines of code, and development effort prediction: a software science validation. IEEE Trans. Softw. Eng. **9**(6), 639–648 (1983)
11. Spring framework: http://projects.spring.io/spring-framework
12. Hayes, F.: The Reality of Object Reuse. Computer World, p. 62, 6 May 1996
13. Adhikari, R.: Adopting OO languages check your mindset at the door. Softw. Mag. **15**(12), 49–59 (1995)
14. Uehara, S., Mizuno, O., Kikuno, T.: A new approach to estimate effort to update object-oriented programs in incremental development. IEICE Trans. Inf. Syst. **E85-D**(1), 233–242 (1999)
15. Cost estimation methods comparison: http://www.computing.dcu.ie/∼renaat/ca421/LWu1.html
16. Boehm, B.W.: Software Engineering Economics. Prentice-Hall, Upper Saddle River (1981)
17. Kusumoto, S., Mizuno, O., Hirayama, Y., Kikuno, T., Takagi, Y., Sakamoto, K.: A new project simula-tor based on generalized stochastic Petri-Net. In: Proceedings of 19th International Conference on Software Engineering, pp. 293–303 (1997)
18. Yadav, C.S., Singh, R.: Prediction model for object oriented software development effort estimation using one hidden layer feed forward neural network with genetic algorithm. Adv. Softw. Eng. **2004**, 1–6 (2014). Hindawi Publishing Corp
19. Reddy, C.S., Raju, K.V.S.V.N.: An optimal neural network model for software effort estimation. Int. J. Softw. Eng. IJSE **3**(1) (2010)
20. Singh, J., Sahoo, B.: UML Based Object Oriented Software Development Effort Estimation Using ANN (2012)
21. Mizuno, O., Kikuno, T., Inagaki, K., Takagi, Y., Sakamoto, K.: Analyzing effects of cost estimation accuracy on quality and productivity. In: Proceedings of 20th International Conference on Software Engineering, pp. 410–419 (1998)

Populational Algorithm for Influence Maximization

Carolina Ribeiro Xavier[1(✉)], Vinícius da Fonseca Vieira[1],
and Alexandre Gonçalves Evsukoff[2]

[1] UFSJ/DCOMP - Federal University of São João del Rei,
São João del Rei, MG, Brazil
carolrx@gmail.com

[2] UFRJ/COPPE - Federal University of Rio de Janeiro, Rio de Janeiro, RJ, Brazil

Abstract. Influence maximization is one of the most challenging tasks in network and consists in finding a set of the k seeder nodes which maximize the number of reached nodes, considering a propagation model. This work presents a Genetic Algorithm for influence maximization in networks considering Spreading Activation model for influence propagation. Four strategies for contructing the initial population were explored: a random strategy, a PageRank based strategy and two strategies which considers the community structure and the communities to which the seeders belong. The results show that GA was able to significantly improve the quality of the seeders, increasing the number of reached nodes in about 25 %.

1 Introduction

The influence maximization problem consists in finding a set of the k most influential nodes in a networks, such as the aggregate influence is maximized. One of the earliest works in this subject was performed by Kempe *et al.* [17] and shows a systematic study of influence maximization as a discrete optimization problem.

As influence maximization is a NP-hard problem [24], an optimal solution can not be obtained and heuristic methods, such as genetic algorithms, can be applied. Genetic algorithms are global optimization algorithms which use a direct analogy to species evolution and work with a population of candidate solutions. Genetic operations, such as crossover and mutation, are applied to the candidate solutions in order to find the best individual, which is considered the solution for the optimization problem.

Influence maximization is of great interest in marketing, specially due to the success in the use of social networks, where people and organizations can promote product, services and ideas through word-of-mouth propagation among friends and acquaintances. This is a great motivation for the study of the different aspects of the influence maximization problem.

In order to be able to study influence maximization computationally, it is important to define a model that simulates its propagation. Many works found

© Springer International Publishing Switzerland 2016
O. Gervasi et al. (Eds.): ICCSA 2016, Part IV, LNCS 9789, pp. 346–357, 2016.
DOI: 10.1007/978-3-319-42089-9_25

in the literature study the simulation of influence propagation in social networks. A class of models is based on epidemiological phenomena [19]. Another class is known, from sociology, as diffusion of innovation, which is the case of Information Cascade Model [13] and Threshold Model [15].

The model explored in this work, *Spreading Activation* (**SPA**), derives from cognitive psychology and represents the semantic processing of human memory. SPA represents behaviour of the brain to retrieve specific information from the memory [1]. This process occurs through a network of concepts, that are activated by their neighbours (related concepts), and the activation is cascaded until the information is retrieved.

Xavier, in a previous work [25], investigate the effect of considering local aspects of the network in order to define the set of nodes that will be selected as seeders for influence maximization. Thus, the community structure of the network was considered for the selection of the initial seeders in SPA.

This work presents a Genetic Algorithm for optimizing influence spreading, considering different ways to define the initial population, considering global and local aspects. The results show that Genetic Algorithm is able to improve the quality of the solution at about 25 % compared to the quality of the initial population.

2 Influence

Social influence, as stated by Rashotte [22], is the modification of thought, sentiment, attitude or behaviour of an individual caused by the interaction with other person or group. Social influence has many forms and it can be easily noticed in online social networks. Currently, influence can be perceived in shares, likes and re-tweet cascades, suggesting that the great connectivity in social networks exposes their users to different kinds of influence from several sources. In large scale data analysis, many applications, such as viral marketing, recommendation systems and diffusion of information are involved in the study of social influence.

Assuming that preferences and knowledge of an individual can influence others, a company interested in promoting a new product could identify and choose the most influential individual as target, potentially reaching a large portion of the network.

The problem arose with the question: "which individuals must be chosen as promising seeders, in order to maximize the influence spreading in a network?".

2.1 Spreading Activation Model

Spreading Activation was conceived cognitive psychology as a model of human memory, representing the way that a brain uses a network of concepts to relate information in memory [21]. Concepts are presented as sequences of cognitive units, which can be represented as nodes connected by edges when the concepts are directly associated.

Spreading Activation can be interpreted as a process of energy propagation. The model has been used to model the processing of information in the brain [1,21], for methods of information retrieval in networks [4,8] and the simulation of influence of behaviour in phone call networks [2,10]. The idea of Spreading Activation can be applied to the context of social networks when the hypothesis that individuals can cause a word-to-mouth effect in a network is considered, i.e., individuals in a network can influence their friends to adopt an idea, behaviour or product. Individuals activated by the influencers will now try to influence their friends, even with less intensity. Then, individuals are activated in the next iterations and activate their friends with each time less intensity, until no node is activated in a certain time step.

The process of influence spreading in a network can be described as follows. Considering a weighted graph $\mathbb{G} = (\mathbb{V}, \mathbb{E})$, from which matrix $\mathbf{W}$ represents the weight matrix of $\mathbb{G}$, the activation of the nodes in the network can be represented by a vector $\mathbf{a}$ with the same dimensionality of the nodes set, where each element a_i denotes the amount of activation associated to the node v_i in the time step t.

The SPA process is started by a set of active nodes, the seeders. At the time step t, each node v_i in the network is associated to a certain level of activation, denoted as a_i, that can be interpreted as the element of a received energy vector or a received influence vector, in which the values are propagated to one ore more of its neighbours. It can be also interesting to define the amount of energy that will be transmitted to a node during the diffusion process. Dasgupta *et al.* [10] define a global parameter β, which represents the fraction of energy propagated by a node to its successors. Thus, it is possible to ensure that, at each iterative step, the amount $1 - \beta$ will be retained in the source node. The spreading factor parameter β determines how much importance is given to the fact that a node is close to the seeders. Higher β values enable an easier propagation to nodes which are distant to the seeders. On the other side, lower β values retain the energy in regions close to the seeders.

According to Ghosh *et al.* [12], a diffusion process in a network is a dynamic and stochastic process, which distributes a certain element in the network. If one considers the amount of influence as an element and that an individual proportionally distributes this quantity among its neighbours, SPA can be interpreted as a network diffusion model.

In this work, Spreading Activation was computationally implemented with igraph [9] library to store and manipulate the network structure. The process was modelled as a succession of multiplications of the weight matrix $\mathbf{W}$ to the influence vector $\mathbf{a}(t-1)$ and the spreading factor β summed to $\mathbf{a}(t-1)$ multiplied to the complement of β, $1 - \beta$.

The implemented algorithm is described by Algorithm 1:

---

begin
   | Initialize β, θ, E_θ e $\mathbf{a}(0)$;
   | $\mathbb{V}^a(0) =$ active vertex set;
   | Build a CSR matrix based on the weight matrix $\mathbf{W}$;
   | **while** $|\mathbb{V}^a(t)| > |\mathbb{V}^a(t-1)|$ *and* $\forall v_i \in \mathbb{V}$: $|a_i(t) - a_i(t-1)| > \theta$ **do**
      | $\mathbf{a}(t) = \mathbf{a}(t-1)(1-\beta) + \beta\mathbf{a}(t-1)\mathbf{W}$;
      | **foreach** $v_i \in \mathbb{V}$ **do**
         | **if** $a_i(t) > E_\theta$ **then**
            | $\mathbb{V}^a = \mathbb{V}^a \cup v_i$;
         | **end**
      | **end**
   | **end**
end

Algorithm 1. *Spreading Activation*

3   Influence Maximization

Some works in the literature [11,23] consider the influence maximization problem as a probabilistic interaction model, and the selection of the most influential seeders is based on the global effect that one individual causes to the network. Other works [5,17,18,20] consider the selection of the seeder nodes as a discrete optimization problem.

Formally, influence maximization problem is defined as follows. Given a value k and a social network, represented as a directed network where agents are represented as nodes and edges represent the relationship between them, the aim is to select a set of seeder nodes of size k, which are initially active so that the expected influence (in terms of the number of active nodes) can be maximized.

Kempe et al. [17], in their work, studied this problem focusing on two basic propagation models: Linear Threshold Model and Independent Cascade Model. Many other works propose greedy heuristic methods for the solution of the problem, such as Leskovec et al. [20], Chen et al. [6], Heidari et al. [16] and Goyal et al. [14].

3.1   Genetic Algorithm for the Influence Maximization Problem

In nature, individuals compete for resources as food, water and refuge. Additionally, among individuals of the same species, those who can not succeed, tend to show a reduced number of descendants. As a consequence, there is a fewer probability for their genes to be propagated along successive generations. According to Charles Darwin: "*The better an individual adapts to an environment, the better are his chances to generate descendants.*".

Genetic Algorithms (GA) use a direct analogy for the evolution phenomenon, in which each individual represents a potential solution for an optimization problem. In the specific case of the study performed in this work, the optimization problem to be solved is the selection of seeders for propagation in a network in order to reach the maximum number of nodes.

For each individual, the algorithm assigns a score, associated to its fitness, which is given by a fitness function, associated to the objective function of the optimization problem. To the most fitted individuals, GA gives a greater opportunity to be selected for reproduction, being combined to another individual of the population, generating descendants with properties of both parts.

The objective function in the implemented GA is the *Spreading Activation*, which takes a set of seeder nodes and returns the number of active individuals by the end of the execution. Thus, the Genetic Algorithm performs the maximization of the objective function.

The GA uses a chromosome based representation. Each chromosome has 40 genes and each gene stores an integer number, representing the index of the corresponding node v_i. Thus, each gene of the chromosome represents a node, and each individual may be associated to up to 40 distinct nodes.

The implemented Genetic Algorithm works with 40 individuals (chromosomes) in the population. Each gene in the chromossome is associated to a node in the network. Different versions of the GA were implemented, considering four ways to define the initial population, based on the same strategies for defining the seeder nodes, as presented by Xavier [25].

Two global strategies were investigated. One of the global strategies (called **PageRank** in this work) randomly selects each individual as one of the 20 % nodes with higher PageRank scores among all the nodes in the network. In the other global strategy (called **Random** in this work), the nodes in the initial population are randomly chosen.

The two local strategies considers the community structure of the network in order to define which nodes will be selected for the initial population. A community in a network can be defined as set of nodes with a high internal density and low external density. Many works found in the literature discuss communities in networks and many definitions of communities have been proposed. In this work, we adopt Newman's modularity definition of community [7], which states that a community is a set of nodes with a high number of internal edges compared to a random version of the network with the same degree distribution. Communities were obtained by the widely adopted Louvain method [3].

The first local strategy (called **Community** 100 % in this work) associates each gene to one of the 40 most modular communities, i.e., one of the 40 communities with higher modularity values. Then, one of the 20 % nodes with higher degree of the associated community is randomly chosen and assigned for each gene. The other local strategy (called **Community** 60 % in this work) associates 60 % of the initial population to the most modular communities, in a similar way to the strategy **Community 100 %**. The remaining 40 % of the initial population is randomly chosen.

The selection method implemented in this work is widely known as "roulette". In this selection method, each individual is associated to a portion of the "roulette" and the size of each portion is proportional to the fitness of the corresponding node. The higher the fitness of an individual, the greater his portion in the roulette and consequently, the higher its chance of being selected to compose the population of the next generation. A common problem occurs in "roulette" selection method is the systematic choice of a single individual, when its fitness is much higher than the others. In order to solve this problem, the rank of the nodes is considered in the calculation of the portions of the "roulette", instead of their scores. This solution smoothes the differences between the fitness values and inserts more diversity among the candidate solutions, avoiding local minima.

The selected individuals are combined in an intermediate population (crossover operation), which is mutated, with probability of 10 %, before being evaluated.

Two versions of the mutation operation were implemented, considering the two main different ways to generate the initial population (local and global). With the global strategies (**Random** and **PageRank**), mutation simply replaces the node associated to a gene by another randomly chosen node. With the local strategies (**Community** 100 % and **Community** 60 %), the gene to be mutated is replaced by an individual of another community, which may be or may be not one of the communities already considered in the initial population.

The implementation was performed with python programming language, using the Louvain method [3] implemented in igraph library [9]. Modularities of each of the communities were calculated as performed by Clauset *et al.* [7]. The degrees of the nodes were ranked according to their connections inside the communities.

3.2   Dataset

Four benchmark networks were considered in this work: the collaboration networks **hep**, **phy** and **ca** and the communication network **email**. The networks are undirected and unweighed. Table 1 shows some basic properties of the networks.

The studied networks show different properties and, even network of the same type (such as collaboration networks) show distinct characteristics (**hep**, **phy** and **ca** has 7.7, 12.4 and 11 as mean degrees, respectively).

Degree assortativity of these networks also shows a significant difference. While **hep** shows an assortativity of 0.45, **phy** and **ca** show a higher assortativity: 0.6 and 0.66, respectively. On the other side, **email** network shows a disassortativity of -0.11. Assortativity is an important property in the study of the seeders. If a network is assortative, high degree nodes tend to be connected to high degree nodes, which can be critical in the analysis of how the influence propagated through the network.

Table 1. Basic properties of benchmark networks.

Network	ca[b]	hep[a]	phy[a]	email[b]
# Nodes	5242	15233	37149	36692
# Edges	28980	58891	231507	367662
Max. degree	162	341	286	2766
Mean degree	11	7,70	12,40	20
Clustering coefficient	0,53	0,50	0,75	0,50
Degree assortativity	0,66	0,45	0,60	−0,11

[a] Downloaded from [6].
[b] Downloaded from: http://snap.stanford.edu/data/.

3.3 Results

The Genetic Algorithm for influence maximization was executed and the results are presented in this section. For **ca** network, the four strategies for initial populations were considered, each one with 50 individuals. For the remaining networks, the experiments considered only the better strategy observed for **ca** network. A limit of 150 generations was considered as stop criterion, since the optimal value for the problem is unknown. For each tested configuration, the GA was executed 10 times.

Table 2 shows the mean and the standard deviation of the number of reached nodes obtained in the execution of the GA with the **ca** network. The mean number of involved communities at each execution is also presented.

Table 2. Results of the GA for each strategy for initial population for **ca** network.

Strategy for initial population	Mean	Std. deviation	# communities
Random	1732,4	25,52	24,3
PageRank 20 %	1678,4	44,25	25,1
Community 60 %	1737,3	19,65	28,2
Community 100 %	1714,3	23,22	30,6

Similar values of mean number of reached nodes can be observed for the different strategies, however, when the standard deviation is considered the strategies which take the communities into account where more regular than the others. The worst results can be observed for PageRank strategy (low mean reached nodes and higher standard deviation).

Moreover, Table 2 shows a great diversity of communities found by the GA when any of the strategies is considered. Even in PageRank and Random strategies, the GA tends to place seeders in different communities, which suggests that spreading seeders among communities makes the spreading process to reach more nodes.

Figure 1 shows the evolution of the 10 executions of the GA considering the four explored strategies for the initial population. A clear evolution among the generations can be noticed for all the executions, showing that the genetic operators can, indeed, improve the initial population.

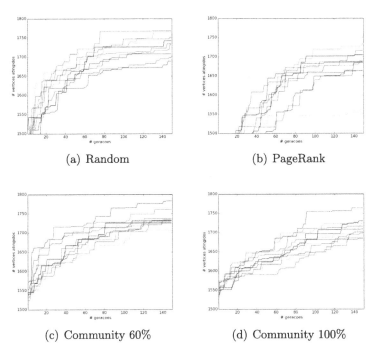

(a) Random

(b) PageRank

(c) Community 60%

(d) Community 100%

Fig. 1. GA evolution for each execution in **ca** network with different kinds of initial population: (a)Random; (b)PageRank; (c)Community 60 %; (d)Community 100 %.

The best individual obtained after the last generation of the GA is the solution of the optimization problem, i.e., the set of seeders that reaches the larger number of nodes. A deeper investigation on these individuals were performed, in order to analyze the properties of good seeders. Figure 2 shows the induced subgraphs in respect to the best seeders, considering the different strategies for initial population. The connectivity between the seeders is clearly very low, which is the consistent with the results in the study about the effect of the selection of seeders presented by Xavier [25]. This result suggests that good seeders tend have a low number of connections among them. In fact, a great variability of connections in the seeders allows them to better explore the different regions of the network in a more efficient way.

The initial population that considers strategy **Community** 60 %, according to Table 2, shows the best results, and only this strategy was explored for the remaining three networks (**hep**, **phy** and **email**). The number of nodes reached

| (a) Random | (b) PageRank | (c) Comm. 60% | (d) Comm. 100% |

Fig. 2. Induced subgraphs in respect to the best seeders obtained by the GA for **ca** network considering different kinds of initial population: (a)Random; (b)PageRank; (c)Community 60 %; (d)Community 100 %.

before the execution of the GA, the mean number of reached nodes and the observed standard deviation after the 10 executions of the GA for each network is presented in Table 3. The mean number of communities involved in each solution is also presented.

Table 3. Result of the GA for the initial population that considers 60 % of the nodes placed in different communities and 40 % randomly placed nodes (Community 60 %) for the networks **hep**, **phy** and **email**.

Network	Before	Mean	Std. Deviation	# communities
hep	1260	1758,1	28,89	31,2
phy	1630	1854,4	36,97	32,8
email	1190	1564,8	51,56	31,2

Table 3 shows that, for **hep**, **phy** and **email**, GA could significantly improve the set of seeders when the strategy of selecting 60 % of the nodes placed in different communities and 40 % randomly placed nodes (Community 60 %) is to construct the initial population. Figure 3 shows the evolution of the GA considering the strategy Community 60 %.

Figure 3 shows a good evolution of the GA among the generations, consistently improving the objective function value for the best individuals of each generation up to 28 % when compared to the initial population.

Again, the best individuals found by the end of the GA execution show a low connection between the seeders. It can also be noticed a high number of communities explored by the seeders. These results combined suggest that good seeders are disperse over the network and have a low number of connections between them.

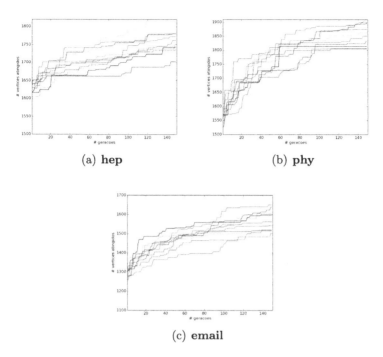

(a) **hep**   (b) **phy**

(c) **email**

Fig. 3. AG Evolution for networks: (a)**hep**; (b)**phy**; (c)**email**.

4   Conclusions and Future Works

This work presents a Genetic Algorithm for influence maximization in real networks, i.e., selecting a set of seeder nodes which maximize the number of reached nodes. Spreading Activation method was considered as a model for influence propagation. Four strategies for constructing the initial population were explored: a random strategy a PageRank based strategy and two community based strategies. The proposed method was computationally implemented and tested with four real networks. The results show that GA was able to significantly improve the quality of the seeders, increasing the number of reached nodes in about 25 %.

The results are consistent with previous works and suggest that a good set of seeders is well diffuse over the network and has a low number of connections between the nodes.

Further studies have to be performed in order to better investigate how the number of random seeders affect the quality of the propagation in the network and how the methodology behaves in other kind of networks, such as innovation and epidemiological networks.

Acknowledgements. The authors thank the financial support agencies: Capes, FAPEMIG and CNPq.

References

1. Anderson, J.R.: A spreading activation theory of memory. J. Verbal Learn. Verbal Behav. **22**, 261–295 (1983)
2. Backstrom, L., Huttenlocher, D., Kleinberg, J., Lan, X.: Group formation in large social networks: membership, growth, and evolution. In: Proceedings of the 12th ACM SIGKDD International Conference on Knowledge Discovery and Data Mining, pp. 44–54 (2006)
3. Blondel, V., Guillaume, J., Lambiotte, R., Mech, E.: Fast unfolding of communities in large networks. J. Stat. Mech. **10**, P10008 (2008)
4. Bollen, J.: A Cognitive Model of Adaptive Web Design and Navigation: A Shared e Knowledge Perspective. Ph.D. thesis, Vrije Universiteit Brussel, Belgium, June 2001
5. Chen, W., Wang, C., Wang, Y.: Scalable influence maximization for prevalent viral marketing in large-scale social networks. In: Proceedings of the 16th ACM SIGKDD International Conference on Knowledge Discovery and Data Mining, pp. 1029–1038. ACM (2010)
6. Chen, W., Wang, Y., Yang, S.: Efficient influence maximization in social networks. In: Proceedings of ACM KDD (2009)
7. Clauset, A., Newman, M., Moore, C.: Finding community structure in very large networks. Phys. Rev. E: Stat., Nonlin., Soft Matter Phys. **70**(6), 066111 (2004)
8. Croft, W.B., Thompson, R.H.: I3R: a new approach to the design of document retrieval systems. JASIS **38**(6), 389–404 (1987)
9. Csardi, G., Nepusz, T.: The igraph software package for complex network research. InterJournal Complex Syst. 1695 (2006). http://igraph.org
10. Dasgupta, K., Singh, R., Viswanathan, B., Chakraborty, D., Mukherjea, S., Nanavati, A.A., Joshi, A.: Social ties and their relevance to churn in mobile telecom networks. In: Kemper, A. (ed.) EDBT. ACM International Conference Proceeding Series, vol. 261, pp. 668–677. ACM (2008). http://dblp.uni-trier.de/db/conf/edbt/edbt2008.html#DasguptaSVCMNJ08
11. Domingos, P., Richardson, M.: Mining the network value of customers. In: Proceedings of the Seventh ACM SIGKDD International Conference on Knowledge Discovery and Data Mining, pp. 57–66. ACM (2001)
12. Ghosh, R., Lerman, K., Surachawala, T., Voevodski, K., Teng, S.H.: Non-conservative diffusion and its application to social network analysis. arXiv preprint (2011). arXiv:1102.4639
13. Goldenberg, J., Libai, B., Muller, E.: Talk of the network: a complex systems look at the underlying process of word-of-mouth. Mark. Lett. **12**, 211–223 (2001)
14. Goyal, A., Lu, W., Lakshmanan, L.V.S.: Celf++: optimizing the greedy algorithm for influence maximization in social networks. In: Proceedings of the 19th International World Wide Web Conference (2011)
15. Granovetter, M.: Threshold models of collective behavior. Am. J. Sociol. **83**(6), 1420–1443 (1978)
16. Heidari, M., Asadpour, M., Faili, H.: SMG: fast scalable greedy algorithm for influence maximization in social networks. Phys. A: Stat. Mech. Appl. **420**, 124–133 (2015). http://www.sciencedirect.com/science/article/pii/S0378437114009431

17. Kempe, D., Kleinberg, J., Tardos, E.: Maximizing the spread of influence through a social network. In: Proceedings of the Ninth ACM SIGKDD International Conference on Knowledge Discovery and Data Mining, KDD 2003, pp. 137–146. ACM Press (2003). http://dx.doi.org/10.1145/956750.956769

18. Kempe, D., Kleinberg, J.M., Tardos, É.: Influential nodes in a diffusion model for social networks. In: Caires, L., Italiano, G.F., Monteiro, L., Palamidessi, C., Yung, M. (eds.) ICALP 2005. LNCS, vol. 3580, pp. 1127–1138. Springer, Heidelberg (2005). http://dx.doi.org/10.1007/11523468_91

19. Khelil, A., Becker, C., Tian, J., Rothermel, K.: An epidemic model for information diffusion in manets. In: Proceedings of the 5th ACM International Workshop on Modeling Analysis and Simulation of Wireless and Mobile Systems, MSWiM 2002, pp. 54–60. ACM, New York (2002). http://doi.acm.org/10.1145/570758.570768

20. Leskovec, J., Krause, A., Guestrin, C., Faloutsos, C., VanBriesen, J., Glance, N.: Cost-effective outbreak detection in networks. In: Proceedings of the 13th ACM SIGKDD International Conference on Knowledge Discovery and Data Mining, pp. 420–429. ACM (2007)

21. Meyer, D.E., Schvaneveldt, R.W.: Facilitation in recognizing pairs of words: evidence of a dependence between retrieval operations. J. Exp. Psychol. 90(2), 227–234 (1971)

22. Rashotte, L.: Social influence. Blackwell Encycl. Soc. Psychol. 9, 562–563 (2007)

23. Richardson, M., Domingos, P.: Mining knowledge-sharing sites for viral marketing. In: Proceedings of the Eighth ACM SIGKDD International Conference on Knowledge Discovery and Data Mining, pp. 61–70. ACM (2002)

24. Wu, J., Wang, Y.: Opportunistic Mobile Social Networks. CRC Press, Boca Raton (2014)

25. Xavier, C.R.: Influence maximization in complex networks for Spreading Activation model: Case Study in a call phone network. Ph.D. thesis, COPPE-UFRJ (2015)

Software Variability Composition and Abstraction in Robot Control Systems

Davide Brugali$^{(\boxtimes)}$ and Mauro Valota

University of Bergamo, Dalmine, Italy
brugali@unibg.it, m.valota@studenti.unibg.it

Abstract. Control systems for autonomous robots are concurrent, distributed, embedded, real-time and data intensive software systems. A real-world robot control system is composed of tens of software components. For each component providing robotic functionality, tens of different implementations may be available.

The difficult challenge in robotic system engineering consists in selecting a coherent set of components, which provide the functionality required by the application requirements, taking into account their mutual dependencies. This challenge is exacerbated by the fact that robotics system integrators and application developers are usually not specifically trained in software engineering.

Current approaches to variability management in complex software systems consists in explicitly modeling variation points and variants in software architectures in terms of Feature Models.

The main contribution of this paper is the definition of a set of models and modeling tools that allow the hierarchical composition of Feature Models, which use specialized vocabularies for robotic experts with different skills and expertise.

1 Introduction

Control systems for autonomous robots are concurrent, distributed, embedded, real-time and data intensive software systems. The computational hardware of an autonomous robot is typically interfaced to a multitude of sensors and actuators, and has severe constraints on computational resources, storage, and power. Computational performance is a major requirement, especially for autonomous robots, which process large volumes of sensory information and have to react in a timely fashion to events occurring in the human environment.

In recent years, a variety of software frameworks have been specifically designed for developing robot control systems that are designed as (logically) distributed component-based systems (see [9] for a survey). Currently, the Robot Operating System [1] is the most widely used robotic framework in research and development. It offers mechanisms for real-time execution, synchronous and asynchronous communication, data flow and control flow management.

A real-world robot control system is composed of tens of components. For each component providing a robot functionality, tens of different implementations may be available. The initial release of ROS in year 2010 already contained

© Springer International Publishing Switzerland 2016
O. Gervasi et al. (Eds.): ICCSA 2016, Part IV, LNCS 9789, pp. 358–373, 2016.
DOI: 10.1007/978-3-319-42089-9_26

hundreds of open source packages (collections of nodes) stored in 15 repositories around the world [1].

Clearly, building complex control applications is a matter of system integration more than of capabilities implementation. The difficult challenge consists in selecting a coherent set of components that provide the required functionality taking into account their mutual dependencies.

In previous papers [14,15] we have presented the HyperFlex Model-driven toolchain and approach for the design of software product lines for autonomous robotic systems.

The key characteristics of HyperFlex are the support to the design and composition of architectural models of component-based functional subsystems, the possibility to symbolically represent the variability of individual functional subsystems using the Feature Models formalism, and the automatic configuration of functional subsystems according to selected features. The HyperFlex approach builds on our experience in developing software architectures for robotic control systems in the context of the EU FP7 BRICS project [7].

The novel contribution of this paper is the description of the new functionality of the HyperFlex toolchain and the definition of a set of guidelines that enable the composition and abstraction of the variability models of individual functional subsystems and of the integrated control system.

Feature models usually don't scale up when the number of variation points and variants becomes substantial, because a single and huge feature model is too complex to maintain and to be understood and processed by humans. Our aim is to simplify the system configuration phase by supporting the definition of feature models at multiple levels of abstraction using specialized vocabularies for each expert involved in system configuration.

System configuration is a crucial phase, which requires to select, integrate, and fine tune the robot functionalities (developed by domain experts) according to the available resources (requiring maintenance by qualified engineers), the environment conditions (often beyond the control of the application engineer), and the task to be performed (often specified by unskilled users).

The paper is structured as follows. Section 1.1 presents the background information about the HyperFlex approach and toolchain by means of a simple example. Section 2 reports on the related works. Section 3 presents the novel contribution of this paper. It extends the previous example with two case studies of variability composition and illustrates the new models and meta-models of the HyperFlex toolchain. The relevant conclusions are presented in Sect. 4.

1.1 The HyperFlex Approach and Toolchain

HyperFlex is a Model-driven engineering (MDE) [21] environment (available open source on GitHub [2,15]) that supports the development of flexible and configurable robotic control systems. It builds on state-of-the-art approaches in software variability modeling and resolution as described in Sect. 2 and consists in a set of Eclipse plugins for the definition and manipulation of three types of software models:

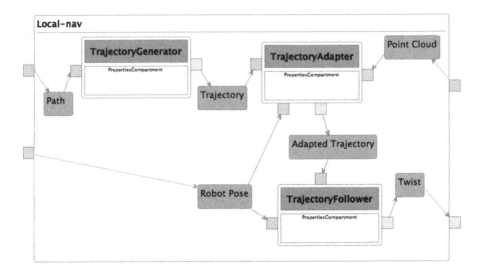

Fig. 1. The architectural model of the Local Navigation system (Color figure online)

- *Architectural Models* represent the structure of control systems in terms of component interfaces, component implementations, and component connectors. The HyperFlex approach promotes the design and composition of domain-specific software architectures for common robotic functionality (e.g. robot navigation), which capture the variability in robotic technologies (e.g. various algorithms for trajectory generation).
- *Feature Models* symbolically represent the variant features [22] of a control system; symbols may indicate individual robot functionality (e.g. marker-based localization) or concepts that are relevant in the application domain, such as the type of items that the robot has to transport (e.g. liquid, fragile, etc.), which affect the configuration of the control system.
- *Resolution Models* define model-to-model transformations, which allow to automatically configuring the architecture and functionality of a control system based on required features. Eventually, the configured architectural model is used to deploy the control system on a specific robotic platform.

As an example, Fig. 1 represents the architectural model of the Local Navigation system of an autonomous mobile robot. Local navigation is the set of functionality that allow the robot to autonomously move from its current position towards a goal position, while avoiding collisions with unexpected obstacles (i.e. moving people) in an indoor environment such as a hospital or a museum.

Architectural components define provided and required interfaces (depicted respectively as yellow and cyan squares), which can be connected by means of registers (green rectangles) according to the topic-based publish/subscriber paradigm supported by the ROS framework.

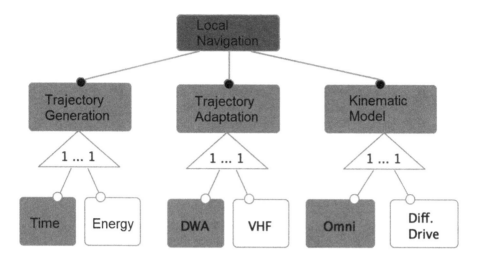

Fig. 2. The feature model of the Local Navigation system (Color figure online)

The *TrajectoryGenerator* receives as input a robot path and produces as output a trajectory, which specifies linear and angular velocity for each one of its waypoints. The *TrajectoryAdapter* receives as input the generated trajectory and produces as output a modified trajectory that avoids unexpected obstacles detected by the robot sensors. The *TrajectoryFollower* receives as input a trajectory and implements a feedback control loop that periodically reads the current robot pose and generates a twist (i.e. linear and angular velocities along the three axis) to minimize the distance to the path.

Figure 2 represents the *Feature Model* of the Local Navigation System, where the selected features are marked in green. The algorithms for *Trajectory Generation* are usually named by the function that is optimized, namely: minimum jerk or minimum time. The algorithms for *Trajectory Adaptation* are classified as reactive, i.e. they use only sensor data to generate robot control commands, such as the Vector Field Histogram (VFH), or as deliberative, i.e. they evaluate alternative paths and choose the best in terms safe distance to obstacles and minimum distance to the original path, such as the Dynamic Windows Approach (DWA).

The robot kinematics (Differential Drive or Omnidirectional) affects the implementation of all the three functionality.

Figure 3 represents the *Resolution Model* for the Local Navigation System. As an example, the selection of feature *DWA* triggers the four transformations that are indicated by the red arrow. They create the connections between the ports of the *TrajectoryAdapter* component that implements the DWA algorithm and the rest of the system.

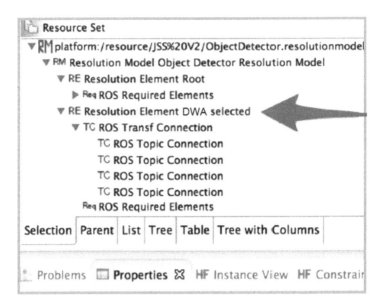

Fig. 3. A portion of the resolution model of the Local Navigation system

2   Related Works

The following subsections illustrates the related works in three areas:
(i) approaches to variability modeling, (ii) approaches to Feature Models composition, and (iii) variability modeling approaches in robotics.

2.1   MDE for Software Variability Management

Hyperflex follows a common approach to model the variability of a software system, which consists in defining four models: (a) the architectural model defines the software architecture of the system in terms of implementation modules (classes, aspects, agents, components) and their interconnections; (b) the variability model describes the functional variability of the system using a symbolic representation (e.g. feature models [11] or the OMG CVL language [17]); (c) the resolution model defines the mapping between the symbols of the variability model and the implementation modules of the architectural model; (d) the configuration model consists in a specific set of variants for the variation points defined in the variability model.

In our approach the first three models are completely orthogonal, i.e. they can vary independently, while the configuration model is an instance of the variability model.

The *Talents* approach [19] aims at modeling and composing reusable functional features for configuring the behavior of a software system. A graphical environment simplifies feature composition. The Talents approach models functional features at the level of instances of a class in an object-oriented programming language. In contrast, HyperFlex models functional features at the level of software components and component-based systems and thus is more adequate to model variability in robotic control systems.

In GenArch [10] the variability model and the configuration model are represented using the same meta-model, while in OMG CVL [17] the variability model and the resolution model are not explicitly separated.

The approach described in [16] defines three modeling categories, i.e. *Commonality*, *Variability*, and *Configuration*. The *Commonality* describes the architecture of a system, in terms of components, sub-components, ports and connectors. These architectural elements can be enriched with variation points, which represent the *Variability* and define how the common parts can be configured. For example, a variant for a component variation point can specify that a new sub-component has to be included in the component. Finally, the *Configuration* describes the selection of variants for all the variation points. The architectural model and the configuration conform to the MontiArc meta-model. Differently from HyperFlex, this approach condenses all the information in a single model.

2.2 Feature Models Composition

A survey of recent papers that propose techniques for Feature Model composition can be found in [5]. The surveyed approaches mostly focus on model composition techniques that are dedicated to support semantics preserving model composition. HyperFlex is a complementary approach, as it focuses on the automatic generation of Feature Model instances in a tree of variability models that are assumed to be semantically coherent and correct.

The approach described in [20] defines a set of composition constraints that specify how the features of the lower level feature models have to be selected according to the configuration of the higher level feature model. Differently from our approach, they don't adopt a component model for the architecture.

The *Compositional Variability* [4] approach supports the hierarchical composition of architectural models and feature models. The associations between a high-level feature model and a low level feature models are defined by means of the so called *Configuration Links*, which are similar to the feature dependencies defined in the HyperFlex *Refinement Model*. Differently from HyperFlex, this approach defines an abstract component model and does not provide the capabilities for modeling domain-specific component-based systems.

2.3 Variability Modeling Approaches in Robotics

In recent years, several model-driven approaches and tools for the development of robotic systems have been proposed, such as OpenRTM [6], Proteus [12], and Smartsoft [18].

In particular, the SmartSoft model-driven approach supports robotics variability management by modeling functional and non-functional properties of robot control system. The approach addresses two orthogonal levels of variability by means of two domain specific languages: (a) the variability related to the operations required for completing a certain task and (b) the variability associated to the quality of service.

These two variability levels are more related to the execution of a specific application (in the paper the example is a robot delivering coffee), while the HyperFlex approach supports modeling the variability of functional systems and the variability of the family of applications resulting from the composition of these functional systems.

3   Variability Composition and Abstraction with HyperFlex

HyperFlex allows structuring a complex control system as a hierarchical composition of functional systems. As an example, Fig. 4 shows the architecture of the *Robot Navigation* composite system, which integrates the *Local-nav* subsystem described in the previous section, with either the *Marker-nav* subsystem or the *Map-nav* subsystem. These components implement two strategies (map-based and marker-based) for generating the robot path between a start position and a goal position that is passed to the *Local-nav* subsystem.

If a geometric map of the environment is available, the robot is able to plan a geometric path in the free space. This strategy requires the robot to estimate its current position with respect to the map reference frame accurately. On the contrary, if artificial visual markers have been placed on the floor or on the walls, a camera mounted on the robot can detect them and the robot can navigate by following a path defined by a specific sequence of visual markers. In this case, the robot needs only to estimate its relative position with respect to the next visual marker. Figure 4 represents the situation where the *Local-nav* subsystem receives the robot path from the *Marker-nav* subsystem.

An interesting challenge that needs to be faced when using feature models to represent the variability of a software product line is the definition of an appropriate vocabulary for naming variation points and variants.

The clear separation of the symbolic representation of the system variability from its architectural model allows the definition of multiple Features Models for the same software system that are meaningful for system integrators with different needs and expertise.

In this context, HyperFlex allows the composition of Feature Models according to two different strategies, that we call *Bottom-up functionality composition* and *Top-down specification refinement*. These two composition strategies are meant for two types of stakeholders in software development for robotics:

– The community of researchers, who keep implementing new algorithms for common robot functionalities as open source libraries, need tools that simplify

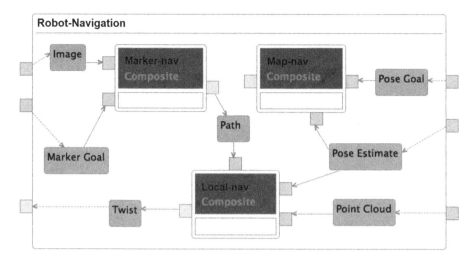

Fig. 4. The architectural model of the composite Navigation system (Color figure online)

the configuration of robotic control systems during test trials in various operational conditions.

- System integrators, who are expert in specific application domains, need tools for the configuration of robot control systems according to specific application requirements.

3.1 Bottom-Up Functionality Composition

Typically, the expert in robotic functionalities is interested in a representation of the control system variability that highlights the different algorithms implemented in the robot control system. For example, in [8] we have analyzed the variability in software library that implement motion planning algorithms. In this context, the relevant features are the type of bounding-box used by the collision-detection algorithm, the sampling strategy, and the type of kinematic model (e.g. single chain, multiple end-effectors).

Feature Models can be hierarchically composed to reflect the composition of functional systems. At each level the feature names abstract the relevant concepts of the corresponding functional system composition level.

For example, Fig. 5 shows three Feature Models that represent the variability of the composite Navigation system depicted in Fig. 4. In particular, the Feature Model of Fig. 5A has two leaf features, namely *Marker Navigation* and *Map Navigation*, that represent two alternative variants of the *Navigation Strategy* variation point.

When the system engineer selects the *Marker Navigation* feature, the Hyper-Flex tool creates a new instance of the corresponding Feature Model depicted in Fig. 5C. Subsequently, the system engineer can select features of lower-level

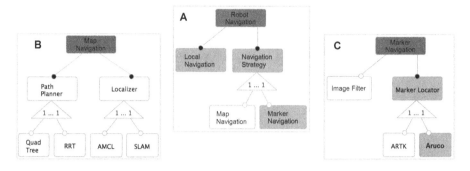

Fig. 5. The feature models of the (A) composite Navigation system, (B) the marker-based Navigation subsystem, and (C) the map-based Navigation system (Color figure online)

Feature Models for specific configuration properties. For example, in Fig. 5C the manual selection of Feature *Aruco* triggers a model-to-model transformation that configures the architectural connectors of the component implementing the Aruco algorithm [13] for marker localization. The approach is not limited to two levels but can be hierarchically extended. Systems made of subsystems can be further composed in order to design more complex systems.

3.2 Top-Down Specification Refinement

The application domain expert is interested in a representation of the system variability that specifies the application requirements supported by the robot control system more than its specific functionality. Figure 6 depicts the feature model of a robot control system for logistics applications. It is structured around three main dimensions of variability in application requirements, namely the type of environment, the type of load that the robot should handle, and the available equipment.

For this purpose, HyperFlex supports the composition of Feature Diagrams representing variability ad different level of abstractions. At each level the feature names abstract the relevant concepts of the specific domain: low-level names represent functional and technical terms while high level names are closer to the application requirements. This approach ensures that the terminology is well known by the system integrators that operates on a specific level.

During the variability resolution process, the application domain expert operates only on the highest-level Feature Model and the selected features trigger the automatic selection of features in the lower levels Feature Models.

For example, the robot operational environment could be a space with narrow passages and only static obstacles (see feature *Warehouse* in Fig. 6) or populated by moving obstacles in crowded areas (see feature *Airport* in Fig. 6). According to the operational environment, the robot should be configured with different algorithms: a slow and complete motion planner is adequate for moving among

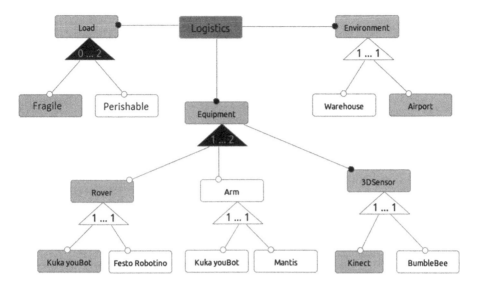

Fig. 6. The feature models of the logistics application (Color figure online)

static obstacles in narrow passages; instead, a fast and approximate motion planner is needed for dynamic environments.

HyperFlex provides a tool that allows to specify that the feature *Warehouse* in the *Logistics* FM is linked to the feature *QuadTree* in the *Map Navigation* FM of Fig. 5B, while the feature *Airport* is linked to feature *RRT*.

If the logistic task consists in transporting objects, the system integrator should select one of the available rovers. Here, the selection of feature *Kuka youbot* in the *Local Navigation* FM will trigger the selection of feature *Omni* in *Logistics* FM of Fig. 2A, which corresponds to the algorithms for omnidirectional rovers. Similarly, if the feature *Fragile* is selected in the FM of Fig. 6 then the feature *Jerk* is automatically selected in the FM of Fig. 2A.

Clearly, the system integrators can focus on the specification of the application requirements and should not be concerned with the functionality that implement them.

3.3   Refinement Model

In this section we illustrate the models, meta-models, and tools that allow the composition of Feature Models and the automatic generation of their instances according to the composition strategies described in the previous section.

The proposed appraoch consists in defining a new transformation model (called *Refinement Model*) that specifies links between the features of a parent Feature Model and the features of its child Feature Models. Figure 7 shows an example, where *FM_A* is a parent Feature Model and *FM_B* and *FM_C* are child Feature Models.

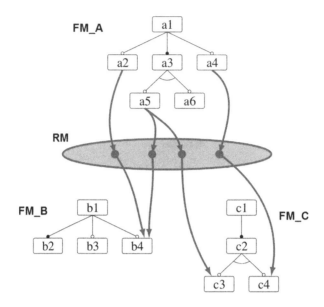

Fig. 7. The links between the features of different feature models

This approach does not require to modify the meta-model defined for Feature Models and thus promotes the reuse of existing Feature Models.

Figure 8 depicts the metamodel that we have defined for creating Refinement Models. The top level class is the *FeatureRefinementModel*, which has a link to the parent Feature Model and encapsulates a list of instances of class *FeatureRefinementPolicy*, one for each child Feature Model.

A Feature refinement policy is a collection of *FeatureRefinementElement*, which store the links between a feature of the associated parent Feature Model and a set of features of the associated child Feature Models. The proposed metamodel imposes the following rules to the definition and use of Feature Refinement Models.

When a new instance of the parent Feature Model is created, the instances of the child Feature Models should be empty, i.e. none of the features is selected. This condition allows to create instances of the child feature models incrementally.

When a feature of the parent FM is selected, all the linked features should be included in the instance of the child FM. This means that it is not possible to define *FeatureRefinementElements* that remove a previously inserted feature.

If a feature of the parent FM (e.g. feature a5 in Fig. 7) needs to be linked to several features of different child FMs (e.g. features b4 and c3), one *FeatureRefinementElement* should be created for each child FM and added to the corresponding *FeatureRefinementPolicy*.

Several features of the parent FM (e.g. features a2 and a5 in Fig. 7) can be linked to the same feature of a child FM (e.g. feature b4) by creating

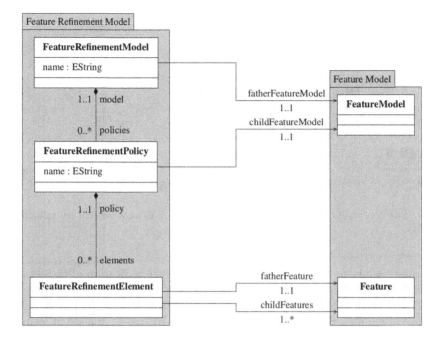

Fig. 8. The feature refinement meta-model

a *FeatureRefinementElement* for each feature of the parent FM and adding all of them to the same *FeatureRefinementPolicy* associated to the child FM.

This set of rules allows to minimize the amount of memory used by the feature refinement tool, which needs to load only two Feature Models at a time (the parent FM and one child FM) during the feature refinement process. The tool takes as input an instance of the parent FM (created with the HyperFlex editor) and a *FeatureRefinementModel* associated to it to generate an instance of each child FM automatically.

It should be noted that some features of the parent FM (e.g. feature a6 in Fig. 7) might not be linked to any feature of the child FMs and vice versa (e.g. feature b3).

The former case corresponds to the situation where the parent FM is used to configure directly some variation points of a functional subsystem as in the example of Sect. 3.2. In this case, feature b3 would be associated to a model to model transformation of the subsystem architecture as described in Sect. 1.1. The latter case requires manual selection of some features of the child FM as in the example of Sect. 3.1.

Feature Models can include constraints that limit the set of possible combinations of selected features. For examples, features $c3$ and $c4$ in Fig. 7 are mutually exclusive. It is not necessary to replicate the constraint in the parent Feature Model (i.e. *FM_A*), because the HyperFlex tool is able to report

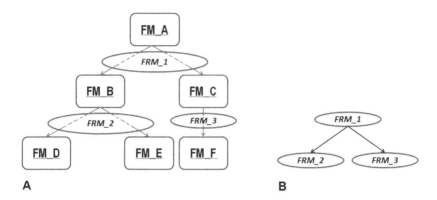

Fig. 9. The composition of feature refinement models (A) and the feature refinement tree (B)

constraint violations in child FM to the user with the indication of the selected features in the parent FM that caused them.

The HyperFlex toolchain includes an Eclipse Wizard that supports the model designer in defining the *FeatureRefinementModels* by means of a set of intuitive Eclipse Forms.

3.4   Refinement Language

The Feature Refinement Model described in the previous section defines a tree structure between a parent Feature Model and a set of child Feature Models. Starting from a manual selection of features in the parent FM, the HyperFlex tool generates instances of the child Feature Models automatically.

This structure can be extended to trees with an arbitrary number of levels by connecting Feature Refinement Models hierarchically. Here, the hierarchy imposes an order according to which the Feature Refinement Models are processed in order to create an instance of each intermediate and leaf Feature Model.

Figure 9A illustrates a simple example with six Feature Models (*FM_A*, ..., *FM_F*) and three Feature Refinement Models (*FRM_1*, ..., *FRM_3*). Figure 9B shows the hierarchical dependencies between Feature Refinement Models.

The refinement process starts when an instance of Feature Model *FM_A* is created manually. The HyperFlex tool processes the *FeatureRefinementPolicy* and the *FeatureRefinementElement* defined in *FRM_1* and generates an instance of *FM_B* and *FM_C*. These instances are then used as input for processing the Feature Refinement Models *FRM_2* and *FRM_3* and generating an instance of the Feature models *FM_D*, *FM_E*, and *FM_F*.

Figure 10 shows the Xtext [3] grammar of the language used to define the tree structure of Feature Refinement Models. The keyword *Node* has an identifier (*ID*), a content, and a list of children nodes. The keyword *Content* indicates that each node of the tree can embed a *Feature Refinement Model*, a sub-tree, or the

```
grammar org.hyperflex.featurerefinementmodels.xtext.editor.FeatureRefinementLanguage
with org.eclipse.xtext.common.Terminals
generate featureRefinementLanguage
"http://www.hyperflex.org/featurerefinementmodels/xtext/editor/FeatureRefinementLanguage"
FRL_Root :
      aliases += (Alias)* 'ROOT' rootTree = Tree trees += Tree*;
Alias : Model | FRL_File;
Model : 'FEATURE_REFINEMENT_MODEL' name = ID importURI = STRING;
FRL_File :
      'FEATURE_REFINEMENT_LANGUAGE' name = ID importURI = STRING;
Tree :
      'TREE' name = ID ':' mode = ('BFS' | 'DFS' | 'SUB') '=' (rootNode = Node |
      multiNode ?= '{''ROOT' rootNode = Node nodes += (Node)* '}');
Child : '=>' node = [Node] ;

Node :
      'NODE' name = ID '(' ((content = Content ')') | (empty ?= ')')) childNodes += (Child)* ;
Content :
      ( modelContent ?= 'MODEL' ':' model = [Model] | treeContent ?= 'TREE' ':' tree = [Tree]
      | frlContent ?= 'FRL' ':' file = [FRL_File] );
```

Fig. 10. Xtext grammar for the feature refinement language

```
file_structure.frl
FEATURE_REFINEMENT_MODEL model1 "../frm/model1.featurerefinementmodel"

FEATURE_REFINEMENT_MODEL model2 "../frm/model2.featurerefinementmodel"

FEATURE_REFINEMENT_MODEL model3 "../frm/model3.featurerefinementmodel"

FEATURE_REFINEMENT_MODEL model4 "../frm/model4.featurerefinementmodel"

FEATURE_REFINEMENT_LANGUAGE file1 "subfile1.frl"
//Root Tree
ROOT TREE Root : BFS = NODE Node_1 (TREE : First_Tree)

/* Another tree */
TREE First_Tree : SUB = {

    ROOT NODE Node_A (MODEL : model1) => Node_B
                                      => Node_C

    NODE Node_B (MODEL : model2) => Node_D
                                 => Node_E

    NODE Node_C (MODEL : model3)

    NODE Node_D (MODEL : model4)

    NODE Node_E (FRL : file1)
}
```

Fig. 11. An example of feature refinement tree

path (URI) to a file that stores a sub-tree. The keyword *Tree* indicates that there is the possibility to specify the algorithm for traversing its children nodes. Here, *BFS* stands for Breadth-first search and *DFS* stands for Depth-first search. The third modality (i.e. *SUB*) is used for sub-trees and indicates that it should be used the search algorithm of the parent tree. Figure 11 exemplifies the use of the *Feature Refinement Language* to build a tree of five nodes.

4   Conclusions and Future Works

In this paper, we presented the functionality, models, and metamodels of the HyperFlex model-driven toolchain for composing Feature models according to different composition strategies. HyperFlex has been conceived for simplifying configuration and deployment of complex control systems of autonomous robots. Nevertheless, the proposed approach to variability modeling and composition can be applied to any application domain.

Our current work aims at exploiting the approach presented in this paper to develop dynamically adaptive robotic systems. Robotic engineers can define several variation points (resources, algorithms, control strategies, coordination policies, cognitive mechanisms and heuristics, etc.). Depending on the context, the system dynamically chooses suitable variants to realize those variation points. These variants may provide better quality of service (QoS), offer new services that did not make sense in the previous context, or discard some services that are no longer useful.

References

1. ROS: Robot Operating System (2007). http://www.ros.org
2. The HyperFlex Toolchain (2014). http://robotics.unibg.it/hyperflex/
3. Eclipse Xtext (2016). https://eclipse.org/Xtext/
4. Abele, A., Lönn, H., Reiser, M.-O., Weber, M., Glathe, H.: Epm: a prototype tool for variability management in component hierarchies. In: Proceedings of the 16th International Software Product Line Conference, vol. 2, pp. 246–249. ACM (2012)
5. Acher, M., Collet, P., Lahire, P., France, R.: Comparing approaches to implement feature model composition. In: Kühne, T., Selic, B., Gervais, M.-P., Terrier, F. (eds.) ECMFA 2010. LNCS, vol. 6138, pp. 3–19. Springer, Heidelberg (2010)
6. Ando, N., Kurihara, S., Biggs, G., Sakamoto, T., Nakamoto, H., Kotoku, T.: Software deployment infrastructure for component based rt-systems. J. Robot. Mechatron. **23**(3), 350–359 (2011)
7. Bischoff, R., Guhl, T., Prassler, E., Nowak, W., Kraetzschmar, G., Bruyninckx, H., Soetens, P., Haegele, M., Pott, A., Breedveld, P., et al.: Brics-best practice in robotics. In: Robotics (ISR), 2010 41st International Symposium on and 2010 6th German Conference on Robotics (ROBOTIK), pp. 1–8. VDE (2010)
8. Brugali, D., Nowak, W., Gherardi, L., Zakharov, A., Prassler, E.: Component-based refactoring of motion planning libraries. In: 2010 IEEE/RSJ International Conference on Intelligent Robots and Systems (IROS), pp. 4042–4049. IEEE (2010)
9. Brugali, D., Scandurra, P.: Component-based robotic engineering (Part I) [Tutorial]. IEEE Robot. Autom. Mag. **16**(4), 84–96 (2009)
10. Cirilo, E., Kulesza, U., Lucena, C.: A product derivation tool based on model-driven techniques and annotations. J. Univ. Comput. Sci. **14**(8), 1344–1367 (2008)
11. Cirilo, E., Nunes, I., Kulesza, U., Lucena, C.: Automating the product derivation process of multi-agent systems product lines. J. Syst. Softw. **85**(2), 258–276 (2012)
12. Dhouib, S., Kchir, S., Stinckwich, S., Ziadi, T., Ziane, M.: RobotML, a domain-specific language to design, simulate and deploy robotic applications. In: Noda, I., Ando, N., Brugali, D., Kuffner, J.J. (eds.) SIMPAR 2012. LNCS, vol. 7628, pp. 149–160. Springer, Heidelberg (2012)

13. Garrido-Jurado, S., Muoz-Salinas, R., Madrid-Cuevas, F., Marn-Jimnez, M.: Automatic generation and detection of highly reliable fiducial markers under occlusion. Pattern Recogn. **47**(6), 2280–2292 (2014)

14. Gherardi, L., Brugali, D.: An eclipse-based feature models toolchain. In: 6th Italian Workshop on Eclipse Technologies (EclipseIT 2011) (2011)

15. Gherardi, L., Brugali, D.: Modeling and reusing robotic software architectures: the HyperFlex toolchain. In: IEEE International Conference on Robotics and Automation (ICRA 2014), Hong Kong, China, 31 May - 5 June 2014. IEEE (2014)

16. Haber, A., Rendel, H., Rumpe, B., Schaefer, I., Van Der Linden, F.: Hierarchical variability modeling for software architectures. In: Software Product Line Conference (SPLC), 2011 15th International, pp. 150–159. IEEE (2011)

17. Haugen, O., Wąsowski, A., Czarnecki, K.: Cvl: common variability language. In: Proceedings of the 17th International Software Product Line Conference, SPLC 2013, p. 277. ACM, New York (2013)

18. Lotz, A., Inglés-Romero, J.F., Vicente-Chicote, C., Schlegel, C.: Managing run-time variability in robotics software by modeling functional and non-functional behavior. In: Nurcan, S., Proper, H.A., Soffer, P., Krogstie, J., Schmidt, R., Halpin, T., Bider, I. (eds.) BPMDS 2013 and EMMSAD 2013. LNBIP, vol. 147, pp. 441–455. Springer, Heidelberg (2013)

19. Ressia, J., Grba, T., Nierstrasz, O., Perin, F., Renggli, L.: Talents: an environment for dynamically composing units of reuse. Softw. Pract. Experience **44**(4), 413–432 (2014)

20. Rosenmüller, M., Siegmund, N.: Automating the configuration of multi software product lines. In: VaMoS, pp. 123–130 (2010)

21. Schmidt, D.: Guest editor's introduction: model-driven engineering. Computer **39**(2), 25–31 (2006)

22. Svahnberg, M., van Gurp, J., Bosch, J.: A taxonomy of variability realization techniques. Softw. Pract. Experience **35**(8), 705–754 (2005)

A Methodological Approach to Identify Type of Dependency from User Requirements

Anuja Soni[(✉)] and Vibha Gaur

Department of Computer Science, University of Delhi, Delhi, India
30.anuja@gmail.com, 3.vibha@gmail.com

Abstract. Agents exhibit a high degree of inter-agent cooperation to achieve designated goals. The success of Multi-Agent System (MAS) depends on how well these agents cooperate with each other. Modeling cooperation with a thrust in finding an appropriate agent for delegating a task is a challenging area of MAS as it requires thorough analysis of the dependencies that exist among agents. In MAS, an agent may exhibit various dependencies viz goal, task, soft goal or resource that may complicate the final system. To handle the intricacy of such a system, it is crucial to identify various inter-agent dependencies from user requirements during the early phases of requirements engineering. This work employs notion of user story, composed of users' requirements in modeling the type of dependencies in an Agent Oriented System. A methodological approach using Fuzzy set theory, Lexical analysis and Vector Model is used to identify Type of Dependency (ToD) from user requirements well before the development of final system. Lexical analysis is applied to obtain index terms from User Stories that eventually are classified in various categories viz. (i) Quality requirements (ii) Supplementary guidelines and (iii) Want of a Resource or Information. These index terms are analyzed on the basis of their physical occurrences in User Stories as well as the importance as perceived by users to identify inter-agent dependencies. The inter-agent dependencies, if identified at the requirements stage, assist the developers in addressing inter-agent coordination issues to reduce the trivial dependencies and thereby unnecessary communication overheads. A case study using Materials e-Procurement System is presented to illustrate the proposed approach.

Keywords: Type of dependency (ToD) · Multi-agent system (MAS) · Lexical-analysis · Vector-model (VM) · Fuzzy set theory (FST)

1 Introduction

An Agent Oriented Paradigm involves a large number of agents playing different roles, interacting with each other to achieve personal and common goals [1–4, 23–30]. To achieve common goals, tasks are distributed and entrusted to other agents with a purpose to share proficiency and capability; to work in parallel or sequence on common problems.

An agent depends on several agents for goals, tasks and various resources. By depending on others, an agent may be able to accomplish goals that are difficult or infeasible to achieve otherwise. Agent Oriented Requirements analysis necessitates supporting ways for identifying and analyzing dependencies in inter-agent coordination.

© Springer International Publishing Switzerland 2016
O. Gervasi et al. (Eds.): ICCSA 2016, Part IV, LNCS 9789, pp. 374–391, 2016.
DOI: 10.1007/978-3-319-42089-9_27

1.1 Introduction to Type of Dependency (ToD)

The inter-agent dependency signifies a compliance that facilitates a depender agent to accomplish a goal through the Service Provider Agents (SPAs) in a distributed environment. The Type of Dependency (ToD) among agents can be goal, task, soft-goal and resource which is explained below [1–5, 23].

(a) **Goal Dependency.** The dependency between the agents is called the goal dependency, when the depender agent does not provide any guidelines to the SPAs for accomplishing a goal. The SPA is accountable for the achievement of the goal itself.

(b) **Task Dependency.** The task dependency is similar to goal dependency, except that, during the delegation of a goal, the depender agent provides some supplementary guidelines such as "Built-in Supplier List" and "Rules for Evaluating Suppliers" to the SPA for its successful accomplishment.

(c) **Soft-Goal Dependency.** The soft-goal dependency specifies the quality requirements associated with requirements. The terms such as efficiently, effectively, satisfactorily and user friendliness are augmented with the goal to define the soft-goal dependency to ensure high quality of the final system.

(d) **Resource Dependency.** The depender agent may depend on the SPA for some information or physical entities viz. Project reports, Files, Artifacts or Computer resource like printer etc. The reliance of a depender agent on other agents for want of a resource is called resource dependency.

These inter-agent dependencies assist the developers in determining appropriate service providers agents and reduce the complexity of the final system. In addition, to enhance the users' perception in the final system, it is essential to identify inter-agent dependencies from the users' requirements.

To effectively handle cooperation among agents, this work proposes a methodological approach to identify Type of Dependency (ToD) from user requirements. ToD refers to various types of inter-agent dependencies viz. goal, task, soft-goal and resource in a coordinated environment. Various approaches such as Fuzzy set theory, Lexical analysis and Vector Model are used to identify ToD at the requirements stage to assist the developers in mitigating the risk arising from the large number of dependencies.

The organization of the paper is as follows: Sect. 2 presents related work; Sect. 3 presents a background study of Vector Model (VM), Fuzzy Set Theory (FST) and Type of Dependency (ToD); Sect. 4 proposes a framework for identifying ToD from user requirements; Sect. 5 presents a case study; Sect. 6 presents a comparative analysis of RE approaches favoring ToD; Sect. 7 concludes the paper and finally Sect. 8 presents a future discussion.

2 Related Work

Software agents are considered as a next generation model for engineering complex, heterogeneous and open distributed systems. Software engineering community is increasingly revising and restructuring projects in terms of MAS as it is a recent paradigm for conceptualizing, designing, and implementing software systems. Agent-Oriented RE models the requirements of a system in terms of autonomous interactive component agents [1–6].

Researchers have tried various ways to address the requirements of agents. In literature various Agent-Oriented requirements frameworks such as i* [1], Gaia [2], TROPOS [3], Formal Tropos [4], Enhanced model [15], RE from Craft to Discipline [16], NorMAS-RE [5], B-Tropos [6] and Iterative REF [17] are presented. A brief description of these is prescribed in order.

i* [1] is well known being one of the most cited methodologies that focuses on intentional and social aspects of agents in a collaborative environment. It represents the primitive requirements of agents in terms of goals, tasks and inter-actor dependencies using the strategic dependency and rationale model diagrams.

Gaia [2] represents MAS as an organization of agent-roles using role model that is described using responsibilities, permissions, activities and protocols. Responsibilities, permissions and activities define the functionalities, rights and private actions of agents in an open and distributed environment, whereas protocols are specified as an institutionalized pattern of interaction among agents.

Tropos [3] augments various mentalistic notions of agents such as actor, goal, plan, belief, resource, dependency and capabilities with the basic constructs of i* to extend the requirements modeling from early to late phases. This methodology focuses on the comprehensive requirements analysis of various agents using Means end analysis, Contribution analysis and AND-OR decomposition of goals in several modeling diagrams that produce a complete set of system requirements involving both functional and non-functional.

Formal Tropos [4] is based on a specification language that adopts primitive concepts for modeling early requirements (such as actor, goal, and strategic dependency), along with a rich temporal specification language. Model checking is employed for the automatic verification of Formal Tropos specifications.

Enhanced model [15] provides a conceptual agent based modeling approach to drive the requirements gathering and user-oriented requirement analysis process.

RE from Craft to Discipline [16] provides a synergistic blend of incremental model analysis, obstacle analysis, divergence patterns, heuristic rules, conflict links, threat model synthesizer and refinement trees for analysis, conflicts resolution, traceability and documentation of requirements.

NorMAS-RE [5] offers a structured approach to requirements analysis, using graphical notation based on dependence networks and is illustrated in the scenario of virtual organizations based on a Grid network.

B-Tropos [6] framework is delineated as the propagated form of Tropos. In this methodology, agent-oriented requirements have been augmented by business constraints using ConDec graphical language. This framework models various kinds of inter-agent dependencies using abductive logic programming based SCIFF specifications.

Iterative REF [17] offers actors, goals and actor dependencies using formal concept analysis (FCA) characterized by FCA lattice and CK theory. The normalized class models, use cases, conceptual models and expert domain knowledge are employed as the key concepts in the prescribed approach.

These frameworks model inter-agent dependencies using various modeling diagrams, dependence networks and specification languages, but don't provide any support for identification of the dependencies from the user requirements. Delegating a goal without identifying type of dependencies from user requirements may result in excessive communication overheads and affect the quality of MAS. To address this problem, this work presents a framework to analyze Type of Dependency (ToD) from user requirements. Vector Model (VM) is used to represent User Stories as vectors of index terms. As the index terms are vague and uncertain, therefore Fuzzy Set Theory (FST) is used. Representing User Stories in terms of index terms using VM and FST assists the developers in discovering ToD directly from the users' requirements to build MAS of users' perception.

3 Background Study

This section briefly describes fundamental concepts of Vector Model [7] and Fuzzy Set Theory [9, 10] employed in the proposed approach.

3.1 Introduction to Vector Model (VM)

VM is an algebraic model used in information retrieval that represents text documents as vectors of index terms and their weights [7]. The definitions used in VM are given below:

Definition (i): Given $D = \{d_1, d_2, \ldots, d_m\}$, where m: the number of text documents, each document $d_i(1 <= i <= m)$ can be represented as a vector $d_i = (w_1, w_2, \ldots, w_N)$, where N is the number of index terms in a document d_i and w_j $(1 <= j <= N)$ are corresponding weights.

Definition (ii): Weight of the index term in a document is driven by its frequency in the current document and number of documents in which it appears. The total frequency 'Tf_{td}' of term 't' in the collection of documents 'd' is described as:

$$Tf_{td} = \sum_{x \in d} f_t(x) \, where \, f_t(x) = \left\{ \begin{array}{ll} 1 & if \, t \, is \, available \\ 0 & otherwise \end{array} \right\}$$

The VM is employed in various applications viz. Information retrieval [7], Information filtering [8] and Requirements traceability [7].

3.2 Introduction to Fuzzy Set Theory (FST)

Fuzzy Decision-Making deals with the vagueness and uncertainty inherent in human formulation of preferences, constraints and goals that is accomplished using FST [9, 10, 18, 19]. FST utilizes fuzzy numbers to model and simulate human thought and linguistic reasoning in a domain characterized by incomplete, imprecise, uncertain and vague data. A fuzzy number is represented by a fuzzy interval of real numbers, each with a grade of membership between 0 and 1. The preliminary definitions of FST are given below [9, 10]:

Definition i: Fuzzy set: A fuzzy set $\tilde{A}$ is defined by $\tilde{A} = \{(x, \mu_{\tilde{A}}(x)) : x \in A, \mu_{\tilde{A}}(x) \in [0,1]\}$. In the pair $(x, \mu_{\tilde{A}}(x))$, the first element x belongs to the classical set A, the second element $\mu_{\tilde{A}}(x)$ called membership function, belongs to the interval [0, 1].

It is often convenient to work with Triangular Fuzzy Numbers (TFNs) because of their computational simplicity. The definition of a TFN is as follows:

Definition ii: Triangular Fuzzy Numbers: If a fuzzy set $\tilde{A}$ is expressed by the triplets as $\tilde{A} = (l, m, u)$, then $\tilde{A}$ is called a TFN, where the parameters l, m, and u respectively indicate the smallest, the most promising and the largest possible values that describe a fuzzy event [10–12]. The membership function $\mu_{\tilde{A}}(x)$ for the TFN is defined as:

$$\mu_{\tilde{A}}(x) = \begin{cases} 0 & x < l \\ \frac{x-l}{m-l} & l \leq x \leq m \\ \frac{u-x}{u-m} & m \leq x \leq u \\ 0 & x \geq u \end{cases} \tag{1}$$

As fuzzy numbers are difficult to deal with, crisp values may facilitate to reach at some substantial concluding results. An α-cut is viewed as a bridge between fuzzy sets and crisp sets. Klir et al. [10] defines α-cut as:

Definition iii: α-cut: Given a fuzzy set $\tilde{A}$ and threshold level $\alpha \in [0,1]$, the α-cut symbolized by $\alpha \tilde{A}$, is the crisp set that has membership degrees not less than α i.e. $\alpha \tilde{A} = \{x | \mu_{\tilde{A}}(x) \geq \alpha, \ x \in A, \ \mu_{\tilde{A}}(x) \in [0,1]\}$

The value of α indicates the level of certainty of stakeholders in obtaining the crisp value of a fuzzy number. Higher value of α indicates that stakeholders are highly confident about the crisp value. Value of α as 0.5 shows the moderate level of certainty for a crisp value.

The α-cut operation applied on a fuzzy number $\tilde{A} = (l, m, u)$ to obtain the crisp interval denoted as $\alpha \tilde{A}_{CrspIn}$ is computed using the Eq. (2) [13, 14].

$$\alpha \tilde{A}_{CrspIn} = [I_{\alpha l}, I_{\alpha u}] = [(m-l)\alpha + l, -(u-m)\alpha + u] \tag{2}$$

Where $I\alpha_l$ = lower bound of crisp interval, $I\alpha_u$ = upper bound of crisp interval. The crisp value denoted by C_{α}^{μ} can be computed from the crisp interval namely $\alpha \tilde{A}_{CrspIn}$ using the Eq. (3) [13, 14]:

$$C_{\alpha}^{\mu} = \mu I_{\alpha u} + (1 - \mu) I_{\alpha l}, where \ \mu \in [0, 1] \tag{3}$$

Where μ refers to the level of optimism of the analyst as positive, moderate or pessimistic. The value of μ as 0.5 refers to moderate level of optimism. Higher value of μ represents the higher degree of optimism.

4   A Framework to Identify ToD from User Requirements

The inter-agent dependency signifies a conformity that assists a depender agent to carry out a goal through the dependee agent in a distributed environment. A depender agent might exhibit a number of dependencies during the delegation of a goal to the SPA. To resolve the coordination issues among agents at the requirements stage, a framework is proposed to identify ToD as shown in Fig. 1.

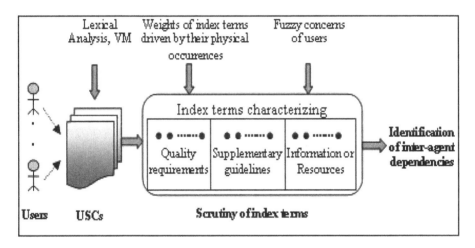

Fig. 1. A Framework to identify ToD from user requirements

To elicit users' requirements directly from users, User Story Cards (USCs) [20] are employed. Preprocessing is carried out on User Stories using Lexical analysis to eliminate unnecessary and redundant words to obtain meaningful index terms [21]. VM is applied to represent the User Stories as vectors of index terms and their weights. The weight of an index term is derived on the basis of its frequency in the current User Story relative to other User Stories. A joint team of developers and domain experts is involved to identify the index terms that characterize (i) Quality requirements (ii) Supplementary guidelines and (iii) Information or Resources.

Developers analyze these index terms on the basis of their physical occurrences in User Stories and the importance given to these User Stories by the users (as described in elicitation phase) to identify inter-agent dependencies. All the index terms might not be equally important to a user. Various users may have their own concerns over the importance of these index terms. As the concerns of the users over the importance of the index terms are vague, uncertain and subjective in nature, therefore the linguistic terms viz. Very Highly Important (VHL), Highly Important (HI), Important (I), Less

Important (LI), Very Less Important (VLI) are utilized to express their concerns that are mapped to TFNs and converted into crisp values using fuzzy α-cut operation.

The procedure for identification of ToDs from the users' requirements is as follows:

(1) *Lexical Analysis is applied to User Stories to obtain index terms.*

Lexical analysis or scanning is the process where the stream of characters forming the User Stories are read from left-to-right and grouped into tokens [21]. Tokens are groups of characters with a specific meaning. It eliminates the words that have no indexing value viz. "and", "or" etc. The remaining words are converted to their meaningful root form.

(2) *VM is applied to User Stories to represent USCs as the vectors of index terms and their weights as shown below:*

$$USC_i = ((Tu_1, w_{i1}) \ldots \ldots (Tu_e, w_{ie}) \ldots (Tu_h, w_{ih}))$$

Where $w_{ie} = w_{term}(Tu_e, USC_i)$ having value in [0,1], signifies the weight of the index term Tu_e in USC_i with 1 <=e <=h and 1 <=i <=N, N represents the total number of USCs and h is the total number of index terms in USC_i.

(3) *The value of wterm(Tue, USC_i) is computed from its physical occurrences in the current User Story relative to other User Stories using the Eq. (4) [7].*

$$w_{term}(Tu_e, USC_i) = FTu_{eUSC_i} * IFTu_{eUSCs} \tag{4}$$

Where FTu_{eUSC_i} defines the frequency of term Tu_e in USC_i and $IFTu_{eUSCs}$ is the inverse frequency of the term Tu_e in the USCs' collection. The inverse frequency is defined as the relative value of the frequency of an index term with respect to total number of USCs. The Inverse frequency $IFTu_{eUSCs}$ is computed using Eq. (5) [7]:

$$IFTu_{eUSCs} = \log_2 \frac{N}{N_{USC_{Tu_e}}} \tag{5}$$

Where $N_{USC_{Tu_e}}$ signifies the number of USCs containing term Tu_e and N denotes total number of USCs. The value of $w_{term}(Tu_e, USC_i)$ is normalized in [0, 1] by dividing its value by the maximum value of w_{term}.

(4) *Domain experts and developers classify the index terms in various categories viz.* *(i) Quality requirements (ii) Supplementary guidelines (iii) Resources or Information.*

The index terms characterizing quality requirements, supplementary information and resources define soft-goal, task and resource dependencies respectively. Absence of supplementary guidelines in a user story indicates the goal dependency.

(5) *The importance of the index terms obtained from the users in linguistic terms at the elicitation phase are mapped to TFNs as shown in Fig. 2.*

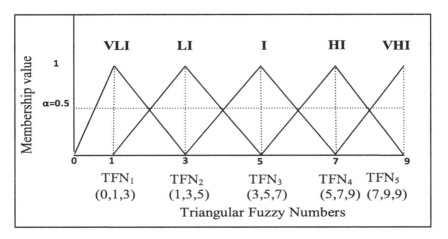

Fig. 2. Linguistic Scale for TFNs

The TFNs are converted to crisp values using the Eqs. (2) and (3). The crisp values obtained are normalized in [0,1] using the Eq. (6).

$$Normalised\ crisp\ value = \frac{Crisp\ value}{Upper\ Limit\ Of\ Scale\ Of\ TFN} \tag{6}$$

Where Upper Limit Of Scale Of TFN signifies the upper limit of the scale of TFNs.

(6) *The weights obtained in step 3 and the normalized crisp values obtained in the last step are aggregated to compute the mean which is treated as the final value of the dependency. The procedure is repeated for various classes of index terms characterizing different inter-agent dependencies.*

The values of the aggregated mean for a class of index terms comes out in [0,1]. The Domain experts and developers determine a threshold value that assists the developers in analyzing a specific dependency corresponding to a class of index terms. However the threshold value may change as per the nature of the application. The aggregated mean greater than the threshold value signifies the presence of a specific dependency.

(7) *The process is repeated for all User Stories.*

The identification of inter-agent dependencies is as an integral part of requirements definition that assists the developers to address inter-agent coordination issues within constrained resources to build MAS of users' perception.

5 Case Study

An agent based Materials e-Procurement System was selected for the case study. Materials e-Procurement is one of the areas of e-commerce technology called Supply Chain Management (SCM). SCM is a network of interconnected businesses that delivers products and services to the customers in the supply chain [22]. Materials e-Procurement is concerned with procuring various materials viz. raw materials, spare parts, packaging materials, consumable items, fuels and miscellaneous items in a distributed environment. Materials e-Procurement System is composed of several agents viz. Purchase Head Agent, Raw Materials Agent, Spares Agent, Packaging Items Agent, Fuel-Energy Agent, Consumable Items Agent and Miscellaneous Items Agent associated with e-procurement of various materials for several projects geographically distributed over various locations.

The experiment was performed with a joint team involving 7 developers, 3 requirements analysts, 5 domain experts and 9 customer representatives. User requirements were identified using User Stories. Following User story was considered for the case study.

*"As a Purchase head I want Raw Materials Incharge **to develop vendors for supplying RawMaterials efficiently, effectively and timely as per the SupplierList and EvaluationRules** So That procurement of RawMaterials can take place successfully"*

To identify ToD from user requirements, an interface was developed in an Integrated Development Environment (IDE) using JSP at front end and MySQL 5.2 as backend on Intel CPU having 1.99 GB of RAM and speed 1.86 GHz. JBoss 4. x was used as the web server.

5.1 Identifying ToDs from Users' Requirements

Various steps undertaken to identify ToDs from the user requirements are presented below step-by-step.

1. Lexical analysis was applied to user story to eliminate unnecessary words and convert it in terms of meaningful index terms. An interface displaying a dictionary of root words with their level of significance was developed as shown in Fig. 3.
2. One more interface shown in Fig. 4. was developed to compute the weights of index terms in various User Story Cards (USCs). Weights of these index terms were computed on the basis of their occurrences in the recent User Story and their occurrences in the remaining User Stories using Eq. (4).
3. The index terms that characterize various inter-agent dependencies were examined with the assistance of customer representatives and developers as shown in Table 1. The index terms "efficiently", "effectively" and "timely" signifying the quality requirements indicated soft-goal dependency, whereas the terms "SupplierList" and "EvaluationRules" signifying the supplementary guidelines defined task dependency.

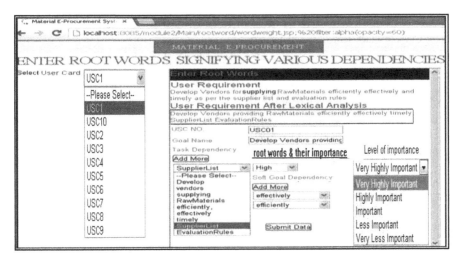

Fig. 3. Interface for lexical analysis and obtaining root words with level of importance

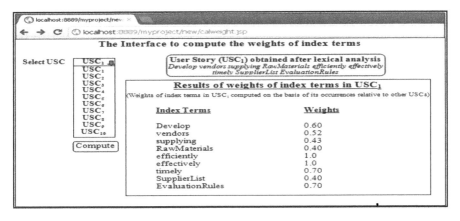

Fig. 4. Interface to compute the weights of index terms

4. The importance of these index terms was obtained from the users as well using linguistic terms viz. VHI, HI and I etc. at the elicitation phase that were mapped to TFNs and converted into crisp values using Eqs. (2), (3), (6).

5. Finally the weights and the crisp values of index terms characterizing the soft-goal and task dependencies were aggregated and values of mean were computed as 0.71 and 0.3 respectively for both the dependencies.

6. The mid-value as 0.5 in [0,1] was considered as a threshold value by the joint team of developers and domain experts. The value of the aggregated mean as 0.71 was found greater than the threshold value that signified the existence of soft-goal dependency, whereas other value of the aggregated mean as 0.3 was found lesser than the threshold value. It signified that though the index terms such as

Table 1. Identification of index terms for various inter-agent dependencies

Class of index terms	Index terms	ToDs
Quality requirements	Efficiently, effectively, timely	Soft-goal dependency
Supplementary guidelines	SupplierList, EvaluationRules	Task dependency

"SupplierList" and "EvaluationRules" at a glance give the impression of task dependency in the goal; however less importance given to these terms nullifies the effect of this dependency in the goal. The experimental results of this case study are presented in Fig. 5.

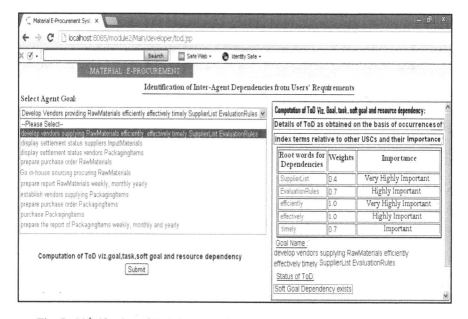

Fig. 5. Identification of ToD in terms of weights and importance of the index terms

The identification of inter-agent dependencies facilitates the developers in resolving the coordination issues during the allocation of the goals to the SPAs. For example, resultant 'soft-goal dependency' for the goal as shown in the Fig. 4, signifies that goal to be delegated requires high quality as indicated by terms 'efficiency' and 'effectiveness' and hence necessitates efficient SPA for its accomplishment.

6   Comparative Analysis

A comparative analysis of existing techniques mentioned in the related work section is provided in Table 2. An evaluation of all these methods was carried out on the basis of a number of parameters viz. year wise publication, support for tool for empirical study, concepts, techniques for modeling inter-agent dependencies, weaknesses and strengths.

Table 2. A Comparative Analysis of RE Approaches favoring ToD

Sr. No.	RE approaches	Year wise publications	Support for Tool for empirical study	Example	Concepts	Techniques for modeling inter-agent dependencies	Weaknesses	Strengths
1.	i*	1997, IEEE	N.A.	Computer based meeting scheduler	Intentional and social aspects of agents	Strategic dependency and rationale model diagrams	Introduces inter-agent social dependencies at the primitive level	Easy to understand due to simple graphical notations
2.	GAIA	2000, Kluwer Academy	N.A.	Agent-based business process management system	Macro-level (societal) and the micro-level (agent) aspects-roles, permissions, responsibilities, protocols, activities, services, acquaintances	Role/Organization based approach	No connection between inter-agent dependencies and user requirements	Responsibilities, permissions and activities define the functionalities, rights and private actions of agents in an open and distributed environment, whereas protocols are specified as an institutionalized pattern of interaction among agents.
3.	TROPOS	2004, Kluwer academy	JACK-Tool	eCulture system	Actor, goal, dependency & plan modeling	Actor diagrams & goal diagrams	Problem of scalability and long term maintenance with large number of AUML diagrams	Formalizes the diagram-based social dependencies in order to model complex inter-agent interactions and their knowledge

(*Continued*)

Table 2. (*Continued*)

Sr. No.	RE approaches	Year wise publications	Support for empirical study		Concepts	Techniques for modeling inter-agent dependencies	Weaknesses	Strengths
			Tool for empirical study	Example				
4.	Formal TROPOS	2004, Springer	T-Tool	Course-exam management	Actor, goal, strategic dependency, dynamic modeling	i*-concepts, temporal specification language, Model Checking	This approach is specific to the primitive requirements of agents	Combines the benefits of i* concepts with rich temporal specification language
5.	Enhanced model for requirements gathering & pre-system analysis	2006, IEEE	N.A.	Agent-based open and adaptive system	Production model, organization model, system user model & cognitive model	Conceptual agent-based modeling	No support for practical implementation of social dependencies	Conceptual agent based modeling appears to be easy to comprehend
6.	RE: From Craft to Discipline	2008, ACM	FAUST (formal analysis suite), animator, SAT-solver Tool	Numerous examples including train control system, phone system on T.V. cable etc.	Agents, objects, operations, behaviors, multi-view modeling, refinement checking, checking operationalization, risk analysis, threat analysis, conflict analysis, goal-oriented animation, reasoning about alternative options	Incremental model analysis, obstacle analysis, divergence pattern, heuristic rules, conflict links, threat model, synthesizer, refinement trees	Key focus is on operations and behaviors of agents, instead of social dependencies	A synergistic blend of multiple approaches appears to be interesting.

(*Continued*)

Table 2. (*Continued*)

Sr. No.	RE approaches	Year wise publications	Support for Tool for empirical study	Example	Concepts	Techniques for modeling inter-agent dependencies	Weaknesses	Strengths
7.	NorMAS-RE	2009, Springer	N.A.	The scenario of virtual organizations based on grid network	Agent view, institutional view, role assignment view, combined view, dynamic & conditional dependency modeling, institution, obligation, sanction & secondary obligation	Graphical notation using dependency networks	No support for inter-agent dependencies from user requirements	Use of dependence networks provides a structural based specification of agent-dependencies
8.	B-TROPOS	2010, Springer	SOCS-SI, a JAVA based tool is suggested	Product development process scenario	Concepts of TROPOS augmented with process-oriented & goal-directed temporal business constraints,	CoDec-graphical language, SCIFF-specifications	Problem of scalability with large sized B-Tropos models, inter-agent dependency requirements found missing during the delegation of goals	Augments business constraints with the concepts of Tropos methodology and provides computational logic-based specification

(*Continued*)

Table 2. (Continued)

Sr. No.	RE approaches	Year wise publications	Support for Tool for empirical study	Example	Concepts	Techniques for modeling inter-agent dependencies	Weaknesses	Strengths
9.	Iterative REF	2012, Springer	N.A.	book Trader System, hotel administration system, ordering system for computer hardware, office material etc. of the universityKULeuven in Belgium, elevator repair system	normalized class models, use cases and conceptual models	formal concept analysis (FCA) using FCA lattice& CK theory	Large number of models for multi-faceted perspectives of agents enhance the complexity of the requirements	The hybrid approach of FCA lattice and CK theory represents expert domain knowledge
10.	Proposed Approach		IDE using JSP at front end and MySQL 5.2 as backend, Web server-JBoss 4. X	An agent based Materials e-Procurement System	ToD viz. Goal, Task, Soft-Goal and Resource dependencies	User Stories, Lexical Analysis, Vector Model Approach and Fuzzy Set Theory	This approach is specific to the role-based architecture of MAS involving various inter-agent dependencies	This approach facilitates identification of various inter-agent dependencies at the requirements stage and assists the developers in addressing inter-agent coordination issues to reduce the trivial dependencies and thereby unnecessary communication overheads.

It was observed that large number of methods exist in literature for modeling inter-agent dependencies. However, only proposed approach is equipped with the potential of identifying inter-agent dependencies from user requirements. The inter-agent dependencies, if discovered at the requirements stage, assist the developers in addressing inter-agent coordination issues to lessen the trivial dependencies and thereby excessive communication overheads.

7   Conclusions

In MAS, tasks are disseminated and delegated to other agents with a rationale of sharing mutual expertise. The inter-agent dependency signifies a compliance that facilitates a depender agent to accomplish a goal through the Service Provider Agents (SPAs) in a distributed environment. In MAS, the large number of inter-agent dependencies would result in a muddled system. To handle the perplexity of such a system, it is crucial to identify various inter-agent dependencies from user requirements well before the development of final system. This work employs a blend of Fuzzy set theory, Lexical analysis and Vector Model to identify Types of Dependencies (ToDs) at the requirements stage that will assist the developers in pre-estimating service provider agents and hence reducing the complexity of the system.

8   Future Discussion

A methodological approach using Fuzzy set theory, Lexical analysis and Vector Model was presented to identify Type of Dependency (ToD) from user requirements well before the development of final system. Future work would also consider a systematic approach for computing the probability of selection of service provider agents driven by ToD to lessen the trivial dependencies and thereby excessive communication overheads of the final system.

References

1. Yu, E.S.K.: Towards modeling and reasoning support for early-phase requirements engineering. In: RE 1997 Proceedings of the 3rd IEEE International Symposium on Requirements Engineering. IEEE, ACM digital library, pp. 226–235 (1997)
2. Wooldridge, M., Jennings, N.R.: The gaia methodology for agent-oriented analysis and design. Journal of Autonomous Agents and Multi-Agent Systems 3(3), 285–312 (2000). Kluwer Academic Publishers, ACM digital library, Hingham, USA
3. Bresciani, P., Perini, A., Giorgini, P., Giunchiglia, F., Mylopoulos, J.: Tropos: an agent-oriented software development methodology. Auton. Agents MAS 8(3), 203–236 (2004). Kluwer Academic Publishers, ACM digital library
4. Fuxman, A., Liu, L., Mylopoulos, J., Pistore, M., Roveri, M., Traverso, P.: Specifying and analyzing early requirements in tropos. J. Requirements Eng. 9(2), 132–150 (2004). Springer

5. Villata, S.: NorMAS-RE: a Normative Multi-Agent Approach to Requirements Engineering. Dagstuhl Seminar Proceedings 09121 on Normative Multi-agent Systems, pp. 1–32 (2009)
6. Montali, M., Torroni, P., Zannone, N., Mello, P., Bryl, V.: Engineering and verifying AOR augmented by business constraints with B-Tropos. J. Auton. Agents Multi-Agent Syst. 23(2), 193–223 (2011). Kluwer Academic Publishers, ACM digital library
7. Hayes, J.H., Dekhtyar, A., Osborne, J.: Improving Requirements Tracing via Information Retrieval. In: IEEE Proceedings on Requirements Engineering, pp. 138–147. ACM Digital Library (2003)
8. Shuda, W., Jiangping, L., Riu, W.: Research of information filtering based on vector space model. In: IWCSE 2009 Proceedings of Second International Workshop on Computer Science and Engineering, vol. 1, pp. 42–46 (2009). IEEE, ACM digital library
9. Zadeh, L.A.: Outline of a new approach to the analysis of complex systems and decision processes. IEEE Trans. Syst. Man Cybern. SMC-3, 28–44 (1973)
10. Klir, G.J., Yuan, B.: Fuzzy Sets and Fuzzy Logic. PHI publications (1995). ISBN:81-203-1136-1
11. Zimmerman, H.J.: Fuzzy Sets, Decision-Making and Expert Systems. International series in Management Science Operations, vol. 10. Kluwer Academic Publishers, London (1987)
12. Yen, J.: Fuzzy Logic Intelligence, Control and Information. Pearson publications, New Delhi (2006). ISBN 978-81-317-0534-6
13. Lee, K.H.: First Course on Fuzzy Theory and Applications. Advances in Intelligent and Soft Computing, vol. 27, p. 149. Springer, Heidelberg (2005)
14. Yang, Y., Wang, X., Wen, X.: Evaluation of english textbook using fuzzy analytic hierarchy process. In: ETTANDGRS 2008 Proceedings of the 2008 International Workshop on Education Technology and Training & 2008 International Workshop on Geoscience and Remote Sensing, vol. 01, pp. 30–33. IEEE, ACM Digital Library (2008)
15. Ranjan, P.: An enhanced model for agent based requirement gathering and pre-system analysis, pp. 187–195. IEEE (2006)
16. van Lamsweerde, A.: Requirements engineering: from craft to discipline. In: ACM-SIGSOFT, pp. 238–249 (2008)
17. Poelmans, J.: An iterative REF based on Formal Concept Analysis and CK theory. Elsevier 2012(39), 8115–8135 (2012)
18. Gaur, V., Soni, A.: A fuzzy traceability vector model for requirements validation. Int. J. Comput. Appl. Technol. (IJCAT) 47(2/3), 172–188 (2013). Inderscience, Special Issue on: "Advanced Software Engineering and Its Applications"
19. Gaur, V., Soni, A.: An integrated approach to prioritize requirements using fuzzy decision making. IACSIT Int. J. Eng. Technol. 2(4), 320–328 (2010)
20. Gaur, V., Soni, A., Bedi, P.: An agent-oriented approach to requirements engineering. In: Proceedings-IEEE 2nd International Advance Computing Conference, pp. 449–454. IEEE Press and IEEE Xplore, USA (2010)
21. Teufl, P., Payer, U., Lackner, G.: From NLP (Natural Language Processing) to MLP (Machine Language Processing). In: Kotenko, I., Skormin, V. (eds.) MMM-ACNS 2010. LNCS, vol. 6258, pp. 256–269. Springer, Heidelberg (2010)
22. Srinivasan, S., Kumar, D., Jaglan, V.: Multi-agent system supply chain management in steel pipe manufacturing. IJCSI Int. J. Comput. Sci. Issues 7(4), 30–34 (2010)
23. Gaur, V., Soni, A.: A novel approach to explore inter agent dependencies from user requirements. J. Procedia Technol. 1, 412–419 (2012). First World Conference on Innovation and Computer Sciences, Antalya. Science Direct, Turkey
24. Gaur, V., Soni, A.: Evaluating degree of dependency from domain knowledge using fuzzy inference system. In: Nagamalai, D., Renault, E., Dhanuskodi, M. (eds.) CCSEIT 2011. CCIS, vol. 204, pp. 101–111. Springer, Heidelberg (2011)

25. Gaur, V., Soni, A.: Analytical inference model for prediction and customization of inter-agent dependency requirements. ACM SIGSOFT Notes **37**(2), 1–11 (2012)

26. Gaur, V., Soni, A.: A knowledge-driven approach for specifying the requirements of multi-agent system. Int. J. Bus. Inf. Syst. **19**(3), 300–323 (2015). Special Issue on: "Requirement Engineering Processes in Information Systems, Inderscience. Print ISSN: 1746-0972, Online ISSN: 1746-0980

27. Soni, A., Gaur, V.: Specifying uncertainties in inter-agent dependencies using rough sets and decision table. Int. J. Comput. Sci. Eng. Technol. (IJCSET) **6**(1), 1–8 (2016). ISSN: 2231-0711

28. Soni, A., Gaur, V.: A novel approach to streamline RE process for multi-agent systems. Int. J. Softw. Eng. Technol. Appl., InderScience (2016, in Press, accepted for publication)

29. Gaur, V., Soni, A., Bedi, P., Muttoo, S.K.: Comparative analysis of ANFIS and ANN for evaluating inter-agent dependency requirements. Int. J. Comput. Inf. Syst. Ind. Manage. Appl. **6**, 23–34 (2014). MIR Labs, ISSN 2150-7988

30. Gaur, V., Soni, A., Muttoo, S.K., Jain, N.: Comparative analysis of mamdani and sugeno inference systems for evaluating inter-agent dependency requirements. In: 12th International Conference on Hybrid Intelligent Systems, Pune, India, pp. 131–136. IEEE Explore, DBLP (2012)

Evolution of XSD Documents and Their Variability During Project Life Cycle: A Preliminary Study

Diego Benincasa Fernandes Cavalcanti de Almeida[✉]
and Eduardo Martins Guerra

National Institute for Space Research, São José Dos Campos, SP, Brazil
contato.inpe@diegobenincasa.com, eduardo.guerra@inpe.br

Abstract. During a software system life cycle, project modifications occur for different reasons. Regarding web services, communication contracts modifications are equally common, which induces the need for adaptation in every system node. To help reduce the contracts changing impact over software source code, it is necessary to understand how these contract changes occur. This paper presents a preliminary study on the evaluation of the change history of different open-source projects that defines XSD documents, specifying metrics for such files, extracting them by software repository mining and analyzing their evolution during the project life cycle. Based on the results, and considering that Web Service Definition Language (WSDL) contracts use XSD, a deeper study focused on web services projects only is further proposed to assess what exactly is changed at each contract revision, possibly revealing changing tendencies to support easy-to-adapt web service development.

Keywords: Software repository mining · XML · XSD · Web service · Contract

1 Introduction

Integration is the new trend in the complex business network. Many companies develop and use their systems to supply different needs, like production control, data management or communication, based on the concept of integrating diverse components into a single solution. To achieve this, one of the alternatives is the use of a Service Oriented Architecture (SOA) to act as the basis for the computational systems creation, where each service provides its particular kind of information to the integrator agent which can be the system itself or an external client. SOA is a reference architecture which aims to create functional modules called services [Con01] with low coupling and favouring code reuse [Sam01].

In distributed computer systems, the services used to provide information over the Internet are known as web services. "A Web service is a software system designed to support interoperable machine-to-machine interaction over a network" [W3C01]. To make this possible, the information exchange should be

© Springer International Publishing Switzerland 2016
O. Gervasi et al. (Eds.): ICCSA 2016, Part IV, LNCS 9789, pp. 392–406, 2016.
DOI: 10.1007/978-3-319-42089-9_28

done by standardized messages, which are made standard by the use of common definitions – structure and vocabulary. These definitions are often referred to as "contracts", as they create the rules for the correct composition and interpretation of messages. In other words, the contract validates the message structure and contents.

Web services messages are created using standard document formats, and a great amount of them is based on a format known as eXtended Markup Language (XML). The contracts that define XML messages vocabulary and structure are also created using a common standard, known as XML Schema Definition (XSD). In fact, the term "contract" in this context is also referred to as "schema". "XML Schema express shared vocabularies and allow machines to carry out rules made by people. They provide means for defining the structure, content and semantics of XML documents" [W3C02]. Likewise, web services operations and endpoints are described using Web Services Definition Language (WSDL) documents, which usually specifies request-response messages structures and involved data types using XSD.

During software system development life cycle, changes occur in source code, even by the introduction of new functionalities, or due to modifications in existing ones. In a system that provides web services, those changes can also happen in contracts, which impacts services providers and consumers. A change in the contract delivers a change in the message structure, and both service nodes must adapt to this in order to maintain the communication. These contract changes can affect different systems, which might be developed by different teams and even by different institutions. Therefore, the extent and frequency of such modifications can influence the adjustment rate that system nodes should meet, since changes in contracts are expensive as it can affect several systems. If adjustment is not performed properly, integration can be compromised.

A recent study revealed that changes in contracts tend to occur real often and they do impact on source code [Fra01]. However, no research has been undertaken to address these changes and classify them according to frequency and location of occurrence. Hence, some questions remain unanswered, like: (Q1) *"How frequent element addition/removal is?"*; (Q2) *"Modifications change document semantics or are merely refactoring?"*; (Q3) *"Is there any changing pattern or tendency in the structural types that experience the most modifications?"*. Despite these – and many other – questions, studies lack on trying to answer them. Thus, this work presents a study focused on analyzing the evolution of XSD documents during the project development life cycle, aiming to capture the most frequent kinds of modifications in XSD files and how often they occur. It should be noted that this initial research is applied to any kind of project that somehow defines XSD documents. To achieve the desired results, three research questions (RQ) were investigated, as follows:

(**RQ1**) What is the frequency of changes in the number of each XSD basic container type (element, attribute and complexType)?
(**RQ2**) Do distinct projects have similar changing frequencies for each XSD basic container type?

(**RQ3**) What is the number of XSD document versions where no changes in XSD basic container types are found?

Understanding the frequency that changes occur in contracts can lead to a redesign of development practices or to a architectural remodeling that eases the adjustment to these changes. If modifications are mapped or measured accordingly, one can undertake a wider contract readjustment to reduce the frequency of changes, ultimately leading to more stable or more adaptable client-server integration.

2   Related Work

Search for similar studies over the literature leads to very few related researches. In fact, only two recent studies are more likely to be linked to this work. This searching was undertaken mainly using the Google Scholar web search engine, with key strings like "evolution of XSD", "contract evolution", "XSD repository mining" and other related expressions.

A first research presents a study over the impact of contracts changing on the software system source code [Fra01]. Through historical data mining and analysis on software repositories, it gathers statistics for XSD files changing frequency, along with information about the most usual modifications and its concentration by developer. Such metrics were defined by the authors and aims to establish the influence of XSD changes on source code, relating XSD modifications quantity with source code modifications quantity. As a final result, it deduces that contracts modifications are frequent and finds evidences of huge impact on source code. That study enforces the need to better understand the contract modifications in order to develop systems prepared to deal with these changes. It is relevant to mention that the comparison between XSD modifications and source code changings were done manually and with few projects, and so is not able to answer the research questions listed at Sect. 3 accordingly.

Another research endeavors a study on the evolution of schemas in relational databases, associating the modifications with the database queries created [Qiu01]. The research classifies the modifications and also gathers some statistics to measure the impact of schema changing on these queries. It is not clear whether the authors use some kind of software to help with the analysis or not, but some proceedings seems to be similar to the ones that are undertaken in the present work, like historical analysis, classification of changes and statistical comparison of modifications. Like the aforementioned research [Fra01], it aims to "measure" the impact of schema changes at code-level, but is related to the present study by the inner-observation of schemas and discrimination of different change types and occurrence frequency.

Despite the aforementioned researches, no others could be found, in the literature, dealing with the changes in contracts and trying to understand how and where these changes occur. With that said, this can be considered an original research. But in the field of software repository mining, however, there are many different publications, as it is an evolving field of research and has numerous works published on the literature. One of them presents a tool that mines components metadata to identify which components available at a repository cannot be installed on the main system, using three project repositories to evaluate the correctness of the tool [Aba01]. Regarding changes in applications, another study defines the concept of unique changes and provides a method for identifying them in software project history [Bai01]. A different publication presents the use of empirical metrics to check what can be inferred from them for projects of the Apache Software Foundation [Gal01]. To enable platform-independent software mining and metric visualization, another publication presents a REST web service implementation directed to such tasks, in order to enable the development of service-oriented applications to mine and visualize software data across the web [Car01]. Nevertheless, there is still no research directed to contracts evaluation and/or web services development guidance, which endorses the originality of the present study.

3   Research Methodology

To evaluate projects, simple metrics were defined, considering the number of three XSD basic container types: `element`, `attribute` and `complexType`. Using these metrics, the quantity of modifications in XSD documents were mapped by their types (in elements, in attributes and in complex types). The documents used for this evaluation were obtained by historical mining over the selected projects repositories, in order to quantify each kind of change. In a future step following the present study, these numbers will help to check if there are changing patterns in contracts during software development life cycle.

To undergo this study, a short number of open-source projects hosted at GitHub were selected, since MetricMiner [Ani01] was chosen as the tool for repository mining and metric extraction and it only works on GitHub repositories. The selection criteria were as follows:

- Projects must have XSD documents in its repository;
- XSD documents must have enough modifications in the repository, represented by the number of commits with XSD files;
- Projects that defines or uses web services are desirable.

MetricMiner [Ani01] is a framework written in Java code that can be used to develop applications that perform deep analysis over GitHub revision control repositories. It runs code written by the researcher over different software revisions in order to capture desired information for further analysis. Benefits from using MetricMiner relies on its multi-thread facility, which can improve complex analysis performance over repositories with huge amounts of files and/or revisions, and also by being highly customisable, as virtually any study can be designed to be performed over source code files. It must be pointed out that MetricMiner does not perform any kind of analysis over the retrieved information, which is an entirely responsibility of the researcher.

To achieve the research proposal, the following information were retrieved from projects, in order to gather the desired metrics and to organize data:

- **HASH** – Hash number of commit;
- **REVCOMMIT** – Serialized number of commit, starting with 0 (most recent, or *"Head"*);
- **FILENAME** – Name of the analyzed file;
- **MODCOUNT** – Serialized number of modification index counter for current file, starting with 0 (most recent);
- **QTY_ELEMENTS** – Quantity of <xs:element> tags in current file;
- **QTY_ATTRIBUTES** – Quantity of <xs:attribute> tags in current file;
- **QTY_CTYPES** – Quantity of <xs:complexType> tags in current file;
- **MOD_ELEMENTS** – Direction of growth in <xs:element> tag quantity related to previous modification index, denoted by -1 for decrease, 0 for no change and 1 for increase;
- **MOD_ATTRIBUTES** – Same as above, but for <xs:attribute> tag;
- **MOD_CTYPES** – Same as above, but for <xs:complexType> tag;

In order to retrieve the aforementioned data, some code development was needed. To extract metrics from XSD files, a simple Java package named XSDMiner was written[1] and coupled to MetricMiner [Ani01] into a processor for historical data mining over projects repositories. Running MetricMiner with XSDMiner outputs a Comma Separated Values (CSV) file with the desired information.

4   Metrics and Analysis Implementation

To evaluate the frequency of changes in XSD documents used in contracts, this work uses the simple metrics defined in Sect. 3, as no related work regarding the evaluation and classification of XSD structural changes could be found in the literature. Also, as said in the previous section, a Java package named XSDMiner was written to perform the study and retrieve metric data. The considered XSD containers at this study were counted for each XSD file found in projects repositories and compared between consecutive revisions (or *commits*) to check if there are evidences of a change.

---

[1] XSDMiner is freely available as an Eclipse project at: http://github.com/diegobenincasa/XSDMiner.

XSDMiner implements the class XSDParser, which parses XSD files and retrieve the counting of container types discriminated before. It uses Java DOM to parse input files and has different methods for retrieving each container type counting. MetricMiner then runs over selected GitHub repositories (downloaded and available locally) and performs user-defined analysis (or "studies") at each revision. MetricMiner Study class was designed so that it can use XSDParser to parse each XSD file in the repository, running its code for every commit. As MetricMiner outputs a CSV file, the Study class was written in such a way that information listed in Sect. 3 could be gathered using this file type. Listing 1.1 illustrates the implemented Study class to be used as a processor by MetricMiner (imports hidden from the code snippet). It shows a working example for a single repository (in this case, the one named "xwiki-platform") to extract information as requested by the class MineXSD for every commit and using four threads at a time.

Listing 1.1. XSDMiner code snippet: the Study class implementation

```
1   package br.inpe.XSDMiner;
2
3   // ... Imports hidden ... //
4
5   public class MyStudy implements Study {
6
7       String project =
8               //"datacite-schema";
9               //"opennms";
10              //"opennms-mirror";
11              //"SOCIETIES-Platform";
12              //"spring-ws";
13              //"XeroAPI-Schemas";
14              "xwiki-platform";
15              //"zanata-server";
16
17      String projectDir = "/home/diego/github/" + project;
18      String output = "/home/diego/Desktop/mm_output/" + project +
                ".csv";
19
20      public static void main(String[] args) {
21          new MetricMiner2().start(new MyStudy());
22      }
23
24      @Override
25      public void execute() {
26          new RepositoryMining()
27                  .in(GitRepository.singleProject(projectDir))
28                  .through(Commits.all())
29                  .withThreads(4)
30                  .process(new MineXSD(), new CSVFile(output))
31                  .mine();
32      }
33  }
```

5   Study Execution

Applying the criteria mentioned in Sect. 3, six projects were selected to be analyzed, as shown in Table 1. It should be highlighted that, for the moment, only one project deals with web services.

After the projects were selected, their repositories were downloaded locally, and MetricMiner with `XSDMiner` was executed over each one of them. In the present study, MetricMiner ran several times, each one for a single project, and CSV files were generated per project. To ease the analysis over these files, they were imported to a PostgreSQL database and a SQL script was written[2] to output metric data separated by XSD container type.

Table 1. Selected projects to perform the study

Project	Description	Address
Datacite/Schema	DataCite Metadata Schema	https://github.com/datacite/schema
OpenNMS	OpenNMS enterprise grade network management application platform	https://github.com/OpenNMS/opennms
SOCIETIES-Platform	SOCIETIES platform software	https://github.com/societies/SOCIETIES-Platform
spring-ws	Spring Web Services	https://github.com/spring-projects/spring-ws
XeroAPI-Schemas	XSD Schemas for api.xero.com	https://github.com/XeroAPI/XeroAPI-Schemas
xwiki-Platform	The XWiki platform	https://github.com/xwiki/xwiki-platform

6   Results

Selected projects have different XSD file quantities, and some projects have a large set of them. Thus, it is difficult to show complete collected data in a summarized manner. Table 2 presents raw metric information that can be gathered together.

The chart presented in Fig. 1 shows the percentage of commits where modifications were found, classified according to each container type (blue: `element`; orange: `attribute`; yellow: `complexType`). On the other hand, the chart illustrated in Fig. 2 shows the ratio between the total number of modifications and the number of commits where these modifications occurred (similar label as before), also classified according to each container type, which represents a mean value of container quantity changes per commit with modifications.

Figure 1 illustrates that there are no visible patterns regarding the percentage of commits with modifications. Project *spring-ws*, the only project in this study that deals with web services, shows numbers close to each other, similar to other three projects, but distinct from the other two. Four of the six selected projects

---

[2] The script is also available at `XSDMiner` repository at GitHub.

Table 2. Raw metric data extracted from projects

Projects →	Datacite/Schema	OpenNMS	SOCIETIES-Platform	spring-ws	XeroAPI-Schemas	xwiki-Platform
XSD files	44	100	78	53	35	4
Total modifications in elements	89	257	268	114	130	33
Total modifications in attributes	97	347	168	109	96	11
Total modifications in complex types	87	240	204	115	78	17
Total commits	200	43614	9761	4286	137	49187
Commits with XSD modifications	34	722	597	60	94	50
Commits with elements modifications	11	192	231	40	64	35
Commits with attributes modifications	19	339	93	36	31	9
Commits with complex types modifications	10	169	141	38	33	15
Commits with no modifications	166	42892	9164	4226	43	49137

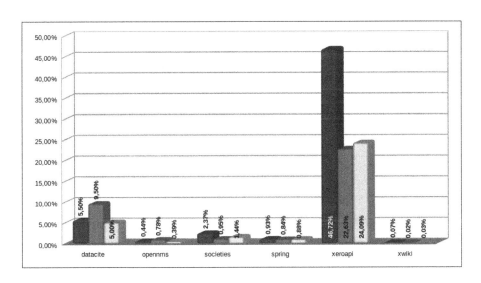

Fig. 1. Percentage of commits where were found changes in each container type (Color figure online)

have greater percentages in `element` containers, and they show a similar tendency: `element` containers have greater percentages, followed by `complexType` and `attribute` containers (in this order). The other two projects also seem to share a common behavior: greater percentages for `attribute` containers, followed by `element` and complexType containers (in this order). One project stands out from the behavior of the others: `XeroAPI-Schemas` changes the number of every container type in a frequency much larger than other projects, particularly the `element` container. Not with the same highlight but also with a much different behavior in respect to the other projects is `Datacite/Schema`, which shows a larger number of commits with modifications in `attribute` containers in comparison to other types.

The chart illustrated in Fig. 2 shows that, visually, no pattern can be inferred, in a similar conclusion as the one found by inspecting Fig. 1. However, the calculated ratios seem to approximate numbers for the three considered container types when analyzing projects individually. Nevertheless, there is not even a "container type ranking" pattern as in Fig. 1, with induces the need to deepen the study and consider more variables and metrics in the analysis to reduce the seemingly heterogeneous behaviour of ratios.

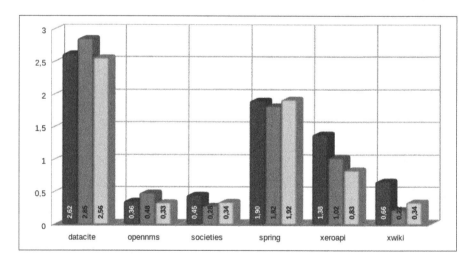

Fig. 2. Ratio between total number of each container type modification and the number of commits where these changes occurred (Color figure online)

No matter how often modifications occur inside XSD documents, they reveal another common characteristic: increase or decrease of containers do not prevail one over another. This establish a new issue to address in future steps after this study, as those metrics derived from collected repository mining information can not be used to sustain an argument for tendencies on XSD documents modifications. Examples of such heterogeneity are shown in Figs. 3, 4 and 5. Blue lines represent `element` containers; red lines represent `attribute` containers; and yellow lines represent `complexType` containers. Left-to-right reading of charts goes from older to recent commits.

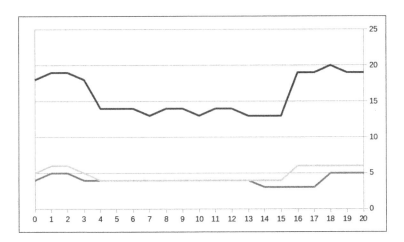

Fig. 3. Quantities of containers through XSD modification commits for project `OpenNMS`, file `users.xsd` (Color figure online)

A different representation of projects behavior heterogeneity is shown in Fig. 6. This illustrates that the amount of commits with contracts modifications (considering the used metrics) do not look similar among distinct projects. Project `XeroAPI-Schemas` is expected to have a large number of contracts modifications, as it is composed only by XSD files, but the other analyzed projects do not seem to follow a common tendency. In Fig. 6, the light blue part of graphs represents the amount of commits without contracts modifications, and the dark blue part represents the opposite.

7 Threats to Validity

The number of projects selected by the criteria specified at Sect. 3 is small when compared to the many existing open-source projects that deals with contracts. The major problem with the selection lies on GitHub search engine: it does not provide an inside-repository search tool that could retrieve projects with specific files in it – like any XSD file (*.xsd filter), for example. This makes the search and selection of projects even more difficult. To improve the quality of the analysis undertaken in this study, it should be continuously executed over other projects, and different ways to search for relevant ones need to be considered.

Another fact that should be taken into account is that some modifications, based on the defined metrics, might become masked by other ones. As an example, in a certain revision a container might have been created, but this can cover

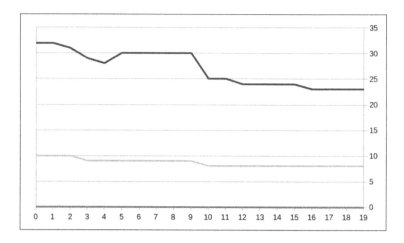

Fig. 4. Quantities of containers through XSD modification commits for project SOCIETIES-Platform, file org.societies.api.internal.schema.privacytrust. privacyprotection.model.privacypolicy.xsd (Color figure online)

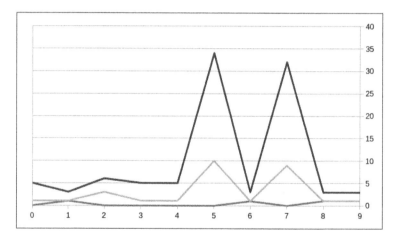

Fig. 5. Quantities of containers through XSD modification commits for project spring-ws, file schema.xsd (Color figure online)

up the removal of another container of the same type, as the current study is mainly based on the number of containers found at each revision of XSD documents. This problem should be addressed in future work, analyzing each modification individually.

Projects have different life cycles, and so have distinct revision quantities and, ultimately, diverse modification rates. This fact is not addressed in the present study, as it only considers the absolute values of the extracted metrics and its percentage portion in total modification count. It does not relate these numbers

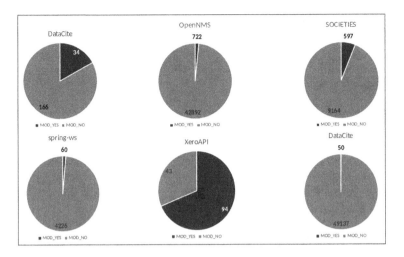

Fig. 6. Amount of commits with XSD modifications per analyzed project (Color figure online)

with the project life cycle, which could hide knowledge not retrieved with this work.

8   Analysis

Over the obtained results presented in Sect. 6, analysis is proceeded by answering the RQs listed in Sect. 1. As so, the answers are as follows.

(RQ1) – The occurrence frequency of changes varies a lot per project and per XSD file. While some projects have more modifications in "element" containers, other ones show this behavior in "complex types". Besides that, different files have distinct modification behaviors, which hampers the task to establish a fixed changing rate for each modification type. Nevertheless, apparently elements modifications are the most frequent and happens several times during projects life cycle.

(RQ2) – Selected projects do not fully share a common behavior regarding changing frequencies, but two groups can be defined: one consisting of four projects in which frequencies of changes follow the sequence `attribute`, `complexType` and `element` (sorted ascending); and another consisting of two projects in which frequencies of changes follow the sequence `complexType`, `element` and `attribute` (sorted ascending). The number of projects selected do not enable the establishment of a unique answer, and an increase in this number might lead to one.

(RQ3) – Three of the six selected projects presents changes in some container type quantities in a frequency not greater than 1 % of total commits. Other two projects show a bit higher values, but do not reach 10 % of total commits, and one project presents results in complete disparity with the others. This reveals that projects that defines XSD documents do not alter these files in a regular basis but to perform many modifications (as seen on Table 2), which sustains the need to deepen the study to understand how and where the changes in XSD files occur, as well as to analyze the impact on the contract semantics.

9 Conclusions

This study aims to evaluate the changes that occur in XSD files used to establish contracts, in order to guide the software development and reduce the impact of modifications over source code. The proposed evaluation tries to understand the behavior of contract changes by quantifying the changes that different types of XSD tags could have, like element, attribute and complex type tags. Regarding this, the present study finds relevant results, but issues are not fully addressed.

It is made clear that modifications in contracts are frequent, but not with the same rate for each selected project. The "element" tag is the most common type of container that experiences greater creation or removal frequencies inside each XSD document, but no pattern could be detected until now. Results shows that while some projects are more likely to have little modifications spread in different revisions, others reveal major changes in lesser commits. Furthermore, there are projects that do not show a significant number of changes in contracts.

The results obtained show that more research needs to be performed, in order to analyze each modification inside XSD documents individually. This can lead to a changing pattern detection, as the present study could not find one using the proposed methodology. Nevertheless, results reveal that changes occur in a regular basis, most of them of specific types and in certain container types, which presents the need for the continuation of this study. This could lead to the detection of changing patterns or at least more accurate knowledge on the behavior of changes in contracts.

10 Future Work

After the results of the presented study, next steps involve dealing with modifications individually for each XSD document and for each project, as well as considering only projects that implements web services contracts. Although the number of selected projects does lead to some valuable inferences, it is not adequate to establish reliable conclusions. As so, new approaches should be considered.

One of the greatest difficulty on selecting a relevant number of projects relies on the fact that GitHub search tool does not provide means to search for projects with the desired characteristics for this research (same as listed at Sect. 3). Thus, at the time this study was undertaken, only few projects were chosen. Despite this, other GitHub search-enabled tool called SearchCode[3] can search over the GitHub database using filters not available natively in it, like searching by code

---

[3] Available at https://searchcode.com.

snippets or particular methods. Using it, a greater number of projects can be selected, considering a search filter that somehow imposes that "must have XSD files in the repository" or "must define WSDL contracts".

Besides selecting and creating a local copy of a relevant amount of projects, code implementation is required to analyze modifications individually. Developed classes and methods should be able to capture modified structures and compare them between consecutive revisions, as well as to check if and count when structure creation or removal results from refactoring or semantic changing. This is the core of an ongoing research following the present study, and data obtained from such analysis can be used to establish answers to new research questions.

As said before, this is an ongoing work. Further steps will try to address the threats listed on Sect. 7 by studying each modification individually, in order to search for semantics modifications and their frequency. With these results, the research will try to detect changing patterns or tendencies in web services contracts. This can lead to a better development of web services by designing them as adaptation-ready systems and reducing the impact of contract changes over source code.

References

[Aba01] Abate, P., Di Cosmo, R., Gesbert, L., Le Fessant, F., Treinen, R., Zacchiroli, S.: Mining component repositories for installability issues. In: Proceedings of the 12th Working Conference on Mining Software Repositories (MSR), pp. 24–33 (2015)

[Ani01] Aniche, M.F., Sokol, F.Z., Gerosa, M.: MetricMiner: supporting researchers in mining software repositories. In: IEEE 13th International Working Conference on Source Code Analysis and Manipulation (SCAM), pp. 142–146 (2013)

[Bai01] Ray, B., Nagappan, M., Bird, C., Nagappan, N., Zimmermann, T.: The uniqueness of changes: characteristics and applications. Microsoft Technical report MSR-TR-2014-149 (2014)

[Car01] Carvalho, L.P., Novais, R., Neto, M.G.M.: VisMinerService: a rest web service for source mining. In: 3rd Workshop on Software Visualization, Evolution, and Maintenance (VEM), pp. 89–96 (2015)

[Con01] Conner, P., Robinson, S.: Service-oriented architecture. US Patent App. 11/388,624 (2007)

[Fra01] França, D.S., Aniche, M., Guerra, E.M.: Como o formato de arquivos XML evolui? Um estudo sobre sua relao com o cdigo-fonte. In: 3rd Workshop on Software Visualization, Evolution, and Maintenance (VEM), pp. 113–120 (2015)

[Gal01] Gala-Pérez, S., Robles, G., González-Barahona, J.M., Herraiz, I.: Intensive metrics for the study of the evolution of open source projects: case studies from apache software foundation projects. In: 10th Working Conference on Mining Software Repositories (MSR), pp. 159–168 (2013)

[Qiu01] Qiu, D., Li, B., Su, Z.: An empirical analysis of the co-evolution of schema and code in database applications. In: Proceedings of the 9th Joint Meeting of the European Software Engineering Conference and the ACM SIGSOFT Symposium on the Foundations of Software Engineering, pp. 125–135 (2013)

[Sam01] Sampaio, C.: SOA e WebServices em Java. Brasport, Rio de Janeiro (2006)

[W3C01] World Wide Web Consortium (W3C): Web Services Glossary. W3C Working Group Note, 11 February 2004

[W3C02] World Wide Web Consortium (W3C): XML Schema. W3C Standard (2004)

Efficient, Scalable and Privacy Preserving Application Attestation in a Multi Stakeholder Scenario

Toqeer Ali[1(✉)], Jawad Ali[2], Tamleek Ali[2], Mohammad Nauman[3], and Shahrulniza Musa[1]

[1] Malaysian Institute of Information Technology,
Universiti Kuala Lumpur, Kuala Lumpur, Malaysia
{toqeer,shahrulniza}@unikl.edu.my
[2] Institute of Management Sciences, Peshawar, Pakistan
jawad2k3@gmail.com, tamleek@imsciences.edu.pk
[3] Max Planck Institute for Software Systems, Kaiserslautern, Germany
mnauman@mpi-sws.org

Abstract. Measurement and reporting of dynamic behavior of a target application is a pressing issue in the Trusted Computing paradigm. Remote attestation is a part of trusted computing, which allows monitoring and verification of a complete operating system or a specific application by a remote party. Several static remote attestation techniques have been proposed in the past but most of the feasible ones are static in nature. However, such techniques cannot cater to dynamic attacks such as the infamous Heartbleed bug. Dynamic attestation offers a solution to this issue but is impractical due to the infeasibility of measurement and reporting of enormous runtime data. To an extent, it is possible to measure and report the dynamic behavior of a single application but not the complete operating system. The contribution of this paper is to provide the design and implementation of a scalable dynamic remote attestation mechanism that can measure and report multiple applications from different stakeholders simultaneously while ensuring privacy of each stakeholder. We have implemented our reference monitor and tested on Linux Kernel. We show through empirical results that this design is high scalable and feasible for a large number of stakeholders.

1 Introduction

Malware detection and its security is a critical concern in todays security research. In this regard, many *Intrusion Detection System* (IDS) techniques have been proposed in the past. Among different IDS techniques, dynamic behavior measurement has at a major advantage [1–3]. The reason is that it does not find a specific pattern in the static executables. Instead it is able to detect 0-Day malware [3] by looking for runtime execution patterns. This ability of dynamic malware detection has pulled the research community towards dynamic behavior-based analysis. However, this approach suffers from several problems that still induce this topic to an open area of research.

© Springer International Publishing Switzerland 2016
O. Gervasi et al. (Eds.): ICCSA 2016, Part IV, LNCS 9789, pp. 407–421, 2016.
DOI: 10.1007/978-3-319-42089-9_29

One major issue with IDSs is that all IDSs either run in user space of operating system or in kernel space. However, recent research shows that there should be a supervision about IDS itself that can give surety that IDS itself is not vulnerable [4]. This is only possible when there is an isolated and shielded hardware module that runs unhindered from the operating system kernel and software stack and keeps track of the security applications. With this need in mind, a hardware root of trust (HOT) concept was introduced by the Trusted Computing Group (TCG) [5]. Another issue is that the IDS runs locally and reports its findings locally. It does not have a capability to report behavior securely to an interested (and authorized) remote computer. For this purpose, TCG introduced the concept of remote attestation.

In remote attestation, a secure measurement agent captures behavior and passes it on to remote party for the purpose of verification. Attestation is divided into static remote attestation and dynamic behavior attestation [6–9]. Static remote attestation techniques take cryptographic hashes of executable from *Core Root of Trust for Measurement* (CRTM) to the application level and store them into *Core Root of Trust for Reporting* (CRTR) [7]. CRTR is typically a coprocessor called *Trusted Platform Module* (TPM) [5]. Once the measurements are complete this system has the capability to securely report stored measurements to a remote party for verification. For more reading on how a static attestation is performed, we refer the reader to the seminal work by Sailer et al. [7].

It has been known since long though, that it is possible to manipulate a benign application into carrying out malicious tasks [10]. As a solution to this problem, dynamic behavior attestation was proposed with the intent to measure and report the *runtime behavior* of the target system for verification. There are some attacks that do not change the code of application, however, they attack system via their vulnerabilities, such as, the recent Heartbleed attack [11]. In that case, standard hash of system does not change. Hence, hash-based attestation techniques cannot tackle with this sort of attacks on the end system. Different dynamic behavior attestation techniques were carried out in past that measure either the behavior of kernel or specific applications [8,12,13]. However, dynamic behavior attestation suffers with the problem of performance. Behavior should be reported first instead of its verification. Previous behavior based techniques focused on measurement and verification instead of an *efficient* measurement and reporting of behavior. This makes all of these techniques infeasible to deploy in practical situations.

One might assert that host based IDS techniques can be adopted straight away for remote attestation. However, there are many distinctions between a host-based IDS and remote attestation. Moreover, remote attestation has its own problems that need to be rectified first to report the behavior of an application to remote party for verification purposes. We list down few dissimilarities so as to clarify this point:

– A traditional host-based IDS runs in the user space while attestation agent must run in kernel space in order to capture the dynamic behavior securely and through a tamper resistant mechanism.

- An IDS runs standalone as software while attestation agent should follow specifications given by the TCG to ensure a chain of trust from an immutable entity. According to the TCG specification, attestation agent should have an un breakable chain-of-trust from a Hardware Root of Trust for Measurement (HRTM) to the attestation agent.
- There is no concept of Static Root of Trust for Measurement (SRTM) and Dynamic Root of Trust for Measurement (DRTM) in an host-based IDS while the attestation agent should have either SRTM or DRTM.
- Host-based IDS captures and verifies behavior locally while in remote attestation, agent measures application and stores that measurement in Hardware Root of Trust for Reporting (HRTR). As per request of the remote party, measurements are reported for verification of behavior.
- Above all, focal point of usual IDS techniques is the detection mechanism while the present-day problems with dynamic behavior is to efficiently measure and report behavior. This problem comes because of the hardware backed security. Since every behavior should be measured and stored in Platform Configuration Register (PCR) that resides in the TPM, a mechanism is required to reduce the behavior and also to detect malicious behavior.

Existing remote attestation techniques deal with a single application's behavior monitoring and reporting. However, for a wide recognition of a technique, the characteristic of scalability is essential. The *behavior* measurements are stored in protected locations of PCR within the TPM. However, the number of PCRs in any TPM are limited. In fact, we cannot assign one separate PCR for each application's behavior since this number is typically limited to 24 PCRs per TPM. Taking care of this limitation of PCRs, [14] proposed an idea to measure multiple applications while storing behavior in a single PCR. Hence, selected many remote attestation techniques that deal with single application's behavior.

The use case they considered in their research is Usage Access Control Model (UCON) [15] and Model Based Behavior Attestation [15]. Finally, they came up with the idea of scaling remote attestation on program execution [8] and remote attestation on function execution [13]. However, according to literature [16] the problem with program execution is that, it is not a feasible technique to measure and report behavior of an application. Program execution measures every system call and stores it into the PCR. Past efforts have shown that behavior generated by an application in program execution is quite high. On this basis we can conclude that the technique is not appropriate for scalability as well.

One highly efficient technique for behavior attestation was given in [16] in which unique behavior of an application is measured and reported based on N-call system call sequences. This technique reduces the measurement and reporting log for verification. This proposed technique is considered to be an optimal solution for behavioral attestation in current scenarios. However, it is not applicable to measure and report the behavior of multiple applications. In fact, they show that it is the limitation of the work that current solution only works for single application.

In this paper we lift this limitation by proposing an extension to the N-Call windows technique [16]. The existing scalability technique is modified from the ground up and adopted for a specific use case [14].

Contributions: The contributions of this paper are as follows,

1. Design and implementation of a scalable and efficient dynamic behavior measurement and reporting architecture. Changes in the current scalability technique to mitigate the privacy issues regarding multiple stakeholders is also included.
2. Design and implementation of a scalable attestation protocol that can attest multiple applications simultaneously.
3. An enhanced algorithm to transform behavior of several applications belonging to different stakeholders into unique behavior log is also provided.

Paper Structure: Next Sect. 2 will provide a brief motivation to the solution given in this paper. In Sect. 3 we will describe some background research in the area. In Sects. 4 and 5 we presented the proposed architecture of the scalable dynamic behavior. Section 6 describes the attestation protocol for the scalable solution, that is, how the attestation is performed of a specific application in a multi-stakeholder environment. Finally, in Sect. 7, short results has been shown that fairly describe the behavior log captured by our solution and the log generated by the previous solutions.

2   Use Case

There are many use cases where organizations require to monitor there own applications running on remote platforms. With the advent of bring your own device, organizations might want to monitor their on applications running on their employees devices. However, this might be possible that an organization *(Org-A)* owns many applications running on end node. Similarly, there is another organization *(Org-B)* and *(Org-C)* having two or more clients running on the same platform (cf. Fig. 1). In that case we need to measure every organization's applications behavior and store them into the PCR accordingly. Apart from that we need to isolate behavior of applications connected to *Org-A* from the behavior of applications associated to *Org-B*. A mechanism is required to monitor dynamic behavior of applications related to different stakeholders while retaining the privacy of each other's applications.

3   Literature Study

Literature study has been performed of remote attestation techniques as well as intrusion detection that depends solely on softwares. IDS methods as mentioned earlier are not appropriate, however, the studies have been performed to find an optimal solution in terms of measurement as well as verification at the remote end.

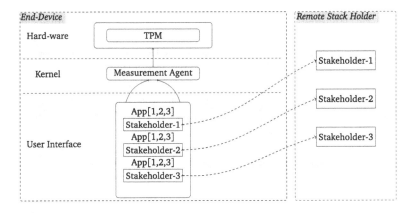

Fig. 1. Motivating use case

Sekar et al. realized the idea of combining strength of misuse detection and anomaly based detection [2]. In this technique, first of all program behavioral specification is developed manually by a number of steps. In first step, a generic specification is formed by grouping system calls according to their similarity. In the second step, the generic specification is further empowered through finding program groups having similar security properties. In the third step, specification is developed more precisely for specific application which are very likely to be attacked frequently i.e. server applications. This application specific specifications are also customized in order to fix various alteration in Operating system. Finally, misuse rules are added to the specification which can be functional to single out features of specific attack. The experimental results on different datasets exhibits that combining anomaly based detection with some misuse rules (for known attacks) can be effective in producing quite short false alarms.

LIN NI et al. proposed an unsupervised clustering method along with simulated annealing algorithm [3]. Unlike supervised machine learning which works on trained label data, unsupervised methods do not need any previous information about the labels of data. In this work, they first divide the data based on similarity into groups of items called clusters. Each cluster gets assigned to a label that classifies whether it is normal or anomalous. After clusters are formed from training data, the system is prepared to detect the intrusions. Due to stochastic nature and better performance of CHAOS, it is used to find high quality optimal solution in terms of efficiency. In experimental phase they consider two metric of performance i.e. detection rate & false positive.

Yunlu Gong et al. [17] uses to combine both the conventional techniques of intrusion detection i.e. misuse & Anomaly. The main purpose of his study is to lessen the false positive rates. To discover data for normal and abnormal behavior class association rule with Genetic network programming is used. Class association rule is basically used to find the relation between different items of data. After finding association rules, they are used to find an effective classifier which can pinpoint the normal, known intrusion and unknown intrusion as well. Thus, the combination of both misuse and anomaly based approaches will ultimately lead to low false positive.

Gideon et al. proposes a new technique for HIDS based on semantic features [18]. The main idea in this technique is to apply semantic features to kernel level system calls in order to achieve full effectiveness in regard of behavior. Apart from this, the techniques based on syntactic analysis required full richness and appropriateness in training for every host. Thus, in case of dynamic nature it will be necessary to retrain the data set periodically. This new method of semantic analysis provided full feature set by analyzing discontinuous system call pattern. The procedure of semantic analysis of system calls is: system calls are represented by 'letters' and combinations of letter are viewed as a 'word'. These words are then transformed to a 'Phrase' by applying CFG (Context Free Grammar). Unlike syntactic methodologies where sometime a retraining is needed for full effectiveness of IDS, the proposed semantic features based approach can easily be transferred between different systems without retraining. However, this technique is based on Extreme Learning Machine (ELM), it requires a longer training period for better accuracy.

Heng Yin et al. proposes a system called Panorama for detecting and analyzing malware such as rootkits, key-loggers and spyware etc. [19]. The procedure for detection is automatic and consists of three steps i.e. test, monitor and analyze. In order to perform an automatic detection and analysis of malware they run a number of tests by test engine. During each test, they generate events that make sensitive information e.g. login or other credentials, in system. This sensitive information is then sent to some trusted application. In parallel, a sample of malware is also running and is not bound for this sensitive data. In the second phase, behavior of sample is monitored by the help of taint engine, which shows influence against sensitive data whether it is affected or not (e.g. send to attacker or any external network). This taint engine uses dynamic taint analysis that monitors propagation of sensitive data inside the whole system to be tested. However, there are some limitations to this approach.

– If the malware is idle and not maliciously behaving while testing.
– Although this approach works in a strong isolation so called an emulated environment but there are possibilities of buffer overflow and integer bugs.

Davide et al. [1] studies about 200 different models on large datasets of system calls, and compares these behavioral models with respect to their effectiveness. The author uses terms atom (system calls), structure (orders of atoms), both are used for behavior specification. They use different set of structures i.e. system calls, system call with arguments etc., and variable number of atoms (1, 2, 3, 5, 10, 20 ...) in order to form specification for behavioral models. And finally they conclude from their experiments that best models are those which depend on few and high-level atoms with their arguments. Because lower number of atoms are more general and have more subsets in terms of behavior which lead to low false alarms and better accuracy. In contrast, higher number of atoms will be more specific in terms of behavior and have more false positive. However, in practical scenario this methodology of finding an optimal approach among different behavioral models is effective, but it required extremely more experimental time and efforts to identify better solution for every particular case.

Tamleek et al. [20] proposed a mechanism for scalable and privacy preserving remote attestation that can measure multiple instances of application running on target platform and get stored in a single PCR through aggregation. The technique stores different application logs in a single system-wide log and also stores their corresponding hashes in single PCR. Now to differentiate the log entries of each application, *app-id* is assigned in order to avoid privacy violation in remote attestation. During remote attestation this technique sends the whole log to challenger, challenger only collects their specific application log based on app-id while hiding the log of other instances or applications.

4 Proposed Architecture

As stated earlier, researchers [14] gave a solution for scalability of applications and virtual machines. However, there are few shortcomings in their work, it needs enhancement. Firstly, the solution given for scalability of dynamic behavior techniques are CPU intensive and it is not feasible to implement. A dynamic behavior attestation technique is required that is appropriate for scalability. Secondly, solution given by them for scalability and privacy is based on limitation of PCRs in TCG's specification 1.2. The number of PCRs are 16 in 1.2 specification while in the new specification, that is, 2.0 given number of PCRs is 24. Based on the aforementioned updated dynamic behavior techniques and technologies, we extended work of [14] and we are proposing a new solution for scalable dynamic behavior attestation.

However, this technique addresses the problem of scarcity in PCR's i.e. single PCR for all stakeholders. In our proposed technique we change the mechanism for scalability by using multiple PCR's for multiple stakeholders. For instance, if there are many stakeholders and each one is running multiple applications, so separate PCR and log are used for each stakeholder. In this case no privacy violation will be occurred during remote attestation, because each PCR has information about a single stake-holder's application(s). As has been discussed the number of PCRs are enhanced in TPM 2.0. It will be a worst case scenario or not possible even, that number of stakeholders are greater than available PCR's. It is possible that a stakeholder is having multiple number of applications running on in parallel.

According to TPM 2.0 [5] specification there are total 24 PCRs. The first 8 PCRs are reserved for initial boot-up such as BIOS, UEFI drivers, MBR etc. The second 8 registers are defined for the use of static OS. PCR [14] is specially reserved for debugging purposes and is not intended in any sealing or attestation because it will not be recorded in event log. The remaining PCRs from 17 to 23 are resettable and can be used any for dynamic kernel module monitoring or its applications.

The size of stored value inside PCRs is determined by the size of a hashing algorithm or digest i.e. SHA-1. Multiple PCRs having association with the same digest (hashing algorithm) is known as PCR bank, and each of PCR in PCR bank is accessed through index - the PCR index. If we have two banks active i.e. SHA-1 and SHA-256, so there is PCR[0] or any other PCR for SHA-1 and PCR[0] for SHA-256. In TPM 2.0 specs there should be at least one PCR bank over 24 PCRs. One of the important operations used with TPM is *pcr_extend* operation. This operation is used during remote attestation. This way the function appends new hash value obtained from hashing algorithm with current value of PCR. Similarly a number of keys are used for different tasks i.e. SRK (Storage Root Key), EK (Endorsement key) and AIK (Attestation Identity Key).

As from the literature study, we have not found out a client node that is running more than four to five stakeholder's applications. However, it is possible that many applications running by a single stakeholder. Keeping this scenario in mind we have designed the architecture that one PCR is allocated to one stakeholder. That means if one stakeholder is having five applications, all behaviors associated to those applications will be stored in that particular PCR. However, the work been done in [14] is that they use single system-wide log for all stakeholders and stored the aggregated measurement in a single PCR. This System-wide log contains all the measurements entries of all the stakeholders applications. In our proposed solution we change the methodology of logging and reporting. We defined a reference monitor that can store a separate Stored Measurement Log (SML) for every stakeholder's application into a single PCR and also aggregate the hash by *pcr_extend* operations in one of PCRs inside TPM. For instance, if we have 3 stakeholders and each runs multiple applications as exhibited in Fig. 2, our reference-monitor records the behavior of each stakeholder's application(s) and make a separate SML file. We specify a separate PCR for storage of each stakeholder's measurement in reference monitor. In Fig. 2 we have shown that stakeholder 1, 2, 3 uses PCR 11, 12, 13 respectively.

At runtime of applications the reference monitor is responsible to distinguish each application from another and identify all applications according to their stakeholders. Afterwards, the applications log details will be properly placed in SML files. As our study depends on application behavior, the author took the system calls as behavior parameter because it is most popular unit of measurement. We specify the reference monitor in kernel space in order to directly interact with operating system.

In our previous work [16], that provided a framework for dynamic behavior attestation based on system calls generated by the application. They developed a behavior monitor in the kernel space that measures a single application and transforms the behavior into sliding windows. When a window is generated it is measured and appended in the PCR. Apart from that the behavior log is also stored in the Stored Measurement Log (SML). The advantage of this type of measurement is that once the behavior is measured it will not be measured again in its life time. However, the drawback of this technique is that it only monitors a single application. In the following section we will elaborate the mechanism for scalable dynamic behavior monitor that measures multiple applications related to one organization. Further, it also monitors and reports the behavior of multiple applications related to multi organization.

As we discussed about the sliding window technique used for remote attestation for single application. We use the same technique for multiple case. Now when multiple applications belong to different stakeholders run on the system, the reference monitor records the behavior (system-calls) of each application and stores the behavior measurement in the form of windows in Log files (SML). Windows are basically a pattern of system calls of fix size. Each SML file contains the behavior details about the application(s) of same stakeholders as shown in Fig. 2.

5   Scalable Dynamic Behavior

The proposed architecture first creates windows of system calls from multiple applications on the same platform. These windows of different stakeholder's applications are stored in log files. The behavior monitor records and measures internal activities and extends the measurements (hashes) within the PCR's. Our behavior monitor can take measurement for multiple applications within and across an organization at the same time. It also maintains a separate log for each stakeholder application(s) and stored the aggregated measurement of log file in a single specified PCR. During an attestation response the log file along with PCR value will be sent to challenger.

5.1   Scalable Measurement

When an application runs, it generates various system calls in order to perform different tasks. For example if an application is needed to open, it will

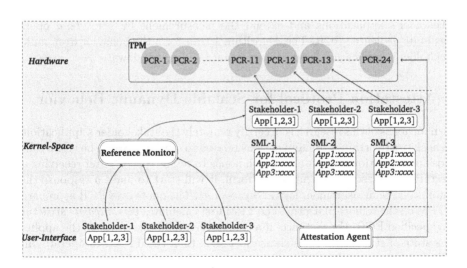

Fig. 2. Proposed scalable behavior measurement architecture

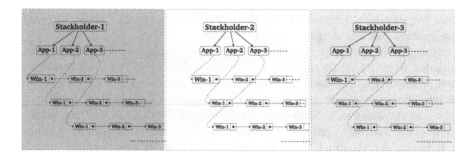

Fig. 3. Link lists representation for system call windows

generate *open* system call. There are numerous system calls for different operations in operating system i.e. *read*, *nmap*, *close* etc. In our reference monitor, we specify the algorithm to identify system calls for each application while multiple applications running on system simultaneously. Each system call is identified by its arguments such as file-access, user-access, Process-Id etc. This is the responsibility of reference monitor to check every system call in respect of Process-id and associate system calls with its required application. During monitoring system calls for applications identification, the system calls are stored in the form of windows. Every application windows is represented in the form of link-list. There is a separate link-list representation for each application windows i.e. if there are three applications so three link-lists will be formed as shown in Fig. 4. The declaration of applications with respect to its stakeholders will be specified explicitly in reference monitor e.g. *mozilla-firefox* is associated with stakeholder-1, send-mail is associated with stakeholder-2 and so on for more applications. Furthermore, as we discussed that separate *PCR* is used for each stakeholder, so reference monitor will make a separate *stored measurement log* SML for each stakeholder's applications and extend the measurement by *pcr_extend* operations in its dedicated PCR. The Algorithm 1 transform the behavior from multi-stackholder's applications and convert in into sliding windows.

6   Attestation Protocol for Scalable Dynamic Behavior

The main focus of this research is to certify remotely the stakeholder's applications. To check trustworthiness of applications related to any corporate, the verifier first sends an attestation request along with nonce to remote party. After receiving an attestation request on a required platform, it will start to make a response that contains all the measurement logs and *pcr-quote*. The quote over PCR is prepared by TPM on the remote platform. The TPM first calculates *pcr-composite* structure on a specified PCR. For instance, if a challenger would like to verify the applications state of stakeholder-1 as shown in Fig. 2, then the TPM will operate only on PCR-12, because PCR-11 is designated only for stakeholder-1 application(s). This way we will not face any privacy violation among different stakeholders.

Algorithm 1. Transform Multiple Stackholder's Application Behavior into Sliding Windows

---

Input : LSM Hooks Associated with Different Applications
$Output : None$
$retrieve\_program\_name(current)$
$select\_associated\_stkholderid(prog\_name)$
$select\_associated\_AppID(prog\_name)$
$get\_prog\_win\_stkholderid\_AppID$
if prog_win_stkholderid_AppID == Null **then**
    $create\_window(prog\_name\_stkholderid\_AppID)$
end if
if $prog\_win\_stkholderid\_AppID! = NULL$ **then**
    if $prog\_win\_stkholderid\_AppID.counter < num\_win\_size - 1$ **then**
        $prog\_win\_stkholderid\_AppID.temp\_window[prog\_win\_stkholderid\_AppID.counter] = num$
        $(prog\_win\_stkholderid\_AppID) + +$
    end if
end if
$prog\_win\_stkholderid\_AppID.temp\_window[prog\_win\_stkholderid\_AppID.counter] = num$
if should_add_window(prog_win_stkholderid_AppID) == 1 **then**
    $inst = (structhg\_inst)kmalloc(sizeof(structhg\_inst), gfp\_kernel);$
    for $i < num\_win\_size$ **do**
        inst.call_val[i] = prog_win_stkholderid_AppID.temp_window[i];
    end for
    INIT_LIST_HEAD(&(inst.list_stkholderid_AppID));
    $list\_add\_tail(\&(inst.list\_stkholderid_{A}ppID)), \&(prog\_win\_stkholderid\_appID.windows);$
    $get\_pcr\_stackholderid();$
    $extend\_prog\_win\_in\_pcr(prog\_win\_stkholderid\_AppID);$
    for $i < num\_win\_size - 1$ **do**
        $prog\_win\_stkholderid\_AppID.temp\_window[i] = prog\_win\_stkholderid\_AppID.temp\_window[i + 1];$
    end for
    $prog\_win\_stkholderid\_AppID.counter = num\_win\_size - 1;$
end if

Afterwards, TPM signs the *pcr-composite* by *attestation identity key* (AIK). Finally, it sends *pcr-composite* in attestation response to the challenger that contains *tpm-quote* along with the behavior log.

Generally, remote attestation is done by two side communication between target and challenger. It is required to define a protocol for secure and trusted attestation. Our protocol is mainly focused on reporting the scalable dynamic behavior while keeping privacy of stakeholders. In paper [14], they used a single system-wide measurement log for applications in which privacy issue is considerable for applications. In our proposed technique for scalability we measured and stored logs for each stakeholder application(s) separately as depicted in Fig. 2.

Now When a challenger sends a verification request for applications of a specific stakeholder, it will send an attestation request. The stakeholder sends all its credentials i.e. *pcr-value* and SML. The challenger first recalculates and extends SML hashes in same order and matches with final *pcr-value* received from stakeholder in order to verify the log integrity. If both the values matches, then log is not changed during communication and considered to be trusted. After successful log integrity, behavior will further classify for normal and malicious through machine learning model. We trained the model by different datasets of application for creating normal and malicious datasets. For further reading about machine learning as how it will be classified, we refer the reader to [21].

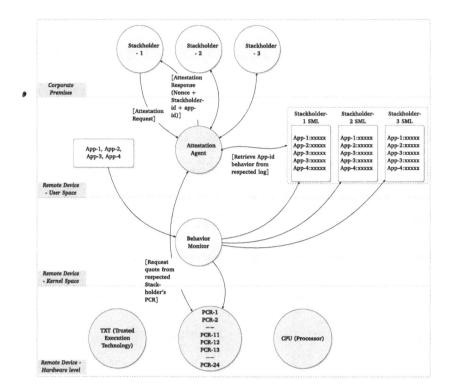

Fig. 4. Scalable dynamic behavior attestation of applications owned by multi stakeholders

7   Results and Discussion

The focus of this research is to show the feasibility of our scalable proposed architecture in real scenarios. Therefore, we implement our reference monitor for scalability in open-source operating system (Linux) and analyze five different applications associated with different stakeholders in parallel. *AMIDE* (Medical Imaging Software) and *FreeMedicalProjectsForms* are two applications belonging to Stakeholder's-1. *Firefox* (web-browser) and *Thunderbird* (email) are associated with stakeholder's-2 and finally *GNUCash* (Accounting Management) is related to stakeholder's-3. For experimental purposes we performed a number of measurements on different windows sizes. i.e. 5, 7 and 10. The variations of these windows sizes will help us to show the trade-off between performance and accuracy of application's behavior.

All the applications were monitored for 15 min with an interval of 5 min. The information about each application with respect to its stakeholder is defined in reference monitor explicitly. At runtime of all the applications reference monitor will take measurement of each application and extend into the dedicated PCR. As we discussed earlier that a separate PCR is used for each stakeholder.

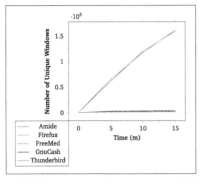
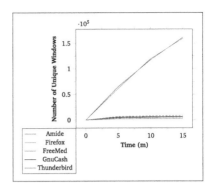
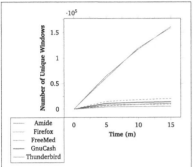

Fig. 5. Measurement comparison of variant of applications between a normal system-call log and the unique behavior log

For instance, stakeholder-1 has two applications i.e. AMIDE & FreeMedicalProjects, it will be extended in PCR-11. Similarly for stakeholder-2 and stakeholder-3 we define PCR-12 and PCR-13 respectively. A log file for each stakeholder applications is also stored in the form of measurement. During the experiments our reference monitor is found to be able to manage three different log file. Furthermore, the graphs show a couple of unique windows for multiple applications at same time. Finally we conclude that the scalability in terms of using separate PCR and separate Log for each stakeholder is possible.

8 Conclusion

Remote attestation is an important and useful technology that can be utilized by corporates to verify health of their own applications running on client devices. Sometime these client devices running multiple applications belongs to different corporates. A generic solution given by TCG to use single SML and single PCR for either complete *integrity* status of the complete operating system or the *dynamic* behavior of a single application. There is no solution given that when there are several applications running and belongs to different stakeholders. In this paper, we designed and implemented a security module in the Linux

Kernel that can achieve the said goal. Our attestation module can measure and report the target application's behavior to the challenger for verification. On top of that the algorithm is designed to store corporates own *application(s)* in a single PCR and SML that ultimately mitigate the privacy issues related to corporates behavior log.

References

1. Canali, D., Lanzi, A., Balzarotti, D., Kruegel, C., Christodorescu, M., Kirda, E.: A quantitative study of accuracy in system call-based malware detection. In: Proceedings of the 2012 International Symposium on Software Testing and Analysis, pp. 122–132. ACM (2012)
2. Uppuluri, P., Sekar, R.: Experiences with specification-based intrusion detection. In: Lee, W., Mé, L., Wespi, A. (eds.) RAID 2001. LNCS, vol. 2212, pp. 172–189. Springer, Heidelberg (2001)
3. Ni, L., Zheng, H.-Y.: An unsupervised intrusion detection method combined clustering with chaos simulated annealing. In: 2007 International Conference on Machine Learning and Cybernetics, vol. 6, pp. 3217–3222. IEEE (2007)
4. Milenković, M., Milenković, A., Jovanov, E.: Hardware support forcode integrity in embedded processors. In: Proceedings of the 2005 International Conference on Compilers, Architectures and Synthesis for Embedded Systems, pp. 55–65. ACM (2005)
5. Trusting Computing Group (2014). http://www.trustedcomputinggroup.org/. Accessed 17 Dec 2015
6. Coker, G., Guttman, J., Loscocco, P., Herzog, A., Millen, J., OHanlon, B., Ramsdell, J., Segall, A., Sheehy, J., Sniffen, B.: Principles of remote attestation. Int. J. Inform. Secur. **10**(2), 63–81 (2011). http://dx.doi.org/10.1007/s10207-011-0124-7
7. Sailer, R., Zhang, X., Jaeger, T., Van Doorn, L.: Design and implementation of a TCG-based integrity measurement architecture
8. Gu, L., Ding, X., Deng, R.H., Xie, B., Mei, H.: Remote attestation on program execution. In: Proceedings of the 3rd ACM Workshop on Scalable Trusted computing, ser. STC 2008, pp. 11–20. ACM, New York (2008). http://doi.acm.org/10.1145/1456455.1456458
9. Kil, C., Sezer, E.C., Azab, A.M., Ning, P., Zhang, X.: Remote attestation to dynamic system properties: towards providing complete system integrity evidence. In: IEEE/IFIP International Conference on Dependable Systems & Networks, DSN 2009, pp. 115–124. IEEE (2009)
10. Prandini, M., Ramilli, M.: Return-oriented programming. IEEE Secur. Priv. **10**(6), 84–87 (2012)
11. Durumeric, Z., Kasten, J., Adrian, D., Halderman, J.A., Bailey, M., Li, F., Weaver, N., Amann, J., Beekman, J., Payer, M., et al.: The matter of heartbleed. In: Proceedings of the 2014 Conference on Internet Measurement Conference, pp. 475–488. ACM (2014)
12. Loscocco, P.A., Wilson, P.W., Pendergrass, J.A., McDonell, C.D.: Linuxkernel integrity measurement using contextual inspection. In: Proceedings of the 2007 ACM Workshop on Scalable Trusted Computing, ser. STC 2007, pp. 21–29. ACM, New York (2007). http://doi.acm.org/10.1145/1314354.1314362
13. Liang, G., Ding, X., Deng, R.H., Xie, B., Mei, H.: Remote attestation on function execution (2009)

14. Tanveer, T.A., Alam, M., Nauman, M.: Scalable remote attestation with privacy protection. In: Chen, L., Yung, M. (eds.) INTRUST 2009. LNCS, vol. 6163, pp. 73–87. Springer, Heidelberg (2010)

15. Alam, M., Zhang, X., Nauman, M., Ali, T., Seifert, J.-P.: Model-basedbehavioral attestation. In: Proceedings of the 13th ACM Symposium on Access Control Models and Technologies, pp. 175–184. ACM (2008)

16. Ismail, R., Syed, T.A., Musa, S.: Design and implementation of an efficient framework for behaviour attestation using n-call slides. In: Proceedings of the 8th International Conference on Ubiquitous Information Management and Communication, p. 36. ACM (2014)

17. Gong, Y., Mabu, S., Chen, C., Wang, Y., Hirasawa, K.: Intrusion detection system combining misuse detection and anomaly detection using genetic network programming. In: ICCAS-SICE 2009, pp. 3463–3467. IEEE (2009)

18. Creech, G., Hu, J.: A semantic approach to host-based intrusion detection systems using contiguous and discontiguous system call patterns. IEEE Trans. Comput. **63**(4), 807–819 (2014)

19. Yin, H., Song, D., Egele, M., Kruegel, C., Kirda, E.: Panorama: capturing system-wide information flow for malware detection and analysis. In: Proceedings of the 14th ACM Conference on Computer and Communications Security, pp. 116–127. ACM (2007)

20. Ali, T., Alam, M., Nauman, M., Ali, T., Ali, M., Anwar, S.: A scalable andprivacy preserving remote attestation mechanism. Inform. Int. Interdisc. J. **14**(4), 1193–1203 (2011)

21. Ismail, R., Syed, T.A., Musa, S.: Design and implementation of an efficient framework for behaviour attestation using n-call slides. In: Proceedings of the 8th International Conference on Ubiquitous Information Management and Communication, ser. ICUIMC 2014, pp. 36:1–36:8. ACM, New York (2014). http://doi.acm.org/10.1145/2557977.2558002

An Approach for Code Annotation Validation with Metadata Location Transparency

José Lázaro de Siqueira Jr.[1]([⊠]), Fábio Fagundes Silveira[1],
and Eduardo Martins Guerra[2]

[1] Federal University of São Paulo – UNIFESP, São José dos Campos, Brazil
joselazarosiqueira@gmail.com, fsilveira@unifesp.br
[2] National Institute for Space Research, São José dos Campos, Brazil
guerraem@gmail.com

Abstract. The use of metadata in software development, specially by code annotations, has emerged to complement some limitations of object-oriented programming. A recent study revealed that a lack of validation on the configured metadata can lead to bugs hard to identify and correct. There are approaches to optimize metadata configuration that add the annotation out of the target code element, such as its definition on the enclosing code element or indirectly inside other annotations. Annotation validation rules that rely on the presence of other annotations are specially hard to perform when it is possible to configure it out of the target element. Available approaches for annotation validation in the literature consider their presence only in the target element. This paper presents a validation of code annotations approach in object-oriented software with location transparency, whereas definitions can occur in different parts of source code related to the target element. An evaluation with a *meta-framework* supports our hypothesis that the approach is capable of decoupling the annotation location from the validation rules.

Keywords: Java · Code annotation · Metadata · Validation · Framework

1 Introduction

Object-orientation has several powerful features, which allow us to model an application to increase reuse enabling development of good quality code. However, there are situations where object-orientation has limitations in its application. For example, it can be hard reusing two similar codes in their functionality, considering that both perform the invocation of a method in a class being encapsulated in the middle of their execution. Thus, the reuse of these components is complicated, as it can implement a different interface, even though the functionality is similar. In this scenario, the use of reusable components in Java with reflection and annotations, for instance, becomes an interesting option to address the limitations of object-oriented programming. Code Annotations are

© Springer International Publishing Switzerland 2016
O. Gervasi et al. (Eds.): ICCSA 2016, Part IV, LNCS 9789, pp. 422–438, 2016.
DOI: 10.1007/978-3-319-42089-9_30

a feature introduced in the Java 5.0 to enable the addition of custom class metadata directly in the source code [1]. Consequently, software that adopted this new feature explores a relatively little-investigated territory.

Frameworks that process their logic based on the metadata of the classes whose instances they are working with are called metadata-based frameworks [2]. They can potentially provide a solution in which the developer can add metadata to the existent classes intuitively. According to Guerra e Fernandes [3], the frameworks, in general, regardless traditional or metadata-based, can bring benefits to the application design, such as a higher reuse level, the coupling reduction, and the increase of productivity of the team, aiming to be more suitably adapted to the application needs.

In the current literature, to the best of our knowledge, there are only three frameworks available for code annotation validation on metadata-based frameworks [4–6]. However, such approaches consider the code annotation configuration only in the target element. Code annotation validation rules that rely on the presence of other annotations are pretty hard to perform [3], especially when it is possible to configure it out of the target element. Furthermore, the use of code annotations practices, such as General Configuration or Annotation Mapping [7], increases the complexity of the validation process due to annotations that can be configured at different levels and locations of the application, such as in packages, classes, methods, attributes, or even into other annotations. Metadata configuration errors, such as a missing property or a misspelled string are not trivial to detect and to find this kind of bugs might be a bottleneck in the team productivity. In this concern, these evidences make metadata validation an important feature for metadata-based frameworks. Errors or warning messages should be designed to help the developer to find a misconfiguration. As a contextualization, Sect. 2.3 reports distinct forms of code annotation configuration in a factual application as a running example.

The aim of this paper is to present an approach for code annotation validation with location transparency on object-oriented software. We have developed a *meta-framework* for code annotation validation that uses a domain specific language (DSL) to evaluate and validate the approach of code annotations configuration usage. Analyses with the *meta-framework* support that our assumption is able to decouple the annotation location from the validation rules. Since code annotations are widely used in mature frameworks (e.g. Hibernate[1], EJB 3[2], and Struts 2[3]) of the software industry, we believe that our tool can be useful for Java developers who use such metadata resources.

2   Metadata Definition

Metadata is an overloaded term in computer science and can be interpreted differently according to the context. In the context of OOP, metadata is information

---

[1] http://www.jcp.org/en/jsr/detail?id=299.
[2] http://www.jcp.org/en/jsr/detail?id=220.
[3] http://www.jcp.org/en/jsr/detail?id=303.

about the program structure itself, such as classes, methods, and attributes. A class, for example, has intrinsic metadata like its name, superclass, interfaces, methods, and attributes. Metadata consumed by frameworks can be defined in different ways [8]: stored in external sources (e.g., XML files and databases), using standard names or in code annotations, which is supported by some programming languages like Java [1] and C# [9]. Some development communities, many recent frameworks developed and APIs are increasingly using metadata resources, such as Hibernate, EJB 3 Struts 2 and JAXB[4]. Most of them define metadata by code annotations, which is the focus of this paper.

2.1 Annotations

Annotation is one of the most interesting features of the Java language, introduced in JDK version 5, which allows metadata insertion directly into the application source code [1]. Such feature allows keeping the source code and metadata in the same place, enhancing application maintenance since there is no need of external references. This resource of the Java language is also offered in other programming languages, such as C# [9], which is called *attributes*. Python language[5] has a similar feature, called *decorator*, which is also used for adding metadata directly into the source code.

Nowadays, much of the frameworks and APIs for Java use annotations. This language feature has been popularizing the use of metadata-based frameworks, through enhancing the definition of metadata to its users, making the use of this approach more feasible.

2.2 Code Annotations Practices

According to Guerra et al. [7], idioms for Code Annotations Practices are a specific language for the usage of annotations in the Java language. These idioms document practices that indicate recurrent solutions used to structure and use code annotations. These practices can help developers of frameworks and components in design metadata definition schema, creating mechanisms that help reducing redundancy and duplication on metadata definition.

General Configuration proposes the use of an annotation that defines a standard configuration for its enclosing code elements. For instance, by using this idiom, the use of an annotation in a class definition configures a default metadata for all its methods and/or its attributes. A single method or attribute can be annotated as well to override the general configuration [7]. It is known that JAX-RS 1.1 and EJB 3 API use this idiom.

Another useful idiom is the Annotation Mapping [7] that proposes creation of a domain-specific annotation that can intermediate the metadata definitions from the framework annotations. This mapping can be dynamic or static, where components search at run-time or compile-time for their annotations inside other

---

[4] http://www.jcp.org/en/jsr/detail?id=311.
[5] https://www.python.org.

annotations of the code element. It is known that the API for Contexts and Dependency Injection - CDI for Java EE 6 and the Bean Validation API use this idiom.

The application of General Configuration and Annotation Mapping practices for code annotations increases the complexity of the validation process since annotations can be configured in different levels and locations of the application, such as in methods, classes, packages or even into another annotation. Moreover, it is difficult when the annotation validation rules rely on the presence of other annotations [3]. Available approaches for code annotation validation in the literature, as suggested by Noguera and Pawlak [4], consider the annotation configuration presence only at the target element and does not support scenarios that use these practices for annotation configuration. According to the previously mentioned examples, we can see that such practices for code annotations are widely used in frameworks of the software industry.

2.3 Motivating Example

In this Section, we describe a motivating example whose goal is to illustrate a scenario for metadata validation, and how this task can be hard when some alternatives for annotation validation are used. To identify which methods are transactional and which ones the execution should be registered, the system uses respectively the annotations *@Transaction* and *@Logging*. Since the registration of the method execution in the system log is required, every method with the *@Logging* annotation should also be associated with a *@Transaction* annotation. Figure 1 (class *OrderProcessing*) presents a code example that illustrates scenarios with valid and invalid annotation configurations.

```
public class OrderProcessing{
  //valid due it has both
      annotations
  @Transaction @Logging
  public void register(){
  //method implementation
  }

  //Invalid due @Logging needs
      to have @Transaction
  @Logging
  public void changeAdress(
      Purchase p, Address a){
  //method implementation
  }
}
```

```
@Transaction
public class
      SaleProcessing{
  //valid due
      @Transaction is
      defined on the
      class
  @Logging
  public void register
      (){
  //method
      implementation
  }
}
```

```
public class
      UserManagement{
  //valid due
      @Transaction is
      defined inside
      @Administration
  @Logging
      @Administration
  public void blockUser(
      Purchase p){
  //method
      implementation
  }
}
```

Fig. 1. Scenarios with valid and invalid annotation configurations.

However, the annotation *@Transaction* implements the patterns General Configuration and Annotation Mapping, meaning that a method can be considered transactional without having a *@Transaction* directly configured on it. The General Annotation pattern indicates that an annotation configured in the class definition can be considered as a metadata defined for all methods. Figure 1 (class *SaleProcessing*) presents a valid example where *@Logging* is defined in a method without *@Transaction* directly on it. According to the Annotation Mapping pattern, the annotation *@Transaction* could be defined inside another annotation. The goal of such practice is to define groups of configurations that make sense together in the application domain.

The use of Annotation Mapping creates a way to indirectly define metadata to a code element. In the example of Fig. 1 (class *UserManagement*), the definition of the annotation *@Logging* should be valid because the method is indirectly configured as transactional by the annotation *@Administration*, which has the annotation *@Transaction* inside of it.

As highlighted by these examples, there are practices that annotations can be used to configure metadata to a code element, without being defined directly on it. These scenarios make difficult the annotation validation rules that rely on the presence of other annotations [3]. All available approaches for annotation validation in the literature, like the one proposed by Noguera and Pawlak [4], consider their presence only in the target element and do not support such scenarios.

3   The Metadata Validation Approach

A recent study revealed that a lack of validation on the configured metadata can lead to bugs hard to identify and correct [3]. There are different approaches, as shown in Sect. 2.2, to optimize metadata definition, called code annotations practices. In such practices, we can add the annotation out of the target code element, such as on the enclosing code element or indirectly inside of other annotations. Annotation validation rules that rely on the presence of other annotations are specially hard to verify when it is possible to configure it out of the target element [3]. As already stated, available approaches for annotation validation in the literature (e.g. [4]) only consider their presence in the target elements.

The goal of this section is to present an approach for annotation validation that supports metadata definitions out of the target element, which also have the constraints definition and implementation independent from the metadata location. For instance, a validation rule that verifies a mandatory presence of other annotation, should not depend on how this other annotation can be located, such as on enclosing elements or inside other annotations. This approach is language independent, and can be implemented in any language that implements a code annotation feature.

Each metadata type represented by an annotation should receive metadata information about the approaches that can be used for its definition and what are the constraints for its valid configuration.

For annotation without metadata about how it can be defined, it is assumed that it can only be defined directly on the target element. Similarly, for annotation without metadata about definition constraints, it is assumed that there are no restrictions on its usage, and it is always considered a valid configuration.

If an annotation can be defined on other elements, it should receive a metadata that configures that. This metadata about the annotation location should be associated with a class that contains the logic that implements such searching. That meta-annotation can contain parameters to be considered by this class, such as a search depth or an exclusion pattern. Based on that structure, new locator's types can be added by creating a new annotation and associating it to a new class that performs its searching logic.

A similar structure is used to define the annotation validation constraints. The annotation receives constraint annotations that define validation rules and each constraint annotation is associated with a class that contains the validation logic. This structure allows extensibility since new rules can be easily introduced by adding an annotation and associating it with a new class.

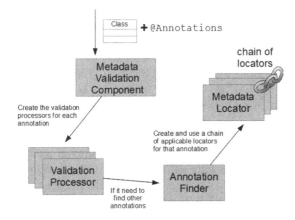

Fig. 2. Annotation validation component.

A representation for a metadata validation component is presented on Fig. 2. A component for metadata validation should receive a class with annotations as a parameter and verify if all the annotations configured on it obey their configured constraints. Based on the metadata structure, the component should go through all code elements, retrieving all its annotations and its respective validation processors associated with the configured validation annotations.

If a validation rule needs to search for other annotation, for instance, to verify if another annotation is present or not, it should use another component responsible for finding annotations. This component should inspect the annotation to search for metadata that configures locators to it. Based on that information, a chain of metadata locators applicable to that annotation will be built and used to locate and return it to the validation rule logic. Considering the described process, the annotation will be searched only in places that its metadata configures that it can be defined.

Based on this proposed solution, it is possible to achieve the desirable property for metadata validation: to define the validation rules independent from the approach for annotation definition. The component for annotation finding removes from the validation rule the responsibility to know where other annotations are defined. That is the main point that differentiates our approach from others, such as Aval [4] where annotations are searched directly on the validation rules implementation. This solution can be used either by components used by frameworks that verify at runtime the validity of the annotations and by developer tools that highlights the annotation definition errors on the code editor. The next section presents an implementation of this approach concepts, named Esfinge$_{\text{METADATA}}$, which aims to be used to evaluate the practicability of the proposed approach.

4   Esfinge$_{\text{METADATA}}$

The Esfinge$_{\text{METADATA}}$[6] is an extensible *meta-framework* to read and validate metadata. The term "meta-framework" is used because the Esfinge$_{\text{METADATA}}$ works with the concept of meta-annotations, aiming to simplify the implementation of metadata-based frameworks. This *meta-framework*, which can also be used on applications that use custom annotations, is an implementation of the proposed approach presented in the previous section. Esfinge$_{\text{METADATA}}$ is used at runtime to evaluate if the annotations configured in a given class obey their validation constraints. We choose to focus on runtime validation, since compile time validation depends on the development environment and validation at runtime is always a good practice for safety, independent whether it was previously validated before the software execution. As a result, Esfinge$_{\text{METADATA}}$ works only with annotations set to *RUNTIME* retention type, i.e. the annotations available to be retrieved at run-time. However, it is important to highlight that the proposed approach can also be implemented to validate annotations at compile or loading time.

In the current implementation, the framework has two modules: Location and Evaluation. Aiming to clarify the concepts pointed out up to here, the following subsections give a general view of it, its internal structure and adopted location strategies.

4.1   General View

This Subsection uses a running example to show a general view of how Esfinge$_{\text{METADATA}}$ can be used to validate annotations. This example illustrates how to create a validation annotation with its respective validator, how the annotations should be used to perform the validation rules with location transparency, and finally how the validation functionality is invoked.

In the example, the validation annotation specification is defined by *@NeedsToHave* annotation (Fig. 3), which configures that one annotation requires another one related to the same element. In the *@NeedsToHave* configuration annotation is added the *@ToValidate* annotation to specify the responsible class to check the validation rules of annotation usage.

---

[6] https://github.com/EsfingeFramework/metadata.

```
1  //@NeedsToHave Annotation Specification
2  @ToValidate(validationClass=
        NeedToHaveAnnotationValidator.class)        1  //Usage of @NeedsToHave by @Logging
3  @Retention(RetentionPolicy.RUNTIME)               2  @NeedsToHave(Transaction.class)
4  @Target(ElementType.ANNOTATION_TYPE)              3  @Retention(RetentionPolicy.RUNTIME)
5  public @interface NeedsToHave {                    4  @Target(ElementType.METHOD)
6      public Class<? extends Annotation> value();    5  public @interface Logging {
7  }                                                  6  }
```

Fig. 3. *@NeedsToHave* annotation specification and usage.

The implementation of the class *NeedToHaveAnnotationValidator* needs to implement the interface *AnnotationValidator*, which defines the method *initialize()* that is responsible to initialize the instance with annotation parameters and *validate()* that contains the validation logic. This class uses the *Annotation-Finder* class to search for other annotation, it will verify the location metadata from this annotation and search on all places that it can be defined.

As the next step, annotations that need to be validated receive the validation and location annotations. For instance, in Fig. 3 (its usage), the annotation *@Logging* uses the annotation *@NeedsToHave* to configure that it is mandatory to have the *@Transaction* annotation configured for the same element.

Finally, to invoke the Esfinge_METADATA validation functionality it is necessary to call the static method *validateMetadataOn()* from *MetadataValidator* class, passing as a parameter the class whose annotations need to be validated. It is recommended for a framework to invoke this functionality right before processing the class annotations, because it can help to ensure that all validation constraints are implemented by the code.

4.2 Domain Specific Language of the *Meta-framework*

The development of frameworks based on metadata, in general, uses a Domain Specific Language (DSL) to configure and specify its components. In this respect, the metadata validation is performed by meta-annotations of validation, which have a specific classification according to the domain to be validated. Table 1 summarizes the meta-annotations that comprise the Esfinge_METADATA DSL and in which application elements these metadata can be applied. They are classified into three types: validation, configuration of validation and location.

Metadata validators presented in Table 1 cover many validation needs. However, there are cases where it is difficult or even impossible to translate a particular domain rule on generic validators. For these cases, the Esfinge_METADATA provides extensibility for the inclusion of new validators to be implemented by the developer, for a particular application domain. New validators necessarily require: (i) a new meta-annotation validation; and (ii) its corresponding implementation, which may optionally extend some available validator. Validators must be annotated with meta-annotations *@ToValidate* or *@ToValidateProperty*, aiming to indicate its corresponding implementation of validation. The implementation of a validator is a class that implements an interface of validation: AnnotationValidator or AnnotationPropertyValidator, validation of structure and property of annotations, respectively.

Table 1. Meta-annotations of Esfinge_METADATA DSL

Meta-annotations	Applicable elements	Validation type
@Unique	Attribute of Annotation	Property
@NotNull	Attribute of Annotation	Property
@MinValue	Attribute of Annotation	Property
@MaxValue	Attribute of Annotation	Property
@RefersTo	Attribute of Annotation	Property
@Prohibits	Annotation	Context
@NeedsToHave	Annotation	Context
@ToValidate	Meta-Annotation	Configuration of Validation
@ToValidateProperty	Meta-Annotation	Configuration of Validation
@SearchOnEnclosingElements	Annotation	Configuration of Location
@SearchInsideAnnotations	Annotation	Configuration of Location

4.3   Internal Structure

The internal structure of Esfinge_METADATA has two main components: *MetadataaLocator* and *MetadataValidator*. They are responsible for locating and validating code annotations, respectively. The first step of the validation process is performed by *MetadataValidator*, which retrieves all class annotations, searching for constraint annotations (annotated with *@ToValidate* or @ToValidateProperty) in each one and invoking the respective implementations of *AnnotationValidator* associated with each annotation.

Whenever a validation logic needs to retrieve another annotation, it should do that through the class *AnnotationFinder*. The *AnnotationFinder* is a facade that provides a simplified interface for annotation searching. It intermediates the interaction between validators and locators aiming to make transparent the search strategy used to locate the annotation. The *AnnotationFinder* creates instances of the *MetadataLocator* interface regarding the annotation metadata. Each locator executes a specific algorithm for searching the metadata represented by the annotation in the code elements. The applicable locators are structured based on the Chain of Responsibility (CoR) pattern. Regarding this structure, each locator tries to find the annotation, and if it is not successful, it delegates the searching to the next one. The chain of locators is created for each annotation based on its metadata. Currently, the locators implemented in the framework are: *RegularLocator*, *LevelLocator*, and *AnnotationLocator*. They are described in detail in the next subsection.

4.4   Location Strategies

As stated in the previous sections, the annotation evaluation process requires a decoupling between the rules implementation and the location where a target annotation is defined. In other words, the annotation might be present in a different element. In this context, the annotation recovering process can be configured for each annotation and is driven by annotation metadata.

The following describes the annotation locators currently implemented: (a) Regular Locator performs the annotation recovery exactly at the target element set as a parameter. This locator is the default one, and it is always included in the locators chain; (b) Level Locator performs the annotation recovery on the enclosing elements of the target element passed as a parameter, and to be included in the locators chain, the annotation needs to have the *@SearchOnEnclosingElements* metadata; and (c) Annotation Locator performs the annotation recovery inside other annotations configured on the target element.

It is important to highlight that the chain structure from the locators is essential when two different strategies for annotation definition are used. For instance, by chaining *LevelLocator* with *AnnotationLocator*, the annotation that is being searched for a method can be found inside an annotation defined in its respective class.

Just to illustrate the possibilities of this structure, a class *ConventionLocator* is under development to perform the recovery of an annotation seeking for name standards or code conventions. For instance, consider an annotation named *@Test* that defines a metadata that can also be defined by starting the method name with *"test"*. By using this locator, it will return *true* if this annotation is present in a method named *testCalculation()*.

5   Evaluation

The proposed approach for annotations validation aims at allowing validation rules to find annotations defined in other code elements, but making this search into other code elements transparent to it. In terms of source code, it means that the validation logic should be decoupled from the annotation location logic.

Table 2. Annotations used in Esfinge_METADATA application.

Annotation	Description	Requirement
@Logging	This annotation is used to register transactions in the system	Requires @Transaction in its configuration for the same element
@Transaction	This annotation configures transactions in the system	Id attribute can not be null
@Administration	This annotation grants administrative privileges in parts of the system	-
@OptimizeExecution	This annotation configures methods which must not be stored in logging	Prohibits its definition with @Logging

This section presents an evaluation performed to verify if the proposed approach is capable of fulfilling these decoupling and functional requirements. The approach evaluation was performed by using the code annotations described in Table 2. Such annotations are commonly used by developers to provide the configuration of characteristics of an application.

5.1 Methodology

The methodology used to evaluate the approach for code annotations validation with location transparency includes both a functional-based evaluation and a modularization analysis.

A careful assessment of the implementation was performed to verify the correctness of the application. In this paper, we use functional testing as the basis to construct test cases to evaluate the correctness of the implemented framework. Two of the most well-known functional-based testing selection criteria were used (equivalence partitioning and boundary-value analysis). The purpose was to test the component as a whole and not just its functions or methods. In order to accomplish that, we implemented functional-based test cases with JUnit[7] framework for the metadata validator following the rules and requirements of the annotation usage. The main goal of the implemented tests was to show that the approach is capable of covering all scenarios properly, including the possibility of annotation configuration in different locations, mainly the annotation configuration out of the target element, which distinguishes our approach from others presented in the literature.

In addition, to analyse the system architecture in terms of the relationships between its constituent components, we performed the modularity analysis through a static Dependency Structure Matrix (DSM) [11], by using IntelliJ[8] IDE. Such model informs system decomposition into subsystems and its integration. The representation and analysis capabilities about partitioning are important to managing system complexity and its dependencies improving the understanding about the system.

By the modularity analysis, we checked the dependencies between metadata validators and locators. The objective of this analysis is to demonstrate that the Metatada validators do not depend on the Metadata location logic. If there was no dependency between validators and locators, it is an evidence that the proposed solution allow the two kinds of components to be freely combined.

5.2 Functional Evaluation

To demonstrate that Esfinge_{METADATA} can check constraints and configuration inconsistencies in annotation definition from several annotation location scenarios, we construct test cases[9] to verify the configuration of all code annotations described in Table 2. In total, 99 test cases were created to evaluate

---

[7] http://junit.org/.

[8] https://www.jetbrains.com/idea/.

[9] Available at: https://github.com/EsfingeFramework/metadata.

Esfinge_METADATA features. However, due to lack of space, it is only shown the test cases for @*Logging* annotation.

As stated, @*Logging* has a specific constraint that requires the metadata defined by the annotation @*Transaction* to be configured for the same element. The strategy used to derive the functional tests considered the @*Logging* annotation defined on a *method* and the validation constraint added to it. Then, we varied the position of @*Transaction* annotation and its location settings through @*SearchOnEnclosingElements* and @*SearchInsideAnnotations* annotations. Furthermore, the scenarios of annotation configuration were based on related code elements including in the class, in a method, in the package, and in another annotation.

For this purpose, several possible annotation configuration usage of @*Logging* with @*Transaction*, presented in Table 3, were successfully verified by the implemented test cases with correct results.

5.3   Modularity Analysis

This subsection reports the modularity analysis by using a Dependency Structure Matrix (DSM). In a nutshell, a DSM can be defined as a square matrix that can be used to represent and navigate across dependencies between components.

According to Baldwin and Clark [11], DSMs are a good tool not only for visualizing dependencies relationships but also for the evaluation of software modularity. Design parameters are placed both in the rows and the columns of a DSM. One parameter depends on another one when there is a mark relating the row to the column.

Costa Neto et al. [12] have analyzed two different types of dependencies named as syntactic and semantic coupling. Syntactic coupling occurs when one component contains a direct reference to some other component, as inheritance, method calls, composition and so on. Semantic coupling is a dependency that is not syntactically defined in the code so that there is no direct reference among the components.

As stated earlier, metadata-based frameworks take decisions based on metadata related to the source code. Such approach represents an improvement in the modularity of the development of frameworks. It is important to notice that there is a difference between a source code that depends on common annotations and those that depend on validation annotations. Common annotations are related to the business and, therefore, allow the source code to remain oblivious of any aspects other than its own. Validation annotations are specifically related to the *MetadataValidator* that performs annotation constraint checking of its correct usage.

Figure 4 presents the dependencies of both modules of the system: *MetadataLocator* and *MetadataValidator*. We have noticed that the classes of each module rely only on its main interface as aforementioned. On the one hand, it is worth mentioning that new annotation location strategies can be added anytime, running independently with the designed validation rules. On the other hand,

Table 3. Test cases derivation for *@Logging*

Id	@Logging	@Transaction	Annotation with @Trasaction	Expected result
CT01	On a Method	Nowhere	@SearchOnEnclosing Elements and @SearchInsideAnnotations	Invalid ✓
CT02	On a Method	On a Method	@SearchOnEnclosing Elements and @SearchInsideAnnotations	Valid ✓
CT03	On a Method	On a Class	@SearchOnEnclosing Elements and @SearchInsideAnnotations	Valid ✓
CT04	On a Method	On a Package	@SearchOnEnclosing Elements and @SearchInsideAnnotations	Valid ✓
CT05	On a Method	Method Annotation	@SearchOnEnclosing Elements and @SearchInsideAnnotations	Valid ✓
CT06	On a Method	Class Annotation	@SearchOnEnclosing Elements and @SearchInsideAnnotations	Valid ✓
CT07	On a Method	Package Annotation	@SearchOnEnclosing Elements and @SearchInsideAnnotations	Valid ✓
CT08	On a Method	Nowhere	@SearchOnEnclosing Elements	Invalid ✓
CT09	On a Method	On a Method	@SearchOnEnclosing Elements	Valid ✓
CT10	On a Method	On a Class	@SearchOnEnclosing Elements	Valid ✓
CT11	On a Method	On a Package	@SearchOnEnclosing Elements	Valid ✓
CT12	On a Method	Method Annotation	@SearchOnEnclosing Elements	Invalid ✓
CT13	On a Method	Class Annotation	@SearchOnEnclosing Elements	Invalid ✓
CT14	On a Method	Package Annotation	@SearchOnEnclosing Elements	Invalid ✓
CT15	On a Method	Nowhere	@SearchInsideAnnotations	Invalid ✓
CT16	On a Method	On a Method	@SearchInsideAnnotations	Valid ✓
CT17	On a Method	On a Class	@SearchInsideAnnotations	Invalid ✓
CT18	On a Method	On a Package	@SearchInsideAnnotations	Invalid ✓
CT19	On a Method	Method Annotation	@SearchInsideAnnotations	Valid ✓
CT20	On a Method	Class Annotation	@SearchInsideAnnotations	Invalid ✓
CT21	On a Method	Package Annotation	@SearchInsideAnnotations	Invalid ✓
CT22	On a Method	Nowhere	-	Invalid ✓
CT23	On a Method	On a Method	-	Valid ✓
CT24	On a Method	On a Class	-	Invalid ✓
CT25	On a Method	On a Package	-	Invalid ✓
CT26	On a Method	Method Annotation	-	Invalid ✓
CT27	On a Method	Class Annotation	-	Invalid ✓
CT28	On a Method	Package Annotation	-	Invalid ✓

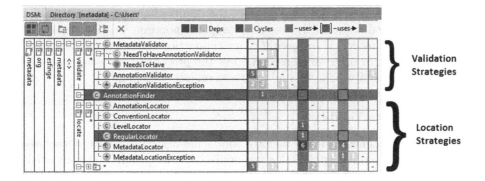

Fig. 4. Esfinge$_{\text{METADATA}}$ dependency structure matrix.

the framework might be extended by adding new validation rules that can also be combined with any location strategy.

As depicted in the DSM (Fig. 4), the classes of the locator package depend only on *MetadataLocator* interface since it has dependencies only on the column number six, which is related to *AnnotationFinder* interface. Such dependency is essential for these classes, so they can call other class of the Chain of Responsibility strategy. On the other hand, the classes of the validator package depend on *MetadataValidator* interface as illustrated by the DSM. It is important to emphasize that any new validator added on the system will depend only on *MetadataValidator*.

5.4 General Analysis

The analysis of the scenarios described by the functional-based derived test cases shown that Esfinge$_{\text{METADATA}}$ is capable of locating and validating code annotations using and combining different location strategies. Obtained results showed that the *meta-framework* for annotation validation can check constraints and configuration inconsistencies properly on annotation definition, independently of its location. This evaluation considered two different location strategies: the definition on enclosing elements and the definition inside other annotations.

By the modularity analyse (Fig. 4) we could identify two main interfaces called *MetadataLocator* and *MetadataValidator*, which handle location and validation code annotations respectively. The *AnnotationFinder* class plays the Facade role among locators and validators. Each locator executes searching in an independent way. For instance, *RegularLocator* depends only on *AnnotationFinder* as we can see highlighted.

Interpreting these results, it is possible to state that the code responsible for implementing the validation rule does not depend on any code that implements a location strategy. Nevertheless, by implementing the proposed approach, Esfinge$_{\text{METADATA}}$ was able to execute correctly test cases that cover all scenarios related to the annotation location and its respective configurations about where it could be searched. Based on that, the results of our assessment showed that the proposed approach is able to decouple annotation location from validation rules. Moreover, it means that any application or framework that uses Esfinge$_{\text{METADATA}}$, or other implementation of the proposed approach, is able to extend locators and validators, and use them on any combination.

5.5   Threats to Validity

Although we recognize that the adopted functional-based evaluation of the proposed approach was not applied in a real application, we believe that code annotations and its definition strategies that have been chosen to assess and validate the Esfinge$_{\text{METADATA}}$ are fairly used in software development industry in Java language. Moreover, it has implemented different scenarios that hardly appear together in a single application. Another limitation of the research is that we consider only two different strategies for annotation configuration, known as General Configuration and Annotation Mapping. However, both of these strategies are extensively used by mature and real frameworks, such as JAX-RS 1.1, EJB 3 API, the API for Contexts and Dependency Injection (CDI) for Java EE 6, and the Bean Validation API. It is important to highlight that the framework provides extensibility for the location strategies, and a new one can easily be incorporated into the framework or in future versions.

6   Related Works

In the context of metadata, relatively little attention has been paid to developing tools that assist in validation of code annotations.

Eichberg et al. [13] and Ruska [14] verify the correctness of applications based on annotations, in terms of their implementation constraints and dependencies that are implied by annotations. Eichberg et al. [13] analyze cases where the verified source code violates constraints and dependencies, which is performed automatically by an extensible tool. Darwin [5] and Córdoba and de Lara [6] suggest a DSL for the design of Java annotations, based on assertions about a set of existing annotations, with a concrete syntax very similar to Java. Noguera and Duchien [10] checked the correctness of the use of annotations, adding to their meta-annotations statements that define restrictions expressed as a constraint language of queries objects, which must be met when declared annotations are used in the program.

Among the related work, we highlight the Aval validator [4] is the most similar approach to our own. Such annotation validator is based on meta-annotations (*Validators*), which provides a set of basic rules of validation as well as means to define more specific directives [4].

Although the relevance and contribution of the studies, none of them perform validation of code annotations with location transparency defined in different contexts. Code annotations can be configured in different elements of a particular application: a method, a class, a package, and even within another annotation.

In that case, code annotations have special constraints of use that are hard to verify with the available approaches in the literature.

7 Conclusion

Metadata validation process, in general, is not a trivial task, taking into account their different ways of definition. In the context of object-orientation, metadata is relevant information about the elements of the source code, especially the configuration resource offered by code annotations, since annotations have been widely used in software development.

In this context, this paper presented Esfinge_{METADATA}, an approach for code annotation validation with metadata location transparency. This research presented the relevance of using metadata and their different ways of definition, the available approaches for annotation validation and the inherent difficulties in the process.

Due to different approaches to optimize metadata definition that add the annotation out of the target code element, one of the current challenges of software engineering is to perform annotation validation. Also, the variability of existing formats and ways of metadata definition increases the complexity in the development of a general-purpose tool for code annotation validation.

The evaluation of the proposed approach met the designed goals and showed interesting results in the context of annotation validation. Analyses with Esfinge_{METADATA} *meta-framework* have demonstrated that the approach is capable of decoupling the annotation location from the validation rules, which could contribute to the software development process. Furthermore, the proposed solution made achievable the definition of validation rules independent from annotation definition. This feature represents a central desirable property for metadata validation. The component for annotation location removes from the validation rules the responsibility to know where annotations are defined. This feature is the main point that distinguishes our approach from others.

Acknowledgments. The authors would like to thank FAPESP (grant 2014/16236-6), CAPES, and CNPq (grants 445562/2014-5 and 455080/2014-3) for financial support.

References

1. JSR175, JSR 175: a metadata facility for the java programming language (2003). http://www.jcp.org/en/jsr/detail?id=175
2. Guerra, E., Souza, J.T., Fernandes, C.: A pattern language for metadata-based frameworks. In: Proceedings of the 16th Conference on Pattern Languages of Programs, pp. 3:1–3:29. ACM (2009)
3. Guerra, E., Fernandes, C.: A qualitative and quantitative analysis on metadata-based frameworks usage. In: Murgante, B., Misra, S., Carlini, M., Torre, C.M., Nguyen, H.-Q., Taniar, D., Apduhan, B.O., Gervasi, O. (eds.) ICCSA 2013, Part II. LNCS, vol. 7972, pp. 375–390. Springer, Heidelberg (2013)

4. Noguera, C., Pawlak, R.: AVal: An extensible attribute-oriented programming validator for java: research articles. J. Softw. Maint. Evol. **19**(4), 253–275 (2007)
5. Darwin, I.: Annabot: a static verifier for java annotation usage. Adv. Softw. Eng. **2010**(540547), 1–7 (2010)
6. Cordoba, I., de Lara, J.: A modelling language for the effective design of java annotations. In: Proceedings of the 30th Annual ACM Symposium on Applied Computing, SAC 2015, pp. 2087–2092. ACM, New York (2015)
7. Guerra, E., Cardoso, M., Silva, J., Fernandes, C.: Idioms for code annotations in the java language. In: Proceedings of the 17th Latin-American Conference on Pattern Languages of Programs, SugarLoafPLoPe, pp. 1–14 (2010)
8. Fernandes, C., Ribeiro, D., Guerra, E., Nakao, E.: Xml, annotations and database: a comparative study of metadata definition strategies for frameworks. In: XML: Aplicaes e Tecnologias Associadas, pp. 115–126, Vila do Conde, Portugal (2010)
9. Miller, J.S., Ragsdale, S.: The Common Language Infrastructure Annotated Standard. Addison-Wesley, Boston (2003)
10. Noguera, C., Duchien, L.: Annotation framework validation using domain models. In: Schieferdecker, I., Hartman, A. (eds.) ECMDA-FA 2008. LNCS, vol. 5095, pp. 48–62. Springer, Heidelberg (2008)
11. Baldwin, C.Y., Clark, K.B.: The Power of Modularity. The MIT Press, Cambridge (2000)
12. Neto, A.C., Ribeiro, M.M., Dsea, D., Bonifcio, R., Borba, P., Soares, S.: Semantic dependencies and modularity of aspect-oriented software. In: Proceedings of the First International Workshop on Assessment of Contemporary Modularization Techniques, pp. 11–16. IEEE Computer Society (2007)
13. Eichberg, M., Schäfer, T., Mezini, M.: Using annotations to check structural properties of classes. In: Cerioli, M. (ed.) FASE 2005. LNCS, vol. 3442, pp. 237–252. Springer, Heidelberg (2005)
14. Ruska, S.: Defining annotation constraints in attribute oriented programming. Acta Electrotechnica et Informatica **10**(4), 89–93 (2010)

Towards a Software Engineering Approach for Cloud and IoT Services in Healthcare

Lisardo Prieto-Gonzalez[(✉)], Gerrit Tamm, and Vladimir Stantchev

SRH Hochschule Berlin, Ernst-Reuter-Platz 10, 10587 Berlin, Germany
{lisardo.gonzalez,gerrit.tamm,
vladimir.stantchev}@srh-hochschule-berlin.de

Abstract. In this work a tailored approach for requirements engineering for risk management solutions is presented. The approach focuses on cross-disciplinary risk paradigms in inpatient healthcare and aims to provide a novel methodology that accounts for specific viewpoints of relevant expert groups. The presented results demonstrate the feasibility and usefulness of the approach in the context of the current evaluation within the German market.

Keywords: Requirements engineering · Risk management · Software engineering · Healthcare

1 Introduction

The usage of information technology services can improve the quality and efficiency of health care and reduce the cost in providing health services in the industrialized countries [1, 2], and the gap between the promises and the existing scientific evidence derived from clinical studies decreases [3, 4]. This is part of a growing trend of technology diffusion into the services sectors of the economy with the knowledge-based sector serving as a prominent example. Specific approaches exist to create mappings between the business and IT side of project management [5]. For this purpose the authors propose the approach of a service-oriented architecture [6]. This allows the assurance of specific service levels [7] by applying standardized reconfigurations such as service replication [8] and represents one of the enablers of Cloud Computing.

Cloud computing is considered nowadays as an important innovation in information technology that can improve the quality of care [9]. Another significant benefit in using information technology in the treatment of patients, is the possibility to enable physicians to remotely monitor their patients' health by means of secure cloud services and IoT devices equipped with sensors that interact with such patients. These are some of the main objectives of the OpSIT project[1], which has led authors to perform this research study. Also, IT improves the quality of healthcare e.g. by using electronic health records instead of traditional paper files [3]. On the other hand, patients are able to manage their health care more easily and reduce the costs of care by spending less time in the hospital [1]. But these new technologies can also bring more risks to the hospitals. As a relevant threat, data security and privacy requirements have to be

---

[1] http://www.opsit.de/.

© Springer International Publishing Switzerland 2016
O. Gervasi et al. (Eds.): ICCSA 2016, Part IV, LNCS 9789, pp. 439–452, 2016.
DOI: 10.1007/978-3-319-42089-9_31

addressed [10]. Privacy means, that the patient retains control even when the data is owned by another party or even after a copy of the data has been provided to another party [1]. Patients are protected by data-protection-law in almost every country, thus, when the hospital is not capable of following applicable regulations in the area of medical confidentiality (a breach of confidentiality is considered as a criminal offense), social data (personally related data that concerns social aspects of the person) and state-specific rules, as a consequence specific risks and crises arise. Since now, issues like data security are in the hospitals' focus e.g. by the introduction of electronic health records. But the more and more daily operations have been managed with information technology especially in hospitals; information systems became a part of the critical infrastructure which maintains the availability of medical care [11, 12]. Over and above this can lead to increased technology dependence, where an unexpected downtime of the information system in hospitals can result in catastrophic consequences for the hospital and the patient [12]. As another point, information technology used in health care has specific requirements in its usability. An American study among 199 child hospitals examined the most common barriers to health care information technology adoption [13]. These are the delivery of services or products to satisfy customers (85.4 %), lack of staffing (82.3 %) and difficulties in achieving end-user acceptance (80.2 %). The three addressed points lead to a new claim of requirement engineering.

The purpose of this contribution is the development of a requirements criteria catalogue to establish a recommender system for the purpose of risk management in hospitals that benefits or plan to get benefit from cloud computing based services and IoT monitoring devices. Based on a growing dataset of risk profiles the system should recommend protective measures that consider the functional areas of hospitals' IT.

1.1   Requirements Engineering

Generating a complete and secure IT systems which will be highly managed implies the usage of well proven development methodologies and frameworks, such as Capability Maturity Model Integration (CMMI) [14], Rational Unified Process (RUP) [15, 16], Microsoft Solutions Framework (MSF) [17], Waterfall [18], Scrum [19] or Extreme Programming (XP) [20]. Some of these frameworks cover more than just the developing process, such as in the case of CMMI and MSF, which can be used to guide process improvement in a project, division, or an entire organization, while other are just focused on the software developing methodologies, like Waterfall or XP. In any case, there exists a common task defined in all of them. This is the requirements capture and analysis phase.

The Rational Unified Process [15] provides the following definition for any requirement: "*A requirement describes a condition or capability to which a system must conform; either derived directly from user needs, or stated in a contract, standard, specification, or other formally imposed document*".

Requirements capture and analysis phase is used in the different methodologies to obtain the customer's needs and translate them in into a set of requirements that define what the system must do and how well it must perform among other characteristics. During this phase, the systems engineer must ensure that the requirements present some specific attributes [21, 22]. In a single statement: requirements must be complete, comprehensive, concise, unambiguous and understandable.

Usually, requirements are documented following a standard classification. Although there are general classification schemes [21], this study will be focused in the categories defined under FURPS + [23] for Information Systems by Hewlett-Packard as it mature and well proven. FURPS + (see Fig. 1) is the acronym for Functional, Usability, Reliability, Performance, Supportability and the extension (+) includes three additional categories: Constraints, Interface Requirements and Business Rules. In order to verify that the different types of system-wide requirements covered by FURPS + have been considered, it seems to be useful to follow an organized strategy during the capture phase and use a checklist [24].

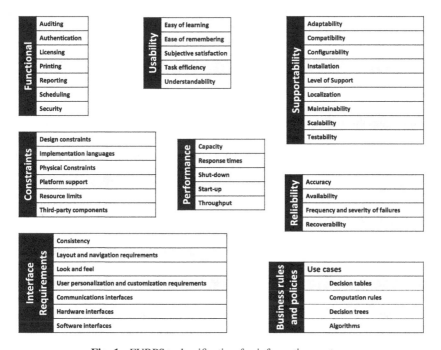

Fig. 1. FURPS+ classification for information systems

1.2 Risk Management for Information Systems

Risk management is defined as the identification of risks that are specific to an organization and respond to them in an appropriate way [25]. Risk management activities are therefore to expose risks that enable recognizing the events that may result in unfortunate or damaging consequences in the future, their severity and how they are controlled [26]. Risks are inherent to any entrepreneurial activity and are as inseparable as the chance of success [27]. In health care they are linked to injured or dead people, as well as financial losses. Hospitals are appreciable risky working environments for patients and staff [28]. For that reason hospital risk management is defined as a systematic program which is designed to reduce preventable injuries and accidents and minimize financial loss to the institution [29]. Risk management in hospitals is used to identify, evaluate and reduce the risk of injuries to patients, staff and visitors and the risk of loss to the organization itself [30]. The goal of the entire risk management process is to prevent or to reduce the frequency, probability and severity of unexpected events, reduce the impact of legal claims, evaluate the acceptability of risks and promote high reliability on performance, system design and the uniqueness of each patient [31]. The core risk management process is often established in different steps, which include the risk identification, the risk assessment and the risk evaluation:

Risk Identification: In hospitals, identifying risks is based on the collection and analysis of information, concerning medical accidents and identification of situations in the hospital that result in financial loss. The institution has to identify processes of the departments of a hospital, which lead to interactions between the departments and affect the corporate environment. Furthermore, a collection and analysis of information concerning medical accidents have to be conducted. The risk identification is a transformation process where experienced personnel generate a series of risks (and opportunities) in a risk register [25]. After the risks have been identified, a quantitative analysis measuring the impact on the enterprise is being conducted.

Risk Assessment: After the risks were identified, the next step is to analyse the likelihood and the impact of the risks and opportunities on the enterprise with a quantitative analysis [25]. Moreover, a causal analysis for every risk is needed in this step for a quick identification of major causes that usually indicate an area for further investigation [25]. For this purpose, the Ishikawa (or fishbone) diagram [32] is an often used method.

Risk Evaluation: The relationship between the individual risks and opportunities and their net effect is calculated when they are combined together [25]. The risk analysis must be checked regularly and needs to be updated.

After the identification, assessment and evaluation of relevant risks and opportunities for the organization, responses and specific action plans must be produced to secure the hospitals' objectives [25]. Besides risk avoidance, risk reduction is one strategy to reduce the probability and/or the impact of an event. Measures of risk reductions are e.g. conferences, continuous and short training courses and workshops on different aspects of risk management systems [27]. Risk reduction is also the avoidance and prevention, where the probability of loss is reduced but not necessarily the severity. Another strategy to deal with the risks is the transfer to another party like an insurance company.

For the risk evaluation of this study, there have been analysed the behaviour of the various key players in the fields of Medical Care and Medicine, supply, HR and IT-systems with regard to the influence of dynamic risk in the context of various simulated scenarios. Information on risks are collected in expert workshops and evaluated in a survey among German hospitals. For the most important hospital risks, identified by literature search and interdisciplinary expert groups, the preparedness of German hospitals are evaluated and adequate management scenarios including IT solutions are developed. These solutions are used for on-site approaches to avoid incidents, to exchange data, as an information source e.g. for guidance documents as well as active training tools.

The rest of this work is structured as follows: In Sect. 2 the advanced model to combine the approaches of requirements engineering and risk management for hospitals is presented. Section 3 gives an overview of how this model can be adopted to identify risks and requirements for information systems in hospitals. In Sect. 4 the results of the evaluation are presented. Finally, in Sect. 5 results and future research activities are discussed.

2 Advanced Model

The conclusions of the previous sections are that designing and developing an information system based system is completely dependent on customer requirements, and a proper capture and analysis, together with a risk evaluation and management plan are fundamental to a successful result. The proposed model focuses on empower the requirement capture in information system based systems oriented to healthcare area. In a former study [33] published in [34], several workshops were conducted among professionals in healthcare in order to evaluate the most common problems and risks in these kind of applications.

As can be seen in the Systems Engineering Process figure [21], the phase corresponding to System Analysis and Control has to do with risk management and as well as the Functional Analysis, provide feedback to Requirements Analysis task. In order to improve this requirements capture in healthcare information systems development, the results from the conducted workshops related to information systems have been analysed, classified and matched to the FURPS + requirements categorization. This way, the most important identified risks can be taken into account when starting with the initial phase of requirements capture, providing insights about possible future risks to be covered and saving time and money preventing possible faulty or missing requirements (Fig. 2).

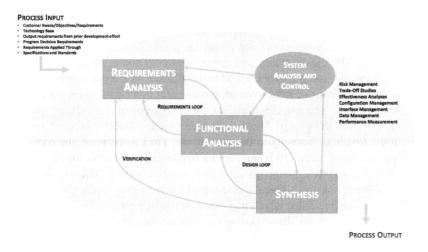

Fig. 2. The systems engineering process

As can be seen in Fig. 3, the requirement capture and analysis when designing an information system solution for health care systems that benefit from cloud based services and also from smart items connected to some of such services, should take emphasis in the requirement categories related to security and communications, and as well in usability.

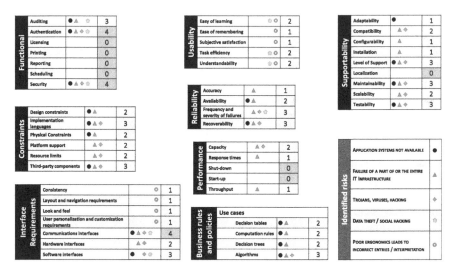

Fig. 3. Requirement priority based on identified risks

3 Applying the Model for Health Care Risks

Due to expert workshops the most relevant risks for hospital information systems were identified. The workshop was moderated by one university professor with expertise in workshop and moderation methods and in the field which the workshop was in. For each five hours workshop a standardized agenda was set. The workshops were structured in five phases: The brainstorming phase, the discussion and precision phase, the evaluation phase, the dyadic phase and the presentation phase. In the brainstorming phase the experts and managers were asked to write down all the risks they could think. Following this, the identified risks were written on cards and clustered on pin boards by the workshop leaders. The workshop leaders for the respective areas shortly presented the risks found from the group. All of them were asked to select the most important risks from the first brainstorming in the third phase. For this purpose, the participants had the opportunity to award 10 points, with a maximum of 5 points for one risk. The assignment had to be based on the question how negatively the participants interpreted the impact of those risks on a hospital and for which risks they wished to be prepared for. Afterwards, the five risks, which had received the highest number of points, were selected. In the fourth phase, the workshop participants were divided into teams of two. Each team of two should choose to edit two risks. The processing was done by dyadic interaction, where the two experts first worked out key features, consequences and costs of a risk and fixed the results on a poster. Experts were paired and given one hour to work together. In the final workshop phase, the results of the teamwork phase were presented. For this purpose, each team introduced their findings to the group and then the results of the dyadic phase were discussed together. In the last step, the participants received the possibility to rate the danger and the probability of occurrence of the risk by setting points to a prepared evaluation scale on the posters. The final assessments served to produce a better ranking of risks. In the last part of the workshop the results for each risk were presented and the other experts had the opportunity to add important aspects to each risk and discuss the findings. The summarized results are presented in Table 1.

Table 1. Identified crises and their main characteristics

Risk	Main characteristics
Trojans, Viruses, Hacking	• A criminal and detrimental attack of the information system of a hospital occurs • Patient data/hospital data can be copied, destroyed or altered/damaged • Confidentiality of the patients is violated • The threat of legal consequences for the hospital present when document/information security cannot be confirmed • Ex.: A new computer virus disables the Picture Archiving and Communication System (PACS) of the hospital. Important results cannot be accessed
Application systems not available	• Application systems are necessary for patient treatment i.e. Hospital information systems must be closed, inaccessible systems • Hospital technical staff are unable to handle the situation within a short period of time • Information required by medical personnel is unavailable (e.g. lab results/imaging) • Patient treatment is limited as is adequate documentation of treatment
Failure of a part of or the entire IT-Infrastructure	• Failure of IT structures interrupts operational processes • Important flow of information within the facility is disrupted • Doctors and health care personnel cannot access important treatment information (e.g. lab results) • Administration cannot access schedules, invoice/pay-roll data • Ex.: Failure of the entire IT-system as a result of flooding in the hospital server room
Data Theft/Social hacking	• Social Hacking results when information is acquired via clever manipulation by a deceptive individual

(Continued)

Table 1. (*Continued*)

Risk	Main characteristics
	• Employees and partners have access to highly-confidential information. Access often occurs by persons directly or through third parties • Carelessness or criminal motives enable confidential information to be publicised • This can damage the reputation of a hospital • Ex.: An unknown individual access and steals 50.000 confidential patient information's
Poor ergonomics leads to incorrect entries/interpretation	• Poor software ergonomics leads to incorrect entries/ interpretation of patient clinical data • It can also increase the number of incorrect entries • Outdated systems increase the possibilities for incorrect entries/interpretations • This can lead to incorrect patient treatment • Ex.: Following the introduction of a new hospital information system there is a subsequent increase in incorrect data entries

In a second step the causes for each crisis were been acquired. This was done through a literature review and two expert interviews. One of the interviewees directed the security management of a hospital, the other worked in the personnel controlling. Furthermore, the participants of the survey had the option to add additional risk which they considered as particularly important for the occurrence of the risk.

Sample. 76 hospitals participated in the online survey. There were 788.5 ($SD = 1327.82$) full-time employees and 376.38 ($SD = 358.45$) patient beds on average. Hospitals were located in 13 of the 16 federal states of Germany. 34.2 % were in public ownership. In 40.8 % of cases institutions were independent non-profit organization (e.g. a church owned hospital) and 25.0 % were privately owned. Employees responding to the online survey were 47.07 ($SD = 8.59$) years old on average and 72.4 % were male. 18.4 % were CEO of a hospital, 10.5 % were chief medical officers and 15.8 % were directors of administration. 17.1 % were head of a department and 31.6 % answered other (e.g. management assistants, quality or risk management representatives).

Procedure. First, participants were welcomed and the survey procedure was presented. Following, demographic information regarding the hospitals and participants was requested. Afterwards, each risk was introduced with the text presented in Table 1. After presenting a risk, the participants had to evaluate the perceived occurrence (past, future – own hospital, future – other hospital) and dangerousness for each risk.

4   Results

Results are displayed in Table 2. In the first columns the occurrence in the past is shown. 61.8 % of the participants indicated that Trojans, viruses and hacking had occurred in the last five years in their hospitals. But in 35.5 % of the hospitals this risk

Table 2. Occurrence (past and future) and the dangerousness of each crisis

Risk	Occurrence (past) in percent					Occurrence (future)		Danger-ousness
	Yes	Nearly c.	Part.	In few div.	No	Own	Other	
Application systems not available	9.2	3.9	23.7	25.0	35.5	39.19 (27.88)	47.67 (27.99)	4.11 (1.56)
Failure of a part of or the entire IT-Infrastructure	7.6	2.6	17.1	14.5	57.9	34.21 (28.78)	44.83 (28.43)	4.42 (1.67)
Trojans, Viruses, Hacking	3.9	0	7.9	7.9	80.3	31.55 (28.21)	42.03 (27.19)	4.44 (1.80)
Data theft/Social hacking	0	0	2.6	1.3	94.7	23.50 (22.45)	36.45 (25.81)	4.13 (1.81)
Poor ergonomics leads to incorrect entries/Interpretation	1.3	1.3	13.2	21.1	63.2	26.68 (24.68)	37.20 (24.68)	3.59 (1.55)
Sum						30.85	41.58	4.16

did not occurred in the last five years in the surveyed hospitals. In contrast, absolutely no data theft and social hacking did occur in 94.7 % of the hospitals in the last five years. The lowest probability was seen for data theft/social hacking (23.50 %) and poor ergonomics leading to incorrect entries/interpretation (26.68 %). In general, the future occurrence for the own hospital was ranked lower (M = 30.85 % vs. 41.58). Failure of a part of or the entire IT-Infrastructure, application systems not available and data theft/social hacking were seen as the most important risk. With the lowest danger-ousness (3.59), poor ergonomics leading to incorrect entries/interpretation was ranked.[2]

Future occurrence multiplied by dangerousness provides an important indicator for the severity of a risk. The multiplication shows the following results (see Fig. 4):

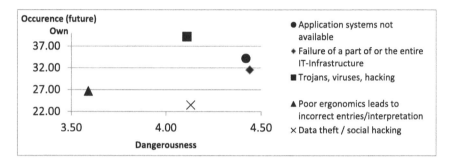

Fig. 4. Estimated future occurrence for own and dangerousness of each risk

The results for the causes are displayed in Table 3. The causes for each risk are organized regarding to their rated importance for the occurrence of the risk. Our results indicate that virtual attacks and failure/disruption of external electrical power are seen as the most important causes for application systems not available. Moreover, defective software and insufficiently qualified staff were added from the participants as relevant causes for the occurrence of the crises. The participants identified the poor security standards and criminal motives as important causes for data theft and social hacking. Incorrect training and difference between programme and treatment processes were identified as the most important cases for poor ergonomics leading to incorrect entries/interpretation. All other evaluations are shown in Table 3.

5   Interdisciplinary Discussion

The aim of this study was to identify and evaluate IT risks in German hospitals and to propose an approach for engineering the software requirements of a recommender system for the purpose of risk management in hospitals that benefits or plan to get benefit from cloud computing based services and IoT monitoring devices. A two-staged study was accomplished. Following a workshop with experts a quantitative online

---

[2] SDs displayed in brackets.

Table 3. Evaluated causes of the crises

Risk	Cause	M	SD
Application systems not available	Virtual Attack	4.36	2,03
	Failure/Disruption of external electrical power	4.13	1.87
	Major fire	3.64	1.68
	Natural event (flood, storm, earthquake)	3.44	1.86
	Attack with conventional explosive device	2.36	1.72
Failure of a part of or the entire IT-Infrastructure	Human/technical error	5.26	1.31
	Failure of external/internal electrical power	4.47	1.70
	Over heating	4.43	1.60
	Destroyed/burned out cables	4.35	1.57
	Short circuiting	4.33	1.60
Additional causes mentioned: Natural catastrophe N = 1, Outdated Hardware N = 1, Failure of radio relay system N = 1, Malware N = 1, (Too) High system complexity N = 1			
Trojans, Viruses, Hacking	Virtual attack on the IT	5.17	1.68
	Criminal and detrimental attack of the information system	4.64	1.93
	Sabotage	4.04	1.87
	Espionage on behalf of competitors	2.70	1.53
Additional causes mentioned: Carelessness N = 1, Irresponsible use of Information media e.g. in social networks N = 1, Human error N = 1, Insufficient separation of Medical technologies and IT N = 1			
Data Theft/Social hacking	Poor security standards	4.72	1.94
	Criminal motives	4.68	2.01
	Sabotage	3.50	2.02
	Espionage on behalf of competition	2.45	1.46
Additional causes mentioned: Hacker antagonism, N = 1, Poor security awareness of hospital staff N = 1			
Poor ergonomics leads to incorrect entries/Interpretation	Incorrect training	5.10	1.61
	Difference between programme and treatment procedures	4.97	1.49
	Error in request definition	4.84	1.61
	Cost saving pressures in IT sector	4.53	1.82
Additional causes mentioned: Poor Usability N = 2			

survey was conducted. Our results show that a crisis as described in the introduction is no isolated event in one hospital. In more than one third of the hospitals questioned application system not available or Trojans, viruses and hacking occurred in the last five years.

In general, the results illustrate that representatives of the hospitals expect the identified crisis to occur in the next five years quite often. The average occurrence probability is estimated to be 30.85 % on average with a rather high dangerousness (M = 4.16). In other hospitals the occurrence probability within the next five years is even perceived over 40 % on average. The gap between the own and other hospitals might be explained with a psychological attribution effect called self-serving bias (the tendency to see oneself better than others [35]. Nevertheless, if the estimated number for own hospital crises is correct, it is still very high. Therefore, results for each crisis should be discussed in detail and linked to prior research and incorporated in future practice. Subsequently, all identified and evaluated crises are discussed.

6 Conclusion and Future Work

In this work a tailored approach for requirements engineering for risk management solutions was presented. The focus of the approach on cross-disciplinary risk paradigms in healthcare provides a novel methodology that accounts for specific viewpoints of relevant expert groups. The presented results demonstrate the feasibility and usefulness of the approach. Nevertheless, the focus of current evaluation on the German market requires a broader interpretation of results and additional evaluations in international settings.

Beside this extended evaluation, future work is oriented towards extending the approach to cover more types of risks, as they are also important when designing an IT system, as well as the inclusion of relevant contingency plans for the detailed risks.

Acknowledgements. This publication is based on the research project "OpSIT" which received financial funding from the ministry of education, science, research and technology (BMBF) under the funding sign 16SV6048 within the programme "IKT 2020 - Forschung für Innovationen". The authors are responsible for the content.

References

1. Avancha, S., Baxi, A., Kotz, D.: Privacy in mobile technology for personal healthcare. ACM Comput. Surv. **45**, 3:1–3:54 (2012)
2. Blumenthal, D.: Stimulating the adoption of health information technology. New Engl. J. Med. **360**, 1477–1479 (2009)
3. Ammenwerth, E., Schnell-Inderst, P., Machan, C., Siebert, U.: The effect of electronic prescribing on medication errors and adverse drug events: a systematic review. J. Am. Med. Inform. Assoc. **15**, 585–600 (2008)
4. Black, A.D., Car, J., Pagliari, C., Anandan, C., Cresswell, K., Bokun, T., McKinstry, B., Procter, R., Majeed, A., Sheikh, A.: The impact of eHealth on the quality and safety of health care: a systematic overview. PLoS Med. **8**, e1000387 (2011)
5. Stantchev, V., Franke, M.R., Discher, A.: Project portfolio management systems: business services and web services. In: Fourth International Conference on Internet and Web Applications and Services, ICIW 2009, pp. 171–176. IEEE (2009)
6. Krallmann, H., Schröpfer, C., Stantchev, V., Offermann, P.: Enabling autonomous self-optimisation in service-oriented systems. In: Mahr, B., Huanye, S. (eds.) Autonomous Systems–Self-organization, Management, and Control, pp. 127–134. Springer, Netherlands (2008)

7. Stantchev, V., Malek, M.: Translucent replication for service level assurance. In: Zhang, L.-J., Paul, R., Dong, J. (eds.) High Assurance Services Computing, pp. 1–18. Springer, USA (2009)

8. Stantchev, V.: Effects of replication on web service performance in WebSphere. International Computer Science, Institute Berkeley, California (2008)

9. Stantchev, V., Hoang, T.D., Schulz, T., Ratchinski, I.: Optimizing clinical processes with position-sensing. IT Prof. **10**, 2–31 (2008)

10. Dzombeta, S., Stantchev, V., Colomo-Palacios, R., Brandis, K., Haufe, K.: Governance of cloud computing services for the life sciences. IT Prof. **16**, 30–37 (2014)

11. Bundesamt für Sicherheit in der Informationstechnik: Schutz Kritischer Infrastrukturen: Risikoanalyse Krankenhaus-IT - Leitfaden - Risikoanalyse_Krankenhaus-IT_Langfassung. pdf (2014). http://www.kritis.bund.de/SharedDocs/Downloads/Kritis/DE/Risikoanalyse_ Krankenhaus-IT_Langfassung.pdf?__blob=publicationFile

12. Lei, J., Guan, P., Gao, K., Lu, X., Chen, Y., Li, Y., Meng, Q., Zhang, J., Sittig, D.F., Zheng, K.: Characteristics of health IT outage and suggested risk management strategies: an analysis of historical incident reports in China. Int. J. Med. Inf. **83**, 122–130 (2014)

13. Menachemi, N., Brooks, R.G., Schwalenstocker, E., Simpson, L.: Use of health information technology by children's hospitals in the United States. Pediatrics **123**(Suppl 2), S80–S84 (2009)

14. CMMI Product Team: CMMI for Development, version 1.2. (2006)

15. Rational Corporation: Rational Unified process: best practices for software development teams. Rational Software Corporation White Paper (1998)

16. Shuja, A.K., Krebs, J.: IBM Rational Unified Process Reference and Certification Guide: Solution Designer (RUP). Pearson Education (2007)

17. Lory, G., Director, G., Carter, J., MSFmentor, U., Rief, J.M., West, T.: Microsoft Solutions Framework version 3.0 overview, June 2003

18. Benington, H.D.: Production of large computer programs. In: ICSE, pp. 299–310 (1987)

19. Takeuchi, H., Nonaka, I.: The new new product development game. Harv. Bus. Rev. **64**, 137–146 (1986)

20. Beck, K.: Extreme Programming Explained: Embrace Change. Addison-Wesley Professional, Reading (2000)

21. Lightsey, B.: Systems engineering fundamentals. DTIC Document (2001)

22. Mazza, C., Fairclough, J., De Pablo, D., Stevens, R., Melton, B., Scheffer, A.: Software Engineering Standards. Prentice-Hall, London (1994)

23. Grady, R.B.: Practical Software Metrics for Project Management and Process Improvement. Prentice-Hall, Inc., Upper Saddle River (1992)

24. Eclipse process framework: checklist: system-wide requirements (FURPS+). http://epf. eclipse.org/wikis/openup/core.tech.common.extend_supp/guidances/checklists/system_ wide_rqmts_furps_3158BF2F.html

25. Chapman, R.J.: Simple Tools and Techniques for Enterprise Risk Management. Wiley, New York (2011)

26. Dickson, G.: Principles of risk management. Qual. Health Care **4**, 75–79 (1995)

27. Krystek, U.: Handbuch Krisen- und Restrukturierungsmanagement: generelle Konzepte, Spezialprobleme. Kohlhammer Verlag, Praxisberichte. W (2006)

28. Smith, M.: The paradox of the risk society state (2004)

29. Morlock, L., Malitz, F.: Do hospital risk management programs make a difference?: relationships between risk management program activities and hospital malpractice claims experience. Law Contemp. Probl. **54**, 1–22 (1991)

30. Joint Commission: Annual Report - Improving America's Hospitals 2013 (2014)

31. Singh, M.P.: Norms as a basis for governing sociotechnical systems. ACM Trans. Intell. Syst. Technol. **5**, 21:1–21:23 (2014)
32. Project Management Institute: Practice Standard for Project Risk Management. Project Management Institute (2009)
33. Glasberg, R., Hartmann, M., Draheim, M., Tamm, G., Hessel, F.: Risks and crises for healthcare providers: the impact of cloud computing. Sci. World J. **2014**, e524659 (2014)
34. Stantchev, V., Colomo-Palacios, R., Niedermayer, M.: Cloud computing based systems for healthcare. Sci. World J. **2014** (2014)
35. Gelfand, M.J., Higgins, M., Nishii, L.H., Raver, J.L., Dominguez, A., Murakami, F., Yamaguchi, S., Toyama, M.: Culture and egocentric perceptions of fairness in conflict and negotiation. J. Appl. Psychol. **87**, 833 (2002)

Smaller to Sharper: Efficient Web Service Composition and Verification Using On-the-fly Model Checking and Logic-Based Clustering

Khai Huynh[✉], Tho Quan, and Thang Bui

Ho Chi Minh City University of Technology, 268 Ly Thuong Kiet Street,
District 10, Ho Chi Minh City, Vietnam
{htkhai,qttho,thang}@cse.hcmut.edu.vn
http://www.cse.hcmut.edu.vn/site/en/

Abstract. Model checking (MC) is an emerging approach recently suggested for the problem of Web Service Composition (WSC), since it can ensure both the soundness and completeness once verifying if an WSC solution fulfills a goal formally described or not. However, as the number of web services to be considered in practice is often very large, the MC-based approach suffers from the state space explosion problem. Clustering has been naturally considered reducing the number of candidates for the WSC problem. However, as typical clustering techniques are mostly semi-formal in terms of cluster representation, it poses a dilemma of maintaining both soundness and completeness. In this paper, we handle this problem by suggesting a logic-based approach for clustering. This work makes twofold contributions. We propose a logic-based similarity between web services, which results in more reasonable clustering results; and we represent the generated clusters as logical formula and enjoy a seamless integration between web service clustering and MC. This approach eventually brings significant improvement of WSC performance when applied on real and relatively large repository of web services.

Keywords: Logic-based web service similarity · Web service clustering · Logic-based web service clustering · Web service composition

1 Introduction

Nowadays, advent of *web service* is considered as a technology bringing revolutionary operations of online B2B (Business to Business) and B2C (Business to Customer) applications. Each web service has functional and non-functional (or Quality of Service – QoS) properties [1]. Functional properties are a set of operations with a certain capability of web services. They represent the functionality of the service using input and output parameters. The user requirement on functional properties is called *hard constraint*. In addition to functional properties, there are many QoS properties of web services, such as the *response time*, *availability*, etc.

© Springer International Publishing Switzerland 2016
O. Gervasi et al. (Eds.): ICCSA 2016, Part IV, LNCS 9789, pp. 453–468, 2016.
DOI: 10.1007/978-3-319-42089-9_32

However, while a web service usually provides a simple functionality, a single service cannot meet the client requirement in many practical cases. It prompts the issue of *Web Service Composition (WSC)*, the process to produce *composite web service*, which is a collection of services that will be executed in a specific order to serve a user requirement, or *goal.*

To ensure the soundness and correctness of a WSC, model checking (MC) [3] is suggested. By treating a WSC solution as a state and the goal a property, a model checker can formally and correctly verify if the WSC solution satisfy the goal or not. In another hand, by exploring all of possible states, the model checker can also find of feasible WSC solutions. Especially, as QoS is concerned, it is significant that the composed web service should satisfy the soft constraints. This is a complex verification task since we need to take into account all of QoS. So far, only MC-based approach is capable of verifying both hard and soft constraints as a single process [6,12].

However, the MC-based approach is suffered from the state space explosion problem in verifying large-scale repository of web services. It is also a big issue in this area, which is commonly addressed by clustering. In this technique, the similar web services will be then grouped into corresponding *clusters.* Thus, when composing web services against a goal, one only needs to consider the candidates in suitable clusters, rather than the whole set of services.

One major cause of this problem is that the typical approach of web service clustering is not suitable w.r.t WSC requirements. Typically, to cluster a dataset of web services, one needs to evaluate the *similarities* or *distance* of the web services. Most of existing works calculate such distances based on the *features* extracted from web service description, such as [2,17,21]. Since two web services which are similar in terms of features may produce two far different results once employed in a same composition situation, the results of such feature-based clustering approaches are not compatible once applied into WSC problem.

In this paper, we propose a novel approach on logic-based clustering for the web service composition and verification. In this approach, web service is represented by a *logical expression.* A dataset of web services is grouped into clusters based on the calculation of the similarity between their logical expressions. Each cluster is also represented by a logical expression so-called *Representative Logical Expression.* Owing to the solid background of mathematical logic, we then prove that our clustering method ensures the soundness and completeness of the composition solution.

Contributions. The contributions of this paper are summarized as follows.

- *From data mining viewpoint:* We propose a logic-based approach for web service clustering, in which the similarities between web services are evaluated based on their logic representation, rather than their typical extracted features. As a result, the clusters are generated more reasonably, bringing better performance of composition.
- *From model checking viewpoint:* We employ clustering to reduce the state space significantly when verifying the correctness of a certain composition solution.

Since the clusters are formally represented, or *abstracted*, as logic formulas, our verification process still maintains both soundness and completeness in a far more compact state space.

Outline. The rest of the paper is organized as follows. Section 2 presents the motivating examples. Section 3 presents the related works. Section 4 presents logic-based web service clustering. Section 5 provides the experiments. The conclusion and future work are presented in Sect. 6.

2   Motivating Examples

The motivation examples presented in this section are based on our previous work in [12]. To be summarized, in this work we adopt the formal definition of *Labelled Transition System* (LTS) [20] to formally represent all of concerned web services as a single LTS-based model, known as *LTS4WS*. In order to compose and verify web services, [12] uses an on-the-fly approach combining with some heuristics allowing early termination to be applied for unfeasible composition. Moreover, the *LTS4WS* is only needed to be generated once and remained unchanged (until the repository changes). It supports various composition purposes, thus we do not need to rebuild the model for every requirement.

   To illustrate how this approach works, suppose that we have a repository with 10 web services as in Table 1, and the user would like to book a tour. The user then provides the place to travel to (*Sightseeing*) and the dates of the travel (*Dates*), and requires the price of some hotels (*Price*) nearby the *Sightseeing* and their reservation (*HotelReservation*). Besides these functional requirements (*hard constraint*), the user also asks for the quality of service (*soft constraint*) that the response time (*respTime*) of web services must not exceed 30 s. In addition, the user also wants to know about *Price* before the *HotelReservation* when the price is the important information to choose the hotel. This requirement can be represented as a *temporal relationship* between web services. These requirements are shown in Table 2.

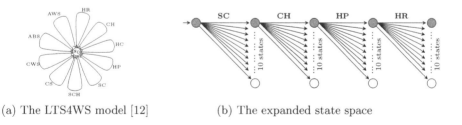

(a) The LTS4WS model [12]   (b) The expanded state space

Fig. 1. The model and expanded state space of unclustered web service repository

   As mentioned in the earlier text, the work of [12] will build the LTS4WS model for the input web services and explore the state space in order to compose

Table 1. The travel booking web service repository

#	Service Name	Input(s)	Output(s)	respTime
1	HotelReserveService (HR)	Dates, Hotel	HotelReservation	5
2	CityHotelService (CH)	City	Hotel	3
3	HotelCityService (HC)	Hotel	City	3
4	HotelPriceService (HP)	Hotel	Price	10
5	SightseeingCityService (SC)	Sightseeing	City	2
6	SightseeingCityHotelService (SCH)	Sightseeing	City, Hotel	16
7	CitySightseeingService (CS)	City	Sightseeing	4
8	ActivityBeachService (ABS)	Activity	Beach	5
9	AreaWeatherService (AWS)	Area	Weather	5
10	CityWeatherService (CWS)	City, Dates	Weather	5

Table 2. Requirements for Travel Booking web service

Kinds of constraint	Value(s)	
Hard constraint:	Input:	$Dates, Sightseeing$
	Output:	$Price, HotelReservation$
Soft constraint:	$respTime \leq 30$	
Temporal relationship:	$\square(\neg HotelReservation \cup Price)$	

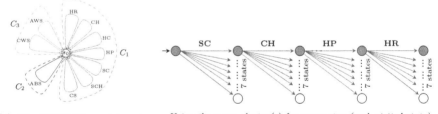

(a) The LTS4WS model and its clusters

Using the same cluster(s) for every step (each visited state)

(b) The expanded state space

Fig. 2. The model and expanded state space of combined feature and semantic-based clustered web service repository [17]

and verify the web services. The LST4WS model of the web services in Table 1 is illustrated in Fig. 1a. At every composing step, all web services are examined, resulting corresponding real states (and their heuristic values) produced in the state space. Then the "best" state (i.e. the current best composition), which is the state that is most likely the goal is chosen. This process is visually represented in Fig. 1b. It is easy to see that, the number of expanded (examined) states is very large, although we can obtain very good composition (that has been verified).

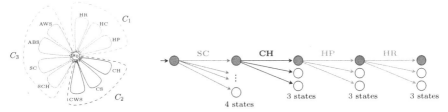

(a) The LTS4WS model and its logic-based clusters

Using different clusters at each step (each visited state)

(b) The expanded state space

Fig. 3. The model and expanded state space of logic-based clustered web service repository

In general, this solution will expand at least $n \times m$ states, where m is the length of composition and n is the number of input services. This approach then still suffers from high memory usage and is time consuming.

To overcome such drawbacks, we extend our work by suggestion of web services clustering and then further enhance it by employing logic-based distance mechanism. The advantages enjoyed by our approach are illustrated as follows.

Example 1. The advantage (and disadvantage) of typical approach on web service clustering.

Intuitively, when one attempts to find a WSC from a repository to achieve a goal, there are a lot of web services in the repository which seem irrelevant to the goal. However, in the approach above mentioned, in order to achieve the goal of $\{SC \bullet CH \bullet HP \bullet HR\}$, we still need to examine all of web services at every step. It is natural to consider clustering the input web services into clusters, and we only find the candidates from the clusters that are regarded as "relevant" to the goal. This approach is suggested in [17], where the web services are clustered, based on their features, into 3 clusters, C_1, C_2, C_3 as represented in Fig. 2a.

Applying this clustering approach to find a suitable composition, [17] uses only 7 web services in cluster C_1, which is find relevant to the goal. Obviously, the number of expanded states at every composition step is reduced to 7, the total expanded states will be only $\sum(n_i) * k$, where $\sum(n_i)$ is the number of web services of the chosen clusters. The expanded state space in this case is shown in Fig. 2b.

Unfortunately, in some cases, the thrown away web services in the unchosen clusters may be needed to compose the answer. For example, suppose that the user also needs the information about the *Weather* at the place where he travels. The suitable solution for this requirement is $\{SC \bullet CH \bullet HP \bullet HR \bullet CWS\}$. In this case, with the clustering approach in Fig. 2a, we have two ways of handling. In the first way, we choose only one cluster which has the highest similarity with the user requirement, which is the cluster C_1 with 7 web services. However, with this approach, we cannot obtain the solution because the web services related to *Weather* is CWS belong to the cluster C_2, which is not selected. In the second way, we choose all of clusters which have certain similarity degrees with the user requirement. In this case, we use two clusters (C_1 and C_2) with 9 services (consider 9 services at each step). That is a redundancy. Moreover, it loses the essence of clustering.

Example 2. The advantage of logic-based distance.

Considering only features may lead to the case that in the same cluster, some web services may be contradictory in term of logical meaning, even though they have the same semantic and features. For example, the service *CityHotelService* (takes *Hotel* and returns *City*) and *HotelCityService* (takes *City* and returns *Hotel*) are in the same cluster. However, at a particular composition step, if we have the *City* and need the *Hotel*, then, obviously, we will consider only the service *CityHotelService*, not *HotelCityService*. They are logically opposite and should not be in the same cluster.

Using our observation on logical aspect, the set of web services in Table 1 can be clustered into three groups consist of 3, 3, and 4 web services, respectively. They are illustrated as C_1, C_2, and C_3 in Fig. 3a. It is easy to observe that, the logical opposite web services *HotelCityService* (HC) and *CityHotelService* (CH) are now groups into two different clusters. So then, there is no contradiction when examining web services of a selected cluster at every composition step.

In this special example, the composition only expanded 13 states in total (see Fig. 3b). It is much better than the previous approach (28 states in Fig. 2b).

3   Related Works

3.1   Web Service Clustering

In this section, we discuss the works related to web service clustering. Kumara et al. [17] proposed a term-similarity approach to calculating the similarity of web services. Firstly, it uses an ontology-learning method. If this fails to calculate the similarities, it then uses an information retrieval (IR)-based method. To address the second issue, [17] proposes an approach that identifies the cluster center as cluster representative by combining the service-similarity value with the term frequency-inverse document frequency (TF-IDF) value of the service name.

Aznag et al. [2] proposed a non-logic-based matchmaking approach that uses correlated topic model to extract topics from service descriptions and search for services in the topic space where heterogeneous service descriptions are all represented as a probability distribution over topics. Xie et al. [21] proposed an ontology-based web service clustering approach. To cluster web services, it calculates the web service similarity based on function similarity. [21] supplies the matching method for inputs and outputs of web service. The web service similarity is based on the concept semantic similarity of the domain ontology.

Although those clustering-based approaches reduce the number of web services considered as candidates for the composition solution, they still face the difficulty on deciding the size and number of the selected clusters. If only few clusters are selected, we may miss some solutions. In contrast, if too many clusters are selected, the problem space may be enlarged unnecessarily. In other words, one cannot ensure the soundness nor the completeness of the solutions suggested from existing clustering approaches reported for the area of web services.

3.2   Web Service Composition and Verification

WSC only involves the functional properties (hard constraints) is the classic problem, which is mostly based on the theory of planning of the artificial intelligence field such as [11,16]. PORSCE II [11] is a framework implementing the WSC based on the requirements on input and output of services. Similarly, OWLS-XPlan [16] also uses web services expressed by OWL-S [4] to transform the problem from WSC domain to planning domain.

Some recent studies are based on abstract models such as Petri net (PN) or Colored Petri net (CPN). Maung et al. [19] proposed an approach to formalize the web service as a CPN, which provides means to observe the behaviors and the relationship of components. The composition mechanism systematically integrates these schemas into a transaction mapping model. Based on this, [19] proposed an algorithm which can verify the behaviors of composition.

WSC methods which combine functional and QoS properties has been proposed in [1]. In [1], the authors applied genetic algorithm (GA) to solve the problem with each possible composition encoded as a gene. However, this study only provides a mechanism to select the best (possible) composition from a set of composition ways (full composition schema) rather than composes from component services.

In the AI planning and workflows approaches, the web service composition is usually only considered on the functional properties. Therefore, these approaches only interested in composition, without verification. In the field of web service verification, most of current approaches verify hard or soft constraints separately. WS-Engineer [9] is a typical work for the web service verification based on functional properties. In [9], authors described a model-based approach to the analysis of service interactions. The composite web service was specified by using the *Finite State Process* algebra notation and the verification was done on this model by using the *Finite State Machine*. Another study that intends to verify combined functional and non-functional requirements is introduced by Chen et al. [5,6]. This approach takes in a full composition schema expressed in BPEL [14], then transforms that schema into a LTS model and uses a model checker to verify (the model). So far, model checking is the only technique which can verifies temporal relationship between web services when composed into composite one.

Recently, Huynh et al. [12] proposed a fast and formalized approach for web service composition and verification. In the proposed approach, authors simultaneously compose and verify web services on both hard, soft, and temporal relationship constraints in an on-the-fly manner. In addition, [12] also proposed some heuristics based on web service characteristics to improve the state space searching performance of the model checker. This approach was implemented as a framework, known as *OnTheFlyWSCV*.

4 Logic-Based Web Service Clustering

4.1 Logic Representation of Web Service

Definition 1 (Web Service). *A **web service** is formally defined by the features on input and output as follows.*

$$WS \equiv f_{in} \to f_{out} \equiv f_{1_{in}} \wedge \ldots \wedge f_{n_{in}} \to f_{1_{out}} \wedge \ldots \wedge f_{n_{out}} \tag{1}$$

Where,

- f_{in}, f_{out} are the logic expression represented the input, output functional properties,
- f_{in_i}, f_{out_j} are the terms.

Example 3. Let $WS1$ be the *HotelReservationService* service. $WS1$ takes *Hotel* and *Dates*, returns information about *HotelReservation*. We have:

- $f1_{in}$: $Hotel \wedge Dates$ $f1_{out}$: $HotelReservation$

or: $WS1 \equiv Hotel \wedge Dates \to HotelReservation$.

4.2 XML-based Web Service Description to Logic Representation Conversion

In practice, to allow being automatic discovered and composed, a web service is usually described by a structured XML-based declarative language like WSDL [7] or OWL-S [4]. Since parameters semantics of those languages are well-defined, we can seamlessly convert from the XML-based description into the logic formulas previously defined. In this section, we illustrate how an OWL-S description can be translated into a formal representation. Other web service description languages can also be translated in a similar manner.

The OWL-S to logic expression translation is the process which extracts the information about *name, input, output, precondition,* and *effect* of web services and converts them to the elements of logic expressions. Let WSP be a profile of web service WS and f is the transformed logic expression, respectively. Each element is transformed as follows.

- The left-hand side of logic expression is the input and precondition of the web service, $f_{in} = \bigcap_{k=1}^{n} WSP_i.hasInput_k \cap \bigcap_{k=1}^{m} WSP_i.hasPrecondition_k$
- The right-hand side of logic expression is the output and effect of the web service, $f_{out} = \bigcap_{k=1}^{n} WSP_i.hasOutput_k \cap \bigcap_{k=1}^{m} WSP_i.hasEffect_k$.

This translation process is described as a mapping in Fig. 4.

4.3 Logic-Based Web Service Similarity

Web service clustering is based on the similarity of web services. In our approach, we propose a novel metric of logic-based similarity between two web services being represented as two logic formulas.

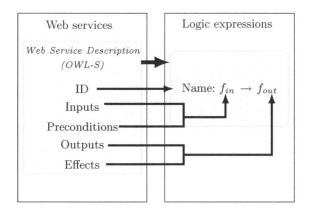

Fig. 4. Web service description to logic expression transformation process

Feature-Based Similarity. To calculate the feature similarity between two logic expressions, we have to determine the *identical part* and the *distinguishable part* of those expressions, which are defined as follows.

Definition 2 (Identical Part of Logic Expressions). *Let f, g be two logic expressions. The **identical part** (Ω) of two expressions is defined as follows.*

$$\Omega = f \cap g = \{\forall f_i \in f\} \cap \{\forall g_j \in g\} \tag{2}$$

Whereas, f_i, g_j are atomic terms in f and g, respectively. Let ω be the identical factor between f and g, then $\omega = |\Omega|$.

Example 4. Let f_1, f_2 be two logic expressions, we have:
$f_1 = Dates \wedge Hotel \rightarrow HotelReservation$
$f_2 = \neg Hotel \vee \neg Dates \vee City \vee HotelReservation$
$\Omega = \{Dates, Hotel, HotelReservation\}$
$\omega = |\Omega| = |\{Dates, Hotel, HotelReservation\}| = 3$.

Definition 3 (Distinguishable Part of Logic Expressions). *Let f, g be two logic expressions. The **distinguishable part** (Ψ) of two expressions is defined as follows.*

$$\Psi = (f \cup g) \setminus \Omega = (\{\forall f_i \in f\} \cup \{\forall g_j \in g\}) \setminus \Omega \tag{3}$$

Whereas, f_i, g_j are atomic terms in f and g respectively. Let φ be the different factor between f and g, then $\varphi = \frac{|\Psi|}{2}$.

Example 5. With f_1, f_2 in Example 4, we have:

$$\Psi = \{City\}, \qquad\qquad \varphi = \frac{|\Psi|}{2} = \frac{|\{City\}|}{2} = \frac{1}{2}.$$

Definition 4 (Feature-based Similarity). *Let f, g be two logic expressions represented of two web services. The* **feature-based similarity** *(Sim_{Fe}) of f and g is calculated based on the following formula.*

$$Sim_{Fe}(f, g) = \frac{\omega}{(\omega + \varphi)} \tag{4}$$

With ω and φ are defined in Definitions 2 and 3.

Example 6. With f_1, f_2 in Example 4 and the values of identical part and distinguishable part are calculated in Examples 4 and 5 respectively, we have the feature similarity of them as

$$Sim_{Fe}(f_1, f_2) = \frac{3}{(3+\frac{1}{2})} = \frac{6}{7} \approx 0.86$$

Logic-Based Similarity. Even though the feature-based similarity can calculate the similarity between two web services, this similarity measure still does not sufficiently reflect the similarity of the logic in data processing of the involved web services. We handle this by introducing the ultimate similarity measure, known as *logic-based similarity of web services*.

Definition 5 (Over-approximation of Two Logic Expressions). *Let f_1 and f_2 are two logic expressions represented of the two web services. The* **over-approximation of two logic expressions** *is defined as follows.*

$$f_{12} = f_1 \oplus f_2 = f_{1_{in}} \wedge f_{2_{in}} \rightarrow f_{1_{out}} \vee f_{2_{out}} \tag{5}$$

Where, $\oplus$ is an *over-approximation operator*. Then, f_{12} will be simplified by a prover, such as Z3 [8].

Example 7. Consider the first three web services in Table 1. Their logic expressions are calculated as follows.

- $f_1 = Dates \wedge Hotel \rightarrow HotelReservation$
- $f_2 = City \rightarrow Hotel$ $f_3 = Hotel \rightarrow City$

  Some over-approximations of these expressions are calculated as follows.

- $f_{12} = f_1 \oplus f_2 = (Dates \wedge Hotel \rightarrow HotelReservation) \oplus (City \rightarrow Hotel) = (Dates \wedge Hotel \wedge City \rightarrow Hotel \vee HotelReservation) = true$
- $f_{23} = f_2 \oplus f_3 = (City \rightarrow Hotel) \oplus (Hotel \rightarrow City) = (City \wedge Hotel \rightarrow Hotel \vee City) = true$
- $f_{13} = f_1 \oplus f_3 = (Dates \wedge Hotel \rightarrow HotelReservation) \oplus (Hotel \rightarrow City) = (Dates \wedge Hotel) \wedge Hotel \rightarrow City \vee HotelReservation = \neg Hotel \vee \neg Dates \vee City \vee HotelReservation.$

Definition 6 (Logic-based Web Service Similarity). *Let W_A, W_B be two web services, f_A, f_B be the logic expressions represented for W_A, W_B, respectively, and f_{AB} be the over-approximation of f_A and f_B as in Definition 5. The* **logic-based web service similarity** *of W_A and W_B is calculated as follows.*

$$Sim_{Lo}(W_A, W_B) = \frac{Sem_{Fe}(f_A, f_{AB}) + Sem_{Fe}(f_B, f_{AB})}{2} \tag{6}$$

Where $Sem_{Fe}(f_1, f_2)$ is the feature similarity of f_1 and f_2.

Example 8. The logic-based similarity of the first three web services in Table 1 is calculated as follows.

The over-approximation expression of each pair of services is calculated as in Example 7.

- $f_{12} = f_1 \oplus f_2 = true$ $\qquad\qquad$ $f_{23} = f_2 \oplus f_3 = true$
- $f_{13} = f_1 \oplus f_3 = \neg Hotel \vee \neg Dates \vee City \vee Hotel Reservation$

The feature-based similarities of them are calculated as follows:

- $Sem_{Fe}(f_1, f_{12}) = 0;$ $\qquad$ $Sem_{Fe}(f_2, f_{12}) = 0$
- $Sem_{Fe}(f_2, f_{23}) = 0;$ $\qquad$ $Sem_{Fe}(f_3, f_{23}) = 0$
- $Sem_{Fe}(f_1, f_{13}) = \frac{6}{7};$ $\qquad$ $Sem_{Fe}(f_3, f_{13}) = \frac{2}{3}$

Eventually, the logic-based similarities of each pair of web services are calculated as follows.

- $Sim(WS_1, WS_2) = \frac{Sem_{Fe}(f_1, f_{12}) + Sem_{Fe}(f_2, f_{12})}{2} = 0$
- $Sim(WS_2, WS_3) = \frac{Sem_{Fe}(f_2, f_{23}) + Sem_{Fe}(f_3, f_{23})}{2} = 0$
- $Sim(WS_1, WS_3) = \frac{Sem_{Fe}(f_1, f_{13}) + Sem_{Fe}(f_3, f_{13})}{2} = \frac{16}{21}.$

4.4 Logic-Based Web Service Clustering

To perform clustering, we apply the k-means algorithm [13], one of the most popular clustering techniques. Determining the number of clusters is an important problem of clustering algorithms in general. In the WSC, suppose that we have N services clustered into k clusters. Each cluster has the average of N/k services. At each composition step, we need k comparisons to select the most appropriate cluster from k clusters. Then, we perform N/k comparisons in examining all web services in the selected cluster. Thus, the number of comparisons in each composition step will be $k + N/k$. As our objective is minimizing this value, we achieve the optimal minimization when $k = \lfloor \sqrt{N} \rfloor$.

Another important problem is how to form the representative element for each cluster. Note again that our web services are represented by logic expressions. Therefore, our representative is also defined using logic expression.

Definition 7 (Representative Element). *Let $F = \{f_1, f_2, \ldots, f_k\}$ is a cluster of logic expressions, the **representative element** (f_r) of F is defined as follows.*

$$f_r = \biguplus f_i = f_{1_{in}} \vee \cdots \vee f_{k_{in}} \rightarrow f_{1_{out}} \wedge \ldots \wedge f_{k_{out}} \tag{7}$$

Where $\biguplus$ is the representative operator.

Proof (of Soundness). Our composition solution is sound, i.e. the final composition correctly satisfies the requirements of the original goal. This soundness is obviously achieved, because the logic-based clustering approach only reduces the number of web service candidates to be taken into consideration at each composition step, and only candidates producing valid composition w.r.t the goal requirements are selected for final composition solution.

Proof (of Completeness). To prove the completeness of our solution, we need to prove that our solution does not miss any suitable candidate when finding a solution. Let us consider a certain web service f_i, which is a suitable candidate for the current composition goal f_g, i.e. the input of f_i can be entailed from the input of f_r, or $f_{g_{in}} \rightarrow f_{i_{in}}$. Hence, if f_i is clustered into a cluster c, whose representative element is f_r, then the input of f_r can also be entailed by $f_{g_{in}}$, or $f_{g_{in}} \rightarrow f_{r_{in}}$.

In other words, we need to prove $f_{g_{in}} \rightarrow f_{i_{in}} \models f_{g_{in}} \rightarrow f_{r_{in}}$. The formal proof is given as follows.

$$\cfrac{\cfrac{\cfrac{\overline{f_{g_{in}}}\ assumption \quad \overline{f_{g_{in}} \rightarrow f_{i_{in}}}\ premise}{f_{i_{in}}} \rightarrow e}{f_{1_{in}} \vee \ldots \vee f_{i_{in}} \vee \ldots \vee f_{k_{in}} \equiv f_{r_{in}}} \vee i}{f_{g_{in}} \rightarrow f_{r_{in}}} \rightarrow i.$$

Example 9. Suppose that two web services *CityHotelService (CH)* and *CitySightseeingService (CS)* in Table 1 are clustered into a cluster. Then, the representative logical expression of this cluster is:

$$f_r = f_{CH} \uplus f_{CS} = City \vee City \rightarrow Hotel \wedge Sightseeing$$
$$= City \rightarrow Hotel \wedge Sightseeing$$

Table 3. Logic-based clustering of Travel Booking repository

Clusters	Web services	The representative logical expression
#1	ABS, AWS, SC, SCH	*Activity* $\vee$ *Area* $\vee$ *Sightseeing* $\rightarrow$ *Beach* $\wedge$ *Weather* $\wedge$ *City* $\wedge$ *Hotel*
#2	CH, CS, CWS	*City* $\rightarrow$ *Hotel* $\wedge$ *Sightseeing* $\wedge$ *Weather*
#3	HR, HC, HP	*Hotel* $\rightarrow$ *City* $\wedge$ *Price* $\wedge$ *HotelReservation*

Table 4. The experimentation datasets

Dataset	No. of web service	Description
Travel Booking (TB)	20	Provide information to serve the travel booking
Medical Services (MS)	50	Support to look up hospital, treatment, medicine, etc.
Education Services (EDS)	100	Supply education services such as scholarship, courses
Economy Services (ECS)	200	Provide information about goods, restaurant, food, etc.
Global	1000	1000 random services from OWLS-TC [15].

After determining the number of clusters and the method of identifying cluster representative element, the k-means algorithm is used to implement. The result of this algorithm is a set of clusters represented by a logic expression.

Example 10. The result of the logic-based clustering for the web services in Table 1 is in Table 3.

5   Experimentation

In this section, we will present the experimental results of our approach. To evaluate, we use the framework OnTheFlyWSCV [12] to compose web services. Our experiments work on the real datasets obtained from the project OWLS-TC [15] with over 1000 web services classified into different domains and described by OWL-S [4]. In this dataset, we select five sub-datasets with the number of web services is varied 20 to 1000 services as shown in Table 4.

We conduct the experiment scenarios based on three approaches. These include *Unclustered*; *Combined feature and semantic based clustering* – the approach was proposed in [17]; and *Logic-based clustering* – the approach is proposed

Table 5. Experimentation results

Dataset	Approach	Expanded states	Visited states	Execution time (s)
TB (20)	Unclustered	1,100	56	0.250
	Combined feature and semantic-based clustering	825	56	0.210
	Logic-based clustering	316	35	0.155
MS (50)	Unclustered	6,251	125	1.008
	Combined feature and semantic-based clustering	4,689	125	0.820
	Logic-based clustering	1,126	75	0.614
EDS (100)	Unclustered	27,400	275	5.570
	Combined feature and semantic-based clustering	20,824	275	4.579
	Logic-based clustering	3,021	151	3.294
ECS (200)	Unclustered	102,800	515	28.198
	Combined feature and semantic-based clustering	76,586	515	23.030
	Logic-based clustering	8,324	287	18.273
Global (1000)	Unclustered	-	-	-
	Combined feature and semantic-based clustering	-	-	-
	Logic-based clustering	91,457	1,429	597.228

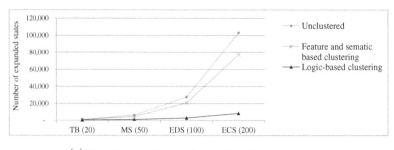

(a) Visual comparison on the number of expanded states

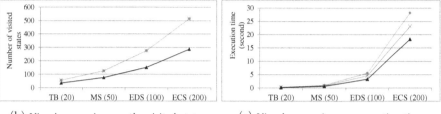

(b) Visual comparison on the visited states   (c) Visual comparison on execution time

Fig. 5. Visual representation of the experimental results

in this paper. The experiments were executed on a PC with core i5-5200 processor (4×2.7 Ghz), 8.0 Gb of RAM, running on Windows 7 64-bit. The result of experiments is evaluated in three aspects: the number of expanded states, the number of visited states, and the execution time. The experiment results are analyzed statistically in Table 5 and Fig. 5.

The experimental results confirm our hypothesis that clustering helps in reducing the number of expanded states (see Fig. 5a). Note that, the heuristic algorithm in the OnTheFlyWSCV tool always chooses the best way to travel in the state space. The number of visited states shows the number of best states that have been chosen during the search. For the original and the second approach, when the "best" states (based only on features and semantic) are the same, then the number of visited states is the same. In our approach, when the contradictory web services cannot be in the same cluster, then the number of "best" states are smaller resulting in the number of visited states obviously smaller (Fig. 5b). As the execution time mostly depends on the number of expanded states, the second approach is faster than the original one and our approach is the best (Fig. 5c).

6   Conclusion

In this paper, we proposed a logic-based web services clustering method. Our method improves the WSC performance when it allows the system to choose a small set of web services which are suitable for user requirement on logical aspect. Moreover, the clustering can be performed beforehand, independently with the requirements. The soundness and completeness of our approach is also proven theoretically.

Also through this paper, we have some contributions about the logical representation of web service; the method to calculate the logic-based similarity between logical expressions represented for web services by means of the over-approximation operator; and the method to build the representative logical expression that represents for web service cluster. Our experiments show that the system parameters such as the number of expanded states, the number of visited states, and the processing time are better than other existing approaches.

In a broader sense, our approach is also used to compose the software components, not just web services. The software components can be described by two elements, inputs and outputs (or precondition and postcondition) are used to compose and verify in our approach. Recently, RESTful services have acquired a great deal of popularity [10]. AS RESTful web services are also described by an XML-based language [18], such as WSDL, we can fully apply our approach for the RESTful web service composition problem.

For the future works, there are several points to be considered. In this work, we have not yet applied the semantic similarity between two concepts to handle ontology alignment issue. It could be investigated in the near future. Also, in this work, we use the k-means clustering algorithm. The experimentation on the pros and cons of using other clustering algorithms has not been done yet. It is another future work.

Acknowledgment. This research is funded by Vietnam National University Ho Chi Minh City (VNU-HCM) under grant number C2015-20-10.

References

1. AllamehAmiri, M., Derhami, V., Ghasemzadeh, M.: QoS-based web service composition based on genetic algorithm. J. AI Data Min. **1**(2), 63–73 (2013)
2. Aznag, M., Quafafou, M., Jarir, Z.: Leveraging formal concept analysis with topic correlation for service clustering and discovery. In: 2014 IEEE International Conference on Web Services (ICWS), pp. 153–160. IEEE (2014)
3. Baier, C., Katoen, J.P., et al.: Principles of Model Checking, vol. 26202649. MIT Press Cambridge, Cambridge (2008)
4. Burstein, M., Hobbs, J., Lassila, O., Mcdermott, D., Mcilraith, S., Narayanan, S., Paolucci, M., Parsia, B., Payne, T., Sirin, E., et al.: OWL-s: Semantic markup for web services. W3C Member Submission (2004)
5. Chen, M., Tan, T.H., Sun, J., Liu, Y., Dong, J.S.: Veriws: a tool for verification of combined functional and non-functional requirements of web service composition. In: Proceedings of the 36th International Conference on Software Engineering, pp. 564–567. ACM (2014)
6. Chen, M., Tan, T.H., Sun, J., Liu, Y., Pang, J., Li, X.: Verification of functional and non-functional requirements of web service composition. In: Groves, L., Sun, J. (eds.) ICFEM 2013. LNCS, vol. 8144, pp. 313–328. Springer, Heidelberg (2013)

7. Chinnici, R., Moreau, J.J., Ryman, A., Weerawarana, S.: Web services description language (WSDL) version 2.0 part 1: Core language. W3C recommendation 26, 19 (2007)

8. de Moura, L., Bjørner, N.S.: Z3: an efficient SMT solver. In: Ramakrishnan, C.R., Rehof, J. (eds.) TACAS 2008. LNCS, vol. 4963, pp. 337–340. Springer, Heidelberg (2008)

9. Foster, H., Uchitel, S., Magee, J., Kramer, J.: Ws-engineer: a model-based approach to engineering web service compositions and choreography. In: Baresi, L., Di Nitto, E. (eds.) Test and Analysis of Web Services, pp. 87–119. Springer, Heidelberg (2007)

10. Garriga, M., Mateos, C., Flores, A., Cechich, A., Zunino, A.: Restful service composition at a glance: a survey. J. Netw. Comput. Appl. **60**, 32–53 (2016)

11. Hatzi, O., Vrakas, D., Bassiliades, N., Vlahavas, I.: The porsce ii framework: using ai planning for automated semantic web service composition. Knowl. Eng. Rev. **28**(02), 137–156 (2013)

12. Huynh, K.T., Quan, T.T., Bui, T.H.: Fast and formalized: heuristics-based on-the-fly web service composition and verification. In: 2nd National Foundation for Science and Technology Development Conference on Information and Computer Science, pp. 174–179. IEEE (2015)

13. Jain, A.K.: Data clustering: 50 years beyond k-means. Pattern Recogn. Lett. **31**(8), 651–666 (2010)

14. Jordan, D., Evdemon, J., Alves, A., Arkin, A., Askary, S., Barreto, C., Bloch, B., Curbera, F., Ford, M., Goland, Y., et al.: Web services business process execution language version 2.0 (2003)

15. Klusch, M.: Owls-tc: Owl-s service retrieval test collection, version 2.1. http://projects.semwebcentral.org/projects/owls-tc/

16. Klusch, M., Gerber, A., Schmidt, M.: Semantic web service composition planning with OWLS-xplan. In: Proceedings of the AAAI Fall Symposium on Semantic Web and Agents, USA. AAAI Press (2005)

17. Kumara, B.T., Paik, I., Chen, W., Ryu, K.H.: Web service clustering using a hybrid term-similarity measure with ontology learning. Int. J. Web Serv. Res. (IJWSR) **11**(2), 24–45 (2014)

18. Mandel, L.: Describe rest web services with wsdl 2.0. Rational Software Developer, IBM (2008)

19. Maung, Y.W.M., Hein, A.A.: Colored petri-nets (CPN) based model for web services composition. IJCCER **2**, 169–172 (2014)

20. Tretmans, J.: Model based testing with labelled transition systems. In: Hierons, R.M., Bowen, J.P., Harman, M. (eds.) FORTEST 2008. LNCS, vol. 4949, pp. 1–38. Springer, Heidelberg (2008)

21. Xie, L.L., Chen, F.Z., Kou, J.S.: Ontology-based semantic web services clustering. In: 2011 IEEE 18th International Conference on Industrial Engineering and Engineering Management (IE&EM), pp. 2075–2079. IEEE (2011)

MindDomain: An Interoperability Tool to Generate Domain Models Through Mind Maps

Alejo Ceballos[1], Fernando Wanderley[2(✉)], Eric Souza[2],
and Gilberto Cysneiros[3]

[1] CESAR - Centro de Estudos e Sistemas Avançados do Recife, Recife, Brazil
alejoceballos75@gmail.com
[2] Universidade Nova de Lisboa, Lisbon, Portugal
fernando.wanderley@gmail.com, eurocha@gmail.com
[3] Universidade Federal Rural de Pernambuco, Recife, Brazil
g.cysneiros@gmail.com

Abstract. Requirements engineering establishes that requirements definition process must be applied to obtain, validate and maintain one or more requirement documents. This process handles different stakeholders expectations and viewpoints, among them, the software designer whose responsibility is to create software models from information provided by domain experts and business specialist. However, due to knowledge differences between stakeholder's technical dialects, communication problems are constant, generating inconsistencies between the conceptual model and the problem to be solved. To help solving these issues an agile and cognitive modeling based approach supported by MDA based tools is proposed promoting better consistency between requirements and the conceptual models, guaranteed by specifying a mind map that serves as the basis for translating requirements to domain models, represented by the UML class diagrams and feature models. Thus, the main contribution of this work is to provide an interoperability tool to generate software models (e.g.: class diagrams and feature models) from mind maps. This tool provides the capability of transformation between different industrial mind map tools (including cloud tools - SaaS) to different domain modelling tools, both class diagrams and for feature models. Finally, a case study was applied to verify this feasibility and check this interoperability assessment. The main contribution is MindDomain can be used in small projects for agile requirements modeling solutions.

Keywords: Agile requirements modelling · Model-driven engineering · Class diagram · Domain model · Mind map

1 Introduction

According to Chaos Report [1], only 32 % of all software development projects were successful completed, 44 % were completed without achieving its goals (late,

© Springer International Publishing Switzerland 2016
O. Gervasi et al. (Eds.): ICCSA 2016, Part IV, LNCS 9789, pp. 469–479, 2016.
DOI: 10.1007/978-3-319-42089-9_33

over budget, or with incomplete functionalities) and 24 % failed (not completed or discarded). The study concluded that the successful projects includes the participation of users with business knowledge and good communication skills, while the other projects typically include users with poor communication skills.

Brooks [2] states that "the most difficult part in building a software is deciding precisely what to build". In the software development process, "what to build" is defined by the requirements that describe how the system should behave, its properties or attributes, and its constraints [3]. Moreover, the requirements must serve as a link between customer needs and the development team. According to Brooks [2], during the requirements elicitation it is necessary to establish the communication between the business expert and the software designer to consolidate the technical details of the business in the same requirements model. Also according to Brooks [2] "if done wrong, no other part of software development process cause more damages to the end result".

With the advent of agile methodologies, more flexible practices of elicitation and documentation were being formalized with a focus on the mutability of the requirements and the improvement in communication between those responsible for the tasks of analysis and software development. From the perspective of software modeling, agile methodologies transformed prescriptive processes of software designers in suggestions to be applied in day to day business so that it becomes more efficient. One of these suggestions, defined by the agile modeling, considers it necessary to reduce the level of detail of the models limiting its function to "understand what is being built, or assist in communication between team member" [4]. Both understanding and assistance in communication occurs with the use of a ubiquitous language, such as Unified Modeling Language (UML) which has "its broad and standardized use within the community of object-oriented software designer" [5]. However "domain experts have a limited understanding of the technical jargon of software development" [6] and these limitations become a crucial factor when the client's wishes are translated in the software to be implemented.

According to the presented context, this paper proposes an agile and interactive approach to fill the gap between the identification of the characteristics of a system (requirements) and how they should be developed (domain models), promoting greater consistency between what is raised by experts domain and what is represented by software designers. Through the use of mind maps, this approach evokes principles of agile modeling as "communication, simplicity and feedback" [4]. Subsequently, the same approach should serve as the basis for translation of several distinct types of models from mind maps.

2   Background

2.1   Agile Modelling

The Agile Modeling is a methodology considered chaordic (chaotic and ordered) that uses various models generated from a collection of practices guided by principles and values without setting any detailed procedure for its creation [4].

Based on the principles and values of the agile manifesto emphasizes communication, simplicity, feedback, courage and humility in software development in
order to improve the activities that allow the application of good modeling teams
using agile processes. But, what does mean an effective modeling? According to
Ambler [4], a model should have the following characteristics:

(i) Serve the purpose of its creation - For example, to communicate the
scope of what should be developed, to understand the problem being presented
or determining a design strategy. The model should be useful for what has
been created; (ii) Be understandable by the target audience. - An architectural
model contains common technical terms to software designers while a requirement model will be written using the domain expert language. The goal is that
the model reaches and serves the public that it was intended; (iii) Be consistent
enough - Inconsistencies can be performed since they do not affect the purpose
of the model. A map that displays the name of a street wrongly cannot be considered useless if the goal is not to use it to be located. (iv) Be as simple as
possible - Simplicity is affected by the level of detail of the model. If a class diagram uses full syntax possible and becomes of difficult to read, a more simplistic
approach to UML can be attempted. Ambler also argues that several models
should be created and worked in parallel as a single model alone cannot convey
the complexity of the problem being analyzed.

2.2   Mind Maps

Mind mapping is a technique used for understanding and memorizing by structuring ideas from their hierarchy and categorization [7]. The mindmap motivation comes from stimulation of storage capacity through associations of ideas,
since memory works through an activation process that spreads from word to
word associated via links.

The entire capacity of the mind maps is applied with an image or key word
and cognitive use of images instead of words [7]. The central idea should be identified and associated complementary subsequent are being placed hierarchically,
always referencing a narrower context from the previous.

Mind Maps is a practical and clear way to organize and prioritize any type
of knowledge, idea, among other activities necessary for learning, memory and
planning [8]. When working with keywords, images numbers, colors and spatial
perception mind maps maximize the stimulus to the cerebral cortex. The association represented ideas in the mind maps used to order the mental freedom and
although often the order is confused with rigidity, and freedom with chaos, mental freedom comes the ability to find and organize this order that exists within
the chaos [7]. This theory is similar to the concept of chaordic defended by Scott
Ambler and his approach to agile modeling.

Figure 1-a illustrates a mind map where its elements are arranged so intuitive,
according to the importance of the concepts related to the domain.

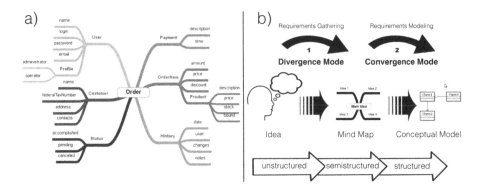

Fig. 1. (a) Mind map from [8] and (b) role to elicit unstructured ideas.

2.3   Agile Requirements Modeling with Mind Maps

The mind map in the context of this paper, is the representation of a ubiquitous language that is a common language that distributes domain knowledge equally among all those involved in the software development process, with the objective to extract and represent the domain using a spatial form.

When we use the mind map as a model semi-structured in order to understand and represent the problem to be solved, we reduced the semantic gap between those involved and create a starting point for a semi-automatic transformation for more specialists models with more functions specific, such as classes diagrams for representing domain entities, can thus generate a closer conceptual model of reality of the problem [8].

Several authors have been making reference to the use of mind maps in the process of obtaining requirements, such as Larman [9,10]. They use mind maps to elicit requirements as well as defining user stories [10]. Mahmud [11] reports empirical evidence on the effectiveness in the use of mind maps to elicit requirements within the context of an agile project.

Figure 1-b illustrates the role of mind map to capture unstructured ideas and turn them into a structured conceptual model closer to the solution of the problem.

3   The MindDomain tool

3.1   The Model Transformation Process

To improve productivity in software development process, the MDE (Model Driven Engineering) emphasizes the need to have models that can be automatically manipulated and transformed into other artifacts [12]. The authors define transformation of models as the generation of a target model from a source model, by defining rules mapping established between the elements of its metamodel.

The Fig. 2-a shows the transformation of a mind map represented in XML, model A (Ma) to domain model represented in XMI, model B (Mb). The Mind Domain execute the transformation from A to B if model A complies with the metamodel A (MMa) and if model B complies with metamodel B (MMb).

This paper presents an approach that uses transformation of model from mind maps to domain models. The transformation is done through a tool called MindDomain. The MindDomain proposed is the "instance of the" or the operationalization of the transformation rules defined by Wanderley et al. [13,14]. The authors defined formally the mapping of elements between source models (mind maps) and target models (class diagrams and feature models), so MindDomain proposed here is the a kind of tool for realizes these transformation rules between industrial modelling tools (e.g.: (i) Freemind and DRichards - for mind map tools; (ii) ArgoUML, jsUML2 and yUML for Class diagrams; and (iii) FeatureModelPlugin for Feature Models). To perform this interoperability capability the MindDomain was design by well-formed interfaces and component-base architecture.

3.2   Mind Domain Architecture

The Mind Domain tool consist of an API (Application Programming Interface) that provides a set of services to allow the transformation from mind maps to conceptual model. The model are represented in XML. Each instance of the model (Ma) complies with the metamodel (MMa) and its transformation correspond to another model (Mb) that complies with the metamodel (MMb).

Figure 2-b depicts the architecture of Mind domain. This was defined as a component architecture to ensure that new models as well as new transformation rules, can be added at no cost in maintaining existing features. Dependencies are represented in this way to illustrate the variability architecture processing functionality. For each source model instance can be multiple instances of target models depending only a translator for this transformation to be implemented. During the development of this tool, various types of sources and targets have been implemented.

During the development of this tool, various types of sources and targets have been implemented. The files generated by Freemind tools and Drichards were used as instances of mind maps source. Possible targets models are compatible with the tools ÁrgoUML, jsuml2, yUML5 and Eclipse Feature Model Plugin (FMP) 6.

Only four transformations were tested during the study: (i) Freemind to ArgoUML, represented by the fm2argouml component; (ii) Freemind to jsuml2, represented by the fm2jsuml2 component; (iii) DRichards to jsuml2, represented by the dr2argouml component; and (iv) Freemind to FMP, represented by the fm2fmp component.

The components of the Domain architecture can be described as:

– Mind Domain API - is the interface definition and publication of services API. Here are all the interfaces whose signatures must be implemented to engage new models and new changes to tool.

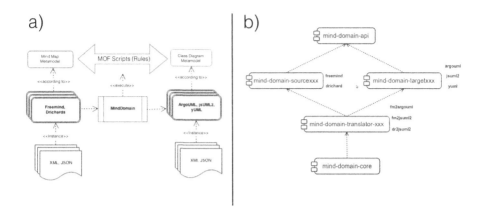

Fig. 2. (a) Transformation model and (b) MindDomain architecture.

- Mind Domain Source - is the set of modules that implement the interfaces that represent the input tool designs. These modules are responsible for reading files and consequent conversion to a model suitable for further processing by Mind Domain. In the context of this study, the entries are summarized instances maps mental (i.e., Freemind and Drichards).
- Mind Domain Target - is the set of modules that implement the interfaces that represent the models resulting from the processing. These modules are responsible for transforming a generated model by Mind Domain in an instance readable by a tool visual. In the context of this study, the outputs may be instances class model (i.e., ArgoUML, jsuml2 and yUML) or feature models (i.e., Eclipse Plugin Feature Model).
- Mind Domain Translator - is the set of modules that implement the transformations from one model to another. Given an input and for each desired output, a change is implemented. This approach was an architectural decision, subject to refactoring, due to factors such as time.
- Mind Domain Core - is the module that manages which instance of each Module should be used and when. Through plants (abstract factory) and other standards projects [15] states that instances of input components, translation and output are created, making thus the possibility of implementing various uncoupled translators independent and cohesive models.

4   Case Study

In order to illustrate the method described in this work will be used as case study: the management product orders for an e-commerce presented originally by Wanderley [8].

In this study, from the perspective of suppliers, the benefits of a constant order cycle creates a continuous flow of capital and avoids situations orders punctual in uncertain times. For this, the development of a logistics operating that meets the customer a satisfactory term is sine qua non condition for success of the process. From this context, the business rules to control Applications must be expressly guaranteed and developed in sufficient space of time.

From the perspective of the domain expert in conceptual terms, a request need to have primarily: (1) a request item that relates directly to the (2) product, which has the description and the price displayed on an invoice, while emphasizing that an order item can have only one associated product, resulting in the existence of a quantity field for the item. The system should be able to manage the information (3) user who generated the request, this user can access the system through your login and password, with (4) operator or administrator profiles. Due to the large flow applications, the expert suggests that the system manage the state (status) of applications. The states range from accomplished, pending and canceled. The system must manage the information as to (6) payment through a brief description and time held, as well as a history (7) reporting the user-recorded changes.

For the case study above, this article defines a process where the specialist domain presents the rules of the business and its conceptual terms and the software designer uses this description of context of the problem to be resolved, and ask question about: (i) The terms and main elements of the problem, (ii) The terms or elements directly associated with the term central, and (iii) What information should be managed by the terms defined previously?

With answers to these questions, experts have resources to extract a mind map representing the elements and associations that consist a solution to the problem to be solved. This semi-structured model serve as input to the process transformation to a well-structured model that should be validated and refined by the domain expert finally serving as artifact to guide the team development. Figure 3 illustrates the mind mapping process and its transformation of a semi-structured model for a structured model.

The mind map shown in Fig. 4 illustrates the resulting artifact of the process questioning, visualized by Freemind tool (for better viewing, is the same structure shown in Fig. 1). The request is the central term, or aggregate root, term advocated by Evans [6] which states that within a domain boundary must always be a root element that holds references to elements associated within that border. Each node within the exposed area can be transformed to an entity or entity attribute, depending on the rule being employed in the transformation.

The mind map is the input artifact for the Mind Domain start transformation process. It is necessary to indicate to the tool which is the type of instance mind map to serve as input and what type of transformation and output instance desired. With that the controller of the transformation, located on the module Mind Domain Core is able to instantiate the correct modules to perform the conversion between the models.

The transformation rules for the scenario presented in this paper are: 1 - The central ideas (aggregate roots) are considered entities, 2 - the ends are considered attributes, and 3 - if an entity to connect to another entity applies an association of aggregation.

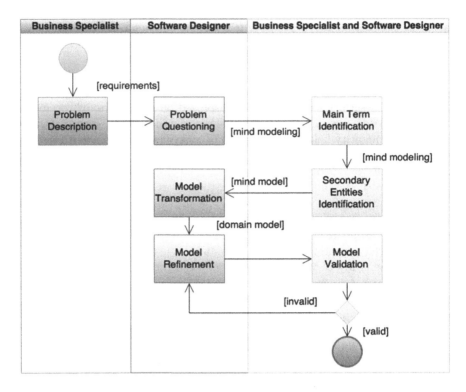

Fig. 3. Mind mapping process.

Figure 5 shows the domain model resulting of the transformation, visualized by ArgoUML tool.

5   Related Works

The Astah-Community [16] a tool that has been growing use and adoption by industry, refers to a feature of generate UML diagram from Mind Maps, but an important and relevant difference from MindDomain proposed here is the reliable transformation and generate software models guaranteed by metamodel mapping and model-driven techniques applied. The Astah-Community proposed the "manually" transformation of each node into a specific class.

Both academy and industry have been considering techniques and mind map based tools for managing requirements elicitation. Mahmud reported [11] interesting results collected during an experiment with experts and non-experts using mind maps. Kof et al. [17] used the concept of map visualization for requirements tracing. Laplante [18] mentioned that mind maps could be used during elicitation discussions with stakeholders aiming to capture the essential parts (business terms) of problem space. Village et al. [19] reports the learnability characteristic of use map visualization in early software engineering design.

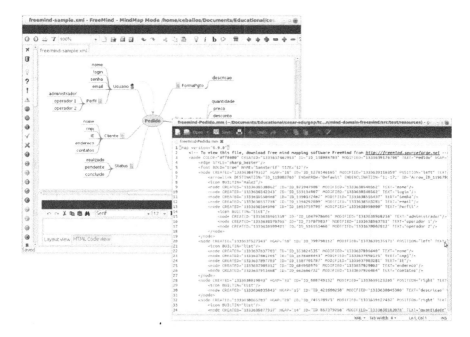

Fig. 4. Mind map artifact and xml file.

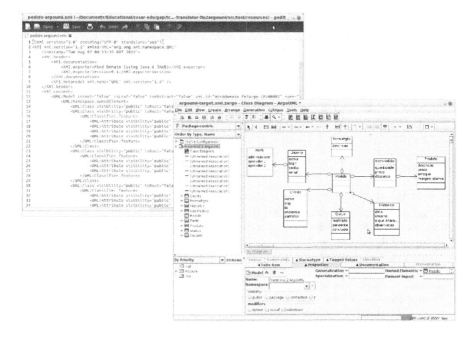

Fig. 5. Domain model visualized by ArgoUML tool.

Alexander et al. [20] reports that mind maps can help requirements engineers to remember and make good holistic visualizations about all system details, using a mixing of words, pictures and other rich resources. Sutcliffe et al. [21] discuss their experience about user-centred requirements engineering, where mind maps were used to encourage epidemiologists (domain experts) to make more use of visualization tools.

6   Conclusion and Future Works

This paper is based on [13, 14] and it presents a tool that from a defined method, and the transformation between models, assists in reducing the existing semantic gap between the domain expert and the software designer. Moreover, the method itself is the tool that allows a real approximation between these two key groups of the software development process.

The proposed method uses adaptive rules that maintains consistency of models created relying on the interaction between those involved, one of the values of Agile Manifesto [22]. The use of multiple models (e.g., mind map and domain model) is also one of the principles advocated by Ambler [4]. In short, mind maps are great tools for memorizing and brainstorming, used within an agile and interactive context allow the formalization of a common language between domain experts and software designers.

Together with MDE, it becomes a powerful resource in the automation of generation more consistent models that reflect more accurately the problem domain They are faced. The Mind Domain tool, together with the process presented here is an example of how this level of automation and agility can be achieved. It brings evidence that mind maps can be a good starting point for Agile Modeling making the participation of the domain expert more active in the software development process.

Currently, we are executing an experimental analysis aims to collect real evidence of the level of productivity that can be achieved with the use of the tool within the context agile software development.

References

1. Chaos Manifesto, The laws of chaos and the chaos 100 best pm practices, The Standish Group International (2011)
2. Brooks Jr., F.P.: No silver bullet essence and accidents of software engineering. Computer **4**, 10–19 (1987)
3. Sommerville, I., Sawyer, P.: Requirements Engineering: A Good Practice Guide. Wiley, New York (1997)
4. Ambler, S.: Agile Modeling: Effective Practices for Extreme Programming and the Unified Process. Wiley, New York (2002)
5. Fowler, M.: UML Distilled: A Brief Guide to the Standard Object Modeling Language. Addison-Wesley Professional, Reading (2004)
6. Evans, E.: Domain-Driven Design: Tackling Complexity in the Heart of Software. Addison-Wesley Professional, Reading (2004)

7. Buzan, T., Buzan, B.: The Mind Map Book How to Use Radiant Thinking to Maximise Your Brain's Untapped Potential. Plume, New York (1993)
8. Wanderley, F.: Uma Infraestrutura para Diminuir o Gap entre o Especialista de Negcio e o Projetista de Software, Ph.D. dissertation, June 2012
9. Larman, C.: Agile, Iterative Development: A Manager's Guide. Addison-Wesley Professional, Reading (2004)
10. Requirements management, user stories, mind-maps, story-trees. https://epistemologic.com/2007/04/08/requirements-management-user-stories-mind-maps-and-story-trees/. Accessed 18 May 2016
11. Mahmud, I., Veneziano, V.: Mind-mapping: an effective technique to facilitate requirements engineering in agile software development. In: 2011 14th International Conference on Computer and Information Technology (ICCIT), pp. 157–162. IEEE (2011)
12. Kleppe, A.G., Warmer, J.B., Bast, W.: MDA Explained: The Model Driven Architecture: Practice and Promise. Addison-Wesley Professional, Reading (2003)
13. Wanderley, F., da Silveira, D.S., Araujo, J., Moreira, A.: Transforming creative requirements into conceptual models. In: 2013 IEEE Seventh International Conference on Research Challenges in Information Science (RCIS), pp. 1–10. IEEE (2013)
14. Wanderley, F., da Silveira, D.S., Araujo, J., Lencastre, M.: Generating feature model from creative requirements using model driven design. In: Proceedings of the 16th International Software Product Line Conference, vol. 2, pp. 18–25. ACM (2012)
15. Gamma, E.: Design Patterns: Elements of Reusable Object-Oriented Software. Pearson Education India, New Delhi (1995)
16. Astah community. http://astah.net/editions/community. Accessed 18 May 2016
17. Kof, L., Gacitua, R., Rouncefield, M., Sawyer, P.: Concept mapping as a means of requirements tracing. In: Third International Workshop on Managing Requirements Knowledge (MARK), pp. 22–31. IEEE (2010)
18. Laplante, P.A.: What Every Engineer Should Know about Software Engineering. CRC Press, Boca Raton (2007)
19. Village, J., Salustri, F.A., Neumann, W.P.: Cognitive mapping: revealing the links between human factors and strategic goals in organizations. Int. J. Ind. Ergonomics **43**(4), 304–313 (2013)
20. Alexander, I.F., Maiden, N.: Scenarios, Â Stories, Use Cases: Through the Systems Development Life-Cycle. Wiley, New York (2005)
21. Sutcliffe, A., Thew, S., Jarvis, P.: Experience with user-centred requirements engineering. Requirements Eng. **16**(4), 267–280 (2011)
22. Beck, K., Beedle, M., Van Bennekum, A., Cockburn, A., Cunningham, W., Fowler, M., Grenning, J., Highsmith, J., Hunt, A., Jeffries, R., et al.: Manifesto for agile software development (2001)

Using Scrum Together with UML Models: A Collaborative University-Industry R&D Software Project

Nuno Santos[1(✉)], João M. Fernandes[1,2], M. Sameiro Carvalho[1,3], Pedro V. Silva[4],
Fábio A. Fernandes[1], Márcio P. Rebelo[1], Diogo Barbosa[1], Paulo Maia[1],
Marco Couto[4], and Ricardo J. Machado[1,5]

[1] ALGORITMI Research Center, School of Engineering, University of Minho,
Guimarães, Portugal
nuno.a.santos@algoritmi.uminho.pt
[2] Department of Informatics, University of Minho, Braga, Portugal
[3] Department of Production and Systems, University of Minho, Guimarães, Portugal
[4] Bosch Car Multimedia Portugal S.A., Braga, Portugal
[5] Department of Information Systems, University of Minho, Guimarães, Portugal

Abstract. Conducting research and development (R&D) software projects, in an environment where both industry and university collaborate, is challenging due to many factors. In fact, industrial companies and universities have generally different interests and objectives whenever they collaborate. For this reason, it is not easy to manage and negotiate the industrial companies' interests, namely schedules and their expectations. Conducting such projects in an agile framework is expected to decrease these risks, since partners have the opportunity to frequently interact with the development team in short iterations and are constantly aware of the characteristics of the system under development. However, in this type of collaborative R&D projects, it is often advantageous to include some waterfall practices, like upfront requirements modeling using UML models, which are not commonly used in agile processes like Scrum, in order to better prepare the implementation phase of the project. This paper presents some lessons learned that result from experience of the authors in adopting some Scrum practices in a R&D project, like short iterations, backlogs, and product increments, and simultaneously using UML models, namely use cases and components.

Keywords: Agile · Scrum · UML · Research projects

1 Introduction

When research and development (R&D) projects are executed within industrial environments, project management commonly follow plan-centric processes, like waterfall [1], the spiral [2] or the Rational Unified Process (RUP) [3]. All these processes allow research to performed and clearly refine the requirements before moving to the implementation phase.

However, requirements are not always clear at the beginning of projects from R&D nature, which, when using plan-centric processes, makes very difficult to define an early schedule to any software development project due to the high uncertainty that

© Springer International Publishing Switzerland 2016
O. Gervasi et al. (Eds.): ICCSA 2016, Part IV, LNCS 9789, pp. 480–495, 2016.
DOI: 10.1007/978-3-319-42089-9_34

characterizes these types of projects. This schedule uncertainty typically occur in early stages of the project when a high effort in domain and technological research tasks is required. When R&D projects are executed in an industrial environment, projects face many challenges as timeliness, addressing the needs of stakeholders, rigor and access [4]. In such industrial environment, agile software development processes bring many advantages, since they are characterized by frequent interactions and collaboration with clients [5]. Agile processes are based in self-organized teams that perform the development tasks. These processes are divided in small iterations, where software systems are periodically assessed, in order to detect/solve possible problems as soon as they emerge. This approach works fine when future requirements are largely unpredictable [6]. Within these processes, eXtreme Programming (XP) [7] and Scrum [8] are among the most popular methodologies [9].

This paper describes how Scrum was adapted by a newly-formed team to develop a software system in a R&D project, called Inbound Logistics Tracking System project (hereafter referred as iFloW project) [10], which is described in this paper as a demonstration case. The Scrum process adaptation was based by performing upfront requirements modelling before the implementation phase (more common in waterfall approaches), instead of starting directly in implementing the software within the small iterations. For the requirements modelling, UML diagrams were modelled, namely use cases and component diagrams. The choice of using UML models was based by their wide acceptance, but any other models that bring upfront knowledge of software requirements could be used. The iFloW project, as its name refers, relates to logistics domain, and was mainly focused in integration with third party logistics (3PL) service providers and integrating Radio Frequency Identification (RFID) technology [11, 12], Global Positioning System (GPS) technologies [13], and an integrated web-based RFID-Electronic Product Code (EPC) compliant logistics information system [14].

The R&D context required domain research tasks in an initialization phase, as well as technological-related research and third-party collaboration during the implementation phase. This phase was performed in form of Sprints, which are small cross-functional development cycles and are used in agile frameworks like XP and Scrum. Software development projects with important integration and interoperability issues require additional concerns when compared with typical agile processes, since implementation requires prior studies related to the technologies involved (except if the team has solid experience with those technologies) and collaboration with third-party entities.

This paper is structured as follows: Sect. 2 presents the related work; Sect. 3 describes the collaborative University-Industry context; Sect. 4 describes our proposed Scrum adaptation, which integrates typical agile practices together with UML models; in Sect. 5 is presented the lessons learned of the process adoption; conclusions are presented in Sect. 6.

2   Related Work

The adoption of agile frameworks commonly found in literature is basically grounded in practical experience. There has been an increased interest in performing some empirical studies in agile software development [15]. Agile principles (namely those from XP) are introduced within a research work related to enterprise architecture development projects and software development projects [16]. Abrahamsson [17] notes that while more research was being done, agile methods were still driven by consultants and practitioners, that there was a lack of research rigour, and that researchers needed to address core questions such as what constitutes agility, how agile methods can be extended, and how mature teams use agile methods. Barroca *et al.* [4] state that collaboration between industry and research in agile methods allowed building trust and regular feedbacks, appropriate contracts at the beginning of the project, and learning experience for both teams. However, to the best of our knowledge, there are no research works in literature that describe the adoption of agile frameworks within R&D projects.

At first glance, agile processes seem to be only suitable for small teams operating in a local environment. The iFloW software system (the main deliverable of the iFloW project) was designed to perform in an environment where the main inputs are provided by integrated systems from suppliers or forwarders. Therefore, agile processes must consider implementations that refer specifically to integration, like the research from Niemelä and Vaskivuo for developing a middleware [18]. Välimäki and Kääriäinen propose an organizational pattern for these cases [19].

Projects where significant interoperability issues strongly rely upon dependency and communication with third-parties, like the case of the iFloW project, commonly involve development effort by distributed teams, which can be highly challenging. In fact, he coordination between different teams in agile development is one of the top concerns identified in [20]. In agile development, some techniques for adapting events, actors and artifacts arise so that both distributed and dispersed teams may work in the same product development [21]. Some examples within Scrum are Isolated Scrums, Scrum of Scrums [22] and Totally Integrated Scrum [23]. In these distributed environments, good requirements modeling, including using UML models, can be advantageous. It is the case of [24], where the identification of contact points where there is a need for synchronizing efforts within distributed Scrums and effort dependencies can be performed using artifacts like an architecture. Managing the distributed work is not an exclusive concern within Scrum framework. An example for these practices within the XP framework is Industrial XP [25].

3   The Collaborative University-Industry R&D Software Project

3.1   The HMI-Excel Program

The Human-Machine Interface Excellence (HMIExcel) project [26] was sponsored by the consortium between University of Minho (UMinho) and Bosch Car Multimedia Portugal (Bosch). It was designed in order to tackle scientific and technological challenges of Bosch Braga and to obtain recognition of this unit as an International Competence Centre in Human-Machine Interface (HMI). The project was 28 months long, starting in March 2013 and ending in June 2015. Although HMIExcel for the funding body is seen as a project, its complexity and uncertainty led the consortium (UMinho and Bosch) to manage it as a program, *i.e.*, a set of projects that are somehow related and contribute to the same goal [27]. The main goal for the HMIExcel was to promote the investment in R&D, for developing and producing future mobility concepts in the automotive domain, where academia and industry worked together aiming innovative products and processes. These solutions were developed to meet production needs and preparing Bosch to respond to the challenges from the Fourth Industrial Revolution (Industry 4.0).

The HMIExcel project divided itself into thirteen research lines, each one tackling one specific challenge. Each research line followed an independent course, as all projects were coordinated by one team focused on the objectives and deliverables expected in each individual line. One of those research lines, iFloW, is used in this paper as a case study analysis for purpose of validation of our approach.

3.2   The iFloW Project

iFloW is an R&D project that aims at developing an integrated logistics software system for inbound supply chain traceability. iFloW is a real-time tracking software system of freights in transit from the suppliers to the Bosch plant, located in Braga. The main goal of the project is to develop a tracking platform that by integrating information from freight forwarders and on-vehicle GPS devices allows to control the raw material flow from remote (Asian) and local (European) suppliers to the Bosch's warehouse, alerts users in case of any deviation to the Estimated Time of Arrival (ETA) and anticipates deviations of the delivery time window.

Figure 1 illustrates the architecture of the iFloW software system, namely the integration of the iFloW main server with other systems. The architecture allows depicting the significant interface protocols and systems that were involved. The system uses information that is provided by different sources, like GPS, Personal Digital Assistant

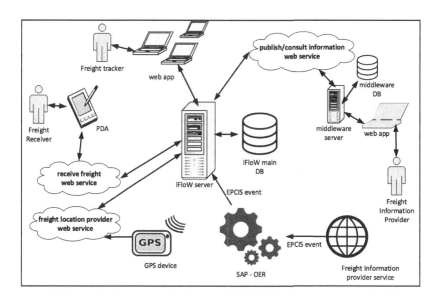

Fig. 1. The architecture of the iFlow software system

(PDA) devices or SAP-OER (Object Event Repository). The '*middleware server*' was developed aiming to standardize the way information between Bosch and providers is exchanged. Then, the '*iFloW server*' executes all business logic, where Bosch employees access all features via a web-based user-interface (web app).

In the case of the iFloW project, this collaboration UMinho and Bosch was based in the premise where Bosch mainly performed as a software customer and UMinho as a contracted software development entity. The core iFloW team was composed of nine collaborators with multidisciplinary backgrounds:

- Bosch:
    - one Product Owner, that was representing other eight elements from the Logistics department, that formally dictated the requirements.
    - one member of the IT department, responsible for validating that each developed product increment could be easily integrated within Bosch information system;
- UMinho:
    - three R&D coordinators, with the role of assuring that the scientific rigor (from both the system and the software development process) and deadlines of the project are met;
    - four software developers with methodological and technological competences (like analysis, requirements, design, database modeling, programming, testing, deployment, etc.).

The entire software development was performed within Bosch's premises, where the iFloW team elements (in exception of R&D Coordinators) were located on a daily basis. The elements from UMinho had no previous knowledge of the domain (in this case, logistics), so the team decided that the project kicked-off by gathering and documenting requirements in a waterfall-based approach.

After the requirements engineering was performed, and since iFloW aimed developing a software system for an industrial context, the team decided to follow the Scrum framework as the iterative approach for the implementation phase. This phase was performed by development iterative cycles in form of Scrum sprints. Based in incremental software deliveries, both UMinho and Bosch could manage their project's expectations.

As a collaborative University-Industry R&D software project, the previously presented roles are slightly different from the roles defined by the Scrum framework (namely, Product Owner, Scrum Master and Development Team) [28], however easily mapped, as depicted in Table 1.

The Product Owner (PO) was the element responsible for dictating the requirements, as previously stated, but also participated in meetings with UMinho R&D coordinators and Software Developers for project monitoring (the Project Manager role is not included within a Scrum team). Thus, it is directly mapped with a typical Scrum PO. Additionally, the PO performed as a contac point for distributed development with third-party service provider entities that composed the ecosystem. The fact that Bosch's information system (not only from Braga plant, but from all Bosch Car Multimedia worldwide plants) is managed by the Bosch Car Multimedia Group, in Germany, and that the system under development will be included in Bosch's information system, also includes Bosch's IT department from Germany in the project's ecossystem. Thus, the PO also performed as a contact point in the negotiation between the iFloW team and the IT department form Germany, regarding the compliance requirements from security and policy issues of Bosch Car Multimedia Group.

The Bosch IT element was responsible for validating the code developed during sprints, in order to be integrated within the existing information systems. He interacts directly with the software developers from UMinho. For this reason, he is considered as part of the development team. However, he was responsible for representing the IT department from Braga during the negotiations with IT department form Germany and with third-party service provider entities. So, within these tasks he also performed the role of a typical Scrum PO.

The R&D Coordination was composed by three professors from UMinho, from the fields of software engineering and also logistics. They were responsible for assuring that the academic concerns of these types of R&D projects were met, where their main concerns was to assure development beyond the state of the art, and to coordinate and monitor the development of the project's deliverables and scientific papers. This responsibility is not mapped to any role in Scrum. Since Bosch had never adopted agile software development processes, they were responsible for training and control the adoption of Scrum practices by the project team, when the project entered the implementation phase. For this responsibility in this phase, these elements are mapped as typical Scrum Masters.

The Software Developers performed tasks related with software engineering (e.g., analysis, requirements, architecture, layout design, coding, testing, deployment, etc.) thus, their responsibility was directly mapped with a Scrum development team.

Table 1. Mapping between iFloW roles and typical Scrum roles

Scrum Role / iFloW Role	Product Owner	Scrum Master	Development Team
Bosch			
Product Owner	✔		
Bosch IT	✔		✔
UMinho			
R&D Coordinators		✔	
Software Developers			✔

4 Using Scrum with UML Models

This section presents the software process and the required adaptations that were due to the research nature of the project and the required integration with third-party services. An overview of the process is depicted in Fig. 2. The process is composed of three phases: Initialization, Implementation, and Deployment.

The initialization phase includes typical activities from domain engineering, requirements engineering and design. In this phase, use cases diagrams and the component diagram were modelled. The implementation phase was performed in small iterations

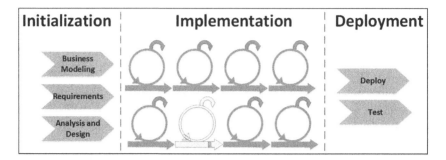

Fig. 2. Overview of the process executed in the iFloW project

and incremental releases in the form of Scrum sprints. This phase is similar to typical Scrum process. The UML models from the previous phase were just verified for changes, except for some new use cases referring newly discovered requirements. It should be noted that Fig. 2 depicts seven cycles performed like typical Scrum sprints (circles with filled border), and one cycle that rather addressed architectural refactoring (circle with dashed border). This different cycle may not be required in other R&D projects if the architecture is stable during the whole project. Finally, the Deployment phase is similar to the Transition phase of RUP. This phase included modeling a UML deployment diagram. This phase is not detailed in this paper due to size reasons.

4.1   Initialization Phase

Within the initialization phase, the objective was to develop a product backlog artifact in order to start the development phase in the form of sprints that, due to the perceived complexity of the project, was delivered together with widely accepted forms of requirements documentation (the overall process is depicted in Fig. 3).

Since iFlow is a R&D software project, some research activities were conducted throughout the project and the research outputs had to be documented. This project kicked–off as a typical waterfall process, and initial tasks were conducted to specify the project scope, to characterize the domain and the organization's (logistics–related) activities, to define terms, and to analyze flows, legacy software and data.

The organization's logistics–related processes and current gaps were documented in a report designated as '*As-Is report*'. Then, the requirements were elicited, formally specified in the form of UML use cases (see Fig. 4), a list of quality (non-functional) requirements and in a first version of the logical architecture (UML component diagram). This set of requirements was documented in a report designated as '*To–Be report*' (which was constantly updated as the implementation went along). Both use case models (especially the '*To-Be*') were used as basis to define a '*Product Backlog*'. This differs from other agile frameworks where, for instance, although complementary and able to be used together [8], in Scrum backlogs are composed of user stories. A user story is a customer-centric characterization of a requirement, containing only the information needed for the project developers to see clearly what is required to implement [29]. Use cases are used in backlogs in [30, 31].

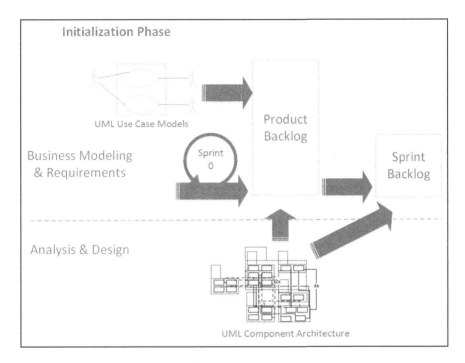

Fig. 3. Initialization phase

The use case diagram illustrated in Fig. 4 shows the overall use case model of the iFloW project. For the purpose of this paper, the use cases are not described, as the diagram is used only for demonstration purposes. Each of the use cases were functionally decomposed, which resulted in a total of 90 lower level use cases.

The Initialization phase ends with a Sprint 0 ceremony. Most of the technological research was performed during this ceremony, prior to the implementation in the following sprints. Like in a typical Sprint 0, each item (use case) was prioritized by its perceived value from stakeholders, in this case by using MoSCoW prioritization technique [32]. For this task, the PO provided input on the perceived business value of the requirement. On the other hand, R&D Coordinators identified critical implementation issues.

Also, each use case was estimated related to a quantitative effort for its implementation, where it was defined that the effort for each sprint corresponds to a total of 20 points (which resulted in approximately five points per week) as basis for distribution of these points per use case. This was a decision made by PO and R&D Coordinators. A commonly used technique is use case points [33], however in this project this technique was not used. Rather, and following a comparative technique similar to a planning poker [34]. Finally, based in efforts and prioritizations, the remaining step to define the '*Sprint Backlogs*' (which use cases from the '*Product Backlog*' to implement during the sprint) and to plan them.

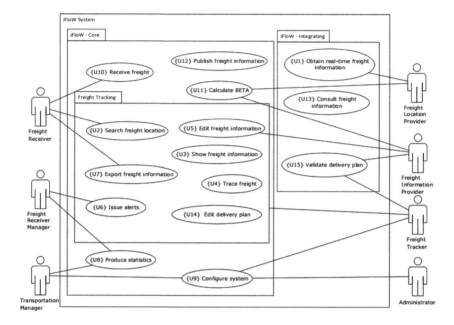

Fig. 4. Use case diagram

4.2 Implementation Phase

Within the implementation phase, the use cases from the '*Product Backlog*' were implemented iteratively and incrementally during eight four-week Scrum sprints. In this phase, typical Scrum iterations were performed, where each '*Sprint Backlog*' is a selected subset from the '*Product Backlog*' (see Fig. 5).

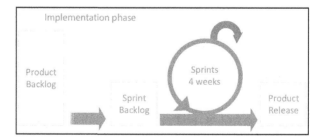

Fig. 5. The implementation phase

In Fig. 6 is depicted an example of a '*Sprint Backlog*' tracking sheet, which allowed monitoring through a breakdown chart. In this sheet, use cases were included until the total effort reached 20 points. Then, the use case implementation was monitored throughout each of the four weeks that compose the Sprint, by registering the remaining units (from the initial estimation) that were implemented during that week. This way, the tracking sheet allowed monitoring the implemented functionalities (and thus, the value delivered to the organization), but also if the Sprint's initial planned target (column SPx Target) was obtained.

Sprint #4Tracking Sheet								
	Project		iDPW					
	Sprint #		4					
	Start date		05/01/14					
					week			
					1	2	3	4
Task ID	Description	SP4 Target (%)	Initial estimate (units)		Units Left	Units Left	Units Left	Units Left
(U9.5)	Configure delivery plan	100%	6		5	3	1	1(b)
(U14)	Edit delivery plan	100%	3		2	1	1	1(b)
(U15)	Validate delivery plan	100%	2		2	2	1	1(b)
(U10)	Receive Freight (TS_SP3_U10_04)	100%	1		1	1	1	1
(U1.1.1)	Obtain freight information from *Forwarder A* from European port	25%	2		2	2	1	0
(U1.1.2)	Obtain freight information from *Forwarder A* from Asian port	25%	2		2	2	1	0
(U12)	Publish freight information	25%	2		2	2	1	0
(U13)	Consult freight information	25%	2		2	2	1	0
	Total estimate units		20					
	Remaining units (actual)				18	15	8	1
	Remaining units (ideal)				15,0	10,0	5,0	0,0

Fig. 6. Example of a sprint backlog based in use cases

Each sprint has a standard planning and structure consisting of several milestones, previously negotiated by the project members:

- **Sprint development**: lasts four weeks, and is allocated to the development of the items from the '*Sprint Backlog*';
- **Sprint Monitoring meeting**: takes place in second week to show sprint progress and monitor sprint tasks. The attendees are the Product Owner, R&D coordination and development team;
- **Sprint Verification and Validation (V + V) meeting**: takes place in the fourth (*i.e.*, last) week and the goal is to test and validate the requirements implemented by the development team. The attendees are the Product Owner, the development team, a member of the Bosch IT department, and an assigned Product User from Bosch. In each Sprint V + V meeting, the Product User was assigned a different user from

Logistics department so the performed tests could encompass different insights from the organization. During the sprint, if any requirement (use case) is moved to a next sprint due to a given constraint and will not be presented in this meeting, the team is notified;

- **Sprint Closure and Planning meeting**: takes place at most two days after the Sprint V + V meeting, and the attendees are the Product Owner, the R&D coordination, a member of the Bosch IT department and the development team. It is similar to a Sprint Retrospective and a Sprint Planning meeting from typical Scrum, performed within the same meeting. The main goal is to analyze the progress of the implementation phase, by assessing the percentage and completion of the use case implementation and thus updating the burndown chart. If applicable, short rework actions (depicted from the Sprint V + V) are approved to perform until the end of the sprint. Additionally, the next Sprint is planned, resulting in the construction of the 'Sprint Backlog' artifact;

- **Sprint Rework meeting**: takes place the day after the Sprint Closure meeting. After Sprint V + V, some rework actions can arise due to a suggestion by the verification and validation team. If applicable, the development team has to implement these rework actions until the end of the sprint. The Sprint Rework meetings are used to validate the rework actions performed. The attendees are the assigned Product Users, Product Owner, a member of the Bosch IT department and the development team.

For implementing each use case, the team performed tasks involving several software engineering disciplines. In this paper, we use the terminology from RUP's disciplines (only for demonstration purposes) to depicts the type of effort involved within the Sprints.

Occasionally, the team performed spikes (originally defined within XP), a technique used for activities such as research, innovation, design, investigation and prototyping. With spikes, one can properly estimate the development effort associated with a requirement or even to better understand a requirement. The use of spikes in the iFloW project justifies the inclusion of the Requirements discipline in the each Sprint, as shown in Fig. 7.

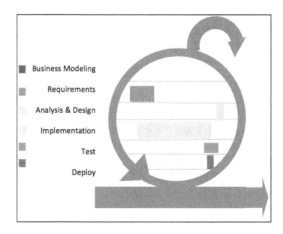

Fig. 7. The performed disciplines within the sprints (Color figure online)

In all sprints, the need for updates to the logical architecture was assessed (within the Analysis & Design). Afterwards, the typical disciplines were carried out within the sprints: Implementation, Testing and Deploy. These spikes were, in their majority, originated from middleware-based use cases (for instance, related to integration with third-party service providers, GPS, EPCIS or SAP-OER). Within the remaining use cases, the Requirements discipline was not required. Thus, in comparison with the disciplines included in Fig. 7, the Sprint performed the remaining disciplines like illustrated with exception of Requirements. In fact, it is what indeed occurs in typical Scrum process (where almost every requirements-related effort is performed before Sprint cycles, like Sprint 0 or similar).

Additionally, the implementation of some middleware-related use cases involved third-party collaboration. The implementation of those use cases must be managed from a distributed team perspective. It is required that the implementation effort by third–parties is properly aligned with the team's development process. In the case of the iFloW project, the integration with third-party service provider entities required that both teams worked together and their work aligned.

At a given point in time, both Bosch and UMinho identified the need for refactoring the code and the architecture of the system, namely to cope with security and standardization issues. Such refactoring led to a pause in the implementation tasks. The software logical architecture was revisited and the impacts were analyzed. Some design-oriented spikes (similar to architectural spikes from XP) were conducted, which then followed the re-design of the architecture. In this case, there was a focus in Analysis & Design instead of Implementation (see Fig. 8, where it is detailed the sixth Sprint included in Fig. 2). Similarly to the other sprints, this effort also lasted four weeks. This effort was required in this case but it may occur, or not, in any project.

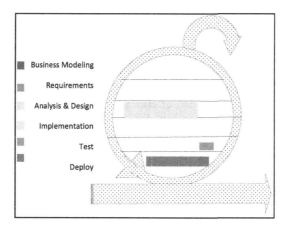

Fig. 8. The performed disciplines within the architectural spike sprint (Color figure online)

5   Lessons Learned

Defining a hybrid approach (waterfall-based during initialization and Scrum-based during implementation) in a collaborative Industry-University context arose many challenges. We believe that the inclusion of artifacts modeling and documentation strengthened the adoption of a Scrum process in these contexts. However, the entire adoption was a learning process, with advantages and disadvantages, which are detailed in this section.

5.1   Advantages

Industrial and scientific interests – typically, these kinds of collaborative Industry-University software projects aim timeliness, addressing the needs of stakeholders, rigor and access. In terms of university interests, the initial approach on documenting requirements allowed the team to gain domain knowledge, to assure the academic rigor of R&D projects. In terms of industry interests, the use of an iterative development process facilitated negotiation concerning schedules and their expectations.

Requirements documentation waterfall-based – the fact that the Product Backlog was composed of 90 use cases led to a shared perception of the system complexity that originated the need to perform proper efforts in documenting the requirements. Thus, consuming efforts in almost exclusively for requirements engineering typically performed in waterfall approaches, in the initialization phase, allowed the project team to gain the required knowledge to properly implement a system of such complexity.

Implementation Scrum-based – within a customer perspective, Bosch was always aware of the system's current state of development. The iterative development, in form of Scrum sprints, was crucial to manage Bosch's expectations, due to the periodical meetings and the incremental delivery of working software.

Use of a logical architecture – to enforce a proper organization on the set of components. The relationships among components suggest dependencies that may impact the implementation of functionalities and their inclusion in the Sprint Backlog.

Assess the logical architecture – the software logical architecture was revisited and the impacts were analyzed at the end of each iteration, in order to predict refactoring efforts. Additionally, when a change was identified, the logical architecture representation allowed to analyze which components are targeted with impacts from those changes.

5.2   Disadvantages

Effort estimation for use cases – the fact that it was a completely new development team (thus team velocity was unknown) and the need to frequently perform research spikes in order to overcome technological issues (for instance, related to GPS, EPCIS or SAP-OER) were the main obstacles for the estimation. In Scrum, estimation is performed using techniques such as planning poker, where user stories are estimated based in comparing efforts between other user stories. Due to the inexperience of the team, estimating the required effort for implementing use cases by comparing with other was itself a learning process. Such approach resulted in sprint backlogs where use cases had not been implemented due to error in estimating and required conclusion in further sprints, and where the effort estimating of the remaining use cases (as well as rework, whenever was required, and the spikes that were performed within almost every sprints) required constant updates on every Sprint Closure and Planning meeting.

Dependence on negotiation for middleware use cases – collaborative coding among iFloW team members and service provider team members was required to implement middleware-related use cases. Most of the times the implementation required previous negotiation and agreements and the implementation did not progress at the desired velocity. The team's work reached a point where they had to pause and wait for those agreements, which resulted in the extension of use cases (and use cases with dependencies with them) through several sprints.

6   Conclusions and Outlook

The iFloW project is a collaborative R&D software project, where University researchers were contracted for developing an industrial software system. Since the project aimed at delivering a software product, the R&D coordination elements decided to use many practices available in agile process, namely in Scrum (and, to a lesser extent, in XP). During this project, there was always the concern to fulfill both industry and university needs. Thus, this project had to face key challenges as timeliness, addressing the needs of stakeholders, rigor and access. It should be noticed that the nature and context of the project created the need for firstly proposing an initialization phase (similar to what happens in waterfall model) and afterwards performing sprints throughout the implementation phase. The project ended with a deployment phase.

This paper presents how a R&D project for developing a software system was conducted by combining some Scrum practices with UML models dedicated to document requirements and architecture. Agile processes are commonly used among practitioners but not much in R&D projects. The main advantages that result from using Scrum practices within the implementation phase are related with the facilitated negotiation with the stakeholders concerning deadlines and their expectations regarding the system. Thus, stakeholders have the opportunity to frequently interact with the development team in short iterations, allowing them to adjust their ideas about the system.

Some issues stated as lessons learned are seen as opportunities for improvement. In order to prevent the stated errors within the sprint backlog definition in future projects, effort estimation techniques as use case points should be considered. Additionally, architecture design (and re-design) can also be improved. By using an architecture derivation method (like the 4SRS method [35]), through traceability mechanisms, requirements change during the implementation can be supported.

Acknowledgements. This research is sponsored by the Portugal Incentive System for Research and Technological Development PEst-UID/CEC/00319/2013 and by project in co–promotion n° 36265/2013 (Project HMIExcel - 2013-2015).

References

1. Royce, W.W.: Managing the development of large software systems. In: IEEE WESCON. Los Angeles (1970)
2. Boehm, B.W.: A spiral model of software development and enhancement. Computer (Long. Beach. Calif) **21**, 61–72 (1988)
3. Kruchten, P.: The rational unified process: an introduction. Addison-Wesley Professional, Boston (2004)
4. Barroca, L., Sharp, H., Salah, D., Taylor, K., Gregory, P.: Bridging the gap between research and agile practice: An evolutionary model. Int. J. Syst. Assur. Eng, Manag. 1–12 (2015)
5. Cho, J.: A hybrid software development method for large-scale projects: rational unified process with scrum. Issues Inf. Syst. **10** (2009)
6. Boehm, B.: Get ready for agile methods, with care. Computer (Long. Beach. Calif) **35**, 64–69 (2002)
7. Beck, K., Andres, C.: Extreme Programming Explained: Embrace Change. Addison-Wesley Professional, Boston (2004)
8. Schwaber, K.: Scrum development process. In: Sutherland, J., Casanave, C., Miller, J., Patel, P., Hollowell, G. (eds.) Business Object Design and Implementation, pp. 117–134. Springer, Heidelberg (1997)
9. VersionOne Inc: 8th Annual State of Agile Survey (2013) http://www.versionone.com/pdf/2013-state-of-agile-survey.pdf
10. Santos, N., Barbosa, D., Maia, P., Fernandes, F., Rebelo, M., Silva, P.V., Carvalho, S.M., Fernandes, J.M., Machado, R.J.: iFloW: an integrated logistics software system for inbound supply chain traceability. In: Mendonça, J.P., Fensterbank, S.-A., Barthet, E. (eds.) Enterprise Interoperability, Proceedings of 8th International Conference on Interoperability for Enterprise Systems and Applications (I-ESA). (in-press). Springer, Guimarães, Portugal (2016)
11. Choy, K.L., Ng, S.W.K., So, S.C.K., Liu, J.J., Lau, H.: Improving supply chain traceability with the integration of logistics information system and RFID technology. Materials Science Forum, pp. 135–155. Trans Tech Publ (2006)
12. Choy, K.L., So, S.C.K., Liu, J.J., Lau, H.: Improving logistics visibility in a supply chain: an integrated approach with radio frequency identification technology. Int. J. Integr. Supply Manag. **3**, 135–155 (2007)
13. Kandel, C., Klumpp, M., Keusgen, T.: GPS based track and trace for transparent and sustainable global supply chains. In: 17th International Conference on Concurrent Enterprising (ICE), pp. 1–8. IEEE (2011)
14. Doukidis, G.I., Chow, H.K.H., Choy, K.L., Lee, W.B., Chan, F.T.S.: Integration of web-based and RFID technology in visualizing logistics operations-a case study. Supply Chain Manag. Int. J. **12**, 221–234 (2007)
15. Dybå, T., Dingsøyr, T.: Empirical studies of agile software development: a systematic review. Inf. Softw. Technol. **50**, 833–859 (2008)
16. Ramos, H., Vasconcelos, A.: eXtreme enterprise architecture planning. In: 29th Annual ACM Symposium on Applied Computing (SAC), pp. 1417–1419. ACM (2014)
17. Abrahamsson, P., Conboy, K., Wang, X.: Lots done, more to do: the current state of agile systems development research. Eur. J. Inf. Syst. **18**, 281–284 (2009)
18. Niemelä, E., Vaskivuo, T.: Agile middleware of pervasive computing environments. In: Second IEEE Annual Conference on Pervasive Computing and Communications Workshops, pp. 192–197. IEEE (2004)

19. Välimäki, A., Kääriäinen, J.: Patterns for distributed scrum—a case study. In: Mertins, K., Ruggaber, R., Popplewell, K., Xiaofei, X. (eds.) Enterprise interoperability III, pp. 85–97. Springer, Heidelberg (2008)

20. Dingsøyr, T., Moe, N.B.: Towards Principles of Large-Scale Agile Development: A Summary of the Workshop at XP2014 and a revised research agenda (2014)

21. Eckstein, J.: Agile Software Development With Distributed Teams: Staying Agile in a Global World. Addison-Wesley, Boston (2013)

22. Sutherland, J., Viktorov, A., Blount, J.: Adaptive engineering of large software projects with distributed/outsourced teams. In: Proceedings of the International Conference on Complex Systems, Boston, MA, USA, pp. 25–30 (2006)

23. Cristal, M., Wildt, D., Prikladnicki, R.: Usage of scrum practices within a global company. In: IEEE International Conference on Global Software Engineering (ICGSE), pp. 222–226. IEEE (2008)

24. Costa, N., Santos, N., Ferreira, N., Machado, R.J.: Delivering user stories for implementing logical software architectures by multiple scrum teams. In: Murgante, B., Misra, S., Rocha, A.M.A., Torre, C., Rocha, J.G., Falcão, M.I., Taniar, D., Apduhan, B.O., Gervasi, O. (eds.) ICCSA 2014, Part III. LNCS, vol. 8581, pp. 747–762. Springer, Heidelberg (2014)

25. Kerievsky, J.: Industrial XP: Making XP work in large organizations. Exec. Report. Cut. Consort. 6 (2005)

26. Fernandes, G., Pinto, E.B., Machado, R.J., Araújo, M., Pontes, A.: A program and project management approach for collaborative university-industry R&D funded contracts. Procedia Comput. Sci. 64, 1065–1074 (2015)

27. Pellegrinelli, S.: What's in a name: Project or programme? Int. J. Proj. Manag. 29(2), 232–240 (2011)

28. Schwaber, K., Beedle, M.: Agile Software Development with Scrum, 1st edn. Prentice Hall PTR, Upper Saddle River (2001). ISBN: 0130676349

29. Ambler, S., Lines, M.: Disciplined Agile Delivery: A Practitioner's Guide to Agile Software Delivery in the Enterprise. IBM Press, Boston (2012)

30. Kroll, P., MacIsaac, B.: Agility and Discipline Made Easy: Practices from OpenUP and RUP. Pearson Education, Boston (2006)

31. Jacobson, I., Spence, I., Bittner, K.: Use case 2.0: The Definite Guide. Ivar Jacobson International (2011)

32. Waters, K.: Prioritization using moscow. Agil. Plan. (2009)

33. Anda, B., Dreiem, H., Jørgensen, M.: Estimating software development effort based on use cases-experiences from industry. In: Gogolla, M., Kobryn, C. (eds.) UML 2001. LNCS, vol. 2185, pp. 487–502. Springer, Heidelberg (2001)

34. Grenning, J.: Planning poker or how to avoid analysis paralysis while release planning. Hawthorn Woods Renaiss. Softw. Consult. 3, 1–3 (2002)

35. Ferreira, N., Santos, N., Machado, R., Fernandes, J.E., Gasević, D.: A V-model approach for business process requirements elicitation in cloud design. In: Bouguettaya, A., Sheng, Q.Z., Daniel, F. (eds.) Advanced Web Services, pp. 551–578. Springer, New York (2014)

Rockfall Hazard Assessment in an Area of the "Parco Archeologico Storico-Naturale Delle Chiese Rupestri" of Matera (Basilicata Southern–Italy)

Lucia Losasso, Stefania Pascale[✉], and Francesco Sdao

School of Engineering, University of Basilicata, 85100 Potenza, PZ, Italy
pascalestefania@gmail.com

Abstract. This work contains the results of the methodology used for the evaluation of the possible trajectories of the unstable blocks and their location along the slope that could destroy valuable rupestrian testimonies. The software used is the Rockfall, managed by RocScience (2002). It is an important tool allowing the rockfalls risk assessment. In the study area various simulations have been performed; they have led to the evaluation of the different parameters of the blocks movements: trajectories, maximum heights of bounce, propagation distances and energies of the blocks, to obtain a mapping of areas with different susceptibility to the transit and to the invasion of the blocks. In order to describe the blocks movement, the RocFall Software apply the parabolic equation of a corps motion in free fall and the principle of total energy conservation. This work leds to the result that the southern side of the site "Belvedere delle Chiese Rupestri" presents a big criticality to the collapse phenomena that reverberates on archaeological assets therein.

Keywords: Hazard · Rockfall · Rupestrian church · Matera · (Basilicata, Italy)

1 Introduction

Basilicata Region (Southern Italy) is considered one of the regions in the Mediterranean basin most subject to slope instability because of its geological, geomorpfological, climatic and seismic characteristics. In some areas of Basilicata, the landslides of different types are so intense that they sometimes generate serious damages to people and properties. In recent years, geomorphological studies conducted by the authors showed that many archaeological sites in Basilicata, are the sign of a widespread and intense landslide activity, which is affecting and damaging the valuable testimony of historic and archaeological representations. Examples of similar situations can be found in the archaeological areas of "Parco Archeologico Storico Naturale delle Chiese Rupestri" of Matera, the object of this paper. It, in fact, as such as Sassi di Matera, is part of the UNESCO World Heritage site since 1993.

The studied area, the "Belvedere delle Chiese Rupestri", located on the left side of the Torrente Gravina, is an area of great historical and archaeological interest. In fact, along these slopes, one of the most valuable and rich archaeological sites and historical heritage has been generated and consequently developed: the so-called Rupestrian

© Springer International Publishing Switzerland 2016
O. Gervasi et al. (Eds.): ICCSA 2016, Part IV, LNCS 9789, pp. 496–511, 2016.
DOI: 10.1007/978-3-319-42089-9_35

civilization. Genesis and subsequent architectural evolution of the original underground cave (in many cases of prehistoric times), were favored by the peculiar geostructural and geomorphologic habitat of Gravina di Matera that fully characterized the socio-economic, religious and strategic aspects of medieval Lucan populations. However, such a geo-morphostructural setting, prepares, at the same time, the steep slope of Gravina di Matera to a widespread and significant geomorphological fragility manifested by rapid and extensive mass movements: collapse, overturning and sliding of rock blocks. This state of hydrogeological instability, accompanied by situations of neglect and vandalism, reverberates, damaging them in varying degrees, on the prestigious rupestrian testimonies in both the Sassi di Matera mentioned above and Historic and Natural Archaeological Park of the rupestrian Churches of Matera, with resulting in serious and deep structural instability. The ultimate purpose is the analysis of the possible trajectories of the individual blocks detached from the rock walls, using integrated software as the Rockfall (Software for the Risk Analysis of Falling Rocks on steep slopes), in order to develop an instrument of real preventive protection, aiming at the preservation of historical and cultural heritage, as well as the monitoring of empirical factors undermining the existence of censused goods.

2 Geological and Geomorphological Framework of the Study Area

The territory of Matera is situated in an area between Apulian foreland and Bradanic fore deep; the "Calcarenite di Gravina" Formation (Lower Pleistocene – Upper Pliocene), is located in the site, at the top of fore deep Pliocene - Pleistocene succession and lies in discordance on foreland "Calcare di Altamura" Formation (Upper Cretaceous) (Fig. 1).

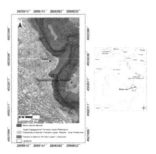

Fig. 1. Geology and geomorphology map of the site Rupestrian Heritage Rich Area of Matera (Basilicata Region, Southern Italy) - (Sdao et al. 2013)

The "Calcare di Altamura" Fm., that outcrops in the lower part of the slopes, is characterized by depositional surfaces, SW-dipping with a dip angle of about $5° – 10°$. It is constituted by a monotonous succession of micritic limestone, packstone and wackestone with aboundant remains of Rudiste. These terrains are generally well stratified and usually affected by a pervasive fracturing. The "Calcarenite di Gravina" Formation (Upper Pliocene - Lower Pleistocene), has a medium thickness of about 40 meters and it is composed of bioclasts and terrigenous limestone fragment (Pomar and Tropeano 2001). In this Formation, it is possible to distinguish two members: one lithoclastic, with a terrigeneous origin, and one bioclastic. These calcarenites are composed of an intense and widespread fracturing, generated by families of discontinuity variously oriented and often intersected between them. In fact, Fig. 2 shows the geomorphological effects generated by fast rock mass movements, due to rock falls, toppling and sliding blocks. In many cases, along the entire edge of the ravine, it is possible to observe open discontinuity crests, irrefutable signs of ongoing morphogenetic dynamics generated by fast landslides in rock (Fig. 3), due to collapse, overturning direct and slipping blocks.

Fig. 2. Exemple of rock slide

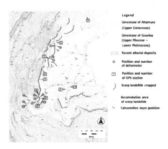

Fig. 3. Geological and geomorphological map of the study site "Belvedere delle Chiese Rupestri"

Recent studies have shown that:

1. the entire site is significantly predisposed to landslides of rock wedges and direct tipping, controlled by the main families of discontinuities, with particular reference to the break with Attitude 51°/68°, 209/69°, 272°/63°;
2. the areas showing the most marked fall in safety hazards in the southern portion of the site investigated.

Such conditions of potential instability of these areas are, however, confirmed by GPS surveys carried out on a network of 10 markers distributed along the edge of the ravine and conducted over the period 2002–2004. The major movements were recorded at stations 2, 3 and 4, falling in the southern portion of the investigated site (Fig. 3).

3   Survey Field and Engineering Properties of Rocks: Landslides Analysis in the Study Area "Belvedere Delle Chiese Rupestri" of Matera

The peculiar geo-structural and geomorphologic aspect of the slopes and the presence of highly fissured rocks, predispose the area to widespread and pervasive landslides. Rocks are widely and intensely affected by tracks and effects (niches of detachment and escarpments of landslide, several rock blocks collapsed and debris accumulations resting in condition of instability on the bottom slopes) due to an active morphological evolution expressed through the activation of mass movements such as rapid collapse, overturning and sliding calcarenitic clusters; rock falls are the most common typology and affect extremely variable massing, from a few to hundreds m^3. Depending on the morphology of the slope (steeper, or more or less inclined) the stores of collapse can stop at the foot of the detachment zone or suffer a transport to the bottom of the ravine. The separation takes place along existing discontinuities in the rock or formed after the release of stress involving the outer portions of the rock in correspondence of the rovine edge. The geomorphological features of the ravine previously outlined, make it a site particularly susceptible to the phenomena described above, related not only to the cracking of the rock walls, but also to the role of the karst process tending to widen the discontinuities in the rock mass and to create real cavity. Other evidence of instability are reported in relation to the presence of natural cavities and/or artificial walls of the ravine. The widespread use of cavities, widened and deepened towards the interior side of the ravine in different historical periods, constitutes another element of weakness of the rock mass (Sdao et al. 2009; Sdao et al. 2013). Both the roof and the walls of the cavities are frequently subject to alteration phenomena that reduce significantly the characteristics of resistance, causing the progressive detachment of the most superficial and altered portions. This process develops sometimes even to cause the fall of large volumes of rock.

4   Historical-Cultural Importance of the Site "Belvedere Delle Chiese Rupestri"

The "Parco archeologico storico-naturale delle Chiese rupestri del Materano" districts, and the neighboring "Belvedere delle Chiese Rupestri" is one of the most clear cases of the persistence of rock cave dwelling practices up to modern times. The birth and the following profuse architectonic evolution of the pristine rupestrian ipogei (several of preistorical age) have been helped along by the odd geostructural and geomorphological habitat of the Matera's gravina. This habitat was perfect for the socio-economical, religious and strategic needs of the medieval Lucan population (Sdao et al. 2013). The people can make house excavating the friable Calcarenite of Gravina (the main lithological type outcropping at the top of Matera's Gravina) using its natural holes and predisposition to form a nether net of pipes (Fig. 4). The medieval population in this habitat was favored by: the easy water availability thanks to engineering hydraulic works (cisterns, catch basins, etc.) and to the low superficial calcarenitic permeability;

the scarcely detecting sites helpful for hiding themselves to the periodic invasions of Arabian or Byzantine people; the satisfaction to make hermitages or rupestrian churches (especially from the VIII and the XIII sec. a. C) for praying thank to the steep slopes or the winding trend of the Gravina. In fact in same of these cult places there are valuable and unique wall painting representing holy images as live testimony of monastic Latin and Greek-Byzantine cultures. The disruptions of the investigated area, greatly influence the state of archaeological heritage. The precarious states of conservation of churches, relating to deterioration of the frescoes, are joined to the risk of destruction because of the phenomena of collapse that may occur following the consequent detachment of rock masses.

Fig. 4. Rupestrian churches of Site "Belvedere delle Chiese Rupestri" of Matera

5   Rockfall Analysis and Simulation

The In the study area, rockfalls are more frequent and they are influenced by discontinuities, wind erosion, freeze-thaw process and water. However, rockfall hazard depends also on the lithology of the rock, the topography, inclination of the slope, size/shape of the boulder and the slope surface properties (Schweigl et al. 2003). During the last decade, various programs have been developed to simulate fall of a boulder down a slope and to compute rockfall trajectories (Bassato et al. 1985; Falcetta 1985; Bozzolo and Pamini 1982; Hoek 1987; Pleiffer and Bowen 1989; Azzoni and de Freitas 1995; Jones et al. 2000; Guzzetti et al. 2002). To simulate the boulders falling movements at a slope scale, it is possibleto identify three different process-based models:

1. 2D slope scale models (movement in a vertical plane);
2. The trajectory of rockfall is defined as a composite of connected a straight lines;
3. Motions are simulated as a succession of flying phases and contact phases (Dorren 2003).

In conclusion, it is possible to say that the simulations models are based on the different falling boulders method: the method considering the falling boulder either with no mass (kinematic method) or with the mass concentrated in a point (lumped mass method) and those that consider the falling boulder as a body with its own shape and volume (i.e. Colorado RockFall Simulation Program – CRSP). The used "RocFall" software is a statistical analysis program that allows the risk assessment of slopes at risk of rockfalls. Computer programs for the rockfall simulation (in particular for the analysis of the block trajectories) require specific parameters, such as the slope profile, necessary because it influences the height of bounce (Azzoni et al. 1995) and the runout distance of the blocks. Another important parameter is the surface roughness that is linked to the material properties: for example a slope face free from vegetation cover, having a value of roughness lower then a surface covered by vegetation, is very dangerous because the movement of falling or rolling rock is not retarded. Finally, another important parameter to consider is the initial velocity to move the boulder: even if the initial velocity of falling boulder is equals to zero, a value of 1–3 m/s is generally recommended (Azzoni et al. 1995; Jones et al. 2000).

6   Simulation of the Trajectory in the Study Area

Rockfalls mainly triggered by climatic or biological events: pore pressure increases, fractures, earthquakes, erosion of surrounding material, freeze-thaw process, chemical degradation or weathering of the rock, root growth and hard winds (Hoek 2007); they represent the main cause of forces changing on a rock. Next there is a general representation of the site "Belvedere delle Chiese Rupestri" (Fig. 5) followed by a representative map containing all profiles detected in the analysis of the rock falls (Fig. 6). In this article it was shown the trend of the most representative profiles.

Fig. 5. The site: "Belvedere delle Chiese Rupestri" of Matera

Fig. 6. Representation of analyzed profiles

6.1 Profile AA' Near "Madonna Degli Angeli" Church Located in the Archaeological Park of the Rupestrian Churches of Matera"

First of all, it should be noted that the church ("Madonna degli Angeli" - Lat 40.6699380° and Long 16.613344°) today is presented without the facade in masonry. The walking surface is lowered compared to the entrance. The profile in the proximity of the Church (Fig. 6) has been chosen in function of the blocks already collapsed or those potentially unstable. Input parameters useful for the simulation are: 1. Elevation profile; 2. Block sizes; 3. Horizontal speed (m /s); 4. Vertical speed (m/s). The size of the blocks utilized in the simulation of rock falls, has been calculated in reference to the average of unstable volume near the church evaluated using the following formula:

$$V = L1 * L2 * L3 \tag{1}$$

Since the volume is known, it has been possible to calculate the mass: assuming that the density of the rocks is equal to 2700 kg/m3, the mass is given by the product between the density of the rock and the average volume:

$$M\,[kg] = \rho\,[kg/m3] * V\,[m3]. \tag{2}$$

In the Table 1 there are the input data:

Table 1. Input data used in the rockfall analysis

Parameters	Value
Density	2700 kg/m^3
Average Volume	5.14 m^3
Block Mass	13878 kg

The volume of the blocks measured in correspondence of the "Madonna degli Angeli" Church near the Archaeological Park of the rupestrian churches of Matera (Sdao et al. 2012). The volume considered in the simulation is the value corresponding to the mathematical average of the various calculated volumes for the different blocks. It is calculated by detecting the three dimensions of the blocks: length, width and height. As far as, instead, the horizontal and vertical speed (which are the same) is concerned, the value is 3.5 m/s obtained by the literature. The "RocFall" program, instead, directly provides output data after entering the values listed above:

– Minimum cut-off speed (slipping) → 0.1 m/s
– Number of horizontal rebounds to analyze → 50.

Program also uses as value of the friction angle the value related to the editor of the outcropping material (the above profile is achieved by asserting the bare rock along the entire route of the boulder).

Next two simulations follow: one concerns the number of horizontal bounces equal to 50 and to the other concerns a number of horizontal bounces equal to 5.

After that, the various diagrams are reported; they allow the analysis of: height of rebounds, total kinetic energy, translation kinetic energy, rotational kinetic energy, translation speed and rotation speed.

The magnitude scale used for the realization of the profile is 2: 2.

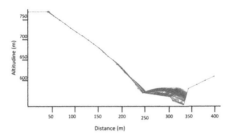

Fig. 7. Rockfall trajectory (profile AA' near "Madonna degli Angeli" Chuch)

The analyzed profile (Fig. 8) has a rather constant trend up to 277858 meters. After that, it is possible to reveal the presence of river terraces influencing the trend. Following the detachment of storage, beginning with a rotatory movement because of the initial speed of 3.5 m/s, it is possible to detect a movement of pure translation until the inclination slightly varies and therefore to highlight rotatory movements with low rebound velocity. Arrived in the valley, the cluster rolls up to reach and then to stop in down in the valley.

6.2   Rockfall Analysis and Simulation

The kinetic energy is associated with the mass and velocity of moving object and can be expressed mathematically as:

$$E_k = \frac{1}{2}mv^2 \tag{3}$$

Kinetic energy is the sum of the rotational and translational term. The importance of knowing the maximum height reached by the boulder after each bounce is another way to investigate the variation of the energy between a rebound and the other one.

- After each bounce the total energy decreases;
- Between a rebound and the other the sum of the kinetic energy and the potential one remains constant;
- The instant of the rebound: all the energy after boulder is kinetic energy of the boulder → the module of the velocity is maximum;
- In the inversion point of the motion, all the energy of the boulder is potential energy → maximum height has been logged.

In the point of maximum height, the boulder stops for a while. Consequently its kinetic energy (E_{TOT}) is void and all the mechanical energy is transformed into potential energy ($E_{P, max}$). So, in the point of maximum height:

$$E_{tot} = E_{p,max} = mgh_{max} \tag{4}$$

Next there are the diagrams trend (Fig. 8):

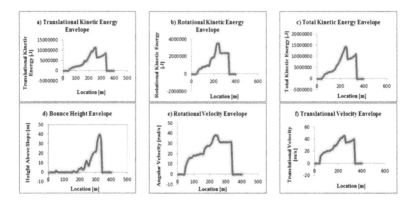

Fig. 8. The analyses results applied on the block: (a) Translational Kinetic Energy, (b) Rotational kinetic Energy, (c) Total Kinetic Energy, (d) Bounce Height, (e) Rotational Velocity, (f) Translational Velocity

6.3 Profile BB' in Proximity of the "Madonna Degli Angeli" Church Located in Archaeological Park of Rupestrian Churches of Matera

In "Madonna degli Angeli" Church location, a second profile (Figs. 6, 7, 8 and 9) for the calculation of the possible trajectories has been realized and the used data are the same as the previous profile.

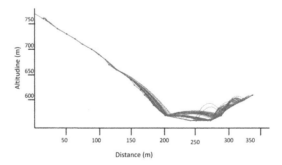

Fig. 9. Rockfall trajectory (Profile BB' in proximity of the "Madonna degli Angeli" church)

The movement of the cluster is characterized by a predominantly translational movement in the first part. Subsequently the mass continues to move with a rotational movement, characterized by continuous and successive bounces until it stops on the opposite side of the slope. Next there are the diagrams trend (Fig. 10):

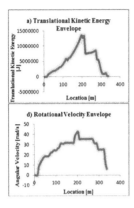
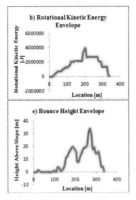
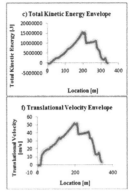

Fig. 10. The analyses results applied on the block: (a) Translational Kinetic Energy, (b) Rotational Kinetic Energy, (c) Total Kinetic Energy, (d) Bounce Height, (e) Rotational Velocity, (f) Translational Velocity

6.4 Profile DD'

The analyzed profile (Fig. 11) is characterized by the presence of many terraces; this is evident from the large number of bounces of the boulder before arriving to the valley and continuing on the other side. Initially, after the boulder is broken away from the wall, it begins its movement with three big rebounds. Subsequently it continues its rotation and translation motion with great prevalence of rotational motion, rather than translational one. Coming downstream, it continues its rotary movement reaching the opposite side on which, after a few rebounds, stops itself. Next there are the diagrams trend (Fig. 12):

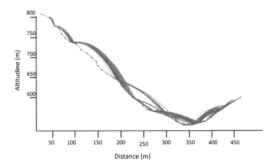

Fig. 11. Rockfall trajectory (Profile DD')

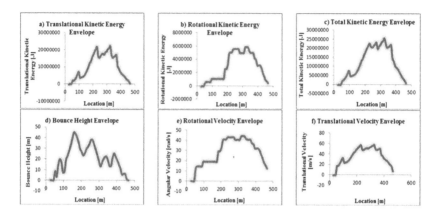

Fig. 12. The analyses results applied on the block: (a) Translational Kinetic Energy, (b) Rotational Kinetic Energy, (c) Total Kinetic Energy, (d) Bounce Height, (e) Rotational Velocity, (f) Translational Velocity

6.5   Profile FF' Near GPS Location Number 3 (Fig. 3)

The profile (Fig. 13) is characterized by an initial rolling movement of the boulder due to the inclination of the first part of the profile, after which it continues its path with a translational movement. Later, because of the inclination of the profile, the clutter continues with a great leap rolling up to fall on the ground. Thus, it advances its path along the side, alternating rotation and translation. It stops just on the other side. Next there are the diagrams trend (Fig. 14):

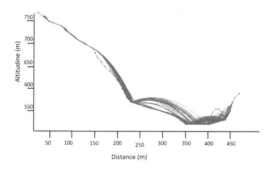

Fig. 13.   One of a rockfall trajectory (Profile FF' near GPS location number 3 - Fig. 3)

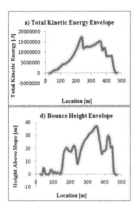 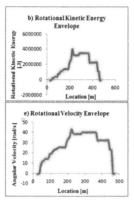 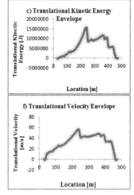

Fig. 14. The analyses results applied on the block: (a) Translational Kinetic Energy, (b) Rotational Kinetic Energy, (c) Total Kinetic Energy, (d) Bounce Height, (e) Rotational Velocity, (f) Translational Velocity

7   Environmental Impacts and Possible Preventive Measures

To reduce the risk associated with the detachment of stone elements from rocky slopes, two types of interventions are taken into consideration: active interventions, ensuring the prevent of the detachment of the stone elements from the slope, and the passive ones, to intercept, divert or stop the boulders on the move.

The rock fall barriers are among the interventions of passive defense. These can be further subdivided into two categories according to their behavior during the impact with the stone elements: the barriers with reduced deformability (also called "rigid") designed and constructed to stop the blocks with small deformations (behavior in field predominantly elastic); the barriers with large deformation (generally defined as "deformable"), designed and built to absorb the kinetic energy from the falling boulders in the regime of large deformations. Because of the establishment of large permanent deformation after an impact of the blocks in motion on a slope, the barrier must obviously be able to stop the boulders without breakthrough or creation of passages. The "stopping capacity" of rock barriers at the network is determined by the characteristics of strength and deformability of the elements that constitute it (nets, ropes, anchors, energy dissipators) and the connections between them; nevertheless structural organization of the device in its entirety is fundamental. The rockfall barriers are generally classified on the basis of limit energy absorbed during the impact, that is the work (elasto-plastic and attractive) connected to the stop of the mass. As evident, the problem of the calculation of limit energy presents many difficulties. On the one hand, in fact, there are marked geometrical non-linearity due to the considerable deformation; on the other hand mechanical non linearities are related both to the plastic deformation of steel elements, as to the energy dissipation due to friction observed in correspondence of the connections (between different elements, but also within the same net panels). On the basis of these considerations, it is clear that the definition of the behavior of these structures can not be carried out with the only theoretical modeling, and the design can not guarantee a correct sizing in the absence of an appropriate experimentation. Anyway, due to the complexity of the problem, any theoretical model still requires an appropriate calibration and validation based on experimental tests. Because of the interactions among the elements that make up the barrier and the different modes of energy dissipation (plastic deformation and friction), it is necessary to resort to evidence full scale (full scale tests), providing experimental results both on the whole device and its components. Information about the kinetic energy and location of impact on a barrier can help determine the capacity, size and location of barriers.

7.1 Profile FF' Passing Through the GPS Location Number 3 (Fig. 3): Example of a Function of Rockfall Barriers

Using the same data of the GPS station number 3 studied above, it has been analyzed below a rolling boulders blocked by a rockfall barrier (Fig. 15).

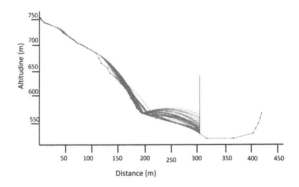

Fig. 15. One of the typical example of a rockfall trajectory with barrier (Profile FF' near GPS location number 3 - Fig. 3)

The initial mass movement is a rolling one, then it falls on the ground and starts a movement of roto-translation. Therefore it continues with a sliding on the surface of the slope until it encounters a part of the steep profile having steep precipice, in correspondence of which the cluster rolls then, falling on the surface, bounces again but is stopped by the rock fall barrier placed in front of it. Next there are the diagrams trend (Fig. 16):

7.2 Profile CC' Passing Through the "Madonna Di Monteverde" Church

In order to protect the churches from phenomena of collapse, it is also important to design protective structures such as "rock barriers." An example that showing very well the importance of rock fall barriers as a method of protection of rupestrian testimonies, follows. The examined Church is "Monteverde" Church (Figs. 17 and 18). Assuming the detachment of a cluster of about 13500 kg it is possible to analyze its path and therefore emphasize the importance of rock fall barrier placed before the Church.

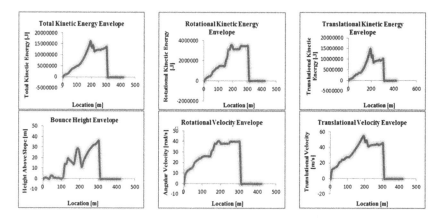

Fig. 16. The analyses results applied on the block: (a) Translational Kinetic Energy, (b) Rotational Kinetic Energy, (c) Total Kinetic Energy, (d) Bounce Height, (e) Rotational Velocity, (f) Translational Velocity

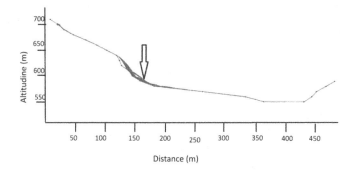

Fig. 17. Rockfall trajectory without barrier (Profile CC' passing through the "Monteverde" Church)

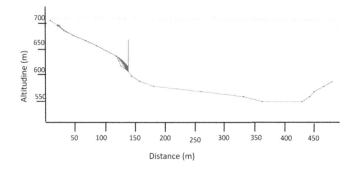

Fig. 18. Rockfall trajectory with barrier (Profile CC' passing through the "Monteverde" Church)

The rock fall barrier is placed immediately before the allocation of "Monteverde" Church. Assuming the movement with a mass of 13 500 kg, in fact it is noted that it would arrive directly on the Church, destroying it completely. Placing the rock fall barrier, the cluster crashes, thus protecting the church below.

8 Analysis of Results and Conclusions

To describe the rockfall hazard around a cultural and natural heritage, an extensive field study including determination of detached blocks, location and dimension of fallen blocks, discontinuity survey and sampling were performed. The obtained results from the simulations have shown that the trajectories followed neglecting a first rolling phase, are characterized mainly by kinematic bounce and free fall because of the morphology of the slope. The presence of terraces causes the continuous shooting of the movement of the boulders but the trajectories stop mostly in the valley. In particular, as far as the kinetic energies, it is possible to observe that the developed maximum values of energy are around 24529 kJ (F-F' profile in Fig. 17) and minimum values are 4396 kJ (profile CC') considering a single boulder or more boulders with a total volume of up to 5 m³; relatively to the heights of impact on the barrier, instead it is noted that the maximum height of 54.4 m is achieved by simulation FF 'and a minimum height of 12.6 m is achieved along the profile CC'.

The results of the performed simulations along the slope profiles of the site "Belvedere delle Chiese Rupestri" are summarized in Table 2:

Table 2. Results of the rockfall analyses along analyzed profiles

Analysis of results	Total kinetic energy [J]	Translational kinetic energy [J]	Rotational kinetic energy [J]	Bounce height [m]	Translational velocity [m/s]	Rotational velocity [rad/s]
Profile A-A'	14506300	11303200	3532020	39,655	45,9647	35,3202
Profile B-B'	15649400	13577500	3906580	33,877	52,1105	42,3467
Profile C-C'	4396360	3580700	843178	12,683	23,0320	16,6591
Profile D-D'	22376900	22376900	5890910	44,812	57,5769	44,0334
Profile E-E'	22217217	17682793	5234291	53,966	51,1577	47,1396
Profile F-F'	24549221	21495158	5914110	54,446	56,4311	44,1200

This work shows that the southern side of the site "Belvedere delle Chiese Rupestri" is the one that has a higher critical phenomena of collapse that can potentially damage the archaeological heritage.

Thanks to the application of the above methodology, based on the possible trajectories of the blocks and the values of associated energy and speed, can be suggested the installation of rock barriers to protect the valuable rupestrian testimonies.

References

Azzoni, A., La Barbera, G., Zaninetti, A.: Analysis and prediction of rockfalls using a mathematical model. Int. J. Rock Mech. Mining Sci. **32**, 709–724 (1995)

Bassato, G., Cocco, S., Silvano, S.: Programma di simulazione per lo scoscendimento di blocchi rocciosi. Dendronatura **6**(2), 34–36 (1985)

Bozzolo, D., Pamini, R.: Modello matematico per lo studio della caduta dei massi. Laboratorio di FisicaTerrestre ICTS. Dipartimento della pubblica Educazione, Cantone Ticino (1982)

Dorren, L.K.A.: A review of rockfall mechanics and modelling approaches. Prog. Phys. Geogr. **27**, 69–87 (2003)

Falcetta, J.L.: Un nouveau modele de calcul de trajectoires de blocs rocheux. Rev. Fr Geotech **30**, 11–17 (1985)

Guzzetti, F., Crosta, G., Detti, R., Agliardi, F.: STONE: a computer program for the three-dimensional simulation of rockfalls. Comput. Geosci. **28**, 1079–1093 (2002)

Hoek, E.: Rockfall - a program in BASIC for the analysis of rockfall from slopes. Unpublished note, Golder Associates/University of Toronto, Canada (1987). Hoek, E: Practical rock engineering. Course note - http://www.rocscience.com/hoek/PracticalRockEngineering.asp (2007)

Jones, C.L., Higgins, J.D., Andrew, R.D.: Colorado rockfall simulation program version 4.0. Colorado Department of Transportation, Colorado Geological Survey (2000)

Pleiffer, T., Bowen, T.: Computer simulation of rockfalls. Assoc. Eng. Geol. Bull. **XXVI**(1), 117–126 (1989)

Pomar, L., Tropeano, M.: The calcarenite di gravina formation in Matera (Southern Italy): new insight for coarse–grained, large–scale, cross–bedded bodies encades in offshore deposits. AAPG Bullettin **85**(4), 661–689 (2001). Pubblica Educazione, Lugano-Trevano

Schweigl, J., Ferretti, C., Nossing, L.: Geotechnical characterization and rockfall simulation of slope: a practical case study from South Tyrol (Italy). Eng. Geol. **67**, 281–296 (2003)

Sdao, F., Pascale, S., Rutigliano, P.: Instabilità dei versanti e controllo, mediante tecniche integrate di monitoraggio, delle frane presenti in due siti sacri del Parco Archeologico Storico Naturale delle Chiese Rupestri di Matera. SIRIS, vol. 9 (2008). Studi e ricerche della Scuola di Specializzazione in Archeologia di Matera, pp. 87–100 (2009)

Sdao, F., Lioi, D.S., Pascale, S., Caniani, D., Mancini, I.M.: Landslide susceptibility assessment by using a neuro-fuzzy model: a case study in the Rupestrian heritage rich area of Matera. Natural hazards and earth system sciences **13**, 395–407 (2013). doi:10.5194/nhess-13-1-2013. ISSN: 1561-8633

Viegas, J.M., Martinez, L.M., Silva, E.A.: Effects of the modifiable areal unit problem on the delineation of traffic analysis zones. Environ. Plan. **36**(4), 625–643 (2009)

Wardrop, J.C.: Some theoretical aspects of road traffic research. Proc. Inst. Civil Eng. Part **2**(9), 325–378 (1952)

Williams, I., Lindsay, C.: An efficient design for very large transport models on PCs. In: European Transport Conference, p. 18. Cambridge (2002)

Integrating Computing to STEM Curriculum via CodeBoard

Hongmei Chi$^{(\boxtimes)}$, Clement Allen, and Edward Jones

Department of Computer and Information Sciences,
Florida A&M University, Tallahassee, FL 32307, USA
hchi@cis.famu.edu

Abstract. Introductory programming has always suffered from low performance rates. These low performance rates are closely tied to high failure rates and low retention in introductory programming classes. The goal of this research is to develop models and instrumentation capable of giving insight into STEM student performance, learning patterns and behavior. This insight is expected to shed some light on low performance rates and also pave the way for formative measures to be taken. CodeBoard is a programming platform capable of managing and assesse student programming via using a functional test-driven approach. Instructors develop programming assignments along with corresponding test cases, which are then used as grading templates to evaluate student programs. The second phase of this research involves developing models for measuring and capturing events relevant to student performance over time. The preliminary results show that this CodeBoard is promising.

Keywords: Cloud computing · Programming environments · Introductory programming · Active learning

1 Introduction

Computation is playing ever increasing role in the conduct of modern scientific inquiry and experimentation. Computational thinking will be a fundamental skill used by everyone in the world by the middle of the 21st everyone in the world by the middle of the 21 Century. There is a growing concern that the United States is not preparing adequate numbers of students, teachers, and practitioners in the area of science, technology, engineering, and mathematics (STEM).

Introductory programming is the basic course for STEM students to master computational thinking skill. However, introductory programming has always suffered from low performance rates [4, 5]. These low performance rates are closely tied to high failure rates and low retention in introductory programming classes. The goal of this research was to develop models and instrumentation capable of giving insight into student performance and behavior. This insight was expected to shed some light on low performance rates and also pave the way for formative measures to be taken.

In many fields of study, knowledge of programming and technology has become fundamental parts of the fields. Most STEM students are required to take one or more programming courses. Programming can be very challenging to grasp, partly because it

© Springer International Publishing Switzerland 2016
O. Gervasi et al. (Eds.): ICCSA 2016, Part IV, LNCS 9789, pp. 512–529, 2016.
DOI: 10.1007/978-3-319-42089-9_36

involves a complex mix of knowledge, skill and creativity that must be mastered concurrently. These three dimensions of learning, taken together, can become overwhelming. Because of this, it is very important to insure that students' introductory programming experiences are bearable, enjoyable, practical and, to the greatest extent possible, successful [6, 7].

There are two main phases and contributions from this paper. The first is CodeBoard, a programming platform capable of managing and assesse student programming using a functional test-driven approach. Instructors, using the system, develop programming assignments along with corresponding test cases which are then used as grading templates to assess student programs. The second phase of this research involved developing models for measuring and capturing events relevant to student performance over time. These models form the Performance Event Space (PES) of collected information that is actively used to interpolate notions and visualize student's performance and behavior. These tools provide versatile, simple and presentable templates that function well with varying depths of data and increasing complexity.

Ideally, programming environments should enhance user experiences by providing on-demand feedback, grades and analysis to provide maximum utility to students and instructors. Reduced analysis time can support functionality for preemptive intervention measures similar to active learning and apprenticeship [2]. Similar tools and models give valid approaches to analyzing performance and learning; however, all analysis occurs after the fact, and can be considered summative assessment approaches [1]. By contrast, CodeBoard provides analytics to assess students formatively by providing immediate feedback that can be used diagnostically and predictively.

The goal of this project was to develop instrumentation that can provide insight into student performance and behavior in introductory programming course. CodeBoard, a fully functional online programming platform for introductory courses has been developed which includes instrumentation capable of capturing data that addresses questions on student performance and behavior. The data captured by CodeBoard is stored in a database for analysis and historical purposes. CodeBoard analytics tools produce reports, graphs, and charts for students and instructors to characterize and visualize student performance and behavior.

This paper is organized as following. In Sect. 2, related work is discussed. Design and implementation of this system is provided in Sects. 3 and 4. The specific modules and case study are presented in Sect. 5. How to use data mining technique to help each student is described in Sect. 6. The survey results and the preliminary feedbacks from students are presented in Sect. 7. Conclusion and future work are presented in Sect. 8.

2 Related Work

CodeBoard use functional testing to advantage by allowing *reuse of test cases*. Assignments derived can a Python assignments can be configured to function the same way in another language. Grading of these assignments will produce identical results to each other based of a single set of test cases. This abstracts away language specific goals and puts the emphasis on the concepts and process of programming.

Unlike FURPS (**F**unctionality, **U**sability, **R**eliability, **P**erformance, and **S**upportability) and CodeBoard, which use functional testing exclusively, other systems like ProgTest [9] and Web-CAT [10] employ [1]unit testing and [2]coverage testing, which both depend heavily on the specifics of particular programming languages. Unit tests written for a C++ assignment with CppUnit[3] are not portable to other languages. Most coverage tools, like JaBUTiService [11] and GCOV[4], used in ProgTest are also language specific. These tools tend to require the student to also develop a high degree of expertise in these testing tools. Users within the system need to get accustomed to writing Unit and Coverage test cases. For introductory programming classes where programs are small to medium sized, the additional requirement to learn a testing tool may be heavy-handed and serve no real value.

As a programming environment, Codecademy presents users with friendly and highly functional interfaces. Each assignment window presents different panes including the description, programming and live preview pane. This is illustrated in Fig. 2. The interface is designed for increased workflow and is meant to host all the requirements of an assignment in a single location. Because only a limited amount of information can effectively be put on a screen at a time, this interface can also restrict assignments to small bite-size chunks of code.

Validation tools are built into each assignment that checks the user's code snippets and returns immediate feedback. If the assignment fails, an error message is displayed and the system allows the user to revise the answer. These validation tools, however, do not give any details on where the user went astray.

The mission of Udacity (http://www.udacity.com/) is to provide accessible high education to the masses across multiple disciplines. Udacity originated at Stanford University and has enrolled students from over 190 countries. Similarly, Coursera (https://www.coursera.org/) offers full-fledged courses online similar to Udacity, but on a much larger scale. Unlike Udacity, Coursera spans beyond science and technology areas and diverges into other areas like business and art. With over 100 partners, six million users and 400 courses, it is clear that there mission is to cater to the masses across any discipline. Many of the courses hosted in these systems are created by academic scholars and by industry professionals from large companies like Reddit and Google. The primary focuses are in new technology, mathematics, science and critical thinking, ranging between beginners to advanced levels.

Khan Academy offers an unprecedented set of educational material for science and math education, in the form of short, free, publicly available video clips. Some of video clips are teaching people how to code. The main purpose of this platform is targeted to individuals not a classroom.

---

[1] **Unit testing** takes the smallest piece of testable software in the application, isolates it from the remainder of the code, and determine whether it behaves exactly as you expect.

[2] Coverage testing measures the degree to which the source code of a program is tested by a particular test suite.

[3] http://cunit.sourceforge.net/doc/index.html.

[4] http://gcc.gnu.org/onlinedocs/gcc/Gcov.html.

Codio [1] is a cloud-based computer programming platform for teaching programming in schools & universities. Codio is a relatively new online IDE that has recently come out of beta. Like other online integrated development tools Codio runs in the browser. The idea behind Codio is close to CodeBoard.

3 Design

CodeBoard is built to provide insight into students' performance progression and programming behavior. Its core infrastructure in built on four pillars: (1) data-centric design; (2) usability; (3) lug-n-play extensibility; and (4) multi-linguality.

The CodeBoard infrastructure is data-centric: all functionality is enabled and driven by data stored in the CodeBoard database. The database also captures data as students use the system. This design promotes efficiency and ensures the capture of data needed for on-demand analytics. Data is abstracted as two main layers. The first involves operation of CodeBoard. The second involves the analysis of this data and presenting on-demand results and interpretations to users.

Usability can play a significant role in improving student performance. In Tilden's design goals for Python [3], he brings up a number of features that improve usability for both students and instructors:

- eliminating the need for software setup
- functioning like a normal website accessible to a wider variety of devices
- providing cloud storage and automatic organization of class assignments
- improving feedback regarding programming errors and behavior.
- Facilitating system modification and extension

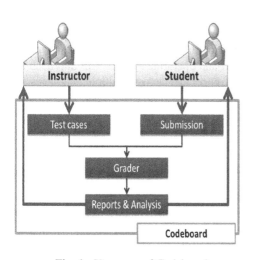

Fig. 1. User case of Codeboard

CodeBoard's design incorporates similar design goals. The goals allow for greater usability of the system (Fig. 1).

CodeBoard is a re-engineered version of FURPS: the driving goal was to improve maintainability and allow future development and easy evolution of the system. To achieve this, a plug-n-play architecture is designed and implemented so that components can be added with minimal impact to the rest of the system. The Model-View-Controller (MVC) architecture is used to separate core functionalities and roles.

Different programming languages are used in programming courses. At Florida A&M University C, C++, C# and Java are the languages used in programming courses. At other universities scripting languages like Python are used to introduce programming concepts to students. A goal of CodeBoard is to support multiple programming languages and, ultimately, to support the sharing of assignments across languages. The support for multiple languages is enabled by the CodeBoard plug-n-play architecture.

At Florida A&M University, recitation laboratories are used to help students in introductory programming courses gain hands-on practice developing and debugging programs. Worldwide, many tools are used to support the hands-on component of programming courses. These range from basic course management to automated support of aspects of program development, grading, and tutoring and intervention. Since 2010, FAMU has used of FURPS, a UNIX based programming management platform. CodeBoard has been used for these recitation labs since the fall of 2013. It is currently used with three recitation labs with 25 to 35 students each. Each lab occurs once each week for a 2.5 h session, and is staffed by two graduate teaching assistants who create the programming problems and graders, and who provide assistance as needed during the lab.

The purpose of recitations is to provide students with a platform where they can explore programming concepts, experience different programming principles, and receive guidance during those recitations.

There are a number of approaches that have been used to grade student programs. Web-CAT grades students based upon how well they test their own programs. Watson [12] takes a different approach by grading students based on the number of errors that they produce within their programs. Watson's model advocates that some degree of the student's final score should be inversely proportional to the frequency of errors the student makes. Both these methods have their benefits, but some common drawbacks include:

Restrictive Development of the System: Watson's approach demands that system functionality include the ability to check each student's errors. Because errors vary across languages, compilers and operating systems and highly reliable generalizations may not be practical. The management and programming environment would need specific knowledge for each language, compiler and operating system. This approach is rigid and is counter to the CodeBoard goal of generality across multiple languages.

The Necessity of Additional Training for Students: Web-CAT forces students to test their own code and this requires students to learn how to test, maybe even before they know to write codes.

The CodeBoard grading system is based on functional testing concepts. Following fundamental software engineering principals, test cases are used to quantify how well a program runs. Each test case contains a specific set of inputs along with a corresponding output set. By checking actual program output against expected output specified in the test case and an overall test case score is assigned (no points if match fails). This approach is a very basic but effective approach to evaluating student's assignments. In software testing terms this approach is called black-box testing, with the added twist of computing a score based on passed and failed test cases.

4   Implementation

When developing CodeBoard a number infrastructure and framework options were considered. In order to ensure CodeBoard's viability, components and requisites had to be established that would guide the direction of CodeBoard's development. From past experiences using FUPRS, CodeBoard design goals were established including ease of use for both the instructor and student, use of modern platforms, and extendibility [13]. Technologies capable of meeting these goals were explored. The most viable direction that stood out was the use of web applications and services like Pythy [14], ProgTest and Web-CAT.

Any web application runs in the browser and is created in browser-supported languages like HTML, JS and CSS. Web-browsers are popular and readily available, and they allow content and applications to be run on a client machine or device without the need to install any additional software. This technology enables an application like CodeBoard to be created and maintained without the need to worry about software distribution. Some applications, like compilers, may not run on some devices like mobile phones, or require third party extensions for students to use them. Web applications solve this problem and are consequently attractive options for development.

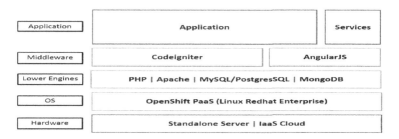

Fig. 2. Structure of CodeBoard

In addition to web applications, the concept of web services has been considered. Web services are a form of internet communication that enables devices and application other than browsers to communicate. Using services, web Application Programming Interfaces (APIs) can be built to provide functionality beyond browser applications. APIs enable the development of connected external applications built on their respective native devices. Android and iOS applications can be built that are optimized for specific devices.

CodeBoard has a back-end which serves and houses all learning material, and associated data. Front-end clients using web-browsers or services can access this content conveniently with any modern web-browsers. This method is similar to other web-bases programming environments mentioned earlier. With today's vast pool of technology, it is worth discussing CodeBoard's current architecture and viable alternative architectures. Figure 2 shows how CodeBoard is structured and the different layers.

Codeboard is built natively in *NIX-based environments, a systems consist of UNIX and Linux based distributions. Its early trial release was built mostly in bash (Bourne Shell) scripts to carry out routine tasks like compiling and executing grades. The majority of data was stored and managed as plain-text files using the RCS revision control system, while grade management was handled by the MySQL database.

Codeboard bash scripts, together with the MySQL database, were integrated through PHP. PHP serves as an interface between the software components. Standard HTML5 and CSS pages are used to display information to the user. This method has some major challenges: the Shell-PHP structure had limitations which defeated the purpose of using web technologies to their full potential. Performance was drastically reduced. Grading could possibly take anywhere between five seconds to minutes to complete depending on program size and the number of test cases. This limitation led to design choices that avoided the limitation while laying a foundation for future improvement and expansion.

The next iteration of the Codeboard system was built entirely from the ground up with performance, modularity and extendibility. The UNIX environment was retained, to enable UNIX based build tools (g++, gcc, python and java) to integrate seamlessly into the grader. Another reason is that UNIX based systems are mostly free, well documented, modifiable and easily maintainable. Many cloud based systems also utilize *NIX based systems, permitting easier deployment to the cloud.

Codeboard has been briefly tested with Windows-based UNIX tools like Cygwin which provides functionality similar to Linux distributions. MinGW can also be used to in lieu of Cygwin; however, some native Linux/UNIX applications may not run as expected. It is advised not to use Cygwin and MinGW tools as replacements to *UNIX because their reliability and performance have not yet been determined.

Codeboard is currently hosted in on the OpenShift (https://www.openshift.com) PaaS public cloud infrastructure. OpenShift runs on a Red Hat Enterprise UNIX system. When demand is low, the PaaS platform limits its usage to only resources that it needs. When demand rises and increased compute power is required, OpenShift's load balancer can scale the PaaS's resources to match the increased demand. Using the cloud has facilitated the rapid development and deployment of Codeboard, without the need to purchase and set up a dedicated server in the CIS Department. As demand within Codeboard has yet to surpass its default usage, its performance from elastic resource scaling is yet to be tested and optimized. For a server with inelastic resources, drastic increase in demand would result in reduced system performance or eventual failure.

A number of cloud technologies have emerged, each with slightly different features and strengths. Though cloud computing usage and boundaries continually expand, the three main categories of Cloud computing include:

IaaS (Infrastructure as a Service). This is a provisioning model where companies outsource equipment for their operations, which usually includes hardware like servers, storage and networking components.

PaaS (Platform as a Service). Much like IaaS, with PaaS various entities can rent out equipment over the Internet. The service delivery model allows the customer to rent virtualized servers and associated services for running existing applications or developing and testing new ones.

Saas (Software as a Service). Saas is a software distribution model in which applications are hosted by a vendor or service provider and made available to customers over a network, typically the Internet.

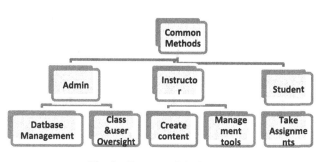

Fig. 3. Features of CodeBoard

CodeBoard utilizes Platform as a Service (Paas) infrastructures like Openshift, Google App Engine (developers.google.com/appengine) or Heroku (www.heroku.com). At FAMU, Openshift's Paas enables rapid deployment. An administrator with good knowledge of the Paas service model and the system configuration (e.g. MQL and database credentials, database initialization scripts etc.), can deploy a functioning instance of CodeBoard (see Fig. 3) in as little as ten minutes. Beyond the initial configuration, further configurations are handled internally and automatically as site and class administrators interact with the system. Although an IaaS set-up can be used, some extra effort is needed to configure equipment and deploy the application.

Stand-alone servers and IaaS cloud infrastructures can be configured in a similar manner. When building a stand-alone server, standard hardware capable of supporting the following software will be required:

- *a UNIX/Linux based system, including Windows Cygwin/MinGW;*
- *Apache 2.4.x;*
- *PHP 5.3+*
- *MySQL 5.1 + (Google Cloud SQL, PostgresSQL and MariaDB can be substituted); and*
- Mongodb.

Apart from the required core software, consideration must be made for user demand levels and the estimated server load. In a cloud environment like Openshift's Paas, applications can be set to scale up automatically, based on load. In a stand-alone instance, physical restrictions on storage, networking capability and processing power sufficient to handle peak loads, should be considered and tested. Failure to plan for peak load could result in server failure during peak loads.

Traditional HTML documents are usually static documents that cannot be changed unless external libraries are used to modify the page's data. When dealing with constantly changing data, external libraries like JQuery (http://jquery.com/) can be used to bind the HTML to defined program logic via the DOM (Document Object Model). The difficulty of implementing the binding depends on the levels of interaction and integration between the elements on the page, the complexity of which can grow exponentially with size. AngularJS transforms static HTML into dynamic web-pages by extending the HTML vocabulary especially for the application. The resulting environment becomes expressive, readable and easier to use.

In CodeBoard, AngularJS is used to enhance individual pages, transforming them into dynamic, personalized and responsive pages. Within the class administrative module, AngularJS is used in many of the creation and management wizards. For students, the online IDE and primary developmental environments are built in AngularJS. The IDEs customize environments to provide a lightweight simplified alternative to UNIX command-line programming and Windows hosted IDEs. AngularJS is also used to construct service driven pages permitting asynchronous loading of pages, an operation that PHP generally cannot do. When heavy data loads are demanded, AngularJS functionality liberates pages from depending on how fast data can load, and enables pages to present partial data fragments as it becomes available.

5 Modules

The CodeBoard authentication and security module manages user groups and permissions associated with user roles. CodeBoard user roles are site administrator, class administrator, and student. Site administrator accounts can only be created by another site administrator account; one generic admin account is created at installation. Once a server has been fully configured, instructors/class administrators are capable of requesting class administrator accounts via the registration page. These accounts must then be approved by a site administrator. Anyone can create a student account and immediately access any publicly available information and services. In order to access material given in private courses student account holders must acquire a registration code from the instructor of the course. This mechanism provides instructors and universities security for with intellectual property they need to protect.

The site administration module manages core objects in CodeBoard. This account type does not have direct access to individual and personal class material for other users. Administrators manage user permissions, whereas course instructors create and manage materials and process related to course delivery and management.

The Class Administrator Modules is the largest module in CodeBoard. The majority of the functionality supports class administrators, who are responsible for creating, using and maintaining their own learning materials.

Upon registering for an account, instructors may initially have no course material of their own. Creating a class entails providing its name and a pass code for students to use when enrolling. Once a class has been created, instructors are free to add in course material into CodeBoard through the use of labs.

A lab is simply a container that holds one or more assignments, quizzes or posted material that must be completed during the 2.5 h session, or prior to the following lab session. When creating a lab, the instructor provides a start date and end date. An optional late submission date can be provided, which enables students to turn in late work for a reduced grade. Lab containers allow different pieces work to be grouped into a single time period without duplicating or redefining dates or metadata. An example of where these containers are useful in FAMU's programming labs. Each 2.5 h lab consists of about three programming assignments with occasional quizzes, tests and tutorials (as blog posts). Labs simplify the process of managing each task (metadata and student work products). CodeBoard also facilitates the importing and sharing of assignments from other sources, including other CodeBoard classes.

"Hands-on" programming assignments are critical for students to master programming skills. CodeBoard provides a wizard that expedites and guides instructors through the process of creating a programming assignment. The wizard consists of the following seven steps:

1. *Gather general assignment information.* The instructor must specify the point values to be used to grade the assignment, and the number of test cases to be used during the grading process, and the number of input/output files required for the assignment, if any.

2. *Define file names and file types.* The instructor defines the deliverables the student should submit, including one or more source code files, and zero or more input files.

3. *Provide solution key source code or executable.* The CodeBoard test Oracle is designed to generate output given some sample output and desired input. This stage enables the instructor to provide the basis for generating test cases to be used in the grading process.

4. *Provide test cases (program inputs which produce expected outputs).* Input sources can be a combination of standard input and data files. The wizard determines what content the instructor must provide based on the number of test cases needed and the data sources and sinks.

5. *Refine output from test cases to be used for grading.* Once the solution key is executed on the input data from step 4, the resulting output is presented so the instructor can select specific output values to be checked during the grading process. For example, an instructor can check for program prompts in one test case, expected data for correct input in the next and finally expected data for wrong data in the final test case.

6. *Create the assignment description.* A WYSIWYG HTML editor is provided to enable the instructor to construct narrative assignment descriptions. The description can be as simple as a short question, or as detailed as narratives and use case examples for students to review.

7. *Conduct a final review.* CodeBoard presents the instructor an overview of the newly created assignment. If acceptable the instructor can officially publish the assignment, which becomes available and visible to students when the lab session opens.

Based on the two semesters that CodeBoard has been in use at FAMU, the completion of the entire creation process took about 3 to 10 min to be completed for small to medium sized programs. This time is based on the assumption that all necessary materials, like the solution key program and program description, have been prepared prior to starting the test case creation process. Once the assignment is activated, no other work needs to be done by the instructor as everything past this phase is automated.

The blog post functionality is extended from the blog helper module. Blogs are a fundamental part of teaching as any material not directly associated with a scored assignment can be added as needed. To date this feature has been used primarily to provide students tutorials or embedded video tutorials.

A student, who creates an account, must first register for a course. Courses can be made public, or private, such that students can only register after obtaining the pass code from the instructor. This pass code mechanism is built in to ensure that only authorized students enroll into classes. Once enrolled, the student is granted access to any material made available by the courses instructor, including labs, the online IDE, class documents and student reporting tools.

The Online IDE provides the programming interface to CodeBoard. The IDE also provides an on-demand program development environment in which students can practice programming. Generally students must download and install a C++ compiler in order to run and execute programs natively on their device. Alternative options include gaining access to remote servers where the server handles the technicalities of compiling the students program and executing it. The CodeBoard IDE allows similar functionality providing students a single point of contact for them to submit and practice their work. The CodeBoard IDE accommodates students with devices incapable of installing software (e.g. mobile devices) or with restricted access.

Behind the scenes, the IDE module makes use of the compile helper module. The compile module provides the online IDE a simple interface between the user and languages supported by the system.

This function is extended from the file management module and provides instructors the ability to manage downloadable documents. Documents may contain tutorials, sample code, output and input files, and other information students may need.

Reports play an essential part in assisting students when instructors may not be able to. These reports are generated automatically when students submit their work and give the student a contrasting view between what is expected of their programs and how their programs actually work. When a student submits an assigned program, the program is compiled and executed against predefined test cases, and a final score is tallied. A detailed grading report is produced detailing discrepancies between expected outputs and the output actually produced by the student's program. The grading report is displayed to the student. This report shows students where they went wrong, and should assist them in correcting the deficiencies. In a lab setting, where a single lab instructor may be outnumbered by the number of students, this feedback has value to both parties. Students get the opportunity to find and fix bugs by themselves and at their own pace. The feedback also reinforces the concept of program correctness based on its execution output.

The modular design of CodeBoard facilitates agile development and on-going maintenance. Modules can be developed and integrated into the system without the need to affect other modules. Through this feature additional modules have been developed that play an important role in extending the systems functionality.

Storing bulk data from files and reports in relational databases like MySQL becomes exponentially difficult to manage, design and maintain as time passes by. Given this, the file manager operates independently from any other modules, and uses an entirely separate MongoDB database to store files. MongoDB is a popular NoSQL[5] database that deviates away from traditional SQL based data manipulation. MongoDB stores its data as indexed documents that can be queried with a relatively more flexibility than standard SQL. These documents can be indexed in any manner defined by the user and queried arbitrarily. MongoDB also includes the GridFS engine, which is built to handle larger document sizes. This engine takes files larger than 16 MB and splits it into 16 MB document chunks. The indexed nature of these chunks allow for easier querying and more efficient manipulation and transfer of documents. CodeBoard uses these capabilities to manage uploaded files reducing complexity and the risk of long-term failure.

The email module has been designed to facilitate an efficient way to contact an entire class roster. In the time span CodeBoard has been operational, there has been a need to send quick messages to students. Instructors who desire this functionality may elect to enable this feature. In the current version of CodeBoard, the email module only provides trivial methods, but the modular system architecture will support future modification and extension of the Email module (Table 1).

Table 1. lists languages supported within the system

Language	Supported	Where it is used	Status
C	Yes	Grader, IDE	In use
C++	Yes	Grader, IDE	In use
Python	Yes	IDE	In use
Java	Yes	Not active yet	Not

In a typical online class setting, instructors have utilized blog functionality to provide a social networking interface to students through which they have access to instructor-provided materials. The Blog module is designed to support instructors in posting announcements and course content including, text, video and images. With labs that have used this system, material that was previously provided to lab students as Word documents are now being translated to HTML and posted as web content.

A large part of the grader constitutes building generalized components capable of spanning multiple languages. In order to achieve this generality, compilation patterns are abstracted from the automated graders and packaged separately in the service oriented compile module. This module provides provide methods for compiling and executing running code across multiple languages.

One important benefit to having programming specific functionality abstracted out is that it enables CodeBoard to add new languages without impacting the source code for the automated graders.

The CodeBoard Analytics module is designed to provide users with a suite of tools that provide deeper understanding of student performance and behavior. Many introductory programming courses lack tools capable of supplying useful insights when needed. The CodeBoard Analytics module implements logic to parse mine and manipulate data, and to display results visually.

---

[5] NoSQL databases provides a mechanism for storage and retrieval of data that is modeled in means other than the tabular relations used in relational databases. This allows for simplicity of design, horizontal scaling and finer control over availability.

6 Data Mining

A fundamental example where separation of concerns has been exploited is the separation of Student, Administrative and Instructor functionality. Figure 4 shows the separation between these fundamental modules in the system. Each module shown in the diagram can run independently of each other, given that these modules inherit the common properties and methods each needs, like security. These three roles, student, administrative and instructor are fundamentally separate modules, but, each can easily import and adopt new functionalities as needed.

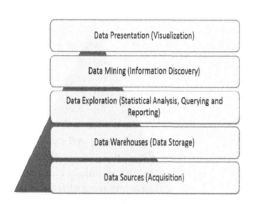

Fig. 4. Data mining process.

Automatic grading at submission is one of CodeBoard's defining features. This has the benefit and ability to capture valuable data at significant points in a student's program development life cycle (see Fig. 4). For each submission, snapshots are taken which, when combined, provide a complete timeline of the student's progression the performance event space. Most components of the PES are captured when programs are submitted. The program is subjected to a range of checkers that generate PES metadata. Measures including time stamps, lines of code, complexity, raw scores, documentation and other rich components are recorded or computed during this submission window.

In addition to submission data, CodeBoard contains instrumentation to capture additional data representing universe events. Embedding triggers in the CodeBoard user interface and software interfaces record non-submission actions taken by the student. These data relate to student behavior – what the student does. Some triggers that are currently used include:

Login Timing: These events allow instructors to see how often students access the system. Instructors will eventually have the ability to correlate how often and how soon students access the system and how well they perform in class.

Opening Assignments: Every time students open an assignment page that time is logged. This logged time in conjunction with submission times is used to estimate the duration of how long students work on assignments.

Checking Grade Reports: Grade reports are allowed students to gain some understanding their progresses. Some reasoning can possibly be established by correlating how often students view their reports against their grades. The underlying research question is whether a student's tendency to check reports affects the grade earned.

'TAB' Clicks: Triggers are built into user interface that students click on – these give the instructor a chance to see what the student was doing, including which CodeBoard features are being utilized by students.

6.1 The Measurable Scope

Unless data are collected in a perfectly controlled environment, observations will have some degree of noise included. Because of this, the observation scope is defined as the reliable boundary and granularity for information collection. In a typical lab setting, various forms of data can be collected. Some information, like submission data, represents the foundation of performance, and therefore becomes a default requirement for data collection. Non-essential information as minute as the number of clicks while looking at an assignment can possibly be analyzed; however, in an uncontrolled environment this data attracts a lot of noise. One might also question the relevance and usefulness of certain minute data relative to the cost it takes to collect, store, and analyze. Too much unnecessary data can possibly get in the way of effectively processing it.

The five categories of CodeBoard analytics tools are defined in the following chapter. Each category defines a subset that represents a grouping of variables that have similar granularity. Categories range from looking at individualized information on individual assignments to overall and multi-class data. This scale shifts from "raw" data to compressed and aggregated data. This is further elaborated in section.

Captured data are separated into categories. Each category is meant to maintain clarity in the way information is presented applying different resolutions to the data. This resolution can scale from atomic sizes, e.g. individual student assignments, to aggregated sizes, e.g. class averages. These define these categories as:

A large part of understanding patterns involves the interpretation of the underlying data. The copious data stream generated by the each of these systems eventually becomes impractical to view as plain and raw data. In order to find a solution to this problem data has to be reduced to understandable measures that depict a sensible image to the user.

Questions in Tables 2 and 3 and course requirements guide the data exploration and data mining process of the data. These performances provide a basis for us to start defining different roles and relationships within the system. These performances determine how the PES relates to performance and behavior and make connections that allow for objective interpretation of data. The relationships have the potential to represent trends in performance, progression and behavior. It is up to CodeBoard to interpolate these measures from captured information and submission data and visualizes these in the form of charts and graphs.

The same information is used to determine performance though these are depicted primarily through the relationships between important measures. The following two subsections make a few of many possible connections that can be made within the system.

6.2   PES and Performance

Performance measures are the first and most direct information to determine and display. These measures are directly tied to student's final results. In order to determine performance measures links need to be defined and collected measures. Common links that research suggest between performance questions and measures are described in Table 2.

Table 2. Performance questions with related measures

#	Questions	Measures
P1	Can students solve a problem?	Score
P2	Can students solve a problem quickly?	Sores, start time, end time
P3	What % of students can pass within time set?	Score, start time, end time

Behavior is a more challenging concept to determine than performance. Behavior is highly subjective and unique to each student. Despite this, because performance measures are in part dependent upon the user's behavior, conjectures about the relationship between behavior and performance can be made and explored. Patterns in behavior can be characterized or described independently of performance but enhanced by performance data collected. Behavior may hold deeper insight that can be used to address student needs through activities like debugging drills.

Question-measure relationships are also established for behavior similar to those described for performance. Measures that have been identified to influence and represent behavior and learning include code growth, change in score and times and durations for programming assignments. Table 3 describes which measures commonly connect to each behavioral question.

6.3   Visualization of Analytics

Visualization of analytics plays a large part in representing and understanding data. Visualization tends to produce levels of simplicity that can make a world of difference in how information reaches and affects users. Today's extensive use of smarter, more visual technology proves how attractive aesthetics empower users and improves utility. For example, a few users have expressed how CodeBoard's pleasantly styled and color coded html reports have been much easier to analyze that regular text file reports.

Table 3. Behavioral questions with related measures

#	Questions	Measures
B1	Do the students work incrementally?	LOC, Code complexity
B2	Are students getting stuck?	Score, LOC, # of submits
B3	Do students start working early?	Effort(score/LOC), Submit time, deadline
B4	What is the source of them getting stuck?	Score, LOC, # of submits, Grading Report

CodeBoard strives to exploit the power of visualization to present analytics for performance and behavior in well translated, compressed and simplified form. Multiple visual options that include textual, tabular or graphical objects provide users with multiple angles to view the data in different light. It is up to users to decide which form suits them well and how best to use that information.

At each submission event, CodeBoard's grading module returns a grading report to the student. Reports are accessible immediately to instructors and can also be accessed at any time thereafter through the grade-book. This feature is very useful for students to review their last submission at any time with relative ease (compared to the legacy system). For each assignment, up to five most recent grading reports are presented for viewing by the student; these snapshots provide a partial timeline of submitted work.

Visualization tools utilized by CodeBoard include charts and graphs. These give users a visual expression of their data and these tools aim to:

- Provide information to the user that is not already given in the grading report.
- Provide timelines of data that tell a program development cycle's story.
- Compress copious data, into understandable measures or visuals.

Web analytical tools like Google Analytics give some brilliant examples of how powerful visual tools can be when tracking data. Our tools specialize and are driven by programming performance and give insight into programming behavior and learning [13].

7 Feedbacks

Since CodeBoard is adopted in our introductory programming, the passing rate is improving 5–10 % for those courses. More promising is that students more likely to do programming than before. Students have more confident in writing codes. We collect responses for each course that adopts CodeBoard at the end of each semester. Several responses are presented below to the following question: What did you like the best about CodeBoard?

"I like to do assignments on CodeBoard and it helps to find my bugs easily."

"These assignments really opened my eyes to mobile security and how to protect my privacy"

"I can do my coding assignment anywhere by using CodeBoard and the most interesting is that I can check my coding assignments."

8 Conclusions

CodeBoard is a web-based programming management environment system capable of creating, scheduling and managing programming assignments. Courses hosted in CodeBoard are semi-automated, where the system tries to handle as much overhead as possible through automated grading and score management. In addition, CodeBoard's analytics platform, which provides on-demand analytics that instructors and students can access. CodeBoard analytics address questions about student performance and behavior by presenting data in a fast and understandable manner.

As for Future Work, New Generalized Analytics Panel will be Added into This System. Currently, analytics provided to students and instructors are all predefined and tailored to the questions asked in this research [8]. A new panel capable of exposing a greater portion of data to instructors would enable instructors to ask their own questions and derive matching interpretations as they desire. The goal of such a tool is to expose as much student data as possible to instructors while maintaining a high level of simplicity.

Acknowledgments. This work has been supported in part by U.S. Department of Education grant P120A080094 and by NSF CPATH grant CNS-0939138. The opinions expressed in this paper do not necessarily reflect those of these funding agencies.

References

1. Codio – taking the pain out of programming for teachers and students. http://www. eschoolnews.com/2014/09/11/codio-taking-pain-programming-teachers-students/
2. Dewan, P.: How a language-based GUI generator can influence the teaching of object-oriented programming. In: Proceedings of the 43rd ACM Technical Symposium on Computer Science Education, SIGCSE 2012, pp. 69–74 (2012)
3. Ilinkin, I.: Opportunities for android projects in a CS1 course. In: Proceedings of the 45th ACM Technical Symposium on Computer Science Education, SIGCSE 2014, pp. 615–620 (2014)
4. Sorva, J., Karavirta, V., Malmi, L.: A review of generic program visualization systems for introductory programming education. Trans. Comput. Educ. **13**(4), 64 (2013). Article 15
5. Bennedsen, J., Caspersen, M.E.: Failure rates in introductory programming. SIGCSE Bull. **39**(2), 32–36 (2007)
6. Tillmann, N., Moskal, M., de Halleux, J., Fahndrich, M., Bishop, J., Samuel, A., Xie, T.: The future of teaching programming is on mobile devices. In: Proceedings of the 17th ACM Annual Conference on Innovation and Technology in Computer Science Education, pp. 156–161. ACM, July 2012
7. Watson, C., Li, F.W.B.: Failure rates in introductory programming revisited. In: Proceedings of the 2014 Conference on Innovation & Technology in Computer Science Education, ITiCSE 2014, pp. 39–44. ACM, New York (2014)
8. Dijksman, J.A., Khan, S.: Khan Academy: the world's free virtual school. APS Meet. Abs. **1**, 14006 (2011)
9. De Souza, D.M., Maldonado, J.C., Barbosa, E.F.: ProgTest: an environment for the submission and evaluation of programming assignments based on testing activities. In: 2011 24th IEEE-CS Conference on Software Engineering Education and Training (CSEE&T), pp. 1–10 (2011)
10. Edwards, S.H., Perez-Quinones, M.A.: Web-CAT: automatically grading programming assignments. ACM SIGCSE Bull. **40**(3), 328 (2008). ACM

11. Eler, M.M., et al.: JaBUTiService: a web service for structural testing of java programs. In: 2009 33rd Annual IEEE Software Engineering Workshop (SEW), pp. 69–76. IEEE, October 2009
12. Watson, C., Li, F.W.: Failure rates in introductory programming revisited. In: Proceedings of the 2014 Conference on Innovation & Technology in Computer Science Education, pp. 39–44. ACM, June 2014
13. Edwards, S.H., Tilden, D.S., Allevato, A.: Pythy: improving the introductory Python programming experience. In: Proceedings of the 45th ACM Technical Symposium on Computer Science Education, pp. 641–646. ACM, March 2014

Trusted Social Node: Evaluating the Effect of Trust and Trust Variance to Maximize Social Influence in a Multilevel Social Node Influential Diffusion Model

Hock-Yeow Yap[✉] and Tong-Ming Lim

Faculty of Science and Technology, Sunway University, 5 Jalan University,
Bandar Sunway, Petaling Jaya, Malaysia
kenny.yap92@gmail.com, tongmingl@sunway.edu.my

Abstract. The use of social networking sites has been very successful on large-scale information sharing. Hence, a vast proposed application possibilities for different people and organizations emerged. Although the use of social networking sites nowadays for large scale information sharing and the spreading of messages on these platforms is considerably effective, this research hypothesizes that trust is able to increase the rate of successfully influenced social nodes. Trust is the fundamental motivation that people cooperates towards a common purpose. This paper discusses trust - a measure of belief and disbelief using experimental simulation to evaluate and compare on the rate of successfully influenced social nodes based on the Trusted Social Node (TSN). This paper considers trust variance and social node impact factor in the Genetic Algorithm Diffusion Model (GADM) to analyze on its successful influential rate with and without the presence of trust in the algorithm. Results produced are a set of influential diffusion time graph where the graph shows there are incremental rate of successfully influenced social nodes with the presence of trust metrics.

Keywords: Trusted social node · Influential strength · Trusted influence · Multilevel influential diffusion · Belief and disbelief

1 Introduction

Social networking sites have tremendously changed the way people around the globe connect to each other. Social networking sites play an important and fundamental role at spreading information, news or ideas to all the connected nodes. This can be achieved by influencing current and new nodes within the social networking environment. Many readers and consumers today rely extensively on information obtained from social networking sites which significantly influence one's decision. Such issue has presented a great concern to business operators because the information spreading around social networking sites may change the viewpoint of an individual if the influences spreading from one social node is not what is desired. Contents on the social networking sites influence one's decision in numerous ways. This include reviews, guides and word-of-mouth. Social networking sites create an endless source of

© Springer International Publishing Switzerland 2016
O. Gervasi et al. (Eds.): ICCSA 2016, Part IV, LNCS 9789, pp. 530–542, 2016.
DOI: 10.1007/978-3-319-42089-9_37

information that is readily available to the world of research. With most of the social networking sites' users have accessed to enormous amount of material, it is acknowledged that not all contents are necessarily reliable. Trust has always been investigated in this context. Various studies were conducted by many researchers [1–4], their works showed that trust played a key role in affecting one's decision. By simulating influence diffusion within a social networking site, they found that there are still much work to be studied on online user generated contents and their credibility assessment.

Motivated by the recent works in identifying trusted social node [5, 6] highlighted that the methodology discussed in paper is able to compute the trust variance and influential impact factor of each social node within the social networking environment. This research commences by developing the base social network algorithm that simulates a hypothetical social networking site. This developed algorithm simulates a working social network with its Virtual Social Node (VSN). The algorithm will be responsible of diffusing influence, accepting and rejecting an influence based on a set of criteria that involves trust variance of each source social node and logical reasoning on the measure of belief and disbelief and the certainty factor of a social node. A set of social network profile data that contain relationship links is used. Trust variance for each social node is generated using the Trusted Social Node (TSN) algorithm [6]. Trust Variance is incorporated into the Genetics Algorithm Diffusion Model (GADM) with VSN to simulate influential diffusion processes.

This article reported results generated from the findings of this research. In this research, social nodes interaction behavior among other social nodes from the social network site are considered in the building of VSN for a more resilient algorithm to simulate the influential diffusion process. This research evaluates the VSN enabled GADM, with and without trusted social node on a simulated social network site based on a data set of 100 million Amazon profiles and relationships. Some data preprocessing tasks are required before the simulation phase and the detail process will be discussed and illustrated in Sect. 3. Finally, this article discuss, compare and debate results from the research and future research will be presented.

2 Related Works

The study of influential maximization focuses on maximizing successful spread of information in any networked environment. Influential maximization is commonly applied in business and marketing to study, analyze and identify market trend to improve profit. Over the years, countless studies have been conducted to identify gaps and to enhance influential maximization algorithm in terms of social node selection, message propagation, influence flow and social triggering. It was found that by selecting an appropriate social node with the appropriate influential trait for a specific influential environment is able to increase the influential capability of the social node thus returning a higher rate of successfully influenced peers. Different influential maximization techniques and models have been developed and experimented. These include Logical Representation extrapolating message instance, Semantic Network Model, Decreasing Cascade Model [7], Genetic Algorithm Diffusion Model [8],

Mathematical Probabilistic Model, and Linear Threshold Model [9] to address gaps in those research where some showed very successful results. David Kempe from Cornell University [7, 9] is one of the pioneer researchers that study influence maximization where they showed that their diffusion model uses probability and general thresholding is able to maximize the influences of a seed node in a network. Kempe [7, 9] defined the activation function as an influential probability

$$P_v(u, S) = \frac{f_v(S \cup \{u\}) - f_v(S)}{1 - f_v(S)}$$

It was found that many researchers also conducted works on influential maximization but very few have considered mutual trust in their algorithm. Trust is one of the key motivation that move people to work towards a common goal. In order to apply trust into Genetic Algorithm Diffusion Modal [7], it is necessary to conduct a study on trust. The concept, definition and value of trust varies from different application thus it is significantly hard to define. Some trust definitions can be found in [10–16]. From past works reviewed, various methods, application and trust profiling approaches were found however no work was found that consider message integrity as a key condition in trust identification. Some of the trust profiling approaches are general rating [4, 12], profile strength [2, 17], reputation system [1, 4, 18] and repetitive action [2, 17, 19]. This research utilizes trust variance generated by the Trusted Social Node [6] algorithm from previous research to uncover trusted social node by applying social interaction pattern analysis. The algorithm takes user generated text contents within a specific timeframe to analyze for message discrepancy hence calculating its objective score, deriving trust variance for each social node and uncovering the most trusted social nodes within the dataset. The purpose of constricting these user generated text contents within a shorter timeframe ensures that the algorithm to detect discrepancy of an opinion from a source node within a short period of time instead of what has been generated in the longer past since every subject in discuss may change or being consistently enhanced and improved over time. This research uses a timeframe that is set to 7 days starting from Sunday 00:00 h to Saturday 23:59 h.

3   Algorithm Design

The key objective of this research is to hypothesize that trust has positive effect on social influential diffusion by increasing the rate of successfully influenced social nodes in a social network. Evaluating the hypothesis involve two (2) stages: the initial stage performs a dry run ensuring the Genetic Algorithm Diffusion Model algorithm is working as expected. Some necessary improvements and customizations are required to ensure the algorithm is working properly. Improvements entail correcting program bugs to mitigate unnecessary computational errors. Results obtained in stage one will be used to compare the influential rate between algorithms that incorporated with and without trust element. The second stage is to design, develop and integrate trust into the Genetic Algorithm Diffusion Model. VSN handles all the trust related calculations. Three of the salient features of VSN are:

- *Direct node trust* – VSN assumes that each node within the simulated social network has a direct relationship from the source node to the recipient node hence only consider direct node trust.
- *Node trust variance* – VSN relies on trust value and node acceptance rate represented by probability score to determine the acceptance of node. Trust variance is a numerical representation of a social node's reliability.
- *BFS tree traverse* – Similar to most social networks available, each social node consists of many social relationship links. The algorithm traverses to the nearest reachable relationship link and attempt to influence that particular node and this process is repeated until all possible reachable links are exhausted.

The Genetics Algorithm Trusted Social Node Diffusion Model (GA-TSNDM) consists of two (2) key components: GADM and the VSN. The algorithm requires two (2) sets of data as inputs – Trusted Social Nodes array and the social node relationship links array. Influential cycle begins by first selecting a node from the social node relationship links array pool along with its connections. Each linked connection has an edge node and each node consists of some personality data. After the process of stripping the personality data from the edge node completes, the personality data and the source trust values are handed to VSN for trusted influential calculation involving the use of certainty factor that measures belief and disbelief. Details of this calculation is presented in Proposition 2. The output consists of an array of binary string carrying the social node's influence decision information. An additional algorithm is designed and developed in this article to translate the output into csv and text files so that it can be illustrated visually into charts.

Definition 1. The GA-TSNDM accepts input from two (2) files: one consists of all the social nodes within the simulated social network along with its trust variance and the other file consists of all the relationship links for each node and their respective acceptance variance. All functions and algorithms in this research are designed to process data in 512 bits chunks to maximize parallel processing. Chunks are divided and distributed automatically by the algorithm.

Definition 2. The social network simulated in this research (Fig. 1) consists of local and global relationship links. A is the highest superset of all nodes in the social network and a, b, and c are subsets of A, which is represented as {a,b,c,...} $\subset$ A. It is also acknowledged that in social networks some nodes share one or more similar relationship links. This is represented a $\cup$ b or c $\cap$ (a $\cup$ b). Each set of relationship links belong to a node and all nodes belong to the superset A. The influence diffusion occurs from the lowest subset all the way to the highest superset.

Proposition 1. The Holland's Hyperplane-Defined Function (HDF) [8] is used to calculate the objective score. However, it is not suitable in this research as it may sometimes result into false-negative or false-positive objective scores.

While performing the baseline analysis on the GADM algorithm paired with HDF with and without the presence of trust variance, it was observed that there are a significant drop on the rate of successful influenced social nodes as trust variance is introduced into the HDF objective scoring method. From the analysis in Fig. 2, it is

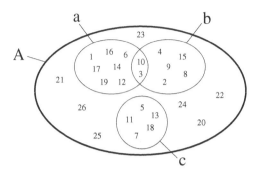

Fig. 1. Social relationship diagram

found that some of the outcomes from HDF are represented as an asterisk '*' which denotes as the "don't care" personality. For "don't care" personality, no influential propagating will occur from a particular node because "don't care" node produces an objective score of zero (0). When the result is applied into the mathematical calculation, some multiplications produces zero. From the analysis, it also showed that HDF wasn't designed to accept additional values directly into its objective score function. Since HDF's binary string evaluation involves simple addition and subtraction, however, the trust variance entails multiplication and division. In cases when a zero (0) is yielded due to division calculation, the calculation don't work; hence the outcomes yield false-negative or false-positive objective scores (Fig. 3).

```
 1  11*0110*01011111**1110**1*11*11*1111*1*11*1111*1*011*11111*1111*
 2  1111111**01100*0**10****11*001*011*111**101111*111011**01**1**1*
 3  0**1111111101*00*1110110***11*01*01**1011101111***1*1*11*00*101*
 4  1**1****1**1*011**0101110*00101*111111**1*011*0*01*1*0*11110101*
 5  1110111111*11111**11111*11*1111111111111*1*1111*1*1100*00111**11
 6  1111*101*110001001****011*111011*111111***1*1*******1****111*1*111
 7  101*11*101*0*10*10***111*11*10**0*1****1**1110111*11**11*1111101
 8  001010111******0000***111**0*0*****1**1111111101***111**11*1*0*1
 9  *1111**1***11****010*10111***10***1*10111*11****1*111001*01*****
10  111**11111*1*100**1**0*1*00**1**100**1*101110*01**11*0*1**1011**
11  *1*1**1**0*1111**00*1111**1*0*1*01**001111**111*1***1*1*10111101
12  1*1*****011111110***11**0*1*111*11*01*011**111*01*011*11*1*1110*
13  **0*1***111*1*110101**11**11**1****11**111*01**1100**01***1**11*
14  **011*1*111**0*****11010*11*****111*11**1*110*10*111*101*011***1
15  ***1***1111*111*1*1100**1110*1**1**0**1010*10*11110*11110*1*****
```

Fig. 2. Samples of HDF objective score snipped during the HDF process

Proposition 2. The probability of a social node to accept or reject an influence from a source node using certainty factor that measures the belief or disbelief and by calculating the occurrence of an influential process given that the source node is trusted using conditional probabilities. In this research, the accuracy of the result uses four (4) decimal points.

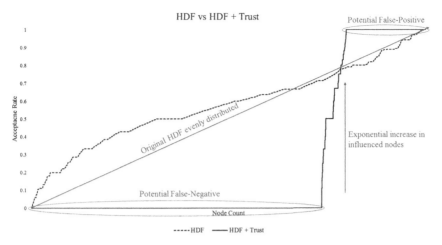

Fig. 3. HDF vs HDF + Trust

To identify the acceptance probability of a selected social node given the source node is trusted; it is denoted as Pr (A | T) where it is then defined as:

$$Pr(A \mid T) = \frac{[Pr(Ra) * Pr(St)] + Pr(St)}{\lceil Pr(Ra) + Pr(St) \rceil}$$

where Pr (Ra) is the probability of the recipient accepting an influence and Pr (St) is the trust variance of the source node. The trust variance of a source node is calculated using the Trusted Source Node algorithm [6] with the recipient acceptance variance value.

Once the probability of the recipient node that accept the influence is obtained, and with the trust variance of the source available, the algorithm is able to measure the belief and disbelief of the recipient that is successfully influenced (**Hypothesis**: accepting the influence) given the trusted acceptance probability of the source (**Evidence**: trusted acceptance probability) denoted by Measure-of-Belief ($H_{accept} \mid E_{trust}$).

$$MB(H_{accept} \mid E_{trust}) = \frac{\max[Pr(A \mid T), Pr(\alpha)] - Pr(\alpha)}{1 - Pr(\alpha)}$$

and the Measure-of-Disbelief ($H_{accept} \mid E_{trust}$)

$$MD(H_{accept} \mid E_{trust}) = \frac{\min[Pr(A \mid T), Pr(\alpha)] - Pr(\alpha)}{1 - Pr(\alpha)}$$

The measurement of beliefs and disbeliefs always require a probability constant for the hypothesis to be evaluated on that is denoted as Pr (H). The hypothesis in the discussion is the acceptance of an influence. By applying acceptance constant into the equation where it can be re-written as Pr(H) = Pr(acceptance constant) = Pr (α). Pr (α). In this experiment, the constant has been set to 0.14 [20].

Definition 3. The function Pr (α), Pr (Ra), and Pr (St) have a range of values between 0.001 and 0.999 in this research.

To compute the certainty factor, the recipient node will accept an the influence from the source node given that the recipient node acknowledges the trust variance of the source node, the certainty factor combines the total strength of belief and disbelief such that

$$CF = \frac{MB(H_{accept}|E_{trust}) - MD(H_{accept}|E_{trust})}{1 - \min\left[MB(H_{accept}|E_{trust}), MD(H_{accept}|E_{trust})\right]}$$

The final result generated from the CF equation measures the recipient's belief of either it has successfully accepted an influence or rejected an influence, where the value falls into the range of $-1.0 \leq CF \leq +1.0$.

Definition 4. Human knowledge is characterized by uncertainty, in contrast to the binary logic of computers. The purpose of a certainty factor is to quantify the degree of confidence in a rule or proposition. Certainty factors are used in cases where the probabilities are too difficult or expensive to identify. Benefits of applying certainty factor allows the algorithm to perform evidential reasoning base on the hypothesis and the available evidences.

4   Analysis of the Experiments and Discussions

This section assesses the performance of the proposed GA-TSNDM algorithm to study its behavior in practice and to ensure that the proposed algorithm performs as designed. Initiating the algorithm requires two (2) inputs where one consists of all the social nodes and trust variance and the other file consists of all the relationship links and acceptance variance. Trust variance and acceptance variance has 2 states - high or low state; therefore the total number of possible outcome for each set of inputs $C = 2^2$ is 4 such that

- C_1: ↑ Trust Variance & ↑ Acceptance Variance
- C_2: ↑ Trust Variance & ↓ Acceptance Variance
- C_3: ↓ Trust Variance & ↑ Acceptance Variance
- C_4: ↓ Trust Variance & ↓ Acceptance Variance

Based on the description above, the outcomes for nodes that fall into category C_1 are absolute and definite influence acceptance, nodes for C_2 and C_3 are weaker or less influence acceptance while nodes in C_4 are absolute and definite influence rejection. Evaluating the outcomes from the experiment on the values in Table 1, the results in Table 2 show that the algorithm is working as expected.

Disclaimer 1. The values used showed in Table 2 are not the actual values in the actual influential maximization. These values are used to demonstrate that the original performance of the calculation algorithm.

Table 1. Dummy values used to assess the performance of the algorithm

	Trust Variance	Acceptance Variance
↑	0.94	0.7817
↓	0.14	0.1817

Table 2. Algorithm assessment results

```
Trust Variance = 0.94
Acceptance Variance = 0.7817
Pr(A|T) = 0.8374
MB(Ha|Et) = 0.8109
MD(Ha|Et) = 0
cf = 0.8109
Influence = Definitely being influenced.
```
↑ T ↑ A

```
Trust Variance = 0.94
Acceptance Variance = 0.1817
Pr(A|T) = 0.5554
MB(Ha|Et) = 0.483
MD(Ha|Et) = 0
cf = 0.483
Influence = Probably being influenced.
```
↑ T ↓ A

```
Trust Variance = 0.14
Acceptance Variance = 0.7817
Pr(A|T) = 0.1247
MB(Ha|Et) = 0
MD(Ha|Et) = 0.1093
cf = -0.1093
Influence = Too weak to define being influenced.
```
↓ T ↑ A

```
Trust Variance = 0.14
Acceptance Variance = 0.1817
Pr(A|T) = 0.0827
MB(Ha|Et) = 0
MD(Ha|Et) = 0.4093
cf = -0.4093
Influence = Probably not being influenced.
```
↓ T ↓ A

For evaluation purposes, it is desirable to use a network dataset to exhibit many structural features of large-scale social networks including relationship links and social interaction patterns as these are the core characteristics and properties that will be evaluated by the algorithm. Database used in this research consists of 93600 unique social nodes. Each social node made up of 5 to 100 relationship links. Overall there are about 1 million unique entities to be processed. In the initial pass, it is found that there are many mistakes and syntax errors in the database such as invalid data structure and invalid csv formatting. These errors are corrected to ensure a clean dataset available for the actual influential maximization experiment. In Definition 2, the social relationship structure has a superset A that consists of a number of subset relationships R = {Ra, Rb, Rc, ... Rn} and for each relationship, it has multiple links Rn = {U1, U2, ... Un}. For every influential cycle at timestamp t, the GA-TSNDM starts at node level $l = 0$:

1. At time t, level l, the pointer traverses to the next closest reachable node via a relationship link and attempt to ingest influence to that node.
2. At time $t + 1$, the pointer checks for next available adjacent node breathy. If adjacent node is reachable, then the influence process repeats until all nodes at the same level has exhausted. The pointer then traverses to $l + 1$ level until it reaches l_n level.
3. The algorithm generates outputs by calculating the following parameters:
   (a) **PrAT:** The probability of acceptance given the trust variance of a source node.
   (b) **MB_HaEt**: The strength (measure) of belief the node will accept the influence.

(c) **MD_HaEt:** The strength (measure) of disbelief the node will accept the influence
(d) **cf:** Certainty factor grouping of the current evaluating node
(e) **success:** Describes the successfulness of the node being influenced. This criterion only applies if the threshold value is set prior to the evaluation. There is no specification on what the threshold value should be used. The threshold value is chosen based on a user's preferred influential range. Changing the threshold value also changes the rate of successfully influenced social nodes.

The algorithm is also used to evaluate the best-case and worst-case scenario of the influential maximizing process yielding the results in Fig. 4. In Definition 3, it states that the values for the GA-TSNDM ranges between 0.001 and 0.999 therefore both of the values are used to evaluate the best-case and worst-case scenarios. The results showed that almost if not all social nodes have been successfully influenced given the source is at its most trusted (0.999) while almost if not all social nodes rejected the influence when the source is at its least trust (0.001).

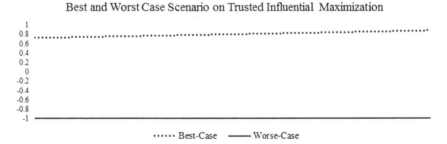

Fig. 4. Best and worst case scenarios

After completed identifying the best and worst case scenario, the actual influence maximization process commences. This process is conducted three (3) times with different settings: highest n trusted nodes, lowest n trusted nodes and adaptive selection. For the highest n trusted nodes, the algorithm searches for n number of nodes with the highest trust variance and use them to perform the influential diffusion, for the lowest n trusted nodes, the algorithm searches for n number of nodes with the lowest trust variance and use them to perform influential diffusion and finally the adaptive selection allows the algorithm to decide for itself to choose the most suitable n number of nodes to perform influential diffusion. All the three (3) processes are mutually exclusive but the same dataset is used for these processes. The value of n is set to 10000 nodes in this experiment. The purpose of doing such is to demonstrate the differences in the result between directed node selection and automated node selection base on the programmed criteria. The results of all three (3) processes are shown in Fig. 5.

From the results in Fig. 5, it is shown that selecting the most trusted node to simulate the influential process has demonstrated significant increase on the rate of successfully

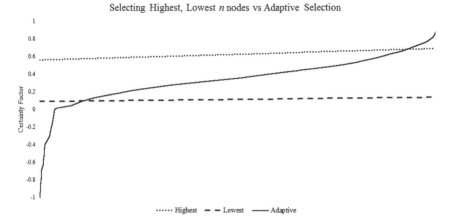

Fig. 5. Results of the three (3) processes from the highest, the lowest trust nodes and the adaptive selection nodes

where influenced nodes has reached 80 per cent depending on the threshold value defined for the influencing environment. In the previous section, we pointed out that there is no predefined threshold value and it is always subject to the user's preferred influential range. Therefore, changing and experimenting few threshold values to ana-lyze the outcomes of the successful influence rate (as shown in Fig. 6) is the usual approach taken by researchers. In this research, the threshold value used is 0.6 (Fig. 6).

From the analysis, the social nodes with the highest trust variance are able to maximize the rate of successfully influenced node but the results shown in Fig. 7 suggest this may not be an ideal solution when the threshold value is increased to 0.7.

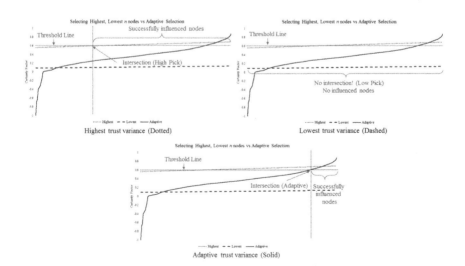

Fig. 6. Three processes with influence rate set at threshold = 0.6

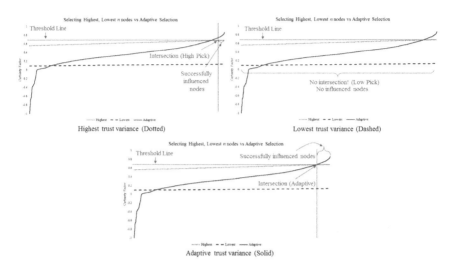

Fig. 7. Three process influence rate with threshold = 0.7

From the outcomes and analysis of the results shown in Figs. 5, 6 and 7, it can be concluded that by applying adaptive node selection algorithm is able to achieve maximum influential capability with the best universally distributed influence diffusion and best rate of successful influenced nodes regardless of the threshold value ranges. Subsequently we select the adaptive node selection for influential maximization in certainty factoring incorporated with trust (TCF) and compare with the results generated from the base Hyperplane Defined Function (HDF) and the enhanced HDF which has incorporated with trust metrics (THDF) (shown in Fig. 8). Considering that the base algorithm that is used to compare with do not support thresholding therefore it is assumed in the comparison threshold value = neutral = 0.

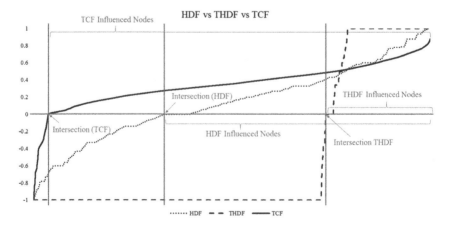

Fig. 8. Algorithm comparison showing results on the influence maximization acceptance rate between HDF, THDF and TCF

Based on the findings, it can be concluded that the TCF algorithm has outperformed HDF by about 25 percent in this research. Although TCF algorithm also significantly outperformed the THDF algorithm by about 60 % but it has also strongly shown that one should not just emphasize on TCF vs THDF because based on Proposition 1, the HDF objective scoring equation does not accept additional values into its calculation therefore by introducing an additional value directly into its objective scoring function resulted into false-negative or false-positive objective scores.

5   Conclusion

From the results and discussions presented in the last section, it can be concluded that it is possible to increase the rate of successfully influenced social nodes within a social network. The rate of successfully influenced nodes may fluctuate depending on the threshold value range being set prior to the influential diffusion process. Generally speaking, the higher the threshold value range, the lower the rate of successfully influenced nodes. Threshold value represents the level of certainty both the source node and the recipient node must commit to. Increasing the threshold value directly increasing the requirements on the quality of influencing messages propagated from the source hence narrowing the scope of potential recipient nodes. Profiling trust and maximizing influence with trust within an online social environment is merely about finding a needle in the world largest haystack and trying to convince the hay to become a needle. These processes are time and resource expensive not to mention the constant changes when human interaction pattern and psychological mentality change dynamically in live data set. While the results do not necessary reflect exactly to the actual outcome in a real world situation but it is an encouraging effort where simulated result shows that the presence of trust has produced considerable improvement on maximizing social influences between social nodes. Future works conducted shall emphasize on improving the calculation function of TCF algorithm specifically on the trust and acceptance probability equation $Pr(A|T)$, introduce heuristic node analysis and learning algorithm that learns and predict social node influential capability, analyze node influential impact and consistently switching trusted social nodes while performing influence maximization, and finally ensuring peak operational efficiency.

References

1. Josang, A., Ismail, R., Byod, C.: A survey of trust and reputation systems for online service provision. Decis. Support Syst. **43**, 618–644 (2007). Amsterdam
2. Caverlee, J., Liu, L., Webb, S.: The SocialTrust framework for trusted social information management: Architecture and algorithms. J. Inf. Sci. **180**(1), 95–112 (2010)
3. Hargittai, E., Fullerton, L., Menchen-Trevino, E., Thomas, K.: Trust online: young adults' evaluation of web content. Int. J. Commun. **4**(1), 468–494 (2010)
4. Resnick, P., Zeckhauser, R.: Trust among strangers in internet transactions: Empirical analysis of ebay's reputation system. Adv. Appl. Microeconomics **11**(2), 127–157 (2002)

5. Hock-Yeow, Y., Tong-Ming, L.: An evaluation and enhancement of the sentiment oriented opinion mining technique using opinion scoring. In: 2015 International Conference on Innovation and Sustainability, Chiang Mai, vol. 1, pp. 42–52 (2015)

6. Hock-Yeow, Y., Tong-Ming, L.: An analysis of opinion variation of social text in the trusted social node identification. In: American Scientific Publishers Advance Science Letters, Sabah, Malaysia (2016)

7. Kempe, D., Kleinberg, J.M., Tardos, É.: Influential nodes in a diffusion model for social networks. In: Caires, L., Italiano, G.F., Monteiro, L., Palamidessi, C., Yung, M. (eds.) ICALP 2005. LNCS, vol. 3580, pp. 1127–1138. Springer, Heidelberg (2005)

8. Lahiri, M., Cebrian, M.: The genetic algorithm as a general diffusion model for social networks. In: AAAI 2010 Proceedings of the Twenty-Fourth AAAI Conference on Artificial Intelligence, United States, pp. 494–499 (2010)

9. Kempe, D., Kleinberg, J., Tardos, E.: Maximizing the spread of influence through a social network. In: Proceedings of the Ninth ACM SIGKDD International Conference on Knowledge Discovery and Data Mining, New York, pp. 137–146 (2003)

10. Morgan, R., Hunt, S.: The commitment-trust theory of relationship marketing. J. Mark. **58** (1), 20–38 (1994)

11. Gefen, D., Karahanna, E., Straub, D.: Trust and TAM in online shopping: An integrated model. MIS Q. **27**(1), 51–90 (2003)

12. Yildirim, E.: The effects of user comments on e-Trust: an application on consumer electronics. J. Econ. Bus. Manage. **1**(4), 360–364 (2013)

13. Weiseberg, J., Te'eni, D., Arman, L.: Past purchase and intention to purchase in e-commerce: the mediation of social presence and trust. Internet Res. **21**(1), 82–96 (2011)

14. Gustavsson, M., Johansson, A.-M.: Consumer Trust in E-commerce. Kristianstad University, Sweden (2006)

15. Gambetta, D.: Can We Trust Trust. Basil Blackwell, Oxford (1988)

16. Falcone, R., Castelfranchi, C.: Social Trust: A Cognitive Approach. In: Castelfranchi, C., Tan, Y.-H. (eds.) Trust and Deception in Virtual Societies. Springer, Netherlands (2001)

17. Massa, P., Avesani, M.: Controversial users demand local trust metrics: An experimental study on epinions.com community. In: AAAI 2005 Proceedings of the 20th National Conference on Artificial Intelligence, United States, vol. 1, pp. 121–126 (2005)

18. Minaxi, G., Paul, J., Mostafar, A.: A reputation system for peer-to-peer networks. In: International Workshop on Network and Operating Systems Support for Digital Audio and Video, New York, pp. 144–152 (2003)

19. Page, L., Brin, S., Motwani, R., Winograd, T.: The Pagerank Citation Ranking: Bringing Order to the Web. Stanford Digital Library Technologies, United States (1999)

20. Sibona, C., Walczak, S.: Unfriending on Facebook: Friend request and online/offline behavior analysis. In: Proceedings of the 44th Hawaii International Conference on System Sciences, Hawaii (2011)

Enhanced Metaheuristics with the Multilevel Paradigm for MAX-CSPs

Noureddine Bouhmala$^{(\boxtimes)}$, Mikkel Syse Groesland, and Vetle Volden-Freberg

Department of Maritime Technology and Innovation,
University College of Southeast, Notodden, Norway
`noureddine.bouhmala@hbv.no`

Abstract. As many real-world optimization problems become increasingly complex and hard to solve, better optimization algorithms are always needed. Nature inspired algorithms such as genetic algorithms and simulated annealing which belongs to the class of evolutionary algorithms are regarded as highly successful algorithms when applied to a broad range of discrete as well continuous optimization problems. This paper introduces the multilevel paradigm combined with genetic algorithm and simulated annealing for solving the maximum constraint satisfaction problem. The promising performances achieved by the proposed approach is demonstrated by comparisons made to solve conventional random benchmark problems.

Keywords: Genetic algorithms · Simulated annealing · Multilevel paradigm · Constraint satisfaction problem

1 Introduction

Many problems in the field of artificial intelligence can be modeled as constraint satisfaction problems (CSP). A constraint satisfaction problem (or CSP) is a tuple $\langle X, D, C \rangle$ where, $X = \{x_1, x_2, ...x_n\}$ is a finite set of variables, $D = \{D_{x_1}, D_{x_2}, ...D_{x_n}\}$ is a finite set of domains. Thus each variable $x \in X$ has a corresponding discrete domain D_x from which it can be instantiated, and $C = \{C_{i1}, C_{i2}, ...C_{ik}\}$ is a finite set of constraints. Each k-ary constraint restricts a k-tuple of variables, $(x_{i1}, x_{i2}, ...x_{ik})$ and specifies a subset of $D_{i1} \times ... \times D_{ik}$, each element of which are values that the variables can not take simultaneously. A solution to a CSP requires the assignment of values to each of the variables from their domains such that all the constraints on the variables are satisfied. The maximum constraint satisfaction problem (Max-CSP) aims at finding an assignment so as to maximize the number of satisfied constraints. Max-CSP can be regarded as the generalization of CSP; the solution maximizes the number of satisfied constraints. In this paper, attention is focused on binary CSPs, where all constraints are binary, i.e., they are based on the cartesian product of the domains of two variables. However, any non-binary CSP can theoretically be converted to a binary CSP [6]. Algorithms for solving CSPs apply the so-called

© Springer International Publishing Switzerland 2016
O. Gervasi et al. (Eds.): ICCSA 2016, Part IV, LNCS 9789, pp. 543–553, 2016.
DOI: 10.1007/978-3-319-42089-9_38

1-exchange neighborhood under which two solutions are direct neighbors if, and only if, they differ at most in the value assigned to one variable. Examples include the minimum conflict heuristic MCH [15], the break method for escaping from local minima [14], various enhanced MCH (e.g., randomized iterative improvement of MCH called WMCH [19], MCH with tabu search [18]), evolutionary algorithms [1,2,4,21]. Weights-based algorithms are techniques that work by introducing weights on variables or constraints in order to avoid local minima. Methods belonging to this category include genet [5], the exponentiated sub-gradient [17], the scaling and probabilistic smoothing [10], evolutionary algorithms combined with stepwise adaptation of weights [11], methods based on dynamically adapting weights on variables [16], or both (i.e., variables and constraints) [8]. Other methods rely on a larger neighborhood [3,13] where a sequence of moves is to be performed and the algorithm will select the move that returns the best solution.

2   Metaheuristics

2.1   Simulated Annealing Algorithm (SA)

The Simulated Annealing algorithm (SA) [12] is a general optimization technique which has been successfully applied to solve combinatorial optimization. An iteration of SA starts by proposing a random perturbation to a state s_0 leading to a new state s_1. The resultant change in the objective function δC is computed. If the change is negative, corresponding to a downhill move, the perturbation is accepted and the new lower cost state s_1 becomes the starting point for the next perturbation. If the cost change is positive, corresponding to an uphill move, the proposed perturbation s_1 may be accepted with a probability $p = exp^{-\delta C/T}$, where T is a control parameter called the temperature. A random number generator that generates numbers distributed uniformly on the interval (0,1) is sampled, and if the sample is less than p, the move is accepted. The higher the temperature, the greater the chance of a solution with a worse cost function value is accepted as the new state. A temperature reduction function is used to lower the temperature. The function together with the initial temperature also describes the annealing schedule which determines when and how the temperature is to be reduced.

2.2   Genetic Algorithms (GAs)

Genetic Algorithms (GAs) [9] are stochastic methods for global search and optimization and belong to the group of nature inspired meta-heuristics leading to the so-called natural computing. GAs operate on a population of artificial chromosomes. Each chromosome can be thought of as a point in the search space of candidate solution. Each individual is assigned a score (fitness) value that allows assessing its quality. Starting with a randomly generated population of chromosomes, GA carries out a process of fitness-based selection and recombination to produce a successor population, the next generation. During recombination, parent chromosomes are selected and their genetic material is recombined to produce child chromosomes. As this process is iterated, individuals from the set of solutions which is called population will evolve from generation to generation by repeated applications of an evaluation procedure that is based on genetic operators. Over many generations, the population becomes increasingly uniform until it ultimately converges to optimal or near-optimal solutions.

3   Combining SA and GA with the Multilevel Paradigm

– **construction of levels (coarsening):** Let $G_0 = (V_0, E_0)$ be an undirected graph of vertices V and edges E. The set V denotes variables and each edge $(x_i, x_j) \in E$ implies a constraint joining the variables x_i and x_j. Given the initial graph G_0, the graph is repeatedly transformed into smaller and smaller graphs G_1, G_2, ..., G_m such that $|V_0| > |V_1| > ... > |V_m|$. To coarsen a graph from G_j to G_{j+1}, a number of different techniques may be used. In this paper, when combining a set of variables into clusters, the variables are visited in a random order. If a variable x_i has not been matched yet, then the algorithm randomly selects one of its neighboring unmatched variable x_j, and a new cluster consisting of these two variables is created. Its neighbors are the combined neighbors of the merged variables x_i and x_j.
– **initial assignment:** The process of constructing a hierarchy of graphs ceases as soon as the size of the coarsest graph reaches some desired threshold. A random initial population is generated at the lowest level ($G_k = (V_k, E_k)$). The chromosomes which are assignments of values to the variables are encoded as strings of bits, the length of which is the number of variables. At the lowest level, the length of the chromosome is equal to the number of clusters. The initial solution is simply constructed by assigning to each variable x_i in a cluster, a random value v_i from D_{x_i} In this work it is assumed that all variables have the same domain (i.e., same set of values), otherwise different random values should be assigned to each variable in the cluster. All the individuals of the initial population are evaluated and assigned a fitness expressed in Eq. 1 which counts the sum of constraint violations where $< (x_i, s_i), (x_j, s_j) >$ denotes the constraint between the variables x_i and x_j when x_i is assigned the value s_i from D_{x_i} and x_j is assigned the value s_j from D_{x_j}.

$$Fitness = \sum_{i=1}^{n-1} \sum_{j=i+1}^{n} Violation(< (x_i, s_i), (x_j, s_j) >) \tag{1}$$

– **parent selection:** During the refinement phase, new solutions are created by combining pairs of individuals in the population and then applying a crossover operator to each chosen pair. Combining pairs of individuals can be viewed as a matching process. In the version of GA used in this work, the individuals are visited in random order. An unmatched individual I_k is matched randomly with an unmatched individual I_l.

- **genetic operators:** The task of the crossover operator is to reach regions of the search space with higher average quality. The two-point crossover operator is applied to each matched pair of individuals. The two-point crossover selects two randomly points within a chromosome and then interchanges the two parent chromosomes between these points to generate two new offspring. The mutation operator is then applied in order to introduce new features in the population. By mutation, the alleles of the produced child have a chance to be modified, which enables further exploration of the search space. The mutation operator takes a single parameter p_m, which specifies the probability of performing a possible mutation. Let $C = vl_1, vl_2, \ldots\ldots vl_m$ be a chromosome represented by a chain where each of whose gene vl_i is a value from the feasible domain (i.e., all the variable have the same domain). In our mutation operator, a chosen gene vl_i is mutated through modifying this gene's allele current value to a new random value from the feasible domain, if the probability test is passed. For coarser graphs ($G_k = (V_k, E_k)$) when ($k > 1$), the mutation is applied to a cluster of variables (i.e., all the variables within the chosen cluster undergo the same mutation).
- **survivor selection:** The selection acts on individuals in the current population. Based on each individual quality (fitness), it determines the next population. In the roulette method, the selection is stochastic and biased toward the best individuals. The first step is to calculate the cumulative fitness of the whole population through the sum of the fitness of all individuals. After that, the probability of selection is calculated for each individual as being $P_{Selection_{I_k}} = f_{I_k} / \sum_1^N f_{I_k}$, where f_{I_k} is the fitness of individual I_k.
- **uncoarsening:** Once GA or SA has reached the convergence criterion with respect to a child level graph $G_k = (V_k, E_k)$, the assignment reached on that level must be projected on its parent graph $G_{k-1} = (V_{k-1}, E_{k-1})$. The projection algorithm is simple; if a cluster belongs to $G_k = (V_k, E_k)$ is assigned the value vl_i, the merged pair of clusters that it represents belonging to $G_{k-1} = (V_{k-1}, E_{k-1})$ are also assigned the value vl_i.
- **refinement:** Having computed an initial solution at the coarsest graph, and depending on the algorithm used (SA or GA), the search process begins from the coarsest level $G_k = (V_k, E_k)$ and continues to move towards smaller levels. The motivation behind this strategy is that the order in which the levels are traversed offers a better mechanism for performing diversification and intensification. The coarsest level allows GA and SA to view any cluster of variables as a single entity leading the search to become guided in far away regions of the solution space and restricted to only those configurations in the solution space in which the variables grouped within a cluster are assigned the same value. As the switch from one level to another implies a decrease in the size of the neighborhood, the search is intensified around solutions from previous levels in order to reach better ones.

4 Experimental Results

4.1 Experimental Setup

The benchmark instances were generated using model A [20] as follows: Each instance is defined by the 4-tuple n, m, p_d, p_t, where n is the number of variables; m is the size of each variable's domain; p_d, the constraint density, is the proportion of pairs of variables which have a constraint between them; and p_t, the constraint tightness, is the probability that a pair of values is inconsistent. From the $(n \times (n-1)/2)$ possible constraints, each one is independently chosen to be added in the constraint graph with the probability p_d. Given a constraint, we select with the probability p_t which value pairs become no-goods. The model A will on average have $p_d \times (n-1)/2$ constraints, each of which having on average $p_t \times m^2$ inconsistent pairs of values. For each pair of density tightness, we generate one soluble instance (i.e., at least one solution exists). Because of the stochastic nature of both GA and SA, we let each algorithm do 100 independent runs, each run with a different random seed. Many NP-complete or NP-hard problems show a phase transition point that marks the spot where we go from problems that are under-constrained and so relatively easy to solve, to problems that are over-constrained and so relatively easy to prove insoluble. Problems that are on average harder to solve occur between these two types of relatively easy problem. The values of p_d and p_t are chosen in such a way that the instances generated are within the phase transition. In order to predict the phase transition region, a formula for the constrainedness [7] of binary CSPs was defined by:

$$\kappa = \frac{n-1}{2}p_d log_m(\frac{1}{1-pt}).$$ (2)

The tests were carried out on a DELL machine with 800 MHz CPU and 2 GB of memory. The code was written in C and compiled with the GNU C compiler version 4.6. The following parameters have been fixed experimentally and are listed below:

- Crossover probability = 0.9
- Mutation probability = 0.1
- Population size = 50
- Stopping criteria for the coarsening phase: The reduction process stops as soon as the number of levels reaches 3. At this level, MLV-GA generates an initial population.
- Convergence during the optimization phase: If there is no observable improvement of the fitness function of the best individual during 5 consecutive generations, MLV-GA is assumed to have reached convergence and moves to a higher level. The MLV-SA is supposed to have reached convergence if the best solution has not been improved during 100 consecutives iterations. The objective function of SA is the same as the fitness used by GA (Eq. 1).

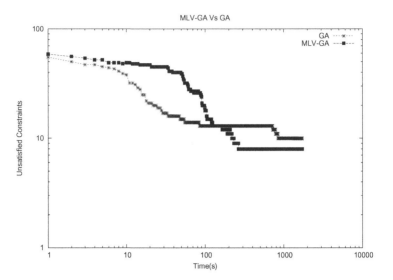

Fig. 1. MLV-GA Vs GA: evolution of the mean unsatisfied constraints as a function of time for csp-N25-DS40-C237 (number of variables 25, domain size is set to 40, number of constraints 237).

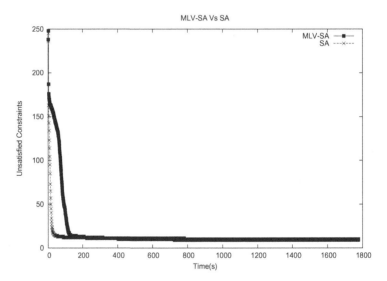

Fig. 2. MLV-SA Vs SA: evolution of the mean unsatisfied constraints as a function of time for csp-N90-DS40-C524 (number of variables 90, domain size is set to 40, number of constraints 524).

Table 1. MLV-SA Vs SA

	MLV-SA				SA			
Instance	Min	Max	Mean	RE_{av}	Min	Max	Mean	RE_{av}
N25-DS20-C225-cd078-ct027	0	1	0.26	0.001	0	1	0.13	**0.0005**
N25-DS20-C229-cd072-ct029	1	3	1.7	0.007	1	3	1.7	0.007
N25-DS20-C242-cd086-ct025	0	1	0.09	0.0003	0	0	0	**0.000**
N25-DS20-C269-cd086-ct025	2	2	2.09	0.007	2	2	2.09	0.007
N30-DS40-C121-cd026-ct0.63	3	6	5.5	0.04	4	6	5.74	0.04
N30-DS40-C173-cd044-ct045	1	2	1.74	0.01	0	2	1.7	**0.009**
N30-DS40-C312-cd070-ct031	4	6	5.26	0.01	3	6	4.83	0.01
N30-DS40-C328-cd076-ct029	2	6	4.09	0.01	2	5	3.91	0.01
N35-DS20-C551-cd094-ct017	1	2	1.74	0.003	0	2	1.65	**002**
N35-DS20-C558-cd094-ct017	1	2	1.87	0.003	1	2	1.74	0.003
N35-DS20-C562-cd094-ct017	1	3	2.13	0.003	1	3	2.13	0.003
N35-DS40-C149-cd024-ct059	3	6	4.83	0.003	3	6	5	0.003
N35-DS40-C157-cd024-ct059	5	8	6.87	0.04	6	8	6.83	0.04
N35-DS40-C486-cd082-ct023	2	5	4.13	0.008	2	5	4.04	0.008
N40-DS20-C653-cd084-t017	3	4	3.48	0.005	1	4	3.3	0.005
N40-DS20-C751-cd096-ct015	2	4	3.09	0.004	2	4	3.26	0.004
N40-DS40-C658-cd082-ct021	5	9	7.48	0.01	6	9	7.61	0.01
N40-DS40-C719.cd092-ct019	4	8	6.87	0.009	6	8	7.04	0.009
N50-DS20-C365-cd030-ct034	2	5	3.78	0.01	2	5	3.78	0.01
N50-DS40-C1141-cd094-ct015	5	9	7.87	0.006	5	9	7.83	0.006
N70-DS20-C474-cd019-ct037	6	9	8.17	0.01	7	10	8.65	0.01
N70-DS40-C2238-cd091-ct011	11	14	12.38	0.005	9	14	12.52	0.005
N90-DS20-C689-cd017-ct033	8	12	10.61	0.01	7	11	10.39	0.01
N90-Ds40-C3173-cd097-ct010	13	18	16.52	0.005	14	18	16.57	0.005

4.2 Analysis of Results

The plot in Fig. 1 compares GA with its multilevel variant MLV-GA. The improvement in quality imparted by the multilevel context is immediately clear. Both GA and MLV-GA exhibit what is called a plateau region. A plateau region spans a region in the search space where cross-over and mutation operators leave the best solution or the mean solution unchanged. However, the length of this region is shorter with MLV-GA compared to that of GA. The multilevel context uses the projected solution obtained at $G_{m+1}(V_{m+1}, E_{m+1})$ as the initial solution for $G_m(V_m, E_m)$ for further refinement. Even though the solution at $G_{m+1}(V_{m+1}, E_{m+1})$ is at a local minimum, the projected solution may not be at a local optimum with respect to $G_m(V_m, E_m)$. The projected assignment is already a good solution leading GA to converge quicker within few generations to a better solution. The plots in Figs. 3 and 4 show the mean relative error $(1 - (n_c - n_s)/n_c)$, where n_c is the number of constraints, n_s is the number of satisfied constraints) as a function of the number of variables for problems having 25 variables and 30 variables. MLV-GA outperforms GA in 36 out of 38 cases. Looking at the average relative error for both algorithms, MLV-GA gives a relative error ranging between 1 % and 18 % compared to 2 % and 22 % for N25 and between 2 % and 9 % compared to 2 % and 15 % for N30. Table 1 shows a comparison between SA and MLV-SA. For each algorithm, the best (Min) and the worst (Max) results are given, while mean represents the average solution and RE_{av} represents the mean relative error. For example, the instance N25-DS20-C225-cd078-ct027 represents a problem with 25 variables, domain size 40, number of constraints is set to 225, constraint density is equal to 0.78 and constraint tightness is 0.27. The algorithm with the lowest average relative error is indicated in bold. SA gives better results compared to MLV-SA in 4 cases out of 24 while similar results are achieved by both algorithms in the remaining cases. The broad conclusions that we draw from these results are that the multilevel framework does not appear to offer any improvement to the convergence of SA and it may indeed hinder its convergence. Figure 2 shows the evolution of the number of unsatisfied constraints as a function of time for SA and MLV-SA. The two algorithms show no cross-over during the early stages of the search. This provides evidence for the superiority of SA compared MLV-SA as it gives consistently better solutions given the same running time. As the time increases, both algorithms reach a premature convergence and return similar results.

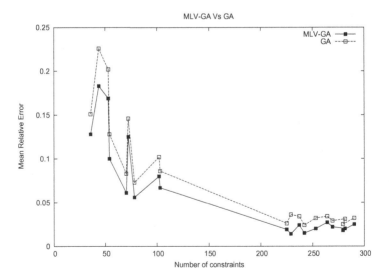

Fig. 3. MLV-GA Vs GA: mean relative error as a function of number of constraint: number of variables 25.

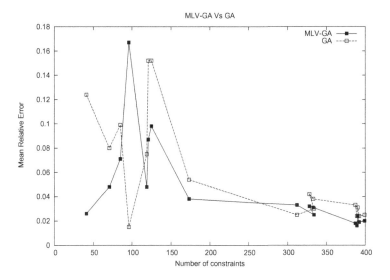

Fig. 4. MLV-GA Vs GA: mean relative error as a function of number of constraint: number of variables 30.

5   Conclusion

In this work, GA and SA are used in the multilevel context for solving MAX-CSP. The multilevel paradigm proved to be an excellent mechanism to enhance GA in order to achieve a suitable trade-off between intensification and diversification. The results have shown that the multilevel paradigm improves GA and returns a better solution for the equivalent run-time for most cases. The multilevel paradigm offers a better mechanism for performing diversification in-intensification. This is achieved by allowing GA to view a cluster of variables as a single entity thereby leading the search becoming guided and restricted to only those assignments in the solution space in which the variables grouped within a cluster are assigned the same value. As the size of the clusters gets smaller from one level to another, the size of the neighborhood becomes adaptive and allows the possibility of exploring different regions in the search space while intensifying the search by exploiting the solutions from previous levels in order to reach better solutions. However, it is disappointing that the multilevel paradigm was unable to enhance the convergence of SA. The reasons behind this sort of convergence behavior are for the time being not clear.

References

1. Bacanin, N., Tuba, M.: Artificial Bee Colony (ABC) algorithm for constrained optimization improved with genetic operators. Stud. Inf. Control **21**(2), 137–146 (2012)

2. Bonyadi, M., Li, X., Michalewicz, Z.: A hybrid particle swarm with velocity mutation for constraint optimization problems. In: Genetic and Evolutionary Computation Conference, pp. 1–8. ACM, New York (2013). ISBN 978-1-4503-1963-8

3. Bouhmala, N.: A variable depth search algorithm for binary constraint satisfaction problems. Math. Probl. Eng. **2015** (2015). doi:10.1155/2015/637809, Article ID 637809

4. Curran, D., Freuder, E., Jansen., T.: Incremental evolution of local search heuristics. In: Proceedings of the 12th Annual Conference on Genetic and Evolutionary Computation, GECCO 2010, pp. 981–982. ACM, New York (2010)

5. Davenport, A., Tsang, E., Wang, C.J., Zhu, K.: GENET: a connectionist architecture for solving constraint satisfaction problems by iterative improvement. In: Proceedings of the Twelfth National Conference on Artificial Intelligence (1994)

6. Dechter, R., Pearl, J.: Tree clustering for constraint networks. Artif. Intell. **38**, 353–366 (1989)

7. Gent, I.P., MacIntyre, E., Prosser, P., Walsh, T.: The constrainedness of search. In: Proceedings of the AAAI-96, pp. 246–252 (1996)

8. Fang, Z., Chu, Y., Qiao, K., Feng, X., Xu, K.: Combining edge weight and vertex weight for minimum vertex cover problem. In: Chen, J., Hopcroft, J.E., Wang, J. (eds.) FAW 2014. LNCS, vol. 8497, pp. 71–81. Springer, Heidelberg (2014)

9. Holland, J.H.: Adaptation in Natural and Artificial Systems. The University of Michigan Press, Ann Arbor (1975)

10. Hutter, F., Tompkins, D.A.D., Hoos, H.H.: Scaling and probabilistic smoothing: efficient dynamic local search for SAT. In: Van Hentenryck, P. (ed.) CP 2002. LNCS, vol. 2470, pp. 233–248. Springer, Heidelberg (2002)

11. Karim, M.R.: A new approach to constraint weight learning for variable ordering in CSPs. In: Proceedings of the IEEE Congress on Evolutionary Computation (CEC 2014), Beijing, China, pp. 2716–2723 (2014)

12. Kirkpatrick, S., Gelatt, C., Vecci, M.: Optimization by simulated annealing. Science **220**(4589), 671–680 (1983)

13. Lee, H.-J., Cha, S.-J., Yu, Y.-H., Jo, G.-S.: Large neighborhood search using constraint satisfaction techniques in vehicle routing problem. In: Gao, Y., Japkowicz, N. (eds.) AI 2009. LNCS, vol. 5549, pp. 229–232. Springer, Heidelberg (2009)

14. Morris, P.: The breakout method for escaping from local minima. In: Proceeding AAAI 1993, Proceedings of the Eleventh National Conference on Artificial Intelligence, pp. 40–45 (1993)

15. Minton, S., Johnson, M., Philips, A., Laird, P.: Minimizing conflicts: a heuristic repair method for constraint satisfaction and scheduling problems. Artif. Intell. **58**, 161–205 (1992)

16. Pullan, W., Mascia, F., Brunato, M.: Cooperating local search for the maximum clique problems. J. Heuristics **17**, 181–199 (2011)

17. Schuurmans, D., Southey, F., Holte, R.: The exponentiated subgradient algorithm for heuristic Boolean programming. In: 17th International Joint Conference on Artificial Intelligence, pp. 334–341. Morgan Kaufmann Publishers, San Francisco (2001)

18. Stützle, T.: Local Search Algorithms for Combinatorial Problems Analysis, Improvements, and New Applications. Ph.D. thesis, TU Darmstadt, FB Informatics, Darmstadt, Germany (1998)

19. Wallace, R.J., Freuder, E.C.: Heuristic methods for over-constrained constraint satisfaction problems. In: Jampel, M., Maher, M.J., Freuder, E.C. (eds.) CP-WS 1995. LNCS, vol. 1106, pp. 207–216. Springer, Heidelberg (1996)

20. Xu, W.: Satisfiability transition and experiments on a random constraint satisfaction problem model. Int. J. Hybrid Inf. Technol. **7**(2), 191–202 (2014)
21. Zhou, Y., Zhou, G., Zhang, J.: A hybrid glowworm swarm optimization algorithm for constrained engineering design problems. Appl. Math. Inf. Sci. **7**(1), 379–388 (2013)

Processing of a New Task in Conditions of the Agile Management with Using of Programmable Queues

P. Sosnin[✉]

Ulyanovsk State Technical University, Severny Venetc street 32, 432027 Ulyanovsk, Russia
sosnin@ulstu.ru

Abstract. The paper deals with an unpredictable appearance of a new project task in real-time designing a system with the software. Usually, this occurs at a time when the designer works in a multitasking mode and other members of the designers' team also work with delegated tasks. For regulating such activity, the use of the Agile management is not sufficient. In described case study, Agile means are combined with means of an interruption management which is adjusted on processing the interruption reason "New task". In the processing of the new task, the designer applies a framework "model of precedent" and its iterative filling by content with using a figuratively semantic support. Combining the indicated means is implemented in the toolkit OwnWIQA.

Keywords: Agile management · Human-computer interruption · A model of precedent · Multitasking · Question-answering · Semantic memory

1 Introduction

Several years ago, in software engineering, a group of prominent scholars and practitioners had claimed the necessity of significant changes in the foundations of this subject area. To date, their initiative "SEMAT" has already received its normative expression in some documents among which "Essence – Kernel and Language for Software Engineering" occupies the central place (http://semat.org/documents-papers).

This document covers the basics of proposed changes and, in particular, the answer to the question "How will project success be affected by changing the way of working?" Here "way-of-working" is understood as "the tailored set of practices and tools used by the team to guide and support their work" [1]. The answer also implicitly supposes that any used way-of-working specifies and restricts the possibilities of interactions of designers with an operational environment of designing.

One more feature of the proposed changing is the orientation on an agile managing the work being implemented with the use of visualized cards reflecting the states of completed pieces of the work.

It should be noted that, in designing, a typical piece of the work is a task that is started in the definite moment of time, and later it will be finished with the definite useful result. So, any task has an own life that happens in the definite interval of time where the task can be presented by its states and steps between states.

O. Gervasi et al. (Eds.): ICCSA 2016, Part IV, LNCS 9789, pp. 554–569, 2016.
DOI: 10.1007/978-3-319-42089-9_39

In a process of designing, life-cycles of tasks in their solutions are crossed that should be managed. There are two basic ways of managing the work with tasks crossed in time the first of which is the agile management when a group of designers solves tasks. The second way is the use of a (human-computer) interruption management that opens the possibility for controlled crossing the work with tasks in intermediate points of their life cycles. The interruption management helps not only to solve the task by the group of designers in parallel, but it also allows supporting the pseudo-parallel work of the designer with a set of delegated tasks. By other words, rational human-computer interruptions help to organize the personal work in conditions of multitasking [2].

Below we describe an approach the use of which supports the work with a new task Z_j that is unpredictable discovered by the designer who is working with the task Z_i inside its life cycle. Such situation is typical for a creative activity of the designer at the conceptual stage of designing. In our approach, discovering the new task is interpreted as a reason of self-interruptions of the designer who must process the appeared reason in conditions of the agile management with using of programmable queues [3].

The offered version of processing is based on the use of reflections of the operational space of designing on the semantic memory of the toolkit WIQA (Working In Questions and Answers) that has been created for conceptual designing of Software Intensive Systems (SISs) [4].

We start the rest of this paper with highlighting the features of the approach in conditions of multitasking. The third section introduces related works used as a source of comparing and inheritances for the offered approach. The fourth section opens features of figuratively semantic support. The fifth section presents some rules of managing the work of the designer with statements of new tasks.

2   Preliminary Bases

When a person mind is active, it implicitly and explicitly creates signs, the definite composition of which can be estimated by thinking as discovering a situation indicating n appearance of a new task. If the new task is important, then such situation plays a role of a reason that initiates the interruption of the current activity of the person for switching on the discovered task. By another word, the reason initiates self-interruption by the person for switching on the work with the new task. However, decision-making about the switching, the designer must take into account the negatives caused by interrupting the current task Z_i' and its relations with the new task Z_j'.

So, in designing, for its effectiveness, the described process, the essential part of which is flown in the mind of the designer, must be managed especially in conditions of multitasking. For the general case, such situation is schematically shown in Fig. 1.

The scheme demonstrates that at the definite interval of time, the designer (as a member of the team developing the definite project) usually works with a set of tasks. Moreover, in the current moment of time, the designer works only with the one of these tasks and this active task Z_i' is chosen by the designer from the definite queue of tasks extracted from the tree of tasks in accordance with agile management rules.

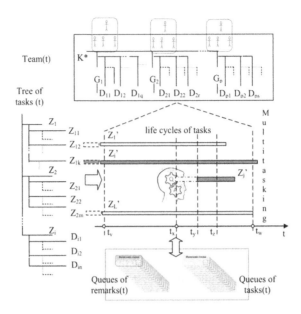

Fig. 1. Pseudo-parallel activity of the designer in a multitasking mode

The scheme also indicates that, at the moment of time t_x, the designer discovers the appearance of the new task Z_j' that, let us suppose, should be included in the process of designing. As told above, the new task starts an own life cycle in the mind of the designer. More definitely, it explicitly begins with a composition of signs (signals), for example, as a set of words $(w_1, w_2,..., w_k)$ that appeared in the left hemisphere of designer's brains.

Beginning with the moment of time t_x till a point-in-time t_y, the designer must decide to continue the work with the task Z_i' or switch the own attention to the new task Z_j'. At the interval (t_x, t_y), thinking of the designer estimates signals $w_1, w_2,..., w_k$ in the context of the tasks Z_i' for resolving the following consequences:

- if the work with task Z_i' will be continued without transferring and registering the signals (keys) $w_1, w_2,..., w_k$ out of brains then these keys to the task Z_j' may be lost, probably without re-return;
- if the attention of the designer will be fully paid for the task Z_j' then a temporary break from the work with the task Z_i' can have a negative impact on the solution of this task. Moreover, the degree of negatives is usually growing when the interval of the break is increased [5].

In our study, we suppose, that, at the interval of time (t_x, t_y), thinking is implemented in conditions that are schematically shown in Fig. 2. After that the task Z_i' is interrupted by the designer on a short interval of time (t_y, t_z) for explicit registering the keywords $(w_1, w_2,..., w_k)$ in specialized queues of remarks. Then the designer activates the process of human-computer interruptions in which the remark about the new task is interpreted as the reason on the self-interruption that should be processed.

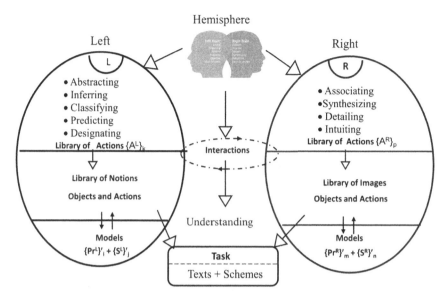

Fig. 2. Conditions of thinking

The scheme clarifies basic responsibilities of the left- and right hemispheres the actions of which can touch "notions" and "images" that are reflected in brain structures. In the work with the task these actions are intertwined and interacted for achieving the different aims the one of which is "understanding", first of all, understanding the statement of the task. Controlled forms of supporting the indicated actions and interactions will be presented below.

Finishing this section we mark that processing the reason of the self-interruption caused by the new task consists of following actions:

- Including the new task in a tasks' structure of the designing;
- Implementing the steps of the task life cycle with using the framework that specifies the task as a model of a precedent [4].

In iterative creating the example of the framework, the designer uses question-answering in the step-wise refinement of the statement of the task, figuratively semantic support for understanding the statement and its components and conceptual experimenting in simulating the task solution [6].

3   Related Works

For our study, the nearest group of related works concerns the subject area of Human-Computer Interruptions in the personal creative activity. In this group, we mark the paper [7] that develops a typology of self-interruptions based on the Multitasking integration of Flow Theory and Self-regulation Theory. The typology is based on negative (Frustration, exhaustion, and obstruction) and positive (stimulation, reorganization, and exploration) triggers. The very useful paper [5] also suggest a typology of self-interruptions including seven basic types ("adjustment", "break", "inquiry", "recollection," "trigger" and "wait") each of which generates for increasing the effectiveness of multitasking. Our approach is aimed at evolving this subject area by combining the interruption management with the agile management and programmable queues of tasks.

The next group combines the papers that present the subject area "Conceptual Development and Experimentation (CD&E)". The name CD&E often use for the corresponding technology and method. For example in [8] CD&E is defined as "A method which allows us to explore and predict, by way of experimentation, whether new concepts that may impact people, organization, process and systems will contribute to transformation objectives and will fit in a larger context".

The specificity of processes in this subject area is presented in the report [8] where the role and place of conceptual experimentation in military applications are indicated in detail. The following publication [9] defines the version of the occupational maturity of the CD&E-process. The publication [10] demonstrates some specialized solutions that are focused on a behavioral side of experimenting. In this area, we additionally suggest the use of conceptual experimenting that defines as "automated thought experiments, the content, and process of which are operatively reflected on a semantic memory, and results of reflections are applied in the process of experimenting with useful purposes". Moreover, conceptual experimenting is conducted with orientation on interactions of the designer with the task presented by the precedent model that is adjusted by the use of the natural experience and its models accumulated in the Base of Experience [6].

One more group of related tasks deals with the explicit and implicit programming of a human behavior caused by an interaction potential embedded in screen interfaces. Registering the human activity in program forms has been offered and specified constructively for Human Model Processor (MH-processor) in the paper [11]. The good example of implicit programming of the human behavior is described in [12] where multitasking and interruptions are estimated for a web-oriented activity. Both these examples indicate the usefulness of implementing the human behavior in program forms.

It should be noted, all papers indicated in this section were used as sources of requirements for developing the set of instrumental means provided the offered version of the interruption management. Any of these papers concerns only a part of the offered approach to interrupting in conditions of multitasking.

4   Precedent-Oriented Approach to Processing the New Task

4.1   Steps of the Interruption Caused by the New Task

In the second section of this paper, the life cycle of the new task was described for its initial steps. Other steps in the context of including the new task in the process of designing are shown in Fig. 3.

This scheme indicates that the designer D_{ps} implements the delegated task Z_i that has been chosen, for example, from the corresponding queue of the Kanban board. (label one on the scheme). In accordance with Kanban-rules, any task of any queue has been extracted from the task tree of the corresponding project. In a management process, the

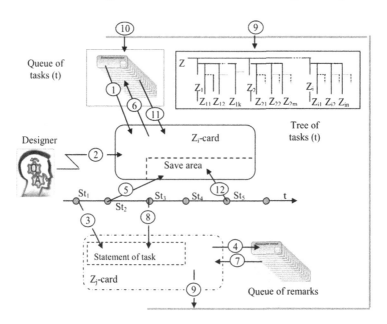

Fig. 3. Steps of processing the new task

necessary information describes the task Z_i in the corresponding visualized card (Z_i-card). Such cards have a normative structure and can be used as units of data in pseudocode programming of interactions with any card and/or queues of cards.

When the designer implements a current task Z_i, a reason of a self-interruption (label 2) appears in brain structures where this reason must get its expression in the consciousness of the designer for explicit initiating the process of the corresponding interruption. In the described study this reason is caused by the appearance of the new task Z_j. In this case, the first step St_1 of such initiating should aim at registering the arisen reason outside of brains with using the reason description (the list of keywords (w_1, w_2,..., w_k)). In the described research, the special utility activated by the designer creates the card (for the potential task Z_j) where it registers the keywords in the field "statement of task" (label 3). After registering, the Z_j-card relocates to the queue of temporal remarks (label 4).

Only after that, it is better to start the second step St_2 of preparing the task Z_i for the future resumption. Actions of this step are similar to actions of the computer interruptions when the system of interruption registers the necessary information in the save area (label 5) that extends the informational structure of the card used in the Kanban management. It should be noted, the point between the first and the second steps can be the point of the next self-interruption if it is necessary or useful to the designer. Therefore, for ordering, the Z_i-card relocates to the corresponding queue of tasks (label 6) after the second step.

At the third step St_3, extracting the Z_j-card from the queue of temporal remarks (label 7), the designer defines the task Z_j in a form that adequately reflects the interruption reason for its successful processing. Such work is creative, and it requires interactions with accessible experience or, in other words, it requires intellectual actions of the designer. More definitely, the designer creates the statement of the task (label 8) by interacting with the Experience Base and Ontology as its part.

In this step, the task Z_j is included in the tree of project tasks (label 9) and in the set of queues of tasks (label 10) that are solved by the designer. The choice of the appropriate queue is caused by the priority of the new task Z_j and its type. If the priority is low then the task Z_i then the task Z_j embeds in a backlog of the used agile managing. In another case, the task Z_j is located in the definite queue of tasks for its solving in multitasking mode.

The fourth step St_4 suggests processing of the reason when the designer creatively solves the task Z_j. This step will begin and fulfill by normative rules of the described management that are applied to an ongoing set of tasks. One part of these rules corresponds to rules of the applied version of the agile management. Some rules define the use of programmable queues [3]. Very important part combines a set of rules that specify the use of interactions with the experience and its models. Solving the tasks corresponding the interruption reasons, the designer can use conceptual experiments any of which helps to understand the reason, outline the way of its processing and check the preliminary solution of the task Z_j. All of these actions include using the interactions with experience. They also are kinds of intellectual processing of human-computer interruptions. After conceptual experimenting (if it was necessary and if processing the reason should be continued), the task Z_j will be a member of a multitasking set, and it will be managed similarly as other members of this set.

At the fifth step St_5, after activating Z_i-card (label 11), actions of the designer provide the recovery of the task Z_i from the point of its interruption. Actions of this step use data from the corresponding save area (label 12).

It should be noted that the third, fourth and fifth steps can be interrupted as many times as necessary. Moreover, each of these steps is activated by rules of the agile management. The toolkit WIQA supports Kanban, Scrum and Scrumban versions that are integrated into the complex of means shown in Fig. 4.

The scheme of the workspace consists of following components:

1. The organizational structure that reflects relations among team K^*, its groups $\{G_p(\{D_{ps}\})\}$, their members $\{D_{ps}\}$ and project tasks $\{Z_i\}$ in real-time.
2. Some means that can be fitted on agile project management in Kanban or Scrum or Scrum-ban version.
3. Some means, including Editor, Interpreter, Compiler and Library, which support conceptually algorithmic programming the work with queues of tasks.
4. Subsystem "Controlling of assignment" that allows appointing the estimated characteristics of time for planned work with each task Z_i included to the front of tasks (backlog of Kanban or Scrum or Scrum-ban).
5. The subsystem of interruption that supports the work with tasks in parallel and in pseudo-parallel mode.

Four positions on this list were presented in our previous publications [3, 4] where the interruption subsystem was only mentioned. At that moment, the necessity of this component was envisaged, but it was specified in general without deep delving and without the possibility of using in practice. Developing the second stage of our approach

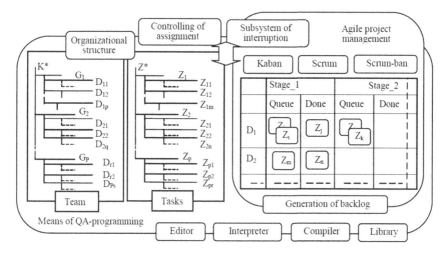

Fig. 4. The workspace of the flexible management

is aimed at an extension of the complex by instruments supporting of the personal multitasking that is implemented in coordination with the agile management by taking into account the human-computer interruptions. Moreover, the main attention is paid on self-interruptions in conditions of appearing the new project task the work with which can be inevitable without interactions with natural experience and its models.

It should be noted, the complex presented in Fig. 4 is realized and successively applied for two years in the project organization (about 2,000 employees). This fact indicates that developed means of the agile management was tested in the real designing.

4.2 Instrumental Environment of Processing the New Task

In the offered approach, the multitasking activity of designers is implemented in a semantic memory of the toolkit WIQA. The intention of this toolkit that supports following basic functions [6]:

- reflecting the operational space of the work with task and groups of tasks onto the semantic memory cells of which are oriented on question-answer specifying the mapped objects;
- pseudocode programming the behavioral actions with such objects and their compositions in the context of interactions of the designer with the natural experience and its models;
- conceptual experimenting with declarative and algorithmic constructions reflected onto the semantic memory;
- agile managing of implementing the flows of tasks and sets of tasks in multitasking mode, based on human-computer interruptions.

It should be noted that indicated functionalities are accessible in conditions that are shown in the scheme (Fig. 5) where the designer can use the Experience Base with embedded Ontology also mapped onto the semantic memory of the question-answer type (or, shortly, QA-memory).

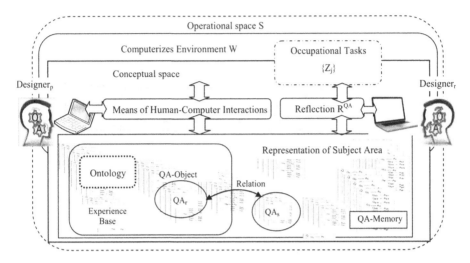

Fig. 5. Operational space

The scheme indicates that the operational space S includes a conceptual space (C-space) as an activity area where designers fulfill the automated part of own conceptual thinking when they work with conceptual artifacts. It should be noted, the C-space is a kind of the actuality, that inherits definite regularity from the operational space, and it has owned regularities expressed the behavioral nature of the human activity.

In solutions of project tasks $\{Z_j\}$, any result of the reflection (or shortly R^{QA}) is accessible for any designer as an interactive object (QA-object). Conceptually thinking, designers work in the C-space, for example, conducting the necessary experiments. In such actions, they interact with necessary concepts, units of the Experience Base and other components of the C-space. Thus, the C-space is the system of conceptual artifacts that are created and used by designers in processes of conceptual thinking when they conceptually solve the project tasks. In this kind of the activity, the C-space models entities and processes of designing, and such modeling is oriented on their reflections in the consciousness of designers. For example, reasoning, the designer combines corresponding dialog processes in consciousness with their models registered in cells of the QA-memory.

4.3 Precedent Oriented Modeling of the New Task

In accordance with our position, any new task Z_j appears as the phenomenon in the brains' structure of the designer. This phenomenon signalizes about its existence by a set of keywords on the background of the active task being solved. If signals lead to the self-interruption, then the successive step of processing the new task are oriented on creating the precedent model for this task. In this work, the designer uses the normative structure (or framework, F(P)) of the precedent model that is shown in Fig. 6.

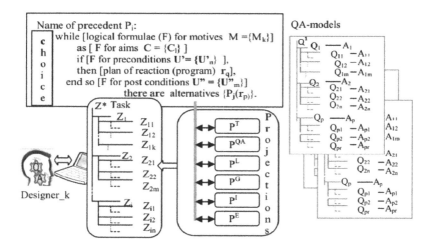

Fig. 6. The structure of the precedent model

The logical scheme of the precedent occupies the central place in the framework F(P). The scheme explicitly formulates "cause-effect regularity" of the simulated behavior of the designer. Framework F(P) includes following components: textual model P^T of the solved task, its model P^{QA} in the form of registered QA-reasoning, logical formulae P^L of modeled regularity, graphical (diagram) representation P^G of precedent, pseudo-code model P^I in QA-program form and executable code P^E. Any component or any their group can be interpreted as projections of F(P), the use of which allow building the precedent model in view of the precedent specificity. However, in any case, the built precedent model should be understandable to its users.

All built models of precedents are divided into two classes one of which includes models embedded in Experience Base of WIQA used by the team of designers not only in the current project. The second class includes models only for the current project of the SIS.

5 Features of Processing the Reason "New Task"

5.1 An Initial Statement of a New Task

The proposed approach assumes that a life cycle of a new task begins with the registration of such a list of keywords further interactions with which helps to the designer to create an adequate statement of this task. At the second step of the life cycle, the designer creates an initial state of the statement.

In our profound conviction, the initial statement must be formulated by the short text that most abstractly (but insufficient measure) expresses the essence of the task. This setting can be confirmed by following substantiations:

6. **The short text.** The statement must be perceived as the understandable wholeness when attention is focused on it. As the "area" focusing is limited the text $S(Z, t_k)$ should be as short as possible.
7. **Most abstractly.** The statement $S(Z, t)$ is a dynamic construction that is evolved step by step (or state $S(Z, t_k)$ by state $S(Z_j, t_{k+1})$) during the lifecycle of the task. Any step decreases the definite "volume" of uncertainty $\nabla U(Z_j, t_k)$ existed in $S(Z_j, t_k)$ by the inclusion to it the next increment $\Delta S(Z_j, t_k)$. The sequence of increments and their contents reflects the sequence of decision-making acts and their results each of which can lead to restrictions on future acts. Accordingly, at the initial state $S(Z_j, t_0)$ of the statement the uncertainty $\nabla U(Z_j, t_0)$ must be so indeterminate as possible. Therefore, the text $S(Z_j, t_0)$ must be formulated as abstract as possible.
8. **The level of abstraction in sufficient measure.** The text $S(Z_j, t_0)$ with its corresponding uncertainty $\nabla U(Z_j, t_0)$ are the initial source for generating the sequence of increments $\Delta S(Z_j, t_1)$, $\Delta S(Z_j, t_2)$, ..., $\Delta S(Z_j, t_{K-1})$ that evolve the initial text till the necessary state $S(Z_j, t_K)$. For such generating, the text can be interpreted as a source of initial axioms that are included in this text. These axioms must be sufficiently for inferring the state $S(Z_j, t_K)$.

In the offered approach, indicated requirements have led us to the pattern of the initial state the text of which combines following three sentences:

1. The first sentence reflects the orientation of its content to a target of a task. This offer includes pointers not only to the target but also on the ability to achieve it. An accompanied uncertainty can be hidden or registered by appropriate signs.
2. The second sentence prompts a feature of an idea which can help to solve this task. The feature implicitly indicates the way of achieving the target.
3. The third sentence concerns an environment in which the designer will work with the task.

Thus, at the second step of the task life cycle, the designer translates the list of keywords to the initial statement of the task with applying the described template. This work is based on the creative use of the naturally-occupational language in conditions of interacting the designer with the own experience, Experience Base, and Ontology.

5.2 Stepwise Refinement in Processing the New Task

After creating the initial statement of the task, the designer turns to an analysis of its text and implementing the other normative actions of the used technology. During these actions, step by step, the statement $S(Z, t)$ will be enriched while its uncertainty will decrease.

The enrichment will be caused by generating increments $\Delta S(Z_j, t_1)$, $\Delta S(Z_j, t_2)$, ..., $\Delta S(Z_j, t_{K-1})$ prioritization of which essentially determines the characteristics of the task being solved. This feature indicates the necessity of managing the development of the task statement.

For managing, the offered approach uses following solutions:

- Using the stepwise refinement way for decreasing the uncertainty of the task statement $S(Z_j, t)$;
- Coordinating the statement development with the process of creating the corresponding model of the precedent.

The first of these solutions is oriented on the use of question-answering in its application for discovering of uncertainty portions, their coding by appropriate questions and decreasing by corresponding answers.

This way is used for any of versions of the statement $S^m(Z_j, t)$ that presents the task as a wholeness at the definite level of abstraction (level L^{m+1}) for its definite applications, for example, for creating the architectural models or for checking the conformity to requirements or for other purposes. As a result, of stepwise refinement, the uncertainty $\nabla U^m(Z_j, t_k)$ is decreased till $\nabla U^{m+1}(Z_j, t_k)$ due to creating the adequate increment $\Delta S^{m+1}(Z_j, t_1)$ that combines question-answer detalization built at the level L^{m+1}. So the increment $\Delta S^{m+1}(Z_j, t_1)$ combines s number of question-answer pairs each of which includes an additional textual unit. Therefore, beginning with the initial task statement, level by level, the textual part of the statement $S(Z_j, t)$ will grow and this process must be controlled.

In the offered approach, the described way of the stepwise refinement is used in the question-answer analysis of the statement $S(Z_j, t)$. This analysis leads to creating the question-answer component P^{QA} of the corresponding precedent model.

The feature of the second solution is the use of iterative creating the precedent model in the real-time work of the designer with the new task. The designer creates any component of this model on the base of the task statement in its current state $S(Zj, t)$. It may happen that steps of such creation will be sources of information for enriching the state $S(Zj, t)$ or its corrections. This new informational unit can have not only the textual form. They can also be diagrams, pictures, tables, formula and algorithmic units. However, in any case, the statement $S(Zj, t)$ includes the textual part that combines the text T_0 of the initial statement with a set of textual units $\{T_k\}$ formulated during the life cycle of the task. Each of these texts should be controlled.

So, in its turn, any changing the state $S(Zj, t)$ can be a reason for corrections any of the components $P^L, P^{QA}, P^G, P^I, P^E$. Thus, components $P^T, P^L, P^{QA}, P^G, P^I, P^E$ are results of the iterative development in coordination with the feedback.

5.3 Figuratively-Semantic Support in Interactions with Textual Units

As told above, after discovering the necessity to include the new task Z_j in the executed work, the designer activates own thinking aimed at creating the initial model of the task in the textual form. This process includes two parts the first of which happens in brains of the designer while the second part provides transferring the results of thinking in their expression using the naturally-occupational language. In this form, initial results are mapped on the QA-memory. So, the second part of the process is what is named above "doing" which is intertwined with thinking. It should be noted, in this part of the work, the intertwining of doing and thinking should be coordinated and controlled. In the offered approach, called requirements are provided by the use of the specialized graphical support.

Switching on a graphics explicitly initiates the process in the right hemisphere of the brain. Drawing helps to extract and visualize only essential semantic components (of the generating symbolic specification of the task) and the scheme of their relations (for example, in a form of a block-and-line scheme). Nonessential semantic components make it difficult for understanding. On the one hand, the use of only significant components allows reducing the text description, and with another side, opens an opportunity for a deeper analysis of the text from the viewpoint of important contradictions.

Features of the used graphical support are caused by a reflection of any textual unit (of the task statement) on its figuratively semantic scheme (FS-scheme). Results of the reflection lead to additional opportunities for checks of the text on «integrity of the informational content», or, in other words, in checks of understanding. Means for creating and using the FS-schemes are embedded in the toolkit WIQA. These means include following components:

- Linguistic processor that provides automated translation of the investigated textual unit in its prolog-like description;
- Converter transforming such description in versions of the FS-scheme in iterative process of its development;
- The graphical editor that supports some transformations of the created FS-scheme from its initial state till the understandable version that is needed for the designer.

The first transformation (by linguistic processor) helps to translate the textual units T_i in the list of simple clause $T^*_i(\{C_j\})$ any of which has one of following structures:

$$
\begin{aligned}
C_j &= P_j(Ob_q),\, AC_j \\
C_k &= P_j(Ob_r,\, Ob_s),\, AC_k
\end{aligned}
\tag{1}
$$

where P – predicate of the clause, Ob – name of the definite object, AC – associative component that is located out of the predicate form. This component is included in the sentence of the text for the necessary utilization of the described meaning.

The second transformation fulfilled by the designer with the help of the converter allows creating the initial state of the corresponding block-and-line FS-scheme. In this work, the description $T^*_i(\{C_j\})$ is processed as a program of the declarative type. Such understanding is inherited from the logical programming that uses prolog-like operators.

The third transformation has following varieties:

- Translation in the pseudo code program that helps to draw the FS-scheme;
- Corrections of the FS-scheme on the base of information from associative components;
- Enrichment of components of the FS-scheme and their relations with new information that is registered in the current state of the project ontology.

For implementing the graphical support, the description $T^*_i(\{C_j\})$ must be stored in the semantic memory of the toolkit WIQA, and it will be visually accessible to the designer in interface environment where the designer has the possibility of correcting the description. After that, the Prolog-like description will be interpreted as a declarative program that is written in an extension of the pseudo code language L^{WIQA}. Operators of this extension have the syntax that corresponds to the expressions (1). Such transformations are iteratively implemented in an operational surrounding that is presented in Fig. 7.

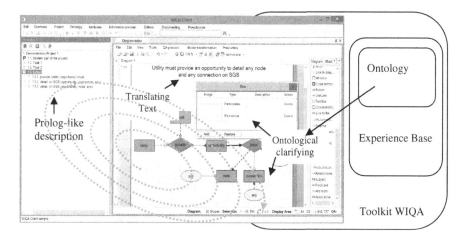

Fig. 7. Iterative coordinating the text and SG-scheme

This figure includes clarifying labels. The FS-scheme can be plugged to the ontology for the scheme specification and its informational enriching. Interactions with the ontology help to open the actual variants of the systematization registered in the ontology for nodes of the built scheme.

The scheme also reflects (by the spiral) the dynamics of iterative coordinating the text unit with its FS-scheme. The basis of such coordination is the feedback of FS-scheme with corresponding prolog-like description (based on forms presented in Fig. 2). Any correction at the scheme leads to the correction of the text equivalent. This mechanism is useful when translating complex sentences into simple clauses described at their prolog-like versions. Note, that the aggregation of the text T, its prolog-like version, and the coordinated graphical model express the understanding that the creator of these constructs embeds in the result of transformations.

6 Conclusion

The following conclusions can be drawn from the described study focusing on an unpredictable appearance of a new task at a time when a team of designers develops the system with the software:

1. The appearance of new tasks is typical for real-time designing of such systems, and therefore, works of designers with new tasks should be managed.

2. New tasks appear in conditions of the collaborative work of designers when they solve delegated tasks in coordination. For the management of such activity, Agile means are widely used, but they are not sufficient for the regulation of multitasking work at the workplace of the designer.
3. The use of programmable queues of tasks and the system of human-computer interruptions leads to the additional possibility of managing that support the multitasking mode and work with new tasks.

Applying the described approach in the operational space of the toolkit WIQA leads to additional positive effects that are caused by iterative creating the precedent model for a new task, conducting the necessary conceptual experiments, using stepwise refinement in the development of the task statement and coordinating the graphical schemes with corresponding textual units.

In coordinating, the textual unit and its graphical model are intertwined in the interactive process of their collaborative developing. The main aim of this process is to attain the necessary level of understanding that finds its expression in the created statement of the task. In its turn, the achieved understanding fixates that the developed statement reflects a concrete wholeness of requirements and restrictions in conditions corresponding the described task. It should be noted that interactions with textual and graphical parts of the statement activate the processes in the left and right hemispheres of any member of the designer team. The reuse of these processes also leads to the version of understanding which can be compared with the first version. With this point of view, understanding fulfills the controlled function.

Acknowledgement. This work was supported by the Russian Fund for Basic Research (RFBR), Grant #15-07-04809a.

References

1. Jacobson, I., Ng, P.-W., McMahon, P., Spence, I., Lidman, S.: The essence of software engineering: the SEMAT kernel. Queue **10**(10), 1–12 (2012)
2. Adler, R.F., Benbunan-Fich, R.: Self-interruptions in discretionary multitasking. Comput. Hum. Behav. **29**(4), 1441–1449 (2013)
3. Sosnin, P., Lapshov, Y., Svyatov, K.: Programmable managing of workflows in development of software-intensive systems. In: Ali, M., Pan, J.-S., Chen, S.-M., Horng, M.-F. (eds.) IEA/AIE 2014, Part I. LNCS(LNAI), vol. 8481, pp. 138–147. Springer, Heidelberg (2014)
4. Sosnin, P.: Combining of kanban and scrum means with programmable queues in designing of software intensive systems. In: Fujita, H., Guizzi, G. (eds.) SoMeT 2015. CCIS, vol. 532, pp. 367–377. Springer, Heidelberg (2015)
5. Jin, J., Dabbish, L.A.: Self-interruption on the computer: a typology of discretionary task interleaving. In: Proceedings of the SIGCHI Conference on Human Factors in Computing Systems, CHI 2009, pp. 1799–1808. ACM, New York (2009)
6. Sosnin, P.: Computerized environment for conceptual experimenting in automated designing. In: Proceedings of the 9th International Conference on Application of Information and Communication Technologies (AICT), Rostov on Done, Russia, pp. 451–457 (2015)

7. Darmoul, S., Ahmad, A., Ghaleb, M., Alkahtani, M.: Interruption management in human multitasking environment. In: Proceedings of the 15th IFAC Symposium on Information Control Problems in Manufacturing, Ottawa, Canada, vol. 48(3), pp. 1179–1185. IFAC-PapersOnLine, 11–13 May 2015 (2015)
8. MCM-0056-2010: NATO Concept Development and Experimentation (CD&E) Process. Brussels: NATO HQ (2010)
9. Wiel, W.M., Hasberg, M.P., Weima, I., Huiskamp, W.: Concept maturity levels bringing structure to the CD&E process. In: Proceedings of the I/ITSEC 2010, Interservice Industry Training, Simulation and Education Conference, Orlando, pp. 2547–2555 (2010)
10. Jaros, V.: Concept development and experimentation in project STRUCTURE. In: Cybernetic Letters. Expertia, o.p.s., Brno, vol. 2, pp. 36–41 (2011)
11. Karray, F., Alemzadeh, M., Saleh, J.A., Arab, M.N.: Human-computer interaction: overview on state of the art. Smart Sens Intell. Syst. **1**, 138–159 (2008)
12. Leiva, L. A.: MouseHints: easing task switching in parallel browsing. In: CHI 2011 Extended Abstracts on Human Factors in Computing Systems, pp. 1957–1962. ACM, New York (2011)

Using CQA History to Improve Q&A Experience

Cleyton Souza[1,2(✉)], Franck Aragão[1], José Remígio[1], Evandro Costa[3], and Joseana Fechine[2]

[1] Federal Institute of Education, Science and Technology of Paraíba - IFPB, Monteiro, Paraíba, Brazil
cleyton.caetano.souza@gmail.com
[2] Federal University of Campina Grande - UFCG, Campina Grande, Paraíba, Brazil
[3] Federal University of Alagoas - UFAL, Maceió, Alagoas, Brazil

Abstract. Social query is the practice of sharing questions through collaborative environments. In order to receive help, askers usually broadcast their request to the entire community. However, the prerequisite to receive help is to have the problem noticed by someone able and available to answer. Some works found a correlation between the characteristics of the questions and the outcome of receiving or not an answer. These findings suggest that there are some characteristics that are more likely to attract the attention of helpers. Our proposal is to analyse CQA history to identify the similar characteristics of previously asked questions that were answered. We believe that adding these characteristics in new questions will impact the receiving of answers. We evaluate our proposal using real world data and a real world experiment. Our results indicate that including "good characteristics" in the question reduce time for first response and improve answer quality.

Keywords: Social query · Community question answering site · Question redesign · Response rate

1 Introduction

The practice of sharing questions through social media is known as social query. Sharing questions on the Web is an ancient way to find information that emulates the Village Paradigm [1]. It originates in forums during early years of internet use [1]. Community Question and Answering sites (CQA) are collaborative environments entirely dedicated to asking and answering questions practice [2].

The most common sharing strategy is broadcasting the problem to everyone in the community. The prerequisite to receive help is that someone able and available notices the questions [3]. However, there is no guarantee if this will happen neither when. Thus, to facilitate this process, researches have been using query routing, i.e., connecting questions and answerers [4]. This could means recommending questions to potential answerers [5] or recommending experts to questioners directing their requests to [3]. Question routing is an effective way to attract the attention of someone [6].

© Springer International Publishing Switzerland 2016
O. Gervasi et al. (Eds.): ICCSA 2016, Part IV, LNCS 9789, pp. 570–580, 2016.
DOI: 10.1007/978-3-319-42089-9_40

However, there are questions that are broadcasted that still receive answers Thus, directing questions is not the only way to find help. Some studies found a correlation between the characteristics of the question and the outcome of receiving or not receiving answers [7–9]. Analyzing answered questions history, it is possible to identify common characteristics among answered questions. We believe that, if users knew which characteristics attract others' attention, they could use this information to improve their chances of finding help.

Thus, in this work, we investigate how adding certain "good" characteristics affects the performance of questions shared through CQA. Our goal with this study is to verify the following claims:

- C_1 – *Questions that receive answers have good characteristics.*
- C_2 – *Questions that do not receive answers do not have good characteristics.*
- C_3 – *Questions with good characteristics will attract more attention than questions with the opposite characteristics.*
- C_4 – *Questions with good characteristics will receive more answers than questions with the opposite characteristics.*
- C_5 – *Questions with good characteristics will receive answers earlier than questions with the opposite characteristics.*
- C_6 – *Questions with good characteristics will receive answers with higher quality than questions with the opposite characteristics.*

To check our claims, we performed two case studies. The first study consisted in sharing the so called "good" questions on real CQA. The second study consisted in analyzing a sample of question regarding the presence and absence of these "good" characteristics. Our findings suggests that questions with good characteristics are answered earlier, receive high quality answers and less requests for clarifications. We aim to use these results to design an interface that computing students can use to share "good" and "attractive" Programming related questions.

The remainder of this paper is organized as follows: Sect. 2 presents Related Work; Sect. 3 describes the context of our study, as well the results of our preliminary investigation through its history; Sect. 4 is about our Evaluation presenting Methodology, Results, Discussion and Threats to Validity; finally, Sect. 5 ends with Conclusion and Future Work.

2   Related Work

The usual social query strategy is broadcasting the question to everyone. However, this is not the best way of taking advantage of the architecture of the environment. After posting a question that will be visible to everyone, there are some struggle scenarios: (1) receiving several responses, (2) receiving wrong or contradictory responses, and (3) to keep receiving responses when no longer needed. Moreover, there is the possibility of receiving no answers because potential responders may never see the question [4].

The prerequisite to receive an answer is someone able and available to help notices the request [4]. The researches about social query usually propose the query routing as the only way to secure help. The process of directing questions to appropriate helpers is known in literature as query routing (this could means (1) recommending questions to answerers or (2) recommending answerers to the questioner). Thus, directing questions is an attempt to attract someone's attention [3]. Nichols and Kang [6] con-firmed that directing questions significantly increases the response rate. However, there are questions that are directed to nobody that still receive answer.

Some studies found a correlation between the characteristics of the question itself and the fact of receiving or not an answer. Burke et al. [10], for instance, found that, in Usenet groups, introductions referencing lurking and a personal connection to the topic of discussion increase the likelihood of getting a reply. In Yahoo! Answers[1], Yang et al. [11] found that medium length questions are less likely to get answered, as well questions from "other" category or with low similarity with their assigned category. According Asaduzzaman et al. [12], the top five reasons to question remain unanswered in Stack Overflow[2] are: "Fails to attract an expert member", "Too short, unclear, vague or hard to follow", "A duplicate question", "Impatient, irregular or inconsiderate members" and "Too hard, too specific or too time consuming".

Regarding personal social networks' studies, Teevan, Morris and Panovich [9] found that, in Facebook, a concise style of question-asking, a predefined audience, and the inclusion of a question mark were associated with more and higher quality responses within shorter periods of time. In [13], they also found that young people and people with larger social networks are more likely to receive answers. In addition, they established a correlation between the length of the questions and the received response: questions with extra sentence are less likely to receive "yes/no" answers or requests for clarification. Lampe et al. [14] found that the question type affect the performance of questions shared on Facebook[3]. Recommendation posts receive more responses than any other question type; while Favor requests usually take a long time until receive a first response. Comarela et al. [15] conducted a study to understanding factors that affect response rate in Twitter[4] and found that tweets with hashtags and URL are more likely to be retweeted and tweets with mentions are more likely to receive a reply. This last result supports Nichols and Kang's claim that directing questions is more effective than broadcasting.

All these findings could be used to improve the likelihood of one getting answers [5]. Imagine that a user is preparing to broadcast a question in a social context. If he had this knowledge, about which factors can affect response rate, he could shape his request to fit these factors and theoretically improve his chances of finding help [5]. In addition, this could be used to improve questions

---

[1] http://answers.yahoo.com.
[2] http://stackoverflow.com.
[3] http://facebook.com.
[4] http://twitter.com.

quality and consequently answer quality [16]. The goal of teaching students to ask better questions was explored in [17]. Results revealed a significant difference in the quality of questions generated on the post-test as a function of condition (participants in the question training condition asked significantly more "deep" questions on the post-test than did the participants in the control condition).

These results open interesting research opportunities like if it is possible to improve Q&A experience through the investigation of CQA history. Through the analysis of question history, we could identify common characteristics among answered questions. While users are phrasing new questions, we could suggest to them to add these characteristics into their request, improving both question quality and question attractiveness. And, finally, these redesigned questions could be easier to respond, whether just for being clearer or for having a specific characteristic that attracts the community.

3 Investigating CQA History

We used a Brazilian CQA about programming called GUJ[5] as the context of our study. This is the larger programming community in Brazil, with almost 200 thousand users. GUJ means Java User Group, in Brazilian Portuguese. The website was created in 2001 and it works like a forum. Users access GUJ and publish questions like a new thread. When other users access GUJ, they are presented to the list of most recent threads. They can access the thread and reply to its author. Since its beginning, it has been made more than 300 thousand questions and it has been exchanged almost 2 million messages. Figure 1 shows an example of a question shared through GUJ.

We split Fig. 1 in areas: (A) the question title; (B) the question tags; (C) the questioner identification; (D) the publishing time; (E) the question description; (F) the social functions buttons (like, share, flag and favorite); (G) the reply

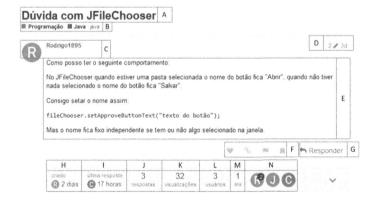

Fig. 1. Example of question thread shared on GUJ

---

[5] http://www.guj.com.br.

button; (H) the questioner identification and publishing time; (I) the publishing time of last answer and answerer identification; (J) the amount of answers received; (K) the amount of views received; (L) the number of users attracted by the question (including questioner); (M) the number of links published through answers; and (N) the list of users that interacted and the amount of contributions of each one.

We conducted a qualitative study in order to identify which are the most common characteristics in answered questions. We started gathering a sample of questions from GUJ containing both answered and unanswered threads. This study included the analysis of this sample. We described questions using attributes like: question length, title length, question and title coherency, code presence, greetings presence, question topic, difficulty level, etc. After outline a list of characteristics, we conducted a literature review searching for articles about asking good programming questions. Since GUJ is strongly popular among Brazilian students, there is a lot of material to help newcomers to ask "good" questions. We confronted and combined both analysis and it resulted in the following list of characteristics which a question can have to attract more responses.

– **Title related characteristics** – The title is the first contact of potential responders with the question. The title should be a summary of the problem and cannot be too short or too long. Regarding with the good title characteristics, users should prioritize: (I) understandable title; (II) medium size title; and (III) a title coherent with the question description subject.
– **Description related characteristics** – After he has been attracted by the title, the potential helper will read the problem description. We observed that some users do not want "waste their time" looking a long code or following a link. Thus, questioners should keep the description with enough information that anyone can answer without additional reading effort. However, we are aware that be concise and clear is not always an easy task. Regarding with good description characteristics, users should prioritize: (IV) understandable description; (V) avoid too long description; (VI) showing an example, but avoiding too much code; (VII) avoid description with code only; and, (VIII) when including links, combining them with partial content.
– **Behaviour related characteristics** – Helpers will be more willing to answer questions from "good" users. We identify that users who follow a community normative sense have more chances of receiving answers. These "good" users are relatively polite, grateful for receiving help, and aware of a correct way to behave that is not written anywhere, but it is unconsciously followed. Regarding this matter, we identify the following good practices: (IX) use of proper language; (X) including greetings; (XI) avoid be impolite; (XII) avoid be demanding; (XIII) restricting the question to a single problem; (XIV) avoid creating duplicated questions (this can be reached by searching in the community for a similar question, before create a new one); and (XV) avoid create factoid questions (this kind of problem is well solved through search engine use).

Since this characteristics' list emerged from a literature review, we are assuming that they are, at least, good characteristics that questions should have, while the fact of their presence be related with question attractiveness and responsiveness will be verified in next section.

4 Evaluation

We believe that adding an "Assistance Phase" to help users, before they disclose their problem in a social environment, can improve question quality and response rate. To validate our approach we test the performance of some "enhanced" questions shared on CQAs. In addition, we compared data from a sample of answered and unanswered questions regarding the presence of the good characteristics. Our results indicate that following the suggestions improve response rate, time for the first response, question quality, response quality and question attractiveness.

4.1 Methodology

To test our claims, we planned two studies. The first study consists in the sharing of questions with these good characteristics and with the opposite characteristics and comparing the performance of both groups. The second study consists in the comparison between the data about answered and answered questions on GUJ, regarding the presence and absence of these characteristics.

Our first study works as a concept proof of our belief that adding certain characteristics will impact what happens to the question after being broadcasted. Basically, we formulate 5 questions with the "good characteristics" and 5 questions with the "bad characteristics" (this means do the opposite of the suggestions). Then, we shared these questions during a week in GUJ and analyzed what succeeded: how much people was attracted to the questions, how much people answered the questions, how long take for them to receive answers, the quality of these answers, etc.

All these questions were based on the main problems faced by students during the classes of "Programming I" and "Data Structure and Algorithms". According [12], too hard questions are more likely to remain unanswered. Thus, while the question topics have a wide range, question difficulties were only low and medium. Regarding answer's quality, although there are many researches in this area [18], since we shared few questions and we would individually define the answers quality, to compare the performance of our questions regarding the answers they received, we used a scale considering all types of answers that we received in ascending order of utility: -2 means that "respondent acted aggressively with the questioner"; -1 means that "respondent did not comprehend the question and asked for more information"; $+1$ means that "respondent only suggested to consult a link"; $+2$ means that "respondent offered a partial/incomplete solution"; $+3$ means that "respondent offer a complete and correct solution". This scale also considers respondent's effort to provide an answer.

The second study consisted in analyzing a sample of question regarding the presence and absence of these "good" characteristics. Our goal was to investigate if questions answered have the characteristics that we believe attract answers and if questions unanswered miss these characteristics. We sample 100 questions of each type and describe then for the presence and absence of these characteristics.

4.2   Results

Table 1 shows the performance of the questions that we shared for the first study. We also add a question index column to refer these questions later–GQ stands for good question and BQ stands for bad question.

Table 1. Performance summary of questions shared on GUJ

Question Index	Difficulty	Topic	Publishing Time	Time for First Response	Time for the Best Response	People Attracted	Amount of Responses	Responses Quality
GQ1	Low	String API	06:37 PM	20 min	157 min	38	2	[2,3]
GQ2	Low	File API	04:18 PM	13 min	13 min	22	1	[1]
GQ3	Medium	Reflection API	11:48 PM	22 min	22 min	18	1	[2]
GQ4	Medium	Theory	10:49 AM	15 min	350 min	16	2	[-1,3]
GQ5	Medium	Generics	11:09 AM	16 min	16 min	8	1	[-1]
BQ1	Low	Theory	01:47 AM	2032 min	2032 min	47	1	[-2]
BQ2	Medium	XML exception	02:52 PM	-	-	-	-	-
BQ3	Medium	Algorithm	11:43 PM	-	-	-	-	-
BQ4	Medium	Algorithm	06:41 PM	248 min	248 min	25	2	[-1,1]
BQ5	Medium	Swing API	03:34 PM	493 min	493 min	19	1	[2]

These results will be discussed in next section, but we can notice that "bad" questions clearly wait longer for a first response. In addition, answer quality of "good" questions is usually higher.

Figure 2 describes our sample of questions, for the second study, regarding the presence and absence of the "good" characteristics. We used the same Roman numerals of Sect. 3.

We further discuss all these data in next section. But, what we can observe is that real users try to add to their questions good "characteristics", without even realizing it. In addition, we can only perceive a slightly higher presence of the good characteristics in answered questions group.

4.3   Discussion

Unfortunately, we still do not have data to statically verify our claims; however, observing the results of both studies, we were able to realize interesting patterns that give us directions to qualitatively check them.

Our first claim (C_1) was that *"Questions that receive answers have good characteristics"*. Although, this seems intuitive, data from questions gathered,

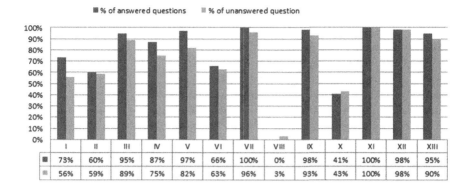

Fig. 2. Data from a sample of answered and unanswered questions

summarized on Table 1, show that both answered and answered questions usually have similar characteristics. However, CQAs are environment with high flow of new threads and GUJ is not different. Our first study shows that, when we have good and bad questions mixed, the first type usually receives answer quickly and has less chance of been lost in the thread flow, as it happens with BQ2 and BQ3. Thus, we believe that good characteristics are correlated with the receiving of good answers.

Our second claim (C_2) was that *"Questions that do not receive answers do not have good characteristics"*. This claim is not true, since Fig. 2 shows that even unanswered questions have good characteristics too. However, what we perceived is that, when a question has bad characteristics, it is likely to not receive many answers. In addition, questions that are hard to comprehend have more chances of receiving answers asking for clarifications.

It was not possible to say something about our third claim (C_3) that *"Questions with good characteristics will attract more attention than questions with the opposite characteristics"*. Since GUJ allows us to see how many times people saw the question, from the first study, we were able to perceive that all questions attract almost the same number of users. However, BQ1, one of the first questions published attract a lot of users and it received just a single answer. We believe that this happens due the characteristics of the question that probably annoy most users, including the user who answered. He said "If you break your question and search here or on Google[6] you will find great material about it. Stop being lazy and search by yourself. When you have a specific question of how to do something you ask here". The question broke the social contract of the community, when asking factoid questions that have easy answers though search engines, thus, attracting a lot of users, but none useful answer. In addition, questions usually will stop attract people, as times goes by, since they are losing positions on the most recent topics. Unless people keep engage in helping each other in the thread.

---

[6] http://www.google.com.

Regarding our forth claim (C_4), *"Questions with good characteristics will receive more answers than questions with the opposite characteristics"*, it was not possible to verify that either. In our first study, most questions received almost the same amount of answers, one or two. Although BQ2 and BQ3 did not receive answers and, probably, this was related to the poor and to long description, respectively, in the second study, answered and unanswered questions had similar characteristics. In addition, we believe that questions will stop receive answers when the community realizes that one already offer a satisfactory solution.

The fifth claim (C_5) that *"Questions with good characteristics will receive answers earlier than questions with the opposite characteristics"* was considered true. Seeing the performance of good and bad questions in Table 1, regarding to the time of the first response, it is clear that the first group receives answers earlier. This, probably, could be explained for the additional effort to answer a "bad question". The first study showed that poorly written questions usually receives request for clarifications, before receiving a satisfactory answer.

The sixth claim (C_6), that *"Questions with good characteristics will receive answers with higher quality than questions with the opposite characteristics"*, was also considered to be truth. When we have good and bad questions mixed, the first type usually receives answer quickly and has less chance of receiving request for clarification. In addition, as BQ1 shows us, good questions have less chance of receiving negative responses from the community.

4.4   Threats to Validity

This section, briefly, discuss about limitations in our work. To analyze the validity of the results we consider the four kinds of threats: external, internal, conclusion and construction.

The external validity is related to the approximated truth of conclusions and generalization to the real world; and the internal validity corresponds to check if the results are a consequence of the manipulation that was done and no others factors. The conclusion validity refers to the correct correlation between what was verified (measured) and the conclusions reached; and, finally, construct validity regards to problems in the design and control of the experiment.

Related to the external validity, a threat to the conclusions reached is that we tested our claims in the context of a CQA about programming. Although, we are confident that our proposal fits any collaborative environment, we cannot guarantee that the same observations will happen outside GUJ.

Related to internal validity, it is true that our observations are highly connected with the questions that we asked, the time when they were released by moderators and who was online on that moment. We tried to use a real world experiment, in order to obtain real world feedback. Sure this brings random factors to the table that could affect the performance of questions shared. However, experiments with historical data from CQA lack these types of "noise", that are important in real world situations.

Related to conclusion validity, problems may occur on the findings if late answers would come or if we establish the wrong "good characteristics".

However, we believe that this is very unlikely, since the flow of new threads on GUJ is intensive, as time goes by, lesser chances a question thread has to receive more replies; and, regarding to the set of "good characteristics", it was based in a mixed study involving literature review and classification task.

Related to construct validity, to our first study, we used ten questions asked by students through Java and data structure classes. We enhance these questions with good and bad characteristics. We are aware that maybe most questions shared on CQAs has mixed characteristics and a wide range of topics and complexities, but, as we was trying to analyze the impact of good characteristics only, we could not use too hard questions or mixed characteristics questions.

5   Conclusion and Future Work

In this work, we propose the inclusion of "Assistance Phase" in query routing process. The goal of this new phase is including in the questions characteristics which were present on previously asked answered questions. In addition, it could be used to reduce scope of Expert Search. We used two studies to evaluate the impact of our proposal, while broadcasting a question.

Our findings suggests that questions with good characteristics are answered earlier, receive high quality answers and less requests for clarifications. In addition, we find interesting patterns related to the presence of "opposite" characteristics. This work represents a first step towards the goal of assisting askers in posting attractive questions, by helping them redesign their question, aiming to improve response rate and time for first response.

To future work, we propose analyzing the individual impact of each good characteristic in the receiving of answers, use statistical analysis to check our claims and also test if questions with bad characteristics remain answered for a long time.

Acknowledgements. We want to thank the support of TIPS (http://www.grouptips. org) and IFPB–Campus Monteiro regarding the conduction of this study.

References

1. Mansilla, A., Esteva, J.: Survey of social search from the perspectives of the village paradigm and online social networks. J. Inform. Sci. **39**(5), 688–707 (2013)
2. Raban, D., Harper, F.: Motivations for answering questions online. New Media and Innovative Technologies (2008)
3. De Souza, C.C., De Magalhães, J.J., De Costa, E.B., Fechine, J.M.: Predicting potential responders in twitter: a query routing algorithm. In: Proceedings of the 12th International Conference on Computational Science and Its Applications, pp. 714–729 (2012)
4. Furlan, B., Nikolic, B., Milutinovic, V.A.: Survey of intelligent question routing systems. In: Proceedings of the 6th IEEE International Conference Intelligent Systems, pp. 014–020 (2012)

5. Dror, G., Koren, Y., Maarek, Y., Szpektor, I.: I want to answer, who has a question? Yahoo! answers recommender system. In: Proceedings of the 17th ACM SIGKDD International Conference on Knowledge Discovery and Data Mining, pp. 1109–1117 (2010)

6. Nichols, J., Kang, J.: Asking questions of targeted strangers on social networks. In: Proceedings of the ACM Conference on Computer Supported Cooperative Work, pp. 999–1002 (2012)

7. Dror, G., Maarek, Y., Szpektor, I.: Will my question be answered? predicting "question answerability" in community question-answering sites. In: Proceedings of the Machine Learning and Knowledge Discovery in Databases, pp. 499–514 (2013)

8. Comarela, G., Crovella, M., Almeida, V., Benevenuto, F.: Understanding factors that affect response rates in twitter. In: Proceedings of the ACM SIGWEB Conference on Hypertext and Social Media, pp. 123–132 (2012)

9. Teevan, J., Morris, M., Panovich, K.: Factors affecting response quantity, quality, speed for questions asked via social network status messages. In: Proceedings of the International Conference of Webblogs and Social Media, pp. 630–633 (2011)

10. Burke, M., Joyce, E., Kim, T., Anand, V., Kraut, R.: Introductions requests: Rhetorical strategies that elicit response in online communities. In: Proceedings of the 3rd Communities and Technologies Conference, pp. 21–39 (2007)

11. Yang, L., Bao, S., Lin, Q., Wu, X., Han, D., Su, Z., Yu, Y.: Analyzing predicting not-answered questions in community-dased question answering services. In: Proceedings of the 25th AAAI Conference on Artificial Intelligence, pp. 1273–1278 (2011)

12. Asaduzzaman, M., Mashiyat, A.S., Roy, C.K., Schneider, K.A.: Answering questions about unanswered questions of stack overflow. In: Proceedings of the 10th Working Conference on Mining Software Repositories, pp. 97–100 (2013)

13. Teevan, J., Morris, M.R., Panovich, K.: Does anyone know how to get good answers? how social network questions shape replies. Technical Report, No: MSR-TR-2013-62 (2013)

14. Lampe, C., Gray, R., Fiore, A., Ellison, N.: Help is on the way: patterns of responses to resource requests on facebook. In: Proceedings of the 2014 Conference on Computer Supported Cooperative Work, pp. 3–15 (2014)

15. Comarela, G., Crovella, M., Almeida, V., Benevenuto, F.: Understanding factors that affect response rates in twitter. In: Proceedings of the ACM SIGWEB Conference on Hypertext and Social Media (HT 2012), pp. 123–132 (2012)

16. Baltadzhieva, A., Chrupała, G.: Question quality in community question answering forums. In: Proceedings of the ACM SIGKDD Explorations Newsletter, pp. 8–13 (2015)

17. Sullins, J., Mcnamara, D., Acuff, S., Neely, D., Hildebrand, E., Stewart, G., Hu, X.: Are you asking the right questions? The use of animated agents to teach learners to become better question askers. In: Proceedings of The 28th International Flairs Conference, pp. 479–481 (2015)

18. Furtado, A., Andrade, N., Oliveira, N., Brasileiro, F.: Contributor profiles, their dynamics, their importance in five qa sites. In: Proceedings of the 2013 Conference on Computer Supported Cooperative Work, pp. 1237–1252 (2013)

The Use of Computer Technology as a Way to Increase Efficiency of Teaching Physics and Other Natural Sciences

E.N. Stankova[1,2(✉)], A.V. Barmasov[1,3,4], N.V. Dyachenko[3], M.N. Bukina[1],
A.M. Barmasova[3], and T.Yu. Yakovleva[3]

[1] St. Petersburg State University, 7–9, Universitetskaya nab., St. Petersburg 199034, Russia
`lena@csa.ru`, `{a.barmasov,m.bukina}@spbu.ru`
[2] St. Petersburg Electrotechnical University "LETI" (SPbETU), 5, ul. Professora Popova,
St. Petersburg 197376, Russia
[3] Russian State Hydrometeorological University, 98, Malookhtinsky pr.,
St. Petersburg 195196, Russia
`{nat230209,abarmasova,yakovtat}@yandex.ru`
[4] St. Petersburg State Pediatric Medical University, 2, Litovskaya ul.,
St. Petersburg 194100, Russia

Abstract. The use of computer technologies for increasing the efficiency of physics teaching is discussed. Proposed approach includes the development of professionally oriented teaching, the use of multimedia lecture courses, the use of Learning Management Systems (such as Blackboard Learn and MOODLE), webinars, etc. It is proposed to organize distributed educational system in a form of Grid infrastructure joining educational facilities of 4 St. Petersburg universities. This Grid infrastructure can provide equal access to all kinds of resources for the students and lecturers in order to teach and study physics more efficiently.

Keywords: Computer technology · Teaching physics · General physics · The concept of modern science · Multimedia lecture courses · Computer learning management system · Blackboard learn · MOODLE · Webinars

1 Introduction

Learning physics is the basis for formation of the scientific worldview for students of natural-science specialties. Nobody doubts it. Physics is the basis of the future student's profession in the subject area and in the area of application of different methods in measurement practices.

Some people think that teaching professional knowledge of physics is not necessary for humanitarians. However, the world of our culture is full of things inextricably linked with the achievements of physics. All incomprehensible things cause human alertness and fear. There is a process of alienation of things from people, as a result – world carries the threat. This must be overcome at the mental level. Basic knowledge of physics should be as natural to the humanities, as the knowledge in the humanities to "techies".

© Springer International Publishing Switzerland 2016
O. Gervasi et al. (Eds.): ICCSA 2016, Part IV, LNCS 9789, pp. 581–594, 2016.
DOI: 10.1007/978-3-319-42089-9_41

In addition, learning physics allows the students of humane specialties to get acquainted with specific natural-scientific way of thinking, allows to enrich their intellectual sphere and also to form the materialist worldview.

Improving the efficiency of teaching physics to various categories of students the authors follow the concepts of physics as an empirical science. This means that the physics is formed by the certain natural phenomena, and while the science develops, the phenomena get the explanation in the system of causal relationships existing in the nature.

Physical experiment plays the exceptional role in the search for such explanations. It presents specially organized investigation process allowing identifying these relationships most definitely.

The key statement of the purpose of scientific research belongs to Galileo: *"...to measure everything that can be measured and make measurable what is not measured..."*. This idea is consistent with the task of developing a system of concepts in physics, which has quantitative measure and characterizes the observed phenomena, and then, allows developing mathematical theories of different degrees of generality based on the abstraction of the concepts.

Our teaching experience has shown that the most difficult matter in learning physics is to create a clear distinction in the learning phenomenon, concepts describing this phenomenon, its models in these concepts and mathematical description of these phenomena and, most importantly, the practical need for such a distinction.

It is very difficult for a student to abstract himself/herself from a phenomenon, to solve the task with the help of the theory and to come back to the reality. This lack of distinction hinders the solution of the tasks that, as a rule contains a description of the situation close to the real circumstances and makes students and in the future, experts, essentially helpless in any unusual situation requiring the simple knowledge of physics.

However, it is very difficult to overcome this gap in the normal course of teaching. Indeed, at first it is necessary to show the phenomenon in its entirety, then to show the process of forming a model of the phenomenon, show the transition-model theory and to show by the "feedback" the effectiveness of the theory and also its limitations.

Computer technologies are able to solve the problem now. A large number of easily controlled multimedia subjects can be used during a lecture: pictures, videos, sounds, design in real-time, etc. Educational programs in interactive mode can be widely used during practical studies.

The possibilities of computer technology are limited only by the experience and creative activity of a lecturer. The exchange of such an experience is of great importance at present when computer technologies are not sufficiently implemented in teaching process.

For several years, the authors have been developing a specific approach to the teaching physics, the concept of modern science, laboratory workshop, as well as special courses for students of natural specialities and humanities taking into account the specific educational programs [1–9].

Teaching physics for different universities, authors make sure that in order to maximize the efficiency of teaching physics it is necessary to organize adequate environment, which should comprise all kinds of learning materials and computational resources of different universities in order to provide equal and the most comfortable access to them. Distributed educational system can be organized as Grid infrastructure joining the best educational courses and computational facilities in the field of physics elaborated in Saint Petersburg State University, Saint Petersburg State Pediatric Medical University, Russian State Hydrometeorological University and Saint Petersburg Electrotechnical University "LETI". This Grid infrastructure can provide equal access to all kinds of resources for the students and lecturers.

2 The Features of the Proposed Approach to Teaching Physics and Concepts of Modern Natural Science for Students of Science and Humanities

The specificity of the method proposed by the authors encloses the integrated use of multiple approaches to the teaching physics, other natural sciences and the concept of modern science.

Firstly, teaching discipline should not deter by its complexity at least at the stage of the first acquaintance. This provides effective assimilation of any material by students. Therefore, the authors aim to make physics accessible to all by seeking to set out a full course of general physics by clear and accessible language. Of course, teaching the course of the general physics implies using of the adequate mathematical apparatus. The distinctive feature of the presentation of the material is the constant matching of physical phenomena with similar appearances described in different sections of the course. Some of the most important provisions are considered from the different points of view and are presented in the different sections of the course. This approach makes it easier to understand and to remember the material [9–12].

The authors also find it possible to follow the ideas expressed many years ago by the outstanding Russian physicists Nikolay Umov (1846–1915) and Yakov Frenkel (1894–1952) about teaching physics at higher school.

Great scientists and authors of the fundamental tutorials emphasize that a good printing course composed by lecturer makes it unnecessary to present all the material in the form of oral lectures. This leads to irrational waste of time by students. The role of professors and lecturers in this case is firstly in the organization of consultations (individual and group), in which the discussion on difficult course issues arising in the process of consultation can evolve into a full lecture [13].

Later (June, 1963) similar thoughts were expressed by the famous American physicist Richard Feynman (1918–1988) in his preface to the printed Lectures on Physics: *«I think, however, that there isn't any solution to this problem of education other than to realize that the best teaching can be done only when there is a direct individual relationship between a student and a good teacher—a situation in which the student discusses the ideas, thinks about the things, and talks about the things. It's impossible to learn very much by simply sitting in a lecture, or even by simply doing problems that are assigned»* [14].

In order to provide such form of education, it is necessary of organize special educational environment, providing equal access to teaching facilities (textbooks, training manuals, virtual laboratories) for students and lecturers. It is necessary to join resources of different universities each having their own unique facilities to teach physics the most efficiently. The environment should be organized as Grid infrastructure functioning. The functioning Grid infrastructure will be selection of the best educational courses in the field of physics elaborated for St. Petersburg State University, St. Petersburg State Pediatric Medical University, Russian State Hydrometeorological University and St. Petersburg Electrotechnical University "LETI". Effective feedback should be organized in order to improve the courses and organize them into the joint educational program.

3   The Use of Multimedia Lecture Courses

At present it is not enough to have only "good printed course". This printed publication can and should be supplemented with the advanced multimedia materials [5, 7, 15].

Training courses developed under this approach contain examples that may be of particular interest for the young reader. Imaginative visualization obtained by using multimedia materials is also widely used in order to bring the "theoretical" science in everyday life.

The important feature of the teaching courses is the combination of fundamental quality and profilization, appearing in the choice of priorities and examples of the application of physics in geology, biology, soil science, meteorology, ecology, etc.

The main feature of these courses is the degree of coverage of practical issues and their relevance. Another feature of the courses is that a large amount of background material allows the use of this knowledge during laboratory and practical studies. Links to Internet sites are provided where students can find additional and more detailed background information [9–12].

At present developers of e-learning resources focus on the new opportunities to work remotely, provided by the network and multimedia technology. Educational materials are developed now as Java-applets, characterizing by the exceptionally high level of interactivity, allowing a remote user to not only change the quantitative parameters of the modelled system, but also to determine its qualitative composition, essentially creating own original computer experiment.

4   Teaching Physics to Medical Students

Medical specialties along with the specialties which indirectly use medical and biological knowledge occupy a special place among the many natural sciences and humanities, which contain physics and the concept of modern science in the learning process.

All physiological processes occurring in the human body are governed by the general physical laws. Furthermore, most diagnostic methods as well as many methods of medical treatment are based on the use of physical phenomena and processes.

Nearly all medical systems are based on physical principles and, in fact, they represent physical devices. Medicine uses the results of theoretical and experimental achievements in physics.

Training an athlete and a coach requires knowledge of the functioning of the human body mechanics, the forces and moments acting on the body and its parts during the performance of various sports movements. Knowledge of physics is widely used in the training of specialists in sports medicine.

Medicine uses the results of theoretical and experimental achievements in physics. Thus, physics is of exceptional importance for medicine as a whole and for the formation of the future physician especially. The necessity of medical biophysics, together with elements of general physics used in physical methods of diagnosis and treatment, the principles of appropriate equipment – are the contents of physics taught in medical schools [16].

The authors have accumulated experience of teaching physics to medical students in St. Petersburg State University (Department of Medicine, Department of Dentistry and Medical Technologies), and in St. Petersburg State Pediatric Medical University, and have wrote several training manuals, taking into account this specificity [17, 18].

It is possible to highlight professionally oriented questions in each of the sections of general physics – the basic methods of definition of physical quantities in medicine; the specificity of physical phenomena and processes in medical practice; the use in medicine of physical phenomena, processes, devices (used in diagnostics for research use in clinical practice); the description of the principle of operation of medical devices.

Use of professionally oriented physical problems in the teaching of future doctors positively affects learning outcomes of students, promotes the development of creative personality of future specialists, the formation of student's valuable attitude to the medical profession. Thus future physicians and biologists need to know the basic details of vision [16–18], proteins [19], photosynthesis [20], effect of physical factors on biosphere [21, 22], etc. Professionally oriented physical tasks can be used in the classroom when new material, revision, consolidation and generalization are studied, in the classroom and in extracurricular organization of independent work of students. The use of professionally oriented physical tasks gives one the ability to individualize the learning process [23, 26, 27].

Now it is very important to give medical students a basic knowledge about telemedicine. The main purpose of telemedicine is medical service of the remote patients who have appeared far from the medical centres and have the limited access to medical services. The technical base of a telemedicine is modern technologies of transfer and reproduction of data that promotes videoconferences and transfer of high-quality digital images at a distance in an effort to render more perfect reanimation aid, the fastest transportation of the patient and accelerated medical decisions. Telemedicine is impossible without the experts having skills of work in remote locations. Medical school students can acquire such skills in the process of training within our grid segment, performing virtual labs, studying the educational materials in remote access.

5   The Use of Learning Management Systems

Under the proposed approach, the authors have developed multimedia presentations for different specializations (biological, medical, geological), including illustrations, animations, movies, as well as corresponding printed manuals [9–12] and webinars. Blackboard Learn-Learning Management System (LMS Blackboard Learn) [28] is used as a platform for creation, storing and presentation of these educational materials. Blackboard has been installed in the St. Petersburg State University from late 2011. Nowadays it is used for the registration of all student' written works and all University teachers must possess the skills to work with Blackboard Learn.

Blackboard represents a virtual learning environment and course management system developed by Blackboard Inc. It is Web-based server software which features course management, customizable open architecture, and scalable design that allows integration with student information systems and authentication protocols. It may be installed on local servers or hosted by Blackboard ASP Solutions. Its main purposes are to add online elements to courses traditionally delivered face-to-face and to develop completely online courses with few or no face-to-face meetings.

Learn distance learning system – MOODLE (acronym for Modular Object-Oriented Dynamic Learning Environment) [29], which is similar to Blackboard, is used in the Russian State Hydrometeorological University.

MOODLE is free and open-source software learning management system written in PHP and distributed under the GNU General Public License. Developed on pedagogical principles, MOODLE is used for blended learning, distance education, flipped classroom and other e-learning projects in schools, universities, workplaces and other sectors. With customizable management features, it is used to create private websites with online courses for educators and trainers to achieve learning goals. MOODLE allows extending and tailoring learning environments using community sourced plugins.

Both LMSs (Blackboard and MOODLE) have much in common (the main differences are listed in Table 1) and according to LISTedTech, both dominate the market for current LMS usage. But the main difference is that Blackboard Learn is a fully commercial product and MOODLE is open-source software using the "freemium" payment model (one gets the basic elements for free, but must pay for extra options). So, this is for you to decide what is better: MOODLE (powerful free solution, but tailoring it for one's specific needs will require extra programming and extra expenses) or Blackboard (fully loaded, but closed-source expensive product that cannot be easily tailored).

Table 1. Blackboard Learn vs MOODLE comparison

	Blackboard Learn	MOODLE
Payment model	Proprietary	Freely licensed open-source software
Course assignment	Weekly activities tab	Activity list on a single page
Managing content	Manual	Automatic
Features	Extra features are included in price	Limited options, free plugins database
Vendor model	One company	Many supporting companies and vendors
Help options	Forums, Knowledge Base, Live Tech Support	Forums, Knowledge Base

Let us compare the main features of these two LMSs.
The main features of Blackboard:

- administrative reporting;
- course catalogue;
- custom branding, fields and functionality;
- data import/export;
- exam engine;
- goal setting;
- grading;
- individual plans;
- multiple delivery formats;
- skills tracking;
- student portal.

Blackboard Learn is considered to be a great value for money, especially for large institutions with a lot of resources.
The main features of MOODLE:

- assessment implementation;
- collaboration management;
- discussion forums;
- file exchange;
- grade management;
- internal messaging, live chat, wikis;
- lesson planner;
- student roster/attendance management.

MOODLE is one of the most popular LMSs, mostly used by institutions with between 1,000 and 2,000 full-time students.

So, if one needs a powerful system and has the resources to pay for it (as St. Petersburg State University with more than 30,000 full-time students) Blackboard will be a good choice. If the budget is limited, but the ambitions are still high (the case of Russian State Hydrometeorological University with about 5,000 students) MOODLE will be a good choice.

In the process of study of a discipline not only traditional technologies, methods and forms of learning are used, but also innovative technologies, active and interactive forms of classes: lectures, practical classes, consultations, independent work, lectures with elements of problem representation, testing, etc. In the educational process active and interactive forms of conducting classes are widely used (conversation, interview, testing, discussion (with "brainstorming session" and without it), etc.). The independent creative work of students is also the important factor is. Good results are obtained by carrying out part of the lectures in the format of conferences, when the students after listening to another section of the general physics course, are preparing reports on the basic physics of natural phenomena and processes occurring in living systems [1–6, 8, 15].

A feature of these courses is the large number of examples that illustrate not just the proposed material, but are taken from actual tasks which the future specialist will regularly encounter in their daily practice [4].

Direct physical demonstrations during lectures on general physics or the concept of modern science would be, of course, ideal (for presenting physics not as some abstract from the reality, but as a science on fundamental laws of nature), but absolutely unrealistic in modern conditions of physics teaching. These demonstrations are impractical not only due to their high financial costs and low mobility (demos require expensive equipment, specialized personnel, storage conditions, etc.), but also a significant amount of time (preparation and display of demonstrations consume a considerable part of lecture time). The obvious way out of this situation is the active use of lectures, videos, interactive computer demonstrations, etc. (as with the use of pre-prepared programs, and access to the Internet and the use of specialized web-sites) [5, 7, 15].

LMS Blackboard Learn and LMS MOODLE promote effective independent work of students by providing access to electronic tutorials, manuals, by realization of effective feedback to lecturer via interactive testing, video conferences, on-line discussions.

The independent work of students requires not only training manuals on general physics (in print or digital formats) [9–12, 24, 25, 30–32], but also handbooks [33], and more in-depth studies of some sections of the course [33–45].

6   The Basis of Special Educational Environment for Teaching Physics

Infrastructural prototype of the distributed educational system which can be considered as Grid segment, consists of 4 sites located in St. Petersburg State University, St. Petersburg State Pediatric Medical University, Russian State Hydrometeorological University and St. Petersburg Electrotechnical University "LETI".

Each Grid-site base configuration consists of minimum 5 computer systems: Configuration server, Computing Element (CE), Storage Element (SE), User Interface (UI) and Working Node (WN).

User Interface element provides access to the Grid segment resources. User logins into UI computer in order to choose Grid resources, install the task for execution, get output data and transfer data if necessary.

Configuration server provides semiautomatic installation and configuration (both initial and secondary) of all base control elements.

Computing Element is the main working point on the local site. CE provides common interface for the computational resources lying beneath. Among its functions is task launching and task scheduling.

Working nodes provide user task implementation. A site can contain several working nodes. Storage Element provides the user with the universal access to the available databases.

The Globus code will be used as the basic software for Educational Grid System. The main function of the Grid infrastructure is to provide access to the educational resources located in different universities. Educational resources can be of different types: electronic textbook, virtual laboratory, data base, intelligent system. User should not care about the place where the resource is located and how to find it. In future the Grid education environment should possess advanced means such as GUI for applications-experiments composition, API to create modules for meta applications.

The first step of the educational Grid infrastructure functioning will be selection of the best educational courses in the field of physics elaborated in the 4 selected universities.

Thus there is the collection of audio tracks fragments of real lectures, read to the students at the Faculty of Physics of St. Petersburg State University and the videos of real experiments and natural phenomena which cannot be reproduced in lecture rooms [46]. A series of electronic media collections "Physics: model, experiment, reality" has been developed, each collection dedicated to one of the relatively isolated areas of modern physics.

Besides educational electronic courses St. Petersburg State University possesses essential computational resources located in Resource Centre "SPbU Computer Centre" which provides access to expensive research equipment and associated software that can be used for installation of virtual laboratories, visualization facilities and data warehouses.

Russian State Hydrometeorological University has gained substantial experience with the virtual laboratories. The university constantly substitutes old laboratory equipment with appropriate software for imitation laboratory workshops in mechanics, optics, thermodynamics and other sections of physics. Working with the software is much more familiar for modern students thus their interest to practical studies increased greatly. So in the section "Molecular physics and thermodynamics" students can perform following virtual laboratory operations: "Determination of dew point at various absolute humidity", "Determination of warmth of transpiration of fluid on pressure of saturated steams", "Determination of a specific heat capacity of a solid body", "Determination of a specific heat capacity of gas at constant pressure by a method of a flowing heating", "Determination of adiabatic index at the adiabatic expansion of gas", "Determination of adiabatic index on velocity of a sound in gas", "Determination of a thermal conduction of gases by a method of heated lines", "Determination of a thermal conduction of a solid body (plate)".

St. Petersburg Electrotechnical University "LETI" is famous for the work in microelectronics. So its staff can provide electronic textbook and manuals in the field of solid phase physics and electronic chips creation.

Lecturers and students of St. Petersburg State Pediatric Medical University can be the acceptors of the Educational Grid Segment facilities. Though being very important general physics is not profiling discipline at the University. Thus their laboratories and visualization facilities are of inferior quality in comparison with the related equipment at other universities. But this advanced equipment and software becomes available at the remote access. Such educational regime allows to train students skills needed in the future to work with telemedicine systems.

7 Conclusions

Thus, the use of modern computer technology for the development of modern multimedia interactive courses and webinars can significantly improve the effectiveness of teaching and learning of general physics and other natural sciences by the students of scientific, technical and humane specialities. We propose to organize special educational environment in the form of Grid segment, which should comprise all kinds of distributed learning materials and provide equal and the most comfortable access to them. Grid infrastructure functioning will be the selection of the best educational facilities in the field of physics, including textbooks, virtual laboratories and databases elaborated in St. Petersburg State University, St. Petersburg State Pediatric Medical University, Russian State Hydrometeorological University and St. Petersburg Electrotechnical University "LETI".

Acknowledgment. This research was sponsored by the The Russian Foundation for Basic Research under the projects: 16-07-01113 "Virtual supercomputer as a tool for solving complex problems", 16-04-00494 "Research of functioning of rhodopsin as the canonical representative of the class A receptors, which are the G-protein, by the methods of the local selective NMR, optical spectroscopy and numerical simulation" and the Contract № 02.G25.31.0149 dated 01.12.2015 (Board of Education of Russia).

References

1. Barmasova, A.M., Barmasov, A.V., Skoblikova, A.L., et al.: Features of teaching general physics to the students-ecologists. In: The Problems of Theoretical and Applied Ecology, 267 p., 15 p., pp. 226–241. Publishing House RSHU, St. Petersburg (2005) (In Russian)
2. Barmasova, A.M., Barmasov, A.V., Bobrovsky, A.P., Yakovleva, T.Y.: To the question about teaching general physics to the students-ecologists. In: Abstracts. Meeting of the Heads of Departments of Physics of Technical Universities in Russia, pp. 46–48. AVIAIZDAT, Moscow (2006) (In Russian)
3. Barmasova, A.M., Yakovleva, T.Y., Barmasov, A.V., et al.: An integrated approach to he teaching physics to students-nature managers. In: Spirin, G.G. (ed.) Abstracts of Scientific-Methodical Workshop on "The Physics in the Engineering Education System of the EurAsEC Member States" and the Meeting of Heads of Physics Departments of Technical Universities of Russia. The Scientific Seminar was Held from 25 to 27 June 2007, 344 p., pp. 40–41. Zhukovsky Air Force Engineering Academy, Moscow (2007) (In Russian)
4. Yakovleva, T.Y., Barmasova, A.M., Barmasov, A.V.: Interdisciplinary connections in teaching general physics to students of science and engineering. In: Spirin, G.G. (ed.) Abstracts of Scientific-Methodical Workshop on "The Physics in the System of Engineering and Pedagogical Education of the EurAsEC Member States". The Scientific Seminar was Held in 2008, 364 p., pp. 355–357. Zhukovsky Air Force Engineering Academy, Moscow (2008) (In Russian)

5. Bukina, M.N., Barmasov, A.V., Ivanov, A.S.: Modern teaching methods for the teaching general physics and mathematical processing of results of measurements of physical quantities. In: Modern Educational Technology in the Teaching Natural Sciences and the Humanities: Proceedings of the International Scientific-Methodical Conference, 27–29 May 2014, 562 p., pp. 408–414. Mining University, St. Petersburg (2014)

6. Bukina, M.N., Barmasov, A.V., Ivanov, A.S.: Some aspects of teaching physics in high school. In: VIII St. Petersburg Congress "Education, Science, Innovation in the Twenty-First Century", Collection of Works, 24–25 October 2014, 414 p., pp. 47–49. Mining University, St. Petersburg (2014) (In Russian)

7. Bukina, M.N., Barmasov, A.V., Lisachenko D.A., Ivanov, A.S.: Modern methods of teaching physics and the concepts of modern natural science. In: Proceedings of the II International Scientific-Methodical Conference on Modern Educational Technologies in Teaching Natural-Scientific and Humane Disciplines, 09–10 April 2015, 732 p., pp. 516–520. Mining University. St. Petersburg (2015) (In Russian)

8. Bukina, M.N., Barmasov, A.V., Ivanov, A.S.: Features of general physics teaching to students of natural science specialties in modern conditions. In: Proceedings of the XIII International Conference on the Physics in the System of Modern Education (FSSO-2015), St. Petersburg, 1–4 June 2015, vol. 2, 393 p., pp. 3–6 (2015) (In Russian)

9. Barmasov, A.V., Kholmogorov, V.E.: Course of general physics for nature managers. In: Chirtsov, A.S. (ed.) Mechanics, 2012, 416 p. BHV-St. Petersburg, St. Petersburg (2008) (In Russian)

10. Barmasov, A.V., Kholmogorov, V.E.: Course of general physics for nature managers. In: Bobrovsky, A.P. (ed.) Oscillations and Waves, 2012, 256 p. BHV-St. Petersburg, Petersburg (2009) (In Russian)

11. Barmasov, A.V., Kholmogorov, V.E.: Course of general physics for nature managers. In: Bobrovsky, A.P. (ed.)Molecular Physics and Thermodynamics, 2012, 512 p. BHV-St. Petersburg, St. Petersburg (2009) (In Russian)

12. Barmasov, A.V., Kholmogorov, V.E.: Course of general physics for nature managers. In: Bobrovsky, A.P. (ed.) Electricity, 2013, 448 p. BHV-St. Petersburg, St. Petersburg (2010) (In Russian)

13. Frenkel Y.I.: The system of teaching in higher education requires revision. J. High. Sch. 4, 35 (1945) (In Russian)

14. Feynman, RP., Leighton, R.B., Sands. M.: The Feynman Lectures on Physics, vol. 1 (1964)

15. Barmasova, A.M., Yakovleva, T.Y., Barmasov, A.V., et al.: Multimedia lecture course on processing of results of measurements of physical quantities for students users. In: Spirin, G.G. (ed.) Abstracts of Scientific-Methodical Workshop on "The Physics in the Engineering Education System of the EurAsEC Member States" and the Meeting of Heads of Physics Departments of Technical Universities of Russia. The Scientific Seminar was Held from 25 to 27 June 2007, 344 p., p. 42. Zhukovsky Air Force Engineering Academy, Moscow (2007) (In Russian)

16. Struts, A.V., Barmasov, A.V., Brown, M.F.: Methods for studying photoreceptors and photoactive molecules in biological and model systems: rhodopsin as a canonical representative of the seven-transmembrane helix receptors. Bulletin of St. Petersburg University. Series 4. Physics, Chemistry, no. 2, pp. 191–202 (2014) (In Russian)

17. Struts, A.V., Barmasov, A.V., Brown, M.F.: Spectral methods for study of the G-protein-coupled receptor rhodopsin: I. Vibrational and Electronic Spectroscopy. Opt. Spectrosc. **118**(5), 711–717 (2015)

18. Struts, A.V., Barmasov, A.V., Brown, M.F.: Spectral methods for study of the G-protein-coupled receptor rhodopsin: II. Magnetic resonance methods. Opt. Spectrosc. **120**(2), 286–293 (2016)
19. Bukina, M.N., Bakulev, V.M., Barmasov, A.V., Zhakhov, A.V., Ischenko, A.M.: Luminescence diagnostics of conformational changes of the Hsp70 protein in the course of thermal denaturation. Opt. Spectrosc. **118**(6), 899–901 (2015)
20. Barmasov, A.V., Korotkov, V.I., Kholmogorov, V.Y.: Model photosynthetic system with charge transfer for transforming solar energy. Biophysics **39**(2), 227–231 (1994)
21. Kholmogorov, V.Y., Barmasov, A.V.: The biosphere and physical factors. Electromagnetic fields and life. In: The Problems of Theoretical and Applied Ecology, 267 p., pp. 27–47. RSHMU, St. Petersburg (2005) (In Russian)
22. Barmasov, A.V., Barmasova, A.M., Yakovleva, T.Y.: The biosphere and the physical factors. Light pollution of the environment. In: Proceedings of the Russian State Hydrometeorological University, no. 33, pp. 84–101 (2014) (In Russian)
23. Biryukova, A.N. : Physics as professionally oriented course in Medical Higher School. Humanitarian Vector **25**(1), 86–89 (2011) (In Russian)
24. Barmasov, A.V., Barmasova, A.M., Struts, A.V., Yakovleva, T.Y.: Dynamics of Rigid Body. Elements of the Theory and the Collection of Tasks, 28 p. Publishing House of St. Petersburg State Pediatric Medical University, St. Petersburg (2012) (In Russian)
25. Barmasov, A.V., Barmasova, A.M., Struts, A.V., Yakovleva, T.Y.: Processing of Results of Measurements of Physical Quantities, 92 p. Publishing House of St. Petersburg State Pediatric Medical University, St. Petersburg (2012) (In Russian)
26. Remizov, A.N., Antonov. V.F., Vladimirov, Yu.: A. Program in Medical and Biological Physics for Medical Students. VUNMC, Moscow (2000) (In Russian)
27. Kholmogorov, V.E.; Krutitskaya, T.K.: The Demonstration Experiments for the General Physics Course. Tutorial, 90 p. Publishing House of St. Petersburg State University, St. Petersburg (2002) (In Russian)
28. Blackboard Learn. www.blackboard.com/platforms/learn/overview.aspx
29. Moodle. https://moodle.org
30. Barmasova, A.M., Yakovleva, T.Y., Barmasov, A.V., Bobrovsky, A.P.: Independent work of students in the conditions of introduction of profile training in high school. In: Schools and Universities: Achievements and Challenges of Continuous Physical Education: Book of Abstracts of the V Russian Scientific-Methodical Conference of Teachers of Universities and School Teachers, 252 p., p. 65. STU-UPI, Yekaterinburg (2008) (In Russian)
31. Nordling, C., Österman, J.: Physics Handbook for Science and Engineering. In: Barmasov, A.V. (ed.) Authorized Translation from the English Language Edition, 528 p. BHV-St. Petersburg, St. Petersburg (2011) (In Russian)
32. Yakovleva, T.Y., Barmasova, A.M., Barmasov, A.V.: Problems of pre-university training of students in physics. In: Spirin, G.G. (ed.) Abstracts of Scientific-Methodical Workshop on "The Physics in the Engineering Education System of the EurAsEC Member States" and the Meeting of Heads of Physics Departments of Technical Universities of Russia. The Scientific Seminar was Held from 25 to 27 June 2007, 344 p., pp. 239–241. Zhukovsky Air Force Engineering Academy, Moscow (2007) (In Russian)
33. Barmasov, A.V., Barmasova, A.M., Yakovleva, T.Y.: The accuracy of definitions in the course of general physics. 1. Material point. In: Spirin, G.G. (ed.) Abstracts of the Meeting of Heads of Physics Departments of Universities of Russia, Moscow, 344 p., pp. 53–55. APR, Moscow (2009) (In Russian)

34. Barmasov, A.V., Barmasova, A.M., Yakovleva, T.Y.: The accuracy of definitions in the course of general physics. 2. Simple and physical pendulums. In: Spirin, G.G. (ed.) Abstracts of the Meeting of Heads of Physics Departments of Universities of Russia, Moscow, 344 p., pp. 55–56. APR, Moscow (2009) (In Russian)
35. Barmasov, A.V., Barmasova, A.M., Yakovleva, T.Y.: The accuracy of definitions in the course of general physics. 3. Ideal and real gases. In: Spirin, G.G. (ed.) Abstracts of the Meeting of Heads of Physics Departments of Universities of Russia, Moscow, 344 p., pp. 56–58. APR, Moscow (2009) (In Russian)
36. Barmasov, A.V., Barmasova, A.M., Yakovleva, T.Y.: The accuracy of definitions in the course of general physics. 4. Vector and vector variable. In: Barmin, A.V. (ed.) Proceedings of All-Russian Scientific-Practical Internet-Conference on School and University: Innovation in Education. Interdisciplinary Connections of Natural Sciences, 180 p., pp. 18–19. OryolSTU, Oryol (2009) (In Russian)
37. Barmasov, A.V., Barmasova, A.M., Yakovleva, T.Y.: The accuracy of definitions in the course of general physics. 5. Gravitational force, gravity force and weight. In: Barmin, A.V. (ed.) Proceedings Of All-Russian Scientific-Practical Internet-Conference on School and University: Innovation in Education. Interdisciplinary Connections of Natural Sciences, 180 p., pp. 20–21. OryolSTU, Oryol (2009) (In Russian)
38. Barmasov, A.V., Barmasova, A.M., Yakovleva, T.Y.: The accuracy of definitions in the course of general physics. 6. Point charge and electric dipole. In: Spirin, G.G. (ed.) Abstracts of the International School-Seminar "Physics in Higher and Secondary Education of Russia, Moscow, 328 p., pp. 65–66. APR, Moscow (2010) (In Russian)
39. Barmasov, A.V., Barmasova, A.M., Yakovleva, T.Y.: The accuracy of definitions in the course of general physics. 7. Quasi-elastic forces. In: Spirin, G.G. (ed.) Actual Problems of Teaching Physics in Universities and Schools Of Post-Soviet Countries. Proceedings of the International School-Seminar "Physics in Higher and Secondary Education", Moscow, June 2011, 280 p., pp. 46–47. APR, Moscow (2011) (In Russian)
40. Barmasov, A.V., Barmasova, A.M., Yakovleva, T.Y.: The accuracy of definitions in the course of general physics. 8. Doppler effect. In: Spirin, G.G. (ed.) Actual Problems of Teaching Physics in Universities and Schools of Post-Soviet Countries. Proceedings of the International School-Seminar "Physics in Higher and Secondary Education", Moscow, June 2011, 280 p., pp. 47–49. APR, Moscow (2011) (In Russian)
41. Barmasov, A.V., Barmasova, A.M., Yakovleva, T.Y.: The accuracy of definitions in the course of general physics. 9. Free electrons. In: Spirin, G.G. (ed.) Actual Problems of Teaching Physics in Universities and Schools of Post-Soviet Countries. Proceedings of the International School-Seminar "Physics in Higher and Secondary Education", Moscow, pp. 38–40. APR, Moscow (2012) (In Russian)
42. Barmasov, A.V., Barmasova, A.M., Yakovleva, T.Y.: The accuracy of definitions in the course of general physics. 10. The equations of state of an ideal gas. In: Spirin, G.G. (ed.) Proceedings of the International School-Seminar "Physics in Higher and Secondary Education", Moscow, 278 p., pp. 43–44. APR, Moscow (2014) (In Russian)
43. Barmasov, A.V., Barmasova, A.M., Yakovleva, T.Y.: The accuracy of definitions in the course of general physics. 11. Electric potential, potential difference and voltage. In: Proceedings of the XIII International Conference on Physics in the Modern Education System (FSSO-2015), St. Petersburg, 1–4 June 2015, vol. 1, pp. 46–48. Publishing House "Fora-print", St. Petersburg (2015) (In Russian)

44. Barmasov, A.V., Barmasova, A.M., Yakovleva, T.Y.: The accuracy of definitions in the course of general physics. 12. Ferroelectrics and antiferroelectrics. In: Science and Education in XXI Century: Collection of Scientific Papers on Materials of International Correspondence Scientific-Practical Conference on 30 January 2015: in 5 Parts, Part III, 153 p., pp. 91–94. AR-Consalt, Moscow (2015) (In Russian)
45. Barmasov, A.V., Barmasova, A.M., Yakovleva, T.Y.: The accuracy of definitions in the course of general physics. 13. Isolated, closed and conservative systems. In: Spirin G.G. (ed.) Proceedings of the International school-seminar "Physics in Higher and Secondary Education". International School-Seminar was Held in 2015, 278 p., pp. 40–41. APR, Moscow (2015) (In Russian)
46. Chirtsov, A.: Computer support in professional education in physics: virtual physical laboratories and real experiments in multimedia. In: International Conference CoLoS 1999: New Ideas in Computer Based Education, 16–18 September 1999, St. Petersburg, Russia (1999)

A Nonlinear Multicriteria Model
for Team Effectiveness

Isabel Dórdio Dimas[1,2]([⊠]), Humberto Rocha[3,4], Teresa Rebelo[5,6],
and Paulo Renato Lourenço[5,6]

[1] ESTGA, Universidade de Aveiro, 3750-127 Águeda, Portugal
[2] GOVCOPP, Universidade de Aveiro, 3810-193 Aveiro, Portugal
idimas@ua.pt
[3] FEUC, Universidade de Coimbra, 3004-512 Coimbra, Portugal
[4] INESC-Coimbra, 3030-290 Coimbra, Portugal
hrocha@mat.uc.pt
[5] IPCDVS, Universidade de Coimbra, 3001-802 Coimbra, Portugal
[6] FPCEUC, Universidade de Coimbra, 3000-115 Coimbra, Portugal
{terebelo,prenato}@fpce.uc.pt

Abstract. The study of team effectiveness has received significant
attention in recent years. Team effectiveness is an important subject since
teams play an increasingly decisive role on modern organizations. This
study is inherently a multicriteria problem as different criteria are typically required to assess team effectiveness. Among the different aspects of
interest on the study of team effectiveness one of the utmost importance
is to acknowledge, as accurately as possible, the relationships that team
resources and team processes establish with team effectiveness. Typically, these relationships are studied using linear models which fail to
explain the complexity inherent to group phenomena. In this study we
propose a novel approach using radial basis functions to construct a multicriteria nonlinear model to more accurately capture the relationships
between the team resources/processes and team effectiveness. By combining principal component analysis, radial basis functions interpolation,
and cross-validation for model parameter tuning, we obtained a data fitting method that generated an approximate response with reliable trend
predictions between the given data points.

Keywords: Team effectiveness · Multicriteria · Radial basis functions ·
Cross-validation

1 Introduction

Teams of individuals working together to achieve a common goal are, nowadays,
a central part of daily life on modern organizations [24]. Hence, over the last
four decades, the use of teams as a way of structuring activities has grown
enormously [29]. This derives in part from the fact that teamwork seems to
be superior in many situations, namely when the tasks and the problems are

© Springer International Publishing Switzerland 2016
O. Gervasi et al. (Eds.): ICCSA 2016, Part IV, LNCS 9789, pp. 595–609, 2016.
DOI: 10.1007/978-3-319-42089-9_42

complex [23]. Moreover, groups appear to be an effective answer to the challenges posed by the actual uncertain and complex environments [29].

Given the fact that teams are created with the purpose of generating value for the organizations, the study of team effectiveness has received a significant attention in recent years [19]. The literature is consensual about the need to consider different criteria to assess effectiveness [16,19] that can be integrated into five dimensions: (i) economic – integrates efficiency and productivity and is related to the team goals achievement; (ii) social – relates to the extend to which the group experience contributes to members' well-being; (iii) political – concerns reputation and legitimacy as assessed by the teams' stakeholders; (iv) systemic – relates to the willingness of members to remain in the team in the future; (v) innovation – concerns the teams' ability to rethink on current processes and to develop innovative solutions [2].

The conceptualization of team effectiveness that is dominant in the literature is the Input–Process–Output (I–P–O) model formulated by McGrath [25]: *inputs* refer to the composition of the team in terms of the individual, team and organizational resources; *processes* refer to activities that team members engage in, combining their resources to manage the tasks; *outputs* concern team results as conceptualized above. Team dimension and team autonomy are examples of team inputs. Examples of processes include team resilience and team learning.

It is of the utmost importance to acknowledge, as accurately as possible, the relationships that team resources and team processes establish with team effectiveness. Typically, these relationships are treated as linear and studied accordingly. However, it has been reported that the relationship between some of the team characteristics/processes and team effectiveness might not be linear. E.g., Bunderson and Sutcliffe reported that too much emphasis on learning can compromise efficiency because it detracts the team from results [8]. Using multiple linear regression models, it is possible to capture the previous study findings for lower levels of team learning, for which increasing the team learning levels would correspond to an increase of team effectiveness. However, for higher levels of team learning, a linear model will fail to capture the previous findings since it will continue to display an increase of team effectiveness when increasing the values of learning levels.

In this study we propose a novel approach using radial basis functions to construct a multicriteria nonlinear model attempting to more accurately capture the relationships between the team resources/processes and team effectiveness. Radial basis function (RBF) methods are interpolation methods, i.e. they exactly fit each data point. There are many different mathematical models that can easily fit a data set exactly no matter how the data points are distributed.However, building a response by using a scarce number of poorly distributed data points is very unreliable, yet necessary in many problems. There is a wide range of applications where RBF interpolation methods were successfully applied, including aeronautics [30,33], radiotherapy [32,34] and meteorology [7]. In most of the applications, RBF models are used as predictive tools. Their good predictive ability underlies their capacity to serve as surrogates that mimic well the unknown responses. RBFs surrogate features are used in this study to capture the trends between the team resources/processes and team effectiveness.

2   Materials and Methods

2.1   Problem Features

In this study, the assessment of team effectiveness is based on four criteria – performance (economic dimension), quality of the group experience (social dimension), viability (systemic dimension) and team process improvement (innovation dimension). Data concerning the political dimension was not available for this study. It is straightforward to formulate a multicriteria mathematical optimization model by considering team effectiveness as a weighted sum of these four criteria. Equal weights were considered for the different criteria.

Team effectiveness will be considered as the result of the presence of six variables: three of them can be conceived as inputs in I-P-O Model, i.e., team dimension, transformational leadership and team autonomy, and the remaining three, team resilience, supportive behaviors and team learning, as processes. Each variable will be briefly described as follows.

Team size corresponds to the number of elements that a team has. In accordance to the literature on team composition, well-composed teams are as small and diverse as possible [17]. Hence, coordinating and integrating individual contributions in large size teams is harder than on small ones, resulting, as a consequence, in negative outcomes.

Transformational leadership can be defined as a leadership style that encourages followers to do more than they originally expected, broadening and changing their interests and leading to conscientiousness and acceptance of the group's purposes [3]. Carless, Wearing and Mann [10] described transformational leaders as those who exhibit the following seven behaviours: they (1) communicate a vision; (2) develop staff; (3) provide support for them to work towards their objectives through coordinated team work; (4) empower staff; (5) are innovative by using non-conventional strategies to achieve their goals; (6) lead by example; (7) are charismatic. A positive association of transformational leadership with team results is suggested in literature (e.g., [4,18]).

Team autonomy can be defined as the level of freedom and independence that a team has in deciding how to carry out the tasks [21] and has been conceived as a critical element of team performance (e.g., [12,21]).

Team resilience is the ability of the team not only to recover from stressful events but also to grow and learn from the adversity [39]. It is an adaptive process, which enables teams to manage difficulties in a positive way, without jeopardizing cohesion and team results. Given that work environments are becoming more and more challenging, team resilience has been related with positive consequences for teams [37].

Supportive behaviors can be defined as the extend to which team members provide voluntary assistance to each other [2]. This concept encompasses both instrumental (tangible help that members may provide to each other) and emotional (members' actions that make other members feel appreciated and that bolster their selfworth) supports. It has has been related to positive team outcomes (e.g., [2, 11]).

Finally, team learning can be conceived as a continuous process of reflection and action, characterized by behaviors like seeking feedback, exploring, experimenting, reflecting, and discussing errors and unexpected outcomes [15]. Previous research presented team learning as a crucial process of adaptation of teams to their environment and highlighted its importance in goals achievement (e.g., [14, 15]).

2.2   Sample

A quantitative study with a cross-sectional design was conducted in which we surveyed teams from different companies, sectors (e.g., industrial, services) and geographical areas (north and center of Portugal). In line with Cohen and Bailey's definition of group [12], teams had to meet the following criteria to be included in the sample: teams must consist of at least 3 members (1), who are perceived by themselves and others as a team (2), and who interact regularly and interdependently to accomplish a common goal (3). In each company, we had to collect two types of information: the team members' questionnaires and the team leaders' questionnaires. Team members were surveyed about transformational leadership, team autonomy, team resilience, supportive behaviors, team learning and quality of the group experience, whereas team leaders were surveyed about team size, team viability, team performance and team process improvement.

Data was collected using two different strategies. In the majority of the organizations, the questionnaires were collected by a person of the organization, with a strategic relationship with the employees, previously instructed by a research team member. For the organizations where this strategy was not viable, the questionnaires were filled online via an electronic platform, with the link being provided to the participants. In both cases, the anonymity and the confidentiality of the answers were guaranteed.

Surveys were collected from 653 members of 117 workgroups and their respective leaders from nine Portuguese organizations. Teams were composed of 9.0 members on average (SD = 9.15). Questionnaires with more than 10 % answers missing were eliminated [5], as well as teams in which less than 60 % of the members delivered their surveys. In consequence, the final sample includes 86 teams.

2.3   Measures

To obtain *team size*, leaders were asked about the number of elements of their teams. To measure *transformational leadership* we used the *Global Transformational Leadership* (GTL) scale developed by Carless et al. [10] and adapted to the Portuguese language by van Beveren [40]. This scale is composed of seven items (each item measures one of the seven characteristics of transformational leaders in accordance with Carless et al. [10]) that are measured on a 5-point Likert scale from 1 = "almost doesn't apply" to 5 = "almost totally applies". A sample item is "My team leader encourages thinking about problems in new ways and questions assumptions". The Cronbach α for this scale is .96.

To measure *team autonomy* we used the *Team-Level Autonomy scale* (TLA) developed by Langfred [21] and adapted to the Portuguese language by van Beveren [40]. This scale is composed of seven items that are evaluated on a 5-point Likert scale from 1 = "almost doesn't apply" to 5 = "almost totally applies". A sample item is "The team is free to choose the method(s) to use in carrying out work". The Cronbach α for this scale is .90.

To measure *team resilience* a three items scale developed by Stephens et al. [37] and adapted to the Portuguese language by Albuquerque [1] was used. Statements are evaluated on a 5-point Likert scale ranging from 1 = "almost doesn't apply" to 5 = "almost totally applies". A sample item is "Team members know how to handle difficult situations when we face them". The Cronbach α for this scale is .92.

To measure *supportive behaviors* a scale developed by Aubé and Rousseau [2] and adapted to the Portuguese language by Pessoa [28] was used. This scale is composed of 5 items that are measured on a 5-point Likert scale from 1 = "almost doesn't apply" to 5 = "almost totally applies". A sample item is "We help each other out if someone falls behind in his/her work". The Cronbach α for this scale is .93.

To measure *team learning* we used the *Team Learning Behaviors' Instrument* developed by Savelsbergh et al. [36] and adapted to the Portuguese language by Dimas et al. [13]. The Portuguese adaptation is composed of 25 items that are measured on 5-point Likert scales from 1 = "almost doesn't apply" to 5 = "almost totally applies". It has five dimensions, which correspond to the five learning behaviors proposed by Edmondson [15] (exploring and co- construction of meaning, collective reflection, error management, feedback behavior, and experimenting). A sample item is "If something has gone wrong, the team takes the time to think it". The Cronbach αs for the five dimensions of this scale are above .88. In the present study, since the intercorrelations between the five team learning dimensions were very high (between .63 and .84), the presence of a second order factor was tested through a Confirmatory Factor Analysis and, as result, a global score of team learning will be used in the following analyses.

To measure team effectiveness, as explained above, four different criteria were used: team performance, team viability, team process improvement and quality of the group experience. All scales used to measure these variables were developed by Aubé and Rousseau [2] and Rousseau and Aubé [35] and were adapted to the Portuguese language by Albuquerque [1].

Team performance scale is composed of five items that are rated on a 5-point Likert scale from 1 = "very low" to 5 = "very high". A sample item is "Achievement of performance goals". The Cronbach α for this scale is .83.

Quality of the group experience scale is composed of 3 items. Each sentence is measured on 5 point-Likert scales from 1 = "strongly disagree" to 5 = "strongly agree". A sample item is "The social climate in our work team is good". The Cronbach α for this scale is .94.

Team viability scale is composed of four items that are measured on a 5-point Likert scale from 1 = "almost doesn't apply" to 5 = "almost totally applies". A sample item is "The members of this team could work together for a long time". The Cronbach α for this scale is .84.

Finally, *team process improvement* scale is constituted by 5 items rated on 5 point-Likert scales from 1 = "almost doesn't apply" to 5 = "almost totally applies". A sample item is "Team members have successfully implemented new ways of working to facilitate achievement of performance goals". The Cronbach α for this scale is .86.

2.4 RBF Interpolation Models

For any finite data set, radial basis functions (RBFs) can provide excellent inter-polants. However, a RBF model may exhibit undesirable trends if the data points are scarce and/or irregularly distributed in a high dimensional space. Therefore, a principal component analysis (PCA) is recommended to detect any collinearity of the attributes of the data points and to transform correlated variables into uncorrelated ones. PCA can also be applied as a data reduction strategy but that is not the goal of this study. Thus, PCA is used here as a structure detection and correction method.

Principal Component Analysis. Given a set of N data points, $\mathbf{p}^1, \ldots, \mathbf{p}^N$, with n components each (in $\mathbb{R}^n$) where $\mathbf{p}_i^j$ represents the ith component of $\mathbf{p}^j$, the PCA is done as follows. First, scale each component by its estimated standard deviation, $\hat{\mathbf{p}}_i^j = \mathbf{p}_i^j / \sigma_i$, where

$$\sigma_i = \frac{1}{N-1} \sqrt{\sum_{j=1}^{N} \left(\mathbf{p}_i^j - \text{ave}(\mathbf{p}_i) \right)^2},$$

with $\text{ave}(\mathbf{p}_i) = \frac{1}{N} \sum_{k=1}^{N} \mathbf{p}_i^k$. Then, compute the covariance matrix $\mathbf{C}$ of the scaled data points $\hat{\mathbf{p}}^1, \ldots, \hat{\mathbf{p}}^N$,

$$\mathbf{C} = \frac{1}{N-1} \sum_{j=1}^{N} \left[\hat{\mathbf{p}}^j - \text{ave}(\hat{\mathbf{p}}) \right] \left[\hat{\mathbf{p}}^j - \text{ave}(\hat{\mathbf{p}}) \right]^T.$$

The collinearity of the variables can be assessed using the spectral decomposition of $\mathbf{C}$, $\mathbf{C} = \sum_{j=1}^{n} \lambda_j \mathbf{u}^j (\mathbf{u}^j)^T$, where $\lambda_1 \geq \lambda_2 \geq \cdots \geq \lambda_n \geq 0$ are the eigenvalues of $\mathbf{C}$, and $\mathbf{u}^1, \ldots, \mathbf{u}^n$ are the corresponding unit eigenvectors. The unit vector $\mathbf{u}^j$ is the jth feature vector of the scaled data set $\hat{\mathbf{p}}^1, \ldots, \hat{\mathbf{p}}^N$ and the scalar $v_j = \hat{\mathbf{p}}^T \mathbf{u}^j$ is the jth principal component of $\hat{\mathbf{p}}$.

The value of each eigenvalue of $\mathbf{C}$, λ_j, indicates the significance the jth principal component of $\hat{\mathbf{p}}$ in representing the variance in the scaled data set.

Thus, the PCA procedure can be seen as an ordination technique for describing the variation in a multivariate data set. The first principal component (first axis) corresponds to the most significant direction of variance in the scaled data set, the second principal component corresponds to the second most significant direction of variance in the principal component, and so forth, with each direction orthogonal to the preceding ones. We can write each data point $\hat{\mathbf{p}}^k$ as a linear combination of the feature vectors $\mathbf{u}^1, \ldots, \mathbf{u}^n$:

$$\hat{\mathbf{p}}^k = \text{ave}(\hat{\mathbf{p}}) + \sum_{j=1}^{n} \left[(\hat{\mathbf{p}}^k - \text{ave}(\hat{\mathbf{p}}))^T \mathbf{u}^j \right] \mathbf{u}^j. \tag{1}$$

By applying PCA to a given set of data points, we can treat the response as a function defined on a feature space, solve the approximation problem by fitting the transformed data in the feature space, and then recover the approximate response in the original input space using Eq. (1). For a more thoroughly description of PCA see, e.g., [22, sect. 3.6].

RBF Interpolation in the Feature Space. Given a set of N data points, $\mathbf{p}^1, \ldots, \mathbf{p}^N$, if the true responses, $f(\mathbf{p}^j)$, $j = 1, \ldots, N$, are known, the goal is to construct a model $g(\mathbf{p})$, using a RBF, $\varphi(x)$, such that $g(\mathbf{p}^j) = f(\mathbf{p}^j), j = 1, \ldots, N$. For the reasons stated above, the RBF interpolant may overfit the data and exhibit undesirable trends between the data points. Thus, using PCA we transform the data fitting problem into a problem in the feature space and find there a RBF interpolant $g(\mathbf{v})$ such that

$$g(\mathbf{v}^j) = f(\mathbf{p}^j), j = 1, \ldots, N, \tag{2}$$

where $\mathbf{v}^j = v_i^j = (\hat{\mathbf{p}}^j - \text{ave}(\hat{\mathbf{p}}))^T \mathbf{u}^i, i = 1, \ldots, n, j = 1, \ldots, N$.
The interpolation model $g(\mathbf{v})$ can be represented as

$$g(\mathbf{v}) = \sum_{j=1}^{N} \alpha_j \varphi(\|\mathbf{v} - \mathbf{v}^j\|), \tag{3}$$

where α_j are the coefficients to be determined by interpolation conditions (2), $\|\mathbf{v} - \mathbf{v}^j\|$ corresponds to the parameterized distance between $\mathbf{v}$ and $\mathbf{v}^j$,

$$\|\mathbf{v} - \mathbf{v}^j\| = \sqrt{\sum_{i=1}^{n} |\theta_i| (v_i - v_i^j)^2}, \tag{4}$$

and $\theta_1, \ldots, \theta_n$ are the model tuning parameters that need to be optimized for obtaining the best prediction model of the given data.

For fixed parameters θ_i, the coefficients $\alpha_1, \ldots, \alpha_N$ in Eq. (3) can be computed by solving the following linear system of interpolation equations:

$$\sum_{j=1}^{N} \alpha_j \varphi(||\mathbf{v}^k - \mathbf{v}^j||) = f(\mathbf{p}^k), \quad \text{for } k = 1, \ldots, N. \tag{5}$$

For multiquadric and Gaussian RBFs, the interpolation matrix of the linear system (5) is nonsingular, provided that all data points are different, which guarantees the existence of a unique interpolant. However, the interpolation matrix of the linear system (5) can be nonsingular for cubic and thin plate spline RBFs. In such case, adding a low-degree polynomial to the interpolation functions in Eq. (3) solves the problem [31].

The most commonly used RBF [31] are multiquadric $\varphi(x) = \sqrt{1 + x^2}$, thin plate spline $\varphi(x) = x^2 \ln x$, cubic spline $\varphi(x) = x^3$, and Gaussian $\varphi(x) = \exp(-x^2)$. These RBFs can be used to model almost linear, almost quadratic, and cubic growth rates, as well as exponential decay of the response for trend predictions. The constructed interpolant $g(\mathbf{v})$ in Eq. (3) depends on "subjective" choice of $\varphi(x)$, and model parameters $\theta_1, \ldots, \theta_n$. While one can try all the possible choices of $\varphi(x)$ in search of a desirable interpolant, there are infinitely many choices for $\theta_1, \ldots, \theta_n$. Instead, cross-validation is used to determine the optimal value of $\theta_1, \ldots, \theta_n$ that yield an interpolant $g(\mathbf{v})$ with the most accurate trend prediction.

Model Parameter Tuning by Cross-Validation. RBF interpolation models use the parameterized distance of Eq. (4). Model parameter tuning for RBF interpolation consists in obtain a set of parameters $\theta_1, \ldots, \theta_n$ that leads to the best prediction model of the unknown response based on the available data. Other metrics instead of fitting errors must be used to compute the optimal scaling parameters θ_i and determine which basis function $\varphi(x)$ are most appropriate to model the response function $f(\mathbf{p})$, because a RBF interpolant, $g(\mathbf{p})$, exactly fits $f(\mathbf{p})$ for $\mathbf{p} = \mathbf{p}^k, k = 1, \ldots, N$. Cross-validation (CV) [38] was proposed to find $\varphi(x)$ and θ_i that lead to an approximate response model $g(\mathbf{p})$ with optimal prediction capability [31]. The leave-one-out CV procedure is usually used in model parameter tuning for RBF interpolation (see [31]).

Leave-One-Out Cross-Validation for RBF Interpolation:

- Fix a set of parameters $\theta_1, \ldots, \theta_n$.
- For $j = 1, \ldots, N$, construct the RBF interpolant $g_{-j}(\mathbf{p})$ of the data points $(\mathbf{p}^k, f(\mathbf{p}^k))$ for $1 \leq k \leq N, k \neq j$.
- Use the following CV root mean square error as the prediction error:

$$E^{CV}(\theta_1, \ldots, \theta_n) = \sqrt{\frac{1}{N} \sum_{j=1}^{N} (g_{-j}(\mathbf{p}^j) - f(\mathbf{p}^j))^2}. \tag{6}$$

The goal of model parameter tuning by CV is to minimize the CV error $E^{CV}(\theta_1, \ldots, \theta_n)$ by finding optimal $\theta_1, \ldots, \theta_n$ so that the interpolation model has the highest prediction accuracy when CV error is the measure. It should be highlighted that, most of the time, this optimization problem is very difficult as the $E^{CV}(\theta_1, \ldots, \theta_n)$ function is highly nonlinear and nonconvex [30]. A straightforward simplification of this problem is to consider $\theta_1 = \cdots = \theta_n$, which reduces the problem to a simple unconstrained minimization of a univariate function. Despite the fact that this simplification has the benefit of dealing with a simple unidimensional optimization problem, it has the disadvantage of not using all different θ_i which allows the model parameter tuning to scale each variable p_i based on its significance in modeling the variance in the response. Thus, considering all θ_i different have the benefit of implicit variable screening built in the model parameter tuning.

3   Results

In this study, the unit of analysis was the group rather than the individual and, as a result, members' responses were aggregated to the team level. To examine whether the data justified aggregation the Average Deviation Index (AD_M Index) developed by Burke, Finkelstein, and Dusig [9] was performed. Following the authors' recommendations, we used the criterion $AD_M \leq 0.83$ to aggregate, with confidence, individual responses to the team level. The average AD_M values obtained for each variable were below the upper-limit criterion of 0.83 revealing that the level of within-team agreement was sufficient to aggregate team members' scores.

A correlation analysis was performed to assure that the variables to include in the models are correlated with the outcome. Significant and negative correlation was found between effectiveness and team size while correlations found between effectiveness and the remaining variables were significant and positive. Supportive behaviors presented the strongest correlation between variables and outcome ($r = .622, p < .01$). The correlation between variables is significant and positive except for team size that has a significant and negative correlation with the remaining variables (Table 1).

MATLAB code *fminsearch*, an implementation of the Nelder-Mead [26] multidimensional search algorithm, was used to minimize the CV error in Eq. (6) and to find the best model parameters $\theta_1, \ldots, \theta_n$. The local optimal solution generated by MATLAB code *fminsearch* for minimization of the CV error is very reliable but also very sensitive to the initial guess. Multiple initial guesses were used for searching a global minimizer of the CV error by *fminsearch*.

The CV error of a constructed approximation can be used as an objective tool to help analysts on the difficult task of deciding which RBF model is better for the problem at hand. Table 2 displays the CV errors of the various RBF interpolation models considering different basis functions. The approximation model that, among all the approximation models, better capture the information "buried" in the data set usually corresponds to the smallest value of minimized CV errors [31]. By simple inspection of Table 2 we can verify that multiquadric RBF lead to the model with smallest CV error. Thus, multiquadric RBF model was selected to study the relationship between the variables and the outcome.

Table 1. Correlation analysis.

	Effectiveness	Team size	Leadership	Learning	Resilience	Supportive behaviors	Autonomy
Effectiveness	1	−.276*	.492**	.558**	.525**	.622**	.445**
Team size		1	−.309**	−.436**	−.297**	−.367**	−.276*
Leadership			1	.784**	.596**	.678**	.557**
Learning				1	.676**	.709**	.691**
Resilience					1	.769**	.496**
Supportive behaviors						1	.555**
Autonomy							1

Note: ** $p < .01$; * $p < .05$.

Table 2. Optimal CV errors for the data set.

Multiquadric CV error	Thin plate CV error	Cubic CV error	Gaussian CV error
0.78	1.13	1.57	1.28

Figure 1 display the relationships between effectiveness and each of the six variables considered. Data points were added to the plots to give an indication of the scatter in the data. The baseline data point, i.e. the data point for which the remaining variables are kept constant is also plotted for perspective. In order to benchmark the multiquadric RBF model results, the following multiple linear regression model, obtained using SPSS, was also added:

$$\text{Effectiveness} = 1.84 + 0.00 \times (\text{Team size}) + 0.00 \times (\text{Leadership})$$
$$+ 0.15 \times (\text{Learning}) + 0.04 \times (\text{Resilience})$$
$$+ 0.34 \times (\text{Supportive behaviors}) + 0.04 \times (\text{Autonomy}).$$

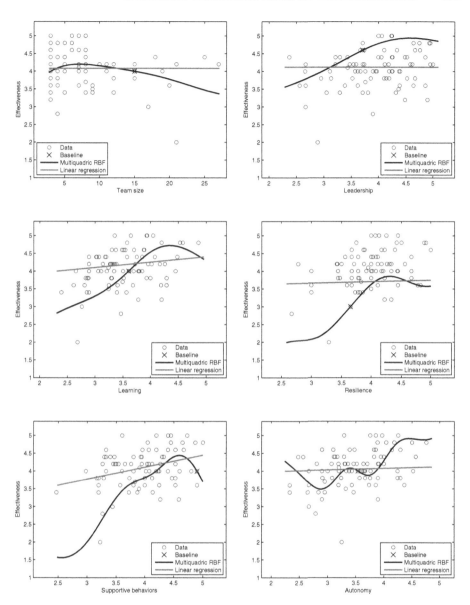

Fig. 1. Two-dimensional plots of linear and multiquadric RBF models.

4   Discussion and Conclusions

In general, the results obtained are quite satisfactory. The trends found by the multiquadric RBF model concerning the relationships between the team resources/processes and team effectiveness are promising. Indeed, the nature of

the results found put forward the need to consider methodologies beyond the linear widespread approach to better capture the complexity of team functioning.
The first result to discuss concerns the relationship between team size and team effectiveness. Literature on groups has shown that team size is a key variable influencing group dynamics and performance. As the size of a team increases, so does the quantity of resources available, but also the need for coordination [29]. In this way, the questions concerning the "optimal" team size are complex and are yet to be fully answered [20]. In fact, whereas some studies suggest that smaller size is better [42], other studies show that increasing team size improves performance [11]. The present study, through a nonlinear approach, and with a large sample composed of different types of organizational teams, gives a contribution to this debate. Hence, our results, which are in line with Nieva, Fleishman and Reick [27], show that the relationship between team size and team effectiveness is nonlinear: teams with five to ten members outperform smaller teams, where resources are lacking, and also larger teams, where coordination becomes difficult.

Concerning transformational leadership, team learning, team resilience, supportive behaviors and team autonomy, nonlinear patterns are also shown. Increasing trends up to a certain threshold are displayed, followed by a deflation for the highest values of the respective variables. Hence, our results show that more is not always better, suggesting that at higher values of the variables considered an inversion is reached, opposing to linear results. These results are in line with those of Bunderson and Sutcliffe [8] on the relationship between team learning and performance. The authors found that too much emphasis on learning can compromise efficiency because detract the team from results, and this is particular salient for teams that have been performing well. The present study extends those results by showing that this pattern is identified not only when performance is considered but also when a more embracing conception of effectiveness is adopted (that integrates four criteria). Moreover, our results highlight that this trend is identified also with other variables important for team functioning. A comprehensive explanation can be presented about the nature of the relationships between transformational leadership, team resilience, supportive behaviors with team effectiveness: when team members gives too much instrumental and emotional support to other members, time is consumed without assurance of results, and that might therefore reduce effectiveness; leaders that present too much transformational leadership behaviors might be to active in stimulating the team to develop, interfering excessively with team functioning and, then, effectiveness might suffers [6]; and finally, too much team resilience could lead to overconfidence, increasing the chance of committing errors and thus affecting effectiveness [41].

It is worth to highlight that, apart from team autonomy, all remaining variables do not present minor trend changes (oscillations). Despite the oscillations displayed for smaller values of team autonomy, the overall trend is similar to the remaining variables. By combining principal component analysis, RBF interpolation, and cross-validation for model parameter tuning, we obtain a data fitting method that generates an approximate response with more desirable trend predictions between the given data points, less likely to overfit the data, i.e. to display unreliable minor trend changes that might lead to undesirable characteristics such as oscillations.

Acknowledgements. This work is supported by FEDER through COMPETE under the funding given by FCT to GOVCOPP (POCI-01-0145-FEDER-008540).

References

1. Albuquerque, L.B.: Team resilience and team effectiveness: adaptation of measuring instruments (Master thesis). Faculdade de Psicologia e de Ciências da Educação da Universidade de Coimbra (2016)
2. Aubé, C., Rousseau, V.: Team goal commitment and team effectiveness: the role of task interdependence and supportive behaviors. Group Din. Theor. Res. **9**, 189–204 (2005)
3. Bass, B.M.: Leadership and Performance Beyond Expectations. Free Press, New York (1985)
4. Braun, S., Peus, C., Weisweiler, S., Frey, D.: Transformational leadership, job satisfaction, and team performance: a multilevel mediation model of trust. Leadersh. Quart. **24**, 270–283 (2013)
5. Bryman, A., Cramer, D.: Quantitative Data Analysis for Social Scientists, rev edn. Routledge, Florence (1994)
6. Buljac-Samardzic, M., van Woerkom, M.: Can managers coach their teams too much? J. Manag. Psychol. **30**, 280–296 (2015)
7. Buhmann, M.: Radial Basis Functions: Theory and Implementations. Cambridge University Press, Cambridge (2003)
8. Bunderson, J.S., Sutcliffe, K.M.: Management team learning orientation and business unit performance. J. Appl. Psychol. **88**, 552–560 (2003)
9. Burke, M.J., Finkelstein, L.M., Dusig, M.S.: On average deviation indices for estimating interrater agreement. Organ. Res. Methods **2**, 49–68 (1999)
10. Carless, S., Wearing, L., Mann, L.: A short measure of transformational leadership. J. Bus. Psychol. **14**, 389–405 (2000)
11. Campion, A.C., Medsker, G.J., Higgs, A.C.: Relations between work group characteristics and effectiveness: implications for designing effective work groups. Pers. Psychol. **46**, 823–850 (1993)
12. Cohen, S.G., Bailey, D.E.: What makes teams work: group effectiveness research from the shop floor to the executive suite. J. Manage. **23**, 239–290 (1997)
13. Dimas, I.D., Alves, M., Lourenço, P.R., Rebelo, T.: Instrumentos de avaliação de equipas de trabalho. Edições Sílabo, Lisboa (2016)
14. Decuyper, S., Dochy, F., Bossche, P.V.: Grasping the dynamic complexity of team learning: an integrative model for effective team learning in organizations. Educ. Res. Rev. **5**, 111–133 (2010)
15. Edmondson, A.C.: Psychological safety and learning behavior in work teams. Admin. Sci. Quart. **44**, 350–383 (1999)
16. Hackman, J.R.: The design of work teams. In: Lorsch, J. (ed.) Handbook of Organizational Behavior, pp. 315–342. Prentice-Hall, Englewood Cliffs (1987)
17. Hackman, J.R.: From causes to conditions in group research. J. Organ. Behav. **33**, 428–444 (2012)

18. Jung, D.I., Sosik, J.J.: Transformational leadership in work groups the role of empowerment, cohesiveness, and collective-efficacy on perceived group performance. Small Group Res. **33**, 313–336 (2002)
19. Kozlowski, S.W.J., Ilgen, D.R.: Enhancing the effectiveness of work groups and teams. PSPI **7**, 77–124 (2006)
20. Kozlowski, S.W.J., Bell, B.: Work groups and teams in organizations. In: Industrial and Organizational Psychology, pp. 333–375. John Wiley & Sons, Chichester (2003)
21. Langfred, C.W.: Autonomy and performance in teams: the multilevel moderating effect of task interdependence. J. Manage. **31**, 513–529 (2005)
22. Li, W., Padula, S.: Approximation methods for conceptual design of complexsystems. In: Chui, C., Neamtu, M., Schumaker, L. (eds.) Approximation Theory XI (Gatlinburg 2004). Nashboro Press, Brentwood, pp. 241–278 (2005)
23. Lourenço, P.R., Dimas, I.D., Rebelo, T.: Effective workgroups: the role of diversity and culture. Eur. J. Work Organ. Psy. **30**, 123–132 (2014)
24. Mathieu, J.E., Tannenbaum, S.I., Donsbach, J.S., Alliger, G.M.: A review and integration of team composition models moving toward a dynamic and temporal framework. J. Manage. **40**, 130–160 (2014)
25. McGrath, J.E.: Social Psychology: A Brief Introduction. Holt, Rinehart, & Winston, New York (1964)
26. Nelder, J., Mead, R.: A simplex method for function minimization. Comput. J. **7**, 308–313 (1965)
27. Nieva, V.F., Fleishman, E.A., Reick, A.: Team Dimensions: Their Identity, Their Measurement, and Their Relationships. U. S. Army, Research Institute for the Behavioral and Social Sciences, Washington, D.C. (1985)
28. Pessoa, C.: Transformational Leadership and Team Effectiveness: The Mediator Role of Resilience and Supportive Behaviors (Master thesis). Faculdade de Psicologia e de Ciências da Educação da Universidade de Coimbra (2016)
29. Rico, R., Maria, C., De, A., Tabernero, C.: Efectividad de los Equipos de Trabajo, una Revisión de la Última Década de Investigación (1999–2009). Revista de Psicologia del Trabajo y de las Organizaciones **26**, 47–71 (2010)
30. Rocha, H.: Model parameter tuning by cross validation and global optimization: application to the wing weight fitting problem. Struct. Multidiscip. Optim. **37**, 197–202 (2008)
31. Rocha, H.: On the selection of the most adequate radial basis function. Appl. Math. Model. **33**, 1573–1583 (2009)
32. Rocha, H., Dias, J.M., Ferreira, B.C., Lopes, M.C.: Selection of intensity modulated radiation therapy treatment beam directions using radial basis functions within a pattern search methods framework. J. Global Optim. **57**, 1065–1089 (2013)
33. Rocha, H., Li, W., Hahn, A.: Principal component regression for fitting wing weight data of subsonic transports. J. Aircr. **43**, 1925–1936 (2006)
34. Rocha, H., Dias, J.M., Ferreira, B.C., Lopes, M.C.: Beam angle optimization for intensity-modulated radiation therapy using a guided pattern search method. Phys. Med. Biol. **58**, 2939 (2013)
35. Rousseau, V., Aubé, C.: Team self-managing behaviors and team effectiveness: the moderating effect of task routineness. Group Organ. Manage. **35**, 751–781 (2010)
36. Savelsbergh, C.M.J.H., van der Heijden, B.I.J.M.: The development and empirical validation of a multidimensional measurement instrument for team learning behaviors. Small Group Res. **40**, 578–607 (2009)
37. Stephens, J., Heaphy, E.D., Carmeli, A., Spreitzer, G.M., Dutton, J.E.: Relationship quality and virtuousness: emotional carrying capacity as a source of individual and team resilience. J. Appl. Behav. Sci. **49**, 13–41 (2013)

38. Stone, M.: Cross-validatory choice and assessment of statistical predictions. J. R. Stat. Soc. **36**, 111–147 (1974)
39. Sutcliffe, K.M., Vogus, T.: Organizing for resilience. In: Positive Organizational Scholarship, pp. 94–121. Berrett-Koehler, San Francisco (2003)
40. van Beveren, P.Q.F.: Liderança transformacional e autonomia grupal: Adaptação de instrumentos de medida (Master thesis). Faculdade de Psicologia e de Ciências da Educação da Universidade de Coimbra (2015)
41. Vancouver, J.B., Thompson, C.M., Williams, A.A.: The changing signs in the relationships among self-efficacy, personal goals, and performance. J. Appl. Psychol. **8**, 605–620 (2001)
42. Wheelan, S.A.: Group size, group development, and group productivity. Small Group Res. **40**, 247–262 (2009)

Assessment of the Code Refactoring Dataset Regarding the Maintainability of Methods

István Kádár, Péter Hegedűs$^{(\boxtimes)}$, Rudolf Ferenc, and Tibor Gyimóthy

University of Szeged, Szeged, Hungary
{ikadar,hpeter,ferenc,gyimothy}@inf.u-szeged.hu

Abstract. Code refactoring has a solid theoretical background while being used in development practice at the same time. However, previous works found controversial results on the nature of code refactoring activities in practice. Both their application context and impact on code quality needs further examination.

Our paper encourages the investigation of code refactorings in practice by providing an excessive open dataset of source code metrics and applied refactorings through several releases of 7 open-source systems. We already demonstrated the practical value of the dataset by analyzing the quality attributes of the refactored source code classes and the values of source code metrics improved by those refactorings.

In this paper, we have gone one step deeper and explored the effect of code refactorings at the level of methods. We found that similarly to class level, lower maintainability indeed triggers more code refactorings in practice at the level of methods and these refactorings significantly decrease size, coupling and clone metrics.

Keywords: Code refactoring · Software maintainability · Empirical study · Refactoring dataset

1 Introduction

Source code refactoring is a very powerful technique to improve the internal quality of software systems. Since its introduction by Fowler [6] it become more and more popular and nowadays IT practitioners think of it as an essential part of the development processes. Despite the high acceptance of refactoring techniques by the software industry, there are some aspects that software companies should take into consideration which may affect the practical application of such techniques; for example, time constraint, cost effectiveness, or return on investment. Due to this shift of priorities between industry and research, we should also explore how developers tend to use refactoring in practice and not just focus on the theoretical concepts of code refactoring. There are evidences in the literature [21] that engineers are aware of code smells, but are not very concerned with their impact as refactoring activity is not focused on them. But as Fowler et al. suggested, code smells should be the primary technique for identifying

© Springer International Publishing Switzerland 2016
O. Gervasi et al. (Eds.): ICCSA 2016, Part IV, LNCS 9789, pp. 610–624, 2016.
DOI: 10.1007/978-3-319-42089-9_43

refactoring opportunities in the code and a lot of research effort [4,5] has been put into examining them. A similar contradictory result by Bavota et al. [2] suggests that only 7 % of the refactoring operations actually remove the code smells from the affected class.

All these seemingly negative results only indicate that although some concepts might be very effective in theory, they may not be applied in industry due to practical reasons. So to be able to elaborate new techniques and methods that better suit the practitioners' needs we should further examine how they apply refactorings. To help addressing this goal, we proposed a publicly available refactoring dataset [11] that we assembled using the *RefFinder* [12] tool for refactoring extraction and the *SourceMeter*[1] static source code analyzer tool for source code metric calculation. The dataset consists of refactoring and source code metrics for 37 releases of 7 open-source Java systems. Every refactoring is bound to the source code elements at the level of methods and classes on which the refactoring was performed. We also store exact version and line information in the dataset that supports reproducibility. Additionally to the source code metrics, the dataset includes the relative maintainability indices of source code elements, calculated by the *QualityGate*[2] tool, an implementation of the *ColumbusQM quality model* [1]. This makes it possible to directly analyze the connection between source code maintainability and code refactoring.

Our first results in utilizing the proposed dataset [11] showed that classes with poor maintainability are subject to more refactorings in practice than classes with higher technical quality. Considering metrics, number of clone instances, complexity, coupling, and size metrics have improved, although comment related metrics decreased. In this paper, we focus on a similar empirical investigation, but not at class level, but at the level of individual methods. The literature lacks such studies on the evolution of methods in systems due to refactorings, which we can examine now by using the proposed dataset.

With the help of the assembled dataset, in this paper we examine the connection between refactorings and practical maintainability of the code by investigating the following research questions:

RQ1. *Are methods with lower maintainability subject to more refactorings in practice?*

RQ2. *Which quality attributes (source code metrics) are affected the most by refactoring methods and to what extent?*

By applying statistical methods on the refactoring data contained in our dataset we found that lower maintainability indeed triggers more code refactorings in practice at the level of methods and these refactorings significantly decrease code lines, coupling, and clone metrics.

The rest of the paper is organized as follows. In Sect. 2 we summarize the empirical works in connection with code refactorings. In Sect. 3 the process of assembling the refactoring dataset and its utilization in this paper is described.

---

[1] http://www.sourcemeter.com/.

[2] http://www.quality-gate.com/.

We present the results of our empirical investigations in Sect. 4. Last, we describe the threats to the validity of our research in Sect. 5 and conclude the paper in Sect. 6.

2   Related Work

In this section we present some relevant works that investigate the relationship between practical refactoring activities and software quality similarly to us. Lot of the below mentioned papers also use the RefFinder tool [12,22] to find refactorings in software systems.

Murgia et al. [17] investigated if highly coupled classes are more likely to be targeted by refactorings than less coupled ones. Classes with high fan-out (and relatively low fan-in) metric was frequently targeted by refactorings, which indicates that developers may prefer to refactor classes with high outgoing rather than high incoming coupling.

Kataoka et al. [10] also examined the coupling metrics and showed that those are quite effective in quantifying the impact of refactoring and helped them to choose the appropriate refactoring types to apply on the source code.

In contrast to the above two works, we did not apply a particular set of metrics to assess the effect of refactorings, but rather performed statistical tests to find those metrics that change significantly upon refactorings and analyzed the way they changed. We could identify that not only coupling, but size and clone metrics also play an important role when doing code refactoring.

Measuring clones and studying how refactoring affects them is also a very popular research topic. The dataset we proposed also includes clone metrics, so code clone related refactoring examinations can also be performed. Choi et al. found [3] that merged code clone token sequences and differences in token sequence lengths vary for each refactoring pattern. They also showed that extract method and replace method with method object refactorings are the most popular choices of the developers performing clone refactoring.

An automated approach presented by Wang et al. [23] recommends clones for refactoring. The built decision tree-based classifier helps the developers to determine if a clone is worth the effort to be refactored or not. The approach achieves a precision of around 80 %, and similarly good precision is achieved in cross-project evaluation. By recommending which clones are appropriate for refactoring, the approach allows better resource allocation for refactoring itself after obtaining clone detection results.

Bavota et al. [2] investigated the relationship between code smells and refactoring activities. They mined the evolution history of 2 open-source Java projects and found that refactoring operations are mainly focused on code components for which quality metrics do not suggest there might be a need for refactoring operations. Contrary to their work, by considering maintainability instead of code smells, we found significant, but not very strong connection with refactoring activities. Bavota et al. also propose a refactoring dataset with 15,008 refactoring operations, but it contains file level data only without precise line information. Our open dataset contains method level information as well and refactoring instances are completely traceable.

The approach presented by Hoque et al. [9] investigates the refactoring activity as part of the software engineering process and not its effect on code quality. The authors found that it is not always true that there are more refactoring activities before major project release dates than after. The authors were able to confirm that software developers perform different types of refactoring operations on test code and production code, specific developers are responsible for refactorings in the project and refactoring edits are not very well tested.

In another work [7] an automatic reviewing tool was developed with the purpose of helping the code review activities by determining which changes in the change set are the results of refactorings. Correctly performed refactorings, by definition, preserve the behavior of the program so cannot introduce bugs. Thus, spending effort on reviewing refactored changes is undesirable because it is more likely that one finds bugs in non-refactoring changes.

The paper by Parsai et al. [20] proposes to adopt mutation testing as a means to verify if the behavior of the test code is preserved after refactoring. Their experiments indicate that mutation testing is suitable for identifying changes on the external behavior of a refactored test and can also be used to detect those parts of the test that was refactored improperly.

An extensive survey on the field of software refactoring is created by Mens and Tourwe [15]. Among others, refactoring activities, specific techniques and formalisms that are used for supporting these activities, important issues that need to be taken into account when building refactoring tools and the effect of refactoring on the software process were taken into consideration in this paper. One activity in the refactoring process is identifying where to apply refactorings. According to the survey, one of the most widespread approach to detect program parts that require refactoring is the identification of bad smells (especially code clones), which decrease maintainability. Our study identifies similar things in practice, because we found that methods with poor maintainability are subject to higher number of refactorings during their lifetime.

Similarly to us, Murphy-Hill et al. [18] empirically investigated how developers refactor in practice. They found that refactoring tools are rarely used: 11 % by Eclipse developers and 9 % by Mylyn developers. Unlike in this work, we do not focus on how refactorings are introduced (i.e. manually or using a tool), but on their effect on source code.

3   Approach

In order to support the further researches on source code refactorings we built a dataset of source code metrics and the applied refactorings between the releases of the investigated projects. Utilizing the data set, we investigated the effect of refactorings applied on the methods of the programs on their various metrics and quality properties.

3.1 Dataset Construction

The basic methodology of the construction of the dataset is described in our previous paper [11], here we emphasize the method-level specific details. The prepared dataset contains data of release versions of 7 open-source Java systems available on GitHub. Table 1 provides details about the projects, their names, URLs, number of analyzed releases and the covered time interval by the releases.

Table 1. The systems included in the refactoring dataset

System	Github URL	# Rel.	Time interval
antlr4	https://github.com/antlr/antlr4	5	21/01/13–22/01/15
junit	https://github.com/junit-team/junit	8	13/04/12–28/12/14
mapdb	https://github.com/jankotek/MapDB	6	01/04/13–20/06/15
mcMMO	https://github.com/mcMMO-Dev/mcMMO	5	24/06/12–29/03/14
mct	https://github.com/nasa/mct	3	30/06/12–27/09/13
oryx	https://github.com/cloudera/oryx	4	11/11/13–10/06/15
titan	https://github.com/thinkaurelius/titan	6	07/09/12–13/02/15

These projects were found ideal for our research purposes because of the adequate number of release versions and the amount of the code modifications between two adjacent releases. We selected 3 to 8 releases of each project. When selecting the releases to include in the data set, we considered the amount of code modifications between two adjacent releases of a project. As long as there is not enough code modification between two adjacent releases, the number of revealed refactorings is rather low, which can mislead statistic or machine learning algorithms. On the other hand, in order to support researches in this topic there should be a large enough number of releases which allows the investigation of the change of refactoring numbers and source code metrics in time. We found that about a half-year time interval between two releases provides sufficient amount of code modifications which proves to be appropriate for our research goals. Thus, in case of every project, we dropped those release versions that were too close to each other in time.

For every selected release version of every project, class and method level metrics and the number of refactorings grouped by refactoring types (e.g. extract method, remove parameter) are available in the dataset. The refactoring types are different in class and method level: there are 23 refactoring types on class level, and 19 on method level. For a complete list of method and class-level refactorings refer to Table 6 in the appendix. In Table 2 we provide an overview of the total number of classes, methods, and refactorings contained in the dataset.

To reveal refactorings between two adjacent release versions we used the RefFinder refactoring reconstruction tool [12]. We note that according to its authors the precision of the tool is 79 % [22]. RefFinder is implemented as an Eclipse plug-in and is able to reveal refactorings between two Eclipse projects.

Table 2. Total number of classes, methods, and refactorings in the dataset

System	# Classes	# Methods	# Refactorings
antlr4	622	5,280	248
junit	1,267	4,124	553
mapdb	850	6,180	2,973
mcMMO	505	4,767	62
mct	2,175	11,765	763
oryx	551	2,592	121
titan	2,429	14,214	3,152
Total	**8,399**	**48,922**	**7,872**

In order to avoid the manual importation of every release of every project into Eclipse and starting the analysis by hand, we extended RefFinder with batch execution support that enables the automatic analysis of all the specified releases. We also implemented an export feature which writes the found refactorings and all of their attributes into CSV files for each refactoring type which makes the later analysis of the refactorings with external tools possible.

To set up the dataset we mapped all of the refactorings to those methods that were affected by them and then counted their numbers. More specifically, if any of the attributes of a refactoring referred to a method, the refactoring was counted to that method. The reference to a method by a refactoring attribute is defined by method name, and because in Java a method name does not specify the method unambiguously, by source code position too. However, we realized that source code positions in refactoring attributes cannot be always determined precisely by RefFinder and the abstract syntax tree of Eclipse, thus there can be roughness in the mapping of refactorings to methods. In the dataset, for every release version, the accounted refactoring numbers indicate how many refactorings of various types were performed that affected the considered method between the current release and the previous one.

Besides code refactoring numbers, the dataset contains more than 50 types of static source code metrics for every method (and class) of the considered projects which were calculated using the SourceMeter static code analysis tool. Beyond these metrics we added the so-called *relative maintainability index* (RMI) which was measured by QualityGate SourceAudit for each method (and class) of the systems. RMI, similarly to the well-known maintainability index [19], reflects the maintainability of a code element, but it is calculated using dynamic thresholds from a benchmark database, not by a fixed formula. Thus, RMI expresses the maintainability of a code element compared to the maintainability of other elements in the system. The technical details of the RMI can be found in our earlier work [8].

The assembled dataset is published on the *tera-PROMISE* repository [16]: http://openscience.us/repo/refactoring/refact.html.

3.2   Data Analysis Methodology

To answer our research questions we utilized the constructed dataset in the following way. For RQ1, we did a correlation analysis between the RMI values of methods and the number of refactorings affecting these methods. In more detail, we took the maintainability indices of methods of revision x_i and the refactoring numbers of revision x_{i+1}. This way we investigated whether poor quality methods are targets of more refactoring operations or not. We performed Spearman rank correlation analysis because we cannot assume anything about the distribution of the maintainability indices nor the number of refactorings.

To answer RQ2, first we calculated the differences of the static metric values of two consecutive releases. Negative values in differences are indicating improvement, because lower metric values (e.g. lower complexity or coupling) are better in most of the cases. We performed a Mann-Whitney U test to determine whether there is a significant difference among the metric decreases in the refactored and non-refactored methods, which indicates which are those metrics that are changed significantly upon refactoring. To investigate the volume of the changes in metric values, we calculated the Cliff's delta (δ) effect size measure as well.

4   Results

In this section we summarize the assessment results on the connection between refactoring activity and maintainability of methods. First, we describe the results of the analysis on the maintainability of refactored methods to answer RQ1. Afterwards, we present the findings on the effect of refactorings on method-level source code metrics to answer RQ2.

4.1   The Maintainability of Refactored Methods

To answer RQ1, we performed a correlation analysis between the number of refactorings affecting the methods of the subject systems and their maintainability indices in the previous release (as described in Sect. 3). Figure 1 depicts the Spearman correlation coefficients between the RMI values in release x_i and the number of refactorings affecting the corresponding methods in release x_{i+1}. For the sake of easy comparison with our previous results on classes [11] we included the class-level maintainability correlations as well.

As can be seen, all the values are negative. Although the coefficients are not particularly high, they are consistently negative and significant at the level of 0.05. The negative values simply mean an inverse proportionality, namely that the worse the maintainability of a method or class is (the lower its RMI value) the more refactorings touch it (the higher the number of refactorings affecting it). There are less correlation coefficients than releases for some systems because we were unable to calculate them when RefFinder found no refactorings between two releases, which happened a couple of times. Table 3 summarizes the mean correlation coefficients both for method and class-level, their deviation

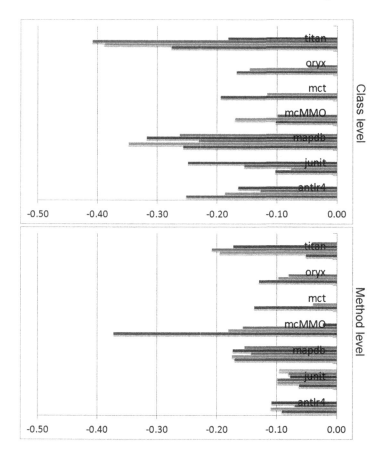

Fig. 1. Correlation of maintainability and refactorings in classes and methods

Table 3. Average Spearman correlation coefficients between RMI and number of refactorings at method and class level

System	Method level			Class level		
	Mean corr.	Deviation	Intervals	Mean corr.	Deviation	Intervals
antlr4	−0.096	0.018	4	−0.183	0.052	4
junit	−0.086	0.014	6	−0.146	0.076	4
mapdb	−0.163	0.014	5	−0.283	0.048	5
mcMMO	−0.183	0.144	4	−0.124	0.040	3
mct	−0.089	0.069	2	−0.156	0.054	2
oryx	−0.103	0.025	3	−0.121	0.063	3
titan	−0.134	0.081	5	−0.314	0.106	4

and the number of evaluated intervals between releases. It can be noticed that the correlation coefficients and the deviations are somewhat larger in the case of classes, but the differences are negligible.

Answer to RQ1: Based on the findings on our dataset we can conclude that methods with poor maintainability are subject to higher number of refactorings during their lifetime compared to those with better maintainability.

4.2 The Effect of Refactorings on Method-Level Source Code Metrics

We found that refactorings affect poorly maintainable code more (i.e. methods), so the question arises whether applying refactorings really improves the internal quality of the code? Furthermore, what are the method level source code metrics that show the highest improvement (i.e. decrease significantly) upon refactoring?

According to the process described in Sect. 3, we first calculated the metric value differences for every method between the adjacent releases. Then, we grouped these metric difference values into two groups: in the first group we put the metric differences of methods affected by at least one refactoring, and in the second group the metric differences of non-refactored methods. Finally, we analyzed which method level metrics show significant differences between the values of the two groups with the help of the Mann-Whitney U test [14].

Table 4. The results of the Mann-Whitney U Test (p-values) for method-level metrics

System name	CC	LLOC	NOS	NOI
antlr4	**0.049**	**0.000**	**0.002**	**0.001**
junit	0.058	0.923	0.667	0.403
mapdb	**0.010**	**0.003**	0.965	**0.002**
mcMMO	0.815	0.824	0.516	0.251
mct	0.703	0.924	0.547	0.660
oryx	0.654	0.555	0.306	1.000
titan	0.601	**0.016**	**0.003**	**0.000**

Out of 50+ source code metrics, the ones listed in Table 4 had the lowest p-values, meaning that the differences in the metric value changes for refactored and non-refactored methods are the most significant for these metrics. We observed that the Number of Outgoing Invocations (NOI), which can be considered as a coupling metric indeed decreases significantly upon refactoring in accordance with the previous findings of other studies [10,17].

But besides NOI, we found a significant decrease in size metrics as well, namely in the case of Logical Lines of Code (LLOC) and Number of Statements (NOS). These can be explained by the fact that typical refactorings, like extract method and pull up method, often have a side effect of reducing the amount of source code. This phenomena is clearly observable on these pure size related metrics.

While this finding is not really surprising, the fact that McCabe's cyclomatic complexity [13] did not show a significant correlation with the number of refactorings applied on methods is just the opposite of what we were expecting. Our perception was that using better code structures will lead to less complex code, but we could not confirm this hypothesis. It is an even more interesting finding in the light of our previous results [11] on the effect of refactorings on the Weighted Method Complexity (WMC) metric of classes, which shows a significant reduction upon refactorings. However, this is not a contradiction. Consider the *Extract method* refactoring for example. In this case duplicated methods in the child classes are extracted and put into their parent class, leading to the removal of the method from several classes and inserting it to their parent. On one hand, this yields to reduction in the average WMC metric as the complexity of child classes decrease, while only the complexity of their parent class increases. On the other hand, at method level the average McCabe's complexity values do not change. So the above results might indicate that refactoring operations tend to decrease complexity at class-level, but not really at the level of methods.

It is interesting that the Clone Coverage (CC) metric also decreased, thus refactoring activity seems to remove copy-paste code parts in practice. This phenomena is similar to the code size reduction, e.g. by extracting common code snippets into a method reduces the copy-pasted code parts, too.

Table 5. Cliff δ effect size measures for method-level metrics

System name	CC	LLOC	NOS	NOI
antlr4	**0.70**	**0.63**	**0.48**	**0.71**
junit	−0.68	**0.01**	−0.08	**0.14**
mapdb	−0.34	**0.27**	0.00	**0.28**
mcMMO	**0.10**	**0.05**	**0.17**	**0.27**
mct	−0.15	−0.03	−0.18	−0.15
oryx	−0.18	−0.14	**0.31**	−0.02
titan	−0.09	**0.16**	**0.21**	**0.33**
Average	−0.09	**0.14**	**0.13**	**0.22**

To quantify the magnitude of the differences between the metric value decreases of the refactored and non-refactored methods, we calculated the Cliff's delta (δ) effect size measure. The detailed results are presented in Table 5. Cliff's δ measures how often the values in one distribution are larger than the values in a second distribution. It ranges from −1 to 1 and is linearly related to the Mann-Whitney U statistic, however it captures the direction of the difference in its sign as well. Simply speaking, if Cliff's δ is a positive number, the metric value differences (thus the metric value decreases) are higher in the refactored methods, while negative value means that the metric value differences are higher in the non-refactored methods. The closer the δ is to $|1|$, the more values are larger in one group than the values in the other group. Generally, the Cliff's δ values are quite hectic; however, the average δ values are positive for every metric except for CC. While in case of LLOC, NOS and NOI the majority of values are positive, only two projects have positive δ values for CC. This might suggest that cloned code is decreased by other targeted changes that are not refactorings, while refactorings often have a side effect to remove code clones as well (e.g. extract method). However, this phenomenon needs further investigation.

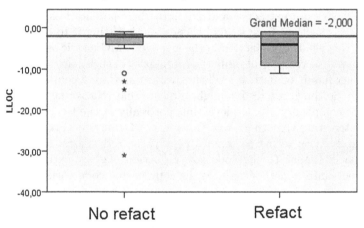

Fig. 2. Boxplot of the LLOC metric decreases in the refactored and non-refactored methods

To have a better overview of the above explained phenomena, we visualized the average size metric differences for the refactored and non-refactored methods in Fig. 2. This boxplot clearly shows that the maximum, minimum, and average numbers of code line reduction are far smaller in case of methods that are not refactored than in the case of refactored methods. While the median of LLOC decrease is 2 in case of non-refactored methods, it is two times larger (around 4) for refactored methods. Based on these findings, we can now conclude RQ2.

Answer to RQ2: We found that size (LLOC, NOS), coupling (NOI), and clone (CC) related metrics decrease the most in refactored methods. Regarding the volumes of the differences, we can say that for these metrics the average Cliff's δ values are mostly positive suggesting a small to medium effect size on the metric decreases in the refactored methods compared to the non-refactored ones.

5   Threats to Validity, Limitations

In this section, we summarize the limitations and threats to validity of our study.

First of all, we note that RefFinder, the tool we used to mine refactorings from the selected projects, is not perfect. According to its authors the precision of the tool is 79 % [22], which means there might be false positive refactorings included in our dataset. Moreover, we have no data about recall at all; however, it is fairly possible that RefFinder does not find all of the refactorings no matter whether they were committed intentionally by the developers or not. To mitigate this threat we already started to manually validate our entire dataset and eliminate false positive instances.

Another threat occurs during the construction of the dataset when the found refactoring instances are mapped to the methods that they affect. As we described in Sect. 3.1, if any of the attributes of a refactoring matches with the name of a method and its source code position, the refactoring will be mapped to that method. Nevertheless, we noticed that the source code positions in refactoring attributes determined by RefFinder using Eclipse AST are inaccurate sometimes, which implies that in case of method overloading, where name does not necessarily identify the method, there might be roughness in the mapping and therefore in the final dataset. However, the number of such mappings is very low.

The key attribute in the dataset is the fully qualified name of the method with parameter descriptions. If a method is renamed between two consecutive releases we do not track it and its metrics, and handle it as a new method in the next release.

Finally, another threat to our results is that we investigated only seven Java systems which may not represents correctly the general characteristic of all of the software systems considering refactoring activities in practice. Therefore, we are plan to continuously extend the number of systems in our dataset.

6 Conclusions and Future Work

In this paper, we used our previously proposed public dataset, which is indented to assist the research of refactoring activities in practice, to investigate the relationship between maintainability and refactoring activities, and we also assessed how refactorings affect different source code metrics at the level of individual methods. The dataset contains fine-grained refactoring information and more than 50 types of source code metrics for 37 releases of 7 open-source systems at class and method levels.

We found that methods with poor maintainability are subject to more refactorings in practice than methods with higher technical quality. Considering concrete metrics, the clone coverage, size metrics and number of outgoing invocations decreased the most intensively in the methods subjected to frequent refactorings. This might indicate that doing code refactoring in practice indeed mitigates unwanted code characteristics such as clones, size, or coupling, and result in more maintainable software systems.

As a future work we plan to manually validate the whole dataset to get more meaningful results. We also plan to reveal more complex phenomena in connection with practical refactorings, especially the relationship between bugs and refactoring activities.

Acknowledgment. This work was partially supported by the European Union project "REPARA – Reengineering and Enabling Performance And poweR of Applications", project number: 609666.

Appendix

Table 6. The type of refactorings extracted by RefFinder at class and method level

Refactoring type	Class level	Method level
Add parameter	✓	✓
Consolidate conditional expression	✓	✓
Consolidate duplicate conditional fragments	✓	✓
Extract method	✓	✓
Inline temporary variable	✓	✓
Introduce assertion	✓	✓
Introduce explaining variable	✓	✓
Remove assignment to parameters	✓	✓
Remove parameter	✓	✓
Rename method	✓	✓
Replace magic number with constant	✓	✓
Replace method with method object	✓	✓
Inline method	✓	✓
Introduce null object	✓	✓
Remove control flag	✓	✓
Replace exception with test	✓	✓
Replace nested condition with guard clauses	✓	✓
Hide method	✓	✓
Replace temporary variable with query	✓	✓
Move field	✓	
Extract superclass	✓	
Extract interface	✓	
Introduce local extension	✓	

References

1. Bakota, T., Hegedűs, P., Körtvélyesi, P., Ferenc, R., Gyimóthy, T.: A probabilistic software quality model. In: Proceedings of the 27th IEEE International Conference on Software Maintenance (ICSM), pp. 243–252, September 2011
2. Bavota, G., De Lucia, A., Di Penta, M., Oliveto, R., Palomba, F.: An experimental investigation on the innate relationship between quality and refactoring. J. Syst. Softw. **107**, 1–14 (2015)

3. Choi, E., Yoshida, N., Inoue, K.: An investigation into the characteristics of merged code clones during software evolution. IEICE Trans. Inf. Syst. **97**(5), 1244–1253 (2014)
4. van Emden, E., Moonen, L.: Java quality assurance by detecting code smells. In: Proceedings of the 9th Working Conference on Reverse Engineering, pp. 97–106 (2002)
5. Fontana, F.A., Spinelli, S.: Impact of refactoring on quality code evaluation. In: Proceedings of the 4th Workshop on Refactoring Tools, WRT 2011, pp. 37–40. ACM, New York (2011). http://doi.acm.org/10.1145/1984732.1984741
6. Fowler, M.: Refactoring: Improving the Design of Existing Code. Addison-Wesley Longman Publishing Co., Inc., Boston (1999)
7. Ge, X., Sarkar, S., Murphy-Hill, E.: Towards refactoring-aware code review. In: Proceedings of the 7th International Workshop on Cooperative and Human Aspects of Software Engineering, CHASE 2014, pp. 99–102. ACM, New York (2014)
8. Hegedűs, P., Bakota, T., Ladányi, G., Faragó, C., Ferenc, R.: A drill-down approach for measuring maintainability at source code element level. Electron. Commun. EASST **60**, 20–29 (2013). http://journal.ub.tu-berlin.de/eceasst/article/download/852/846
9. Hoque, M.I., Ranga, V.N., Pedditi, A.R., Srinath, R., Rana, M.A.A., Islam, M.E., Somani, A.: An empirical study on refactoring activity. ACM Computing Research Repository abs/1412.6359 (2014)
10. Kataoka, Y., Imai, T., Andou, H., Fukaya, T.: A quantitative evaluation of maintainability enhancement by refactoring. In: Proceedings of the International Conference on Software Maintenance, pp. 576–585 (2002)
11. Kádár, I., Hegedűs, P., Ferenc, R., Gyimóthy, T.: A code refactoring dataset and its assessment regarding software maintainability. In: Proceedings of the 23rd IEEE International Conference on Software Analysis, Evolution, and Reengineering. IEEE Computer Society (2016, to appear)
12. Kim, M., Gee, M., Loh, A., Rachatasumrit, N.: Ref-Finder: a refactoring reconstruction tool based on logic query templates. In: Proceedings of the 18th ACM SIGSOFT International Symposium on Foundations of Software Engineering (FSE 2010), pp. 371–372 (2010)
13. McCabe, T.J.: A complexity measure. IEEE Trans. Softw. Eng. **2**, 308–320 (1976)
14. McKnight, P.E., Najab, J.: Mann-Whitney U Test. Corsini Encyclopedia of Psychology. Wiley, New York (2010)
15. Mens, T., Tourwe, T.: A survey of software refactoring. IEEE Trans. Softw. Eng. **30**(2), 126–139 (2004)
16. Menzies, T., Krishna, R., Pryor, D.: The Promise Repository of Empirical Software Engineering Data (2015). http://openscience.us/repo
17. Murgia, A., Tonelli, R., Marchesi, M., Concas, G., Counsell, S., McFall, J., Swift, S.: Refactoring and its relationship with fan-in and fan-out: an empirical study. In: Proceedings of the 16th European Conference on Software Maintenance and Reengineering (CSMR), pp. 63–72, March 2012
18. Murphy-Hill, E., Parnin, C., Black, A.P.: How we refactor, and how we know it. IEEE Trans. Softw. Eng. **38**(1), 5–18 (2012)
19. Oman, P., Hagemeister, J.: Metrics for assessing a software system's maintainability. In: Proceedings of the International Conference on Software Maintenance, pp. 337–344. IEEE Computer Society Press (1992)
20. Parsai, A., Murgia, A., Soetens, Q.D., Demeyer, S.: Mutation testing as a safety net for test code refactoring. CoRR abs/1506.07330 (2015)

21. Peters, R., Zaidman, A.: Evaluating the lifespan of code smells using software repository mining. In: Proceedings of the 16th European Conference on Software Maintenance and Reengineering (CSMR), pp. 411–416, March 2012
22. Prete, K., Rachatasumrit, N., Sudan, N., Kim, M.: Template-based reconstruction of complex refactorings. In: IEEE International Conference on Software Maintenance (ICSM), pp. 1–10, September 2010
23. Wang, W., Godfrey, M.W.: Recommending clones for refactoring using design, context, and history. In: 2014 IEEE International Conference on Software Maintenance and Evolution (ICSME), pp. 331–340. IEEE (2014)

A Public Bug Database of GitHub Projects and Its Application in Bug Prediction

Zoltán Tóth[✉], Péter Gyimesi, and Rudolf Ferenc

Department of Software Engineering, University of Szeged, Szeged, Hungary
{zizo,pgyimesi,ferenc}@inf.u-szeged.hu

Abstract. Detecting defects in software systems is an evergreen topic, since there is no real world software without bugs. Many different bug locating algorithms have been presented recently that can help to detect hidden and newly occurred bugs in software. Papers trying to predict the faulty source code elements or code segments in the system always use experience from the past. In most of the cases these studies construct a database for their own purposes and do not make the gathered data publicly available. Public datasets are rare; however, a well constructed dataset could serve as a benchmark test input. Furthermore, open-source software development is rapidly increasing that also gives an opportunity to work with public data.

In this study we selected 15 Java projects from GitHub to construct a public bug database from. We matched the already known and fixed bugs with the corresponding source code elements (classes and files) and calculated a wide set of product metrics on these elements. After creating the desired bug database, we investigated whether the built database is usable for bug prediction. We used 13 machine learning algorithms to address this research question and finally we achieved F-measure values between 0.7 and 0.8. Beside the F-measure values we calculated the bug coverage ratio on every project for every machine learning algorithm. We obtained very high and promising bug coverage values (up to 100 %).

Keywords: Bug prediction · Bug database

1 Introduction

Software systems are likely to fail occasionally that is obviously unwanted both for the end users and for the software developers. Keeping the software quality at high-level is more important than ever, since customers define the reputation of the used subject system. Open-source software development paved its way, and has become a corner stone in the domain of evaluating research ideas and techniques dealing with computer science [19]. These publicly available systems gather a huge amount of historical data stored for example in version control systems or bug tracking systems. Researches have been using the opportunity given by these public information sets for a long time to prove the power of their approaches [1,14,24]. In spite of this fact, only a few publicly available

© Springer International Publishing Switzerland 2016
O. Gervasi et al. (Eds.): ICCSA 2016, Part IV, LNCS 9789, pp. 625–638, 2016.
DOI: 10.1007/978-3-319-42089-9_44

bug databases are presented to take role as a basis for further investigations. Many authors do not make the corpus used in their studies public, thus the experiments are not repeatable [12].

Our study tries to endorse the use of public databases for addressing different research questions such as bug prediction related ones by showing the power of our automatically generated bug database in bug prediction domain. We have developed a toolchain that automatically gathers different information about publicly available projects to build a bug database. We selected 15 Java projects from different domains to ensure the generality of the constructed database. The characteristics of these open-source projects were extracted from **GitHub**[1] that hosts millions of projects and using a static source code analyzer tool called **SourceMeter**.[2] We analyzed 15 projects with more than 3.5 million lines of code, and more than 114 thousand of commits in total. From the analyzed commit set we detected almost 6 thousand commits that referenced at least one bug (inducing a bug fix intention) according to the SZZ algorithm [22]. We used release versions of the systems and created bug databases for six-months-long intervals approximately.

To show the usefulness of the contained information we experimented with 13 machine learning algorithms and achieved quiet good results. For class level, the best algorithms resulted in higher than 0.7 F-measure values. For file level we achieved similar values, however a little lower ones. Almost full bug coverage can be reached by using these models by tagging only 30 % of the source code elements as buggy. We defined two research questions, which are the following:

RQ 1: Is the constructed database usable for bug prediction? Which algorithms or algorithm families perform the best in bug prediction?
RQ 2: Which machine learning algorithms or algorithm families perform the best in bug coverage?

The remainder of the paper is organized as follows. Section 2 enumerates the most important research papers dealing with public and private bug databases, and bug prediction techniques. In Sect. 3, we propose our approach and show how the database is constructed, and what kind of data entries are stored in the dataset. Section 5 presents the power of the constructed database by evaluating different results of the applied machine learning algorithms. Finally, we summarize and conclude the paper.

2   Related Work

Publishing databases as public resources for the scientific community is not a new idea [13,24]. Many papers have dealt with bug databases and used some kind of bug prediction approaches as demonstrations [9]. Notwithstanding the

---

[1] https://github.com.
[2] https://www.sourcemeter.com.

numerous studies dealing with bug prediction, the number of publicly available bug databases are incredibly low and neglected. Researchers often use a database created for their own purposes but these datasets are not published for the community.

Many research studies deal with bug prediction and they use a database created for specific purposes. We tried to create a database that is publicly available and general enough to test different bug prediction methods [3,15,17,20]. We gathered a wide range of software product metrics to characterize the known bugs, amongst others the classic object-oriented metrics [4,23].

In our research work, we only found four publicly available bug databases. These four datasets mainly operate with classic C&K [4] metrics and contain accumulated information about bugs at a pre-release or post-release time. Granularity is usually file level or class level that means the database contains bug characteristics for files or classes, consequently bug prediction is limited to this granularity. None of these databases consist data obtained from GitHub, they mostly gathered them from Bugzilla and Jira. We conducted an experiment using GitHub as the source of information (both for version control and for bug tracking).

Out of these databases *Terapromise* is the most up to date and has also a coding rule violation [18] database. Based on the capability of the tool we used for static source code analysis, we gathered C&K metrics, rule violations, and software code clone related metrics such as number of clone instances located in the given source code elements.

The *Bug prediction dataset* [6] contains data extracted from 5 Java projects by using inFusion and Moose to calculate the classic C&K metrics for class level. The source of information was mainly CVS, SVN, Bugzilla and Jira from which the number of pre- and post-release defects were calculated.

The *Zimmermann Eclipse* [24] database is still publicly available, however the last extension/modification was applied on March 25, 2010. Zimmermann et al. gathered complexity metrics and metrics describing the structure of the built AST for file level to detect pre- and post-release defects. The dataset is created by using the public information stored in Bugzilla.

Bugcatchers [10] operates with bad smells (solely), and found that coding rule violations have a small but significant effect on the occurrence of faults at file level. Bugcatchers used Bugzilla and Jira as the sources of information.

Many other papers used bug databases to extract some additional data, but these databases have never been published. Such databases amongst others are the following: IBugs [5], Mozilla [9], and Eclipse [2].

In this paper, we present an approach that uses GitHub, and collects a wide set of metrics for approximately six-months-long time intervals. This database is suitable for bug prediction purposes, and can be easily extended to involve more open-source projects.

3   Approach

In this section we briefly summarize our previous work [8] that also dealt with bug databases, however in the approach some major differences are present. We will highlight the hot spots where the two approaches differ from each other. In our previous work, first we downloaded the data from GitHub, then we processed the raw data to obtain statistical measurements on the projects. At this point we selected the relevant software versions to be analyzed by the static source code analyzer. After the source code analysis, we performed the database building that results in a dataset that stores entries in pairs such that a source code element that used to have at least one bug in it is present with the source code metrics calculated before the bug(s) was/were fixed (buggy state) and with the state when the bugs were already fixed. In this process, for each issue, we determined the following important source code versions:

- the last version that contains the untouched bug (the version before the first commit that references the issue),
- the first version that contains the fixed source code (the version after the last commit that references the issue),
- the versions that also contain the bug (versions after the issue reported and before the first fix was made).

We detected the references between the commits and the bugs by using the SZZ algorithm [22]. GitHub also provides the linkage between issues and commits. These links are determined from the message of the commits. With the use of these links, we accumulated the bug related source code elements (faulty classes) on issue level. A source code element is bug related, if it was modified in a commit that references the issue. Then we marked the buggy source code elements in the versions listed above. The database was constructed from the last version that contains the untouched bug and from the first version that contains the fixed source code as mentioned above.

In this current study, we followed another approach that rather follows the usual methods described in Sect. 2 [6,13,24]. Let us consider a few bugs that were later fixed (consider Fig. 1). There are 3 versions of the system: A, B, and C, and we have 3 bugs in the software. We fixed bug A before version B that means bug A is present in the system in version A. The same is true for bug B, however bug B was finally fixed after version B, thus bug B appears also in the output of version B. At this point bug A is already fixed that causes it not to appear in version B. Bug C is similar to bug A.

Since the faulty elements are determined from the viewpoint of reported issues, and the issues are independent from the selected release versions, this means that the bug information is scattered in time. If a bug was reported after a specific release version and fixed before the subsequent selected version then the bug does not appear in either of the databases. To solve this issue, a common solution is to aggregate the bug information to the selected release versions. For every issue, we determined the preceding release version and marked the buggy source code elements.

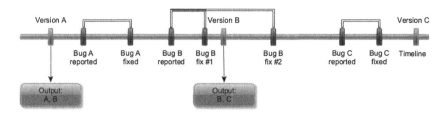

Fig. 1. The relationship between the bugs and release versions (Color figure online)

In addition to the previous list, we determined

– the versions that partially contain the bug (versions after the first fix and before the last fix).

For the construction of the database we used the so-called traditional approach that means we collected release versions with approximately six-months-long time intervals for every project. We used six-month-long intervals since enough bugs and versions are present for such long time interval. Based on the age of a project, the number of selected release versions could differ for each project. We selected the release versions manually from the list of releases located on the projects GitHub pages. It is a common practice that projects use the release tag on a newly branched (from master) version of the source code. Since we use only the master branch as the main source of information, we had to perform a mapping when the hash id of the selected release is not representing a commit located in the master. Developers usually branch from master and then tag the branched version as release version, so our mapping algorithm detects when (time stamp) the release tag was applied on a version and searches for the last commit in the master branch that was made right before this time stamp.

We created a database for each of the release versions. Since bug tracking was not always used from the beginning of the projects, we could not assign any bug information to some of these earlier release versions. Also, the changing developer activity could result in lack of bug reports and consequently bug fixing commits are rare. All of these factors play roles in that the created databases vary in the number of bugs.

Similarly as in our earlier study, we computed some process metrics on file level from the data gathered from GitHub. This extra information is based on the actions performed on files by the developers. This means that if a file was not modified since it was uploaded with the initial commit then these extra metrics are zero. To avoid the misleading rows, we removed these files from the final database.

4 Chosen Projects and the Created Databases

To select projects for the database construction, we examined many projects on GitHub. The main aspects were similar as in our previous paper. We chose 15

Table 1. The selected projects

Project	Domain	kLOC	NC	NBR	Class	File	DB files
Android Universal I. L.[a]	Android library	13	996	89	639	478	12
ANTLR v4[b]	Language processing	85	3, 276	111	2, 353	2, 029	10
Elasticsearch[c]	Search engine	677	13, 778	2, 108	54, 562	23, 252	24
jUnit[d]	Test framework	36	2, 053	74	5, 432	2, 266	16
MapDB[e]	Database engine	83	1, 345	175	2, 740	962	12
mcMMO[f]	Game	42	4, 552	657	1, 393	1, 348	12
Mission Control T.[g]	Monitoring platform	204	975	37	6, 091	1, 904	6
Neo4j[h]	Database engine	648	32, 883	439	32, 156	18, 306	18
Netty[i]	Networking framework	282	6, 780	1, 039	11, 528	8, 349	18
OrientDB[j]	Database engine	380	10, 197	174	11, 643	9, 475	12
Oryx[k]	Machine learning	47	363	36	2, 157	1, 400	8
Titan[l]	Database engine	119	3, 830	121	5, 312	3, 713	12
Eclipse p. for Ceylon[m]	IDE	165	6, 847	666	4, 512	2, 129	1
Hazelcast[n]	Computing platform	515	16, 854	2, 354	25, 130	14, 791	18
Broadleaf Commerce[o]	E-commerce framework	283	9, 292	652	17, 433	14, 703	22
Total		3, 579	114, 021	8, 732	183, 078	105, 105	210

[a] https://github.com/nostra13/Android-Universal-Image-Loader
[b] https://github.com/antlr/antlr4
[c] https://github.com/elasticsearch/elasticsearch
[d] https://github.com/junit-team/junit
[e] https://github.com/jankotek/MapDB
[f] https://github.com/mcMMO-Dev/mcMMO
[g] https://github.com/nasa/mct
[h] https://github.com/neo4j/neo4j
[i] https://github.com/netty/netty
[j] https://github.com/orientechnologies/orientdb
[k] https://github.com/cloudera/oryx
[l] https://github.com/thinkaurelius/titan
[m] https://github.com/ceylon/ceylon-ide-eclipse
[n] https://github.com/hazelcast/hazelcast
[o] https://github.com/BroadleafCommerce/BroadleafCommerce

projects as data source. The selected software systems are listed in Table 1, together with some statistics. The first column contains the name of the projects with links to the GitHub repository in the footnote. The next column is the main domain of these systems. We can see that there is a large variance between the projects regarding the domain that strengthens the generality of the constructed database. The next three columns is the thousand Lines of Code, the Number of Commits and the Number of Bug Reports, respectively, on the master branch measured in May of 2015.

We constructed separate databases for class and file level. These databases are in CSV form (comma separated values). The first row in the CSV files contains header information such as unique identifier, source code position, source name, metric names, rule violation groups and number of bugs. The data in the rest of the lines follows this order. Each line represents a source code element (class, file). In total we selected 105 release versions for the 15 projects and created 210 database files for six-months-long intervals. The last three columns in Table 1 present the number of entries constructed for each project.

Figure 2 depicts the above mentioned entry numbers on a bar chart. Some projects have an outstanding number of class and file entries, however we are going to present results on every project one-by-one by evaluating the best machine learning algorithms for different release versions. Out of the total 183,078 class level entries, Elasticsearch has 54,562 in 12 databases that is not surprising if we consider the size of the project (677 kLOC). However, Neo4J has the most commits (twice as much as the second project which is Hazelcast), it has considerably less bug reports that results in a smaller database. In general, the bigger the project and the more bug reports a project has the bigger database it results in.

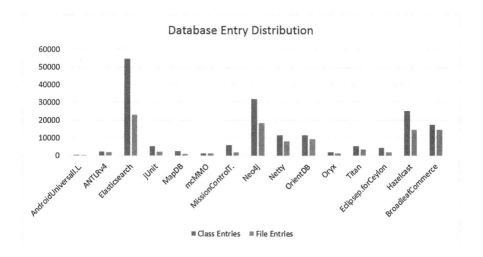

Fig. 2. Number of entries distribution (Color figure online)

5   Evaluation

In this section we give exhaustive answers for the research questions by presenting our final results and achievements we made.

RQ 1: Is the constructed database usable for bug prediction? Which algorithms or algorithm families perform the best in bug prediction?

We evaluated our database by applying machine learning algorithms for all of the constructed data sets. The bug information in our database is present as number of bugs. To apply machine learning (classification), first we grouped the source code elements into two classes based on the occurrence of bugs in them. Instances with non-zero bug cardinality form a class (defective elements) and instances with zero bug number constitute the second separate class (non-defective elements).

If we look at the ratio between the number of defective and the number of non-defective elements, we may notice that there are way more non-defective elements in a software version than defective. Considering that we are planning to apply machine learning algorithms, it could distort the results, because the non-buggy instances get more emphasis. To deal with this issue, we applied random under sampling method to equalize the learning corpus [11,21]. We randomly selected elements from the non-buggy class to match the size of the buggy category. This way we got a training set with the same number of positive and negative instances. We repeated this kind of learning 10 times and calculated an average. For the training, we used 10-fold cross validation and compared the results based on precision, recall, and F-measure metrics where these metrics are defined in the following way:

$$precision = \frac{TP}{TP + FP}$$

$$recall = \frac{TP}{TP + FN}$$

$$F - measure = 2 \cdot \frac{precision \cdot recall}{precision + recall}$$

where TP (True Positive) is the number of classes/files that were predicted as faulty and observed as faulty, FP (False Positive) is the number of classes/files that were predicted as faulty but observed as not faulty, FN (False Negative) is the number of classes/files that were predicted as non-faulty but observed as faulty. We carried out the training with the popular machine learning library called Weka.[3] It contains algorithms from different categories, for instance Bayesian methods, support vector machines, and decision trees.

We used the following algorithms:

– NaiveBayes
– NaiveBayesMultinomial
– Logistic
– SGD
– SimpleLogistic
– SMO
– VotedPerceptron [7]
– DecisionTable
– OneR
– PART
– J48 (C4.5) [16]
– RandomForest
– RandomTree

---

[3] http://www.cs.waikato.ac.nz/ml/weka/.

We analyzed software versions with six-month intervals from 15 projects. In total, we selected 105 release versions. 80 of these versions contain bug information due to the reasons mentioned in Sect. 3. 5 of the 80 versions contain too few buggy elements to apply machine learning. We ended up with 75 suitable versions for the training on class level. On file level, we got only 72, because in one buggy file there could be more than one buggy class, thus the size of the training set for a specific version could differ based on the granularity of the database.

Class Level. First we investigated whether the class level databases are suitable for bug prediction purposes. Presenting the results for 15 projects (105 release versions) using all 13 machine learning algorithms would end up in a giant table that human eyes could not process, or at least can not focus on the most relevant parts. Consequently, we only present the best algorithms here to make it more easy to overview and find the best ones. Furthermore, for each project we selected the interval which has the most database entries to ensure the suitable size of the training corpus. Then, we used 10-fold cross-validation for that interval as described earlier. We chose the algorithms simply by calculating the averages on F-measure values and considered the best 5 algorithms. Table 2 presents the F-measure values for these 5 algorithms at class level.

As one can observe, values can be highly different by projects which can be caused by various reasons (size of the constructed dataset). For instance, let us consider the Android Universal Image Loader and Broadleaf Commerce projects.

Table 2. F-measures at class level

Project	SGD	SimpleLogistic	SMO	PART	RandomForest
Android Universal I. L.	0.6258	0.5794	0.5435	0.6188	0.7474
ANTLR v4	0.7586	0.7234	0.7379	0.7104	0.8066
Elasticsearch	0.7197	0.7304	0.7070	0.7171	0.7755
jUnit	0.7506	0.7649	0.7560	0.7262	0.7939
MapDB	0.7352	0.7667	0.7332	0.7421	0.7773
mcMMO	0.7192	0.6987	0.7203	0.6958	0.7418
Mission Control T.	0.7819	0.7355	0.7863	0.6862	0.8161
Neo4j	0.6911	0.7156	0.6835	0.6731	0.6767
Netty	0.7295	0.7437	0.7066	0.7521	0.7937
OrientDB	0.7485	0.7359	0.7310	0.7194	0.7823
Oryx	0.8012	0.7842	0.8109	0.7754	0.8059
Titan	0.7540	0.7558	0.7632	0.7301	0.7830
Eclipse p. for Ceylon	0.6891	0.7078	0.6876	0.7283	0.7503
Hazelcast	0.7128	0.7189	0.6965	0.7267	0.7659
Broadleaf Commerce	0.8019	0.8084	0.8081	0.7813	0.8210

The Android project is the smallest one in size, Broadleaf is one of the middle-sized projects. Android has 639 class level entries in total (6 DB files), however Broadleaf has 17,433 entries (11 DB files) that is more suitable for being a training corpus. Nevertheless, if we take a closer look on the results we can see that the best F-measure values occurred also in small projects such as in Oryx or MCT, ergo we cannot generalize this conjecture to be true; however, further investigations should to be done to prove that. Tree-, function- and rule-based models performed the best in this scenario. F-measure values are up to 0.8210 that is a promising result. Before answering the first research question let us investigate the results at file level as well.

File Level. File level is different in some aspects from class level. For example, a completely distinct set of metrics (and also fewer) are calculated for file level entries. The best file level machine learning results are shown in Table 3. At first sight one can see that the results are in a wider range than in the case of class level. RandomForest has the highest F-measure values in case of files too. Furthermore, another tree based algorithm (J48) also performs nicely in this case. Two function-based (Logistic and SimpleLogistic) and one rule-based algorithm are in the top. Considering these results we can answer our research question.

Table 3. F-measures at file level

Project	Logistic	SimpleLogistic	PART	J48	RandomForest
Android Universal I. L.	0.5983	0.6230	0.6632	0.6215	0.6214
ANTLR v4	0.7638	0.7941	0.7443	0.8267	0.7645
Elasticsearch	0.6303	0.6280	0.6718	0.7025	0.7169
jUnit	0.6950	0.6530	0.6142	0.6613	0.6591
MapDB	0.7466	0.7337	0.7702	0.7790	0.8158
mcMMO	0.6864	0.6717	0.6583	0.6509	0.6951
Mission Control T.	0.7039	0.6700	0.6287	0.6573	0.7049
Neo4j	0.6621	0.7154	0.6766	0.6504	0.7150
Netty	0.6483	0.6549	0.6646	0.6823	0.7120
OrientDB	0.6868	0.6772	0.7157	0.7182	0.7234
Oryx	0.5537	0.5687	0.6500	0.6569	0.7331
Titan	0.6590	0.6813	0.6595	0.6407	0.6919
Eclipse p. for Ceylon	0.6664	0.6403	0.7141	0.7026	0.6837
Hazelcast	0.6883	0.6980	0.6742	0.6790	0.6946
Broadleaf Commerce	0.7244	0.7206	0.7736	0.7797	0.7875

> **Answering RQ 1:** *Considering F-measure values for the chosen releases we can state that such databases are suitable for bug prediction by using machine learning algorithms to build prediction models. In bug prediction domain the RandomForest performed best in addition to function and rule based machine learning algorithms thus one should consider these first to build prediction models using our databases.*

After having insight in the bug prediction results, another question is put in words since the algorithms could perform better if they mark more classes/files buggy. It is an important aspect to see how many bugs are covered by the marked classes/files and what proportion of classes/files were marked as buggy.

> **RQ 2:** Which machine learning algorithms or algorithm families perform the best in bug coverage?

Contrary to the investigation for the previous research question, in this context we cannot perform the same evaluation since we used random under sampling to equalize the number of buggy and non-buggy source code elements for the learning corpus, thus not all entries are included in the evaluation. For bug coverage we use the previously built 10 models (for the equalized training sets - with random under sampling) and evaluate it on the whole training set (without random under sampling). During the evaluation we use majority voting for an element (if more than five models predict the element as faulty then we tag it as faulty otherwise we tag it as non-faulty).

Tables 4 and 5 show bug coverage values (ratio of covered bugs) and the ratio of how many classes or files have been tagged as faulty to obtain the bug coverage. Trees are performing best if considering only the bug coverage, however they tagged more than 31 % of the source code elements as buggy in average. NaiveBayes is the other end of the story, since it has the lowest average of bug coverage, but tags the smallest amount of entries as buggy. Same results occurred at file level but here we present some other algorithms (not the best five) to show the differences in machine learning algorithms. We can state that our database is useful for finding bugs in software with high bug coverage.

Since we are in lack of space to introduce wide tables here we present our whole set of results as online appendix together with the full bug database at the following URL:

http://www.inf.u-szeged.hu/~ferenc/papers/GitHubBugDataSet/

We can now answer our second research question.

> **Answering RQ 2:** *Tree based machine learning algorithms performed best in this scenario, with the highest bug coverage ratio. At class level circa 31% of the elements were tagged as buggy, but the F-measure values are still high (higher than 0.71). For file level the values are lower but in total the results are very similar to class level.*

Table 4. Bug coverage at class level

Project	NaiveBayes	PART	J48	RandomForest	RandomTree
Android Universal I. L.	0.71 (0.21)	1.00 (0.39)	1.00 (0.47)	1.00 (0.42)	1.00 (0.42)
ANTLR v4	0.93 (0.20)	1.00 (0.35)	1.00 (0.26)	1.00 (0.27)	1.00 (0.27)
Elasticsearch	0.86 (0.14)	1.00 (0.33)	1.00 (0.32)	1.00 (0.32)	1.00 (0.32)
jUnit	0.82 (0.15)	1.00 (0.26)	1.00 (0.29)	1.00 (0.27)	1.00 (0.24)
MapDB	1.00 (0.25)	1.00 (0.29)	1.00 (0.21)	1.00 (0.26)	1.00 (0.26)
mcMMO	0.72 (0.18)	1.00 (0.40)	1.00 (0.39)	1.00 (0.41)	1.00 (0.36)
Mission Control T.	0.80 (0.21)	1.00 (0.22)	1.00 (0.32)	1.00 (0.18)	1.00 (0.17)
Neo4j	1.00 (0.14)	1.00 (0.34)	1.00 (0.27)	1.00 (0.36)	1.00 (0.39)
Netty	0.82 (0.18)	0.98 (0.34)	0.98 (0.32)	0.98 (0.35)	0.98 (0.33)
OrientDB	0.83 (0.18)	1.00 (0.31)	1.00 (0.32)	1.00 (0.31)	1.00 (0.33)
Oryx	0.92 (0.26)	1.00 (0.30)	0.93 (0.25)	1.00 (0.28)	1.00 (0.30)
Titan	0.66 (0.11)	0.94 (0.29)	0.94 (0.29)	0.94 (0.29)	0.94 (0.32)
Eclipse p. for Ceylon	0.79 (0.14)	1.00 (0.34)	0.98 (0.27)	1.00 (0.32)	1.00 (0.36)
Hazelcast	0.85 (0.14)	0.99 (0.32)	0.99 (0.31)	1.00 (0.31)	1.00 (0.32)
Broadleaf Commerce	0.60 (0.19)	1.00 (0.30)	0.94 (0.29)	1.00 (0.28)	1.00 (0.31)
Average	0.82 (0.18)	0.99 (0.32)	0.99 (0.31)	1.00 (0.31)	1.00 (0.31)

Table 5. Bug coverage at file level

Project	RandomForest	DecisionTable	SGD	Logistic	NaiveBayes
Android Universal I. L.	1.00 (0.46)	1.00 (0.68)	0.46 (0.10)	0.81 (0.27)	0.81 (0.33)
ANTLR v4	1.00 (0.32)	0.91 (0.30)	0.91 (0.20)	0.91 (0.24)	0.82 (0.18)
Elasticsearch	1.00 (0.39)	0.94 (0.35)	0.82 (0.19)	0.83 (0.24)	0.73 (0.16)
jUnit	1.00 (0.44)	0.94 (0.30)	0.83 (0.25)	0.83 (0.31)	0.89 (0.20)
MapDB	1.00 (0.33)	1.00 (0.36)	0.93 (0.19)	0.97 (0.28)	0.90 (0.25)
mcMMO	1.00 (0.42)	0.93 (0.44)	0.81 (0.27)	0.82 (0.29)	0.75 (0.21)
Mission Control T.	1.00 (0.25)	1.00 (0.38)	1.00 (0.24)	1.00 (0.24)	1.00 (0.19)
Neo4j	1.00 (0.30)	1.00 (0.38)	1.00 (0.24)	1.00 (0.25)	0.80 (0.12)
Netty	1.00 (0.44)	0.99 (0.60)	0.85 (0.34)	0.88 (0.36)	0.73 (0.14)
OrientDB	1.00 (0.41)	0.97 (0.49)	0.95 (0.42)	0.92 (0.38)	0.79 (0.14)
Oryx	1.00 (0.43)	1.00 (0.66)	0.64 (0.17)	0.73 (0.32)	0.36 (0.09)
Titan	1.00 (0.37)	1.00 (0.45)	1.00 (0.64)	0.89 (0.38)	0.72 (0.11)
Eclipse p. for Ceylon	1.00 (0.39)	1.00 (0.42)	0.76 (0.21)	0.83 (0.25)	0.65 (0.11)
Hazelcast	1.00 (0.38)	0.95 (0.37)	0.87 (0.21)	0.89 (0.28)	0.80 (0.12)
Broadleaf Commerce	1.00 (0.33)	0.88 (0.34)	0.78 (0.21)	0.80 (0.24)	0.69 (0.14)
Average	1.00 (0.38)	0.97 (0.43)	0.84 (0.26)	0.87 (0.29)	0.76 (0.17)

6   Conclusion and Future Work

In this paper we proposed an approach for creating bug databases for selected release versions in an automatic way using the popular source code hosting system named GitHub. We gathered 15 Java projects from different domains to fulfill the need of generality. After constructing six-months-long release intervals we gathered bugs and the corresponding source code elements and organized them into databases.

We applied 13 machine learning algorithms on them to investigate whether the database is usable for bug prediction purposes. We experienced quite good results for tree based algorithms (Random Forest, J48, Random Tree) with respect of F-measure values and bug coverage ratios.

In the future, we are planning to make our tool open-source, thus anybody can use or even improve our method. We plan to do more experiments with our models on other projects. We will try to identify (with statistical methods) connection between the usefulness of the database and other descriptors such as size of the projects or amount of the reported bugs.

Acknowledgment. This work was partially supported by the European Union project "REPARA – Reengineering and Enabling Performance And poweR of Applications", project number: 609666.

References

1. Arisholm, E., Briand, L.C.: Predicting fault-prone components in a java legacy system. In: Proceedings of the ACM/IEEE International Symposium on Empirical Software Engineering, pp. 8–17. ACM (2006)
2. Bangcharoensap, P., Ihara, A., Kamei, Y., Matsumoto, K.: Locating source code to be fixed based on initial bug reports - a case study on the eclipse project. In: 2012 Fourth International Workshop on Empirical Software Engineering in Practice (IWESEP), pp. 10–15, October 2012
3. Catal, C., Diri, B.: A systematic review of software fault prediction studies. Expert Syst. Appl. **36**(4), 7346–7354 (2009)
4. Chidamber, S.R., Kemerer, C.F.: A metrics suite for object oriented design. IEEE Trans. Softw. Eng. **20**(6), 476–493 (1994)
5. Dallmeier, V., Zimmermann, T.: Extraction of bug localization benchmarks from history. In: Proceedings of the Twenty-Second IEEE/ACM International Conference on Automated Software Engineering, pp. 433–436. ACM (2007)
6. D'Ambros, M., Lanza, M., Robbes, R.: An extensive comparison of bug prediction approaches. In: 2010 7th IEEE Working Conference on Mining Software Repositories (MSR), pp. 31–41. IEEE (2010)
7. Freund, Y., Schapire, R.E.: Large margin classification using the perceptron algorithm. In: 11th Annual Conference on Computational Learning Theory, pp. 209–217. ACM Press, New York (1998)
8. Gyimesi, P., Gyimesi, G., Tóth, Z., Ferenc, R.: Characterization of source code defects by data mining conducted on GitHub. In: Gervasi, O., Murgante, B., Misra, S., Gavrilova, M.L., Rocha, A.M.A.C., Torre, C., Taniar, D., Apduhan, B.O. (eds.) ICCSA 2015. LNCS, vol. 9159, pp. 47–62. Springer, Heidelberg (2015)
9. Gyimothy, T., Ferenc, R., Siket, I.: Empirical validation of object-oriented metrics on open source software for fault prediction. IEEE Trans. Softw. Eng. **31**(10), 897–910 (2005)

10. Hall, T., Zhang, M., Bowes, D., Sun, Y.: Some code smells have a significant but small effect on faults. ACM Trans. Softw. Eng. Methodol. (TOSEM) **23**(4), 33 (2014)

11. He, H., Garcia, E., et al.: Learning from imbalanced data. IEEE Trans. Knowl. Data Eng. **21**(9), 1263–1284 (2009)

12. Kamei, Y., Shihab, E.: Defect prediction: Accomplishments and future challenges

13. Menzies, T., Caglayan, B., He, Z., Kocaguneli, E., Krall, J., Peters, F., Turhan, B.: The promise repository of empirical software engineering data, June 2012

14. Nagappan, N., Ball, T., Zeller, A.: Mining metrics to predict component failures. In: Proceedings of the 28th International Conference on Software Engineering, pp. 452–461. ACM (2006)

15. Ostrand, T.J., Weyuker, E.J., Bell, R.M.: Automating algorithms for the identification of fault-prone files. In: Proceedings of the 2007 International Symposium on Software Testing and Analysis, pp. 219–227. ACM (2007)

16. Quinlan, R.: C4.5: Programs for Machine Learning. Morgan Kaufmann Publishers, San Mateo (1993)

17. Śliwerski, J., Zimmermann, T., Zeller, A.: When do changes induce fixes? ACM SIGSOFT Softw. Eng. Notes **30**(4), 1–5 (2005)

18. Tufano, M., Palomba, F., Bavota, G., Oliveto, R., Di Penta, M., De Lucia, A., Poshyvanyk, D.: When and why your code starts to smell bad. In: 37th IEEE/ACM International Conference on Software Engineering, ICSE 2015, Florence, Italy, 16–24 May 2015, vol. 1, pp. 403–414 (2015)

19. Von Krogh, G., Von Hippel, E.: The promise of research on open source software. Manage. Sci. **52**(7), 975–983 (2006)

20. Wang, S., Lo, D.: Version history, similar report, structure: putting them together for improved bug localization. In: Proceedings of the 22nd International Conference on Program Comprehension, pp. 53–63. ACM (2014)

21. Wang, S., Yao, X.: Using class imbalance learning for software defect prediction. IEEE Trans. Reliab. **62**(2), 434–443 (2013)

22. Williams, C., Spacco, J.: Szz revisited: verifying when changes induce fixes. In: Proceedings of the Workshop on Defects in Large Software Systems, pp. 32–36. ACM (2008)

23. Zhou, Y., Leung, H.: Empirical analysis of object-oriented design metrics for predicting high and low severity faults. IEEE Trans. Softw. Eng. **32**(10), 771–789 (2006)

24. Zimmermann, T., Premraj, R., Zeller, A.: Predicting defects for eclipse. In: International Workshop on Predictor Models in Software Engineering, PROMISE 2007: ICSE Workshopps 2007, p. 9. IEEE (2007)

Towards an Intelligent System for Monitoring Health Complaints

André Oliveira[1], Filipe Portela[1,2(✉)], Manuel Filipe Santos[1], and José Neves[1]

[1] Algoritmi Research Centre, University of Minho, Braga, Portugal
a61561@alumni.uminho.pt, {cfp,mfs}@dsi.uminho.pt,
jneves@di.uminho.pt
[2] ESEIG, Porto Polytechnic, Vila do Conde, Portugal

Abstract. Users' dissatisfaction, related to healthcare institutions in Portugal, has increased in the recent years. This fact can be seen through the increase of formal complaints, which the responsible regulatory entity has been receiving daily, in this country. More and more technical efforts have been done in order to understand and analyze this tendency. In this paper, the authors pretend to present a group of studies about the distribution and use of words in the formulation of a formal complaint, which aims to find the existence of a characteristic pattern. This paper presents an analysis of the words used in complaints and data analysis platform with dashboards, to offer the opportunity of visualizing the information in different ways.

Keywords: Complaints management · Healthcare · Information system · Information visualization · Data analysis

1 Introduction

Nowadays, the entity responsible for complaints management in the healthcare area, in Portugal, collects a large volume of data every day. A very large part of the data collected states real problems, most of them are associated to services provided to the users of healthcare institutions. This fact results in an increasing necessity to understand the basis of the problems recorded by the development of data analysis, in order to find an adequate answer to the complaints and to solve the problems, improving the quality and efficiency of the services provided in healthcare units.

This article aims to explore the Data Science concept through a first study in the health complaints area. In this work, a large group of analysis and studies were performed. These studies were focused on information contained in complaints of healthcare establishments produced online. In this way, a study of complaints is presented through the exploration of text mining techniques, information extraction techniques and information visualization with the future goal of developing an automatic system of classification and recommendation of complaints focusing mainly in the words which compose complaints and their frequency. In order to better understand the results of the study, a few analysis about the distribution of the words by typology and classification were developed. All work was performed using real data and it was

© Springer International Publishing Switzerland 2016
O. Gervasi et al. (Eds.): ICCSA 2016, Part IV, LNCS 9789, pp. 639–649, 2016.
DOI: 10.1007/978-3-319-42089-9_45

developed with collaboration of the Portuguese regulatory institution of healthcare. After the introduction, in the second section, it was made a presentation of the work background. In the third section, designated as "solution development", there is a total description of the work executed in "Knime" and "QlikView" software. This article ends with the discussion of the results and a section with the conclusion and future work.

2 Background

As stated before, this article aims to offer a study about the complaints produced in healthcare establishments, with all the information data, worked on this project, being collected in the Portuguese regulatory institution. It is important to say that, at first, it was performed a data understanding task, in a previous article, developed by the same authors, denominated "Developing an Ontology for Health Complaints Management" [1]. The development of this paper and this entire study were realized with two main goals. At first, the opportunity to offer a contribution to the scientific community by showing a way to develop analysis and studies about complaints, mainly on the healthcare area. The second goal is to offer a different view of the data and information collected to the professionals that have to deal with it in order to produce actions that effectively reduce the production of complaints to the Portuguese health system, improving services quality and patients satisfaction.

2.1 Complaints Management Entity

The entity involved in this project is a Portuguese regulatory entity, with the mission of regulating all the activity developed by healthcare providers establishments [2]. This institution has the skills to cover the national network of public healthcare and private services. Primarily concerned with the protection of the rights of the user basic principles of public service, namely the universality and equality of service access, safety and quality of provision [3]. In recent years, as part of its function, this institution has targeted the exercise of mainly regulation in two dimensions: economic (to fix prices and control the production in health establishments) and on a social dimension (humanizing services and monitoring the compliance of users rights) [4].

2.2 Complaints

According to the current legislation in Portugal, all users of healthcare establishments have the constitutional right to pursue and formally complain about the services provided, when they are not adequately treated [4]. This system can also be used to make a praise. The formal complaint is not only used to presenting negative aspects, but also can be used to expose the most positive aspects. All citizens, as users of healthcare facilities, have the constitutional right to complain (negative or positive) about the service or the way that they had been treated. Effectively, this right gives a clear and perceptible response in health matters. It is one of the majors concerns of the regulator entity. It is very important to have a proper processing of all complaints and expositions of displeasure by the users. The main goal is allowing a better assessment of the weak points of the national health system and identify areas requiring further intervention. In 2009, the regulator entity received a total of 7848 complaints related to health services, and most of them were related to waiting times and quality of administrative assistance and healthcare. Therefore, the claims are a good indicator of the response that the people expect receive in health care facilities. The complaints handling process should reflect the needs and expectation of users, as well it shall be in accordance with the goals of organizations [5]. A correct management of complaints clearly affects the sense of justice and satisfaction of users, strengthening the bonds of loyalty of users to the institution concerned. An effective and efficient management of complaints could result in organizational success on a highly competitive market of direct competition as is observable in the health sector.

3 Solution Development

The development of the solution presented in this paper was performed with the contribution and acquire knowledge in the previous article developed by the same authors. The solution presented focus an analysis of the distribution of words that compose complaints in health sector and a visual perspective of that distribution and information.

With the development of this research, the authors intend to focus the attention of the reader to a real world problem and to find areas of intervention. This work aims to solve some problems in the Portuguese health sector and find departments of this sector that need to improve their quality of assistance by the opinion of the patients.

3.1 Data Analysis

For the development of this research informational data were collected from the responsible entity for the regulation of health complaints. The dataset used was only referring to the complaints produced online. This decision was supported by the greater consistency of the data collected in the online platform. It was also supported by the fact that complaints written in the traditional book of complaints have to be transcribed or scanned into the computer system of the regulatory institution by a designated agent. In fact, this action can provide some changes to the complaint originally produced.

In this way, the online complaints dataset is composed by five informational fields which have the description field, composed by the complaint produced by the protestant; the year field, respecting to the year that the complaint is produced; a numeric field, that can be translated as the identification number of the complaint; the Classification field, which contains the classification assigned to the complaint and the Typology field which regards to the type assigned to the complaint. In order to fully understand these two last fields, the classification field corresponds to the several groups of complaints assigned by the regulatory agent based on the complaints description text and the matter focused; and the typology field focus on the type of complaint, in order to distinct the negative ones to the praises and thanks produced in this formal document.

The online complaints dataset is composed by 889 complaints formed by 15201 words. All the complaints were from the year of 2014.

3.2 Knime Analysis

On the online complaints data analysis, it was defined the use of open-source software Knime, in order to develop an evaluation of the information existing in the documents collected.

Before starting the work with this software, it was decided that only would be interesting to work with three informational fields of the online complaints dataset: Description of the complaint, its respective Classification and Typology. Inside the Knime software, the dataset was loaded and analyzed by techniques and models of text mining, but unfortunately it was completely impossible for the software to assimilate the data involved, because of the linguistic barrier. The system did not have an incorporated Portuguese technical dictionary and, for this reason, this goal was suspended. In order to achieve some results with the used software, all the attention was focused on data analysis process. The most relevant analysis was the study of frequency of words in complaints for each classification assigned using the model below.

Analyzing the Knime model (Fig. 1), it was used two types of filters for data treatment, which one is responsible for filtering the columns of the dataset and which the other filter namely "Row Filter" is responsible for controlling the rows to be analyzed. In "Column Filter", the information contained in Typology is going to be excluded, remaining Description and Classification.

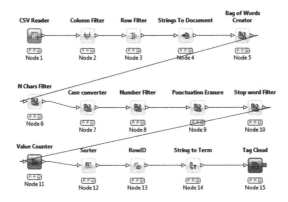

Fig. 1. Knime model (Color figure online)

When analyzing the model, it is possible to see the use of some algorithms related to the data treatment. This use is related to the final goal of words count and to obtain good and interesting results with scientific value. The tool named "bag of words" is one of the most important for the study, due to their capability in individualizing the complaints in terms, i.e., occurs an individualization and isolation of the words that compose the complaints text. Another tool used in the model was "punctuation erasure". The objective is deleting all the symbols of punctuation from the texts.

After the data processing phase, the use of the node "value counter" can be verified. This node is dedicated to calculate the occurrence of the terms early identified in the complaints. The model ends with the use of the node "tag cloud", which offers to the user an alternative mode of visualizing the study results.

3.3   Complaints Studies by Knime

With the model presented in the previous section, an analysis of the frequency of words contained in the complaints was performed. This analysis was realized on the total of complaints and subsequently to the most relevant types of classification of complaints.

To study the complaints, the group of 20 more frequent and most used words in each group of complaints was analyzed. Later, this frequency was submitted to two mathematical formulas that calculated the appearance percentage in the dataset:

$$\% \, Frequency \, by \, Complaints = \left(\frac{Frequency \, of \, Words}{Number \, of \, Complaints} \right) \times 100$$

$$\% \, Frequency \, by \, Total \, of \, Words = \left(\frac{Frequency \, of \, Words}{Number \, of \, Words} \right) \times 100$$

After a quick analysis, it was obtained a cloud of words composed by the most used words in the complaints (Table 1).

Table 1. Knime model to complaints

Words	Frequency	% Frequency by complaints	% Frequency by total of words
Health	404	45 %	3 %
Hospital	398	45 %	3 %
Appointment	366	41 %	2 %
Situation	350	39 %	2 %
Service	291	33 %	2 %
Doctor	289	33 %	2 %
Establishment	273	31 %	2 %
Complaint	272	31 %	2 %
Answer	272	31 %	2 %

3.3.1   Application to the Group of All Complaints

In this first study (Table 1), it was observed that the word most used in complaints is the word "health", and this word appears 404 times with a percentage of appearance of 45 %. 3 % of the words used in complaints is actually "health". Following the analysis, it is possible to note that the words "hospital" and "appointment" complete the podium of the most frequent words, having both the rate above 40 %, not being this result a complete surprise, by the fact of both words are closely related with healthcare. Near 40 % margin is also the word "situation", excluding the word "have" for its natural formation and for being an auxiliary verb, having this way a reduced importance. In order to complete the analysis, remains to be said that the words figured on the Table 1 perform a total of 5240 frequencies of words and represent 34 % of the words commonly used in the formulation of the complaints.

3.3.2   Application of the Model to Different Groups of Classification

In the sequence of the study of words frequencies, the model presented was executed to the 10 classifications that have larger number of complaints registered in their type. So, with the realization of this work, it is visible, in the Tables 2 and 3, the group of 10 most frequent words for every classification previously selected. For example in the Waiting times complaints the most used work was appointment and waiting. This means that most of the complaints of this type was related to questions with the appointment and the waiting time.

Table 2. Knime model to restrict classification

"Other issues"	"Quality in healthcare"	"Patient rights"	"Waiting times"	"Complaints book"
Hospital	Hospital	Health	Appointment	Complaints
Health	Situation	Situation	Waiting	Appointment
Situation	Health	Hospital	Hospital	Doctor
Appointment	Service	Appointment	Time	Answer
Establishment	Same	Service	Doctor	Hospital
Same	Urgency	Establishment	Health	Book
Answer	Doctor	Doctor	Service	Medical
Any	Complaint	Right	Situation	Health
Doctor	Answer	Same	Any	Same
Service	Medical	Complaint	Answer	Doctors

Table 3. Knime model to restrict classification 2

"Health fees"	"Access"	"Unfounded rejection"	"Quality"	"Billing"
Fees	Health	Health	Health	Situation
Moderators	Appointment	Establishment	Hospital	Value
Health	Hospital	Appointment	Service	Payment
Exemption	Establishment	Medical	Any	Annexes
Establishment	Situation	Doctor	Services	Same
Answer	Wait	Situation	Stay	Any
Hospital	People	Family	Complaint	Appointment
Complaint	Service	Hospital	Doctor	Hospital
Moderator	Access	Appointments	Patients	Some
Situation	Family	Same	Medical	Complaint

3.4   Dashboards

Currently, the development of dashboards based on informational data and business indicators is offering to users an innovative experience of information visualization, improving the development of new points of analysis and study, supporting the decision making process, in order to achieve goals and identify problems within organizations. The representation of information through dashboards can be done in various ways, in order to support decision making process within an organization.

The development of dashboards in this work offers a better opportunity to have a greater understanding of the raw data. Remains to say that all graphics created belong to a Pivot-Table and the data used is always the same for all analyzes only differing the fields used on every dashboards. Due to the condition of development by a dynamic table, when we select a particular group of data in a graphic, the page assumes the selection for all graphics on the same page and, for this reason, the information presented can instantly change, improving the experience of information visualization.

At this point of work, the software used was the QlikView solution, due to the possibility of easy understanding the data involved in the work and for the large diversity of options for representation and visualization of information.

3.4.1   Study of the Most Frequent Words on Complaints

In order to obtain a more attractive experience of information visualization about the distribution of words frequencies by classification classes, some dashboards and graphics were developed. For the correct development of this study, it was created a new dataset, taking into account mainly three dimensions: a field with the words obtained in the previous study with Knime, another field with the respective frequency of each word collected and another one, respecting to the classification associated with the word.

Two graphics for analysis were designed taking into account the group of data used and the information contained in the dataset. In this way, the Fig. 2 represents a general overview of the dashboards created.

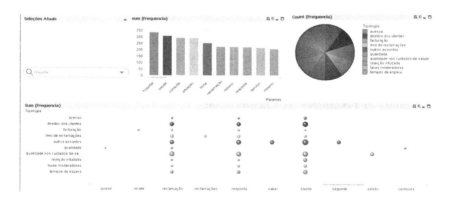

Fig. 2. General aspect of dashboards (Color figure online)

3.4.2   Study of the Words by Frequency

In this first study, only the words and frequency of words were analyzed. It was developed a graphic with the sum of the frequency of words for each word involved in the process. The main reason that led to this decision was the fact that the same word can have different frequencies in different types of classification. In the graphic presented in Fig. 3, it is possible to see a bar chart that represents the words most frequents. Inside the QlikView software all the words can be checked by sliding a horizontal bar displayed in the Fig. 3. From the analysis of the graphic, the most used word in the formulation of complaints in healthcare establishments is undoubtedly the word "Hospital" and this word has a frequency of 333 appearances. It is also possible to observe that the next group of most frequent words is formed by the words "health" "appointment" and "Situation" being each one of them equally used and, because of this fact, the difference of frequency is not very accentuated. In a different point of view, we can realize that the words which record a lesser frequency in this analysis are the words "clarification", "euros", "declaration" and "attendance", having all a frequency value of 8. This chart can also analyze the percentage of use.

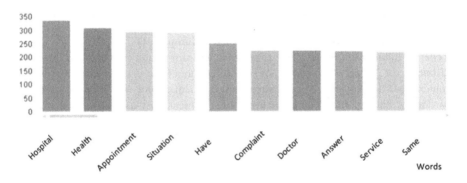

Fig. 3.   Dashboard words by frequency (Color figure online)

3.4.3   Study of the Words by Classification Types

This study presented in the Fig. 4 considers just two fields of informational data: the classification and the number of words involved.

Easily, it is possible to check that the main purpose of development of this graphic is to have an overview of the words by classification. It allows the study of percentage of words by classification. By clicking inside of the chart, it is possible to see the words related to each class.

3.4.4   Study of Frequency by Classification

This study was developed taking into account all dataset fields. In this way, we can see, by the graphic in the Fig. 5, the relation of a determinate word with the several classifications. In the plot graphic, we can see the occurrence of each word in the classifications analyzed. By a certain example, we can visualize that the word "appointment" occurs and it has a considerable frequency in every classification study, being the classes with bigger frequency "patient rights", "other issues" and "waiting

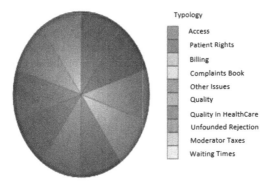

Fig. 4. Dashboard words by classification (Color figure online)

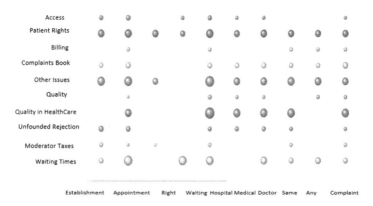

Fig. 5. Dashboard frequency by classification (Color figure online)

times", as much bigger the dimension of a plot, greater is the frequency of the word. In another point, we can check that some words are restricted to a unique class, as the case of the word "access" that only occurs in the classification "access" and the case of the word "waiting" that only occurs on the classification "waiting times".

4 Discussion

The collection of complaints is an important source of information for the Portuguese regulatory institution, although sometimes information cannot be observable easily and for that reason is extremely necessary the use of technology. For that purpose, the studies presented in this paper reflects the actual distribution of complaints in the Portuguese health system by classification and distribution of words in this particular type of documents.

Although it was impossible in this stage to apply text mining techniques, the authors think that this paper presents several information that, in the past, no one paid attention and with this contribution the Portuguese regulatory institution of health sector can now lead a path to further and better discoveries.

All the analyses were developed in a real context and this is the strongest point of the entire study, because this improves the accuracy of the dashboards and analyses presented. Another strong point of this work is its capability of performance with newer and bigger collection of data.

5 Conclusion and Future Work

This article presents analyses and studies developed about complaints produced in healthcare units basing it on real facts. All information analyzed was collected near the Portuguese regulatory institution of healthcare units. For the development of this work, it was necessary a deep comprehension of the complaints management process combined with a higher sensitivity to analyze the information collected resulting in the development of dashboards.

In this paper, we can see the distribution of words in the group from the complaints collected. We can also see the frequency of the most used words and its respective percentage. In this way, we can see that the words "health", "hospital" and "appointment" are the most used words with a percentage near to 50 %. This result can be analyzed in other way: from every two complaints produced, one of them have these words. In the sequence of the study, we can also see an example of information visualization by the words cloud presented. With this work, it is also possible to see a table with the most used words for the ten most important classifications applied. By developing the dashboards, we can have an interesting and exciting way to represent information and, with this tool, the reader can have an instantaneous perception and assimilation of the information.

As a result of the partnership around this project, all work developed and all the results achieved are going to be sent to the Portuguese regulatory institution with the purpose of the information be interpreted for their experts. Then they will decide if these dashboards are good enough to be applied in this field. All studies can easily be adopted and adapted to evaluate the elaboration of complaints in several hospitals and departments. Taking as example *Centro Hospitalar do Porto*, this study can be used to evaluate complaints in maternity care [6], emergency [7], intensive care units [8, 9], dermatology [10] and others [11–14].

Finally, it is important to note that this work is the first phase of a big goal. In the future a set of algorithms will be used in order to facilitate the automatic classification of complaints by using the words presented in the text.

Acknowledgments. This work has been supported by FCT - Fundação para a Ciência e Tecnologia within the Project Scope UID/CEC/00319/2013.

References

1. Oliveira, A., Portela, C.F., Santos, M.F., Machado, J., Abelha, A., Neves, J.M., Vaz, S., Silva, Á.: Developing an ontology for health complaints management. In: KMIS Conference, Braga, Portugal (2015)
2. Lobo, A.F.: Entre o direito a reclamar e o direito à saúde. Serviço social em gabinetes do cidadão, SNS. Instituto Superior Miguel Torga, Escola Superior de Altos Estudos, Coimbra (2004)
3. Simões, J.A.: As parcerias público-privadas no sector da saúde em Portugal. Revista Portuguesa de Saúde Pública, vol. 4, no. Saúde Pública, pp. 79–90 (2004)
4. Entidade Reguladora da Saúde: Relatório sobre "A carta dos Direitos dos Utentes". Entidade Reguladora da Saúde, Porto (2011)
5. Firmino, C.F.: Reclamações: A sua importância nas Unidades de Saúde Privadas. Mestrado em Gestão da Saúde, Universidade Nova de Lisboa, Lisboa (2011)
6. Abelha, A., Pereira, E., Brandão, A., Portela, F., Santos, M.F., Machado, J., Braga, J.: Improving quality of services in maternity care triage system. Int. J. E-Health Med. Commun. (IJEHMC) **6**(2), 10–26 (2015)
7. Portela, F., Vilas-Boas, M., Santos, M.F., Abelha, A., Machado, J., Cabral, A., Aragão, I.: Electronic health records in the emergency room. In: 9th IEEE/ACIS International Conference on Computer and Information Science (2010)
8. Portela, F., Vilas-Boas, M., Santos, M.F.: Improvements in data quality for decision support in intensive care. In: Szomszor, M., Kostkova, P. (eds.) e-Health. LNICST, vol. 69, pp. 86–94. Springer, Heidelberg (2011)
9. Portela, F., Veloso, R., Oliveira, S., Santos, M.F., Abelha, A., Machado, J., Silva, Á., Rua, F.: Predict hourly patient discharge probability in intensive care units using data mining. Indian J. Sci. Technol. **8**(32) (2015). ISSN: 0974–6846. Indian Society for Educat.
10. Duarte, J., Portela, C.F., Abelha, A., Machado, J., Santos, M.F.: Electronic health record in dermatology service. In: Cruz-Cunha, M.M., Varajão, J., Powell, P., Martinho, R. (eds.) CENTERIS 2011, Part III. CCIS, vol. 221, pp. 156–164. Springer, Heidelberg (2011)
11. Portela, F., Santos, M.F., Machado, J., Abelha, A., Silva, Á., Rua, F.: Pervasive and intelligent decision support in intensive medicine – the complete picture. In: Bursa, M., Khuri, S., Renda, M. (eds.) ITBAM 2014. LNCS, vol. 8649, pp. 87–102. Springer, Heidelberg (2014)
12. Braga, P., Portela, F., Santos, M.F.: Data mining models to predict patient's readmission in intensive care units. In: ICAART - International Conference on Agents and Artificial Intelligence, Angers, France (2014)
13. Silva, E., Cardoso, L., Portela, F., Abelha, A., Santos, M.F., Machado, J.: Predicting nosocomial infection by using data mining technologies. In: Rocha, A., Correia, A.M., Costanzo, S., Reis, L.P. (eds.) New Contributions in Information Systems and Technologies, Advances in Intelligent Systems and Computing 354. AISC, vol. 354, pp. 189–198. Springer, Heidelberg (2015)
14. Cabral, A., et al.: Data acquisition process for an intelligent decision support in gynecology and obstetrics emergency triage. In: Cruz-Cunha, M.M., Varajão, J., Powell, P., Martinho, R. (eds.) CENTERIS 2011, Part III. CCIS, vol. 221, pp. 223–232. Springer, Heidelberg (2011)

Proposal to Reduce Natural Risks: Analytic Network Process to Evaluate Efficiency of City Planning Strategies

Roberto De Lotto[✉], Veronica Gazzola, Silvia Gossenberg,
Cecilia Morelli di Popolo, and Elisabetta Maria Venco

DICAr – University of Pavia, via Ferrata 3, 27100 Pavia, Italy
{uplab,roberto.delotto}@unipv.it

Abstract. Natural hazards have greater social and economic impact in urban areas because urbanization and economic development increase people and assets' concentration in high-risk prone areas: hazards generate risks in relation to population's exposure and its physical and economic assets.

Advanced urban planning is one of the involved disciplines in the process of human exposure and risks reduction: defining strategic actions, it can reduce losses following natural disasters and, in the same time, ensure a flexible design able to absorb external impacts, to transform and to adapt itself, increasing urban resilience.

The paper defines two possible strategies of intervention in city as risk mitigation methods: Areal Change and Functional Change.

Authors describe the use of scenario planning and Multicriteria evaluation (ANP method) to deepen suitable strategies at urban level. Moreover, authors introduce the first instruments useful for the future application of the presented method: ideal city modeling and evaluation criteria definition.

Keywords: Natural risk assessment · Exposure reduction · Multicriteria evaluation · City planning · Qualitative and quantitative criteria

1 Natural Hazards, Disasters, Risks

The exponential growth of urbanization depending on human presence and the intensification and diversification of land-use practices and patterns (especially in low-income and developing countries) lead to cities to be exposed to multiple effects of different types of natural hazards. Effectively, a natural event is a potential threat to human life and property. It could be an extreme phenomenon that may not involve damage to persons or property or it may become disaster[1] or catastrophe[2] when human beings (and their physical and economic assets) live or work in its path.

The probability of occurrence that an undesired hazard event contributions to a potential disaster configures over time a risk [1].

---

[1] Disaster is a hazardous event that occurs over a limited time in a defined area.
[2] Catastrophe is a massive disaster that requires significant expenditure of money and a long time for recovery.

© Springer International Publishing Switzerland 2016
O. Gervasi et al. (Eds.): ICCSA 2016, Part IV, LNCS 9789, pp. 650–664, 2016.
DOI: 10.1007/978-3-319-42089-9_46

The field of risk assessment is rapidly growing with contributors from many natural and social sciences areas (Table 1). From Blaikie et al. (1994) [2] on, risk is considered as a product of natural hazard (shaped by the magnitude and frequency of a physical event) and human vulnerability (result of the interaction of physical and human elements). In the last years the concept of vulnerability is further divided into other components [1]:

- Exposure: product of physical location and the character of surrounding built and natural environment;
- Resistance: capacity of an individual or group of people to withstand the impact of a hazard on the basis of economic, psychological and physical health and their system of maintenance;
- Resilience: ability of actors to cope with or adapt to hazard stress, intended as coping capacity.

Therefore, disaster risk becomes function of hazard (H), vulnerability (V), exposure (E) and adaptation (A) [3].

The identification of all risk factors provides the basis to create an order of priorities among initiatives that will contribute to reduce disaster risk.

2 Strategic Approaches to Reduce Risk

Adverse impacts of hazardous natural events on natural environment, buildings, human beings, economy and society often cannot be fully prevented.

In the last years, a worldwide attention for the need to implement effective policies for disaster risk management is developed [5]. Advanced urban and city planning is one of the involved disciplines that can be considered to pursue disaster risk reduction as part of the everyday development process. In particular, city planning strategic actions can reduce losses following natural disasters and in the same time, they can ensure a flexible design able to absorb external impacts, to transform and to adapt itself, increasing urban resilience. Two strategic approaches together can reduce the possibility that a disaster may produce irreversible disruptions on a global ecosystem:

- Mitigation action: structural and non-structural measures undertaken to limit the adverse impact of natural hazards, environmental degradation and technological hazards [2] or lessening of the potential adverse impacts of physical hazards (including human-induced) through actions that reduce hazard, exposure, and vulnerability [6]. Considered as indirect damage prevention, it includes actions that reduce the severity of future disasters with longer-term global benefits [7];
- Adaptation action: adjustment in natural or human systems to a new or changing environment [6]. Considered as direct damage prevention, it gives immediate localized benefits trying to moderate harm or exploit beneficial opportunities [7].

Since that hazardous events are often repetitive and it can measure the probability they come back (employing scientific method, using statistical and probabilistic tools, studying natural events' histories) [8], longer-term mitigation actions, acting directly on

Table 1. Risk analysis: different formula.
Source: Tiepolo 2014a [4], modified by authors

Formula	Origins	Organizations manuals
R = H * V	Blakie et al. (1994) [2] Birkmann 2006	GTZ 2002 UNDP 2005
R = Pr * Co	Jones et al. (2003)	
R = H, V, E	Dwyer et al. (2004) [32] Dilley (2005)	ADRC 2005
R = H * V * LP	Villagran de Leon (2004)	
R = H * V * Co	Kaynia 2008 [33]	
R = (H * V * Va)/P	De La Cruz-Reyna (1996) [34]	
R = H + V + E - Cc	Davidson (1997) [35]	IADB-Hahn 2003 [36]
R = (H * V)/C		UN-ISDR 2009 [1] USAID 2011
R = (H * V * E)/A	Gotangco et al. (2010) [3]	

A = adaptation; C = capability (of adaptation); Cc = capability to cope;
Co = consequences; LP = lack of preparation; E = exposure; H = hazard;
P = preparation; Pr = probability; S = sensibility; V = vulnerability; Va = value
at risk

reducing risk components, represent the valid solution to prepare city and its elements
to a possible hazardous events.

2.1 Mitigation Actions to Reduce Exposure

Since, in an urban area, human beings (and buildings) are the most exposed elements to risks, the assessment of the exposure of the different urban functions (residential, commercial and tertiary buildings, critical facilities, social and economic infrastructures) is one of the essential features of disaster risk reduction. Mitigate risks acting on exposure means remove, reduce and control the quantity and quality of exposed items presented in territorial sensitive areas. To reach these goals it is essential to know the localization and distribution of people, goods and human activities.

The exposure analysis has to be involved in urban planning at regional (with particular reference to infrastructure and strategic facilities) and city level (referring to expansion's areas and to interventions on existing urban fabric).

Responsible and advanced urban planning means that the pre and post hazardous event issues are taken into proper consideration and for human settlements, industrial activities and infrastructure (main roads, railways, electrical lines, gas pipelines), particular areas are chosen in order to minimize possible damage. Therefore, it is necessary to develop strategies of risk reduction, to create guidelines for trans-scalar strategic urban planning and to define multidisciplinary instruments in order to respond to the city needs in short term.

As mitigation tools, two possible strategies of intervention to reduce exposure are the relocation [9–11] of a specific urban function from high-risk to low-risk area (Areal change) and the replacement of high exposure functions with lower exposure ones in the same area (Functional change) [12] (Fig. 1).

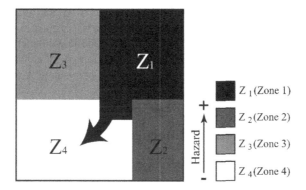

Fig. 1. Areal change framework. Source: authors

- Areal change acts directly on physical relocation of buildings in areas less hazardous: the function keeps unchanged and planners define the new use of the abandoned building that should have a lower exposure and the policies to build in the new area (Fig. 1).
- Functional change acts on the function by reducing exposure that implies a reduction of physical and economic losses. There are two possible solutions: change of function in the same building (the functional changes ensure the exposure and crowding's reduction, i.e. from residential to tertiary or from facilities to retail); change of function with significant changes of building (new safer construction replaces the previous in the same area) (Fig. 2).

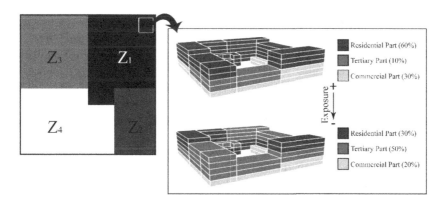

Fig. 2. Functional change framework Source: authors

In any risk prevention project, it is possible to recognize direct/indirect and private/public costs (social and economic). In particular, they derive from: significant modifications of inner structural organization and their relations with cities; increase of land use; relocation studies, project design and expertise advice; public support activities; buildings acquisition; disused buildings demolition; renaturalization of abandoned areas; definition of anthropological value of assets; private owners' acceptance and will to pay (due to the highly fragmented system of private property) and so on.

Beyond risks reduction, it is possible to recognize additional benefits: improvement of environmental quality in abandoned areas; reduction of other disaster mitigation expenses; reduction of reconstruction costs; reduction of systemic losses; reduction of costs (social and economic) related to emergency.

3   Efficiency Evaluation Instruments

After the definition of Area or Functional Changes, it is crucial to understand, supposing the equal level of risk reduction, which is the most suitable strategy to adopt at urban level and how to develop it.

3.1   Scenario Planning

In order to identify the best solution to risk reduction, the study of possible scenarios and the way to realize them become essential and it is necessary to analyze the key elements (type of hazard, behavior of citizens, urban fabric with its functions, economic feasibility, anthropological value of places and cultural heritage) of the purpose. This process is well known as Planning Scenario [13], a strategic planning method suitable to spatial planning because it allows to simplify huge amounts of data in few possible configuration and the integration of quantitative data with qualitative expressions and disciplines.

The process to develop scenario is well described by Shoemaker [13]:

1. Define the scope: set the time frame and the analysis scope;
2. Identify major stakeholders: who have a strong interest in these issues;
3. Identify basic trends: which political, economic, societal trends effect the issues;
4. Identify key uncertainties: which uncertain outcomes of events affect the issues and which relationship among uncertainties may occur;
5. Construct initial scenario themes: there are different ways to define the initial scenario depending on the elements involved and the main objective;
6. Check for consistency and plausibility: there are, at least, three different test to verify the inner consistency. "Are the trends compatible within the chosen time frame? [...] Do the scenarios combine outcomes of uncertainties that indeed go together? [...] Are the major stakeholders placed in positions they do not like and can change?" [13, pp. 29];
7. Develop learning scenarios: identification of strategically relevant themes and organization of related outcomes and trends;
8. Identify research needs to understand uncertainties and trends;
9. Develop quantitative methods: quantification of different scenarios' consequences;

10. Evolve toward decision scenarios with an iterative process that uses all the previous steps.

Each defined "scenarios should cover a wide range of possibilities and highlight competing perspectives while focusing on interlinkages and internal logic within each future" [13, pp. 30].

Given the nature of the complex assessment problem, it is necessary to use Multicriteria evaluation methods as decision support system. The evaluation process is very close to scenario planning: the inner coherence of the presented method is kept.

3.2   Evaluation Process

Among the various methods available in literature, the use of Analytic Network Process (ANP, proposed by Saaty in 2005 as a development of Analytic Hierarchy Process, AHP, 1980) [14–19] allows defining a priority ranking among alternatives through pairwise comparisons between qualitative and quantitative criteria. In the ANP method, the dependencies among evaluation criteria is clearly represented by a network structure. It is based on the idea that not only the importance of criteria determines the importance of alternatives, but also the alternatives determine the importance of criteria. It is this theoretical assumption that allowed the evolution from a multilevel hierarchy structure to a network structure (from AHP to ANP), from independence among elements of each level to a dense network of interdependencies and feedback: so, this structure better reflects the complex interactions among different components of a real system.

The application of method starts from the structure definition of the decision problem identifying nodes, clusters (main goal, alternatives and evaluation criteria) and the choice of assessment procedure network (Fig. 3).

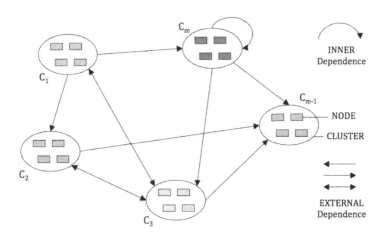

Fig. 3. Dependencies representation. Source: authors

There are two possible structures:

- Simple Network, based on a single network of dependencies;
- Complex Network, which presupposes the existence of a control hierarchy that creates subnets organized as a simple network made up of clusters, nodes and networks of influence. Saaty suggests the use of logical framework comparable to SWOT analysis, called BOCR that uses a control hierarchy based on the so called Merit nodes: "Benefits (B), the good things that would result from taking the decision; Opportunities (O), the potentially good things that can result in the future from taking the decision; Costs (C), the pains and disappointments that would result from taking the decision; and Risks (R), the potential pains and disappointments that can result from taking the decision" [16].

In a city planning process with complex interventions on territory, the Simple Network model is able to represent a greater interaction among elements, but is much more complicated to build and applied because of the huge amount of criteria.

The BOCR model allows a better understanding of technical aspects, simplifying the assessment and defining the interaction of nodes in the four control categories (Benefits, Opportunities, Risks, and Costs). It also allows buyers' greater understanding of evaluation and a more appropriate presentation and consideration of stakeholders' different points of view [20–22].

For these reasons, the use of Complex Network as evaluation support is chosen.

4 Methodology

After the definition of main goal (natural risk reductions), the presented method can be described as follows:

1. Scenarios definition;
2. Criteria definition: individuation and explicit criteria formulation is a fundamental assumption for method's applications because they express a measurable matter of judgment. Each criterion is characterized by semantics, metrics, intelligibility, consensus, coherence, completeness, no redundancy, expression of judgment's mode, a preferred structure and a scale of ordinal or cardinal values [23–26]. They could be:

   - Quantitative Criteria: they are mathematically evaluated because relate to objectively available and comparable data; they present units of measure and numerical ratios;
   - Qualitative Criteria: they indicate the quality of a given element; they represent evaluations of not mathematically calculable data, but easily measurable and comparable with other data.

3. Problem modeling: the BOCR method defines the interaction of nodes into the four control categories. In this way, the multiplicity of factors can be simplified considering the decision problem characterized by certain aspects, evaluated in a short-middle period, are positive (Benefits) or negative (Costs); uncertain aspects, evaluated in long term period, have positive (Opportunity) or negative connotation (Risks) and are more difficult to predict [16, 20]. Each merit node has an attached

sub-network that is analyzed separately and produces an alternatives ranking that will be synthesized and integrated with the other subnets in order to reach the final selection options list, taking into account the external and inner dependencies among all elements. The identification of influence relationships is a crucial step because it determines the final evaluation outcome.

4. Compilation of pairwise matrices: the ANP method differs from the classic Multicriteria analysis (MCDA) [24, 27] because of the pairwise comparisons made, here, with the fundamental scale of Saaty. It allows converting the subjective opinions expressed by evaluators on a 9-point scale: each score represents how many times the greater of the two elements is dominant on the lower one, related to a particular characteristic. In this way, the subjectivity is reduced.

   The procedure consider that every network element is taken in rotation as the "parent" and is used as reference to create multiple comparisons among all the elements previously connected to it, called "children". So, during evaluation, the question to ask is: given a parent node and two children nodes to compare, which of the two children nodes influence more the parent node? By how much? The procedure is repeated for each element and each cluster. Finally, it determines the consistency of judgments of each matrix (consistency ratio less than 0.1);

5. To implement the method, authors use SuperDecisions software, created by R.W. Saaty and W.J. Adams 1999 [28][3]

6. Definition of alternatives' ranking: to reach the final priority vector and then the final ranking of alternatives, it is necessary combine the four priority vectors, resulting from each subnet, using mathematical combinatorial formula;

7. Sensitivity analysis: the last step is useful to assess the robustness of decision model. It evaluates the stability of the final ranking (outputs) varying the weights assigned to the control criteria (input): if the output variation, related to inputs' changes, is not significant, then the decision method is robust and the result stable.

4.1   Ideal City Modeling

Application of studied methodology takes place through the creation of ideal model of flexible city [29] characterized by some typical elements of European historical cities: urban fabric (high, medium and low density), morphology (block size and shape) and urban historical evolution (historic core, post-war expansion and new development).

The ideal town[4], is structured with several blocks (with evocative roman block dimensions: 80 m × 80 m) characterized by different building density[5] and functional

---

[3] For each pairwise matrix, it defines: principle normalized eigenvalues; non-weighted supermatrix; weighted supermatrix, limit supermatrix and the alternatives' ranking.

[4] Authors choose Pavia (medium sized city located in Northern Italy) to apply the ideal city modeling because it is characterized by the typical elements of European historical cities and it has a database with detailed urban information that authors can easily consult. [30].

[5] High-density ≥ 6 $[m^3/m^2]$; Medium-density between 2.5 and 6 $[m^3/m^2]$; Low-density ≤ 2.5 $[m^3/m^2]$.

mix: low density block (opened urban fabric with only residential function), medium density block (mixed urban fabric with residential and commercial functional mix), high density block (closed urban fabric of historical city center with functional mix of residence, commerce and tertiary) and expansion area (part of city aim of urban modifications). In the definition of ideal city, strategic facilities (i.e. hospitals and schools) may not be considered since that, in the field of prevention, specific regulations exist for them [30].

It is supposed that the territory of ideal city is characterized by different values of general risk (high, medium and low risk) (Fig. 4).

For each considered function, many data have to be collected. In particular the analysis of the people's exposure represents the main step but also the most difficult to be performed. The population (and its main features such as density, individual characteristics, behavior during night and daytime) is not a static entity but dynamic: significant presence of people varies in a cyclic manner over time for each considered urban function. Several areas are particularly exposed during day (shopping malls, buildings with tertiary and offices activities), some other during night (residential buildings). Therefore, it is important to consider time and spatial dimensions of risk exposure jointly; in particular, for each urban block of ideal city, data that have to be collected are:

- Ground floor surface (GFS) of each urban function and Percentage of each urban function (residence, commerce and tertiary);
- Number of building floors;
- Number of people that are in each urban function compared to total people number of the ideal city;
- Number of hours of occupation and use for each urban function;

These collected data become instrument of scenario planning and Multicriteria evaluation (through criteria definition) applied to the two urban strategies of risk mitigation presented by the authors in the paper (Areal Change and Functional Change).

4.2   Criteria Definition

Quantitative criteria:

- Intended use changes [sqm]. It evaluates the quantity of surfaces that change intended use. In case of the functional change, the variation is related to small portion of territory and to portions/floors of buildings. In case of areal changes, it represent also the abandonment area with the high exposure and consequently high risk.
- Urban loads changes [number of inhabitants over services' needs]. It represents the quantity generated by the services and infrastructure needs on a given territory. In case of functional changes, the variation depends on which function is moved: infrastructure mostly remain unchanged, while other facilities depend on each scenario. In case of areal changes, the variation of urban loads is maximum: it is

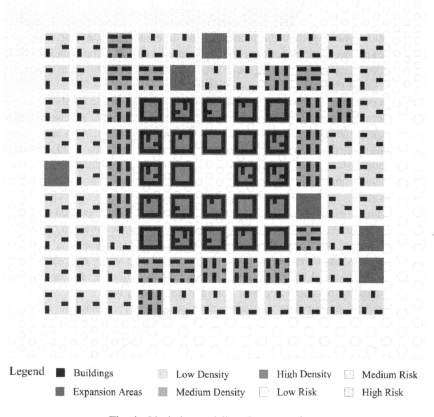

Legend   ■ Buildings   Low Density   ■ High Density   Medium Risk
          ■ Expansion Areas   Medium Density   □ Low Risk   High Risk

Fig. 4. Ideal city modeling. Source: authors

necessary to highlight the relationship between citizens and the need of services and infrastructure and to recreate the maximum efficiency with the minimum land consumption.

- Land use [sqm]. Intended as a subtraction between the whole area and the coverage index. The intensive use of soil (in particular, permeability reduction and urban sprawl) increase the vulnerability of a given territory.
- Economic feasibility [€/sqm or €/m] involves different disciplines considering the new areas and the abandoned ones. The public administration (at local, regional or national level) must carefully evaluate which is the most effective action to reach risk reduction (taking into account also valuable buildings and anthropological elements). In case of functional changes, the economic impact can be more limited, because the relocation influenced a smaller territory or punctual situations. In case of areal changes, the economic impact is huge because the relocation of large portions of land involves significant structural and infrastructural investments: expropriation or equalization are useful instrument to reach the main goal.
- Realization time [unit of time]. It is the amount of time useful to realize the mitigation actions; it can change significantly in relation to each scenario.

Qualitative criteria:

– Spatial organization of urban fabric and morphological characteristics. In a city, its forms, its morphology and the relations b its different parts are crucial for city's structure and the relations between the city itself and citizens. The capacity to adapt the interventions to an existing built area (functional change) or to a non-built territory (areal change) is the basis of its success.

– Social acceptability. The decision to modify the function or the inhabited territory itself represents a citizens' remarkable effort. With functional change, the most significant social impact occur on inner communities' relations. With areal change, the habits related to territory and community change: citizens could oppose the modification of a whole portion of city. It is necessary to find a balance among intervention's opportunity, the economic factor and the possibility to recreate new communities.

– Maintaining of balance and function relations. The relationship between the existing functions within the urban core are well demonstrated by the diffusion of functional mix. In case of functional change, the modification of functions can lead to the deterioration or improvement of an existing territory. The ability of planners will be linked to the essential improvement of existing balance. In case of areal change the balance must be recreated in a new core and, later, improved.

– Environmental impact. Interventions, acting on a territory, change its characteristics. In case of functional change, that represents a less significant impact, the quantity of actors involved and of the dimension of functions to be replace are reduced: the global impact is not necessarily inferior because it depends on what should be relocated. In case of areal change, that represents a significant impact, it is necessary to verify the impact both on the place of departure (risker area) and on the place of arrival.

– Functional and territorial compatibility. It is necessary to analyze the territory and its inner relations in order to adapt to the new territory and to existing functions.

The criteria defined above are divided in the four Merit nodes:

– Benefits:
  • Urban aspects:
    – Intended use changes;
    – Spatial organization of urban fabric and morphological characteristics;
– Costs (Fig. 5):
  • Economic aspects:
    – Economic feasibility;
    – Realization time;
  • Environmental aspects:
    – Environmental impact;
– Opportunities:
  • Urban aspects:
    – Urban loads changes;
    – Maintaining of balance and function relations;

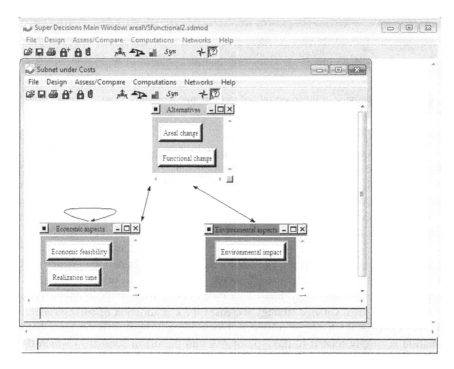

Fig. 5. Subnet costs of BOCR method. Source: SuperDecisions software and authors

- Risks:
  - Environmental aspect:
    - Land use;
  - Social aspects:
    - Social acceptability;
  - Urban aspect:
    - Functional and territorial compatibility.

5   Results and Discussion

The following section presents the results of the evaluations carried out on the ideal city modeling, taking as example the pairwise comparison between the two urban strategies (Areal Change and Functional Change) related to the quantitative criteria Economic Feasibility, included in Merit node "Costs" (Fig. 6):

The evaluator expresses a judgment between 1 and 9 (using the fundamental scale of Saaty) answering the question: "From an economic point of view, it is more feasible realize a Functional or an Areal Change?" In the ideal block of 80 m × 80 m, the Areal Change provides a displacement of 19500 sqm of GFS (residence, commerce and

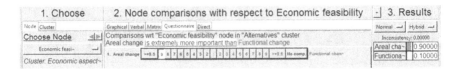

Fig. 6. Detail of pairwise comparison related to economic feasibility. Source: Super Decisions software and authors

tertiary) with a total cost[6] (including reclaim and renaturalization costs) of about €24 million while the Functional Change, with an exchange of 4746 sqm of commercial and tertiary GFS, creates a total cost of about €3 million. It is clear that the Areal Change's dominance is maximum: it obtains 9 as evaluation. Therefore, it is important to check out the consistency ratio (less than 0.1): from a methodological point of view, the assessment is consistent.

Finally, Fig. 7 shows the results of the assessment: the Functional Change exceeds the Areal Change of about 60 %. The results' analysis demonstrate the greater weight of economic and environmental data: from the economic point of view, the costs of Functional Change is definitely limited so more sustainable; from an environmental perspective, the Areal Change strategy increases the consumption of soil considerably while the Functional Change is related to blocks already built. It is obvious that the evaluation is strictly related to the type of risk and the urban context in which these interventions have to take place[7].

Name	Graphic	Ideals	Normals	Raw
Areal change	▬▬▬▬	0.380433	0.275590	0.300794
Functional change	▬▬▬▬▬▬▬	1.000000	0.724410	0.790661

Fig. 7. Final evaluation's results. Source: Super Decisions software and authors

6 Conclusion

In presented paper authors focused on a comprehensive procedure able to reduce risk in urban settlement throughout city planning instruments. To reach this goal, authors proposed an integration of three different methodologies belonging diverse disciplinary fields: risk analysis (centering on mitigation and exposure reduction); scenario planning (as organizational model for the entire planning process); Multicriteria analysis (in particular ANP, as tool to assess the efficiency of different scenario). The study of risk analysis carried to measure how exposure influences the value of risk, and how it

---

[6] Data from Pavia Camera di Commercio price list [31].

[7] Ideals: final normalized judgments in the 0–1 gap; Normals: final normalized judgments related to their sum; Raw: values from priority vector of limit supermatrix.

depends on the probability to have a certain density of people in specific urban functions (evaluated as percentage in 24/7/356 h/day/year estimation). In this context, authors outlined two main city planning strategies, both aimed to reduce exposure. One is related to functional modifications in a predefined area, the other is the relocation of predefined functions in different risk area. These two strategies are assessable throughout qualitative and quantitative criteria and indicators; ANP seems to be the best tool to solve this kind of complex problems.

In the next step, authors will verify in some real contexts the efficiency of the methodology, considering real available data and the (related) definition of criteria and indicators.

References

1. UNISDR (United Nations International Strategy for Disaster Reduction): terminology on disaster risk reduction. In: International Strategy for Disaster Reduction Secretariat, Ginevra (2009). http://www.unisdr.org/eng/terminology/terminology-2009-eng.html
2. Blaikie, P., Cannon, T., Davis, I., et al.: At Risk: Natural Hazards, People's Vulnerability and Disasters. Routledge, Londra (1994)
3. Gotangco, K., Perez, R.: Understanding vulnerability ad risk: the CCA-DRM nexus. Klima, Climate Change Center, Manila (2010)
4. Macchi, S., Tiepolo, M. (eds.): Climate Change Vulnerability in Southern African Cities. Springer Climate. Springer International Publishing, Switzerland (2014a)
5. DRM (Disaster Risk Management). http://www.worldbank.org/en/topic/disasterriskmanagement
6. IPCC (Intergovernmental Panel on Climate Change): Climate Change 2001: Impacts, Adaptation and Vulnerability. Contribution of Working Group II to the Fourth Assessment Report of the Intergovernmental Panel on Climate Change. Cambridge University Press, Cambridge, UK (2001)
7. Verheyen, R.: Climate Change Damage and International Law: Prevention Duties and State Responsibility. Martinus Nijhoff, Boston (2005)
8. Renn, O.: Social theories of risk. In: Krimsky, S., Golding, D. (eds.) Social Theories of Risk. Praeger, Westport (1992)
9. Mileti, D.S., Passerini, E.: A social explanation of urban relocation after earthquakes. Int. J. Mass Emerg. Disaster **14**, 97–110 (1996)
10. Perry, R.W., Lindell, M.K.: Principles for managing community relocation as a hazard mitigation measure. J. Contingencies Crisis Manag. **5**, 49–59 (1997)
11. Menoni, S., Pesaro, G.: Is relocation a good answer to prevent risk? Criteria to help decision makers choose candidates for relocation in areas exposed to high hydrogeological hazards. Disaster Prev. Manag. **17**, 33–53 (2008)
12. De Lotto, R., Morelli di Popolo, C., Morettini, S., Venco, E.M.: La valutazione di scenari flessibili per la riduzione del rischio naturale. Planum. J. Urbanism 27, Atelier 10 (2013)
13. Schoemaker, P.J.H.: Scenario planning: a tool for strategic thinking. Sloan Manag. Rev. **36**, 25–40 (1995)
14. Saaty, T.L.: A scaling method for priorities in hierarchical structures. J. Math. Psycol., Washington (1977)
15. Saaty, T.: How to make a decision: the analytic hierarchy process. Eur. J. Oper. Res. **48**, 9–26 (1990)
16. Saaty, T.: The Analytic Network Process. RWS Publications, Pittsburgh (1997)

17. Saaty, T.: Decision Making with Dependance and Feedback: The Analytic Network Process. RWS Publications, Pittsburgh (2005)
18. Saaty, T.: The analytic hierarchy and analytic network processes for measurement of intangible criteria and for decision-making. In: Figueira, J., Greco, S., Ehrgott, M. (eds.) Multiple Criteria Decision Analysis: State of the Art Surveys. Springer, Berlin (2005)
19. Ferretti, V.: Multicriteria Analysis. Interview with Thomas Saaty. In: Valori e Valutazioni, Siev 13, 25–32 (2014)
20. Saaty, T., Vargas, L.G.: Decision Making with the Analytic Network Process: Economic, Political, Social and Technological Applications with Benefits, Opportunities, Costs and Risks. Springer, New York (2006)
21. Bottero, M., Lami, I., Lombardi, P.: Analytic Network Process. La valutazione di scenari di trasformazione urbana e territoriale. Alinea Editrice, Firenze (2008)
22. Bottero, M., Lami, I., Abastante, F., Greco, S.: The role of analytic network process and dominance-based rough set approach in strategic decisions for territorial transformations. Int. J. Multicriteria Decis. Making 3(2/3), 212–235 (2012)
23. Meadows, D.: Indicators and information systems for suitable development. The Sustainability Institute (1998). http://sustainer.org/
24. Nijkamp, P., Rietveld, P., Voogd, H.: Multicriteria Evaluation in Physical Planning. North Holland Publishing Company, Amsterdam (1990)
25. Lee, M.-C.: The analytic hierarchy and the network process in multicriteria decision making: performance evaluation and selection key performance indicators based on ANP model. In: Crisan, M.: Convergence and Hybrid Information Technologies, InTech (2010)
26. Bottero, M., et al.: Integrating multicriteria evaluation and data visualization as a problem structuring approach to support territorial transformation projects. EJDC (Euro J. Decis. Process.). Springer (2014)
27. Voogd, H.: Multicriteria Evaluation for Urban and Regional Planning. Pion Limites, London (1983)
28. Adams, W.J., Saaty, R.W.: Software super decisions (1993–2003). http://www.superdecisions.com/
29. De Lotto, R., Morelli di Popolo, C.: Opportunità e limiti nella dimensione fisica della città flessibile. Planum. The Journal of Urbanism 25, Atelier 7 (2012)
30. De Lotto, R.: Città e pianificazione. La tradizione di Pavia e le opportunità per il futuro. Maggioli Editore, Politecnica, Santarcangelo di Romagna, RN (2008)
31. Camera di Commercio industria, artigianato e agricoltura di Pavia: Prezziario opere edili della Provincia di Pavia, Pavia (2015)
32. Dwyer, A., Zoppou, C., Day, S., Nielsen, O., Roberts, S.: Quantifying Social Vulnerability: A methodology for identifying those at risk to natural hazards, Geoscience Australia Technical Record, Canberra (2004)
33. Kaynia, A.M., Uzielli, M., Nadim, F., Lacasse S.: A conceptual frame work for quantitative estimation of physical vulnerability to landslides. Eng. Geol. 102, 251–256 (2008)
34. De La Cruz-Reyna, S.: Long term probabilistic analysis of future explosive eruptions. In: Scarpa, R., Tilling, R.I. (eds.) Monitoring and Mitigation of Volcano Hazards, pp. 599–629. Springer, Berlin (1996)
35. Davidson, R.: An urban earthquake disaster risk index, the John A Blume earthquake engineering center, department of civil engineering, Report no. 121. Stanford University, Stanford (1997)
36. IADB, Hahn, H., Villagrán de León, J.C.: Comprehensive risk management by communities and local governments. Indicators and other disaster risk management instruments for communities and local governments (2003)

Data Processing for a Water Quality Detection System on Colombian Rio Piedras Basin

Edwin Castillo[1], David Camilo Corrales[1,2(✉)], Emmanuel Lasso[1],
Agapito Ledezma[2], and Juan Carlos Corrales[1]

[1] Grupo de Ingeniería Telemática, Universidad del Cauca,
Campus Tulcán, Popayán, Colombia
{efcastillo,dcorrales,eglasso,
jcorral}@unicauca.edu.co
[2] Departamento de Ciencias de la Computación e Ingeniería, Universidad
Carlos III de Madrid, Avenida de la Universidad 30, 28911 Leganés, Spain
davidcamilo.corrales@alumnos.uc3m.es,
ledezma@inf.uc3m.es

Abstract. Freshwater is considered one of the most important of planet's renewable natural resources. In this sense, it is vital to study and evaluate the water quality in rivers and basins. A study area is Rio Piedras Basin, which is the main water supplier source of 9 rural communities in Colombia. Nevertheless, these communities do not make a water quality control. Different research has been conducted to develop water quality detection systems through supervised learning algorithms. However, these research approaches set aside the data processing for improve the outcomes of supervised learning algorithms. This paper presents an improvement of data processing techniques for a water quality detection system based on supervised learning and data quality techniques for Rio Piedras Basin.

Keywords: Water quality data · Lotic ecosystem · Dimensionality reduction · Imbalanced classes · Classifier · DT · ANN · BN · K-NN · SVM · PCA · Boosting and SMOTE

1 Introduction

Freshwater is considered one of the most important of planet's renewable natural resources. In this sense, it is vital to study and evaluate the water quality in lotic ecosystems, which represent water ecosystems in constant motion and in the same direction, such as rivers and basins [1]. A problem occurs in Rio Piedras Basin, which is located on the western slope of the Central Cordillera, west of Popayán (Colombia) and it is the main water supplier source of 9 rural communities: Huacas, Laureles, Canelo, Quintana, San Juan, Santa Teresa, Laguna, San Ignacio, and San Isidrio [2, 3]. Nevertheless, these communities do not make a control water quality.

A significant amount of research has been conducted to develop water quality detection systems, which allow monitoring activities of water quality on different basins [4–7] and lotic ecosystems [8–10] around the world [5, 11–17], through supervised

© Springer International Publishing Switzerland 2016
O. Gervasi et al. (Eds.): ICCSA 2016, Part IV, LNCS 9789, pp. 665–683, 2016.
DOI: 10.1007/978-3-319-42089-9_47

learning (SL) algorithms. SL tasks predict or classify a new input data from examples (instances), commonly called training data (composed of attributes and a target variable), through algorithms such as decision trees (DT), Bayesian networks (BN), Artificial Neural Networks (ANN), K-Nearest Neighbor (K-NN) and Support Vector Machines (SVM) [18]. However, these research approaches set aside the data quality verification (i.e. redundant attributes, duplicate instances, imbalanced dataset, etc.).

Therefore, an improvement of data processing techniques for water quality detection system for *Rio Piedras Basin*, based on supervised learning techniques, which considered the issues founded on a data quality verification phase. The remainder of this paper is organized as follows: Sect. 2 describes the study area description, the data quality issues addressed and supervised learning algorithms used; Sect. 3 refers the data processing for water quality detection system proposed; Sect. 4 presents results and discussion and Sect. 5 relates conclusions and future work.

2 Background

2.1 Study Area Description

The data used in this study were collected quarterly in the *Rio Piedras* Basin, located in Cauca department, Colombia (source: 76° 31' 10" west of Greenwich and 2° 21' 45" north latitude, mouth of river: 76° 23' 45" west longitude and 2° 25' 40" north latitude), by the Environmental Studies Group (ESG) from the University of Cauca, between 2011 and 2013, taking into account the methodology followed in [19]. Captured samples contain biological (macroinvertebrates) and physicochemical variables, at three points of the basin: Puente Alto, Puente Carretera and Bocatoma Diviso, in different precipitation periods: high (October-November), average (June-July) and low (August September).

Thus, there were captured 10 physicochemical indicators, 5 biological indicators and 3 precipitation periods. In total, the built dataset consists of 645 records, and 3 values to classify (classes) [17], as set forth in Table 1.

According to the latest work, the three (3) values to classify are denoted by the numbers 1, 2 and 3, which represent a high water quality (very clean water), good (slightly polluted water) and Regular (moderately polluted), respectively.

2.2 Dimensionality Reduction

The dimensionality reduction is the transformation of high-dimensional data in a meaningful representation of smaller dimensions. This reduced representation must have the minimum number of parameters required for expressing the observed data properties [20, 21]. The mentioned task is primarily oriented towards two objectives: instances and attributes reduction techniques.

Attributes Reduction (AR). The attribute reduction decreases the dimension of the attributes within a dataset [21–24]. AR methods are grouped into two categories: attribute selection and extraction; the first looks the best subset of features according to

Table 1. Dataset *Rio Las Piedras* attributes

Category	Attribute	Unit of measurement	Range	Class
Physicochemical indicators	Temperature	°C	13.0 - 17.8	High Water Quality (1)
	Conductivity	μs/cm	35.2 - 89.0	
	Total dissolved solids	mg/L	16.5-42.1	
	Dissolved oxygen	mg/L	7.17-8.23	
	pH	mg/L	6.62-8.17	
	Ammonium	mg/L	0.01-0.04	
	Nitrates	mg/L	0.01-0.09	
	Nitrites	mg/L	0.01-0.06	Good Water Quality (2)
	Phosphates	mg/L	0.08-0.24	
	Turbidity	mg/L	1.0-9.8	
Biological indicators	Class	-	-	
	Order	-	-	
	Family	-	-	Low Water Quality (3)
	Taxon	-	-	
	Number of individuals	-	-	
Precipitation periods	Month	-	-	
	Year	-	-	
	Sampling point	-	-	

certain criteria (choice of attributes and/or number of attributes to be selected), discarding redundant, inconsistent and irrelevant attributes, while the second transforms the high dimension attributes set in a space of smaller dimension [25, 26].

Noteworthy is the importance of AR mechanisms, in optimizing a dataset. However, when applying techniques for selecting attributes, information loss is generated [22], which is a problem in small datasets, such as the dataset *Rio Las Piedras*. Therefore, this paper chose to use attribute extraction tactics.

At the same time, it was conducted a systematic review of 44 published researches from 2004 to the present, based on the guidelines set forth in [27], taking as search sources: IEEE Xplore (35 items), ScienceDirect (9 items), focused in application domains such as: intruder detection, medicine, biometrics, facial recognition, satellite images classification, among others, obtaining that the algorithm of Principal Component Analysis (PCA) is the most used (33 items), and also overcomes the capabilities of information viewing and understanding than other extraction techniques. Therefore, this algorithm is taken as a starting point to reduce attributes that allow a water quality evaluation task. The explanation of the chosen techniques is presented in Sect. 3.

Instances Reduction. Instances Reduction (IR) decreases the number of irrelevant instances within a dataset [28–30]. Several authors propose classifications of IR methods, as is the case of [31], in which group IR techniques: Noise Filters, Condensation Algorithms and Prototype Algorithms, while in [32] are classified in:

Wrapper and Filter. Furthermore, in [33–35] ensemble methods as Cascading, stacking, Bagging, Boosting, Random Forest are used for IR tasks.

To select the appropriate algorithm for the instances extraction, 34 studies published from 2006 to the present were reviewed, using as search sources: IEEE Xplore (12 items), ScienceDirect (10 items), Springer Link (6 items) and Google scholar (6 items) and focusing the search in application areas such as: intruder detection, security, classifiers building, time series, text recognition, among others. In this review it was found that the most commonly used techniques are Ensemble and Wrapper with 16 and 12 papers respectively, while "Filter" methods are referenced only 6 times.

It is important to indicate that the Wrapper algorithms tend to be over-trained (overfitting) due to frequent use of cross-validation as evaluation technique on a single dataset, which tends to be adjusted to very specific training data features that do not have no causal relationship with the objective function [36]. Moreover, within the Ensemble methods, Boosting is the IR technique most often used [34] and it has a greater capacity of data generalization. Based on the above reasons, this research will take as its starting point the Boosting algorithm for IR tasks.

2.3   Imbalanced Classes

The classes imbalance problem occurs when the number of class label instances is greater (majority or negative class "C−") on the number of instances that have other labels class (minority or positive class "C+") [37, 38], and whose unbalance degree can be measured using *IL (Imbalance level) reason* [39–41].

In this scenario, the classifiers have a tendency classification to the majority class, thereby minimizing the classification error and correctly classifying majority class instances detriment of minority class instances.

To solve the imbalance class problem, there have been two main approaches [38, 42, 43]: the external method (at data level) and the internal method (at classification algorithms level). The first consists of achieving a balance between classes by eliminating instances of the majority class (sub-sampling) or the inclusion of instances in the minority class (over-sampling); while the internal method adjusts the classifiers to favor the class minority. In this comparison the external method is the most versatile (because they do not require any change in the algorithm and also can be used in different application domains), the most widely used in the last decade and whose most representative algorithm is SMOTE (Synthetic Minority Over-Sampling Technique), which adds to the minority class examples by creating new instances (which called synthetic data) obtained from an interpolation process.

2.4   Classifiers for Water Quality Detection

For classifiers selection, there were taken 4 researches as a starting point [18, 44–46], in which performing a literature review and theoretically evaluate supervised learning algorithms most commonly used as the case of Decision Trees (DT), Artificial Neural

Networks (ANN), Bayesian Networks (BN), K-Nearest Neighbor (K-NN) and Support Vector Machines (SVM) considering metrics as: accuracy, noise tolerance, ability of explanation, learning speed and classification speed.

3 Mechanism for Detecting Water Quality in *Rio Piedras Basin*

The water quality detection system in *Rio Piedras Basin*, contains various components to perform a pre-processing data and supervised learning algorithms based on data mining tool: KNIME® Analytics Platform, all transparent to the end user. In Fig. 1 the architecture of the proposed mechanism is exposed, which is made up of data processing and classification modules. The first module consists of two components for dimensionality reduction (attributes and instances) to address the redundant values problem and a component that balances the classes. The second module consists of some supervised learning algorithms: DT, ANN, BN, K-NN and SVM.

Below, the proposed mechanism components are briefly described:

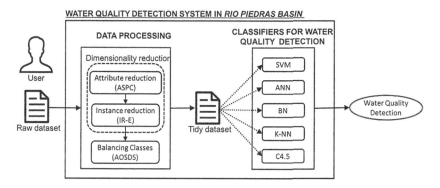

Fig. 1. Mechanism architecture

3.1 Automatic Selection of Principal Components (ASPC)

The Automatic Selection of Principal Component (ASPC) is based on the PCA algorithm definition, and proposes a new approach to Principal Components (PC) selection.

First, PCA is a multivariate technique from data exploratory analysis (statistical), which transforms a large number of attributes, correlated with each other, in a number of attributes smaller uncorrelated (linearly independent or orthogonal), called principal components (PC). The main components are a linear combination of the original attributes, which describe the greatest amount of dataset information and are sorted by the amount of information that they contain (high to low) [47–49].

Currently, there is no definite rule on the exact number of PC to be used. However, to select the optimal PC subset, the authors in [50] exhibit various techniques such as:

B1-Backward, B1-Forward, B2 and B4; obtaining better results with B4 method. B4 intends to use the PC whose accumulated explained variance exceeds 60 % of the information.

On the other hand, ASPC is the mechanism, which consists of building data subsets $X'_{n*p'}$, for $p' = 1, 2, \ldots, p$ and $p = EV$ (PCA determines eigenvalues (EV) and attributes p within the dataset) with p' variables and n observations, from each of the eigenvalues and their eigenvectors.

From here, data subsets are constructed based on the accumulated $\{1\}, \{1+2\}, \{1+2+3\}, \ldots, \{1+2+3+m\}$ PC. Subsequently, each data subset is evaluated by means of classification accuracy obtained by some supervised learning techniques: SVM, BN, K-NN, DT and ANN. Finally, it is selected the data subset with better accuracy and thus the number of PC associated with this.

It is worth noting, for the main components analysis should be observed the relationship between PC and the initial attributes from the feature matrix taking into account the correlations sign and magnitude.

3.2   Instances Reduction (IR-E)

The mechanism defined for Instances Reduction (IR-E) takes as a starting point the model proposed in [34], which is based on the Boosting Classifier definition to create a model that selects redundant and irrelevant instances within a dataset and is called BIS (Boosting Instance Selection). This algorithm replaces the classification models for IR algorithms as: Decremental Reduction Optimization Procedure 3 (Drop3), IB3, Iterative Case Filtering (ICF), Modified Selective Subset (MSS), Reduced Nearest Neighbor (RNN), Condensed Nearest Neighbor Rule (CNN).

BIS performs a process M times (M is defined a priori) and, in each iteration, a voting process is executed, which consists of assigning one vote (v_i) to each selected instance (x_i) through a particular IR technique randomly chosen. After M iterations, it is obtained as a result a votes vector V, which records the obtained votes by each instance and, from the latter, a set of thresholds is built $(\emptyset_1, \emptyset_2, \ldots, \emptyset_M)$, with the instances which obtained most votes. Now, to get the best threshold it is defined a criteria $J(\emptyset)$, which is represented by an instances subset $S_\emptyset$ belonging to the training set T, so that meet the Eq. 1.

$$\theta : S_\theta = x_i \in T : v_i > \emptyset \tag{1}$$

To asses each $J(\emptyset_i)$ criteria, the K-NN classifier is trained with the subset $S_{\emptyset_i}$ where the quality of each instances subset is selected according to both, the classification performance and the number of removed instances, as seen in the Eq. 2. In fact, any classifier can be used (DT, SVM, BN o ANN); Nevertheless, for this research we will restrict ourselves to use K-NN due to its simplicity, ability to detect wrong results and high speed learning [18, 44–46],

$$J(\theta) = \alpha C + (1 - \alpha)r \qquad (2)$$

Where, C symbolizes the classification performance (precision or AUC), r denotes the deleted instances percentage and α is the parameter used to distinguish the importance of each factor.

On the other hand, each threshold assessment involves a K-NN classifier training with the instances subset that represents it, which implies a high degree of complexity of assessment process $(2M + 2)$ when number of iterations M is high (the maximum number of possible thresholds is equal to M).

To address the problem, optimal threshold selection called $\emptyset_o$, from an approach that works on two levels. The first level consists of constructing a subset of thresholds $(\theta_1, \theta_2, \ldots, \theta_p)$ from the set of possible thresholds $(\emptyset_1, \emptyset_2, \ldots, \emptyset_M)$ for $p \leq M$, discarding the repeated or equivalent thresholds latter, because these get similar assessments in terms of performance and classification codes. Thus a degree of less than or equal complexity $(2p + 2)$ is achieved.

At the second level it seeks to find a threshold value representing both high voting values as low values and thus counteract the drawbacks mentioned above. For this, the arithmetic mean of the thresholds subset resulting from the previous step is determined, which process is represented by the Eq. 3.

$$\emptyset_o = \frac{1}{p} \sum_{i=1}^{p} \theta_i \qquad (3)$$

3.3   Automatic Optimal Synthetic Data Selection (AOSDS)

AOSDS is based on the over-sampling algorithm SMOTE (Synthetic Minority Over-Sampling Technique) definition and proposes a strategy to generate the appropriate number of synthetic data in such a way that minimizes over-training the classifier.

SMOTE currently works with binary classes, a majority and a minority class, denoted as C− and C+ respectively. This algorithm creates synthetic instances or data for the minority class, by interpolating an instance and its closest K neighbors belonging to that class. First, K closest neighbors from the minority class are selected and, subsequently, the synthetic instantiated percentage (denoted by P) is chosen. Then, to generate a new synthetic data, interpolation between the line connecting each minority class instance with any (or all) of their closest K neighbors previously selected is performed. This calculation is made using the Euclidean distance definition and a replacement or overlay (called overlap) function that assigns a value 0 (if both values are equal) or a value 1 (in the case they are different). Since SMOTE only applied to binary classes, this work focused the problem of imbalance dataset Rio Las Piedras on two classes: Class 1 and 3 (C− and C+ respectively) and ignoring the remaining class (class 2). However, it is important to mention that class 2 is only ignored in over-sampling process.

Although this technique generates new instances that allow balancing classes, so far not found a method to indicate the optimal number of instances to be created, since a large number of synthetic instances can over-train the classifier and generate inaccurate results [42, 51, 52].

The proposed strategy involves determining that the original dataset imbalance level (IL) meets a minimum classes threshold unbalance. If the dataset meets the threshold of imbalance, we proceed to apply to the minority class definition SMOTE. For this case, an imbalance level $IL \geq 3.9$ is defined and instances percentage to oversample (P value) equal to 50, 100, 150 and 200 % will be used, thereby preventing the ratio C+ instances exceed those of C−. Furthermore, [53] discloses that to achieve a 200 % of oversampling only required two nearest neighbors, enough for the study case. Therefore, $K = 2$ is taken as the number of nearest neighbors for SMOTE.

Once applied the over-sampling process for a given P, the behavior of each subset generated is evaluated and the data subset that gets the best performance is selected. At the same time, an optimal synthetic data percentage, associated with that dataset, is defined.

3.4 Classifiers to Water Quality Detection

This component consists of classifiers to perform a water quality detection in *Rio Piedras Basin*, which were selected based on the research presented in [18, 44–46] where theoretically evaluate some classifiers: DT, ANN, BN, K-NN and SVM, considering metrics as: precision, noise tolerance, explanatory capacity, learning speed and classification speed. In these researches, is claimed that there is no algorithm that satisfies all evaluation metrics. Also, depending on the dataset used, the algorithm has a different behavior (in metrics of model evaluation such as precision). Therefore, it is necessary to evaluate each of these algorithms in order to obtain a high degree of precision in predictions and easy interpretations. Additionally, it is important to mention that to evaluate the proposed mechanism, cross-validation with $k = 10$ was used.

4 Experimental Results

ASPC, IR-E and AOSDS methods were evaluated individually and collectively, using classifiers as SVM, BN, K-NN, C.4.5 (DT) y ANN, applied to the original and processed datasets. The experimental process can be seen in more detail in the Fig. 2.

4.1 Attributes Reduction

The results of PCA process on the *Rio Las Piedras* dataset are shown in Fig. 3. As shown in Fig. 3(a), 18 principal components (PC) are generated, where first PC represents 21.2 % of the total information, the second PC explains 15.2 % of the original variance, the third explains 11.6 %, and so on until PC 18, which represents 0.1 % of the total information. As discussed above, explained variance data are important to

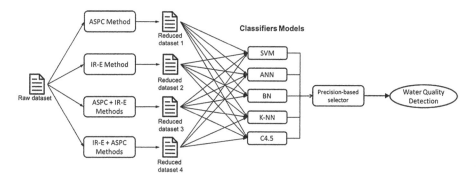

Fig. 2. Experimental process for Dimension Reduction

know the number of PC that will be used in the analysis. If we take B4 methodology as a criterion for components selection, it is considered that the optimal number of PC (reducing the PC amount as much as possible) is 5 components, which variance is 65.2 %. This means that the five (5) first PC represent 65.2 % of the total dataset information. As shown in Fig. 3(b), the other components explain significantly lower percentages comparatively to the first five components.

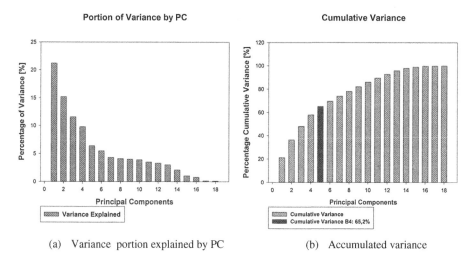

(a) Variance portion explained by PC   (b) Accumulated variance

Fig. 3. PCA results on *Rio Las Piedras* dataset

Once applied the PCA process on the dataset, the ASPC proposed mechanism is applied, which involves as a first step in building a dataset with $\{1\}, \{1+2\}, \{1+2+3\}, \ldots, \{1+2+3+\cdots 18\}$ principal components. Subsequently, the classifiers mentioned above are training with each of these datasets, with a 10-fold cross-validation. Finally, we obtain as a result the data subset that best precision obtained and the number of PC associated with this dataset. These results are shown in Table 2.

Table 2. ASPC results

Dataset	No. attributes	Information representation	Classifiers precision					Average precision
			BN	SVM	C.4.5	K-NN	ANN	
Not processed	18	100 %	80.2 %	92.8 %	77.1 %	73.2 %	96.1 %	83.8 %
ASPC	5	65.2 %	91 %	91.4 %	91.4 %	86.9 %	89 %	89.9 %

As seen, the ASPC mechanism selects the dataset consisting of the first 5 PC as the best data set, because the latter had the best average precision (89.9 %) among all dataset formed. The reduction of 13 components obtained by the ASPC method led to improved classification performance by approximately 6 %.

In practical terms, the problem initially represented in an 18-dimensional cyberspace has been reduced to a smaller hyperspace (5 dimensions), capturing 65.2 % of the original variance. This result implies greater and easy data interpretability, process that, as mentioned above, is performed by analyzing the relationship between PC and initial attributes.

From these results, it can be assumed that the ASPC method can perform a quite strong selection of training data without deteriorating the capacity of classifiers.

Now, to compare the ASPC method with the B4 selection method in the Table 3, the results obtained are reflected.

Table 3. Comparison between ASPC and B4 methods

ASPC			B4		
No. PC	Information representation	Average precision	No. PC	Information representation	Average precision
5	65.2 %	90.1 %	5	65.2 %	90.1 %

The ASPC mechanism applied on the *Rio Las Piedras* dataset was able to reduce the space to five components. These results corresponded with those obtained with the B4 method, which suggests that the new dataset is constituted with five components and whose information amount is within the accumulated variance confidence interval 60 % − 95 %. Moreover, the mechanism increased both the classification process performance as the model interpretability (fewer attributes).

4.2 Instance Reduction

By applying the IR-E method proposed for instance reduction on a water quality dataset, the results presented in Table 4 were obtained.

Table 4. RI-E results

Dataset	No. attributes	No. instances	No. classes	Classifiers precision					Average precision
				RB	SVM	C.4.5	K-NN	RNA	
Not processed	18	645	3	80.2 %	92.8 %	77.1 %	73.2 %	96.1 %	83.8 %
IR-E	18	336	3	50.6 %	57.1 %	62.8 %	55.2 %	55.2 %	56.1 %

As noted, the IR-E technique achieves to reduce to 309 the instances within the *Rio Piedras* dataset. However, IR-E decreased the classifiers precision in 27.7 %, indicating that important dataset instances were removed, thus losing important information.

4.3 Attributes and Instances Reduction

Under the same context, the Table 5 shows the results product to train classifiers with the newly acquired dataset after applying jointly reduction techniques presented above. As noted, the sequential composition of attribute reduction and subsequent instances reduction is denoted as ASPC + IR-E, while the sequential combination of instances reduction and subsequent attributes reduction is represented as IR-E + ASPC.

Table 5. Attribute and instance reduction methods in conjunction results

Reduction methods	No. attributes	No. instances	Information	Classifier precision					Average precision
				BN	SVM	C.4.5	K-NN	ANN	
ASPC + IR-E	5	493	65.2 %	72.7 %	73.8 %	76 %	72.7 %	74.2 %	73.9 %
IR-E + ASPC	3	225	41.5 %	44.7 %	41 %	60.9 %	65.7 %	59.1 %	54.3 %

It is interesting that the IR-E + ASPC methods sequence reduced in greater quantity both, attributes such as number of instances, in comparison with the technique ASPC + IR-E. The first technique reduced 15 components and 420 instances, while the second reduced 13 components and 152 instances. However, when a classifier is trained with the dataset obtained trough ASPC + IR-E a higher precision (73.9 %) is obtained in comparison when the training process is performed with the dataset obtained trough IR-E + ASPC method (54.3 %). In addition, it may be noted that the dataset reduced by IR-E + ASPC is represented by three characteristics that explain only 41.5 % of the total variance, which is outside the confidence range suggested by the B4 criterion. This means that this method is not appropriate to reduce the *Rio Las Piedras* dataset size.

As a result, we can see that the ASPC + IR-E mechanism is a suitable solution for reducing the water quality dataset size, which allows classifiers provide a similar precision to that obtained with the original dataset (no pre-processing). Accordingly, the IR-E + ASPC method is not appropriate to reduce the dataset size.

4.4   Average Classifiers Training Time

It is noteworthy that another way to evaluate the algorithms performance (efficiency) is by reducing classifiers runtime in dataset processing task. This measure is important when the task of training a classifier for detecting water quality is related to a highly complex scenario (large amounts of data) or a decision support system that requires to define a contingency action in relatively short times.

For this analysis, the arithmetic average time that each classifiers takes to process a dataset is calculated. For this case study, the datasets would be the no pre-processing or original *Rio Las Piedras* dataset and processed or reduced *Rio Las Piedras* dataset. In Fig. 4 these results are summarized.

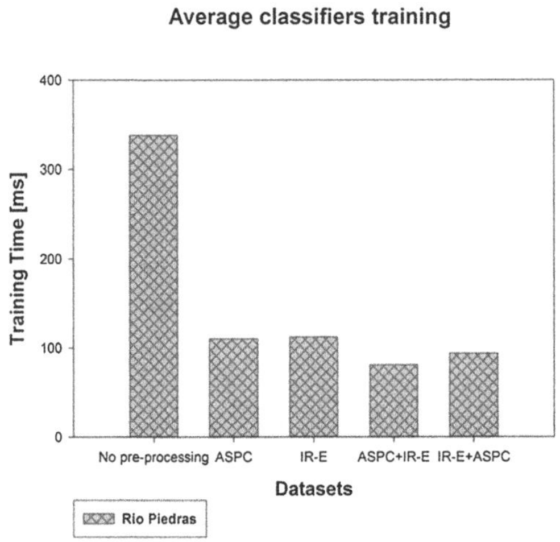

Average classifiers training

Fig. 4. Average classifiers training

In a first step, it is observed that the ASPC technique, in addition to reduce the number of dataset attributes (13), reduces the classifiers training time in 228 ms in comparison with the training time for the original dataset. This indicates that the existence of redundant features in the dataset influences the performance of classifiers.

Following the same behavior, it is observed that the IR-E technique reduce the dataset in 309 instance and the classifiers training time in 226 ms. The result suggests that by minimizing the amount of redundant instances and noise, it is possible to reduce the time necessary to training the classifiers (computational cost).

Now, in the case of reduction methods combined ASPC + IR-E and IR-E + ASPC applied on the *Rio Las Piedras* dataset, we have that in both cases is greatly reduced the time for training the classifiers (244 ms and 257 ms respectively). Additionally, it is noted that the approach ASPC + IR-E allows a classification process significantly faster than individual methods and even the combined method IR-E + ASPC.

These results are to be expected, because such mechanisms remove both datasets characteristics (instances and attributes), which represent a computational cost reduction in data mining tasks (classification and/or grouping).

As a result, the proposed strategy gets similar precision values to the original dataset and greatly reduces classifiers training time, making this proposal the most appropriate to reduce the dataset size for water quality detection.

4.5 Class Balance

To evaluate the proposed mechanism, a classifiers cross-validation was used with k = 10: SVM, BN, K-NN, C.4.5(DT) and ANN. On this approach, an imbalanced dataset *Rio Las Piedras* version (Table 6) and over-sampled datasets with 50 %, 100 %, 150 % and 200 % of synthetic instances (Table 7), was evaluated. As indicated above, the majority and minority class are denoted as C− and C+ respectively.

Table 6. Imbalanced dataset description

No. classes	No. instances	No. instances C−	No. instances C+	IL
3	493	307	54	5.6

The performance of each of these classifiers is obtained in terms of ROC metrics (Receiver Operating Characteristic) and F-Measure (measures in %), considered to evaluate both the original and the over-sampled datasets with different amounts of synthetic instances (50 %, 100 %, 150 % and 200 %). The results are shown in Table 7.

Table 7. Classifiers performance behavior over C− and C+ classes

Class/Dataset	SVM		ANN		C.4.5		BN		K-NN	
Original	F-M	ROC	F-M	ROC	F-M	ROC	F-M	ROC	F-M	ROC
C−	87.4 %	77.8 %	86.8 %	78 %	88 %	75.6 %	86.6 %	76.5 %	80 %	75.8 %
C+	0	66.3 %	0	62 %	0	60.8 %	0	60 %	9 %	51.5 %
SMOTE-50										
C−	84 %	74.4 %	84.8 %	76 %	84.7 %	76 %	83 %	73.6 %	78 %	72.8 %
C+	0	60 %	0	62 %	18.6 %	66 %	0	62.5 %	43.6 %	60.8 %
SMOTE-100										
C−	81 %	71.7 %	80.8 %	75 %	84.6 %	81 %	80 %	73 %	75 %	71 %
C+	0	60 %	0	67.7 %	51.6 %	79 %	0	64.8 %	41 %	64 %
SMOTE-150										
C−	78 %	69 %	76 %	64 %	84.5 %	82 %	72 %	73.5 %	74 %	72.5 %
C+	0	61.4 %	21 %	68 %	59.5 %	79.8 %	29 %	68.5 %	47.7 %	66.5 %
SMOTE-200										
C−	69 %	75.7 %	72 %	72.6 %	79 %	80.7 %	67.7 %	72.5 %	73 %	72 %
C+	0	72 %	21.5	68.6	60 %	81 %	40.6	69 %	53 %	67.7 %

As indicated in the above table, to train the C.4.5 classifier with the original *Rio las Piedras* dataset (*IL* = 5.6), instances belonging to the negative class tend to be correctly classified (M-F = 88 % y ROC = 75.6 %), while those belonging to the positive class tend to be classified incorrectly (M-F = 0 y ROC = 60.8 %). This is because the classification methods tend to favor the negative class.

Similarly, increasing the number of positive class instances with synthetic data, the classification performance of the latter improves, obtaining the best results when synthetic data at 150 % and 200 % are generated, with F-M = 59.5 %, and ROC = 79.8 % and F-M = 60 %, ROC = 81 % respectively. Although there is no significant difference in these results, and considering that what is sought in class balance process is that the classification process can correctly predict C+ instances without affecting significantly the C− instances detections, the best classification results are obtained when C+ is balanced with 150 % of synthetic instances. This oversampling level improvement performance metrics F-Measure and ROC in C+ class at 59.5 % and 19 % respectively, without affecting the C− class detection, moreover, affects only in 3.5 % the F-Measure and increasing the ROC value by 6.4 %.

Now, the K-NN measures shown in Table 7 exposes clearly the same behavior of the previous classifier (C.4.5), where the C+ class is oversampling at 150 % of synthetic instances and this allows the classifier to discriminate it better and detect it correctly (F-Measure and ROC metrics increase their performance to 38.7 % and 15 respectively), at the expense of 6 % of precision and 3.3 % of ROC area reduction for C− class. Meanwhile, when C+ class is oversampled with a sampling level of 200 %, F-Measure and ROC metrics improves at 5.3 % and 1.2 % with respect to the previous process, further reducing the performance of C− classification.

On the other hand, the SVM classifier got for each C+ class a value of F-Measure = 0. Otherwise it occurs in the C− class, where good results are obtained. From here, we can say that SVM classifier is very sensitive to the class imbalance problem, because this algorithm has its separation hyper-plane very close to the minority class, resulting in a low or null performance for examples of this class, compared to those of the majority class [54].

Like the BN, the ANN are more difficult to profit in the metrics when the dataset is oversampled with a percentage of less than 150 % synthetic instances. In fact, like the previous cases, the best results for all cases are obtained when the dataset is oversampled with this value (150 %), and can be seen as the best balance between the number of true and false positives.

From these results, we can see how in spite of balance the dataset trough SMOTE technique, the percentage of correctly classified instances by SVM, BN and ANN algorithms does not significantly increase, and is even lower than the results obtained when classifiers are trained directly with imbalanced dataset (original). Therefore, the fact of using synthetic samples generated trough SMOTE represents a gain in the discriminant capacity of classes, but not equal in all classifiers.

In this vein, the experimental results showed that oversample the imbalanced dataset with levels of 150 % and 200 % improves the detection of classifiers in C+ class without significantly impairing the detection of the C− class. In addition, it is observed that by using a level of imbalance or another, the same results are obtained approximately. However, to create too many synthetic instances produces overtraining, which affects the ability to detect the minority class.

From the foregoing, it is considered that the best level of synthetic instances oversampling is 150 %, and the SMOTE algorithm is considered as a practical method for generating synthetic instances. However, we must be especially careful in the number of generated instances, as it can make poor results in C− class prediction.

Based on the considerations submitted, in the Table 8 balanced dataset (with 150 % of synthetic instances) is described.

As shown in the above table, to oversample the dataset with 150 % of synthetic instances increase in 54 the examples of *Rio Las Piedras* dataset.

4.6   Classification Module

This section presents the classifiers evaluation (DT, ANN, BN, K-NN, SVM) and results analysis, applied on the dataset described in the Table 1 and on the processed dataset after applying the dimensionality reduction and balancing classes approach (Table 8).

Table 8. Overview of processed dataset

Dataset	No. attributes	No. instances	No. classes
Rio Las Piedras	5	547	3

As mentioned above, the performance of each of these classifiers is obtained in terms of some metrics as: precision, recall, F-Measures and ROC, considered to evaluate both the original as the processed dataset (proposed mechanism). The results are presented in Fig. 5.

In the case of the original dataset, the ANN and C.4.5 algorithms obtained the best results among all evaluated classifiers, incorrectly classifying a smaller instances proportion, compared to the other methods, which can be checked with the obtained precision   (C.4.5 = 83 %   and   RNA = 76 %,   compared   to   SVM = 62.6 %, K-NN = 70.1 % and BN = 65 %).

Additionally, it is important to mention that the C.4.5 classifier gets the highest proportion of true positives than other techniques because it has a higher recall (83 %), while other algorithms obtained a recall value not more than 70 % for the case of SVM, BN and K-NN. Regarding to F-Measure, the C.4.5 and ANN followed the same behavior, as they obtain the best values (83.3 % and 76.3 % respectively) among all the evaluated techniques. These results suggest that C.4.5 and ANN supervised algorithms have a best behavior for working with this dataset.

Similarly, the five supervised learning algorithms are trained with the processed dataset (Table 8), as shown in Fig. 5. This graph shows that for all classification models, the number of incorrectly classified instances was reduced, except for C.4.5, where remained approximately constant. This behavior is reflected in the increased

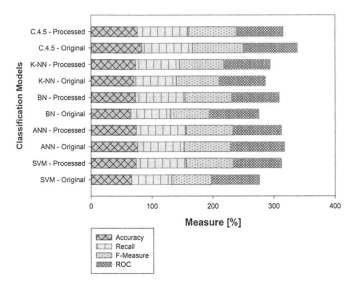

Fig. 5. Result of classifiers evaluation: SVM, ANN, K-NN, BN and C.4.5 applied to *Rio Las Piedras* original and processed datasets

precision of classifiers: BN (72.7 %), SVM (73.8 %) and K- NN (72.7 %) with the exception of ANN which failed to increase its accuracy (74.2 %).

However, the true positives proportion in relation to false positives is high, since the recall values exceed 72 % (SVM = 82.2 %, ANN = 81.7 %, BN = 81.3 %, K-NN = 72.4 % and C.4.5 = 83.6 %). The mentioned results let us to identify that the 5 evaluated models have a good confidence since the number of false positives is low and the number of relevant instances classified is high, as can be contrasted through F-Measure values (SVM = 77.3 %, ANN = 77.2 %, BN = 76.6 %, K-NN = 72.6 % and C.4.5 = 78.8 %).

5   Conclusions and Future Works

In the previous sections, mechanisms for datasets dimensionality reduction and water quality detection were proposed. From the results achieved, it can be inferred that the proposed mechanism (ASPC + IR-E) is an appropriate solution for pre-processing water quality datasets, in order to reduce their size, which allows classifiers to provide a similar precision to that obtained with the original dataset (no pre-processing). Similarly, the execution times to classifiers training tasks were reduced.

Additionally, the evaluated classification algorithms present good results. However, ANN and C.4.5 algorithms obtained the highest precision values, classifying fewer instances wrong and also kept the same behavior in all experiments. Therefore, these supervised learning models are chosen as the base classifiers to detect water quality in *Rio Piedras Basin*, taking into account the dataset characteristics associated with this domain.

As future work, it is intended to deploy the proposed mechanisms in a production environment; address other dataset problems, such as missing values and outliers; and use a similar methodology to generate a water quality prediction, in order to get the ability to construct an early warning system for lotic ecosystems.

Acknowledgments. The authors are grateful to the University of Cauca and its Telematics Engineering Group (GIT) for the technical support and the Environmental Studies Group (GEA) for providing the dataset, Control Learning Systems Optimization Group (CAOS) of the Carlos III University of Madrid, KNIME® Analytics Platform, AgroCloud project of the RIC-CLISA Program, Ministerio de Economía y Competitividad de España (Proyecto TRA2011-29454-C03-03. i-Support: Sistema Inteligente Basado en Agentes de Soporte al Conductor) and Colciencias (Colombia) for PhD scholarship granted to MsC. David Camilo Corrales.

References

1. Álvarez, L.F., Arango, M.C., Arango, G.A., Torres, O.E., de Jesús Monsalve, A.: Calidad Del Agua De Las Quebradas La Cristalina Y La Risaralda, San Luis, Antioquia. In: EIA, pp. 121–141, Julio 2008
2. Marchant, C., Mergili, M., Borsdorf, A.: Agricultura Ecológica y Estrategias de Adaptación al Cambio Climático en la Cuenca del Río Piedras. Cuenca Río Las Piedras (2012)
3. Acosta, M., Devereux, T.: Manual de las medidas de adaptación al cambio climático practicadas por los campesinos de Asocampo de la cuenca Río Las Piedras, Cauca, Colombia: Un resumen visual de las medidas de adaptación local frente al cambio climático y el trabajo y la investigación en campo. Centro Internacional de Agricultura Tropical CIAT 2013
4. Dang, J., Huo, A.-D., Song, J.-X., Chen, X.H., Mao, H.-R.: Simulation modeling for water governance in basins based on surface water and groundwater. Agric. Water Manage. (2016)
5. Sun, W., Liao, H.: Forecasting and evaluating water quality of chao lake based on an improved decision tree method. Procedia Environ. Sci. **2**, 970–979 (2010)
6. Lek, S., Cheng, L., Lek-Ang, S., Li, Z.: Predicting fish assemblages and diversity in shallow lakes in the Yangtze River basin. Limnologica **42**, 127–136 (2012)
7. Zhang, W., Wang, Y., Engel, B.A., Peng, H., Theller, L., Shi, Y., Hu, S.: A fast mobile early warning system for water quality emergency risk in ungauged river basins. Environ. Model Softw. **73**, 76–89 (2015)
8. Yan, J., Tan, G., Gao, C., Yang, S.: Prediction of water quality time series data based on least squares support vector machine. Procedia Eng. **31**, 1194–1199 (2012)
9. Basant, N., Gupta, S., Singha, K.P.: Support vector machines in water quality management. Anal. Chim. Acta **703**, 152–162 (2011)
10. Liong, S.-Y., Tkalich, P., Palani, S.: An ANN application for water quality forecasting. Mar. Pollut. Bull. **56**, 1586–1597 (2008)
11. Xu, J., Liao, Y., Wang, W.: A method of water quality assessment based on biomonitoring and multiclass support vector machine. Procedia Environ. Sci. **10**, 451–457 (2011)
12. Sophatsathit, P., Areerachakul, S., Lursinsap, C.: Integration of unsupervised and supervised neural networks to predict dissolved oxygen concentration in canals. Ecol. Model. **261–262**, 1–7 (2013)
13. Tai, H., Liua, S., Ding, Q., Li, D., Xu, L., Wei, Y.: A hybrid approach of support vector regression with genetic algorithm optimization for aquaculture water quality prediction. Math. Comput. Model. **58**, 458–465 (2012)

14. Bucak, I.O., Karlik, B.: Detection of drinking water quality using CMAC based artificial neural networks. Ekoloji Dergisi **20**, 75–81 (2011)
15. Park, Y.-S., Bae, M.-J.: Biological early warning system based on the responses of aquatic organisms to disturbances: a review. Sci. Total Environ. **466–467**, 635–649 (2014)
16. Gupta, S., Singh, K.P.: Artificial intelligence based modeling for predicting the disinfection by-products in water. Chemometr. Intell. Lab. Syst. **114**, 122–131 (2012)
17. Gonzales, W.F., Castillo, E.F., Corrales, D.C., López, I.D., Hoyos, M.G., Figueroa, A., Corrales, J.C.: Water quality warnings based on cluster analysis in Colombian rivers basins. Sistemas y Telemática (S&T) **13**, 9–26 (2015)
18. Corrales, J.C., Corrales, D.C., Figueroa-Casas, A.: Towards detecting crop diseases and pest by supervised learning. Ing. Univ. **19**, 207–228 (2015)
19. Pérez, G.R.: Bioindicación de la Calidad del Agua en Colombia: Propuesta Para el Uso del Método BMWP Col, Primera ed. vol. 1. Universidad de Antioquia (2003)
20. Fukunaga, K.: Introduction to Statistical Pattern Recognition. School of Electrical Engineering-Purdue University-West Lafa yet te, Indiana
21. Inza, I., Larrañaga, P., Saeys, Y.: A review of feature selection techniques in bioinformatics. Bioinformatics **23**, 2507–2517 (2007)
22. Khalil, K., Nasreen, S., Khalid, S.: A survey of feature selection and feature extraction techniques in machine learning. In: Science and Information Conference (SAI), 27–29 August 2014, pp. 372–378 (2014)
23. Wang, X., Paliwal, K.K.: Feature extraction and dimensionality reduction algorithms and their applications in vowel recognition. Pattern Recogn. **36**, 2429–2439 (2002)
24. Deepa, T., Ladha, L.: Feacture selection methods and algorithms. Int. J. Comput. Sci. Eng. (IJCSE) **3**, 1787–1797 (2011)
25. Popescu, M.C., Sasu, L.M.: Feature extraction, feature selection and machine learning for image classification: a case study. In: IEEE (2014)
26. Paliwal, K.K.: Dimensionality reduction of the enhanced feature set for the HMM-based speech recognizer. Digit. Sig. Process. **2**, 157–173 (1992)
27. Kitchenham, B.: Procedures for performing systematic reviews. Joint Technical report, July 2004
28. Ahmad, S.S.S., Pedrycz, W.: Feature and instance selection via cooperative PSO. In: IEEE, 9–12 October 2011, pp. 2127–2132 (2011)
29. Tsai, C.-F., Chang, C.-W.: SVOIS: support vector oriented instance selection for text classification. Inf. Syst. **38**, 1070–1083 (2013)
30. Chan, Z.-Y., Ke, S.-W., Tsaia, C.-F.: Evolutionary instance selection for text classification. J. Syst. Softw. **90**, 104–113 (2014)
31. Jankowski, N., Grochowski, M.: Comparison of instances seletion algorithms I. Algorithms survey. In: Rutkowski, L., Siekmann, J.H., Tadeusiewicz, R., Zadeh, L.A. (eds.) ICAISC 2004. LNCS (LNAI), vol. 3070, pp. 598–603. Springer, Heidelberg (2004)
32. Ariel Carrasco-Ochoa, J., Arturo Olvera-López, J., Francisco Martínez-Trinidad, J., Kittler, J.: A review of instance selection methods. Artif. Intell. Rev. **34**, 133–143 (2010)
33. Blachnik, M.: Ensembles of instance selection methods based on feature subset. Procedia Comput. Sci. **35**, 388–396 (2014)
34. García-Pedrajas, N., De Haro-García, A.: Boosting instance selection algorithms. Knowl.-Based Syst. **67**, 342–360 (2014)
35. Blachnik, M., Kordos, M.: Bagging of instance selection algorithms. In: Rutkowski, L., Korytkowski, M., Scherer, R., Tadeusiewicz, R., Zadeh, L.A., Zurada, J.M. (eds.) ICAISC 2014, Part II. LNCS, vol. 8468, pp. 40–51. Springer, Heidelberg (2014)

36. Jordan, M.I., Karp, R.M., Xing, E.P.: Feature selection for high-dimensional genomic microarray data. In: ICML 2001 Proceedings of the Eighteenth International Conference on Machine Learning, pp. 601–608 (2001)
37. Hong, X., Gao, M., Chen, S., Harris, C.J.: A combined SMOTE and PSO based RBF classifier for two-class imbalanced problems. Neurocomputing **74**, 3456–3466 (2011)
38. Fernández, A., Garcia, S., Herrera, F.: Enhancing the effectiveness and interpretability of decision tree and rule induction classifiers with evolutionary training set selection over imbalanced problems. Appl. Soft Comput. **9**, 304–1314 (2009)
39. Elrahman, S.M.A., Abraham, A.: A review of class imbalance problem. J. Netw. Innovative Comput. **1**, 332–340 (2013)
40. Satyasree, K.P.N.V., Murthy, J.V.R.: An exhaustive literature review on class imbalance problem. Int. J. Emerg. Trends Technol. Comput. Sci. (IJETTCS) **2**, 109–118 (2013)
41. Verbiest, N., Ramentol, E., Cornelis, C., Herrera, F.: Improving SMOTE with fuzzy rough prototype selection to detect noise in imbalanced classification data. In: Pavón, J., Duque-Méndez, N.D., Fuentes-Fernández, R. (eds.) IBERAMIA 2012. LNCS, vol. 7637, pp. 169–178. Springer, Heidelberg (2012)
42. Cooper, E.W., Nguyen, H.M., Kamei, K.: Borderline over-sampling for imbalanced data classification. In: Fifth International Workshop on Computational Intelligence & Applications (2009)
43. Kamel, M.S., Sun, Y., Wong, A.K.C., Wang, Y.: Cost-sensitive boosting for classification of imbalanced data. Pattern Recogn. **40**, 3358–3378 (2007)
44. Bhavsar, H., Ganatra, A.: A comparative study of training algorithms for supervised machine learning. Int. J. Soft Comput. Eng. (IJSCE) **2**, 74–81 (2012)
45. Kotsiantis, S.B.: Supervised machine learning: a review of classification techniques. Informatica **31**, 249–268 (2007)
46. Zaharakis, I.D., Pintelas, P.E., Kotsiantis, S.B.: Machine learning: a review of classification and combining techniques. Artif. Intell. Rev. **26**, 159–190 (2006). Springer Science
47. Wu, C.-M., Zhang, Y., Luo, Y.: Facial expression feature extraction using hybrid PCA and LBP. J. China Univ. Posts Telecommun. **20**, 120–124 (2013). ScienceDirect
48. Xu, D., Wang, Y.: An automated feature extraction and emboli detection system based on the PCA and fuzzy sets. Comput. Biol. Med. **37**, 861–871 (2007)
49. Xiao, B.: Principal component analysis for feature extraction of image sequence. In: International Conference on Computer and Communication Technologies in Agriculture Engineering, 12–13 June 2010, vol. 1, pp. 250–253 (2010)
50. King, J.R., Jackson, D.A.: Variable selection in large environmental data sets using principal components analysis. Environmetrics **10**, 67–77 (1999)
51. Makond, B., Wang, K.-J., Chen, K.-H., Wang, K.-M.: A hybrid classifier combining SMOTE with PSO to estimate 5-yearsurvivability of breast cancer patients. Appl. Soft Comput. **20**, 15–24 (2014)
52. Sicilia, M.Á., Riquelme, J.C.: SMOTE-I: mejora del algoritmo SMOTE para balanceo de clases minoritarias. Actas de los Talleres de las Jornadas de Ingeniería del Software y Bases de Datos **3**, 73–80 (2009)
53. Bowyer, K.W., Chawla, N.V., Hall, L.O., Philip Kegelmeyer, W.: Synthetic minority over-sampling technique. J. Artif. Intell. Res. **16**, 321–357 (2002)
54. He, H., Ghodsi, A.: Rare class classification by support vector machine. In: 2010 20th International Conference on Pattern Recognition (ICPR), 23–26 August 2010, pp. 548–551 (2010)

Validation of Coffee Rust Warnings Based on Complex Event Processing

Julián Eduardo Plazas[✉], Juan Sebastián Rojas, David Camilo Corrales,
and Juan Carlos Corrales

Grupo de Ingeniería Telemática, Universidad del Cauca, Popayán, Colombia
{jeplazas,jsrojas,dcorrales,jcorral}@unicauca.edu.co

Abstract. The rust is the main coffee crop disease in the world. In the Colombian and Brazilian plantations, the damage leads to a yield reduction of 30 % and 35 % respectively in regions where the meteorological conditions are propitious to the disease. Recently, researchers have focused on detecting the coffee rust disease starting from climate monitoring and parameters of crop control; however most of the monitoring systems lack the ability to process multiple source information and analyse it in order to identify abnormal situations and validate the generated warnings. In this paper, we propose a CEP engine and a prediction system integration for early warning systems applied to the coffee rust detection, capable of analysing multiple incoming events from the monitoring system and validating the warnings detection; evaluating an experimental prototype in a field test with satisfactory results.

Keywords: Complex event processing · Early warning systems · Coffee rust · Climate monitoring

1 Introduction

Coffee rust is a leaf disease caused by the fungus Hemileia Vastatrix. Coffee rust epidemics, with intensities higher than previously observed, have affected a number of countries including: Colombia, from 2008 to 2011; Central America and Mexico, in 2012–2013; and Peru and Ecuador, in 2013. There are many contributing factors to the onset of these epidemics e.g. the state of the economy, crop management decisions and the prevailing weather which develop several impacts on production, on food security and on farmers and workers' income and livelihood. Production has been considerably reduced in Colombia (by 31 % on average during the epidemic years compared with 2007) and Central America (by 16 % in 2013 compared with 2011–12 and by 10 % in 2013–2014 compared with 2012–2013) [1].

On the other hand there is an area focused on building models (algorithms) for regression and classification known as supervised learning. These algorithms learn through examples (data training), aiming to predict or detect a new input data. The outcome of the learning (training) process of the algorithm (such as Bayesian networks, Decision tree, Support vector machine, Artificial neural networks, K-nearest neighbors, Ensemble methods, etc.) the creation of a classifier (hypothesis or model) for a training dataset [2].

© Springer International Publishing Switzerland 2016
O. Gervasi et al. (Eds.): ICCSA 2016, Part IV, LNCS 9789, pp. 684–699, 2016.
DOI: 10.1007/978-3-319-42089-9_48

The idea of using information as training references has inspired the researchers to apply supervised learning algorithms for detecting the coffee rust disease. Colombian researchers [3] made use of meteorological variables (relative humidity average in the last 2 months, temperature variation average in the last month, accumulated precipitation in the last 2 months), physic crop properties (coffee variety, crop age, percentage of shade), and crop management (coffee rust control in the last month, fertilization in the last 4 months) aiming to report the incidence of rust. In the same way, Brazilian researchers [4, 5] sought to notify the appearance of coffee rust considering meteorological variables (daily average of night hours with air relative humidity, wind speed sum average, average daily temperature), and physic crop properties (fruit load of the plantation, spacing between plants).

Despite the efforts made by Colombian and Brazilian researchers to detect and notify the coffee rust, most of the existing rust prediction systems lack the ability to systematically analyse information coming from multiple sources, hence there is an insufficient amount of context information which may provoke significant misleads on the system reports. Therefore, in order to successfully reduce the harmful impacts of the coffee rust disease, the prediction systems can be designed as early warning systems, enabling the assessment of a situation's actual risk based on the emergency related information, performing systematic analysis on the received information, generating accurate warnings in a timely fashion, delivering the warnings to the final user through multiple channels or services and finally defining how the people must prepare and react before a dangerous situation [6].

Considering the previous statements this paper introduces a proposal focused on the integration of a CEP module within a coffee rust prediction system, in order to develop two key elements of an early warning system [6] capable of a properly detection of Coffee Rust warnings: Risk Knowledge, which has been acquired for five years by systematically collecting climate conditions data of coffee plantations and an enhanced Monitoring & Warning Service, capable of validating the warning generation via the CEP component.

This paper is arranged as follows. The next section presents a conceptual base of the different technologies related with this work. The Sect. 3 describes different proposals that focus on predicting the coffee rust incidence probability using supervised learning algorithms and the application of CEP as a monitoring tool in various scopes. On Sect. 4 the proposal is described, focused on integrating a warning validation module based on CEP with a coffee rust prediction system; this section presents the application scenario, the system design, operational example and the experimental prototype. The Sect. 5 explains and discusses the field experiment results. Finally the Sect. 6 presents the conclusions from the proposal evaluation along with the possible future works.

2   Background

2.1   Early Warning Systems

In the disaster management, early warnings are important to reduce the harmful impacts of a disaster, empowering individuals and communities threatened by hazards to act in sufficient time and in an appropriate manner to reduce the possibility of personal injury, loss of life and damage to property and the environment. Therefore an early warning system needs to actively involve the communities at risk, facilitate public education and awareness of risks, effectively disseminate messages and warnings and ensure there is constant state of preparedness [6]. A complete and effective early warning system comprises four key elements: Risk Knowledge, a component which allows the system to assess the actual risk of a situation due to its emergency related collected information; Monitoring & Warning service, a full time operating component capable of performing real time analysis on the received information in order to generate accurate warnings in a timely fashion; Dissemination and communication, this component handles all the procedures needed to deliver the warnings to the final user through multiple channels or services, ensuring a more complete dissemination, avoiding failures and reinforcing the warning message; Response capability, this component represents the knowledge and response mechanisms the population has, i.e. how the people must react before a dangerous situation and which tools they have to face the phenomenon (escape routes, first aid kits, etc.) [6]. It is worth mentioning that this paper proposes the implementation of two of the four components of an early warning system: Risk Knowledge, which has been acquired for five years by systematically collecting climate conditions data of coffee plantations, and Monitoring & Warning Service, via the enhancement of an existing coffee rust prediction system with a CEP component capable of validating the warning generation.

2.2 Complex Event Processing

The recent development of event processing methodologies has allowed building event-driven software systems, which are based on different event analysis tools such as the Complex Event Processing (CEP). This tool is focused on processing information from multiple sources in order to accurately detect event patterns, allowing the system to react and report the occurrence of an incident as soon as it is detected. CEP-based applications are composed by three main components: a monitoring component, responsible for the observation of events and its representation; a transmission mechanism that performs event notification by different means; and a reactive component, which defines the actions the system will take once an incident is detected. CEP is leveraged in this proposal to enhance the Monitoring & Warning Service component of an early warning system.

3 State of the Art

After an extensive literature review about the usage of CEP on coffee rust early warning systems, no relevant projects were found. Hence some of the most important works related with CEP and coffee rust prediction systems will be mentioned as follows, starting with the projects focused on predicting the coffee rust appearance probability and concluding with the works that involved event-driven applications and CEP.

Recent research efforts have focused on detecting the incidence of rust in coffee using supervised learning algorithms. Colombian researchers can detect the incidence rate of rust among 0 % and 15 % with supervised learning algorithms as: decision trees, support vector machines, Bayesian networks, K-NN, artificial neural networks and -novel approach- ensemble methods [2, 3, 7, 8]. In the same way, Brazilian researchers have focused on detecting the incidence of the disease using supervised learning algorithms as decision trees, support vector machines and Bayesian networks [4, 5, 9]. They used numerical values of the infection rates and mapped them into two categories: the first option of the binary infection rate defined value 1 for infection rates equal or greater than 5 percentage points (pp), and 0 otherwise. The second option defined value 1 for infection rates equal or greater than 10 pp, and 0 otherwise.

Event-driven applications and CEP have been implemented in warning systems undertaking different topics. In the health scope, Foley and Churcher [10] integrated CEP in an existing patient monitoring platform known as SAPHE (Smart and Pervasive Healthcare Aware Environment), which gathers information through a group of sensors placed on the patient and subsequently sends the collected information to the medical staff; enabling a surveillance for the welfare of the patients. Kobayashi and Fujita [11] sought to leverage the versatility of ICT to propose the idea of a "human-centric intelligent society". The objective of this proposal is that through the collection and analysis of large amounts of information, the quality of life of people can be monitored, predicted and improved. Miyazawa and Hayashi [12] proposed a distributed performance monitoring, based on event processing, to offload the performance management activities from Network Management Systems (NMS) into multiple network devices, performing a real-time fined-grained monitoring at less than one-second intervals. The proposed system is implemented in a proof of concept, enabling the detection of trivial network variations like burst traffic, but also reducing the processing load in NMS.

The previous related works show that there is a trend towards the improvement of coffee rust prediction systems and the development of warning systems applied in various scopes leveraging event driven applications and CEP; however none of these proposals considered validating the warnings of an existing coffee rust prediction system through CEP tools, restructuring these platforms into early warning systems. Taking this scenario into account, our approach pretends to upgrade and redesign an existing coffee rust prediction system as the implementation of two key elements (Risk Knowledge, Monitoring & Warning) of a completely functional early warning system, leveraging the features of CEP for the warning validation.

4 Proposal

With the aim of remodeling an existing coffee rust prediction system as an early warning system with alert validation, three main stages have been established: the application scenario definition, describing the field experiment context and the warning validating event patterns; the system design, illustrating the integration of the prediction system and the CEP module; and the experimental prototype implementation, based on the system design, conducting a performance evaluation of the validation mechanisms through a field test. Consequently, the implementation of the Risk Knowledge and

Monitoring & Warning Service components of an early warning system architecture are described through the completion of these stages.

4.1 Application Scenario

Coffee production is the main Colombian agricultural activity; more than 350.000 families depend on the coffee harvest for their sole income. Therefore, diseases, pests and even low prices have a negative impact on the economic and social aspects of the main coffee-growing regions. Studies focused on coffee rust have concluded that the spores carrying the infection are spread by climatic elements such as temperature, humidity, wind and rainfall, wind being the vector for long distance spore transport, while precipitation droplets are responsible for vertical propagation from infected leaves or soil [7]. Consequently, this proposal's application scenario is focused on the Colombian coffee production, establishing the Experimental Farm *Los Naranjos* of Supracafé, in Cajibio, Cauca, Colombia $(21°35'08''N, 76°32'53''W)$ as the field experiment context for data acquirement and experimental prototype evaluation.

Risk Knowledge. This component represents all the gathered information about the selected emergency situation, i.e. which parameters enable the prediction of the objective emergency.

A large quantity of projects [4, 5, 7, 13] have provided a broad quantity of information on the coffee rust disease, observing that the fungus Hemileia Vastatrix (the cause of the disease) needs very specific conditions to parasitize the leaves of the coffee plant. Particularly, it requires rain splashes to begin the process of dispersion between leaves and later between plants, as well as the presence of a layer of water on the underside of the leaves to begin its germination, along with temperatures between 16 and 28 °C and low sunshine conditions. Therefore, environments with constant rainfall, especially in the afternoon or evening, with occurrence of cloudy skies, preventing a temperature rise in noon, or with very low temperatures in the early morning, are conducive to the development of strong coffee rust epidemics [13]. The most popular areas for coffee production in Colombia are located in the optimum range of the disease development, with an average temperature of 22 °C; under these conditions the rust is favored to the extent that, at the end of the harvest period, the crops have reduced foliage and show symptoms of incidence and severity.

Based on the obtained results of the aforementioned researches, the selected attributes to be analysed in order to opportunely detect the coffee rust appearance probability are Relative Humidity, Temperature, Wind Speed and Precipitation. Therefore, the Risk Knowledge element from an early warning system must be defined and divided into two types of situations in order to be completed.

Detection of Risky Situations: a risky situation is a serious, unexpected and dangerous occurrence which require an immediate intervention, these threatening events may hold one or various types of hazards for the society, environment and economy, like natural disasters (wildfires, earthquakes, tornados, tsunamis, etc.) or man-made emergencies (explosions, radiological emergencies, chemical threats, etc.). Therefore a system that enables the prevention and early detection of risky situations represent an important advantage in the emergency management, since by establishing a constant surveillance and preparation before a specific risky situation, the negative impact of such situation can be decreased. Most of the risky situations have specific characteristics, which enable their identification through the constant monitoring of a set of parameters arranged as patterns; such is the case in the appearance of the coffee rust in Colombian crops.

This proposal leverages the Corrales et al. [7] research, which enables the early detection of the coffee rust disease in Colombian crops, based on three years (2011–2013) worth of collected information from *Los Naranjos* Experimental Farm; however the warnings generated by this system [7] must be validated in order to avoid false alarms.

Detection of Abnormal Situations: an abnormal situation is a critical, complex, unplanned and ambiguous event which compromises the expected performance of any kind of system. These events usually are all kinds of unexpected situations that leads to the malfunctioning (human, hardware, software, etc.) of a system. An early warning system has as objective the continuous monitoring of a specific kind of emergency, and once a possible risky situation is detected, the system must proceed to warn the population about the emergency. Hence if an abnormal situation provokes a malfunctioning of the early warning system, it could lead to the generation of false alarms, obtain mistaken information about the monitored phenomenon, or even provoke a total absence of the reports the early warning system must be providing. The abnormal situations present some characteristics that can be identified by analysing and comparing the expected and actual performance of the system; therefore, most of the events that indicate the malfunctioning of a system are intrinsic to it. Therefore, in order to detect abnormal situations on the implementation of the experimental prototype, the performance of the coffee crops data collecting tool (meteorological station) must be considered.

Under normal circumstances a meteorological station is powered by the electrical network and has an Uninterruptible Power Supply (UPS) in case of a power failure. The station collects climate information from the sensors network about the four parameters previously described (relative humidity, temperature, wind speed and precipitation), then every fifteen minutes the collected information is sent via HTTP. The provided information is the average relative humidity, the average temperature, the average wind speed and the accumulated precipitation along the day. Therefore the expected failures of the meteorological station are:

– *Sensors or Datalogger damage:* the meteorological station presents an irregular behavior sending mistaken information like out of range data, drastic changes or illogical changes between readings, and repeated information for an extended period of time. Such behavior can be provoked by damages on the sensors, i.e. the sensors take mistaken readings, or damages in the datalogger, i.e. although the sensors gather correct readings, such readings are stored in a wrong way.

– *Transmission system damage:* the transmission system of the meteorological station fails presenting two possible conditions: when the station does not transmits any data, or transmits only null values.

– *Power failure:* when the electrical network presents a blackout, and the UPS runs out of energy, the station stops transmitting information.

The Monitoring & Warning Service component leverages the climate information provided by several meteorological stations, which periodically sends a set of data about the current environment of the coffee crops (precipitation, temperature, wind speed, and relative humidity as humidity); hence it is necessary to take into consideration the different abnormal situations that may affect the system: Null Values, the received data only have null or zero values; Sensor Damage, when the incoming data gets out from the theoretical bounds, has an illogical value, has a repeated value or has drastic changes; and Absence of Incoming Information, when the system does not receive any data. The Table 1 illustrates a set of event patterns related to the defined abnormal situations, which may compromise the performance of the early warning system.

4.2   System Design

This section describes the system design for the integration of the prediction system and a CEP engine in order to implement the two components of an early warning system. The Fig. 1 illustrates the system design, whose modules are explained according to their usage into the Monitoring & Warning Service component.

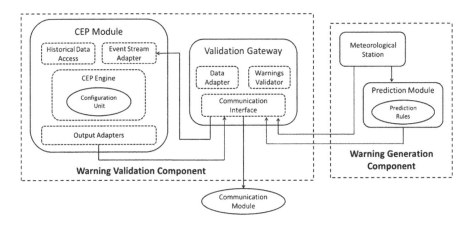

Fig. 1. System design

Monitoring & Warning Service. This component represents all the mechanisms that enable the monitoring of risky and abnormal situations, for the consequent generation and validation of warning. The different mechanisms will be presented as follows:

– *Meteorological station:* the station collects the climate information from a sensors network focused on four meteorological parameters (relative humidity, temperature, wind speed and precipitation), then every fifteen minutes the collected information is sent via HTTP to the Prediction Module and the Validation Gateway.

Table 1. Abnormal situations event patterns

Meteorological station failures	Abnormal situation	Event pattern
Sensor damage	Relative humidity out of range	Humidity < 0 OR Humidity > 100
	Temperature out of range	Temperature < −5 OR Temperature > 45
	Wind Speed or Precipitation out of range	Wind Speed < 0 OR Precipitation < 0
	Precipitation illogical changes	Current Precipitation < Previous Precipitation
	Repeated readings of Humidity, Temperature or Wind Speed during 5 h	Average Deviation Humidity = 0 OR Average Deviation Temperature = 0 OR Average Deviation Wind Speed = 0
	Drastic changes in readings of Humidity, Temperature and Wind Speed	Current Humidity > > Previous Humidity OR Current Humidity < < Previous Humidity OR Current Temperature > > Previous Temperature OR Current Temperature < < Previous Temperature OR Current Wind Speed > > Previous Wind Speed OR Current Wind Speed < < Previous Wind Speed
Transmission system damage	Null values	Humidity = 0 AND Temperature = 0 AND Wind Speed = 0 AND Precipitation = 0
Power failure	Absence of incoming information during 2 h	Number of received events = 0

– *Prediction module:* this module rece ives the climate information from the meteorological station and analyses it in order to predict the appearance of coffee rust based
on the Risk Knowledge component (specifically the risky situations). When this
module detects a real threat it generates a warning with a unique identification code
(warning code) based on the type of warning, timestamp and station ID. Then the
warning is sent to the Validation Gateway via the HTTP protocol.

– *Validation Gateway:* this module is composed by three components:

  – *Communication Interface:* this component enables all the communication needed
  inside the Validation Gateway and with the outside modules (Warning generation
  Component, Dissemination and Communication Component and the CEP
  Module), i.e. the sending and reception of meteorological data, the generated risk
  warnings, the abnormal warnings and the validated risk warnings.

  – *Warnings Validator:* performs the validation process of the generated warnings,
  i.e. this component takes the warning generated by the prediction module (identified by a warning code) and the result obtained in the analysis performed on the
  meteorological information by the CEP Module in order to define if the warning
  is real, or there is an anomaly in the meteorological station performance.

  – *Data Adapter:* this component performs an adaptation of the data received from
  the meteorological station to the format needed in the analysis performed by the
  CEP Module.

– *Communication Module*: this module delivers the different warnings to the relevant
users. The validated coffee rust warnings are sent to the farm managers, the abnormal
situation warnings are sent to the Monitoring System managers, and the invalidated
warnings are discarded.

– *CEP module:* this module represents the Monitoring & Warning Service component
of an early warning system by handling all the complex event processing operations,
i.e. it performs a real time analysis of the incoming information to subsequently
validate the warnings if an abnormal situation is detected. It is composed by:

  – *CEP Engine:* this component processes all the incoming events and analyses them
  in real time, according to the event patterns illustrated in the Table 1, triggering
  different types of reports. Furthermore a set of configuration parameters (analysis
  conditions, number of warnings per day, number of events received, etc.) are set
  by the Risk Knowledge component of an early warning system, and define which
  warnings answer to each event pattern.

  – *Event Stream Adapter:* this adapter communicates with the Validation Gateway
  and receives all the events from the platform so they can be processed by the CEP
  engine; the received events are represented as java objects containing the collected
  climate information from the Meteorological Station.

  – *Historical Data Access:* this module stores the previous received events in order
  to help the CEP Engine to process and analyse the incoming events. In a more
  precise way, this component is intended to maintain a record of the past events,
  in order to compare them with the current events and obtain a better conclusion
  of the stations' condition.

- *Output Adapters:* this module is an output interface for the CEP Engine, formatting its reports or warnings into standardized results, to finally return such reports to the Validation Gateway via HTTP.

4.3   Experimental Prototype

This section presents the experimental prototype of the implemented platform, with the aim of implementing two components of an early warning system with the capabilities of detecting abnormal situations, concerning the coffee rust appearance in Colombian coffee crops, and validating the generated warnings. The experimental prototype is presented as a deployment model, describing all the selected technologies for each component, subsequently showing an operational example, and finally exposing the field experiment design.

Deployment Model. The deployment model of the experimental prototype is illustrated on Fig. 2 and its components are explained as follows.

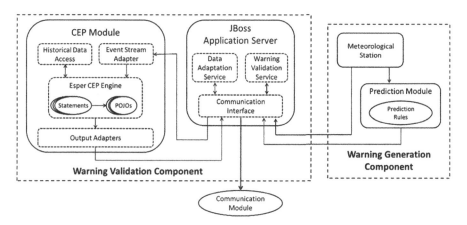

Fig. 2. Deployment model

Warning Validation Component: this module handles all the complex event processing operations, communicating with the converged control module and it is composed by:

- *Esper:* Esper is an open source component designed for complex event processing and event series analysis, it is compatible with different programming languages through its two distributions: Esper for Java platforms and Nesper for .Net platforms [14]. Esper and Nesper enable an agile and efficient development of applications capable of processing large volumes of events, regardless of whether those events have already happened or are being analysed in real time, besides it enables the filtering and analysis of events in multiple ways in order to successfully meet the requirements set by the developer.
- *JBoss Application Server:* JBoss AS is a certified Java platform for developing and deploying enterprise applications, which supports Java EE APIs and includes an

enhanced performance and scalability. It provides a complete Java platform by integrating Apache Tomcat and capabilities for clustering, messaging, transactions that simplify the development of Web services through metadata [15]. It is composed by a HTTP adapter which enables the communication throughout all the components and two services: the Warning Validation Service and the Data Adaptation Service, which are explained in the previous section.

– *Communication Module*: the definition and implementation of this module is not handled in this proposal, since it is a component of the Communication & Dissemination element of an early warning system, therefore it is proposed as future work in Sect. 6.

Warning Generation Component: these components are external to the development of this prototype and a complete description can be found in [3, 7].

Operational Example.   The operation of the experimental prototype can be explained in four different states: Regular Operation, Coffee Rust Warning Validation, Abnormal Situation Detection, and Coffee Rust Warning Invalidation. These states are explained as follows:

– *Regular Operation*: on this state no warning (valid or invalid) is generated by any component of the system.
    The prototype operation starts with the Meteorological Station collecting and sending crop related climate data to the Prediction Module and the JBoss Application Server. The Prediction Module receives and analyse the climate data but does not identify any threat; while the JBoss Application Server receives the meteorological information through the HTTP Adapter and processes it in the Data Adaptation Service. The Data Adaptation Service sends the adapted data as a Java object to Esper, which receives it through the Event Stream Adapter and analyses it in the Esper CEP Engine. Finally, based on the defined Statements and supported by the Historical Data Access, the Esper CEP Engine does not detect any abnormal behavior on the Meteorological Station information.

– *Coffee Rust Warning Validation*: on this state a coffee rust warning is generated, validated and deployed.
    The prototype operation starts with the Meteorological Station collecting and sending crop related climate data to the Prediction Module and the JBoss Application Server. The Prediction Module receives and analyse the climate data and identifies a threat, which is sent to the JBoss Application Server. The JBoss Application Server receives the meteorological information and the coffee rust warning through the HTTP Adapter, processing the climate information in the Data Adaptation Service and the coffee rust warning in the Warning Validation Service.
    The Data Adaptation Service sends the adapted data as a Java object to Esper, which receives it through the Event Stream Adapter and analyses it in the Esper CEP Engine. Based on the defined Statements and supported by the Historical Data Access, the Esper CEP Engine does not detect any abnormal behavior on the Meteorological Station information.

Finally, as Esper has not detected any abnormal situation, the coffee rust warning is validated by the Warning Validation Service, which activates the Communication Module through the HTTP Adapter in order to inform the farm managers there is a valid coffee rust warning.

– *Abnormal Situation Detection*: on this state an abnormal situation warning is detected, stored and deployed.

The prototype operation starts with the Meteorological Station collecting and sending crop related climate data to the Prediction Module and the JBoss Application Server. The Prediction Module receives and analyse the climate data but does not identify any threat; while the JBoss Application Server receives the meteorological information through the HTTP Adapter and processes it in the Data Adaptation Service.

The Data Adaptation Service sends the adapted data as a Java object to Esper, which receives it through the Event Stream Adapter and analyses it in the Esper CEP Engine. Based on the defined Statements and supported by the Historical Data Access, the Esper CEP Engine detects an abnormal behavior on the Meteorological Station information, activating the Output Adapter to send a POJO (Plain Old Java Object) with the abnormal behavior report to the JBoss Application Server.

Finally, the JBoss Application Server receives the abnormal behavior report through the HTTP Adapter, which sends the report to the Warning Validation Service to be stored and activates the Communication Module in order to inform the Monitoring System managers.

– *Coffee Rust Warning Invalidation*: on this state a coffee rust warning is generated and invalidated.

The prototype operation starts with the Meteorological Station collecting and sending crop related climate data to the Prediction Module and the JBoss Application Server. The Prediction Module receives and analyse the climate data and identifies a threat, which is sent to the JBoss Application Server. The JBoss Application Server receives the meteorological information and the coffee rust warning through the HTTP Adapter, processing the climate information in the Data Adaptation Service and the coffee rust warning in the Warning Validation Service.

The Data Adaptation Service sends the adapted data as a Java object to Esper, which receives it through the Event Stream Adapter and analyses it in the Esper CEP Engine. Based on the defined Statements and supported by the Historical Data Access, the Esper CEP Engine detects an abnormal behavior on the Meteorological Station information, activating the Output Adapter to send a POJO with the abnormal behavior report to the JBoss Application Server. The JBoss Application Server receives report through the HTTP Adapter, which sends it to the Warning Validation Service to be stored and activates the Communication Module in order to inform the Monitoring System managers there is an abnormal situation warning.

Finally, as Esper successfully detected an abnormal situation, the coffee rust warning is invalidated by the Warning Validation Service.

Field Experiment Design. The field experiment is a set of tests performed in an almost-real environment (*Los Naranjos* Experimental Farm) [16]. This evaluation design is exposed defining the prototype platform, the tests' objectives, and the methodology.

Prototype Platform: the Warning Validation Component is deployed in a Dell Power-Edge T100 server (Intel Xeon 3000 CPU, 8 GB RAM, Single embedded Gigabit NIC). The Warning Generation Component implementation is specified on [7].

Evaluation Objectives: in order to evaluate the performance of the experimental prototype the following objectives were stated:

- Observe the Latency of the implemented Monitoring & Warning Service, i.e. how much time the prototype takes to finish the analysis of a predefined set of events.
- Observe the Success Rate of the implemented Monitoring & Warning Service i.e. the percentage of successfully analyzed events within a set of information.
- Measure the physical resource compsumtion of the Monitoring & Warning Service in terms of the CPU Load and the Memory Usage of the service, facing an increasing stream of events.

Evaluation Methodology: in order to evaluate the performance of the experimental prototype two main tests were conducted: the first test was the Latency and Pollution Test, the latency by observing how much time the prototype takes to finish the analysis of a predefined set of events measured in seconds (Latency), and the Pollution by observing the percentage of successfully analyzed events (Success Rate). The second test was the Load Test observing the performance of the server measured in the percentage of CPU Load and Memory Usage facing an increasing stream of events. The Fig. 3 illustrates the obtained results of both evaluations.

5   Results and Discussion

This section presents the different tests that were conducted in order to observe the behavior of the experimental prototype, illustrating the obtained results.

In the Fig. 3 two graphics can be observed. The first one (Latency and Pollution Test Results) illustrates that the latency behaves in a linear fashion regarding the number of events, which was expected since a larger quantity of events requires more time to be processed and analyzed by the prototype, on the other hand the success rate represented by a line shows an initial decrease and a later stabilization, where the initial decrease was not due to the triggering of false alarms in the reports, but the absence of expected reports; presumably caused by hardware limitations within the server, i.e. the server could not handle the increasing stream of incoming events, generating the absence of some expected reports; notwithstanding, considering the application scenario the prototype should not handle more than 54 events every 15 min.

The second illustration (Load Test Results) shows two different behaviors, where the CPU Load had a linear increase as expected, since with a larger quantity of events the server must implement more physical resources to process and analyse them, but the

Memory Usage is always around 9 %, showing that JBoss AS has an standardized memory consumption, which supports the previous idea where the server could not handle the increasing amount of events generating the decrease in the obtained success rate.

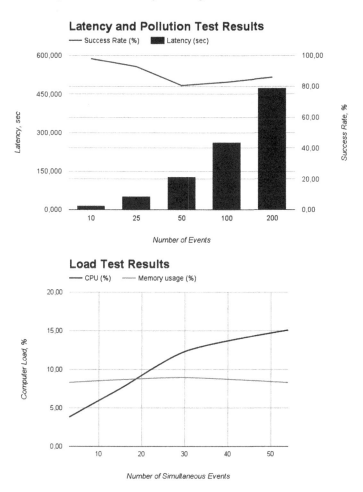

Fig. 3. Latency, pollution and load test results

6   Conclusion and Future Works

This paper presented a redesign and upgrade of a coffee rust prediction system integrating a Complex Event Processing component, capable of analysing and validating the coffee rust incidence predictions. In order to evaluate our proposal, an experimental prototype, with two of four components of an early warning system, was implemented with the Esper CEP engine and subsequently tested through two main evaluations: the latency and pollution test, and the load test. From the obtained results, observed in

Fig. 3 and analyzed in Sect. 5, it is possible to conclude that a successful integration between the CEP component -supported by the Esper engine- and the prediction system was achieved; validating the coffee rust incidence probability predictions, and obtaining satisfactory results in both performed evaluations. Nevertheless, such results could be improved by enhancing the hardware from the implemented server, since most of the presented difficulties were caused by the inability to handle large amounts of events at the same time, generating the absence of some expected reports during the testing of the prototype.

As future works it is proposed the improvement of the CEP prediction capabilities, since it would prevent system failures and data losses, enabling the system to operate in optimal conditions. On the other hand the implementation of the third component of an early warning system (identified in this paper as the Communication Module within the Dissemination & Communication component) within the proposed architecture would allow the system to deliver the coffee rust warnings to the relevant users through multiple channels or services; also a Response Capability component can be designed in order to handle the respective preparations of the population involved with the coffee rust management, establishing a completely functional early warning system aimed for the coffee rust warning.

Acknowledgements. The authors would like to thank University of Cauca, Colciencias, Supracafé and AgroCloud project of the RICCLISA Program for supporting this research.

References

1. Avelino, J., Cristancho, M., Georgiou, S., Imbach, P., Aguilar, L., Bornemann, G., Läderach, P., Anzueto, F., Hruska, A.J., Morales, C.: The coffee rust crises in Colombia and Central America (2008–2013): impacts, plausible causes an d proposed solutions. Food Secur. **7**(2), 303–321 (2015)
2. Corrales, D.C., Carlos, C., Casas, A.F.: Toward detecting crop diseases and pest by supervised learning. Ing. Univ. **19**(1), 207–228 (2015)
3. Corrales, D.C., Ledezma, A., Hoyos, J., Figueroa, A., Corrales, J.C., et al.: A new dataset for coffee rust detection in Colombian crops base on classifiers. Sist. Telemática **12**(29), 9–23 (2014)
4. Luaces, O., Rodrigues, L.H.A., Alves, M., Bahamonde, A.: Using nondeterministic learners to alert on coffee rust disease. Expert Syst. Appl. **38**(11), 14276–14283 (2011)
5. Cintra, M.E., Meira, C.A.A., Monard, M.C., Camargo, H.A., Rodrigues, L.H.A.: The use of fuzzy decision trees for coffee rust warning in Brazilian crops. In: 2011 11th International Conference on Intelligent Systems Design and Applications (ISDA), pp. 1347–1352 (2011)
6. Developing early warning systems: a checklist, III Third International Conference on Early Warning, Bonn, Germany, p. 10 (2006)
7. Corrales, D.C., Peña, A.J.Q., León, C., Casas, A.F., Corrales, J.C.: Early warning system for coffee rust disease based on error correcting output codes: a proposal. Rev. Ing. Univ. Medellín **13**(25), 57–64 (2014)

8. Corrales, D.C., Figueroa, A., Ledezma, A., Corrales, J.C.: An empirical multi-classifier for coffee rust detection in colombian crops. In: Gervasi, O., Murgante, B., Misra, S., Gavrilova, M.L., Rocha, A.M.A.C., Torre, C., Taniar, D., Apduhan, B.O. (eds.) ICCSA 2015. LNCS, vol. 9155, pp. 60–74. Springer, Heidelberg (2015)
9. Neto, C.D.G., Rodrigues, L.H.A., Meira, C.A.A.: Warning models for coffee rust (Hemileia vastatrix Berkeley & Broome) by data mining techniques. Coffee Sci. 9(3), 408–418 (2014)
10. Churcher, G.E., Foley, J.: Applying complex event processing and extending sensor web enablement to a health care sensor network architecture. In: Hailes, S., Sicari, S., Roussos, G. (eds.) Sensor Systems and Software, pp. 1–10. Springer, Berlin, Heidelberg (2010)
11. Kobayashi, G., Fujita, K.: Convergence service platform. FUJITSU Sci. Tech. J. 47(4), 8 (2011)
12. Miyazawa, M., Hayashi, M.: In-network real-time performance monitoring with distributed event processing. In: IEEE Network Operations and Management Symposium, NOMS 2014, pp. 1–5 (2014)
13. Rivillas, C.A., Serna, C.A., Cristancho, M A., Gaitan, A.L.: La roya del cafeto en Colombia: Impacto manejo y costos del control, February 2011
14. Zámečníková, E., Kreslíková, J.: Comparison of platforms for high frequency data processing. In: Proceedings of the 2015 IEEE 13th International Scientific Conference on Informatics, INFORMATICS 2015, pp. 296–301 (2015)
15. Benothman, N., Clere, J.-F., Schiller, E., Kropf, P., Maucherat, R.: Network performance of the JBoss application server. In: Proceedings of the Conference on Local Computer Networks, LCN, 2013, pp. 739–742 (2013)
16. Sampieri, R.H., Collado, C.F., Lucio, B.P., Valencia, S.M., Torres, C.P.M.: Metodología de la investigación (2014)

Author Index

Printed in the United States
By Bookmasters